DYNAMIC RESPONSE OF STRUCTURES

Proceedings of the Third Conference
organized by the Engineering Mechanics Division
of the American Society of Civil Engineers
and sponsored by the
Civil Engineering Department of the
University of California-Los Angeles

March 31-April 2, 1986
University of California
Los Angeles, California

Edited by Gary C. Hart and Richard B. Nelson

Published by the
American Society of Civil Engineers
345 East 47th Street
New York, New York 10017-2398

The Society is not responsible for any statements made or opinions expressed in its publications.

PREFACE

These are the proceedings from the third specialty conference on the Dynamics of Structures sponsored by the Engineering Mechanics Division of the American Society of Civil Engineers including the keynote papers, the invited papers, and contributed technical papers presented at the conference held at UCLA March 31 through April 2, 1986.

The primary goal of the conference is to provide a forum for dialogue between engineers conducting experimental tests and those developing analytical models. Engineers in the area of earthquake, wind, system identification, full-scale structural response, and structural control are represented in these proceedings.

Each of the papers included in the Proceedings has been reviewed and accepted for publication by the Proceedings Editors. All papers are eligible for discussion in the Journal of the Engineering Mechanics Division of ASCE. All papers are eligible for ASCE awards.

The editors wish to express their appreciation for the guidance and assistance provided by the members of the Steering Committee in arranging the technical program of the conference.

Gary C. Hart
Richard B. Nelson
Editors

STEERING COMMITTEE

Gary C. Hart (Co-Chairman)
University of California, Los Angeles

Richard B. Nelson (Co-Chairman)
University of California, Los Angeles

Mihran S. Agbabian
University of Southern California, Los Angeles, California

Charles Culver
National Bureau of Standards, Washington, DC

Barry Goodno
Georgia Institute of Technology, Atlanta

Robert D. Hanson
The University of Michigan, Ann Arbor

Neil M. Hawkins
University of Washington, Seattle

I. M. Idriss
University of California, Los Angeles;
Woodward-Clyde Consultants, Santa Ana, California

W. D. Iwan
California Institute of Technology, Pasadena

Paul C. Jennings
California Institute of Technology, Pasadena

James Noland
Atkinson-Noland Associates, Boulder, Colorado

Gerard Pardoen
University of California, Irvine

R. Parmelee
Alfred Benesch & Company, Chicago, Illinois

Jon D. Raggett
J. D. Raggett, Inc., Carmel, California

Timothy A. Reinhold
Applied Research Associates, Inc., Raleigh, North Carolina

Christopher Rojahn
Applied Technology Council, Palo Alto, California

Rodolpho Saragoni
University of Chile, Santiago

G. I. Schueller
University of Innsbruck, Innsbruck, Austria

Firdaus E. Udwadia
University of Southern California, Los Angeles, California

A. S. Veletsos
Rice University, Houston, Texas

J. T. P. Yao
Purdue University, Lafayette, Indiana

CONFERENCE STAFF

T. A. Sabol
Director, Technical Program

D. Haines
Director, Administrative Arrangements

CONTENTS

Keynote Papers

Invited Papers

Session 1A — Lessons Learned from Recent Earthquakes

*Manuscript not available at time of printing

Session 1B — Full-Scale Structures

Session 2A — Laboratory Test Studies

Session 2B — Full-Scale Structures

*Manuscript not available at time of printing

Session 3A — Laboratory Test Studies

Session 3B — Full-Scale Structures

Session 4A — Strengthening of Existing Buildings

Session 4B — Wind Tunnel Studies

Session 5A — US/Japan Earthquake Research

*Manuscript not available at time of printing

Session 5B — Base Isolation

Session 6A — U.S./Japan Earthquake Research

Session 6B — System Identification

*Manuscript not available at time of printing

Session 7A — Soil/Structure Interaction

Session 7B — Deterministic Structural Dynamics

Session 8A — Soil/Structure Interaction

*Manuscript not available at time of printing

Session 8B — Probabilistic Structural Dynamics

Session 9A — Soil/Structure Interaction

Session 9B — Structural Control

*Manuscript not available at time of printing

xiv

INTERDEPENDENCE OF DYNAMIC ANALYSIS AND EXPERIMENT

by

V.V. Bertero, F.ASCE[1] and Ray W. Clough, F.ASCE[2]

ABSTRACT

This paper emphasizes the importance of experimental research to the development of reliable analytical methods for predicting the behavior of structures subjected to static or dynamic loads. A brief summary is presented first indicating the varying relationship between analysis and experiment during the history of structural mechanics. Emphasis is then placed on the essential part played by experiments in developing analytical procedures for evaluating the nonlinear response of buildings to severe earthquake motions. The special problems associated with earthquake response analysis are mentioned, including the difficult tasks of specifying the earthquake input and the acceptable levels of response performance, and the present state of the art is described. Then the U.S.-Japan Cooperative Earthquake Research Project involving study of a seven story RC shear-wall frame building is presented to illustrate the interrelationship between analysis and experiment in this field. Finally, conclusions are drawn about the expected earthquake performance of code designed RC buildings.

INTRODUCTION

Objective and Scope

The rapid advances that have been made in methods for analysis of the dynamic response of structures since the advent of electronic computation are obvious, and enormous research efforts are being directed toward further improvements in such techniques of computational mechanics. However, these spectacular achievements tend to obscure the fact that the analytical procedures, no matter how powerful and refined, cannot lead to a reliable estimate of the dynamic behavior of a real structure unless the mathematical model used in the calculations truly represents the physical properties of the actual system. Obtaining the same results with a different computer program using a different computer does not constitute a check of the validity of the calculated results if the same assumptions are employed in defining the mathematical model for both cases. In general, the proof that a given analysis procedure and program is producing reliable results can only be obtained by comparison of the analytical predictions with the behavior of a real structure subjected to real dynamic loads, that is, by correlation of analysis with experiment.

[1] Professor of Civil Engineering, University of California, Berkeley.
[2] Nishkian Professor of Structural Engineering, University of California, Berkeley.

   It is the purpose of this paper to discuss this interrelationship
of analysis and experiment with reference to the field of earthquake
engineering, where the limitations of a purely analytical approach became apparent in the early stages of computer analysis. The paper
begins with a review of the changing relationships between analysis
and experiments in the historic development of structural mechanics;
this is followed by an overview of the special problems inherent in the
design and analysis of earthquake-resistant buildings and a summary of
the present state of the art of earthquake-resistant design. The U.S.-
Japan Cooperative Research Project on earthquake behavior of concrete
shear-wall frame buildings then is used to illustrate the inter-
dependence between analysis and experiment in earthquake engineering,
starting with a brief description of the project. Comparisons are
made between the response anticipated by code loads, the response
calculated from dynamic analysis and the measured test results.
Finally these comparisons are used to evaluate the controversial response reduction factors introduced in current design codes.

Historical Perspective

   Advances in the theory and practice of structural engineering
traditionally have been made by combining observations of structural
performance under load with the development of theories that explain
the observed behavior. Indeed the origins of structural mechanics may
be traced back to the experiments of Robert Hooke who announced his law
of proportionality of deformation and force in 1678. Based on this
concept and supported by experimental evidence, many early researchers
made contributions to the theory of bending of beams--including
Bernoulli, Euler, Coulomb and of course Navier. The assumption that
plane sections remain plane during bending was introduced by Bernoulli,
but Navier is given credit for developing the general theory of
flexure for beams of arbitrary cross-section--in a book published more
than 100 years after Bernoulli's death. In that work Navier referred
extensively to the experimental results that were available at that
time.

   It is significant that Navier's hypothesis and Hooke's law were
sufficient to develop essentially the entire theory of framed
structures, so subsequent developments in the analysis of elastic
structures were mainly mathematical in nature and did not depend
heavily on experiment. The force method approach to treatment of
statically indeterminate structures was systematized during the 19th
century by Maxwell, Mohr, Müller-Breslau and Castigliano. The
alternative formulation using displacements or deformations as the
primary variable was first expressed in general form by A. Ostenfeld
in Denmark during the 1920's. With either of these approaches, the
analysis of elastic framed structures of any form or complexity
could be formulated. The only difficulty to be dealt with was the
solution of simultaneous equations, and this led to the first
recognition of the importance of computational mechanics in the field
of structural analysis. The advantages of iteration methods in solving
the simultaneous equations were soon recognized, and the Hardy Cross
Moment-Distribution method became a standard tool for dealing with the
analysis of building frames. The great capability of this technique
for analysis of such systems essentially eliminated the need for

experimental studies of elastic building frames. Even for more general structural systems, such as framed domes for example, the generalized iteration method called the Southwell relaxation procedure was capable of providing solutions. However experimental model studies were sometimes used in analysis of such structures as a matter of convenience, treating the model essentially as an analog computer to avoid the tedious iterative numerical work.

The ultimate stage in the development of methods of analysis of framed structure was the matrix formulation of structural theory. Several researchers participated in this development, including Falkenheimer in France and Langefors in Sweden, but by far the most important contribution was the work of Argyris in England in 1954-55[24] in which he presented an integrated matrix formulation of the force and the displacement methods. The timing of this development was quite fortuitous because the matrix formulation was ideally suited for implementation by computers, and electronic computers were just becoming available at that time. The subsequent generalization of structural theory by the finite element method then opened up the possibility of applying the Argyris matrix formulation to all types of structures--involving two-dimensional continua such as plane stress systems, plates and shells, and three-dimensional solids in addition to the one-dimensional frame members. This finite element explosion quickly displaced the use of experimental methods such as photoelasticity in analysis of elastic solids, and led to significantly reduced interest in experimental mechanics during the 1960's.

With the development of general purpose structural analysis computer programs based on the finite element idealization, the structural engineering profession had a tool that could solve all of their analysis problems, but only within the limits of validity of the mathematical model adopted to represent the physical system; the most important of these limitations was the assumption of linear response behavior. Although the initial applications of those programs were to static problems, their extension to deal with dynamic loadings was straightforward for linear systems, involving only the addition of inertial and damping effects to the static structural resistance mechanisms. Thus, the extended programs provided the capability of dealing with arbitrary dynamic loadings applied to essentially any type of structural system. With this new capability, structural engineers began turning considerable attention to the field of earthquake engineering, especially in the state of California where earthquake-resistant design was recognized as a critical topic. However, it soon was recognized that the linear elastic response problem has little practical importance in earthquake engineering. It became apparent that it is not economically feasible to design all structures so that they would not be damaged by a maximum intensity earthquake motion. But this fact was coupled with the difficult problem to design structures which respond inelastically during the earthquake but suffer only acceptable and predictable amounts of damage. It was at this point, some 20 years ago, that experimental studies again acquired major importance in structural engineering because the true nonlinear physical properties of a structure can only be established by experiment. Hence during the past two decades there has been a close relationship between analysis and experiment in the field of earthquake

engineering. Typically tests have been performed on structural components and assemblages to determine their actual behavior when subjected to dynamic cyclic loads similar to what might be experienced during
an earthquake, and then mathematical models were formulated which could
reproduce analytically the observed behavior.

Techniques for evaluating the nonlinear response of structures,
usually using step-by-step methods, were an early target in the field
of computational mechanics, so as soon as nonlinear mathematical models
of structural components were developed from test data they could be
incorporated into general purpose nonlinear dynamic response analysis
programs. So in principle the key to the earthquake engineering
analysis problem was in hand. However, the close relationship between
analysis and experiment in earthquake engineering continues because
the range of problems to be dealt with is so broad and extensive.
Tests performed by means of shaking tables or by pseudo-dynamic
methods provide data on damages to typical structures induced by
simulated earthquake motions, and correlation of the measured results
with the response behavior calculated from these specified input
motions provides an opportunity to verify all aspects of the analytical
procedure--computational methods as well as the mathematical model
assumptions.

In a similar way, the analysis of damage to real buildings resulting from actual earthquakes provides an important opportunity for
interaction between analysis and experiment because the actual structures include factors that cannot be treated in a laboratory model--
such as soil-structure interaction. Of course, the actual structural
response during an earthquake seldom is measured and recorded, but
reasonable estimates often can be made of the general character of the
earthquake motions and the calculated dynamic response to these
motions can then be correlated with the observed damages caused by
the earthquake.

SPECIAL PROBLEMS OF EARTHQUAKE ENGINEERING ANALYSIS

Preliminary Remarks

The analysis of the dynamic response of a building mentioned above
is only a small part of the problem of designing a building intended
to resist earthquake effects at a selected site. The first requirement is to establish an adequate representation of the most severe
earthquake motions that may possibly occur at the site. Then the
response must be calculated taking account of the entire soil-founda-
tion-superstructure system including participation of non-structural
components and of the building contents at the time when the earthquake
strikes. Moreover, the analysis must deal with the building not as it
was designed but as it actually was constructed and maintained up to
the time of the event.

The building response during an earthquake involves the interaction of DEMAND and SUPPLY, the DEMAND being made on the structure by
the earthquake and the SUPPLY being the resistance mechanisms developed
in the structure. This relationship can be called the General Design
Equation and expressed as follows:

DEMAND $\leq$ SUPPLY

The problem for the designer is to estimate the DEMANDS that may be made by the earthquake and then to create a design that can SUPPLY the necessary resistance. These supplies and demands in general involve the properties of stiffness, strength, stability and capacities for energy absorption or dissipation. Evaluation of the earthquake demands usually is done by numerical analysis of a mathematical model of the building systems, taking account of the complete structure and its foundation system. An important factor in the problem is that the earthquake input is influenced to some extent by the response of the structure, thus it cannot be defined as an independent entity.

In spite of the remarkable advances that have been made in structural analysis as described above, it must be recognized that such analyses often have failed to predict the earthquake behavior of real buildings, particularly at deformation states that approach the ultimate, mainly due to inability to effectively model the non-linear behavior. Consequently, it is apparent that inability to determine the SUPPLY side of the general equation makes it impossible also to establish the earthquake DEMANDS. The three basic elements of the earthquake response problem--earthquake input, demands on the structure and supply capacity of the structure are discussed briefly below, together with comments on proper construction and maintenance.

Design Earthquakes and Criteria

In principle, a design earthquake may be considered as a ground motion which drives the structure to its design limit state; thus the choice of earthquake depends on the degree of damage envisioned by the design criteria. In practice, seismic codes have specified design earthquakes in terms of a seismicity zone, or as a site intensity factor or a peak site acceleration. However, it is difficult to establish that the structure will be driven to its limit state by an input described only by such simple indices. The current trend in code descriptions of earthquakes is to use ground motion spectra (GMS) based on an effective peak acceleration (EPA); this is a great improvement, however, there still is uncertainty in the choice of the EPA and the GMS to be used for a given project.

Earthquake Demands

Major uncertainties in the earthquake response analysis which establishes the demands made on the structure are due to difficulties in establishing: (1) the critical seismic loading (design earthquake), (2) the state of the entire building-foundation-soil system at the time of the earthquake (mathematical model), (3) the internal forces (deformations) and stresses (strains) that will be induced in the model (structural analysis and stress analysis), and (4) the supplies of stiffness, strength, stability, and the energy absorption and dissipation provided by the complete system (structural and nonstructural components).

Building Supplies

The properties supplied by a real building include contributions not only from the bare superstructure, but also the effects of interaction of the structure with the soil and with its nonstructural components. For example, masonry walls and/or partitions tightly infilled into moment-resisting frames introduce significant changes in the dynamic characteristics of the building, as illustrated in Fig. 1. An evaluation of these changes and of their implications on the design of earthquake-resistant buildings is given in Ref. 7. This evaluation demonstrates clearly that the neglect of such interaction between structure and nonstructural components in the dynamic response analysis can lead to completely unrealistic estimates of the demands made on the building and thus to a poor design. In this context it is evident that extrapolation of results from studies of a bare test superstructure must be done with great care. Also, the designer must be cautioned against arbitrary increases on the supply side of the General Design Equation in order to cover uncertainties on the demand side, because supply increases may lead to corresponding increases of demand.

Construction Inspection and Maintenance

Design and construction are intimately interrelated--the achievement of good workmanship depends, to a large degree, on simplicity of detailing of the members, and of their connections and supports. A design is only effective if it can be constructed and maintained. Post-earthquake studies have revealed that a large part of the damage and failure has been due to poor quality control of structural materials and/or poor workmanship--problems which could have been avoided if the building had been carefully inspected during construction [8]. In many other cases, damage may be attributed to improper maintenance of buildings during their service lives. Improper alteration, repair, and/or retrofitting of the structure as well as of the nonstructural components can lead to severe damage in the case of major earthquake shaking [8].

Summary

To summarize, the above review and analysis of the problems encountered in achieving efficient earthquake-resistant construction of buildings clearly point out the need for a comprehensive approach to the solution of these problems, integrating the various disciplines involved in the design, construction and maintenance of earthquake-resistant buildings. The need for such an approach has been discussed in references 6 and 8; its ultimate goal should be the development of a sound seismic design code procedure which should be simple enough to facilitate the preliminary design of the building and its enforcement through capable inspections during all the phases of the design, construction and maintenance (modifications, repair and/or retrofitting) of the building.

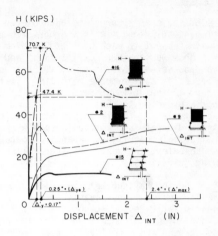

FIG. 1 - EFFECTS OF ADDITION OF INFILLS ON THE LATERAL LOAD-INTERSTORY
DRIFT RELATIONSHIP OF MOMENT RESISTING FRAMES [7].

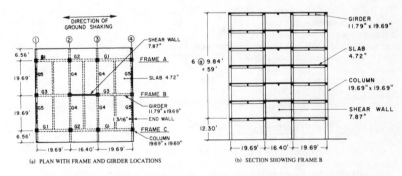

FIG. 2 - US-JAPAN 7-STORY RC FRAME-WALL TEST STRUCTURE: PLAN AND SECTION OF
FULL SCALE MODEL.

STATE OF THE ART OF EARTHQUAKE RESISTANT DESIGN

Selection of Design Earthquakes

Having discussed the special problems encountered in the design of earthquake-resistant buildings, it is appropriate to describe the state of the art with respect to each of these problem areas, following the same sequence. Thus we consider now the first and most difficult step in the design process--establishment of the Design Earthquake (EQ) --giving due consideration to the three main limit states that may be specified with regard to the building response: serviceability level - where the building is expected to continue performing its designated function, damageability level - where the damage is limited to pre-determined amounts, and safety against collapse - where any degree of damage short of collapse is permitted. When the appropriate design criterion has been selected, the design earthquake is defined according to the following guidelines.

Serviceability Limit Design Earthquake: For all practical purposes the building should remain in the linear elastic state. A design EQ based on a smoothed linear elastic design response spectrum (LEDRS) is the most reliable and convenient approach to use for the preliminary design. However, the ground spectrum that is used in the derivation of the LEDRS must be appropriate to the site and not based just on standard values. Values selected for the damping ratio, determination of allowable stresses and computation of natural periods and internal forces must be consistent with the expected behavior. This was the approach followed by ATC3-06 in defining the recommended spectral shapes for deriving LEDRS (Ref. 5, Vol. 2).

Ultimate Limit Design Earthquake: Derivation of a reliable inelastic design response spectrum (IDRS) requires full characterization of the expected severe ground motions at the site as well as the acceptable responses of the structure. However, current IDRS do not account for the duration of strong shaking. Extensive integrated analytical and experimental studies will be required to obtain all the information necessary to establish reliable design earthquakes when ultimate limit states control the design. Until this is done the procedure suggested in references 6 and 10 can be used.

Specifying the Effective Peak Acceleration (EPA)

The concept of EPA was introduced in the development of zoning maps for ATC3-06. At first sight EPA appears to be a sound parameter for application in seismic hazard analysis; however there is no systematic, quantitative definition of this parameter at present. From results obtained in a recent study [11], it has been concluded that "generally EPA depends both on the type of earthquake considered and on the interaction of the dynamic ground motion characteristics and the soil-foundation-structure system. Furthermore, EPA will depend on the limit state under consideration. Although the use of EPA can provide an idea of the relative damage potential of a given ground motion, its use as the sole parameter to define this damage potential can be very misleading".

In developing the ATC design provisions [5], two parameters were used to characterize the intensity of design ground shaking: the EPA (A_a) and the Effective Peak Velocity-Related Acceleration, EPV (A_V). According to ATC, for any specific ground motion the values of these two parameters can be obtained by the following procedure: (a) The 5 percent damped $(\xi=5\%)$ linear elastic pseudo-acceleration spectrum is drawn for the actual given motion, (b) Straight lines are fitted to the actual spectral shape for fundamental building periods, T, in the range of 0.1 and 0.5 second for the EPA and at a period of about 1 second for the EPV to obtain a smoothed spectrum, (c) The ordinates of the smoothed spectrum are divided by 2.5 to obtain the EPA and EPV.

Analysis of the 5 percent damped linear elastic response spectrum for the recorded ground motion at the Pacoima Dam (or for the derived Pacoima Dam record) shows that the maximum ATC values for the EPA and the EPV specified for this earthquake, namely $A_a=0.40$ and $A_V=0.40$, can be significantly exceeded for certain period values in the range used for their derivation. Nevertheless, the recommendation of these values by ATC has been a welcome step towards a more realistic appraisal of the severity of the ground motion that can occur at sites located in the proximity of major active faults (Map Area No. 7 of the ATC Maps).

Response Reduction Factors

The approach followed by ATC3-06 (described above) in recommending design earthquakes in the form of smoothed linear elastic design spectra (LEDS) appears to be appropriate for the design of essential facilities that must remain practically undamaged even under the Maximum Credible Earthquake (MCEQ) shaking. However, except for these essential facilities, it would be unrealistically conservative and uneconomical to design most buildings to respond to MCEQ shaking within the linear elastic range of the structural material, or even in the so-called "effective linear elastic" behavior of the structure (i.e., up to its "Significant Yield Level"). As already recognized by the engineers that developed the original SEAOC Blue Book and by those that developed the ATC3-06 recommendations, in order to attain economical design of buildings that are quite unlikely to be subjected to MCEQ shaking during their service life, it is necessary to allow these buildings to undergo significant but controllable (acceptable) inelastic deformations. These inelastic deformations usually allow reduction of the required strengths at the yield and ultimate states without increasing the maximum resulting deformations significantly. In the ATC recommendations, the beneficial effect of the inelastic deformation in reducing the required strengths is introduced by means of a RESPONSE MODIFICATION FACTOR R which is independent of T. It is not well recognized, however, that even for a given structural system the acceptable decrease in these strengths cannot be constant for the whole range of the fundamental period T, for which the structure might be designed.

The SEAONC Seismology Committee, by proposing the lateral force coefficient $C = \dfrac{0.5S}{T^{2/3}}$ in their present revision of the Blue Book, has adopted similar LEDS to those recommended by ATC. To include the

effect of energy dissipation through inelastic deformation, the SEAONC
Committee proposes to use a factor R_w called the STRUCTURAL QUALITY
FACTOR to reduce C. In discussing the selection of this value it has
been stated that R_w is a measure of the anticipated ductility and
flexibility of the structural system. The only apparent difference in
the definition of R and R_w is the level of the design forces to which
the LEDS is reduced. ATC reduces this spectrum to the "Significant
Yield Level of the Structure", while the SEAONC Committee reduces it
to the "Working or Allowable Force Level". Thus R_w should be somewhat higher than R, (for reinforced concrete $R_w \doteq 1.4$ R).

It is very difficult to judge the rationality and reliability of
the values recommended for these R and R_w factors, because there is no
indication of how they have been derived and what they physically
represent. In reference 5 (Chapter 4 of the Commentary) it is stated
that R "is an empirical response reduction factor intended to account
for both damping and the ductility inherent in the structural system
at displacements great enough to surpass initial yield and approach
the ultimate load displacement of the structural system". In judging
this statement it should be noted that the LEDS selected by ATC is
already based on a 5% damped linear elastic response spectrum. The
equivalent viscous damping expected in typical structures should not
be significantly greater than this, particularly in the case of steel
structural systems, therefore the response reduction factor should not
include a viscous damping contribution.

A possibly better explanation of R is given in Chapter 3 of the
Commentary [5]. "The response modification factor, R, and... have
been established considering that structures generally have additional
overstrength capacity, above that whereby the design loads cause
significant yield." The writers believe that this overstrength together with inherent toughness is a "blessing" which enables structures
that are designed on the basis of presently specified design seismic
forces (UBC or the recommended ATC) to withstand the MCEQ shaking
safely. Properly sized and detailed buildings, when well constructed
and maintained, provide considerably higher force at first "SIGNIFI-
CANT EFFECTIVE YIELDING" than that on which the code design is based
and in addition offer significant overstrength above that first yield
level of the structure. The resulting total strength is usually 2 to
3 times greater than the minimum code specified effective yield
strength. This overstrength has been observed by the authors in the
past; fortunately experimental test data now give specific proof of
its existence.

Because of the above observations some of the statements included
in Section 3-1 of the Commentary of the ATC Recommendations [5] should
be emphasized. "The values of R must be chosen and used with judgement. For example, lower values must be used for structures possessing
a low degree of redundancy wherein all the plastic hinges required for
the formation of a mechanism may be formed essentially simultaneously
and at a force level close to the specified design strength. This
situation can result in considerably more detrimental P-delta effects."
The importance of the above statement will be illustrated and discussed
further in the following section on the U.S.-Japan Research project.
However, the implications of those results can be summarized in the

statement: "if the constructed building has a Real Ultimate Strength just equal to the code minimum specified Effective Yield Strength, its response to the MCEQ shaking will exceed acceptable limits."

Expected Performance of Code Designed Buildings

The performance during recent earthquakes of buildings designed according to U.S. codes has been studied and described in detail in several publications [6,8]. From these studies, the following observations have been made: (1) The maximum lateral strength of code designed structures, especially RC wall structures, is typically higher than that required by code--usually by more than a factor of 2. (2) Despite this observed extra strength, many code designed buildings collapse or suffer serious damage (either structural or nonstructural) to an extent that constitutes failure; accordingly it can be concluded that present seismic code regulations do not guarantee safety against collapse or unacceptable damage. (3) While the collapse of buildings with moment-resisting frames usually involves "pancaking" of one or more stories, this type of failure is not often observed in buildings with shear walls that extend through the foundation.

RESULTS FROM THE U.S.-JAPAN COOPERATIVE PROJECT

General Features of the Project

In order to take full advantage of the major earthquake engineering facilities available in each country, a cooperative research program involving these large scale experimental test facilities was initiated by the United States and Japan. General features of the program are described in References 1 and 2; the specific phase of the research considered in this paper is described in Reference 3. The essential feature of this phase of the project is the test of a full scale seven story RC shear wall-frame building using the pseudo-dynamic test facility of the Building Research Institute in Tsukuba, Japan, and a test of an equivalent one-fifth scale model on the 20 ft square earthquake simulator of the Earthquake Engineering Research Center at the University of California, Berkeley.

There were several principal objectives to this test: to compare the performance of the 1/5 scale model with that of the full scale building, to compare the response developed by pseudo-dynamic testing with that from a true dynamic test, to compare the performance predicted by analysis with measured results, and to evaluate the earthquake resistance of a code-designed building. The dimensions and general arrangement of the full scale structure tested in Japan are shown in Fig. 2. The structure was designed with reference to the 1979 UBC and the Japanese code; strictly speaking, however, the design did not satisfy the minimum UBC detailing requirements.

Although many aspects of the structural response were measured and calculated, in this discussion the gross lateral force-displacement relationship will be described, with emphasis on comparison of measured results with behavior implied by the code requirements and with the results of analyses. All of these types of results are described in the following paragraphs.

U.S. Code Predicted Responses

The relationship between the total lateral force, V, and roof displacement, δ_r, (or roof drift index) according to UBC minimum requirements is shown as curve OABC in Fig. 3. Similar relationships according to the minimum recommended by ATC [5], are shown in Fig. 4. Note that, in order to compare all these relationships directly amongst themselves not only has the demanded strength been plotted, but also the resulting MINIMUM NOMINAL STRENGTHS are shown.

Analytical Response Predictions (Before Testing)

Before testing any of the models, the expected behavior of the test structure was analyzed using available techniques and computer programs; static and dynamic analyses were conducted, based on the specified mechanical characteristics of the materials and using code expressions to estimate the nominal axial-flexural strengths of the critical sections of the structure. In all these analyses it was assumed that the axial-flexural strength would control the behavior, i.e., that shear failure would not occur and that the critical regions have sufficiently large ductility to allow formation of the collapse mechanisms. The results of these analyses also are presented in Figs. 3 and 4, as well as in Fig. 5.

Static Analyses: Two different techniques of static analysis were used: limit analysis, and a step-by-step nonlinear analysis. Each of these types of analysis was conducted based on a pseudo-3 dimensional (3D) model which actually is based on 2 dimensional (2D) behavior of each frame and/or wall component in the loading direction as illustrated in Fig. 6. In other words the 3D behavior due to interaction of these planar frames and wall with the elements framing perpendicular to them is neglected.

To carry out these analyses two different static lateral loading patterns were assumed: inverted triangular (as specified by UBC and ATC) and rectangular.

Limit Analyses. The two limit analyses resulted in a predicted axial-flexural strength significantly higher than the minimum required strength specified by codes (more than 1.8 times), confirming the expected overstrength indicated by other studies. Reasons for this overstrength are given in references 2, 3, 10 and 12. As expected, the total lateral load resistance (base shear) is higher for the case of a rectangular load distribution.

In Figs. 3 and 4 the values of the minimum base shear demanded from the shear wall alone are also given (converted into required nominal strength).

Step-by-Step Nonlinear Analyses. Charney and Bertero [13] calculated the nonlinear responses shown in Fig. 5 using a step-by-step method. When a triangular distribution of lateral force was assumed, the static maximum base shear was 664 kips (0.27W); where a rectangular distribution of lateral force was assumed, the static maximum base shear was 819 kips (0.33W).

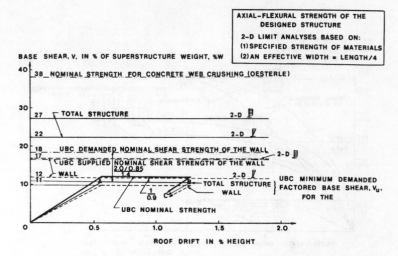

FIG. 3 - COMPARISON OF THE 1979 UBC MINIMUM DEMANDED STIFFNESS, FACTORED
BASE SHEAR, AND NOMINAL STRENGTHS ASSUMING K=0.8 VS. THE PREDICTED
NOMINAL STRENGTHS BASED ON 2D LIMIT ANALYSES OF THE DESIGNED TEST
STRUCTURE OF FIG. 2.

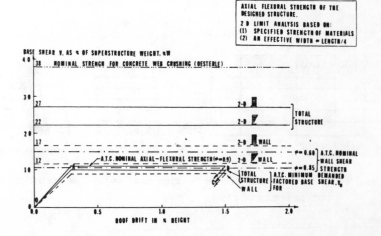

FIG. 4 - COMPARISON OF THE ATC3-06 MINIMUM DEMANDED STIFFNESS, FACTORED BASE
SHEAR, AND NOMINAL STRENGTHS VS. THE PREDICTED NOMINAL STRENGTHS BASED
ON 2D LIMIT ANALYSES OF THE DESIGNED TEST STRUCTURE OF FIG. 2.

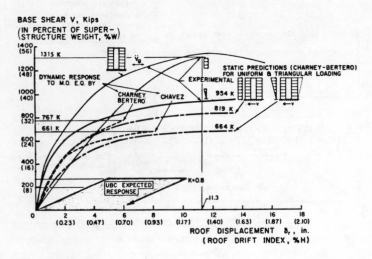

FIG. 5 - COMPARISON OF UBC MINIMUM DEMANDED RESPONSE WITH ANALYTICALLY PREDICTED
         AND EXPERIMENTALLY OBTAINED BEHAVIORS.

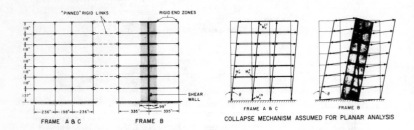

FIG. 6 - ANALYTICAL MODEL USED IN THE STRUCTURAL ANALYSES OF THE FRAME-WALL
         TEST STRUCTURE OF FIG. 2.

Dynamic Analysis. From the nonlinear dynamic step-by-step analysis conducted using the computer program DRAIN-2D, the maximum base shear for the N-S component of the Miyaki-Oki (MO) earthquake record was computed to be 767 kips (0.30W) (Fig. 5). By a different analysis procedure, Chavez [14] predicted a maximum base shear of 661 kips (0.26W) for the MO record normalized to a peak acceleration of 0.36g.

Measured Response

The full-scale model of the building was subjected to two series of pseudo-dynamic tests in Japan [2,15]. In the first series of tests, the model was considered as a single-degree-of-freedom system and tested under an inverted triangular distribution of load. The envelope of hysteretic behavior is shown in Fig. 5. The maximum base shear was 954 kips (0.38W), i.e., three-and-one-half times greater than the UBC-expected lateral resistance as controlled by flexural yielding. This value was 44% and 25% larger than the value predicted analytically by Chavez and by Charney and Bertero, respectively, and clearly indicates that the state of the art in predicting inelastic damage behavior should be improved.

The full-scale model did not fail during the pseudo-dynamic test in which the inverted triangular lateral load distribution was used, although the displacement ductility considering initial yielding of the wall was more than 6 and the effective displacement ductility of the structure as a whole was close to 4. This high observed ductility arises primarily because under the inverted triangular load distribution, the wall and the beams and columns of the frame were subjected to small shear stress. The wall was designed for a unit nominal shear stress, v_u, smaller than $5\sqrt{f_c'(psi)}$ and at the maximum measured V = 954 kips, v_u was smaller than $12\sqrt{f_c'(psi)}$. The experimental results confirm the authors' previous conclusions: If by proper conceptual design it is possible to keep shear stresses low, RC shear wall structures can offer sufficient dissipation of energy (ductility and stable hysteretic behavior) through flexural yielding to resist even the effects of extreme ground shaking.

After being subjected to the first series of tests with the inverted triangular distribution of lateral load, the full-scale model was repaired. 'Nonstructural' spandrel walls and partitions were also added in all stories except the first. The model was then subjected to a series of tests similar to the first series. Finally, the model was loaded statically using a rectangular distribution of lateral load. As illustrated in Fig. 5, the structure under this load was able to resist a maximum base shear of 1315 kips (0.53W) or 4.8 times the UBC-required design ultimate lateral strength. However, when the base shear of 1315 kips was reached at a lateral displacement of 11.3 in., the wall failed in shear at the first story with a sudden drop in resistance. Note that this displacement was about 16% smaller than that obtained in the first series of tests without failure. Comparison of results from the two test series indicates the importance of the type of loading and of the added 'nonstructural' elements to the behavior of frame-wall structures. While resistance may be considerably increased, ductility or deformability may be jeopardized.

The 1/5th scale model of the test building was subjected to several series of tests at the Earthquake Simulator facility at Berkeley, and in Fig. 5 the force-deformation envelope obtained from these tests is compared with those obtained pseudo-dynamically on the full-scale model. As can be seen, the envelope obtained from the 1/5th scale model shows a significant increase in strength when compared with results obtained for the full-scale model using a triangular distribution of lateral force. Although the failure of this 1/5 model was triggered by axial-flexural yielding which led to buckling of the reinforcing bars in the edge of the walls, there were clear indications that the concrete in the web of the wall had started to crush.

Correlation of Analysis and Experiment

Comparison of results presented in Figs. 3-5 clearly shows that the measured responses are not only significantly higher than those expected from the minimum requirements of the UBC and ATC provisions, but that they also are higher than the analytically predicted response of the designed test structure using presently available techniques, i.e., according to the state-of-the-art. The reasons for the discrepancies observed between the analytically predicted and the measured responses are discussed in detail in references 2, 3, 10 and 12, and arise primarily from the way in which frame-wall structures are modelled, from underestimation of the contribution of the floor system to the lateral stiffness and strength of the total structure, and from difficulties of predicting the actual mechanical characteristics of the structure at the time when it was subjected to the earthquake shaking.

Analytical Response Predictions (After Testing)

After the experiments had been performed, it was possible to improve the mathematical model of the test structure, taking account of the observed 3D behavior and making use of material characteristics measured at the time of testing the structure. The improved model developed by Kabeyasawa, et al. is described in Reference 16; the results of the static limit analyses carried out with this mathematical model [17] are presented in Fig. 7 and are discussed briefly here. The two static lateral load patterns mentioned earlier (inverted triangular and rectangular) were used in these analyses, in addition to the force distribution measured in the experiments at the time of the maximum response. The strengths predicted for the total test structure and for the wall alone for each of the assumed load distributions are plotted in Fig. 7; also shown are the measured response and the behavior demanded according to the minimum requirements of UBC and ATC.

From a comparison of the results shown in Fig. 7, it is clear that the predicted strength, based on the measured distribution of inertia force along the height of the structure, correlates very well with the measured strength for the total structure as well as for the wall alone. While a 3D limit analysis using a rectangular distribution of forces resulted in some overestimation of the axial-flexural strength (capacity), which will lead to a conservative estimate of the demanded shear resistance, the analysis based on an inverted triangle loading pattern as specified by present U.S. seismic codes led to an under-

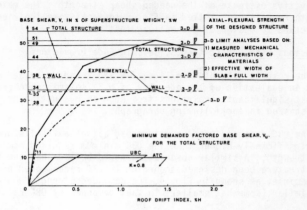

FIG. 7 – COMPARISON OF THE 1979 UBC AND THE ATC3-06 MINIMUM DEMANDED RESPONSES
WITH THE EXPERIMENTALLY OBTAINED BEHAVIOR AND WITH THE ESTIMATED NOMINAL
STRENGTHS BASED ON 3D LIMIT ANALYSES OF THE CONSTRUCTED AND TESTED
STRUCTURE OF FIG. 2.

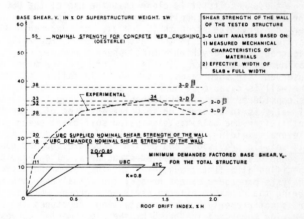

FIG. 8 – COMPARISON OF THE 1979 UBC MINIMUM DEMANDED NOMINAL SHEAR STRENGTH OF
THE WALL WITH THE EXPERIMENTALLY OBTAINED RESISTANCE, WITH THE ESTIMATED
DEMANDED NOMINAL STRENGTHS BASED ON 3D AND 2D LIMIT ANALYSES OF THE TESTED
STRUCTURE OF FIG. 2, AND WITH THE PREDICTED NOMINAL CONCRETE WEB CRUSHING
OF THE WALL.

estimation of the axial-flexural capacity, which could produce a nonconservative estimate of the needed shear strength. The main shear
strength results are plotted in Fig. 8 to facilitate their evaluation
with regard to design of the wall against shear.

Conclusions with Regard to U.S. Seismic Code Provisions

The results presented in Figs. 3-5 and 7 and described above
permit an evaluation of the adequacy of U.S. seismic code provisions.
The most significant conclusions from such an evaluation are presented
and discussed in the following paragraphs.

(1) The UBC designed test structure had actual axial-flexural strength
    significantly higher than the UBC minimum required nominal
    strength. A similar observation could be made had the test
    structure been designed according to ATC recommended provisions
    because, as shown in Fig. 4, the demanded strength would be
    similar (somewhat smaller) to that demanded by UBC requirements.

(2) The shear force developed at the critical region of the wall,
    when its axial-flexural yielding capacity was reached, was signi-
    ficantly higher than that for which the wall was designed (0.34W
    measured vs. 0.18W nominal UBC shear strength). Note that the
    supplied nominal shear strength estimated according to UBC
    expressions is practically the same as the demanded value (Fig. 8).
    From these results it is clear that the use of the UBC expressions
    for estimating the supplied nominal shear capacity resulted in a
    very conservative strength (0.20W vs. the measured 0.34W) for the
    particular wall used in the test structure. For walls that are
    designed using very high nominal shear stresses, close to the
    maximum allowed ($10\sqrt{f'_c}$), this type of behavior might not be
    observed. In such cases, a semi-brittle failure may occur before
    the wall is capable of supplying the required deformations. If it
    is considered that the philosophy for earthquake resistant design
    of walls is to suppress or to delay any type of shear failure
    until after the wall has dissipated sufficient energy through
    flexural yielding, it can be concluded that the UBC as well as
    ATC seismic provisions and design procedures do not guarantee
    a sound design of frame-walled structures. The seismic forces are
    specified at a fictitious level and the recommended linear elastic
    analysis procedure cannot predict the actual response developed
    during the probable extreme earthquake ground motion. The design-
    er is not properly guided in detailing structures for actual
    behavior (depicted in Fig. 5 by the experimental curves), because
    he/she is required to make computations, and therefore to visual-
    ize, only seismic behavior according to the code-conceived linear
    elastic response represented in Figs. 3 and 4 by the line OA. The
    inadequacy of UBC code-procedure is illustrated in Fig. 9.

(3) The failure of the full-scale model when subjected to the rectan-
    gular lateral load distribution, and the clear signs of the
    initiation of concrete crushing at the maximum resistance of the
    1/5-scale model, suggest that the expressions for estimating web
    crushing of RC structural walls presented in reference 18 could be
    quite nonconservative (0.55W vs. 0.34W as indicated in Fig. 8)

when walls are subjected to severe interstory drift (about 2%) and several cycles of full reversals, as was the case in the experiments conducted on the 1/5 scale model.

(4) It is conservative to estimate the required shear strength of the wall on the basis of the axial-flexural-capacity of the structure at the moment that the earthquake occurs. This capacity can be estimated through a static 3D limit analysis using a rectangular distribution of the lateral forces and based on the full contribution of the floor system and on the probable mechanical characteristics of the structural material at the time of the severe earthquake (Fig. 8). The use of a 2D analysis for predicting such axial flexural capacity can be somewhat nonconservative for determining the required shear strength (Fig. 8).

EVALUATION OF RESPONSE REDUCTION FACTORS

Variation with Building Period

Correlation of the observed performance of the test structure when subjected to known base motions with the code specified seismic strength provisions made possible a critical evaluation of the response reduction factors R or R_w used in defining the design force levels. The ATC code proposal [5] implies that the reduction factor is intended to account for the ductility inherent in the structural system. However, as noted earlier it is hard to justify the use of a constant reduction factor over the whole range of periods of vibration that a particular structural system might have. Studies reported in References 19-22 show that the reduction used to derive an inelastic design spectrum from a linear elastic spectrum is function of the damping, ductility and other resistance characteristics of the structure. For any specific values of these properties the reduction varies with the period of the structure--decreasing as T decreases. For a period less than 0.5 sec., the reduction may be half (or even one-fourth) as much as it would be for a period greater than 4 seconds--depending on the degree of ductility developed. The extent to which the reduction varies with period generally increases with increased ductility.

From these comments it is evident that the recommendation of a constant reduction factor (independent of T) cannot be justified by ductility considerations alone. For short period structures, in particular, the recommended reductions appear to be too high if the designer intends to provide only the code required strength. Fortunately, as mentioned earlier, the code design typically results in significantly greater strength than is indicated by the code forces; the implications of this overstrength are summarized below.

To facilitate discussion of the implications of the test results on the response reduction factor R, the data presented in Figs. 3-8 are replotted in Figs. 10-12 in the form of pseudo-acceleration spectra.

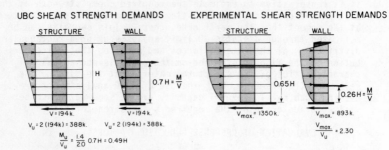

FIG. 9 - COMPARISON OF THE 1979 UBC MINIMUM DEMANDED NOMINAL SHEAR STRENGTH WITH
THE EXPERIMENTALLY DETERMINED DEMANDS FOR THE TOTAL STRUCTURE AND FOR THE
WALL OF FIG. 2.

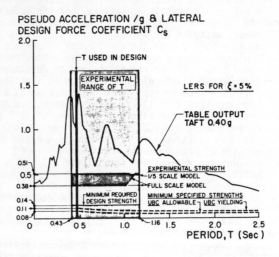

FIG. 10 - COMPARISON OF UBC MINIMUM SPECIFIED LATERAL SEISMIC STRENGTHS WITH
REQUIRED STRENGTH FOR A 5% DAMPED LINEAR ELASTIC RESPONSE TO THE
SHAKING TABLE MOTION AND WITH MEASURED STRENGTHS FOR THE US-JAPAN
7-STORY RC STRUCTURE OF FIG. 2.

Measured Period of Test Structure

It is shown in Figs. 10-12 that the actual measured period of vibration of the test structure varied significantly during the test programs. Although the period considered in its design was 0.48 sec. at the start of testing it was only 0.43 sec; however, after a series of tests at the service limit states it increased to about 0.61 sec (a nearly 50% increase due to the effect of cracking under service loads). After a series of tests in the damageability limit states and just before the final test, the period was about 0.90 sec; at the end of the test program T was 1.16 sec.

Shaking Table Excitation

The shaking table input motion used in the final test to failure was the recorded Taft earthquake ground motion normalized to 0.40g with some modifications. The 5% damped pseudo acceleration spectrum of the shaking table motions is shown in Fig. 10. As can be seen the effective peak acceleration, EPA, does not appear to be sufficiently well defined in the range 0.1 to 0.5 sec to obtain a reliable value according to the procedure suggested by ATC [5]. For values of T around the initial period of the test structure (i.e., about 0.5 sec) the EPA seems to be higher than 0.40g if the ATC procedure is applied. From Fig. 10 it is also obvious that if it had been required to keep the structure in the linear elastic range, it would have been necessary to design it for a lateral seismic design force coefficient, C_s, larger than 1.00. The structure actually was designed according to the UBC and the UBC required minimum yielding strength is equivalent to $(C_s)_y = 0.11$. However, when it is recognized that UBC requires the wall alone to resist the entire code specified lateral force and also requires the Ductile Moment-Resisting Space Frame (DMRSF) to resist at least 25% of the required lateral force, then the combined minimum required design strength is equivalent to $(C_s)_y = 0.14$. This corresponds to a reduction of the 5% damped LERS of the shaking table motions by a factor of more than 8 (Fig. 10) and a reduction of about 7 with respect to the 5% damped LEDRS recommended by ATC (Fig. 11).

The results of the experiments show that the first significant yielding of the wall occurred under a $(C_s)_y \doteq 0.18$ (see Figs. 5 and 7), but that the maximum strength was $(C_s)_{max} \doteq 0.51$(Fig.10).This confirms what was mentioned previously, that "for a well-designed structure the effective yielding of the structure occurs at a significantly higher load than the minimum code required value (0.18/0.11 = 1.64 or 0.18/0.14 = 1.29), and that the actual maximum strength shows a considerable overstrength beyond the first effective yielding (0.51/0.18 = 2.83). This is really a blessing because it allows UBC designed structures to withstand safely the effects of the Maximum Credible Earthquake (MCEQ) shaking. If the actual maximum strength of the designed structure was only the minimum required by UBC,$(C_s)_{max} = 0.14$, it is doubtful that its performance under the Taft 0.40g motion would have been acceptable. To dissipate the input energy from the Taft 0.40g motion it would have been necessary that the interstory drift ratio of a structure with a $(C_s)_{max} = 0.14$ be significantly higher than the 1.7% maximum observed during the tests; this already is larger than the maximum value of 1.5% recommended by ATC.

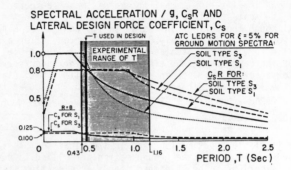

FIG. 11 - COMPARISON OF ATC 5% DAMPED PSEUDO-ACCELERATION SPECTRA FOR FREE
FIELD GROUND MOTIONS AND THE LATERAL SEISMIC DESIGN COEFFICIENTS
FOR THE US-JAPAN 7-STORY RC TEST STRUCTURE OF FIG. 2, R=8.

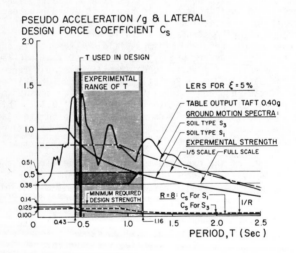

FIG. 12 - COMPARISON OF ATC MINIMUM REQUIRED DESIGN STRENGTHS, THE USED DESIGN
STRENGTH, THE ATC 5% DAMPED LEDRS, THE 5% DAMPED LERS FOR THE SHAKING
TABLE MOTION AND THE MEASURED STRENGTHS.

Figure 11 illustrates the relations between the ATC minimum recommended lateral design force coefficients C_s and the 5% damped LERS from which they have been derived when applied to the 7-story RC test structure for a site located in a seismic region like San Francisco. It is clear that if the structure had been designed according to the C_s recommended by ATC, its $(C_s)_y$ would have been < 0.125. In fact,

$$(C_s)_y = \frac{0.48 \times 1}{8 \times 0.48^{2/3}} \doteq 0.10,$$

which is significantly lower than the UBC required combined $(C_s)_y$ = 0.14. The results plotted in Fig. 12 allow a comparison of the required yielding strength if the structure were to remain linear elastic and damped with ξ = 5% during the actual shaking table motions: with (a) the spectral shape used by ATC for sites when A_a and A_v are 0.40 and the corresponding seismic design coefficients using the recommended R = 8 for the dual system of the test structure; with (b) the yield strength for which the structure was designed, $(C_s)_y$ = 0.14; and with (c) the measured strength, $(C_s)_{max}$ = 0.51. From Fig. 12 it can be seen that if the designed test structure were built on rock or stiff soil (soil profile type S_1) the maximum strength of this structure $(C_s)_{max}$ = 0.51 attained in the final test (where the effective period T of the structure was $\doteq$ 1 sec) would have been equal to the yield strength required by the ATC 5% damped LERS. In spite of this the maximum measured interstory drift ratio (1.7%) exceeded the maximum value recommended by ATC (1.5%).

From the results presented in Fig. 12 it can also be seen that the actual value of R, called R_E, of the structure would have varied depending on the state of the structure (cumulative damage) at the moment that the table output (simulating Taft 0.40g) occurred. If this ground motion had occurred just after construction, T $\doteq$ 0.50 sec, then R_E could have been as large as $\frac{1.4}{0.51} \doteq 2.7$. On the other hand if it had occurred after the structure had already been damaged by moderate earthquake shaking and/or strong winds, so the value of T was increased to about 0.8 sec, then R_E could have been as low as 1.5, and even lower if the damage to the structure had increased T to a value of about 1.0 sec.

Code Spectrum Input

If the 5% LERS specified by ATC for Soil S_3 type is considered instead of the 5% damped LERS of the table motions, R_E is about 1.6 for periods up to about 0.9 sec. It should be noted that the R_E values given above are the effective values for which the 5% LERS could be reduced to attain the "actual maximum shear resistance" that the structure has to possess in order to have good performance when subjected to ground shaking having a 5% damped pseudo acceleration similar to that specified by ATC or to the one corresponding to the shaking table motions for 0.40g Taft.

When the above values of R_E are compared with the value of R recommended by ATC (8 for dual systems based on RC shear wall and RC DMRSF), it is clear that it is not wise to try to provide the structure with just the ATC minimum required significant yielding strength. It is necessary to assure that the resulting ATC design has an actual maximum strength equal to $\frac{R}{R_E}$, i.e., about $\frac{8}{1.6} = 5$ times the ATC minimum required significant yield level. Clearly, therefore, a designer who uses ATC (or UBC) should check that a design he has made based on the minimum specified code forces has a maximum strength of about 5 times that demanded by these code factors. Without such a check, a design based on just the specified code forces might result in a poor earthquake performance. To summarize, it can be concluded that it is very difficult to rationalize (justify) quantitatively the values recommended for R in ATC [5]. The use of the value R = 8 in the design of the RC frame-wall dual system without any other require- ments is not reliable; it should be tied with other conditions. In the present ATC recommendation [5], it is tied with the stringent requirements for detailing RC DMRSF members and structural walls of Appendix A of ACI 318-83 [23]. However, the authors believe that this is not enough. Therefore, it is suggested that the preliminary design obtained using the ATC recommendations (or UBC specifications) should be subjected to a limit analysis for an estimation of its actual maximum resistance to verify that this resistance is about 5 times, say the minimum yielding strength required by ATC (the actual value to be used should be investigated further). In addition, the design (sizing and detailing) against shear of the wall (as well as the DMRSF members) should be based on this maximum resistance.

The SEAONC Seismology Committee has suggested the reduction factor $R_W = 10$ for dual systems composed of RC frames and shear walls. Studies and comparisons similar to those discussed above for the value of R indicate that although the SEAONC factor is a little more conservative than ATC ($\frac{1.4}{10} > \frac{1}{8}$), its value is still significantly higher than the R_E value obtained experimentally (Fig. 12). Therefore, the same observations as made for R apply in the case of R_W.

CONCLUDING COMMENTS

The field of earthquake engineering demonstrates dramatically the importance of maintaining close ties between analysis and experiment. The design process requires that a reasonable prediction be made of the inelastic response behavior of a structure subjected to a critical design earthquake, but only by tests of similar types of structures can the actual nonlinear performance be determined. Developing the mathematical model to be used in a nonlinear earthquake response analysis requires test data from components subjected to cyclic de- formations at the damage level, and the validity of the resulting non- linear analysis program should similarly be demonstrated by correlation of predicted performance with response to actual or simulated earthquake. Regardless of the size and speed of the computer used to carry out the analysis or the refinement and number of degrees of freedom considered in the mathematical model, the analytical results can never be better than the basic assumptions made in formulating the analysis, and the ultimate test of the analysis must be its ability to predict the true behavior of the real physical system under consideration. All of the

preceding discussion leads to the conclusion that the more sophisticated the analyses becomes the more important it is to provide experimental checks of its performance.

ACKNOWLEDGMENT

The U.S.-Japan Cooperative Research discussed here was supported
by the National Science Foundation with Grant Nos. CEE80-09478 and
CEE83-20587. However, the opinions, discussions, findings, conclusions
and recommendations are those of the authors.

The valuable contributions to this work made by Dr. A. Aktan and
by graduate students F. Charney, R. Sause, and S. Moazzami are greatly
appreciated by the authors.

REFERENCES

[1] U.S.-Japan Planning Group, Cooperative Research Program Utilizing
Large-Scale Testing Facilities, "Recommendations for a U.S.-Japan
Cooperative Research Program Utilizing Large-Scale Testing
Facilities," Report No. UCB/EERC-79/26, Earthquake Engineering
Research Center, University of California, Berkeley (1979).

[2] "Earthquake Effects on Reinforced Concrete Structures: U.S.-Japan
Research," J.K. Wight, Editor, Publication SP-84, American
Concrete Institute, Detroit, Michigan (1985).

[3] Bertero, V.V. et al., "U.S.-Japan Cooperative Earthquake Research
Program: Earthquake Simulation Tests and Associated Studies of a
1/5th-Scale Model of a 7-Story Reinforced Concrete Test Structure,"
Report No. UCB/EERC-84/05, Earthquake Engineering Research Center,
University of California, Berkeley, June 1984.

[4] Structural Engineers Association of California, "Recommended
Lateral Force Requirements and Commentary," Seismology Committee,
San Francisco, California.

[5] Building Seismic Safety Council "NEHRP Recommended Provisions for
the Development of Seismic Regulation for New Buildings," Washington, D.C., 1984.

[6] Bertero, V.V., "State of the Art in the Seismic Resistant Construction of Structures," Proceedings of the Third International
Earthquake Microzonation Conference, University of Washington,
Seattle, Washington, USA, June 28-July 1, 1982, Volume II, pp.767-
808.

[7] Brokken, S.T. and Bertero, V.V., "Studies on the Effects of Infills
in Seismic Resistant Construction," Report No. UCB/EERC-81/12,
Earthquake Engineering Research Center, University of California,
Berkeley, October 1981.

[8] Bertero, V.V., "Seismic Performance of Reinforced Concrete
Structures," Anales de la Academia Nacional de Ciencias Exactas,
Fisicas y Naturales, Buenos Aires, Argentina 1978, pp.75-145.

[9]   Uniform Building Code, International Conference on Building
      Officials, Whittier, California, 1982 Edition.

[10] Bertero, V.V., "Implications of Recent Research Results on Pre-
      sent Methods for Seismic-Resistant Design of R/C Frame-Wall
      Building Structures," Proceedings of the 51st Annual Convention
      of SEAOC, Sacramento, Calif., 1982, pp.79-116.

[11] Cartin, J.F., "Performance of School Canopies During the 10th
      October 1980 El-Asnam, Algeria Earthquake," Proceedings of the
      8th World Conference on Earthquake Engineering, San Francisco,
      Calif., 1984, Volume IV, pp.775-782, EERI, Berkeley, California.

[12] Bertero, V.V., "State of the Art and Practice in Seismic Resistant
      Design of RC Frame-Wall Structural Systems," Proceedings of the
      8th World Conference on Earthquake Engineering, San Francisco,
      Calif., 1984, Volume V, pp.613-620, EERI, Berkeley, California.

[13] Charney, F.A. and Bertero, V.V., "An Evaluation of the Design
      and Analytical Seismic Response of a Seven-Story Reinforced
      Concrete Frame-Wall Structure," Report No. UCB/EERC-82/08,
      Earthquake Engineering Research Center, University of California,
      Berkeley (1982).

[14] Chavez, J.W., "Study of the Seismic Behavior of Two-Dimensional
      Frame Buildings: A Computer Program for the Dynamic Analysis--
      INDRA," Bulletin of the Int'l. Inst. of Seismology & Earthq.
      Engrg., Vol. 18, BRI, Tsukuba, Japan (1980).

[15] Okamoto, S. et al., "A Progress Report on the Full Scale Seismic
      Experiment of a Seven-Story R/C Building," Proceedings, 3rd
      Joint Coordinating Comm. Mtg. of the U.S.-Japan Cooperative
      Research Program, BRI, Tsukuba, Japan (1982).

[16] Kabeyasawa, T. et al., "Analysis of the Full Scale Seven-Story
      Reinforced Concrete Test Structure: Test PSD3," Proceedings, 3rd
      Joint Coordinating Comm. Mtg. of the U.S.-Japan Cooperative
      Research Program, BRI, Tsukuba, Japan (1982).

[17] Moazzami, S., "Limit Analyses of a 7-Story Reinforced Concrete
      Frame-Wall Structure," Individual Study Report, SESM Division,
      Civil Engineering Department of the University of California,
      Berkeley, California, October 1984.

[18] Oesterle, R.G. et al., "Web Crushing of Reinforced Concrete
      Structural Walls," ACI Journal, May-June 1984, pp.231-241.

[19] Bertero, V.V. et al., "Establishment of Design Earthquakes -
      Evaluation of Present Methods," Proceedings, International
      Symposium on Earthquake Structural Engineering, St. Louis,
      August 1976, pp.551-580.

[20] Mahin, S.A. and Bertero, V.V., "An Evaluation of Inelastic Seismic
      Design Spectra," Journal Str. Div., ASCE, Vol. 107, No. ST9,
      September 1981, pp.1777-1795.

[21] Newmark, N.M. and Hall, W.J., "Earthquake Spectra and Design," Earthquake Engineering Research Institute, Berkeley, Calif., 1982.

[22] Newmark, N.M. and Riddell, R., "Inelastic Spectra for Seismic Design," Proceedings, 7th World Conference on Earthquake Engineering, Istanbul, Turkey, 1980, Vol. 4, pp.129-136.

[23] "Building Code Requirements for Reinforced Concrete," (ACI 318-83)," American Concrete Institute, Detroit, Michigan (1983).

[24] Argyris, J.H., "Energy Theorems and Structural Analysis," Aircraft Engineering 1954-55, reprinted by Butterworth's Scientific Publications, London, 1960.

Structural Dynamics in Hurricanes and Tornadoes

Kishor C. Mehta, F. ASCE* and
James R. McDonald, M. ASCE*

ABSTRACT

The random fluctuations of wind gusts produce dynamic interaction with structures. The objectives of this paper are to identify distinguishing differences in hurricane and tornado winds and to discuss issues relating to structural dynamics for these windstorms.

INTRODUCTION

Hurricanes and tornadoes cause extensive damage to structures each year. Although both are atmospheric vortices, there are significant differences in their wind characteristics. These characteristics affect structures in different ways. Hurricanes tend to be long duration storms because of the size of the hurricane and the slow moving translational speed of the storm. Tornadoes on the other hand are short-lived and rather narrow in extent. In relatively rare cases tornado winds can exceed 200 mph, whereas hurricane winds at landfall are typically less than 125 mph.

The random nature of wind gusts, their frequency and spatial distribution can cause dynamic interaction with structures. Traditionally, though, wind loads are treated as static loads on structures rather than dynamic loads. In this paper the characteristics of hurricane and tornado winds are described. It is not the purpose of the paper to discuss methods of dynamic wind analysis. Rather, the issues related to structural dynamics are discussed. The need for additional collection of data and research also are identified.

HURRICANES

Hurricanes are tropical storms that affect the Atlantic and Gulf Coast areas of the United States. During the period 1900-75, 126 hurricanes made landfall in U.S. coastal areas--an average of 1.7 hurricanes per year (Simpson and Reihl, 1981). A tropical storm becomes a hurricane when its sustained wind speed exceeds 73 mph. Hurricanes are large-scale vortices. The eye of a hurricane can be 10 to 50 miles in diameter, and winds of hurricane intensity can affect as much as 300 miles of coastline, as illustrated in Figure 1.

In this section, wind characteristics of hurricanes (or typhoons and cyclones) are reviewed. Issues that are related to structural dynamics because of hurricane wind effects are discussed.

* Professor of Civil Engineering, Institute for Disaster Research, Texas Tech University. P ^ B^x 4089, Lubbock, Texas 79409.

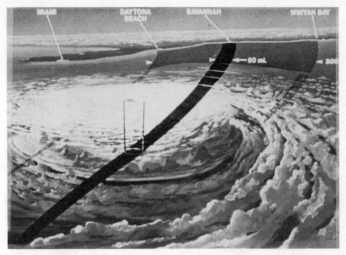

FIGURE 1. SPATIAL EXTENT OF HURRICANE STORM

Wind Characteristics

Measurements made by instruments located in the path of hurricanes provide the basis for understanding engineering related wind characteristics in the storm. Experiments are underway in Japan (Mitsuta, 1974; Ishizaki, 1983), Taiwan (Loh, 1985), and Hong Kong (Lam and Lam, 1985) where instruments are set up to record typhoon winds. In the United States, there are no organized land based experiments to record hurricane wind characteristics; one reason being the relatively low probability of a hurricane crossing a particular location on the U.S. coast. The Hurricane Research Center, NOAA, obtains wind data by flying hurricane hunter airplanes through the storms. These data are generally collected over water at altitudes of more than 1500 m. The data are more directly applicable to hurricane forecast rather than ascertaining the response of land based buildings and structures.

Strip chart records of wind speed and wind direction obtained at an industrial plant during passage of Hurricane Alicia over Galveston, Texas in 1983 is shown in Figure 2. The peak wind speed recorded was 94 mph. The fluctuating nature of wind and the persistent high intensity winds over several hours are evident in the Figure. The fluctuating nature of wind over a long period of time has implications on structural dynamics considerations. The associated wind direction trace in Hurricane Alicia is also shown in Figure 2. The traces cover a period in excess of two hours. The wind direction in a hurricane changes slowly over a period of time. However, wind direction fluctuations of up to 30 degrees over a short time period are evident in the trace.

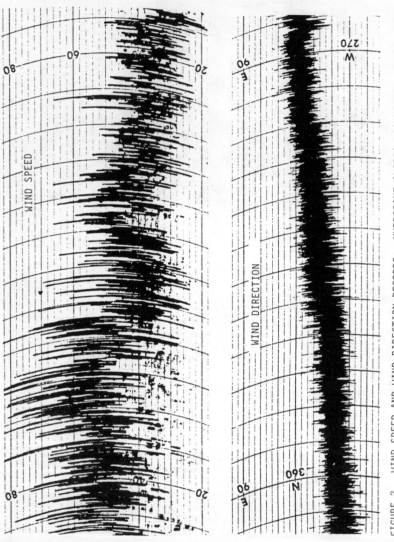

FIGURE 2.   WIND SPEED AND WIND DIRECTION RECORDS: HURRICANE ALICIA, GALVESTON, TEXAS, 1983

Wind characteristics are generally described by a vertical profile of mean wind, turbulence intensity and gust spectrum. These characteristics for non-hurricane wind are reasonably well established. For example, mean wind profiles are modeled by an exponential equation. The power-law exponent varies for different terrain conditions as suggested in the U.S. standard (ANSI, 1982). However, Ishizaki (1983) found that the wind characteristics vary significantly from one typhoon (hurricane) to another and their characteristics are different from extra-tropical winds. Variations in power-law exponents and turbulence intensity in three typhoons as measured by Ishizaki (1983) are shown in Figure 3. Using the ensemble of several typhoon records, Ishizaki proposed values for power-law exponent and turbulence intensity for design purposes as shown in Table 1. The current U.S. standard values for coastal areas, where wind is coming over water, are also shown in Table 1 for comparison. it is evident from the table that the values of the power-law exponent and turbulence intensity in hurricane winds can be quite different from those of non-hurricane winds.

TABLE 1. POWER LAW EXPONENTS AND TURBULENCE INTENSITY
VALUES FOR TYPHOONS (HURRICANES) AND OTHER WINDS

Recommended Parameters for Typhoons (Hurricanes), Coastal Areas, 10 m Height (Minor, 1984)			Parameters in Current U.S. Standard for Coastal Areas, 10 m Height (ANSI, 1982)	
Mean 10 Minute Wind Speed (m/s)	Power Law Exponent	Turbulence Intensity (I)	Power Law Exponent ($1/\alpha$)	Turbulence Intensity (T_z)
20	0.33	0.200	0.10	0.13
30	0.29	0.176		
40	0.27	0.163		
50	0.26	0.153		
60	0.24	0.147		

Wind gust spectra for typhoon winds obtained through measurements in Hong Kong are shown in Figure 4 (Lam and Lam, 1985). The spectra are of similar shape as that for non-hurricane winds; the magnitude of the peak varies at different heights above ground as indicated in Figure 4. It is important to note that the energy level of wind gusts for frequencies above 0.1 Hz is relatively low and that above 1 Hz is very low. These gust energy levels suggests that dynamic interaction between wind and structure is negligible if the fundamental frequency of the structure is above 1 Hz.

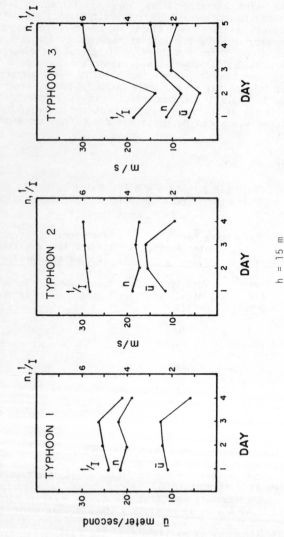

n = 1/power law exponent; I = turbulence intensity

h = 15 m

FIGURE 3. VARIATION IN POWER LAW EXPONENTS AND TURBULENCE INTENSITY IN THREE TYPHOONS (Minor, 1984)

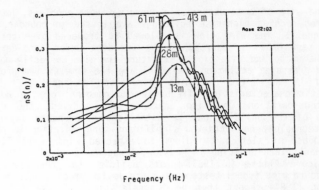

Frequency (Hz)

FIGURE 4. WIND GUST SPECTRA FOR TYPHOON
WINDS (Lam and Lam, 1985)

Structural Dynamics Issues in Hurricanes

Background in wind characteristics of turbulence intensity and gust spectrum provides guidance to the issues related to structural dynamics. In addition, spatial correlation of wind, wind induced vibrations, fatigue due to wind and impacts of windborne debris can lead to structural dynamics problems. These issues are discussed below.

Along Wind Response

As indicated in Figure 2, wind speed fluctuates randomly. For the purpose of an analytical model for structural dynamics, wind speed is considered to be a mean value averaged over a period of time (10 to 60 minutes) plus a component fluctuating about the mean value. Wind speeds in hurricanes and extra-tropical cyclones exhibit relatively steady state characteristics with at least a weakly stationary process. Wind speeds in thunderstorms and tornadoes are not steady state; the mean component of wind speed is likely to have an upward and/or downward trend. The steady state wind speed in hurricanes permits dynamic analysis using the fluctuating component. The mean component of wind can be treated as a static load.

Davenport (1961, 1962, 1963) established analytical procedures to assess dynamic response of a variety of structures using a statistical approach based on the concepts of the stationary random series for wind gusts. These procedures give peak structural responses by using wind speed characteristics of turbulence intensity, gust spectrum, and gust correlation in space and structural properties of natural frequency and damping. Difficulties with mean wind speed profile, turbulence intensity, gust spectrum and gust correlation in space are encountered when a structure is located in an urban environment. Surrounding terrain with man-made structures can alter the approaching wind to an extent where it is not possible to predict wind characteristics. In addition, wind pressures on a structure depend on the

shape and size of a structure. To overcome these problems, important
structures such as high-rise buildings are tested in boundary layer
wind tunnels. The wind flow is allowed to traverse over correctly
modeled terrain as it approaches the building under consideration.
Since the shape and size of the structure are also correctly modeled,
the wind pressures acting on the model are the same as the prototype
building in the field (assuming correction for scaling). If the
aeroelastic characteristics (frequency and damping) of the building
under consideration are also appropriately modeled, the dynamic
response of the building can be established. Boundary layer wind
tunnel technology has achieved significant sophistication to predict
the response of slender buildings to expected wind. The 71-story
Allied Bank Plaza building in Houston, Texas was exposed to the winds
of Hurricane Alicia in 1983. This building had been previously
subjected to wind tunnel tests during its design process. Isyumov and
Halvorson (1984) report that peak accelerations at the 71st floor
level obtained from the aeroelastic model in the wind tunnel and
full-scale measurements compared reasonably well.

Two significant points can be made from the above discussion.
Aeroelastic models based on the fundamental mode predict dynamic
response reasonably accurately. This means that higher modes of
vibration contribute relatively little to the total response of
slender buildings. The second point is that, though it is possible to
use wind tunnel technology to assess response of tall slender structures through appropriate physical modeling, it is not possible at
this time to assess analytically wind characteristics in urban areas
and the transfer function from wind speed to wind load. Thus, wind
characteristics and transfer function pose a major stumbling block to
development of analytical procedures for dynamic analysis of structures subjected to wind loads in urban areas.

Other Wind Induced Vibrations

Steady winds flowing over a structure can cause wind induced
vibrations that are described as vortex oscillations, galloping and
fluttering. These vibrations occur at relatively low wind speeds and
are dependent on characteristics of the structure. When steady winds
flow around a structure, vortices are detached from the structure at
regular intervals. The frequency at which the vortices are detached
depends on the Strouhal number for the characteristic cross section.
If the frequency of vibration of the structure matches the frequency
of vortex detachment, resonance develops in the structure and the
across wind response is amplified. The magnitude of forces generated
by vortex detachment is not yet well known. Hence, an analytical
procedure to ascertain across wind response is still in the development stage. Galloping is the term used to describe large amplitude
oscillations occurring in a direction normal to the wind at frequencies much lower than those of vortex shedding. The alternating
forces are created by the motion of the structure. When the motion
stops, the forces disappear. Fluttering is a high frequency vibration
of flexible, membrane-like surfaces in steady wind flowing across the
surface. The magnitude of displacement can be significant and can

cause damage to the structure. Hurricane winds have not been judged to be causing fluttering. Vortex oscillations, galloping and fluttering are generally not problems in hurricane winds.

Fatigue Problem

Fatigue can be a significant problem for flexible structures in the path of sustained hurricane winds. The extent of damage is related to the number cycles and stress levels produced by the hurricane winds. Building cladding failures and subsequent damage to low-rise buildings in Darwin, Australia was traced to low-cycle fatigue of the cladding material adjacent to the screw fastenings (Walker, et al., 1975). Cyclone Tracy traversed the city of Darwin in 1974 at a speed of 5 mph, exposing buildings to fluctuating winds for more than four hours. The maximum 3-second gust at 10 m above ground was 170 mph in the northern part of the city. Fatigue failures of the cladding elements led to virtual destruction of residential units as shown in Figure 5.

FIGURE 5.   DESTRUCTION OF RESIDENCES BY CYCLONE TRACY

Windborne Debris

Windborne debris is inevitable in hurricanes and has significant effects on buildings and structures in urban areas. Minor (1984) describes effects of windborne debris in five specific hurricane events: Hurricane Celia in 1970, Cyclone Tracy in 1974, Hurricane Frederic in 1979, Hurricane Allen in 1980, and Hurricane Alicia in 1983. Windborne debris was common in all of these events. In each case, windborne debris was found to be a significant building damage mechanism. Dynamic effects of windborne debris impacts have not been explored to any extent.

TORNADOES

Tornadoes are usually spawned by severe thunderstorms. The tornado itself is an atmospheric vortex which extends from within the cloud to the ground. The vortex is sometimes narrow and smooth in appearance, but more often is broad and turbulent. A cloud of condensed moisture (a funnel cloud) may be visible within the vortex, but the vortex (hence the tornado) is much larger in dimension and may extend well beyond the area of visible funnel.

Tornadoes form when vorticity is concentrated into a limited column beneath a thunderstorm. The concentration is produced by converging winds at low levels stimulated by rapid vertical motion in the thunderstorm. The vorticity may originate in the rotation of the thunderstorm and/or the wind difference (shears) between thunderstorm-generated outflow and environmental inflow.

During the past 30 years an average of 687 tornadoes were reported each year, although the values vary significantly from year to year. Tornadoes vary in both size and intensity--the large, intense ones occurring much less frequently than the small, weak ones. In this section, the characteristics of tornadoes that relate to structural damage are reviewed. Issues that are related to dynamic effects on buildings are discussed.

Tornado Winds

A complete understanding of tornado windfields is lacking because of the absence of accurate measurements of the winds. Attempts to measure tornado wind speeds have generally been frustrating. Chase teams from the University of Oklahoma have tried to deposit an indestructible instrument package in the paths of tornadoes, with only a small amount of success (Burgess et al., 1985). Doppler radar holds promise for measurements of tornado windfields. Under the right set of circumstances Doppler radar can detect circulation in the tornado wind flow patterns, but much more research and experience are needed. The nature of tornado wind fields has in general been deduced from indirect methods such as damage analyses, photogrammetric analyses of tornado movies and studies of cycloidal ground marks.

Tornado intensity is rated according to the F-Scale (Fujita, 1971) which has wind speeds associated with each of six damage categories that range from 0 to 5. The intensity of a tornado is determined from appearance of the worst damage observed along the path. The F-Scale has significant limitations in that the system does not take into account strengths and weaknesses of building construction. The absence of damage indicators in open country also make the F-Scale difficult to apply. The wind speed ranges associated with each classification have never been calibrated on a scientific basis. Despite these short shortcomings, the F-Scale, along with the Pearson Path Length and Path Width Scales, are the generally accepted means of characterizing the intensity and size of tornadoes.

On April 29, 1985 the University of Oklahoma chase team was able to place the instrument package in the path of a relatively weak tornado near Ardmore, Oklahoma. Figure 6 shows the wind speed trace. The instrument was placed at about 330 ft from the tornado path center-line. Wind gust of 60 mph were measured for the tornado, which was rated F2 on the F-Scale. The trace is characterized by rapid fluctu-ations about a running mean that can be visualized as varying with time. While this trace probably does not represent the maximum tornado wind speeds of the Ardmore tornado, it does give an indication of the nature of the wind speed fluctuations that occur in a tornado. Both gustiness and more general time variation of wind speed are observed. While analysis of the tornado wind record for turbulence intensity, gust spectrum and gust correlation are possible, only a very limited number of records are available at this time.

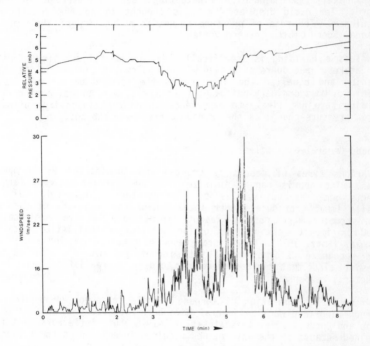

FIGURE 6.   TORNADO WIND SPEED AND ATMOSPHERIC PRESSURE
CHANGE TRACES: ARDMORE, OKLAHOMA (Burgess
et al., 1985)

Atmospheric Pressure Change

The cyclostrophic relationship, which states that the inward radial pressure gradient force and the outward centrifugal force balance one another is thought to be a good approximation outside the boundary layer. By assuming an appropriate variation of tangential velocity as a function of distance from the tornado center, the cyclostrophic equation can be integrated to obtain the total pressure change, which in general is a function of air density and maximum tangential wind speed.

The trace of atmospheric pressure change for the Ardmore, Oklahoma tornado is also shown in Figure 6. The time scale is the same as that for the wind speed trace. The total pressure change at the instrument location was 5.5 mb (0.01 psi), which occurred over a period of approximately two minutes. Fluctuations can be observed in the record. The rapid drop at 4 min. is thought to be the result of turbulence. The rate of pressure change is a function of the translational speed of the tornado vortex.

Unless a building is specifically sealed, natural porosity, fresh air intakes, open doors or broken windows allow air to flow out of the building and equalize the pressure as the tornado passes over the building. Most ordinary buildings (residences, warehouses, commercial buildings) are not likely to experience significant loads from atmospheric pressure change as the tornado passes over the building.

Tornado-Generated Missiles

Various types of debris is created and transported by tornado winds. The debris ranges in size from roof gravel to very large objects such as automobiles or storage tanks (McDonald, 1976). Missile impacts produce either local damage (perforation) or damage from structural response. Present impact formulas are empirical, based on impact test results or on adaption of ballistic missile formulas (Rotz, 1975). Structural response requires complex analyses that recognize deformation of both target and missile. Inelastic behavior also must be considered in order to realistically model structural response.

Attempts of simulation of tornado missile trajectories require a tornado windfield model. Several have been proposed (McDonald, 1975; Simiu and Cordes, 1976; Twisdale et al., 1978) but little agreement is obtained because of the assumptions required in developing the trajectory models.

Structural Dynamics Issues in Tornadoes

Unlike the hurricane winds discussed earlier, tornado winds change rapidly in both magnitude and direction. Gustiness created by both mechanical and atmospheric turbulence affect structures through size effects and frequency effects, as with hurricane winds. In addition the rapidly changing wind velocity with respect to time produces

acceleration effects on the structure that have received little attention from tornado researchers.

Wind Gusts

Knowledge of hurricane (and other "straight") winds have come from
wind measurements and extensive analyses of the collected data. These
types of measurements and analyses of data have not been made for
tornadoes and are not likely to be made in the near future.

Accelerating Winds

A rapid change in velocity of air particles with time results in
acceleration of the air particles and produces inertia forces on a
structure. Wen (1975) addressed this problem by adopting Morrison's
equation to inviscid tornado-like flow. Wen's results are questionable, because of an absence of experimental coefficients needed for
the equations. McDonald and Selvam (1985) applied the Boundary
Element method to the problem of a two-dimensional building immersed
in tornado-like flow and calculated inertia forces by assuming the
fluid to be inviscid. The study is a first step in the solution of a
much more complex problem that must include the effects of viscosity
and turbulence. analytical solutions of the more general problem will
require the use of supercomputers and could take 10 years to accomplish.

Pressure Coefficients

Local variations of wind pressures on structures due to drag
effects are determined in wind tunnels that are designed to model the
wind flow in the boundary layer. Pressure coefficients are then used
to determine both local and overall forces created by the winds on the
structure. Because it is not possible to model the changing directions of the tornado windfield in conventional wind tunnels,
pressure coefficients corresponding to tornado wind flow are not
available.

Laboratory models of tornado vortices have been developed (see
Davies-Jones, 1976 for a critical review of laboratory models), but
the scale of the models is so small that measurement of wind induced
pressures on buildings is not feasible.

Tornado Problem Discussion

From the above discussion two things are clear: there are definite
differences in hurricane and tornado winds and there is a lack of
knowledge of the characteristics of the tornado winds. These are the
primary reasons why only a limited number of dynamic analyses have
been performed on structures subjected to tornado winds. The works of
Wen (1975), Tan (1975), and Seniwongse (1977) all attempted to calculate tornado loads on tall buildings. Time histories of wind loads
are obtained by assuming a tornado windfield model and a translational
speed of the tornado vortex.

Observations of tornado damage to structures made after the Lubbock tornado (Mehta et al., 1970) and after additional investigations (Mehta et al., 1976) suggest that effects of tornado winds on structures and components are not radically different from that produced by nontornadic winds. The same types of damage is observed whether caused by straight winds or tornado winds. The effects of external pressures on windward, leeward and side walls can be seen in both tornado and straight wind damage. Likewise, local effects at wall corners, eaves, ridges and roof corners are observed in both types of storms. The appearance of damage is so similar that one cannot look at a picture of a damaged building and easily determine from the damage itself whether the damage was produced by a tornado or other type of windstorm. There indeed may be some differences in the characteristics of tornado and other straight winds, but their effects on structural damage are difficult, if not impossible to detect in the damage. Pressure coefficients obtained from wind tunnel tests on ordinary buildings appear to be adequate for determining upper bound tornado loads on structures for design purposes (see for example ANSI, 1982).

Effects of gustiness and inertia forces in tornado winds are accounted for with the following rationale: tornado wind speeds are estimated from damage to structures (either by applying the F-Scale or by analysis of the damage). The observed damage was caused by the wind, including the effects of gustiness and inertia pressures. In tornado-resistant design, tornado wind speeds are treated as gust speeds; gust-related factors used in determining wind pressures are taken as unit. While the approach may not be exact, it addresses the problem of gustiness and inertia pressures in an indirect manner.

CONCLUSIONS

The current state of knowledge leads to the following general conclusions regarding dynamic issues associated with hurricane and tornado winds.

·   The long-duration, fluctuating nature of hurricane winds can cause significant dynamic response of structures.

·   The gust spectrum in hurricanes appears to have little energy in the frequency range greater than 1 Hz. This suggests that dynamic interaction is minimal in structures with fundamental frequencies greater than 1 Hz.

·   Wind profiles and turbulence intensity appear to be different in hurricane winds than in non-hurricane winds.

·   A lack of knowledge of how wind characteristics are altered in urban areas has slowed theoretical development of a methodology for dynamic wind-structure interaction.

·   The transient nature of tornado winds acting on a structure for a short duration suggests that a different approach is required than for steady state winds.

- Atmospheric pressure change associated with tornadoes is not likely to induce dynamic loads on ordinary structures. Windborne debris may cause structural dynamic interaction.

- Until additional data on wind characteristics of both hurricane and tornado winds are collected and further research is performed, wind loads likely will continue to be treated as equivalent static loads on ordinary structures.

REFERENCES

ANSI, 1982, "Minimum Design Loads for Buildings and Other Structures," ANSI A58.1-1982, American National Standards Institute, New York, NY.

Burgess, D.W., Vasiloff, S.V., Davies-Jones, R.P., and Zrnic, D.S., 1985: "Recent NSSL Work on Windspeed Measurement in Tornadoes," Proceedings, Fifth U.S. National Conference on Wind Engineering, (Nov.6-8, 1985; Lubbock, Texas), Institute for Disaster Research, Texas Tech University, Lubbock, TX.

Davenport, A.G., 1961: "The Application of Statistical Concepts to the Wind Loading of Structures," Proceedings, Institute for Civil Engineers, Vol. 19, p. 449.

Davenport, A.G., 1962: "The Response of Slender Line-Like Structures to a Gusty Wind," Proceedings, Institute for Civil Engineers, Vol. 23, p. 389.

Davenport, A.G., 1963: "The Buffeting of Structures by Gusts," Proceedings, Symposium on Wind Effects on Buildings and Structures (June 26-28, 1963; Teddington, England), pp. 357-391.

Davies-Jones, R.P., 1976: "Laboratory Simulations of Tornadoes," Proceedings, Symposium on Tornadoes: Assessment of Knowledge and Implications for Man (June 1976; Lubbock, TX), Institute for Disaster Research, Texas Tech University, Lubbock, TX.

Fujita, T.T., 1971: "Proposed Characterization of Tornadoes and Hurricanes by Area and Intensity," SMRP No. 91, Satellite and Mesometeorology Research Project, The University of Chicago, Chicago, IL.

Ishizaki, H., 1983: "Wind Profiles, Turbulence Intensities and Gust Factors for Design in Typhoon-Prone Regions," Journal of Wind Engineering and Industrial Aerodynamics, Vol. 13, Nos. 1-3, pp. 55-66.

Isyumov, N. and Halvorson, R.A., 1984: "Dynamic Response of Allied Bank Plaza During Alicia," Hurricane Alicia: One Year Later, Proceedings of the Specialty Conference (August 1984; Galveston, TX), ASCE, New York, NY, pp. 98-115.

Lam, R.P. and Lam, L.C.H., 1985: "Wind Data for Typhoon-Resistant Design of Structures in the South China Sea Region," Proceedings, Fifth U.S. National Conference on Wind Engineering (Nov. 6-8, 1985; Lubbock, TX), Institute for Disaster Research, Texas Tech University, Lubbock, TX, pp. 1A-25 to 1A-32.

Loh, C-H., 1985: "Spectral Analysis of Typhoon Wind Data and Its Effects on Structures," Proceedings, Fifth U.S. National Conference on Wind Engineering (Nov. 6-8, 1985; Lubbock, TX), Institute for Disaster Research, Texas Tech University, Lubbock, TX, pp. 1A-33 to 1A-40.

McDonald, J.R., 1975: "Flight Characteristics of Tornado Generated Missiles," Institute for Disaster Research, Texas Tech University, Lubbock, Texas.

McDonald, J.R., 1976: "Tornado-Generated Missiles and Their Effects," Proceedings, Symposium on Tornadoes: Assessment of Knowledge and Implications for Man (June 1976; Lubbock, TX), Institute for Disaster Research, Texas Tech University, Lubbock, Texas.

McDonald, J.R. and Selvam, R.P., 1985: "Tornado Forces on Buildings Using the Boundary Element Method," Proceedings, Fifth U.S. National Conference on Wind Engineering (Nov. 6-8, 1985; Lubbock, TX), Institute for Disaster Research, Texas Tech University, Lubbock, TX.

Mehta, K.C., McDonald, J.R., Minor, J.E. and Sanger, A.J., 1970: "Response of a Structural System to the Lubbock Storm," SRR 03, Texas Tech University Storm Research Report, Lubbock, Texas.

Mehta, K.C., McDonald, J.R. and Minor, J.E., 1976: "Wind Speed Analysis of April 3-4, 1974 Tornadoes," Journal of the Structural Division, ASCE, Vol. 103, No. ST9, Proc. Paper 12429.

Minor, J.E., 1984: "Window Glass Performance and Hurricane Effects," Hurricane Alicia: One Year Later, Proceedings of the Specialty Conference (August 1984; Galveston, TX), ASCE, New York, NY, pp. 151-167.

Mitsuta, Y., 1974: "Preliminary Results of Typhoon Wind Observation at Tarama Island, Okinawa," Proceedings, Second U.S.A.-Japan Research Seminar on Wind Effects on Structures, Ishizaki and Chiu, Eds., University of Tokyo Press, Tokyo, Japan, pp. 27-37.

Rotz, J.V., 1975: "Summary of Missile Test Results on Concrete Panels," Vol. 1-A, Proceedings, ASCE Specialty Conference on Structural Design of Nuclear Plant Facilities (New Orleans, LA), ASCE, New York, NY.

Seniwongse, M.S.N., 1977: "Inelastic Response of Multistory Buildings to Tornadoes," a dissertation submitted in partial fulfillment of the requirements for the degree of Doctor of Philosophy, Texas Tech University, Lubbock, TX.

Simiu, E. and Cordes, M., 1976: "Tornado-Borne Missiles Speeds," NBSIR 76-1050, National Bureau of Standards, Washington, DC.

Simpson, R.H. and Reihl, H., 1981: The Hurricane and Its Impact, Louisiana State University Press, Baton Rouge, LA, 1981.

Tan, C.T., 1975: "Inelastic Response of High-Rise Buildings to Tornadoes," a thesis submitted in partial fulfillment of the requirements for the degree of Master of Science in Civil Engineering, Texas Tech University, Lubbock, TX.

Twisdale, L.A., Dunn, W.L. and Chu, J., 1978: "Tornado Missile Risk Analysis," EPRI NP-768, Final Report, Electric Power Research Institute, Palo Alto, California.

Walker, G.R., 1975: "Report on Cyclone Tracy--Effect on Buildings (3 Vols.)," Department of Housing and Construction, Commonwealth of Australia, Melbourne, Victoria (Australia), 1975.

Walker, G.R., Minor, J.E. and Marshall, R.D., 1975: "The Darwin Cyclone: Valuable Lesson in Structural Design," Civil Engineering, Vol. 45, No. 12, December, pp. 82-86.

Wen, Y.K., 1975: "Dynamic Tornadic Loads on Tall Buildings," Journal of the Structural Division, ASCE, Vol. 101, No. ST1, Proc. Paper 11045.

STOCHASTIC FINITE ELEMENT METHODS IN DYNAMICS

M. Shinozuka[1], M. ASCE and G. Dasgupta[2], M. ASCE

ABSTRACT

This paper develops a stochastic finite element solution method utilizing a Neumann expansion of the operator matrix involved and at the same time, devises an efficient Monte Carlo method consistent with the solution method. These analytical and Monte Carlo methods both utilize the successive nature of approximation. The methods demand inversion of the "average" operator matrix only once, achieve successive improvements of the solution by means of a stationary operator, and, unlike conventional perturbation formulations, require no evaluation of the partial derivatives of the operator matrix.

In this paper, the generic system deviator formulation, which is analagous to that of a static case, is illustrated in order to avoid lengthy algebraic expressions. The computational steps for natural extension to steady-state and transient response analyses are then indicated.

1. INTRODUCTION

Most modern structures possess a high degree of structural complexity. Therefore, when structural behavior is to be predicted under various loading and environmental conditions, advanced analytical and numerical techniques, notably the finite element method, supported by experimental collaboration, are required and are indeed extensively used at present. The current investigation evaluates the static and dynamic structural response variabilities when the structural complexity necessitates the use of computer codes such as those based on the finite element method for response calculation, and more specifically, when spatial variations and irregularities of the structural properties are significant to the extent that they must be treated as random functions of spatial coordinates. A typical example of this is the case in which the soil and the structure embedded in it act as a soil-structure interactive system under seismic conditions with the soil properties varying randomly in space, and thus, if a finite element

---

[1] Renwick Professor of Civil Engineering, Department of Civil Engineering and Engineering Mechanics, Columbia University, New York, NY 10027.

[2] Associate Professor, Department of Civil Engineering and Engineering Mechanics, Columbia University, New York, NY 10027.

code is used, a different set of values for the parameters representing the soil properties is assigned to each element.

Astill et al. [1], Shinozuka [10] and Shinozuka and Astill [13]
solved similar problems involving spatial variations of system parameters through a "straightforward" application of the Monte Carlo method, although the Monte Carlo analysis in Ref. 13 was performed for the
purpose of checking the perturbation analysis result derived therein.
These Monte Carlo solutions utilized a technique also developed by
Shinozuka and Jan [14], Shinozuka [11,12] and Yang [17,18] which permits the digital generation of sample functions of multidimensional
and/or multivariate random processes of spatial coordinates. Reference 1, for example, dealt with a stress wave propagating and reflecting within a randomly non-homogeneous cylinder of finite length in
which Young's modulus and the density of the cylinder material vary
along with the cylinder axis as a bivariate random process with a
specified cross-spectral density or equivalently cross-correlation
matrix. A set of one hundred cylinders were conceptually constructed
by generating one hundred sample pairs of Young's modulus and the density variations along the axis with the prescribed correlations. A
finite element analysis was performed for each of these sample cylinders to produce a set of one hundred patterns of the stress wave propagation from which it was possible to extract such useful statistics
as the mean value and standard deviation of the absolute maximum
stress within the cylinder. Such a "straightforward" Monte Carlo approach may not be practical in more complex real-life structural systems primarily because of the high computational costs that the approach usually entails. However, the approach is useful in the sense
that one could obtain valuable numerical insight into response variabilities within complex structural systems from the variabilities
evaluated for less complex systems with its use.

This paper represents an extended version of an earlier study by
Shinozuka and Nomoto [15]. Specifically, it develops a stochastic
finite element solution method utilizing a Neuman expansion of the
operator involved and devises an efficient Monte Carlo method consistent with the solution method. These analytical and Monte Carlo
methods both capitalize on the successive nature of approximation and
do not require the evaluation of the partial derivatives of the operator matrices. As such, the methods provide a highly competitive alternative to the perturbation method widely used for the stochastic
finite element solution by other researchers including Hisada and
Nakagiri (see, e.g., [5]) and Der Kiureghian [4]. The method also has
great potential for application to nonlinear dynamic problems (see for
example, Liu et al. [7]).

In the interest of brevity, only the theoretical development associated with the Neumann expansion method will be illustrated here.
Furthermore, employment of operational notation such as representation
of a time-dependent solution as $D^{-1}F$ is used, where D is the dynamic
system operator (see Dasgupta [3]) given by

$$D \equiv M \frac{d^2}{dt^2} + C \frac{d}{dt} + K \qquad (1a)$$

with M, C and K being the mass, damping and stiffness matrices, respectively. It is henceforth implied that the solution strategy for typical dynamic problems (in time as well as in frequency domains) is available to the potential user of the proposed method. In fact, it can be readily recognized that an important advantage of the Neumann expansion technique via the so-called delta method is that any available finite element computer code can be immediately employed without substantial modification, and straightforward convergence criteria exist based on the stationary nature of the operator involved. On the other hand, in the perturbation method, as in the so-called epsilon method, the derivative matrices are to be calculated in numerical form and rigorous convergence criteria are difficult to develop. Experience indicates that the evaluation of the derivatives of the element stiffness matrices, say for a shell element, with respect to the thickness parameter, may not be a nontrivial numerical chore.

2. SUCCESSIVE EXPANSION METHOD

2.1 The Neumann Expansion Method

Consider the following equation of equilibrium formulated by the standard finite element method:

$$Du = F \qquad\qquad or \qquad\qquad u = D^{-1}F \qquad\qquad (1b)$$

where D is the dynamic system stiffness matrix, u the displacement vector and F the externally applied force vector. Assume that D contains parameters that are subject to spatial variabilities, and then decompose D into two matrices,

$$D = D_0 + \Delta D \qquad\qquad\qquad (2)$$

where D_0 represents the stiffness matrix in which the spatially variable parameters are replaced by their mean values and ΔD (hence the name delta method) consists of the elements representing the deviatoric parts of the corresponding elements in D: indeed, ΔD can, in principle, be constructed by subtracting D_0 from D and its expected value $E\{\Delta D\} = 0$. In what follows, however, Young's modulus is assumed to be a parameter that varies randomly in space.

Using Eq. 2 in Eq. 1,

$$u = D_0^{-1}F - D_0^{-1}\Delta Du \qquad\qquad\qquad (3)$$

which can be modified into the following recursive expression:

$$u_i = u_0 - D_0^{-1}\Delta Du_{i-1} \qquad\qquad (i=1,2,\ldots.) \qquad (4)$$

with

$$u_0 = D_0^{-1}F \tag{5}$$

The repeated use of Eq. 4 produces

$$u = u_0 - D_0^{-1}\Delta D u_0 + D_0^{-1}\Delta D D_0^{-1}D\Delta u_0 - \cdots = u_0 - Pu_0 + P^2 u_0 - \cdots \tag{6}$$

where $P = D_0^{-1}\Delta D$ is a stable operator. The fact that successive improvements of the solution for u as shown in Eq. 6 can be made on the basis of such a stationary operator is a major advantage of the delta method. Equation 6 can also be derived directly from Eqs. 1 and 2, utilizing a Neuman expansion of the following form

$$\begin{aligned} D^{-1} &= \left(D_0 + \Delta D\right)^{-1} = \left(I + D_0^{-1}\Delta D\right)^{-1}D_0^{-1} \\ &= \left(I - D_0^{-1}\Delta D + D_0^{-1}\Delta D D_0^{-1}\Delta D - \cdots\right)D_0^{-1} \end{aligned} \tag{7}$$

where I is an identity matrix of appropriate dimensions and the characteristic values λ_i of $D_0^{-1}\Delta D$ are assumed to be such that $\left|\lambda_i\right| < 1$ for convergence.

In the following subsections (2.2 and 2.3), the formal expressions for the mean values and covariances of the various quantities of interest are given for the record within the framework of the delta method by truncating Eq. 6 after the first two terms (first-order approximation) and after the first three terms (second-order approximation). However, these expressions are not directly used in the present study involving dynamic problems. Instead, an intelligent Monte Carlo method developed at Columbia University is used on the strength of the Neumann expansion (Eq. 6).

2.2 First-Order Approximation

Truncating the right-hand side of Eq. 6 after the second term, the first-order approximation for the displacement is obtained as

$$u = u_0 - D_0^{-1}\Delta D u_0 \tag{8}$$

with mean value

$$E\{u\} = u_0 \tag{9}$$

and covariance matrix

$$\text{Cov}\{u,u\} = \mathbf{E}\{[u - \mathbf{E}\{u\}][u - \mathbf{E}\{u\}]^T\}$$

$$= \mathbf{E}\{(u - u_0)(u - u_0)^T\} = \mathbf{E}\{D_0^{-1} \Delta D u_0 u_0^T \Delta D^T (D_0^{-1})^T\} \tag{10}$$

The covariance matrix given in Eq. 10 can be evaluated since the correlations among the elements of ΔD are assumed to be known or prescribed. Obviously, the diagonal elements of Eq. 10 represent the variances of the displacement components. For further simplicity, the present study deals only with appropriate simplex elements.

Once the first-order displacement vector u_0 is evaluated, the relationship between the strain vector $\varepsilon^{(i)}$ and the displacement vector $u^{(i)}$ associated with a particular finite element i is given by

$$\varepsilon^{(i)} = B^{(i)} u^{(i)} \tag{11}$$

with $B^{(i)}$ being a constant matrix. Using Eq. 8 in Eq. 11,

$$\varepsilon^{(i)} = B^{(i)} \{u_0^{(i)} - (D_0^{-1} \Delta D u_0)^{(i)}\} \tag{12}$$

where $(D_0^{-1} \Delta D u_0)^{(i)}$ indicates a vector consisting of those components of the vector $D_0^{-1} \Delta D u_0$ that pertain to the computation of $\varepsilon^{(i)}$. Then the expected value of $\varepsilon^{(i)}$ is

$$\mathbf{E}\{\varepsilon^{(i)}\} = B^{(i)} u_0^{(i)} \tag{13}$$

while the variance of $\varepsilon^{(i)}$ is evaluated as the diagonal elements of the following covariance matrix

$$\text{Cov}\{\varepsilon^{(i)}, \varepsilon^{(i)}\} = \mathbf{E}\{[\varepsilon^{(i)} - \mathbf{E}\{\varepsilon^{(i)}\}][\varepsilon^{(i)} - \mathbf{E}\{\varepsilon^{(i)}\}]^T\}$$

$$= \mathbf{E}\{[B^{(i)}(D_0^{-1} \Delta D u_0)^{(i)}][B^{(i)}(D_0^{-1} \Delta D u_0)^{(i)}]^T\} \tag{14}$$

Consider that the spatial variation of Young's modulus is random in such a way that the randomness is isotropic and ergodic. It then follows that (1) the average value of the modulus is constant (E_0), whether or not it is taken as an ensemble average or a spatial average, and (2) the correlation between the values of the modulus at two arbitrary points depends only on their separation distance. For further analytical expedience, we assume that the value $E^{(i)}$ of the modulus in element i is uniform throughout the element, although it is random. The implication of this spatial discretization in terms of the finite element procedure adopted here is in essence to convert the random field of the modulus of elasticity into correlated random variables. In the present study, the value of the modulus at the centroid of element i is used for $E^{(i)}$ to be consistent with the assumed use of the simplex elements in performing the finite element analysis. The above assumption is reasonable if the size of each element is much smaller than the correlation distance. In general, the correlation distance indicates a separation distance over which the correlation of values of the random process at two corresponding points is still of some significance. Hence, the validity of such spatial discretization depends on the representative length of the finite element relative to the correlational distance. Further studies on such discretization appear to be warranted. Vanmarcke and Grigoriu shed some light on this issue [16].

The stress vector $\sigma^{(i)}$ in element i is

$$\sigma^{(i)} = E_0\left(1 + e_i\right)\tilde{D}B^{(i)}u^{(i)} \tag{15}$$

where e_i = the non-dimensional deviation of $E^{(i)}$ from E_0 or

$$E^{(i)} = E_0\left(1 + e_i\right) \tag{16}$$

Substituting Eq. 8 into Eq. 15 and disregarding the higher-order term $e_i\left(D_0^{-1}\Delta Du_0\right)^{(i)}$, one obtains the first-order approximation of $\sigma^{(i)}$ as

$$\sigma^{(i)} = E_0\tilde{D}B^{(i)}\left\{u_0^{(i)} - \left(D_0^{-1}\Delta Du_0\right)^{(i)} + e_i u_0^{(i)}\right\} \tag{17}$$

The matrix $\tilde{D}$ in Eqs. 15 and 17 indicates the "stress from strain" matrix for element i when it is multiplied by $E^{(i)}$ and involves only Poisson's ratio. Since Poisson's ratio is assumed to be deterministic and constant throughout the body, matrix $\tilde{D}$ does not depend on element i. The specific form of matrix $\tilde{D}$ for isotropic plane stress, plane strain, axisymmetric and three-dimensional problems can be found in, for example, Ref. 2. The expected value vector and covariance matrix of $\sigma^{(i)}$ are then given by, respectively,

$$\mathbf{E}\left\{\sigma^{(i)}\right\} = E_0\tilde{D}B^{(i)}u_0^{(i)} \tag{18}$$

and

$$\begin{aligned}
\text{Cov}\left\{\sigma^{(i)}, \sigma^{(i)}\right\} &= \mathbf{E}\left\{\left[\sigma^{(i)} - \mathbf{E}\left\{\sigma^{(i)}\right\}\right]\left[\sigma^{(i)} - \mathbf{E}\left\{\sigma^{(i)}\right\}\right]^T\right\} \\
&= E_0^2\mathbf{E}\left\{\left[\tilde{D}B^{(i)}\left\{\left(D_0^{-1}\Delta Du_0\right)^{(i)} - e_i u_0^{(i)}\right\}\right]\left[\tilde{D}B^{(i)}\left\{\left(D_0^{-1}\Delta Du_0\right)^{(i)} - \right.\right.\right. \\
&\quad\left.\left.\left. - e_i u_0^{(i)}\right\}\right]^T\right\}
\end{aligned} \tag{19}$$

The variances of the stress components are then obtained as the diagonal elements of Eq. 19.

2.3 Second-Order Approximation

Consider the following second-order approximation of the displacement vector by truncating Eq. 6 after the third term:

$$u = u_0 - D_0^{-1} \Delta D u_0 + D_0^{-1} \Delta D D_0^{-1} \Delta D u_0 \tag{20}$$

in which the third term on the right-hand side involves the second-order terms $e_i e_j$ of the non-dimensional deviations of Young's modulus. Then, the expected value of u is given by

$$\mathbf{E}\{u\} = u_0 + \mathbf{E}\{D_0^{-1} \Delta D D_0^{-1} \Delta D u_0\} \tag{21}$$

whereas the covariance matrix can be shown to be

$$
\begin{aligned}
\mathrm{Cov}\{u,u\} = &- \mathbf{E}\{D_0^{-1} \Delta D D_0^{-1} \Delta D u_0\}[\mathbf{E}\{D_0^{-1} \Delta D D_0^{-1} \Delta D u_0\}]^T \\
&+ \mathbf{E}\{(D_0^{-1} \Delta D u_0)(D_0^{-1} \Delta D u_0)^T\} - \mathbf{E}\{(D_0^{-1} \Delta D u_0)(D_0^{-1} \Delta D D_0^{-1} \Delta D u_0)^T\} \\
&- \mathbf{E}\{(D_0^{-1} \Delta D D_0^{-1} \Delta D u_0)(D_0^{-1} \Delta D u_0)^T\} + \mathbf{E}\{(D_0^{-1} \Delta D D_0^{-1} \Delta D u_0)(D_0^{-1} \Delta D D_0^{-1} \Delta D u_0)^T\}
\end{aligned} \tag{22}
$$

in which the third and fourth terms of the right-hand side involve the expectations $\mathbf{E}\{e_i e_j e_k\}$ of the third-order terms of e. It is well known [6] that

$$\mathbf{E}\{e_i e_j e_k\} = 0 \tag{23}$$

provided that the e's are mean-zero Gaussian variables. Hence, if Gaussian spatial fields are considered, the third and fourth terms of the right-hand side of Eq. 22 will vanish. On the other hand, the fifth term involves the expected values of the forth-order terms of e. It is also well known [6] that

$$\mathbf{E}\{e_i e_j e_k e_\ell\} = \mathbf{E}\{e_i e_j\}\mathbf{E}\{e_k e_\ell\} + \mathbf{E}\{e_j e_k\}\mathbf{E}\{e_i e_\ell\} + \mathbf{E}\{e_i e_k\}\mathbf{E}\{e_j e_\ell\} \tag{24}$$

provided that the e's are mean-zero Gaussian random variables. Hence, if the cross-correlation matrix of the deviation vector e = $[e_1 \ e_2 \ .... \ e_n]^T$ is specified, the fourth-order expected values in Eq. 24 can be easily evaluated, where n indicates the total number of finite elements.

The expression for the strain vector $\varepsilon^{(i)}$ for the second-order approximation is derived from Eqs. 11 and 20:

$$\varepsilon^{(i)} = B^{(i)}\{u_0^{(i)} - (D_0^{-1}\Delta D u_0)^{(i)} + (D_0^{-1}\Delta D D_0^{-1}\Delta D u_0)^{(i)}\} \qquad (25)$$

with

$$\mathbf{E}\{\varepsilon^{(i)}\} = B^{(i)}u_0^{(i)} + B^{(i)}\mathbf{E}\{(D_0^{-1}\Delta D D_0^{-1}\Delta D u_0)^{(i)}\} \qquad (26)$$

and

$$
\begin{aligned}
\mathrm{Cov}\{\varepsilon^{(i)},\varepsilon^{(i)}\} = {} & \mathbf{E}\{[B^{(i)}(D_0^{-1}\Delta D u_0)^{(i)}][B^{(i)}(D_0^{-1}\Delta D u_0)^{(i)}]^T\} \\
& - \mathbf{E}\{[B^{(i)}(D_0^{-1}\Delta D u_0)^{(i)}][B^{(i)}(D_0^{-1}\Delta D D_0^{-1}\Delta D u_0)^{(i)}]^T\} \\
& + B^{(i)}\mathbf{E}\{(D_0^{-1}\Delta D i_0)^{(i)}\}[B^{(i)}\mathbf{E}\{(D_0^{-1}\Delta D D_0^{-1}\Delta D u_0)^{(i)}\}]^T \\
& - \mathbf{E}\{[B^{(i)}(D_0^{-1}\Delta D D_0^{-1}\Delta D u_0)^{(i)}][B^{(i)}(D_0^{-1}\Delta D u_0)^{(i)}]^T\} \\
& + \mathbf{E}\{[B^{(i)}(D_0^{-1}\Delta D D_0^{-1}\Delta D u_0)^{(i)}][B^{(i)}(D_0^{-1}\Delta D D_0^{-1}\Delta D u_0)^{(i)}]^T\} \qquad (27) \\
& - B^{(i)}\mathbf{E}\{(D_0^{-1}\Delta D D_0^{-1}\Delta D u_0)^{(i)}\}[B^{(i)}\mathbf{E}\{(D_0^{-1}\Delta D D_0^{-1}\Delta D u_0)^{(i)}\}]^T \\
& + B^{(i)}\mathbf{E}\{(D_0^{-1}\Delta D D_0^{-1}\Delta D u_0)^{(i)}\}[B^{(i)}\mathbf{E}\{(D_0^{-1}\Delta D u_0)^{(i)}\}]^T \\
& - B^{(i)}\mathbf{E}\{(D_0^{-1}\Delta D D_0^{-1}\Delta D u_0)^{(i)}\}[B^{(i)}\mathbf{E}\{(D_0^{-1}\Delta D D_0^{-1}\Delta D u_0)^{(i)}\}]^T \\
& + B^{(i)}\mathbf{E}\{(D_0^{-1}\Delta D D_0^{-1}\Delta D u_0)^{(i)}\}[B^{(i)}\mathbf{E}\{(D_0^{-1}\Delta D D_0^{-1}\Delta D u_0)^{(i)}\}]^T
\end{aligned}
$$

in which $(D_0^{-1}\Delta D D_0^{-1}\Delta D u_0)^{(i)}$ indicates a vector consisting of those components of the vector $D_0^{-1}\Delta D D_0^{-1}\Delta D u_0$ that pertian to the computation of $\varepsilon^{(i)}$. Some of the terms in Eq. 27 vanish by virtue of Eq. 23 and others can be evaluated utilizing the relationship in Eq. 24, if the e's are mean-zero Gaussian. The expression for the stress vector $\sigma^{(i)}$ for the second-order approximation is obtained from Eqs. 15 and 20 as

$$
\begin{aligned}
\sigma^{(i)} = {} & E_0\tilde{D}B^{(i)}\{u_0^{(i)} - (D_0^{-1}\Delta D u_0)^{(i)} + (D_0^{-1}\Delta D D_0^{-1}\Delta D u_0)^{(i)} \\
& + e_i u_0^{(i)} - e_i(D_0^{-1}\Delta D u_0)^{(i)}\} \qquad (28)
\end{aligned}
$$

In deriving Eq. 28, the term $E_0\tilde{D}B^{(i)}e_i(D_0^{-1}\Delta D D_0^{-1}\Delta D u_0)^{(i)}$ has been disregarded since it involves the non-dimensional deviations e_i in the form of a third order. The expected value of $\sigma^{(i)}$ is then given by

$$E\{\sigma^{(i)}\} = E_0 \bar{D} B^{(i)} [u_0^{(i)} + E\{(D_0^{-1} \Delta D D_0^{-1} \Delta D u_0)^{(i)}\} - E\{e_i (D_0^{-1} \Delta D u_0)^{(i)}\}]$$

$$(29)$$

The expression for the covariance matrix $\text{Cov}\{\sigma^{(i)}, \sigma^{(i)}\}$ follows in a straightforward manner from its definition, with the aid of Eqs. 28 and 29. However, the expression is even more lengthy than Eq. 27 and hence is not shown here.

3. COMPUTATIONAL STRATEGY

The use of straightforward Monte Carlo techniques in such cases as considered in the present study involving a finite element stress analysis is usually not practical for the following two reasons: (1) In most engineering applications, the finite element analysis is expensive, particularly when dynamic and/or nonlinear responses are sought. Conventional Monte Carlo techniques require many repeated executions of the finite element code to obtain statistical stability. Thus, the associated cost could quickly become prohibitive. (2) The numerical generation of spatially distributed random parameters or numerical generation of sample functions of a (possibly multidimensional [12], multivariate [1], and multidimensional-multivariate [10]) random process is required at a large number of points such as the nodal points of the finite element mesh or the centroids of all the finite elements. Such a generation is expensive especially when two (or more) dimensional processes are involved. The difficulty associated with (1) above is much more serious, however. This is particularly true now that many efficient methods of digital generation of sample functions of stochastic fields is available, including those based on ARMA models (Naganuma et al [8] and Samaras et al [9]).

As mentioned earlier, an intelligent Monte Carlo method developed at Columbia University is used to obtain the mean values and covariances of the resopnse in the present study. It is important to note that porposed enhancement of the conventional Monte Carlo simulation scheme used here takes advantage of the successive Neumann expansion method in Eq. 6.

In this context, it is pointed that numerical implementation of the inversion of the dynamic system matrix D (as introduced in Eq. 1a) does not pose a basic problem. For example, in a frequency domain formulation, for each frequency of excitation ω, the matrix assumes the form:

$$D(\omega) = K - \omega^2 M + i\omega C \qquad (30)$$

In subsequent calculations, $D(\omega)$ essentially behaves similarly to the static stiffness matrix. Similarly, typically implicit schemes in time-domain formulations reduce the dynamic equations via termporal discretization into an algebraic problem resembling the static case. Hence, the aforementioned derivation associated with the system stochasticity remains valid in general (nonlinear) dynamic situations.

The proposed Monte Carlo method still requires repeated generation of the ΔD matrix, involving generation of the appropriate discretized random field, but does not require inversion of the global stiffness matrix D; <u>actually, it requires inversion of D_0 only once</u>, and thus dramatically reduces the difficulty associated with (1) in the preceding paragraph. Furthermore, an efficient algorithm for numerically evaluating $Pu_0 = D_0^{-1} \Delta D u_0$, $P^2 u_0 = D_0^{-1} \Delta D D_0^{-1} u_0$, ... is implemented so that the Neumann expansion of higher orders, say up to the fourth order, can be used with a minimal increase in computer cost. A similar successive approximation in which the inversion of the "average" global stiffness matrix is required only once can also be devised within the framework of the perturbation method. However, it requires the evaluation of the higher-order partial derivatives of the stochastic matrices corresponding to the order of approximation. In some cases, such higher-order partial derivatives are extremely difficult, if not impossible, to evaluate, thus making implementation of the higher-order perturbation approximation impractical in these cases. More importantly, however, the major difficulty associated with the perturbation methods lies in the fact that higher-order computations are practically impossible to perform. This is particularly contrary to the intelligent Monte Carlo scheme based on the delta method. Also, another serious problem stems from convergence issues particularly when perturbation techniques are used for dynamic and transient analyses.

4. ACKNOWLEDGEMENT

This work was partially supported by the National Science Foundation under Grant No. NSF CEE-83-18123 with Dr. M. Gaus and Dr. G. Albright as Program Directors.

REFERENCES

1. Astill, C.J., Nosseir, S.B. and Shinozuka, M., "Impact Loading on Structures with Random Properties," Journal of Structural Mechanics, Vol. 1, No. 1, 1972, pp. 63-77.

2. Bathe, D.-J. and Wilson, E.L., Numerical Methods in Finite Element Analysis, (Englewood Cliffs, NJ: Prentice Hall), 1976.

3. Dasgupta, G., "Stochastic Finite Element Analysis of Soil-Structure Systems," Proceedings of the 4th International Conference on Structural Safety and Reliability, Kobe, Japan, Vol, II, 1985.

4. Der Kiureghian, A., "Finite Element Methods in Structural Safety Studies," Structural Safety Studies, edited by J.T.P. Yao, R. Corotis, C.B. Brown and F. Moses, (NY: American Society of Civil Engineers), 1985.

5. Hisada, T. and Nakagiri, S., "A Note on Stochastic Finite Element Method (Part 2) - Variation of Stress and Strain Caused by Fluctuations of Material Properties and Geometrical Boundary Conditions," Journal of the Institute of Industrial Science, University of Tokyo, Vol. 32, No. 5, May 1980, pp. 262-265.

6. Lin, Y.D., Probabilistic Theory of Structural Dynamics, (NY: McGraw-Hill), 1967.

7. Liu, W-K., Belytschko, T. and Mani, A., "A Computational Method
    for the Determination of the Probabilistic Distribution of the
    Dynamic Response of Structures," ASCE PVP-98-5, 1985, pp. 243-
    248.

8. Naganuma, T., Deodatis, G. and Shinozuka, M., "An ARMA Model for
    Two-Dimensional Processes," Technical Report No. 1 under NSF
    Grant No. CEE-83-18123, October 1984; submitted for publication
    to the Journal of Engineering Mechanics, ASCE.

9. Samaras, E., Shinozuka, M. and Tsurui, A., "ARMA Representation
    of Random Processes," Journal of Engineering Mechanics, ASCE,
    Vol. 111, No. 3, March 1985, pp. 449-461.

10. Shinozuka, M., "Probabilistic Modeling of Concrete Structures,"
    Journal of the Engineering Mechanics Division, ASCE, Vol. 98, No.
    EM6, December 1972, pp. 1433-1451.

11. Shinozuka, M., "Monte Carlo Solution of Structural Dynamics," In-
    ternational Journal of Computers and Structures, Vol. 2, 1972,
    pp. 855-874.

12. Shinozuka, M., "Digital Simulation of Random Processes in Engin-
    eering Mechanics with the Aid of FFT Technique," Stochastic
    Problems in Mechanics, edited by S.T. Ariaratnam and H.H.E.
    Leipholz, (Waterloo: University of Waterloo Press), 1974.

13. Shinozuka, M. and Astill, J., "Random Eigenvalue Problems in
    Structural Mechanics," Journal of the American Institute of
    Aeronautics and Astronautics, Vol. 10, No. 4, April 1972, pp.
    456-462.

14. Shinozuka, M. and Jan, C.-M., "Digital Simulation of Random
    Processes and Its Applications," Journal of Sound and Vibration,
    Vol. 25, No. 1, 1972, pp. 111-128.

15. Shinozuka, M. and Nomoto, T., "Response Variability Due to Spa-
    tial Randomness," Technical Report No. CU-CEEM-MS-80-1, Depart-
    ment of Civil Engineering and Engineering Mechanics, Columbia
    University, August 1980.

16. Vanmarcke, E. and Grigoriu, M., "Stochastic Finite Element Anal-
    ysis of Simple Beams," Journal of Engineering Mechanics, ASCE,
    Vol. 109, No. 5, October 1983, pp. 1203-1214.

17. Yang, J.-N., "Simulation of Random Envelope Processes," Journal
    of Sound and Vibration, Vol. 21, No. 1, 1972, pp. 73-85.

18. Yang, J.-N., "On the Normality and Accuracy of Simulated Random
    Processes," Journal of Sound and Vibration, Vol. 26, No. 3, 1973,
    pp. 417-428.

U.S.-JAPAN COORDINATED PROGRAM ON MASONRY RESEARCH

Shin Okamoto[1], James Noland[2], M.ASCE

ABSTRACT: An overview of the U.S.-Japan Coordinated Program on Masonry Research is presented. The program consists of parallel efforts in Japan and the United States to develop design methods for reinforced masonry buildings located in areas of seismic risk. Both programs rely heavily upon experimental work to produce data and confirm design methods. Both programs also include tests of a full-scale reinforced masonry building specimen to observe and verify system response.

INTRODUCTION

The U.S.-Japan Coordinated Program on Masonry Research is the third in a series of building research programs conducted by Japan and the United States under the auspices of the Panel on Wind and Seismic Effects of the U.S.-Japan Natural Resources Development Program (UJNR). The first two were conducted on reinforced concrete and structural steel buildings.

The joint masonry programs are intended to obtain experimental data and establish design methods pertinent to design and construction of masonry buildings in areas of seismic risk. However, the knowledge and techniques developed in the course of the two programs will also be applicable to design and construction of masonry buildings subjected to other types of loading.

The two research programs have begun. The program in Japan began in 1984 and the U.S. program commenced in 1985. This paper presents a brief overview of each program.

PART I - OVERVIEW OF THE JAPANESE PROGRAM

Background

Masonry structural systems are known worldwide as traditional and efficient with high durability, fire resistance, sound and heat insulation, and high flexibility in construction. The structural system was imported to Japan from Europe with little attention given to earthquake resistance of masonry buildings as then constructed. The masonry buildings constructed by the imported concepts sustained extensive damage in the Great Kanto Earthquake. Since then the construction of masonry structures practically ceased in Japan and little effort was made on the study of the seismic resistance of masonry structural systems.

---

[1] Director, Production Department, Building Research Institute Ministry of Construction, Tsukuba, Japan

[2] Principal, Atkinson-Noland & Associates, Boulder, Colorado, U.S.A.

Because of the large demand for building construction after World War II, Japan developed various construction systems amendable to mass production of buildings. Masonry buildings were generally not built because of the problems then perceived regarding masonry structures, e.g., tedious labor. What is desired in building construction today has changed. Now more value is placed on quality then quantity. Versatility and variety are more appreciated than uniformity.

It was natural, therefore, to reconsider masonry structures. Since the late 1960's various types of new masonry systems have been developed to make construction more reliable and competitive. The reinforced concrete block (RCB) system is one example.

The Building Research Institute has contributed a great deal to the improvement of masonry structural system. One current project is entitled "Application of Small PC units to Urban Housing" in which a high quality masonry unit will be developed and a new earthquake resistant construction technique proposed.

Additional research is required to further improve masonry design and construction techniques for greater seismic safety thus enabling greater utilization of this type of construction. The immediate target of the Japanese part of the U.S.-Japan joint research program will be the development of a masonry construction system applicable to low and medium rise building structures.

Research Team Organization

The organization of the Japanese masonry research effort is shown in Figure 1 and its relationship to the U.S. masonry research team organization. The U.S. organization is presented in more detail in Part 2 of this paper. Note that the Japanese and U.S. research efforts are linked at the governmental level by a coordinated research agreement and at the project level by the Joint Technical Coordinating Committee on Masonry Research (JTCCMAR).

Three major committees are included in the Japanese masonry research organization, i.e., the Technical Coordinating Committee on Masonry Research (PROCMAR), and the Building Construction Committee on Masonry Research (BLDCMAR). The first committee, TECCMAR, coordinates technical research planning on materials, components, assemblies and full-scale experiments and will develop proposals for aseismic design guidelines for low and medium use reinforced masonry building structures in Japan.

PROCMAR was organized under a domestic cooperative research agreement between the Building Research Institute, the Japan Association for Building Research Promotion and the Building Contractor's Society for the purpose of promoting the U.S.-Japan research program in Japan on masonry structures. The members of PROCMAR include government officials, local self-governing body officials and representatives from many industrial organizations as shown in Figure 1.

There are many construction problems to solve in this research program to develop technology for the design and construction of low and medium rise reinforced buildings in Japan. The third committee, BLDCMAR,

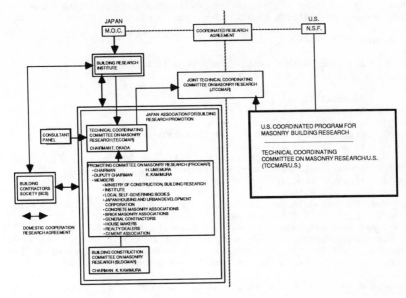

FIG. 1 ORGANIZATION OF U.S.-JAPAN JOINT RESEARCH
ON MASONRY BUILDING STRUCTURES

RESEARCH ITEMS	FISCAL 1984 APR. MAR.	1985 APR. MAR.	1986 APR. MAR.	1987 APR. MAR.	1988 APR. MAR.
1. MATERIAL TEST	●			●	
2. STATIC TEST OF WALLS	●		●		
3. STATIC TEST OF BEAMS		● ●			
4. STATIC TEST OF WALL AND BEAM ASSEMBLIES	●	●			
5. FULL-SCALE PLANAR STATIC TEST		● ●			
6. SHAKING TABLE TEST OF SCALE MODEL SPECIMEN			DESIGN OF SPECIMEN	TEST	ANALYSIS
7. FIVE STORY FULL-SCALE TEST		PRELIMINARY DESIGN	DESIGN & CONSTRUCTION OF SPECIMEN	TEST	TEST AFTER REPAIR AND RETROFITTING, ANALYSIS
8. ASEISMIC DESIGN GUIDLINES			DRAFT OF GUIDELINES		PROPOSAL OF GUIDELINES
9. JOINT TECHNICAL COORDINATING COMMITTEE ON MASONRY RESEARCH (JTCCMAR)	WORK SHOP ●	1ST JTCCMAR ●	2ND ●	3RD ●	4TH ● 5TH ●

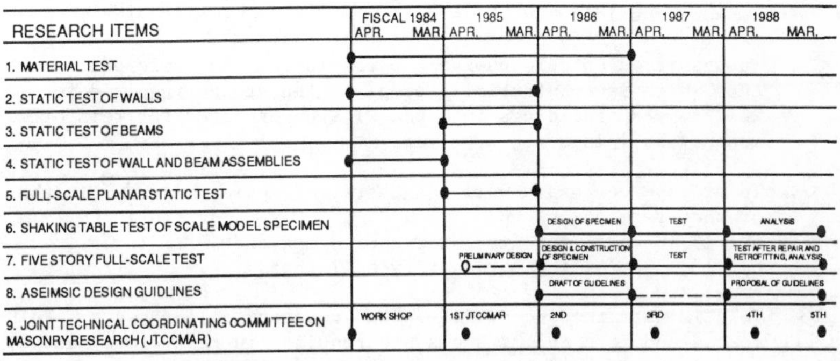

FIG. 2 JAPANESE RESEARCH PLAN ON MASONRY BUILDING STRUCTURES

was organized under PROCMAR to address the problems concerning reinforced masonry building construction techniques.

Research Plan

The Japanese masonry research effort will take approximately five
years and is composed of the following basic components:

1. Material Tests
2. Static Tests of Walls
3. Static Tests of Beams
4. Static Tests of Wall and Beam Assemblies
5. Full-Scale Planar Static Tests
6. Shaking Table Tests of a Scale Model Specimen
7. A Full-Scale Five-Story Structure Test
8. Establishment of Aseismic Design Guidelines

The time schedule for the Japanese research effort is shown in Figure 2. During the first two years, fundamental tests of materials, of
wall and beam components, masonry structural assemblies, and of a full-
scale planar specimen are planned. A great deal of this has been done.
The shaking table tests of a scale model specimen and a full-scale five-
story structure will be done in the later part of the program.

Japanese Research Objectives

The final target for the Japanese side of the joint research program
is to propose aseismic design guidelines for low and medium rise reinforced masonry building structures in Japan. Presently, only reinforced masonry buildings of 12 meters (one to three stories) are permitted. As a result of this program design guidelines for low-to-med-
ium rise reinforced masonry building (up to five stories) will be proposed.

There are also many other objectives related to the structural design of masonry buildings such as:

1. To assure acceptable aseismic performance. Aseismic performance which is substantially equal to that to be required for
   wall type RC buildings in Japan will be expected for reinforced
   masonry buildings.

2. To propose decreased wall length rate. Revise 21 cm/m^2 in the
   present regulations to:
   a) 15-18 cm/m^2 for five story buildings and
   b) 12-15 cm/m^2 for three story buildings.

3. To eliminate the use of reinforced concrete collar beams. Reinforced concrete collar beams are required in reinforced masonry
   buildings under present regulations.

4. To simplify joint works of reinforced bars. Effectiveness of
   lap and mechanical joints of vertical reinforcing bars in walls
   must be established.

Review of Research Done in 1984

The following topics were addressed in 1984:

1. Masonry Units - Standard concrete and clay block units developed for use in the Japanese program are shown in Figures 3 and 4.

2. The predictions of the compressive strength of prisms based on the properties of block units, grout concrete and mortar has been investigated.

3. High lift grouting procedures have been studied using wall specimens 2-1/2 units long (1000 mm or 39.4 in.) by 12 courses high (2400 mm or 94.5 in.).

4. Effectiveness of lap splices of reinforcing bars in masonry has been studied using six types of splice configuration. This included evaluation of spiral reinforcement around lap splices.

5. Anchorage of reinforcing bars in grouted masonry has been studied considering grout cover and proximity of the reinforcement to face shells of units.

6. Static tests of masonry wall and beam components were conducted. The major purpose was to investigate their deformation performance and shear capacity varying area of main reinforcement, shear reinforcement and shear-span ratio. Results are described in a paper to be presented later at this conference.

Conclusions

Masonry buildings of either concrete or clay unit masonry are at present few in Japan compared with the number of buildings of other materials, e.g., wood, concrete, and steel. This is because of the susceptibility to earthquake damage of masonry buildings previously constructed by methods imported from Europe one hundred years ago. It is expected that the five-year masonry research program just described on low to medium rise masonry buildings will allow and encourage greater numbers of such buildings in Japan.

Acknowledgement

The author of Part 1 of this paper wishes to express his sincere thanks to TECCMAR-Japan members who contributed great effort to the technical research planning of the Japanese program and for coordination with the U.S. program. Gratitude is also expressed to PROCMAR members for their contributions to promote the Japanese side of the joint U.S.-Japan masonry research program.

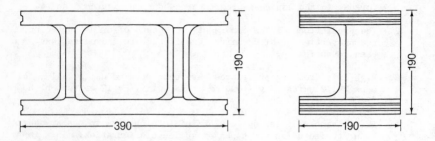

Fig. 3   Standard Concrete Block Unit—Japanese Research Program

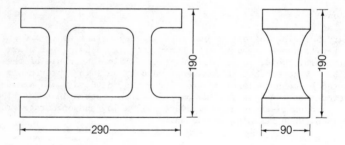

Fig. 4   Standard Clay Block Unit—Japanese Research Program

PART II - OVERVIEW OF THE UNITED STATES PROGRAM

Background

Load-bearing masonry buildings have been built in the United States for many years - nearly since the time European settlers first arrived. Masonry buildings are a significant percentage of buildings built from that time up to the 1930's and are generally unreinforced.

Reinforced concrete and structural steel buildings gradually became a larger and larger part of the construction market. Their use was and is encouraged by the great amount of research and development done and by the gradual improvement in design methods and codes.

In contrast, the use of masonry as the primary structural system declined in comparison to the use of steel and concrete possibly due to the perception that masonry design technology was not at the same level as that for steel and concrete and to the perception that masonry structures perform poorly in earthquakes.

While some U.S. research in structural masonry was done earlier, it began to increase in the mid-to-late sixties and continues to increase. Masonry building code improvement has also recently been emphasized as exemplified by the 1985 Uniform Building Code and by the work of the joint ASCE/ACI Committee 530 on masonry. This author believes that some of the impetus for renewed interest in the structural use of masonry is due to its architectural flexibility and that it can be competitively economic for certain types of building plans. The introduction of reinforcement has been shown to provide the ductility required for adequate performance in earthquake conditions.

However, the design methodology is still basically a working stress approach based on the assumption of linear-elastic material behavior although the latest UBC does permit strength methods for certain applications. While many reinforced masonry structures have been built and have performed successfully, economic considerations and a better ability to predict ultimate behavior would make structural masonry a more viable alternative as a building system particularly for earthquake conditions. A need was recognized for complete strength-method design techniques for reinforced masonry based upon adequate experimental data to bring masonry structural technology to a level consistent with modern needs.

Research Team Organization

A team of masonry researchers known as the Technical Coordinating Committee for Masonry Research/U.S. (TCCMAR/U.S.) was formed to identify the research necessary and to organize in a manner such that the research can be done in an orderly fashion. The organization of TCCMAR/U.S. and the basic research areas are depicted in Figure 1.

Funding is provided by The National Science Foundation (Dr. A. J. Eggenberger) and liason with the UJNR Committee on Wind and Seismic Effects is done by G. R. Fuller (HUD) and Dr. H.S. Lew (NBS). A single coordinator (J. Noland) is responsible for overall operation of the U.S. program with the advice and assistance of the Executive Panel. The Consultants Panel is composed of individuals highly qualified to provide basic guidance on the course and conduct of the program. Two were intimately associated with previous U.S.-Japan joint research programs.

A panel of Industry Observers was formed to provide industry perspective and to relay information about the program to the industry. The purpose of the Industry Participation Panel is to define ways in which the industry can assist the program and arrange for that assistance.

Research Plan

The U.S. masonry research program consists of a group of coordinated research tasks beginning with tasks which address basic material behavior, progressing to tasks addressing behavior of masonry structural components, e.g., walls, and finally to experiments on a full-size segment of a masonry building. Development of analytical methods for stress and strain analysis and system behavior will proceed in parallel with experimental tasks. The final task will be the preparation of recommendations for design procedures and building code provisions.

The research tasks which were identified by TCCMAR/U.S. fell into 10 categories as shown in Figure 1. The following is a more detailed list of research tasks in each category and their objectives.

Category	Task	Title-Purpose
1.0	1.1	Preliminary Material Studies - To establish the range of continuity of masonry behavior to provide a basis for selection of the type or types of masonry to be used. To establish standardized materials test procedures for all the experimental tasks.
1.0	1.2	Material Models - To evaluate K1, K2, and K3 for the flexural stress-block. To determine uniaxial and biaxial material properties for analytical models (Task 2.1 and 2.2) including post-peak behavior. To evaluate non-isotropic behavior.
2.0	2.1	Force-Displacement Models for Masonry Component- To develop force-displacement mathematical models which accurately characterize reinforced masonry components under cyclic loading to permit pretest predictions of experimental results. To develop models suitable for parameter studies and models suitable for design engineering.
2.0	2.2	Strain Analysis Model for Masonry Components- To develop a strain model for reinforced masonry components in conjunction with Task 2.1 to enable regions of large strain to be identified thus assisting in experimental instrumentation planning. To develop a simplified model to be used to provide data for

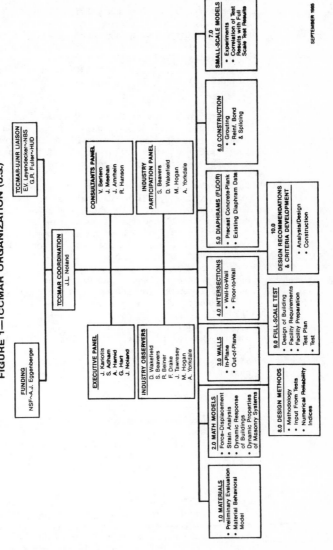

U.S.-JAPAN COORDINATED PROGRAM FOR MASONRY BUILDING RESEARCH

FIGURE 1—TCCMAR ORGANIZATION (U.S.)

SEPTEMBER 1985

strength design rules and in-plane shear design procedures.

2.0	2.3	Dynamic Response of Masonry Buildings - To develop a generalized dynamic response model to predict inter-story displacements using specified time histories. To correlate force displacement models and to investigate force-displacement characteristics of structural components in the near-elastic and inelastic displacement range. To provide data for building test planning.
2.0	2.4	Dynamic Properties of Masonry Systems- To develop consistent, unified, rationale for seismic design of masonry buildings considering elastic and inelastic response of masonry buildings and of the soil/structure interaction and related to seismic hazard zones.
3.0	3.1(a)	Response of Reinforced Masonry Story-Height Walls to Fully Reversed In-Plane Lateral Loads- To establish the behavior of story-height walls subjected to small and large amplitude reversals of in-plane lateral deflection, axial force, and bending moments considering aspect ratios, reinforcement ratios and patterns, and the effect of openings.
3.0	3.1(b)	Development of a Sequential Displacement Analytical and Experimental Methodology for the Response of Multi-Story Walls to In-Plane Loads- develop a reliable methodology for investigating, through integrated analytical and experimental studies, the in-plane behavior of multistory reinforced hollow unit masonry walls. The methodology will be the basis of studying the response of a full-scale masonry research building.
3.0	3.2(a)	Response of Reinforced Masonry Walls to Out-of-Plane Static Loads- To verify the behavior of flexural models developed using material models, to evaluate the influence of mortar joint spacing, unit properties, reinforcement ratios and grouting upon wall behavior. To provide stiffness data for correlation with dynamic wall test results (Task 3-2(b)).
3.0	3.2(b)	Response of Reinforced Masonry Walls to Out-of-Plane Dynamic Excitation- To determine effects of slenderness, reinforcement amounts and ratios, vertical load and grouting on dynamic response, to verify mathematical response models, to develop design coefficients for equivalent static load methods.
4.0	4.1, 4.2	Wall-to-Wall Intersections and Floor-to-Wall Intersections of Masonry Buildings - To determine the effectiveness of intersection details to connect masonry wall components, to construct a nonphenomeno-

logical analytical model of intersection behavior.

5.0   5.1   Concrete Plank Diaphram Characteristics- To investigate experimentally concrete plank diaphram floor diaphrams to determine modes of failure and stiffness
characteristics including yielding capacity in terms
of distortion as needed for masonry building models.

5.0   5.2   Assembly of Existing Diaphram Data- To assemble extensive existing experimental data on various types
of floor diaphrams, to reduce to a form required for
static and dynamic analysis models.

6.0   6.1   Grouting Procedures for Hollow Unit Masonry- To
identify methods of grouting hollow unit masonry such
that the cavity is solidly filled and reinforcement
is completely bonded.

6.0   6.2   Reinforcement Bond and Splices in Grouted Hollow Unit
Masonry- To develop data and behavioral models on the
bond strength and slip characteristics of deformed
bars in grouted hollow unit masonry, to develop data
and behavioral models on the bond strength and slip
characteristics of deformed bar lap splices in grouted hollow unit masonry as needed for building modeling.

7.0   7.1   Small Scale Models- To experimentally evaluate the
use of small-scale modeling for reinforced hollow-
unit masonry walls by correlating test results with
test results of full-scale walls of the same configuration. To determine if tests of small-scale specimens can reveal basic characteristics and failure
modes of full-scale masonry specimens.

8.0   8.1   Limit State Design Methodology for Reinforced Masonry-
To select an appropriate limit state design method-
ology for masonry. To select and document a procedure to compute numerical values for strength reduction factors. To review program experimental research
tasks assure that statistical benefits are maximized
and proper limit states are investigated.

8.0   8.2   Numerical Reliability Indices- To develop numerical
values of statistically –based strength reduction
(i.e., $\emptyset$) factors using program experimentally developed data, other applicable data, and judgement.

9.0   9.1   Design of Reinforced Masonry Research Building, Phase
I- To develop the preliminary designs of the poten-
tial research buildings which reflect a significant
portion of modern U.S. masonry construction. To estimate inter-story displacements using methods developed in Category 2 tasks and the associated load magnitudes and distributions. To select a single

configuration in consultation with TCCMAR which will
be used as a basis for defining equipment and other
laboratory facilities in Task 9.2.

9.0   9.2   Facility Preparation- Define, acquire, install and
check-out equipment required for experiments on a
full-scale masonry research building.

9.0   9.3   Full-Scale Masonry Research Building Test Plan- To
develop a detailed and comprehensive plan for conducting static load-reversal tests on a full-scale
reinforced masonry research building.

9.0   9.4   Full-Scale Test- To conduct experiments on a full-
scale reinforced masonry research building in accordance with the test plan and acquiring data indicated.
To observe building response and adjust test procedures and data measurements as required to establish
building behavior.

10.0   10.1   Design Recommendations and Criteria Development- To
develop and document recommendations for the design
of reinforced masonry building subject to seismic
excitation in a manner conducive to design office
utilization. To develop and document corresponding
recommendations for masonry structural code provisions.

11.0   11.1   Coordination- To fully coordinate the U.S. research
tasks to enhance data transfer among researchers and
timely completion of tasks. To schedule and organize
TCCMAR and Executive Panel meetings. To establish
additional program policies as the need arises. To
stimulate release of progress reports and dissemination of results. To coordinate with industry for the
purposes of informing industry and arranging industry
support. To interface with NSF and UJNR on overall
funding and policy matters.

The names, affiliations and research task arrangements of current
TCCMAR/U.S. members are as follows:

Task	Researcher	Institution
1.1	R. Atkinson	Atkinson-Noland & Assoc.
1.2(a)	A. Hamid	Drexel Univ.
1.2(b)	R. Brown	Clemson Univ.
2.1	R. Englekirk	AKEH Joint Venture
2.2	G. Hart	"
2.3	J. Kariotis	"
2.4	R. Ewing	"
3.1(a)	J. Noland/B. Shing	Univ. of Colo.- Boulder

3.1(b)	G. Hegemier/F. Seible	Univ. of Calif.-San Diego
3.2(a)	A. Hamid	Drexel Univ.
3.2(b1)	S. Adham	Agbabian Assoc.
3.2(b2)	R. Mayes	Computech
4.1	G. Hegemier	Univ. of California-
4.2	F. Seible	San Diego
5.1	M. Porter	Iowa State
5.2	"	University
6.1	L. Tulin	Univ. of Colorado-
6.2	"	Boulder
7.1	A. Hamid	Drexel Univ.
8.1	G. Hart	Univ. of Calif.- Los Angeles
8.2	TCCMAR*	-
9.1	J. Kariotis A. Johnson	Barnes-Kariotis Joint Venture
9.2	G. Hegemier	Univ. of Calif.-San Diego
9.3	TCCMAR*	-
9.4	TCCMAR*	Univ. of Calif.-San Diego
10.1	TCCMAR*	-
11.1	J. Noland	Atkinson-Noland & Assoc.

* Task to be done by TCCMAR/U.S. or subgroup thereof.

Task Coordination

The research tasks are interdependent, i.e., results from a given task may be required for the execution of others and vice-versa. Analytical tasks will generally require interaction with experimental tasks on a fairly continuous basis so that analytical model development may incorporate data as they are obtained. The needs of the analytical tasks will in turn serve to define, in part, the manner in which experimental tasks are designed and conducted and the data to be obtained.

The anticipated intra-task interaction is depicted generally in Figure 2.

Schedule

The schedule for tasks comprising the U.S. program is shown in Figure 3. The total time required to complete the program is estimated to be approximately six years.

Review of 1985 Progress

Task 1.1, Preliminary Studies was completed by June 1985. The results will be reviewed in a paper to be presented later in this conference. Work was begun on Tasks 1.2(a & b), 2.1, 2.2, 2.3, 3.1(a), 3.2(a)

U.S.-JAPAN COORDINATED PROGRAM FOR MASONRY BUILDING RESEARCH

FIGURE 2 — TASK INGREDIENTS-DEPENDENCE CHART (U.S. PROGRAM)

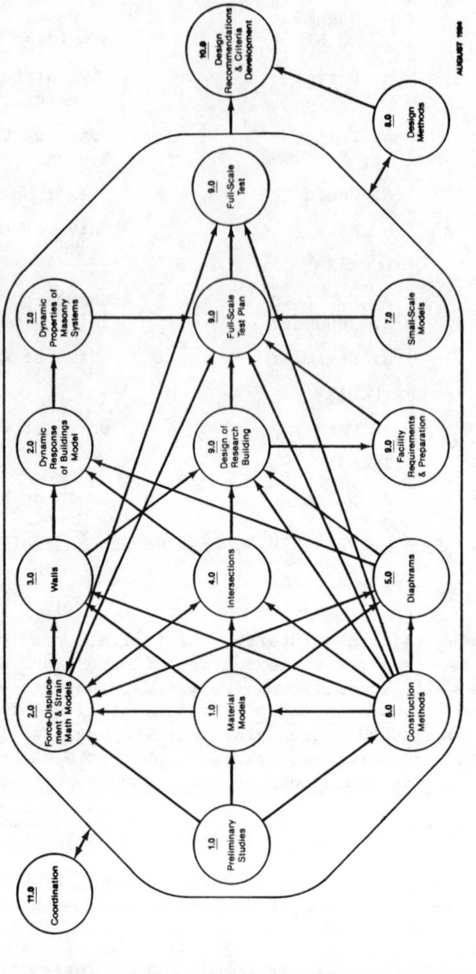

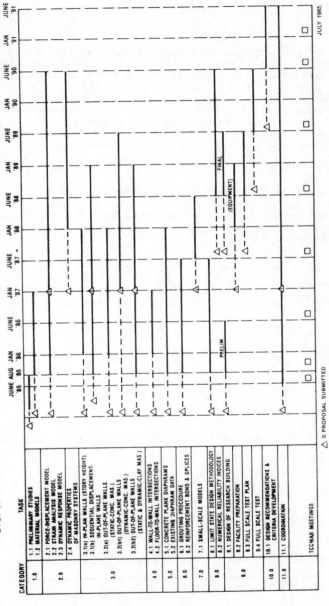

FIGURE 3—TASK SCHEDULE (U.S. PROGRAM)
U.S.-JAPAN COORDINATED PROGRAM FOR MASONRY BUILDING RESEARCH

5.1, 5.2, 6.2, 8.1 and 9.1 in the Fall of 1985 therefore results are
not yet available.

CONCLUSIONS

The U.S. program was designed to provide a strength design methodology for reinforced masonry buildings based upon an integrated experimental and analytical program of specific research tasks. Study of the
task descriptions will reveal that only issues believed to be essential
to this development are addressed. Budget limitations also restrict
the number of parameters which can be evaluated in these tasks. However, the work will provide a unified framework and reference point so
that future research, as funds become available, will be able to effectively build upon the knowledge developed in this program.

ACKNOWLEDGEMENT

The support of the National Science Foundation is greatly acknowledged. The contribution of hollow concrete units by the Concrete
Masonry Association of California-Nevada and of hollow clay units by
the Western States Clay Products Association for specimen construction
is also acknowledged.

A STOCHASTIC SOURCE SPECTRUM FOR
ENGINEERING APPLICATION

R.J.Scherer*   and   G.I.Schuëller**, M.ASCE

ABSTRACT

The solution of a point source (double couple) combined with Aki/Papageorgiou's stochastic source model is used to establish the power spectrum at the acceleration in the vicinity of the earthquake source. The influence of the far-, intermediate- and near field term of the solution of a point source on the acceleration process is discussed. The model incorporates the directivity effect as well as the radiation pattern. Both are seen to be important factors in determining the intensity of the acceleration as a function of the azimuth. The proposed model is based on the evolutionary spectrum theory which provides the simulation of time histories exhibiting quite realistic accelerations.

INTRODUCTION

For evaluation of seismic risk at a specific site - including the effect of soil-structure-interaction analysis, the wave field in the base rock has to be known. Generally, based on measured accelerations on the earth surface, stationary or unimodulated stochastic processes with smooth wide-band power spectra are assumed or estimated. Due to the fact that the spectrum of the wave field is negligibly influenced by path effects, the spectrum at the seismic source can be approximated by the spectrum of the base rock. It is shown in [1,2] that for epicentral distances < 100 km, the apparent time dependency of the seismic process is mainly due to the superposition of unimodulated subprocesses, where each subprocess represents one of the four wave types, i.e. the direct P- and S-waves and the Love- and Rayleigh waves respectively. In the following a time dependent source spectrum for body waves will be discussed.

SPECTRUM OF THE MEAN CRACK

First a short discussion of Aki's barrier model [3] is given. The rupture plane which is a simplification of the actural rupture volumina is strongly inhomogeneous in both the strength of the material and the amount of accumulated strain energy. After the critical strength at a crack tip is reached the nucleation of a potential rupture may start. As long as the amount of mechanical energy supplied by the increase of the crack is greater than the surface energy consumed by the crack (Griffith theory) the rupture

 * Fa. Dyckerhoff & Widmann AG, Munich, FR-Germany
** Institute of Engineering Mechanics, University of Innsbruck, Austria

propagates and sweeps over the idealized rupture plane. The rupture
plane does not break totally because of the roughness effects which
may act as barriers by locally stopping the rupture process or at
least decelerate it. Therefore the whole rupture process consists of
starting and stopping phases, i.e. energy pulses which behave like
shot noise. It is a well known fact that the irregularities in the
time histories are the source of the high frequency content of the
process. Therefore for strong motion modeling it is not only a
question of the strength of the overall earthquake, but also of the
way the seismic energy will be supplied by the rupture process.

Papageorgiou and Aki [4] suggested the following relation between the
seismic energy radiated by a mean idealized circular crack and the
dynamic source parameters v_z, v_r, r_c, $\Delta\sigma$ in terms of the power
spectrum of the acceleration

$$S_{ac}(f) = v_z \ v_r^4 r_c (\frac{\Delta\sigma}{\mu})^2 2C \ (\frac{\mu}{\rho})^2 \ \frac{1}{v^6} \ \frac{1}{r^2} \ A_{FS}^2 \tag{1}$$

which is per unit width

$$S_a^0(f) = [v_z v_r^4 \ (\frac{\Delta\sigma}{\mu})^2 C] \ (\frac{\mu}{\rho} \ \frac{1}{4\pi})^2 \ \frac{1}{v^6} \ \frac{1}{r^2} \ A_{FS}^2 \tag{2}$$

where v_z = mean velocity of propagation of the active rupture zone,
v_r = mean rupture velocity, r_c = mean radius of the cracks, $\Delta\sigma$ = mean
local stress drop, μ = shear modulus, ρ = density, C = a constant,
suggested in [4] with C 100 as a mean value, r = hypocentral distance,
v = wave velocity, A_{FS} = radiation coefficent of the far field term
of the S-wave.

Papageorgiou and Aki derived eq.(1) for the far field term of the S-
wave only and did not consider the time-dependency of the source
spectrum which is due to the variation of the distance between the
active source zone and the station. The constant C is derived for a
mean directivity effect. Because the rupture is not confined to a
plane which means that the different cracks may be distributed randomly with respect to their orientation over the idealized fault plane one
can use eq.(1) as the power spectrum of a mean crack or as the power
spectrum of the stochastic rupture process per unit crack. For
convenience and because of the uncertainties inherent in the derivation of the constant C, eq.(1) is adopted for S-waves as well as for
P-waves using the same value for C. The main difference between the
present approach and the one suggested by Aki and Papgeorgiou [4] is
the way by which the relation between the power spectrum of the
acceleration and the source parameters is used. In [4] eq.(1) is
multiplied by the mean width of the source plane and interpreted as the
power spectrum of the total rupture process, using a constant value for
the distance. Hence the acceleration process is idealized as a
stationary stochastic process. On the contrary in the present approach
eq.(1) is intepreted as the instantaneous power spectrum per unit
width. The distance r between the instantaneous active source plane
and the receiver is a variable value which introduces a time-dependent
geometrical attenuation. Therefore the acceleration process is a
nonstationary stochastic process determined by an evolutionary
spectrum. It is a special case of an unimodulated process because the

time - varying distance does not influence the frequency content. Additionally the directivity effect in the derivation of the power spectrum in [4] is interpreted here as an averaged directivity of the idealized mean crack. Consequently one will deal again with the directivity effect when the evolutionary spectrum of the total acceleration process is derived, because the directivity is a global nonrandom effect for the particular earthquake under consideration and the rupture as a whole.

DETERMINISTIC SOLUTION OF THE POINT SOURCE

It is assumed that the total rupture can be expressed by the solution of a point source integrated over the source area. The solution for the displacement $\vec{w}$ generated by a point source is (see [5])

$$\vec{w}(\vec{x},\omega) = \frac{1}{4\pi\rho} M_O(\omega)\{[-(\frac{r}{v_p} - \frac{1}{i\omega})\frac{1}{i\omega}e^{-i\omega r/v_p} + (\frac{r}{v_s} - \frac{1}{i\omega})\frac{1}{i\omega}e^{-i\omega r/v_s}]\frac{1}{r^4}\vec{A}_N +$$

$$+ e^{-i\omega r/v_p}\frac{1}{v_p^2}\frac{1}{r^2}\vec{A}_{IP} + e^{-i\omega r/v_s}\frac{1}{v_s^2}\frac{1}{r^2}\vec{A}_{IS} +$$

$$+ i\omega e^{-i\omega r/v_p}\frac{1}{v_p^3}\frac{1}{r}\vec{A}_{FP} + i\omega e^{-i\omega r/v_s}\frac{1}{v_p^3}\frac{1}{r}\vec{A}_{FS}\} \tag{3}$$

with
$$\begin{matrix}\vec{A}_N \\ \vec{A}_{IP} \\ \vec{A}_{IS} = \\ \vec{A}_{FP} \\ \vec{A}_{FS}\end{matrix}\begin{bmatrix} 9 & -6 & 6 \\ 4 & -2 & 2 \\ -3 & 3 & -3 \\ 1 & 0 & 0 \\ 0 & 1 & -1 \end{bmatrix}\begin{bmatrix} F_{\hat{r}} \\ F_{\hat{\theta}} \\ F_{\hat{\phi}} \end{bmatrix} \tag{4}$$

where
$$\vec{F} = \begin{bmatrix} F_{\hat{r}} \\ F_{\hat{\theta}} \\ F_{\hat{\phi}} \end{bmatrix} = \begin{bmatrix} \sin2\theta & \cos\phi \\ \cos2\theta & \cos\phi \\ \cos\theta & \sin\phi \end{bmatrix} \tag{5}$$

and v = wave velocity, r = distance between source and receiver, M_O = seismic moment, $\vec{A}$ = radiation coefficients, ω = frequency rad/s, $\vec{x}$ = location of the receiver, $\vec{F}$ = basic coefficients, with the indexes: P = P-wave, S = S-wave, N = near field term, I = intermediate field term, F = far field term, and r,θ,ϕ = spherical polar coordinates.

FORMULATION AS A EVOLUTIONARY SPECTRUM

The exponential term $e^{-i\omega v/r}$ is identical with the time delay which has to be considered during the superposition of the seismic energy radiated by different parts of the source and therefore radiated at different time instances. The interest is less focussed at a time invariant Fourier-Spectrum but at an evolutionary spectrum, where the time lag, i.e. delay has to be expressed explicitely by a time variable. This means that

$$M_O(\omega)e^{-i\omega r/v} \to M_O(\omega,t-\frac{r}{v}) = M_O(\omega)w(t-\frac{r}{v}) \tag{6}$$

Furthermore, if the complex wave number $\hat{n}$ is introduced,

$$\hat{n} = in = i\frac{\omega r}{v} = i2\pi \frac{fr}{v} \tag{7}$$

where $n/2\pi$ is a measure of the number of waves between source and receiver, then eq.(3) with eqs.(6) and (7) yields

$$\vec{w}(\vec{x},\omega,t) = \frac{1}{4\pi\rho r^2}M_o(\omega)\sum_{i=1}^{2}\frac{1}{v_i{}^2}\ w(t-\frac{r}{v_i})\ \cdot$$

$$\cdot\ [-(-1)^i A_N\hat{n}_i{}^{-2}+(-1)^i A_N\hat{n}_i{}^{-1}+A_{Ii}\hat{n}_i{}^0+A_{Fi}\hat{n}_i{}^1];\ i = P,S \tag{8}$$

By taking the second derivative of eq.(8) one obtains the time-dependent Fourier Spectrum of the acceleration process, i.e.

$$\vec{a}(\vec{x},\omega,t)=\frac{1}{4\pi\rho r^4}\ M_o(\omega)\sum_{i=1}^{2}\ w(t-\frac{r}{v_i})\cdot$$

$$\cdot[-(-1)^i A_N\hat{n}_i{}^0+(-1)^i A_N\hat{n}_i{}^1+A_{Ii}\hat{n}_i{}^2+A_{Fi}\hat{n}_i{}^3];\ \ i = P,S \tag{9}$$

From this one can see that the time dependent Fourier Spectrum of the acceleration process can be expressed by a polynomial expression of the complex wave number $\hat{n}$. The contribution of the various terms, i.e. near-, intermediate and far field terms, to the total spectrum may be discussed by setting the three radiation coefficients $F = 1.0$ [6]. It can be seen that the contribution of the near field term decays quite rapidly with increasing distance for the high frequency content, but slowly for the very low frequency content. The frequency range below 0.1 Hz which may be seen as a lower cut-off frequency is of no interest for engineering purposes. In general, structural eigenfrequencies < 0.1 Hz are due to code- and structural requirements not feasable. Therefore for engineering purposes the near field term can be neglected. This assumption is supported by the investigations of the displacements integrated from the uncorrected acceleration recorded in Friuli 1976. In most of the time histories they show a systematic trend with frequencies quite lower than 0.1 Hz during the direct body wave phase which points towards the near field term. The intermediate field term can not be neglected within the first ten to twenty km. Significant contributions up to frequencies of 1 to 2 Hz are to be expected. Assuming that for moderate and high frequencies content respectively the radiated seismic energy is statistically mutually independent for each crack and also that the P- and S-waves are mutually independent, one can immediately express the evolutionary spectrum of the acceleration in the following form:

$$S_{aa}(f,t)=v_r v_z^4(\frac{\Delta\sigma}{\mu})^2 CW(\frac{\mu}{4\pi\rho})^2(\frac{v^2}{r^4 f^3})^2\ \sum_i\ w(t-\frac{r}{v_i})(A_{Ii}\hat{n}_i{}^2+A_{Fi}\hat{n}^3)^2 \tag{10}$$

where W indicates the mean width of the rupture plane. The apparent velocity of the rupture propagation $\hat{v}_z$ implies the directivity effect. Simplifying $v_z\sim v_r$ one can write

$$\hat{v}_z = v_r \cdot \frac{1}{(1-\frac{v_r}{v}\cos\Psi)} \tag{11}$$

with Ψ being the angle between the direction to the rupture propagation and the direction to the receiver. Eq.(10) describes the direct body waves and therefore is valid in a time window with the duration $\hat{T}_r$, the ̄ ̄rent rupture time

$$T_r = T_r(1 - \frac{v_r}{v}\cos\Psi) \tag{12}$$

If it is assumed that the far field and the intermediate field terms are independent, the squared term of eq.(10) can be simplified

$$(A_I\hat{n}^2 + A_F\hat{n}^3)^2 \approx A_I^2\hat{n}^4 + A_F^2\hat{n}^3 \leq 2(A_I^2\hat{n}^4 + A_F^2\hat{n}^3) \geq 0 \tag{13}$$

where the upper and lower bounds are given which are actually reached only in an oscillating way. Utilizing eq.(10) to (13) and rearranging the terms one can write the expression for the evolutionary source spectrum of the acceleration in a clearly arranged form [6] (s.APPENDIX).

The major advantage of the suggested seismic model when simulating synthetic accelerogramms from evolutionary spectra is the fact that the governing parameters are based on physical reasoning and moreover, they can be easiliy calculated - for some cases even by hand [6].

SUMMARY AND CONCLUSIONS

It is shown that the classical linear wave theory combined with stochastic considerations is able to describe the governing features of the acceleration processes which are responsible for the structural damage due to earthquakes. The time-dependent dominant frequencies of the acceleration process are not continuously varying with time but are mainly constant during certain time intervals. The beginning of these time intervals match quite well with the time lags of the P- and S-waves and the surface-waves. The model also reveals that the shape of the source spectrum is of negligible importance because of the resonance amplificiation of the seismic wave field by the local site conditions.

ACKNOWLEDGEMENT

This research has been partially supported by the Deutsche Forschungs-gemeinschaft under contract No. SFB 96/B8 and by the Austrian Fond zur Förderung der wissenschaftlichen Forschung under contract No. FSP S-30/01 which is gratefully acknowledged by the authors.

REFERENCES

[1] SCHERER,R.I.: "Die Beschreibung der Transienten Tragwerksbelastung aus Erdbebenwellen als Zufallsprozess",Diss.,TU Munich,FR Germany, 1984.
[2] SCHUËLLER,G.I.andR.J.SCHERER: "A Stochastic Earthquake Loading Model",Proc.,PRC-US-JAPAN Trilateral Symposium/Workshop on Engineering for Multiple Natural Hazard Mitigation, Beijing, China,Liu,H.(Ed.),Harbin,China, 1985,pp.G-2-1 - G-2-8.
[3] AKI,K.: "Prediction of Strong Ground Motion: Theoretical Formulations",Proc.NATO,Adv.Study,Ankara,Turkey,June,1985.

[4] PAPAGEORGIOU,A.S., and K.AKI: "A Specific Barrier Model for the Quantitative Description of Inhomogeneous Faulting and the Prediction of Strong Ground Motion,I.Description of the Model, II.Application of the Model",Bull.Seism.Soc.Am.,Vol.73,1983,pp. 693-722,pp.953-978.

[5] AKI,K.and RICHARDS,P,G.: "Quantitative Seismology,Theory and Methods",Vol.1,Vol.2,Freeman & Co.,San Francisco, 1980.

[6] SCHERER,R.J.and G.I.SCHUELLER: "A Stochastic Earthquake Load Model and its Effects on Aseismic Design Procedures",Proc.2nd Int. Sem.on Stoch.Math.in Struct.Mech.,F.Casciati u.L.Faravelli (Eds.), Univ.Pavia,1986.

[7] SAVAGE,J.C.: "Relation of Corner Frequency to Fault Dimensions", J.Geophys.Res.,pp.3788,1972.

APPENDIX

REARRANGED EQUATIONS FOR THE SOURCE SPECTRUM

Utilizing eq.(10) to (13) and rearranging the terms one can write the expressions for the evolutionary source spectrum of the acceleration in a clearly arranged form:

- evolutionary source spectrum of the acceleration

$$S_a(f,t) = \sum_j \sum_i \{\vec{S}_{oji} S_j {}^H G_j w_{Ti} w_F\} \quad \begin{array}{l} i = P,S \\ j = F,I \end{array} \tag{A1}$$

- spectral forms

$$\begin{aligned} S_F(f) &= 1 \\ S_I(f) &= 1/f^2 \end{aligned} \tag{A2}$$

- basic intensity

$$S_o = (\frac{1}{4\pi})^2 (\frac{\mu}{\rho})^2 \, v_r{}^5 (\frac{\Delta\sigma}{\mu})^2 C \tag{A3}$$

- basic intensities of the body waves

$$S_{oi} = S_o \frac{1}{v_i{}^4} c_{Di} \quad i = P,S \tag{A4}$$

with

$$c_{Di} = \frac{1}{1- \frac{v_r}{v_i} \cos\Psi} \tag{A5}$$

- intensities of the body waves

$$\vec{S}_{oFi} = S_{oi} W_F \frac{1}{v_i{}^2} \vec{A}_{Fi}^2$$

$$S_{oIi} = S_{oi} W_I \frac{1}{4\pi^2} \vec{A}_{Ii} \quad i = P,S \tag{A6}$$

- time and frequency windows

$$w_{Ti}(t) = \hat{w}_{Tr}(t - \frac{r}{v_i} + \frac{\hat{T}_r}{2}) \quad i = P, S$$

$$w_F(f) = w_F(f + \frac{F}{2}) \tag{A7}$$

with

$$w_X(x) = \{ \begin{matrix} 1 \\ 0 \end{matrix} \quad -\frac{X}{2} \leq x \leq \frac{X}{2}$$

$$\text{elsewhere} \tag{A8}$$

where $F = f_{max} - f_{corner}$ represents the bandwidth in the frequency range

with

$f_{max} = 25$ Hz   after empirical suggestion

$f_{corner} = \frac{1}{T_r}$ [4] and [7]

If one is also interested in velocity or displacements in the low frequency range one can improve the frequency window by changing the idealized cut-off at the corner frequency by a linear decrease of the spectral amplitudes between the corner frequency and zero. This represents the spectrum deduced from far field investigations, i.e. Aki's ω-sqare model

- geometrical attenuation

$$H_{GF}^2 = \frac{1}{r^2(\xi)} = \frac{1}{r_o^2 + (v_z t - \xi_a)^2}$$

$$H_{GI}^2 = H_{GF}^4$$

SEQUENTIAL EARTHQUAKE RESPONSE OF AN EARTH DAM

Ahmed M. Abdel-Ghaffar[*], M. ASCE

ABSTRACT

Sequential earthquake-response records of the well-instrumented Long Valley earth dam (Mammoth Lake area in California) are analyzed and engineering interpretations are made. The dam, instrumented with a multi-channel system having 22 accelerographs tied together with common start, survived strong shakings during a series of nine earthquakes. Dynamic properties of the dam are identified from the spectral and correlation analysis of the earthquake records. The applicability and validity of existing 3-D models in predicting these properties are examined. In addition, travelling seismic wave effects or nonuniformity of ground inputs as well as nonlinear behavior of the dam response were evident from the analysis of these records.

GENERAL DESCRIPTION

The Long Valley Dam [3] is a rolled earthfill, compacted embankment with maximum height of 180 ft (54 m). The dam which is built in a narrow canyon and constructed in the late 1930's, is located in Mono County approximately 22 miles (35 Km) northwest of the city of Bishop and 240 miles (366 Km) north of the city of Los Angeles. The dam has an impervious zone which is a significant portion of the embankment (Fig. 1). The dam lies in the close proximity of an active fault area [3]; both the Sierra Frontal fault and the Hilton Creek fault systems are only 5 miles (8 Km) from the dam.

RECENT SEISMICITY

In 1980, a series of earthquakes occurred in the Mammoth Lakes area in the vicinity of Long Valley Dam [6]. The initial shock occurred May 25, 1980. This $M_L=6.0$ earthquake was followed by numerous aftershocks and another earthquake ($M_L=6.3$) on the same day. Two days later, a third earthquake ($M_L=6.0$) occurred at 7:51 a.m. Minor cracks were discovered in the roadway at the contact between the dam fill and the north abutment (Fig. 1) and extended a short distance down the groin. Investigation showed that these insignificant cracks did not extend into the dam or bedrock. There has been a marked increase in local seismic activity near the dam involving an $M_L=5.5$ on September 30, 1981 [4] and two moderate earthquakes, both of $M_L=5-5.5$ on January 6, 1983 [5].

STRONG-MOTION RECORDS

The dam is instrumented with a multi-channel central-recording accelerograph system (set up by the California Division of Mines and Geology) in which all instruments were tied together with common start. The shaking during each earthquake triggered a total of 22 strong motion accelerographs located on or near the dam (see Fig. 1), providing the most

(*)Associate Professor, Civil Engineering Department, Princeton University, Princeton, New Jersey 08544

78

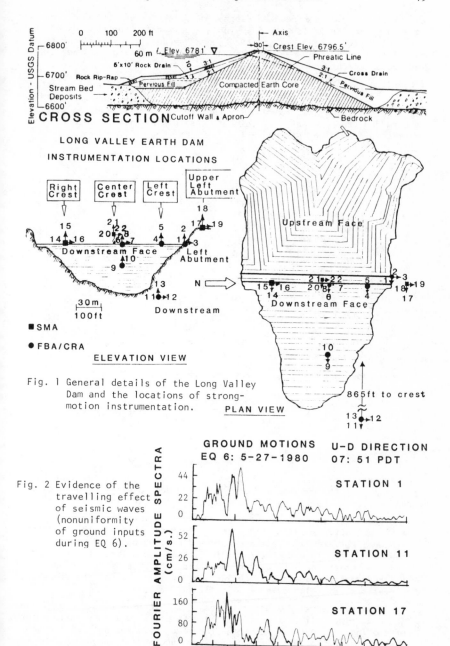

Fig. 1 General details of the Long Valley Dam and the locations of strong-motion instrumentation.

Fig. 2 Evidence of the travelling effect of seismic waves (nonuniformity of ground inputs during EQ 6).

extensive array of earthquake response measurements yet obtained for
earth dams in the United States. Table 1 summarizes the records of the
May 1980 series of earthquakes that shook the dam. It is evident that
all of the six earthquakes vary in the degree of shaking and the maximum
induced acceleration.

SPECTRAL ANALYSIS OF THE EARTHQUAKE RECORDS

The records of the dam were analyzed to obtain the Fourier amplitude
spectra, the cross-spectral density functions, spectral coherence, phase
spectra and the apparent wave velocity (associated with propagation of
seismic waves) which is given by:

$$v_s \equiv v_{mn}(f) = \frac{L}{\tau_{mn}(f)} \quad , \tag{1}$$

where

$$L = \sqrt{(x_m - x_n)^2 + (y_m - y_n)^2 + (z_m - z_n)^2} \tag{2}$$

is the distance between two points m and n, and

$$\tau_{mn}(f) = \frac{\theta_{mn}(f)}{2\pi f} \tag{3}$$

is the delay time with the phase angle:

$$\theta_{mn}(f) = \tan^{-1}\left[\frac{Q_{mn}(f)}{C_{mn}(f)}\right] \tag{4}$$

in which $Q_{mn}(f)$ and $C_{mn}(f)$ represent the cospectrum and quadspectrum
of the cross spectral density function.

The coherence function is defined as

$$\gamma_{xy}^2(f) = \frac{|G_{xy}(f)|^2}{G_{xx}(f)G_{yy}(f)} \quad , \quad 0 \leq \gamma_{xy}^2(f) \leq 1 \tag{5}$$

in which

$$G_{xy}(f) = |G_{xy}(f)|e^{-i\theta_{xy}(f)} \quad \text{and} \quad |G_{xy}(f)| = \sqrt{C_{xy}^2(f) + Q_{xy}^2(f)} \tag{6}$$

The spectral analysis of this unique set of records provided valuable
information on the dynamic properties of the dam material as well as the
ground motions. Following are some of the important findings:

1. The characteristics of the recorded ground motion at the right abutment were different from those of the left abutment and the downstream
   ground site (Fig 2). This nonuniform distribution of ground acceleration along the length of the dam influenced the nature of the dynamic
   response of the dam, as evidenced by Fig. 3; phase difference along
   the boundaries and strong coupling between longitudinal and transverse
   vibrations were also evident (Fig. 3).

2. The analysis revealed substantial changes in the dynamic properties of
   the dam, since all of the six earthquakes of Table 1 vary in the degree
   of shaking. Figure 4 illustrates the changes in the natural frequencies from one shaking to the other (the fundamental frequency from
   EQ 1, 3, 4, and 6 was estimated to be 1.88 Hz, 1.68 Hz, 1.38 Hz, and

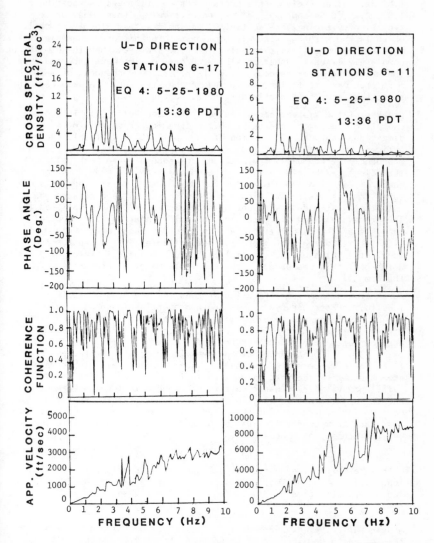

Fig. 3 Spectral analysis of the recorded motions on the dam
crest (St. 6) and the abutment (St. 17) as well as
the downstream site (St. 11); nonsimilarity of the
cross spectral density between the response and the
two inputs is an indication of the effects of
travelling seismic waves.

1.83 Hz, respectively); this corresponds to stiffness degradation of 1.00, 0.79, 0.54 and 0.95 ratios, respectively. This behavior is typical of yielding, or "softening", dynamic systems involving soil, which is known to be extremely nonlinear. Furthermore, possible source of low coherence (less than unity) of Fig. 3 arises when the dam is not completely linear. Other possible sources for low coherence could be the existance of extraneous noise of the measured response or when more than one excitation is applied to the structure (in such case the response cannot be directly attributed to the measured single input).

3. Because of the V-shape narrow canyon, the Long Valley Dam is a good example of an earth dam for which the dynamic response is of a three-dimensional nature. Evaluation of the applicability of existing analytical and numerical 2-D and 3-D analyses in predicting the resonant frequencies (and modes of vibration) corresponding to peak values of the calculated Fourier and correlation spectra were made. The adequacy of existing 3-D procedures for dynamic and seismic analysis of earth dams is verified. Figure 5 shows the comparison between the computed first mode (using both finite element [7] and simplified [2] 3-D procedures and the modal configuration inferred from spectral analysis of the records); higher computed 3-D modes match very well with the measurement of vibration modes (Table 2). In the two 3-D models the value of the low-amplitude shear modulus is taken as 3.39×10^6 psf; this was based on the available material properties of the dam [3].

4. Parametric system identification methods [1] are used to estimate the in-situ, undistrubed dynamic-strength properties (such as shear modulus or stiffness and damping, both as functions of induced dynamic strain) of the dam from its earthquake records. Predicted results were compared with those measured under controlled laboratory conditions [3]. It was found that the dynamic properties extracted from field performance during earthquakes are of relatively higher values than those from laboratory-measured soil properties of the dam materials. However, the general pattern of changes in modulus strength and energy dissipation of (Fig. 6) the dam materials is very similar. Considering the results of this study, laboratory dynamic strength test data from Long Valley Dam should be increased to simulate the undisturbed strength condition.

ACKNOWLEDGMENTS

This research is supported by a Grant (No. ECE84-15256) from the National Science Foundation; this support is gratefully acknowledged. The computer assistance by Ms. C. Zeniou is greatly appreciated.

REFERENCES

1. Abdel-Ghaffar, A. M. and Scott, R. F., "Shear Moduli and Damping Factors of Earth Dams," Journal of Geotech. Engrg Div., ASCE, Vol. 105, No. GT12, Proc. Paper 15034, Dec. 1979, pp. 1405-1426.

2. Abdel-Ghaffar, A. M., and Koh, A.-S., "Three-Dimensional Dynamic Analysis of Nonhomogeneous Earth Dams," Intl. Journal of Soil Dyn. and Earthquake Engrg, Vol. 1, No. 3, 1982.

3. Hoye, W. W., Hengenbart, J. L., and Matsuda, S., "Long Valley Dam: Stability Evaluation," City of Los Angeles, Dept. of Water and Power, Report No. AX203-24, Sept. 1982.

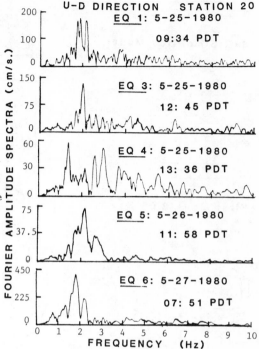

Fig.4 Changes in the resonant frequencies (frequency shift) from several earthquakes vary in degree of shaking.

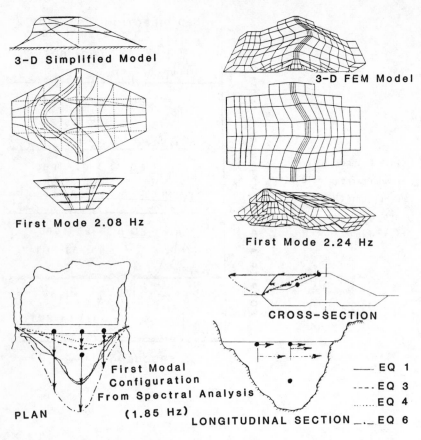

Fig. 5 Comparison between the first computed 3-D mode shape and measured modal configuration.

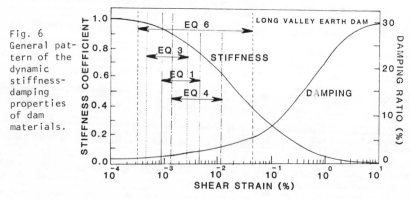

Fig. 6 General pattern of the dynamic stiffness-damping properties of dam materials.

4. McJunkin, R. D., and Kaliakin, N. A., "Strong-Motion Records Recovered from the Mammoth Lakes, CA, Earthquake of Sept. 30, 1981," CDMG Office of Strong-Motion Studies Report OSMS 81-10.1, 22 p., 1981.

5. McJunkin, R. D., Shakal, A. F., and Kaliakin, N. A., "Strong-Motion Records Recovered from the Mammoth Lakes, CA, Earthquakes of January 6, 1983," Calif. Div. of Mines and Geology, Office of Strong Motion Studies, Sacramento, CA.

6. Turpen, C. D., "Strong-Motion Records from the Mammoth Lakes Earthquakes of May 1980," CDMG Preliminary Report, 27, 42 p., 1980.

7. Prevost, J. H., Abdel-Ghaffar, A. M. and Lacy, S. J., "Nonlinear Dynamic Analyses of an Earth Dam," Journal of Geotech. Engrg., ASCE, Vol. 111, No. 7, July 1985, pp. 882-897.

TABLE 1
The May 1980 Series of Earthquakes that Shook The Long Valley Earth Dam, California

ACCELEROGRAPH LOCATION	CHANNEL # STATION #	EQ.1 May 25, 1980 9:34 PDT PEAK ACCELERATION	EQ.2 May 25, 1980 9:49 PDT PEAK ACCELERATION	EQ.3 May 25, 1980 12:45 PDT PEAK ACCELERATION	EQ.4 May 25, 1980 13:36PDT PEAK ACCELERATION	EQ.5 May 26, 1980 11:58 PDT PEAK ACCELERATION	EQ.6 May 27, 1980 7:51 PDT PEAK ACCELERATION
Upper Left Abutment	1 (St. 17)	0.276 g	0.067 g	0.193 g	0.154 g	0.073 g	0.398 g
	2 (St. 18)	0.119 g	0.037 g	0.121 g	0.147 g	0.028 g	0.313 g
	3 (St. 19)	0.427 g	0.194 g	0.508 g	0.292 g	0.110 g	1.024 g
Center Crest	1 (St. 20)	0.225 g		0.121 g	0.094 g	0.118 g	0.472 g
	2 (St. 21)	0.146 g		0.138 g	0.115 g	0.030 g	0.191 g
	3 (St. 22)	0.231 g		0.210 g	0.163 g	0.090 g	0.287 g
Right Crest	1 (St. 14)	0.15 g				0.102 G	0.485 g
	2 (St. 15)	0.131 g				0.029 g	0.248 g
	3 (St. 16)	0.201 g				0.078 g	0.472 g
Left Abutment	1	0.079 g		0.075 g	0.063 g		0.207G
	2	0.112 g		0.73 g	0.055 g		0.119 g
	3	0.125 g		0.088 g	0.099 g		0.208 g
Left Crest	4	0.153 g		0.120 g	0.092 g		0.279 g
	5	0.178 g		0.143 g	0.108 g		0.240 g
Center Crest (of downstream face)	6	0.223 g		0.131 g	0.097 g		0.482 G
	7	0.227 g		0.203 g	0.168 g		0.309 g
	8	0.148 g		0.132 g	0.115 g		0.155 g
Center of Downstream face	9	0.191 g		0.141 g	0.118 g		0.315 g
	10	0.19 g		0.153 g	0.116 g		0.210 g
Downstream on outlet in bedrock	11	0.068 g		0.060 g	0.043 g		0.180 g
	12	0.109 g		0.112 g	0.083 g		0.219 g
	13	0.081 g		0.075 g	0.069 g		0.089 g

Table 2
Comparison Between Natural Frequencies (in Hz) Computed
by the 3-D Models and Those Obtained from Spectral
Analysis of Earthquake Records (EQ 1)

Mode Order	Resonant Frequencies from Spectral Analysis of E/Q 1			3-D Simplified Model [2]	3-D Finite Element Model [7]
	$U-D^{(a)}$	$L^{(b)}$	$V^{(c)}$		
1	1.85	1.85	1.85	2.08	2.24
2.	2.15	2.15	2.15	2.44	2.96
3	2.45	2.05	2.50	2.65	3.00
4	2.65	2.65	-	2.96	3.08
5	2.95	2.85	2.80	3.36	3.39
6	3.25	-	3.10	3.39	3.53
7	3.40	3.45	-	3.50	3.58
8	3.65		3.60	3.52	3.72
9	3.80	3.95	3.85	3.79	3.74
10	4.10	4.25	4.25	4.02	3.77

(a) U-D ≡ Upstream-downstream direction.
(b) L ≡ Longitudinal direction.
(c) V ≡ Vertical direction.

IMPORTANCE OF PHASING OF STRONG GROUND MOTION
IN THE ESTIMATION OF STRUCTURAL RESPONSE

Norman A. Abrahamson[1] and Jogeshwar P. Singh,[2] M. ASCE

ABSTRACT

A ground motion time history and its response spectrum are a
function of its Fourier amplitude and Fourier phase spectrum. For a
given Fourier amplitude spectrum, different Fourier phase spectra can
produce widely different time histories and response spectra. For
proper estimation of time histories or response spectra that preserve
the characteristics of strong ground motion due to the effects of
source, travel path and local soil conditions, it is important to
understand the characteristics of both the Fourier amplitude and the
Fourier phase spectra. In this paper, the influence on Fourier phase
spectrum of seismic source directivity due to rupture propagation along
the earthquake source is investigated. The phasing is described using
frequency-dependent phase difference histograms. Strong motion recordings from the 1979 Imperial Valley earthquake are used to demonstrate
the influence of directivity on phase. Significant differences are observed in the mean and variance of the phase difference histograms due
to directivity for stations equidistant from the fault and located on
similar soil conditions. The synthetic time histories generated for a
given Boore-type Fourier amplitude spectrum, together with phase difference histograms representative of directivity effects, preserve the
observed increase in the long-period spectral velocities, due to
directivity.

INTRODUCTION

Evidence of the nonstationary nature of recorded strong motion
data has increased interest in the nonstationy properties of strong
ground motion. The goal of many studies has been to generate nonstationary time histories for which the response spectrum will approximate a given "target" spectrum. From examination of the characteristics of the Fourier phase of several strong motion recordings, we
have found the phasing to be significantly sensitive to near-source
effects such as directivity. In this paper, we have selected Fourier
phase to describe the nonstationarity of ground motion and we examine
the effect on the response spectrum and time histories of variations
in Fourier phase due to near-source effects.

Although (a) Fourier amplitude spectrum is also affected by
directivity and (b) Fourier amplitude and phase spectra are affected

---

[1] Senior Seismologist, [2]Director, New Technology Group, Harding
Lawson Associates, San Francisco, California

by low Q values for stations located in the fault zone (Singh, 1982), we have considered only the effect on ground motion of the phasing due to directivity.

METHODOLOGY

Generation of the synthetic ground motions is carried out in the frequency domain. This requires specifying both an amplitude and a phase spectrum. A simple model for the Fourier amplitude spectrum of SH-waves based on established seismic theory is used. It consists of the modified Brune spectrum given by Boore (1983). In the Boore model, the acceleration spectrum $A(\omega)$ of SH waves at a distance R from a fault generating an earthquake with moment M_0 is given by

$$A(\omega) = \frac{C M_0}{R} \quad \frac{\omega^2}{1+ (\omega/\omega_c^2)} \quad \frac{1}{[1+(\omega/\omega_m)^8]^{1/2}} \quad \exp -\omega R/2Q\beta$$

where C is a constant, ω_c is the corner frequency, ω_m is a high-frequency cutoff, Q is the coefficient of inelastic attenuation, and β is the shear wave velocity. The constant C is given by

$$C = \frac{R_{\theta\phi} \cdot FS \cdot PRTITN}{4\pi\rho\beta^3}$$

where ρ is the mass density, $R_{\theta\phi}$ is the radiation pattern, FS is the free surface effect, and PRTITN is the factor that accounts for the partitioning of the S-wave energy in the two horizontal components. The corner frequency is related to the moment and dynamic stress drop $\Delta\sigma$ as

$$\omega_c = 2\pi \, 4.9 \times 10^6 \, \beta \, (\Delta\sigma/M_0)^{1/3}.$$

Near-source estimates of the high-frequency cutoff ω_m are discussed in Hanks (1982).

We have adopted parameter values used by Boore (1983): $\rho = 2.7$ gm/km^3, $\beta = 3.2$ km/sec, $R_{\theta\phi} = 0.63$, $Q = 300$, $FS = 2$, $PRTITN = .71$, and $\omega_m = 30\pi$ rad/sec.

The phase spectrum is much more complex than the amplitude spectrum. This is in part due to the phase being modulo 2π with zero mean. To examine the phase characteristics, the phase derivative or unwrapped phase is often used. In this paper, we have used the phase difference to study the phase. Phase differences have been used in many studies for generating nonstationary time histories (e.g., Sawada, 1984; Nigam, 1984; Ohsaki, 1979).

Over a selected frequency band, the phase at near-frequencies is subtracted and a histogram of phase differences is formed. The histogram closely resembles a normal distribution (Nigam, 1984) and therefore can be parameterized by the mean and variance. The phase difference is called the group delay because the mean value of the phase difference controls the arrival time of the energy in the given bandwidth. If a constant Fourier amplitude A and a linear Fourier phase $\phi(\omega)$ is assumed over the frequency bandwidth $\Delta\omega$, then the envelope function is

$$X(t) = \frac{4A}{\phi'+t} \quad \sin \quad \frac{(\phi'+t)\,\Delta\omega}{2}$$

where ϕ' is the mean phase derivative over the frequency band.
Including the variance of the phase difference histogram in the phase
model affects the envelope of the motion. That is, increasing variance causes a decrease in amplitude of the envelope and an increase in
duration of motion, while decreasing in variance causes an increase in
amplitude of the envelope and a decrease of the duration in the motion.
This is consistent with squeezing or stretching of the ground motions
in the near field due to directivity (Singh, 1981). Together, the
mean phase and variance of the phase difference histogram define an
envelope of the ground motion. By estimating the mean and variance of
different frequency bands, frequency-dependent envelopes of the ground
motion can be estimated.

Because the envelope of the acceleration time history is generally
the envelope of the high-frequency energy, the envelopes of velocity
and displacement time history related to long-period motions may be
different from the acceleration envelope. Many time domain methods for
generating nonstationary acceleration time histories apply an envelope
to filtered random noise. The accelerograms generated in this manner
may appear realistic; however, the amplitude and timing of the peaks
of velocities and displacements in the record obtained by integrating
these synthetic accelerograms may be unrealistic. This can lead to
inaccurate estimates of long-period ground motions because of misrepre-
sentation of the effects of source travel paths and local soil conditions.

By defining a frequency-dependent phase difference histogram, we
include envelopes for the velocity and displacement range of the
spectra as well as for the acceleration range.

EMPIRICAL EXAMPLE

We have examined selected recordings from the 1979 Imperial Valley
earthquake to demonstrate the importance of phasing in estimation of
strong ground motion. Other recent earthquakes that exhibit similar
effects of phasing are the 1979 Coyote Lake, 1980 Livermore, and 1984
Morgan Hill earthquakes.

During the 1979 Imperial Valley earthquake, the rupture from the
epicenter propagated primarily northward along the Imperial fault.
Stations El Centro 7 and Bonds Corner, located within 1 and 3 km from
the fault breaks, respectively, and on similar soil conditions, represent the near-source phasing in the forward rupture azimuth and the
near-source phasing in the back rupture azimuth, respectively. The
time histories from the 230° component for these two stations are
shown in Figure 1. The clear difference in the waveforms is due to
directivity and is discussed in detail by Singh (1982).

Figure 2 shows the difference in the phasing of the El Centro 7
and Bonds Corner time histories using the wide-band (1 to 15 Hz) phase
difference histograms. The wide-band histograms have similar shapes,
but significant differences in the frequency-dependent mean and variance are appparent in Figures 3 and 4. The differences are largest at
the long periods because directivity has its strongest influence at
the long periods.

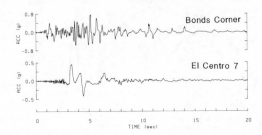

Figure 1. Recorded accelerograms from 1979 Imperial Valley earthquake, 230° component.

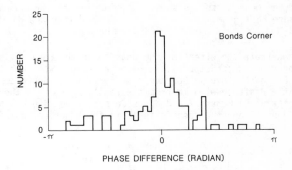

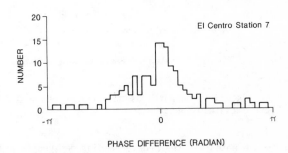

Figure 2. Phase difference histograms of El Centro 7 and Bonds Corner accelerograms for frequency band of 1 to 15 Hz.

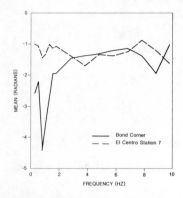

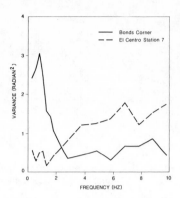

Figure 3. Frequency dependence of the mean of the phase
difference histogram.

Figure 4. Frequency dependence of the variance of the
phase difference distribution.

    To demonstrate the effect of the phasing on the response spectrum
and the shape of time histories, we have generated two synthetic time
histories using the inverse Fourier transform of the Boore amplitude
spectrum at a distance of 3 km from the fault, along with the recorded
phase spectra at El Centro 7 and Bonds Corner. The time histories are
shown in Figure 5 and the response spectra for 5 percent damping are
shown in Figure 6.

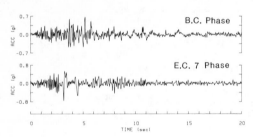

Figure 5.   Time histories generated using recorded phase spectrum with
            Boore amplitude spectrum.

It is important to note that the Fourier amplitude spectra are identical for the two synthetic motions; the differences in the shape of
time histories and response spectra are due only to the difference in
phasing. Furthermore, the spectral velocity over periods between 1
and 4 sec in the forward rupture azimuth increases up to 50 percent.
For a given Fourier amplitude spectrum, this increase is attributed to
the change in phasing in the forward azimuth.
    The above comparison was based on recorded phase spectra. Using
the normal distribution to approximate the phase difference histogram,

a synthetic phase spectrum can be constructed by summing random samples of the phase difference distribution. Here, we have used the frequency-dependent mean and variance of the recorded phase at El Centro 7 and Bonds Corner to demonstrate the method. The time histories generated using the phase sampled from the phase differences distribution, along with the identical Boore amplitude spectra used earlier, are shown in Figure 7. The response spectra for these two synthetic time histories are shown in Figure 8. Comparing these time histories and spectra with those computed using the recorded phase, it is apparent that the shape of time histories and response spectra can be preserved using the random sample from phase difference distributions at different azimuths.

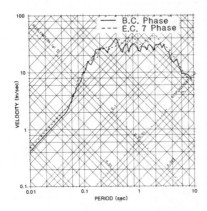

Figure 6. Response spectra for time histories generated using recorded phase spectrum with Boore amplitude spectrum (damping = 5 percent).

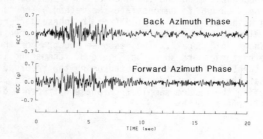

Figure 7.   Synthetic time histories generated using phase sampled from phase difference distribution with Boore amplitude spectrum.

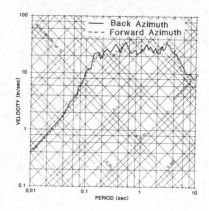

Figure 8.   Response spectra for time histories generated using phase sampled from phase difference distribution with Boore amplitude spectrum (damping = 5 percent).

CONCLUSIONS

We have shown the importance of the phasing in the estimation of ground motions. Phase is strongly affected by near-source effects such as directivity, and the spectral velocity is influenced by the phase. When generating synthetic time histories, it is important to use an appropriate frequency-dependent phase model rather than a filtered noise model based on a single envelope for all frequencies to preserve the influence on ground motions of source, travel path, and local soil conditions. This study demonstrates one method for generating a phase spectrum through the use of the frequency-dependent phase difference histogram. Other parameters influencing the frequency-dependent mean and variance of this distribution require further study. The influence on the structural response of two time histories having comparable peak ground accelerations but different phasing is discussed in detail by Singh (1985).

REFERENCES

Boore, D. M. (1983). Stochastic simulation of high-frequency ground motions based on seismological models of the radiated spectra, Bull. Seism. Soc. Am, vol. 73, 1865-1894.

Hanks, T. C. (1982). f_{max}, Bull. Seism. Soc. Am., vol. 72, 1867-1879.

Nigam, N. C. (1984). Phase properties of earthquake ground acceleration records, Proc. Eighth World Conf. Earthquake Eng., vol. II, 549-556.

Ohsaki, Y. (1979). On the significance of phase content in earthquake ground motion, Earthquake Eng. Struct. Dyn., vol. 7, 427-439.

Sawada, T. (1984). Application of phase differences to the analysis of nonstationarity of earthquake ground motion, Proc. Eighth World Conf. Earthquake Eng., vol. II, 557-564.

Singh, J. P. (1982). Importance of local structure and source characteristics in estimation of nearfield strong ground motions, Proc. Third Int. Earthquake Microzonation Conf., Seattle, Washington.

Singh, J. P. (1981). The influence of seismic source directivity on strong ground motions, Ph.D. dissertation, University of California, Berkeley.

Singh, J. P. (1985). Earthquake ground motions: Implications for designing structures and reconciling structural damage, Earthquake Spectra, vol. 2, November 5, February, 239-270.

EVALUATION OF BUILDING RESPONSE RECORDED
DURING THE MORGAN HILL EARTHQUAKE

Charles A. Kircher[1] (M. ASCE) and I-Kwang Chang[2] (M. ASCE)

Introduction

On April 24, 1984, the Morgan Hill Earthquake (Richter magnitude 6.2) occurred within approximately 15 km of IBM Corporation's computer-equipment manufacturing facility located just south of San Jose, California. This event produced significant response, but only limited damage, to buildings and equipment at the facility. Damage would probably have been much more extensive at the facility had IBM not undertaken a Seismic Program to strengthen older buildings and brace important equipment. As one element of the Seismic Program, three buildings at the facility had been instrumented prior to the Morgan Hill Earthquake and excellent records of building response were obtained for this event.

This paper describes related studies and presents findings of the evaluation of recorded building response for one of the three instrumented buildings, Building 012. Building 012 is of particular interest since it had been extensively tested and analyzed prior to the earthquake, and, unlike most other older buildings at the site, had not been seismically upgraded. Thus, the results of the earthquake-response evaluation provided the basis to confirm previous test and analytical findings regarding the building's capacity to withstand moderate and severe earthquakes.

Description of Building

Building 012 is a five-story, reinforced-concrete, moment-frame structure with a Penthouse. The building was designed in 1967-1968 to meet Seismic Zone 4 requirements of the 1964 Edition of the UBC (Reference 1), the edition of the UBC governing building design for that period. The 1964 Edition of the UBC required five-story, moment-frame structures to be seismically designed for less than 5 percent of the building's weight laterally, and did not include special ductility provisions now considered essential for construction of moment-frame buildings in zones of high seismicity.

---

1. President, Jack Benjamin and Associates, Inc., Mountain View, California

2. Advisory Engineer. IBM Corporation, White Plains, New York

The building measures approximately 288 feet in plan at the 2nd floor (Terrace) level, 225 feet each way at the upper floor levels (3rd floor through 5th floor). Story heights are about 14 feet at all elevations except the 2nd story which is a little over 17 feet in height. The floor plan of the Building at the Terrace level and an elevation view of a section at the perimeter of the building are shown in Figure 1.

All lateral forces are carried between upper floor elevations and the foundation elevation by the interior and perimeter frames. Perimeter frames are located on column lines 2, 10, C and L, and the interior frames are located on column lines 3 through 9, and D through K. Due to larger member sizes, the perimeter frames are much stiffer laterally than the interior frames. The floor slabs are essentially rigid in plane and the bulk of the lateral force, up to yield of the structural system, will be transferred from the interior of the building through the rigid floor slabs to the building's perimeter. From there lateral force is carried by the perimeter frames down to the foundation. Thus, the perimeter columns and girders are designed to primarily carry lateral forces, and the interior columns and floor slabs are designed to primarily carry gravity loads.

Pre-Earthquake Dynamic Tests of the Building

In 1981 an experimental study of Building 012 was performed to measure the dynamic properties of the structure (Reference 2). This study was undertaken to identify the vibrational characteristics of the building as the first step in the development of equipment and piping seismic-design parameters.

Nine dominant modes of vibration, which are the lowest three translational modes in the east-west direction, the lowest three translational modes in the north-south direction, and the lowest three torsional modes, were identified from the analyses of test data. North-south frequencies and mode shapes were found to be virtually identical to east-west modes (due to building symmetry). The fundamental translational-mode frequency for both the north-south and east-west direction was measured at 1.65 Hz.

On the basis of the mode-shape measurements, it was concluded that a soft story exists between the 2nd and 3rd floors of Building 012 and that previous building analyses, which did not fully account for the soft-story effect, may have underestimated seismically induced force in columns at the 2nd-floor level. As a result of an increased concern regarding Building 012's capacity, a seismic analysis, described in the next section, was undertaken.

Pre-Earthquake Seismic Analysis of the Building

A seismic analysis of Building 012 was performed to calculate and evaluate seismic capacity, considering potential soft-story and nonlinear response, for two earthquake levels: the Moderate Level Earthquake (MLE) and the Severe Level Earthquake (SLE). The MLE corresponds to an event which has a reasonable likelihood of occurring one or more times during the life of the structure (i.e., 50 percent probability of being exceeded for a 50-year design life).

The SLE corresponds to an event which is unlikely, but may occur during the life of the structure (i.e., 10 percent probability of being exceeded for a 50-year design life). Ground response spectra corresponding to MLE and SLE events were obtained from an existing seismic-hazard analysis study of the IBM San Jose site.

The seismic analysis found that moderate structural damage would occur for the MLE and that significant cracking at the top and bottom of 2nd-to-3rd floor perimeter columns and, to a lesser degree, cracking of interior columns would occur.

Most importantly, the seismic analysis found that extreme structural damage would occur for the SLE and that all 2nd-to-3rd floor columns would experience complete shear/compression failure. It was concluded, therefore, that Building 012 would be potentially life threatening during a severe earthquake.

Strong-Motion Instrumentation

As part of a 1983 project to instrument the IBM San Jose facility with strong-motion accelerograph units, four units were installed in Building 012 and one unit was installed at the ground level near the base of Building 012. The locations for the DSA-1 units are shown in Figure 1.

The four locations for DSA-1 units in the building were selected to provide general information about changes in building response with elevation and to specifically measure relative displacement between the 2nd and 3rd floors. Two units were located on the 3rd floor to separate torsional and translational response of the building. Due to the limited number of units available for installation, the 5th floor, 4th floor and 1st floor of Building 012 could not be instrumented. All units in Building 012, as well as the nearby ground-level unit, were wired together to ensure coincident triggering and proper synchronization of the recordings.

Damage Sustained as a Result of the Morgan Hill Earthquake

On April 24, 1984, the Morgan Hill Earthquake (Richter magnitude 6.2) occurred approximately 15 kilometers from the IBM San Jose site. This event produced moderately strong ground shaking at the site (i.e., peak ground accelerations of 0.17g near Building 012) of significant duration (i.e., long-period waves continued to be present even after 40 seconds of shaking). In the frequency region corresponding to the fundamental mode of Building 012 (i.e., 1.1 to 1.5 Hz), the ground shaking was only about one-half of the level of shaking expected for the MLE, and thus corresponds to an event which can be expected to occur approximately every 10 years.

A post-earthquake survey of Building 012 discovered earthquake-induced damage, as summarized below.

● Horizontal cracks were observed at the tops of 2nd-floor perimeter columns.

● Vertical separations were observed at the interface between perimeter columns and architectural panels which border the base of windows at the 3rd-, 4th-, and 5th-floor levels.

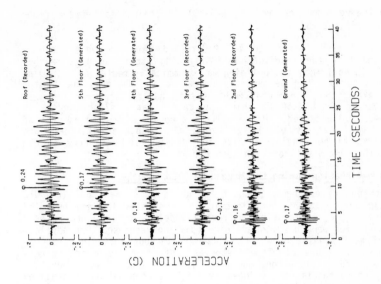

Figure 2   North-South Acceleration Time Histories
Showing Locations of Strong-Motion Insturments

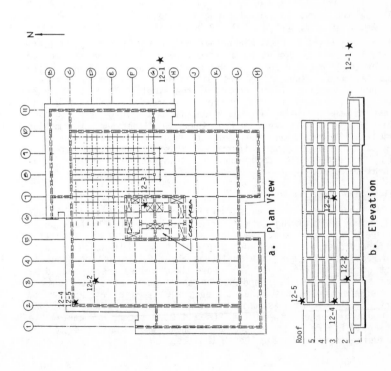

Figure 1   Plan and Elevation Views of Building 012
Used to Evaluate Building Response

● Damage was observed to drywall in stairwells between floor
   levels 2 and 3.

The observed damage to the 2nd-to-3rd floor perimeter columns
indicated that the longitudinal reinforcing steel may have reached
yield. In early May the building was dynamically tested to determine
if the structure had permanently decreased in stiffness due to the
earthquake. The fundamental mode in both horizontal directions was
found to be about 1.50 Hz which indicated that a slight change in
stiffness had occurred. Using stiffness models developed as part of
the seismic analysis of the building, the change in stiffness was
determined to be consistent with the damage which occurred to the
3rd-, 4th-, and 5th-floor architectural panels.

Building-Vibration During the Morgan Hill Earthquake

During the earthquake, the DSA-1 units in or near Building 012
recorded acceleration time histories of ground motion and building
response. After processing to remove baseline errors, etc., these
time histories were used to evaluate building response. The acceleration time histories used to evaluate north-south building response
are shown in Figure 2. Time histories labeled as "Recorded" correspond to measured vibration, and time histories labeled as "Generated"
correspond to functions developed by weighted combination of measured
response. The time history of ground motion is labeled as "Generated"
since only the first 10 seconds of ground motion were recorded, due
to equipment failure, and the balance had to be generated.

The acceleration time histories, shown in Figure 2, indicate
that the building is dominated by strong pulses of the ground motion
for about the first 7 seconds of the event. After this point in time
fundamental-mode response becomes signficant and remains dominant
from about 10 seconds to 30 seconds of the event. After about 30-35
seconds the fundamental-mode response decreases appreciably and
response of the building is again dominated by long-period ground
motion. For both east-west and north-south directions, the motion
recorded at the corner and the center of the building was found to be
within 10 percent of each other; and thus it was concluded that
building response was dominated by translational, rather than
torsional response.

Calculation of Base Shear

Base shear, expressed as a function of earthquake duration, was
calculated by multiplying each floor acceleration time history shown
in Figure 2 by the respective floor mass and summing the resulting
floor forces over all elevations. In either the north-south or east-
west directions the peak base shear was about 6,000 Kips or about 11
percent of the building's weight. Base shear values are dominated by
fundamental mode response, and second and higher modes have only a
slight effect on the base shear. Base shear values at or near 10
percent of the building's weight occur for many cycles of fundamental-mode response between the 9th and 23rd second of earthquake
vibration. Figure 3 shows the base shear which occurred in the
north-south direction during the earthquake.

Figure 3 Base Shear in the North-South Direction

Figure 4 Relative Displacement Between the 2nd and 3rd Floors in the North-South Direction

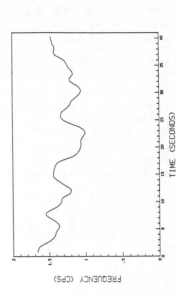

Figure 5 Change in the North-South Fundamental-Mode Frequency During the Earthquake

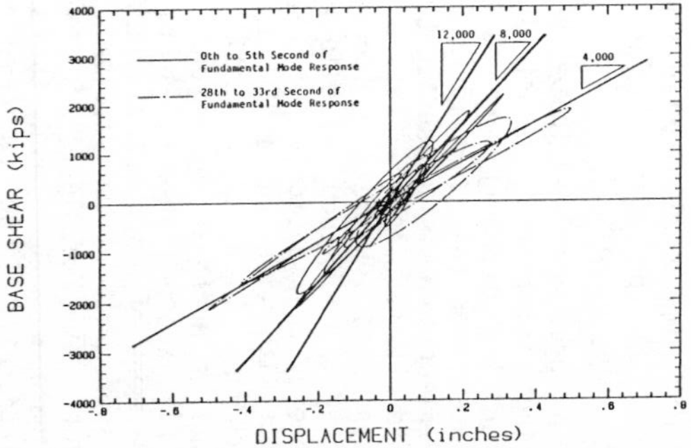

Figure 6 Degradation in North-South Stiffness Frequency During the Earthquake

Evaluation of Frequency and Stiffness Nonlinearity

Nonlinearity of the fundamental mode of vibration was examined by calculating the change in this mode's frequency during the 40 seconds of recorded earthquake shaking. Frequency changes were calculated by measuring the duration of each cycle of fundamental-mode response measured by the relative displacement of the 3rd floor with respect to the 2nd floor. The changes in frequency are similar for both directions of response. Relative displacement of the 3rd floor with respect to the 2nd floor in the north-south direction is shown in Figure 4, and the change in the north-south fundamental-mode frequency is shown in Figure 5.

The initial fundamental-mode frequency of 1.65 Hz (i.e., the measured pre-earthquake frequency) begins to degrade immediately and continues to degrade (with short-term fluctuations) to a value of 1.1 Hz at the point in time of strongest shaking. After the level of shaking has reduced significantly, the fundamental mode frequency begins to gradually increase (again with short-term fluctuations) until it reaches 1.5 Hz (i.e., the measured post-earthquake frequency). In the north-south direction the lowest value in the fluctuation of the fundamental frequency (i.e., 0.9 Hz) occurred about 10 seconds after the point in time of peak fundamental-mode response.

Two trends in the change of fundamental-mode frequency of both the north-south and east-west directions are apparent. First, there are short-term fluctuations in the frequency which are directly correlated with short-term fluctuations in the level of modal response (i.e., the larger the amplitude of vibration, the lower the frequency). Second, there is a gradual decrease in frequency with earthquake duration and build-up of modal response. This decrease in frequency remains even after the level of shaking decreases, which suggests that permanent changes in the stiffness characteristics of the building occurred.

In order to evaluate possible permanent changes in the building's stiffness characteristics, force-deflection characteristics of the building were calculated for each cycle of building response. For this calculation, force was represented by base shear and deflection was represented by displacement of the 5th floor relative to the 1st floor. Plots of base shear versus 5th floor deflection produced a hysteresis loop for each cycle of building response. In general, each loop was eliptical and relatively thin (only moderate energy dissipation). The average slope of the hysteresis loops changed with time and amplitude of response. Figure 6 illustrates the changes in force-deflection characteristics by comparing five seconds of north-south response at the beginning of the earthquake with five seconds of north-south response at the end of the shaking. The segment of response at the end of the shaking was selected to have approximately the same amplitude of base shear as the beginning segment. Average slopes of the hysteresis loops are also indicated in Figure 6. As shown, the stiffness of the building, which is approximately 12,000 Kips/inch initially, degraded to about 8,000 Kips/inch within the first five seconds of shaking. By the end of the shaking the stiffness degraded to approximately 4,000 Kips/inch, or about one-third of the original stiffness. This reduction in stiffness from 8,000 Kips/inch to 4,000 Kips/inch, for the same force level, indicates significant stiffness changes in lateral-force-carrying elements. The stiffness reduction which is apparent during the earthquake is in contrast to pre- and post-earthquake fundamental-mode frequency measurements which indicate only minor changes in stiffness.

Summary and Conclusion

The dramatic decrease in fundamental-mode frequency and degradation of building stiffness which occurred during the earth-quake indicates that logitudinal-steel reinforcement in perimeter columns yielded, and, therefore, was not fully effective in resisting seismic moment. In essence, the perimeter of the building above the 3rd floor rocked on top of the 2nd-floor columns as small horizontal cracks opened and closed at the top (and possibly the bottom) of these columns. Evidence of cracking was most pronounced at corner columns where compressive stress due to gravity load is less than that on other columns (i.e., due to smaller tributary area of the floor). Post-earthquake dynamic tests of fundamental-mode frequency (i.e., 1.50 Hz) did not indicate that the longitudinal reinforcement had yielded, since the level of lateral load (i.e., ambient wind force) during these tests was not large enough to re-open the cracks in perimeter columns.

Although observed damage was not extensive, the results of the building-response evaluation revealed that significant yielding of 2nd-floor perimeter columns occurred as a result of Morgan Hill Earthquake. This finding confirmed the results of an analytical study which had previously predicted the same degree of nonlinear behavior for a comparable event. Since the analytical study had also predicted complete failure of all 2nd-floor columns for a severe earthquake, confirmation of analytical-study validity provided the basis for a decision to vacate and strengthen the building.

Special effort by IBM in terms of dynamic testing, seismic analysis, and earthquake-data evaluation discovered a potentially dangerous condition which would otherwise have gone undetected. As a result of this discovery, Building 012 is now undergoing seismic-upgrade modification.

References

1. International Conference of Building Officials, Uniform Building Code, 1964 Edition.

2. Kircher, C. A., "Dynamic Tests of Three Commercial Buildings, "Recent Advances in Engineering Mechanics and Their Impact on Civil Engineering Practice, W. F. Chen, et al., eds., 1983, p. 368-371.

DYNAMICS OF FENDER AND RACK SYSTEMS FOR DOCKING FERRIES[*]

By Zachary Sherman, Ph.D., P.E., FASCE

ABSTRACT:
    The dynamics of large ferries striking fender systems
and piers is currently being studied in a more systematic
manner than has been done heretofore. A dynamic analysis
of a ferry fender and rack system is developed based on the
conservation of energy principle and Newton's laws of
motion. Starting with some general pattern of wales and
piles and attaching face rubber pads (bumpers) to the wales,
the nearly-elastic system can be analyzed. The dynamics of
the interaction of the ferry, rubber bumpers, wales, and
piles are developed. The equations of motion of the system
elements, coupled with their ability to store energy, can
be reduced to the simultaneous solution of two time-dependent algebraic equations. A rational analysis of such a problem must be based on a set of design criteria and assumptions which would include the maximum design velocity of the
ferry at impact, depth of water at the ferry slip, general
layout of the pile and rack system, depth and fixity of
piles, properties of soil in which piles are embedded, etc.

INTRODUCTION:
    Ferry fender and rack systems, to help berth large
ferries such as ply across the Hudson River between Manhattan Island and Staten Island in New York City, can be
rationally analyzed and designed[8,9]. Traditionally, the design
and construction of such ferry fender and rack systems have
been done empirically. High maintenance and replacement
costs of wood piles and wales has prompted some recent analytical studies of this problem.
    Early in 1978, at the time of this study (1), a major
area of concern for the Department of Marine and Aviation
of the City of New York was the condition of the Staten
Island Ferry docks (7) located at St. George, Staten Island.
Slip No. 5 in particular, had been considered in urgent
need of renovation, and its ferry fenders and associated
racks are the subject of analytical study here (1, 4).

THE SLIPS AND SLIP STRUCTURE:
    The geometry of the slip should be such that a rela-
tively large contact area between the ferry and the slip's
fender is provided for. This will result in better distribution of the impact forces.
    The slips at this site consist of rows of piles embed-
ed in a thin layer of soil and rip-rap overlaying the bedrock. Individual rows of piles are bolted to timber wales

---

[*] Consulting Engineer, Long Beach, New York 11561

made of dense wood. The bulk of the impact energy is stored in the piles, and some of this energy, which is not dissipated by friction and viscous forces, is then returned, causing rebound action.

Our study indicated that wales need to be much stiffer in order to more uniformly distribute the ferry impact forces to the piles. While ferry slips have been in existence a long time, they have typically been built by experience, with little or no attempt at a rational engineering approach, such as has been attempted in this paper or in Ref. (4).

DESIGN CRITERIA:

The design criteria and assumptions, taken from Ref. (1) for the ferry fender/rack system for Slip No. 5, St. George Terminal, Staten Island, are as follows:

1. Water depth at constant elevation.

2. Impact to piles occurs at 27 ft. from assumed point of fixity.

3. The ferry rack must be effective at an impact design velocity of 3 ft./sec. (within 99% of occurrence).

4. Ferry assumed to strike perpendicular to fender at mid-point.

5. Fender system is simply supported at both ends of the slip.

On the basis of these assumptions, the design selected consisted of 5 rows of twelve inch diameter piles spaced 24 inches on centers. Three lines of wales, 15 in. by 16 in., will run along the entire length of the rack. Interlocking pads, of 50 durometer rubber with a spring constant of 15 x 10^4 lb/in., were mounted to the vertical planks (which are fastened to the 16 in. by 15 in. wales) in vertical rows of three along the fender system. The dimensions of the rubber pads are 36 in. by 36 in., 9 in. thick, with a 10 1/2" x 10 1/2" cutout.

Preliminary analysis and design of ferry fender/rack system
In addition to the design criteria, the following basic assumptions are made:

1. At any instant of time, K.E.+P.E.=Constant for the entire system.

2. $K.E._{max}=P.E._{max}$ .... (1)

3. $\frac{1}{2}m_f v_0^2 = P.E._{piles}+P.E._{wales}+P.E._{rubber}$ .... (2)

   +Additional P.E., due to friction and viscous forces— see Sect. B, item g.

4. Ferry rack is simply supported at Points A and B, some distance L apart (see Figure 1).

5. Ferry impacts at L/2, at a 90° angle to the rack, where, K.E.=Kinetic Energy, P.E.=Potential Energy, $\frac{1}{2}m_f v_0^2$= K.E. of ferry at instant of strike, and m_f=ferry mass, v_0=ferry impact velocity.

The system of coordinates used to describe the motion of the ferry and the wales is given in Figures 3 and 4.

$$y = \xi - H + (H - \delta) = \xi - \delta \quad \dots \quad (3)$$

$$\delta = \xi - y \quad \dots \quad (3a)$$

$$\dot{y} = \partial y / \partial t = \dot{\xi} - \dot{\delta} \quad \dots \quad (3b)$$

where δ is the shortening of the rubber spring, and $\dot{y}$ is velocity of the centerline of the wales at any x, $\dot{\xi}$ the velocity of the ferry.

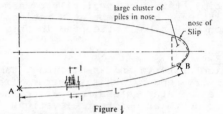

Figure 1

Ferry Rack Assumed Simply Supported at Points A & B
Pier #5 - St. George, Staten Island

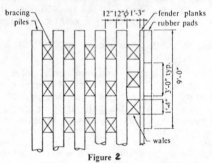

Figure 2

Section 1-1 — Fender & Rack Cross-Section

A.

Coordinate Systems:

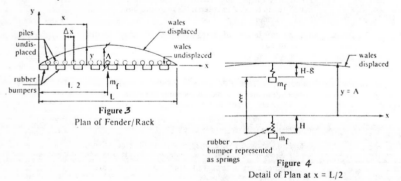

Figure 3
Plan of Fender/Rack

Figure 4
Detail of Plan at x = L/2

The displacement of the Piles and Wales is expressed in the "y" coordinate system. The displacement of the Ferry and Rubber is expressed in the "ξ" coordinate system.

B. Energy Expressions

(a) Ferry (K.E.)—(neglect P.E. stored in ferry hull due to impact).

$$\text{K.E.}_{m_f} = \tfrac{1}{2} m_f \dot{\xi}^2 = \tfrac{1}{2}(w_f/g) \cdot \dot{\xi}^2 \quad \dots \quad (4)$$

From Newton's Law (see Ref.10):

$$m_f \ddot{\xi} = -K_r \delta = -K_r(\xi - y) \quad \dots \quad (5)$$

where K_r is the spring constant of the rubber bumpers, and $\ddot{\xi} = \partial^2 \xi / \partial t^2$, the acceleration of the ferry at $x = L/2$.

(b) Wales (P.E.):

$$\text{P.E.}_{wales} = \tfrac{1}{2} \int \frac{M^2 dx}{EI} = \frac{EI}{2} \int (y'')^2 dx \quad \dots \quad (6)$$

where $1/R \cong y'' = M/EI$, $y'' = d^2y/dx^2$, and E and I are the mod. of elasticity and stiffness of the wales, respectively.

(c) Wales (K.E.):

$$\text{K.E.}_{wales} = \tfrac{1}{2} \int_0^L \dot{y}^2 dm_w \quad \dots \quad (7)$$

where m_w = mass per unit length of wales.

(d) Rubber (P.E.):

$$\text{P.E.}_{rubber} = \tfrac{1}{2} K_r \delta^2 \quad \dots \quad (8)$$

(e) Piles (P.E.):

$$\text{P.E.}_{piles} = \Sigma \tfrac{1}{2} K_p y^2 \cdot \Delta x \quad \dots \quad (9)$$

where K_p = pile spring constant, lb/in/in, and Δx = pile spacing.

(f) Piles (K.E.):

$$\text{K.E.}_{piles} = \tfrac{1}{2} \int_0^h C_1 \dot{y}^2 m_p \cdot dz \quad \dots \quad (10)$$

where m_p = mass/unit length of pile, C_1 a coefficient (see Figure 5).

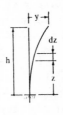

Figure 5
Typical Deflected Pile

(g) Viscous Forces:

(1) On Ferry = $F_f = f(K_{D_f} \cdot \dot{\xi}^2)$

(2) On Piles = $F_p = f(K_{D_p} \cdot \dot{y}^2)$ } Neglect

C. Motion of the Wales

The position of the wales at any time may be described by a Fourier series. As a first approximation, for preliminary design, the first term of this series will be used to describe the shape. Let

$$y = A \sin \frac{\pi x}{L} \cdot \sin \omega_{nw} t \quad \dots \quad (11)$$

where $\omega_{nw} = \pi^2 \sqrt{gEI/wL^4}$, the natural frequency of the wales

w = lb/unit length of wales

A = maximum displacement of wales at $x = L/2$

$$\dot{y} = A \omega_{nw} \sin \frac{\pi x}{L} \cdot \cos \omega_{nw} t \quad \dots \quad (11a)$$

$$\ddot{y} = -\omega_{nw}^2 A \sin \frac{\pi x}{L} \cdot \sin \omega_{nw} t = -\omega_{nw}^2 \cdot y \quad \dots \quad (11b)$$

$$\ddot{y}\big|_{x = \frac{L}{2}} = -\omega_{nw}^2 A \cdot \sin \omega_{nw} t \quad \dots \quad (11c)$$

(a) Evaluation of Potential Energies at $\dot{\xi} = 0$:

Assume that $\dot{\xi} = 0$, K.E. = 0, i.e., the Potential Energy of the system is a maximum. From Eqs. 6, 8, 9, it can be shown that:

$$(P.E._{wales})_{max.} = \frac{EI}{2} \cdot A^2 \left(\frac{\pi}{L}\right)^4 \left(\frac{L}{2}\right) \sin^2_{nw}t \quad \dots\dots\dots\dots \quad (12)$$

$$(P.E._{rubber})_{max.} = \tfrac{1}{2} K_r \delta^2_{max.} = \tfrac{1}{2} K_r (\xi - y)^2 \big|_{\xi=0}. \quad \dots\dots\dots \quad (13)$$

$$(P.E._{piles})_{max.} = \tfrac{1}{2} K_p A^2 \frac{L}{2} \quad \dots\dots\dots\dots\dots\dots\dots\dots \quad (14)$$

D. Solution of Motion Equation (of Ferry) at x=L/2
From Eqs. 5, 11:

$$m_f \ddot{\xi} = -K_r (\xi - A \sin_{nw}t). \quad \dots\dots\dots\dots\dots\dots\dots \quad (15)$$

$$m_f \ddot{\xi} + K_r \xi = K_r A \sin \omega_{nw}t \quad \dots\dots\dots\dots\dots\dots \quad (15a)$$

The general solution is:

$$\xi = \xi_h + \xi_p = \text{complementary solution} +$$
$$\text{particular solution} \quad \dots\dots\dots\dots\dots \quad (16)$$

From the initial conditions, i.e., Eqs. 1, 2, and

$$\dot{\xi}(O^+) = v_o = \text{initial velocity of ferry}, \quad \dots\dots\dots\dots \quad (16a)$$

Eq. 16 can be shown to be:

$$\xi = C \sin \omega_{nr}t + E_1 \sin \omega_{nw}t \quad \dots\dots\dots\dots\dots\dots \quad (17)$$

and,

$$\dot{\xi} = C \omega_{nr} \cos \omega_{nr}t + E_1 \omega_{nw} \cos \omega_{nw}t \quad \dots\dots\dots \quad (17a)$$

$$\ddot{\xi} = -C \omega^2_{nr} \sin \omega_{nr}t - E_1 \omega^2_{nw} \sin \omega_{nw}t \quad \dots\dots\dots\dots \quad (17b)$$

where $C = \dfrac{v_o - \omega_{nw}E_1}{\omega_{nr}}$ $\quad \dots\dots\dots\dots\dots\dots\dots\dots \quad (17c)$

$$E_1 = \frac{K_r A}{K_r - m_f \omega^2_{nw}} \quad \dots\dots\dots\dots\dots\dots\dots\dots \quad (17d)$$

and $\omega_{nr} = \sqrt{K_r/m_f}$, natural frequency of rubber $\dots$ (17e)

The above equations can be reduced to Eqs. 18, 19, and shown to be:

$$\tfrac{1}{2} m_f v_o^2 = \tfrac{1}{2} K_p A^2 \cdot \frac{L}{2} \sin^2 \omega_{nw}t + \frac{EI}{2} \cdot A^2 \cdot \frac{\pi^4}{L^4} \cdot \frac{L}{2} \cdot \sin^2 \omega_{nw}t +$$

$$\tfrac{1}{2} K_r [C \sin \omega_{nr}t + E_1 \sin \omega_{nw}t - A \sin \omega_{nw}t]^2 \quad \dots\dots \quad (18)$$

$$C\omega_{nr} \cos \omega_{nr}t + E_1 \omega_{nw} \cos \omega_{nw}t = 0 \quad \dots\dots\dots\dots\dots\dots \quad (19)$$

Equations 18, 19 can be solved simultaneously in "A" and "t", since "C" and "E_1" can both be expressed in terms of "A".

E. Actual Design
(a) Design Basis:
Based on the parametric studies, and the systems of graphs drawn for Equations 18, 19, the design was selected (and was given under the heading Design Criteria) on the following basis:

(1) Maximum bending stresses in the piles were 7,724 psi, based on an allowable stress of 8000 psi, and the diameter and spacing selected(6).

(2) Wales of 16 inches on face allows for good distribution of the impact force. This yields a reasonable maximum displacement (deflection) of about 15 inches with a corresponding maximum bending stress of only 386 psi.

(3) A 9-inch thickness for rubber appeared most economical, allowing for about 5 percent maximum absorption of the stored energy by the rubber, 94 percent by the piles, and less than 1 percent by the wales. Other combinations of piles and rubber pads gave a range of 80 to 95 percent for the impact energy absorbed by the piles.

(b) Additional Constants

(1) Modulus of Elasticity
    =3×10^6 psi

(2) Density=66 pcf

(3) Mass of Ferry=2,655(lg. tons)
    $\cdot 2240 \cdot \dfrac{1}{32.2 \cdot (12)}$
    $= 15,391 \dfrac{\text{lb-sec}^2}{\text{in}}$

(4) Length of Ferry Rack
    =$L \simeq 245$ ft. (see Figure 1)

(5) Length of piles from point of
    fixity to point of Ferry impact
    =27 ft. (see Figure 6)

(6) K_p=56 lb/in/in, for 5 rows of
    12 inch diameter piles, spaced
    @24 inches o.c.

(7) Moment of inertia of
    3-16"×15" wales=13,500 in^4

(8) K_r=$15 \cdot 10^4$ lb/in, Ref. 2, 3, 5

(9) Natural frequencies:
    ω_{nr} =3.12 rad./sec
    ω_{nw}=0.575 rad./sec

(c) Evaluated Constants ($\dot{\xi}$=0):
    At t=0.88 sec., $\dot{\xi}$=0, i.e., the
    ferry's velocity normal to the fender is
    reduced from 3 fps to zero in 0.88 sec,
    and,

    A=31.05", C=5.62", E_1=32.14"

    ξ=17.74", y=15.05", δ=2.69"

    $\dot{\xi}$=0, $\ddot{\xi}$=−26.26 in/sec^2

    F=4,050 lb/pile, at x=L/2 (see
    Figure 6)

(d) Plots of Ferry Motion:
    Plots showing the deceleration $\ddot{\xi}$,
    velocity $\dot{\xi}$, and displacement ξ, of the
    ferry from the time of impact (t=0) to
    time of zero ferry velocity (t=0.88), are
    shown in Figures 7, 8, and 9, respec-
    tively. A force of 404,000 pounds is ex-
    erted by the fender/rack system on the
    ferry at the instant $\dot{\xi}$=0.

assumed impact

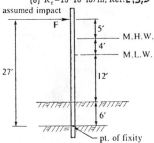

Figure 6

Pile Designed for Maximum
Bending Due to Impact

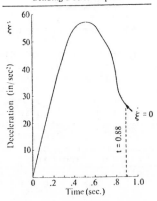

Figure 7

Deceleration of Ferry vs. Time

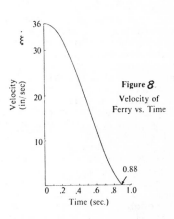

Figure 8

Velocity of
Ferry vs. Time

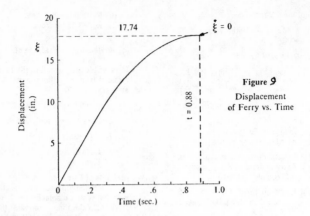

Figure 9
Displacement
of Ferry vs. Time

SUMMARY AND CONCLUSIONS:
        Based on Newton's laws of motion and the conservation
of energy, a ferry fender/rack system can be economically
analyzed/designed. Such an approach would be recommended
for the slips of the Staten Island Ferry, both on the Staten
Island side, or the Manhattan side of the Hudson River, at
least for the type and size (weight) of ferry boats in use
today.
        One assumption made in this analysis, was that all
five piles in a given row normal to the fender, deflect the
same amount. An adequate bracing system, not designed here,
could accomplish this.
        Another assumption made was that the ferry strikes the
fender at the midpoint. If it strikes anywhere else along
the wales, the deflected shape of the wales will be different, causing a different distribution of the maximum stored
potential energy in the rubber pads, piles, and wales,
respectively. The piles and rubber will still absorb nearly
all of this energy, with the piles still carrying most of
it. The distribution of the forces in the piles along the
rack system will be altered according to the new deflected
shape of the wales, and these forces will be proportional
to the displacements. For this reason, there will be no
reduction in the number of rows of piles along the fender,
although in a final design/analysis, the design criteria
would include the ferry striking the fender at least at the
quarter points. If the ferry strikes the slip at its outboard end (nose), the large cluster of piles there will
absorb nearly all of the energy, with the wales contributing
little or nothing to the action. This outboard cluster of
piles must be designed/analyzed separately.

REFERENCES:
1. Arcidiacono, A. and Berkowitz, W., _Ferry Fender Design_, May 5, 1978, (A Civil Engineering term project developed under the guidance of Dr. Zachary Sherman, Consulting Engineer and Adjunct Professor of Civil Engineering, Cooper Union School of Engineering, New York, New York).

2. Gobel, E. F., _Rubber Springs Design_, New York, John Wiley & Sons, 1974.

3. Goodyear, _Handbook of Molded and Extruded Rubber_, Ohio, Goodyear Tire and Rubber Company, 1949.

4. Mueser, Rutledge, Wentworth and Johnston, Consulting Engineers, _Ferry Rack Study_, New York, 1975.

5. Payne, A. R. and Scott, J. R., _Engineering Design with Rubber_, New York, Interscience Publishers, Inc., 1960.

6. Popov, E. P., _Mechanics of Materials_, 2d ed., New Jersey, Prentice-Hall, Inc., 1976.

7. Port Authority of New York and New Jersey, _Staten Island Ferry System Study_, New York 1975.

8. Sherman, Z., _Ferry Fender and Rack Design_, A/E Concepts in Wood Design (AWPI), March/April 1980.

9. Sherman, Z., _Synthesis Analysis and Design of a Ferry Fender and Rack System_, presented at the First National Conference on Bridge and Pier Protection, Stevens Institute of Technology, Hoboken, New Jersey, Dec. 14-16, 1981.

10. Thomson, William T., _Vibration Theory and Application_, New Jersey, Prentice-Hall, Inc., 1965.

A CRITICAL LOOK AT BUILDING INSTRUMENTATION IN

EARTHQUAKE ENGINEERING

Vahid Sotoudeh[1]

Haresh C. Shah[1] M. ASCE

ABSTRACT

Collection of dynamic response data from man-made buildings and structures has been one of the research topics since earthquake engineering was initiated nearly fifty years ago. There are three main tasks related to this topic; instrumentation lay out, selection of appropriate sensor and other hardware, and processing of the collected data.

This paper discusses different issues and requirements existing in the area of building instrumentation. Strategies and types of equipment currently being used by the United States Geological Survey (USGS) and California Division of Mines and Geology (CDMG) is discussed. A brief review of different methodologies which have been, or can be, applied in the processing of structural response data is also presented.

Building Instrumentation Digital Recorder-Analyzer (BIDRA) is introduced in this paper. BIDRA is a new instrumentation system which has been developed at Stanford University to fulfill most of the building instrumentation requirements.

INTRODUCTION

Strong motion accelerograms obtained from instrumented structures are useful and quantitative information about the response of large structures to major earthquakes. Checking the validity of dynamic modeling techniques used by structural engineers in their design practice, and providing guidelines to improve such methodologies are the main goals in the analysis of these data.

Most of the present structural response accelerograms, recorded during moderate and major earthquakes, have been collected from

---

[1] The John A. Blume Earthquake Engineering Center (JABEEC), Stanford University, Stanford, CA. 94305

several buildings (primarily mid- and high-rises) which have been instrumented by either the U.S. Geological Survey (USGS) or the California Division of Mines and Geology (CDMG). Within each of these programs, a building is thoroughly instrumented using 10 to 13 (in few cases more) accelerometers which are remotely hardwired to multi-channel central film recorder(s). The locations of these accelerometers are selected based on the criteria developed by the USGS (EERI, 1984).

The strong motion instrumentation of a structure can also be used to record the low-amplitude ambient vibrations (Hart et al., 1982). Although, structures are known to behave differently during their response to the large-amplitude strong excitations, ambient vibration data can be used in the pre-instrumentation analytical studies, for the purpose of theoretical model confirmation, as well as in the post-earthquake damage assessment studies (Sotoudeh et al., 1985).

Generally, a project of strong motion instrumentation of a structure consists of three major tasks. These tasks are: Design of instrumentation lay-out, selection of appropriate hardware, and processing of the collected data.

INSTRUMENTATION LAYOUT

Design of the instrumentation layout is the first task to be performed in a structural instrumentation project. The related question to be answered can be stated as follows. Given a fixed number of sensors (accelerometers), where should they be installed, within a structure, so that the structural behavior can be optimally identified from the recorded vibrations, during an earthquake?

Mathematical optimization procedures have been introduced in the literature for the selection of optimal sensor locations to provide specific structural parameters with minimal error (Udwadia, 1984; Rojahn et al., 1977).

It has been observed in the records of the earthquake response of structures that the first few modes of vibration are the dominant frequency components in the structural vibrations. To improve the signal-to-noise ratio while recording the structural response, the sensors need to be located as close as possible to the antinodes of all the lower frequency mode shapes. For most regular buildings consideration of the first two bending modes along each of the horizontal directions as well as one or two torsional modes is sufficiently thorough. Except for special structures, vertical vibrations are ignored.

An optimal instrumentation layout can be designed as follows. First, an analytical model of the structure, such as a finite element model, should be made and analyzed to calculate the modal frequencies and mode shapes. The next step is confirmation of these results by experimental ambient testings. Finally, if necessary, modifications in the analytical model should be made to match the experimentally identified structural parameters, and based on the final analytical results the optimal locations of the sensors are selected (Sotoudeh et al., 1985).

The cost and effort involved in such analytical modeling and ambient vibration testing is minimal compared to the value and the cost of the permanent, strong motion instrumentation itself. Therefore, the information provided by these pre-instrumentation activities should always be used in the design of final instrumentation layout.

HARDWARE REQUIREMENTS

The basic requirements for a typical structural instrumentation hardware are: reliability, ease of installation, operation and data communication, and cost effectiveness.

The three main components of a typical seismic instrumentation system, for a structure are: sensors, recorder(s), starter and battery-back-up power supply. These components are usually interconnected by cables to provide synchronization and remote access to the signals. This set-up can also be used for ambient vibration recordings.

Force-balance servo accelerometers have been the most commonly used type of sensor in structural instrumentation. This type of accelerometer has demonstrated reliable performance and high sensitivity in picking both the high-amplitude seismic as well as the low-level ambient vibrations. When good quality and identical sensors are used, the instrumental relative phase distortions in the recorded signals at different locations are minimized.

Recorders are either analog or digital, and are either located centrally or widely scattered at the sensor locations. Common practice with film recorders has been remote recording on multi-channel recorders connected by cables to the sensors installed throughout the structure. Extensive cabling for data collection and for simultaneous triggering takes a noticeable part of the budget and effort allocated to a structural instrumentation project. A feasible alternative for simultaneous triggering is the use of wireless RF devices or transmission through power lines which are capable of reliably and simultaneously triggering scattered recorders.

In some situations digital recording is superior to analog recording, especially for building instrumentation where a hostile environment is not a limiting factor for the equipment electronics. Digital recorders with solid-state memories are, in turn, superior to those which record on magnetic tapes, require mechanical tape drives, and have a much slower rate of data communication.

Cost-effectiveness is another important requirement for the hardware of a building instrumentation system. Presently, the strong motion sensors, recorders and the required cabling are too expensive to be afforable to most of the building owners. Therefore, a low-cost, strong motion recording system which is designed specifically for the purpose of instrumenting buildings is necessary to overcome this problem and to increase the rate of the buildings being instrumented.

DATA PROCESSING

The term "Data Processing" here refers to a series of operations performed on the signals which are obtained from the seismic instrumentation. These operations are usually done in two stages: Preliminary and Secondary processing.

The preliminary stage consists of digitization of analog records, fixing the position of zero amplitudes (or baseline correction), instrument (sensor) correction, low- and high-frequency noise elimination through digital filtering, and finally, integration of acceleration records for calculation of velocities and displacements (Brady, 1980).

Secondary data processing may include estimation of auto- and cross-correlation functions, frequency domain analysis through Fourier transform or other spectral analysis techniques, calculation of elastic response spectrum, time-domain analysis of strong-motion non-stationarities, and structural system identification.

Analysis of ambient vibrations of a structure includes identification of structural dynamic characteristics such as modal frequencies, dampings, and mode shapes. Ambient vibration testings usually can be repeated as many times as necessary to get a good set of data. Therefore, an initial processing of the signals, at the testing site, should be done for data screening and correcting the obvious inefficiencies in the instrumentation layout.

Analysis of earthquake response records has the same system identification purposes as the ambient vibrations case, plus the consideration of time-variations and nonlinearities which normally occur in large-amplitude structural vibrations. Non-stationarities of the earthquake excitations also contribute to this time-dependent feature and need to be identified.

BUILDING INSTRUMENTATION DIGITAL RECORDER-ANALYZER (BIDRA)

BIDRA is a low cost building instrumentation system, designed and developed at the JABEEC at Stanford University. It is designed to fulfill most of the previously discussed requirements. This system is basically a network of multi-channel digital recorders which digitize the analog outputs of their cable-connected accelerometers and record these digitized signals on their battery-backed, solid-state memories. Simultaneous triggering of the recorders is done by RF transmission.

Each recorder is programmable and is equipped with an internal clock. Within each recorder, signal conditioning modules perform anti-aliasing filtering, and can selectively amplify the amplitudes of the analog signals, at several adjustable gain levels. Up to eight channels can be recorded on each of the recorders which corresponds to several minutes of recording time at a sufficiently high sampling rate. Since the high frequency components existing in the structural vibrations are usually low, relatively low sampling rates can be used for the digitization of lowpass filtered signals.

The necessary communications for set-up and data read-out of each recorder is done by a microcomputer through a standard RS-232 connector. Software packages are also developed for a microcomputer, for the purpose of primary and secondary data processing at site.

SUMMARY

The basic issues in the area of strong motion building instrumentation are discussed in this paper. BIDRA, a new building instrumentation network is introduced.

ACKNOWLEDGMENTS

Mr. Shahram Barkhordarian's and Mr. Ali Fotowat Ahmadi's Electrical Engineering assistance, and Mr. Robert Whalen's Mechanical Engineering assistance were the key elements in the development of BIDRA.

REFERENCES

Brady, A.G., Current Strong-Motion Ground Record Processing, Workshop on Interpretation of Strong-Motion Earthquake Records Obtained in and/or Near Buildings, UCLA Report No. 8015, April 1980.

EERI, Evaluation of Instrumenting Buildings for Natural Earthquakes, Experimental Research Needs for Improving Earthquake-Resistant Design of Buildings, Report No. 84-01, Jan. 1984.

Hart, C.G., Estrada, J. and Yoosefi, N., Ambient Vibration Survey of Multistory Buildings with Strong Motion Instruments, UCLA School of Engineering and Applied Science, UCLA ENG 82-22, Feb. 1982.

Rojahn, C. and Matthiesen, R.B., Earthquake Response and Instrumentation of Buildings, Journal of the Technical Councils, ASCE, Vol. 103, No. TC1, 1977, pp. 1-12.

Sotoudeh, V. and Brady, A.G., Dynamic Study of Great Western Building, Berkeley, California, USGS Open file report, in press.

Udwadia, F.E., Optimal Sensor Locations for Geotechnical and Structural Identification, Proceedings of the Eighth World Conference on Earthquake Engineering, San Francisco, 1984.

The Effects of Vertical Earthquake Motions on the Response of Highway Bridges

Douglas A. Foutch, M.ASCE[1]
M. A. Saadeghvaziri[2]

Introduction

Bridges are a vital part of the lifeline systems of a metropolitan area. Immediately after an earthquake it is imperative that emergency vehicles be able to gain access to devastated areas to remove victims and to provide disaster relief. As a city tries to recover from a major disaster, bridges are important links in the transportation system that must be in service if the economic system of the city is to recover and operate efficiently.

The San Fernando Earthquake of February 9, 1971, had a major impact on bridge engineering in the United States. The earthquake caused damage to 62 bridges in the epicentral region at a cost of over 15 million dollars. The bridges that suffered total or partial collapse numbered 15, and 31 suffered major damage (Elliot, et al., 1973).

As a result of this, both analytical and experimental research on the seismic design of highway bridges increased. New and improved seismic design codes were developed by Caltrans (Gates, 1976) and by the Applied Technology Council (ATC) (1981). The latter was recently adapted by the American Association of State Highway and Transportation Officials (AASHTO) (1983).

In spite of all of this activity, relatively little has been said about the effects of vertical earthquake motions on the response of highway bridges. In fact, it is not required to incorporate the effect of the vertical component of ground motions in either the design codes for bridges or for buildings. This probably stems from the results of early building studies where vertical motions were shown to be unimportant.

The purpose of this paper is to demonstrate that vertical vibration of bridges alone or in conjunction with lateral motion can cause serious damage and even lead to collapse. To demonstrate this, the preliminary results of two case studies will be presented.

[1]Associate Professor, Department of Civil Engineering, University of Illinois at Urbana-Champaign, Urbana, Illinois

[2]Graduate Research Assistant, Department of Civil Engineering, University of Illinois at Urbana-Champaign, Urbana, Illinois

The San Fernando Road Overhead Bridge

One of the bridges that suffered collapse during the San Fernando Earthquake was the San Fernando Road Overhead Bridge. It was located about 14 km from the epicenter of the earthquake and about 10 km from Pacoima Dam. The bridge was a two-span (122 feet each) RC box girder bridge with a single column at midspan and with monolithic abutments. This is probably the most common type of bridge currently being built in California. Some details of the bridge are shown in Fig. 1.

The collapse resulted from failure of the pier at its base. The concrete was crushed and spread across the ground. The longitudinal steel buckled into large loops and many ties were fractured. Although the bridge deck did not touch the ground, it lost a few feet of elevation at midspan, and the entire bridge was replaced. The monolithic abutment and the strong deck kept the collapse from being complete. The P-M curve for this column is shown in Fig. 2 along with the plot of ϕ_u/ϕ_y vs. P_u/P_o.

An analytical study of the bridge was conducted in order to determine if vertical vibrations could have been a contributing factor in the collapse. An elastic finite element model was developed. The first five natural frequencies and mode shapes are shown in Fig. 3. Note that the lowest two natural frequencies are associated with vertical modes. The participation factors are also given in Fig. 3. These indicate that the contribution to the axial force comes almost entirely from the second mode, or the first symmetric mode. The first mode, which is asymmetric contributes almost nothing, as expected.

The model was subjected to the vertical component of ground motion recorded at Pacoima Dam. The effective peak acceleration of this component was about 0.7 g. The time history of the axial force in the pier is shown in Fig. 4; and the vertical abutment force is shown in Fig. 5. The dashed horizontal lines in Fig. 4 represent the balance and cracking loads for the column. The column force actually exceeds the balance load three or four times and approaches it several other times.

If the axial force in a RC member is greater than the balance load a brittle failure may result because the concrete crushes before the longitudinal steel yields. Note that even if the balance load is not actually exceeded, the ductility of the column is greatly reduced as P approaches P-balance. This column was severely deficient in hoop ties, so the concrete was surely unconfined. Had enough ties or an adequate spiral been used to confine the core concrete, the expected ductility would have been improved as shown in Fig. 2(b).

The analysis also indicated several other potential problems. The net tension force in the pier exceeded the cracking load. This would cause uniform cracking along the height of the column, greatly reducing its shear capacity and flexural strength. It could also cause serious deterioration of the bond of the longitudinal steel extending from the pier into the pile caps. The large, oscillating vertical abutment reaction could lend to deterioration of the abutment

and its foundation. For other types of the bridges, this force should
be considered in the design of restrainers or hold-down devices.
Currently it is not required that this force be calculated.

The Rio Del Overpass

        The Rio Del Overpass is also a two-span RC box girder bridge with
monolithic abutments. It differs from the San Fernando Overhead
Bridge in that it has two columns and it is severely (39°) skewed.
Both of these aspects have a significant impact on its dynamic
behavior. Plan and elevation views of the bridge are shown in Fig. 6.

        The Rio Del Overpass was instrumented with 20 accelerometers under
the California Strong Motion Instrumentation Program (Shakal, et al.,
1984). It has been subjected to three small earthquakes for which
records are available. During Cape Mendocino Offshore Earthquake of
August 1983, the base of the bridge experienced a peak acceleration of
about 0.028 g. The response spectra for the vertical motion at the
abutment and at centerspan are shown in Fig. 7; and the response
spectra for the transverse motions are shown in Fig. 8. These figures
indicate that the vertical motion was amplified by a factor of about 10
for vertical shaking, but by only about 2 for transverse excitation.
These observations along with the availability of recorded data lead
to the choice of this bridge for the second case study.

        For a skewed bridge, the transverse and torsional modes may be
coupled. Therefore, large vertical accelerations that were measured
at the edge of the deck at centerspan could have resulted from
torsional motion associated with the transverse vibration rather than
from the vertical ground excitation.

        Fourier transform and cross-correlation analysis of the measured
data along with a detailed finite element analysis was conducted.
These revealed that the vertical response and torsional motions
probably contributed about equally to the measured vertical
accelerations.

        An additional analysis of the bridge was conducted to evaluate
the contributions of transverse and vertical excitation to the axial
forces in the piers. The measured transverse and vertical
accelerograms from the Cape Mendocino Offshore Earthquake were used,
but they were scaled upward so that the effective peak acceleration for
each was about 0.7 g. That was about the same level of shaking that
the San Fernando Road Overhead experienced.

        The time histories of the total axial force for column 1 (South)
and for column 2 (North) are shown in Fig. 9 and Fig. 10 respectively.
For both columns the axial compression force exceeded the balance load
and the axial tension force exceeded the cracking load. The plots of
axial force vs. time developed from transverse excitation and from
vertical excitation are shown in Fig. 11 and Fig. 12. These indicate
that each component of excitation contributed about equally to the
total axial force.

The transverse vibration produces these large axial forces because of the torsional motion that develops about the longitudinal axis of the deck. The columns oppose this torsion by developing opposite axial forces. This type of action is not possible for a single-column bent like the San Fernando Road Overhead Bridge.

The vertical contribution to the axial force resulted from response in both the first and second vertical modes, unlike the San Fernando Road Overhead where only the second mode contributed. This difference was the result of the unequal span length and the skewness of the Rio Del Overpass.

Concluding Remarks

Based on the results of these two case studies, the following observations may be made:

1. The response to vertical earthquake excitation should be considered in the design of highway bridges. The axial forces that develop in the piers may reduce their flexural and shear capacities and the ductility.

2. For straight two-span bridges with equal spans the second vertical mode is the only one that contributes significantly to the axial force in the pier.

3. For skewed bridges, both the first and second vertical modes contribute to the axial forces in the piers.

4. For two-column bents, torsional motion associated with transverse modes can generate large axial forces in the columns.

5. In addition to the pier design, the vertical accelerations should be considered in the design of the abutments, pile caps, hold-down devices and connection details.

6. For future bridge instrumentation programs, transducers should be placed such that torsional motion can be identified. This was not possible for the Rio Del Overpass Bridge.

Acknowledgements

The authors greatly acknowledge the help of James Gates of Caltrans who provided the structural drawings of the San Fernando Road Overhead Bridge and the Rio Del Overpass, and A. F. Shakal of the Office of Strong Motion Studies California Division of Mines and Geology who provided the measured earthquake motions at the Rio Del Overpass.

References

1. The American Association of State Highway and Transportation
    Officials (AASHTO), "Guide Specifications for the Design of
    Highway Bridges, 1983," AASHTO, Washington, D.C., 1983.

2. Applied Technology Council, "Seismic Design Guidelines for Highway
    Bridges," Report ATC-6, Applied Technology Council, Berkeley,
    California, October 1981.

3. Gates, James H., "California Seismic Design Criteria for Bridges,"
    Journal of the Structural Division, ASCE, ST12, December 1976,
    2301-2313.

4. Elliot, A. L. and Nagai, I., "Earthquake Damage to Freeway
    Bridges," San Fernando, California, Earthquake of February 9,
    1971, Vol. II, U.S. Department of Commerce, Washington, D.C.,
    1973, 201-234.

5. Shakal, A. F., Ragsdale, J. T., and Sherburne, R. W., "CSMIP
    Strong Motion Instrumentation and Records from Transportation
    Structures-Bridges," Symposium sponsored by Technical Council on
    Lifeline Earthquake Engineering/ASCE, ASCE National Convention,
    San Francisco, California, October 4-5, 1984.

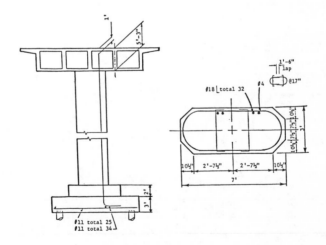

Fig. 1 San Fernando Road Overhead Details

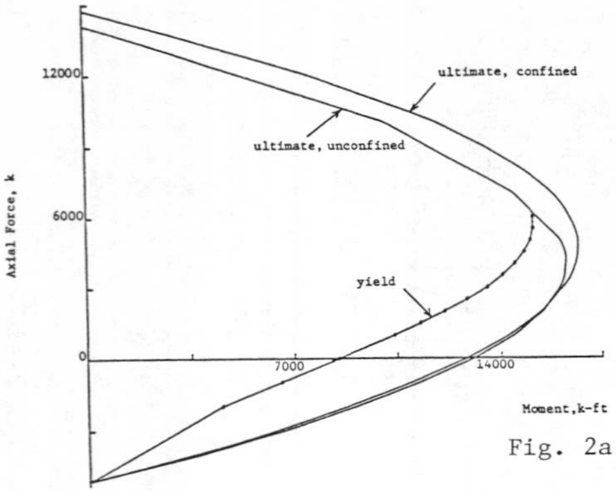

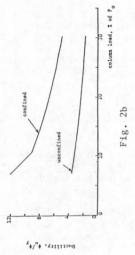

Fig. 2b

Fig. 2   Interaction curvers and ductility
         vs. axial load for the column

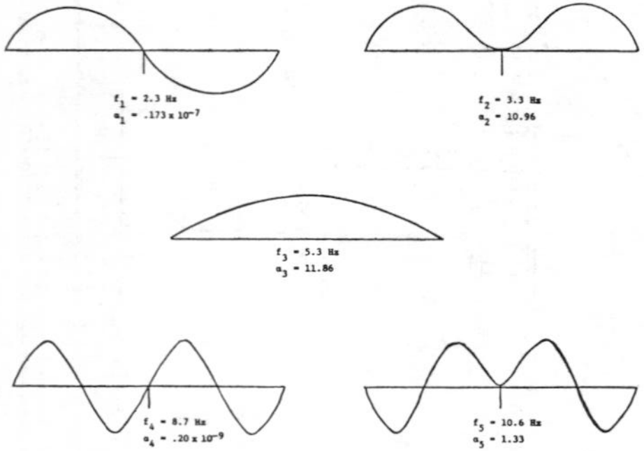

Fig. 2a

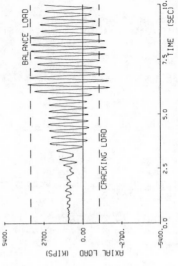

Fig. 4   Time history of the axial load
         due to vertical acceleration

Fig. 3   Mode shapes, natural frequencies
         and participation factors

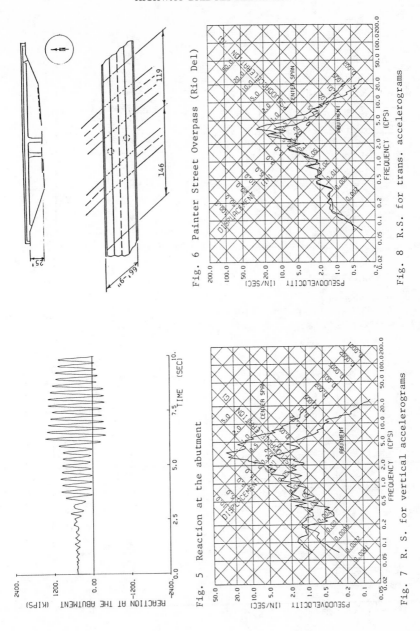

Fig. 5 Reaction at the abutment

Fig. 6 Painter Street Overpass (Rio Del)

Fig. 7 R. S. for vertical accelerograms

Fig. 8 R.S. for trans. accelerograms

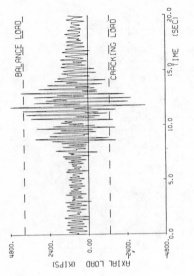

Fig. 9   Total axial load in Column 1

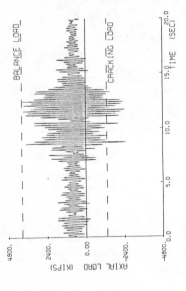

Fig. 11   Axial load due to vertical motion

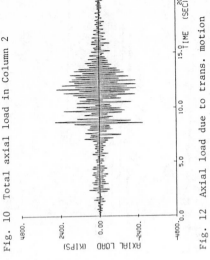

Fig. 10   Total axial load in Column 2

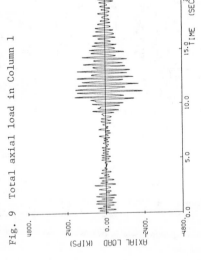

Fig. 12   Axial load due to trans. motion

DYNAMIC TESTING OF STRUCTURES USING THE RESCUE TECHNIQUE

Jeffrey Simons,[1] Albert Lin,[2] and Herbert Lindberg[3]

Abstract

This paper describes the most recent phase of an NSF supported
program at SRI International to develop the technique of repeatable earth
shaking by controlled underground expansion (RESCUE) for testing soil-
structure systems to earthquake-like ground motions. In the most recent
phase, we have upgraded our facility to produce three pulses of ground
motion per test, performed free-field tests to characterize the ground
motion, and tested a one-story 1/11-scale model of Building 180 at JPL,
Pasadena, California. In the structure tests, the highest peak-to-peak
values produced for ground motion were 0.5 g for acceleration, 3 in./s
(7.6 cm/s) for velocity, and 0.1 in. (0.25 cm) for displacement. By
recording the free-field motion during the structure tests, we were able
to investigate soil-structure interaction and found that little soil-
structure interaction occurs at this level of excitation. We performed
computer analyses to determine methods of increasing the ground motion
and structural response. We estimate that by modifying the array design
to maximize the displacements, we can produce displacements of several
inches.

Introduction

SRI International is conducting an on-going program funded by NSF
to develop the technique of repeatable earth shaking by controlled underground expansion (RESCUE) (Abrahamson et al., 1977; Bruce et al., 1979
and 1982; Lindberg et al., 1984; Simons et al., 1984).

The RESCUE technique produces ground motion by simultaneous firing
of a planar array of buried vertical line sources. Each line source
comprises three steel canisters bundled together inside a rubber bladder.
In the current configuration, we have eight line sources spaced at 8-ft
(2.6-m) intervals. Each line source is 39 ft (12 m) long and 1 ft
(0.31 m) in diameter. Small amounts of propellant (rifle powder) are
burned in the canisters to produce a source of high pressure gas. The
gas is vented into the rubber bladder, which expands against the soil.
The process is controlled to produce ground motion at earthquake
frequencies and stress levels. Because the sources are not damaged by
the process, they can be reused for repeat tests and over a range of
amplitudes and frequencies.

---
[1] Research Engineer, SRI International, 333 Ravenswood Avenue, Menlo Park,
CA 94025.
[2] Assistant Professor, Kansas State University, Manhattan, KS 66506.
[3] Aptek, Inc., San Jose, CA 95129.

In the previous phases of the program, each line source contained a single canister. For this phase, we increased the number of canisters to three per line source so that three pulses are produced, one for each canister.

Free-Field Tests

After upgrading the array to produce three pulses, we performed three free-field tests (FF-1, FF-2, and FF-3) to characterize the ground motion produced and to relate the amplitude of the ground motion to the charge density in the canisters. Each test produced three pulses separated by 250 ms. The charge density in pounds per canister for the three pulses was 2/2/2 for test FF-1, 2/3/3 for test FF-2, and 3/4/4 for test FF-3.

Figure 1 shows the surface ground motion in the principal direction (perpendicular to the array) for test FF-1 recorded at station C3 located at the center of the test area, 20 ft (6.1 m) from the center of the array. We obtained velocities and displacements by integrating the acceleration records. The motion is reasonably repeatable from pulse to pulse. As reported earlier, the displacements follow the applied pressure, which indicates that the response is quasi-static (Bruce et al., 1979). The increase in displacement from the first to the third pulse is attributed to compaction of the soil. The Fourier amplitude spectrum of the accelerations shows that most of the energy is low-frequency excitation (less than 20 Hz), and this characteristic is what distinguishes the technique from those that use buried explosions to load structures.

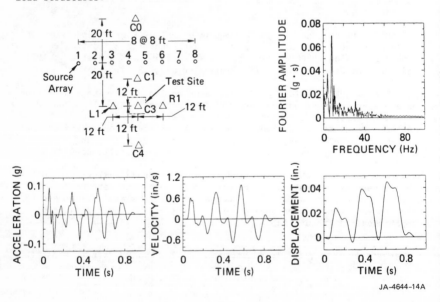

JA-4644-14A

Figure 1. Ground motion from Test FF-1.

Figure 2 shows the ground displacement for test FF-2 at locations C3 and R1 at the 10-ft (3.0-m) depth. There is good agreement between the shape and magnitude of the pulses, indicating that the ground motion is uniform in the direction parallel to the array for a distance that extends beyond the 8-ft (2.4-m) designated test area.

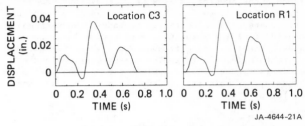

Figure 2. Variation in ground motion along the array.

JA-4644-21A

The variation of motion with distance from the array can be seen in Figure 3, which shows displacement records for test FF-1 obtained at distances of 12, 20, and 32 ft (3.6, 6.1, and 9.8 m) from the array. Interpolating these records gives an attenuation for displacements of about 25% from the front to the back of the designated test area shown in Figure 1. Of the methods we have investigated for reducing this attenuation, the most promising is to dig a trench beyond the test area.

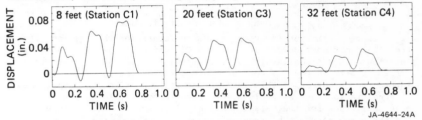

JA-4644-24A

Figure 3. Variation of displacement with distance from the array.

Figure 4 shows the variation in surface displacement with charge density. The displacement at location C3 is shown for the first pulse for the three tests. The charge density was 2 lb/canister in tests FF-1 and FF-2 and 3 lb/canister in test FF-3. The magnitude and shape of the pulses for the first two tests agree very well. As described in the analysis section, because of interaction between the sources, the magnitude of the pulse for the test FF-3 is more than twice as large, which is significantly higher than the 50% increase in charge.

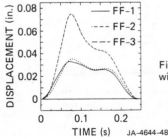

Figure 4. Variation in displacement with charge density.

JA-4644-48

Structure Tests

Following the free-field tests, we designed and constructed a 1/11-scale model [8 ft (2.6 m) square and 9 ft (3.0 m) high] of Building 180 at JPL in Pasadena, California, and performed three tests, ST-1, ST-2, and ST-3. The model had a top concrete mass supported by four steel columns and a concrete base. A large (65,000-lb) lead mass was suspended from the top of the structure mass to correctly reproduce the vertical soil stresses. To characterize the structure, we recorded the response to ambient wind loading. The first mode frequency was about 5.7 Hz. The response of the structure for the test ST-1 measured at the center of the top mass is shown in Figure 5. The maximum peak-to-peak value for acceleration is 0.26 g, for velocity is 2.05 in./s (5.2 cm/s), and for displacement is 0.16 in. (0.41 cm).

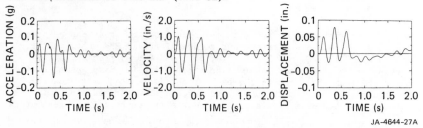

JA-4644-27A

Figure 5. Response of top mass for Test ST-1.

Figure 6 compares the response at the center of the base and the free-field motion recorded at location C0 (which is the corresponding location on the opposite side of the array) for test ST-1. The close agreement between the two records indicates that little soil-structure interaction occurred at this level of excitation. The peak amplitudes of acceleration were slightly higher for the free-field motion, and the peak amplitudes for the velocity and displacement were slightly higher for the base record. The frequencies of the peaks in the Fourier amplitude spectra of the two accelerations, shown in Figure 7(a), show very close agreement. The amplitudes for the low-frequency peaks (0-10 Hz) were greater for the base record, and the amplitude of the higher frequency peaks (20-50 Hz) were greater for the free-field record, a result that is consistent with the differences in time histories.

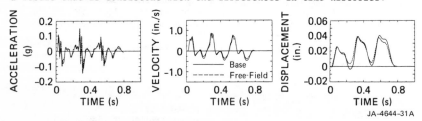

JA-4644-31A

Figure 6. Base response and free-field motion for Test ST-1.

The versatility of the technique was demonstrated in test ST-3 by delaying the third pulse by an extra 150 ms. The resulting Fourier amplitude spectra of the free-field accelerations, shown in Figure 7(b), show a richer harmonic content than in the previous tests.

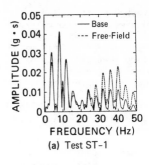
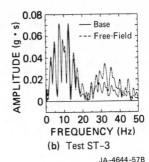

(a) Test ST-1
(b) Test ST-3

Figure 7. Fourier spectra for structure tests.

JA–4644–57B

Analysis

We performed two sets of analyses to determine methods of increasing the ground motion and structural response for this technique. In the first analysis, we investigated how to vary the pulse timing to increase the structure response. In the second analysis, we performed a parameter study to investigate modifying the array design to increase the ground displacement.

We investigated varying the pulse timing to increase the structure response by using a computer model to predict the structure response to free-field ground acceleration. The soil-structure system was modeled using a three-degree-of-freedom linear lumped parameter model, as shown in Figure 8. The resulting first mode frequency of the soil-structure model of 5.3 Hz is about 8% less than the measured ambient frequency of 5.7 Hz, which suggests some softening of the soil under the forced motion. The model reproduced the foundation and top mass response to the free-field motion reasonably well.

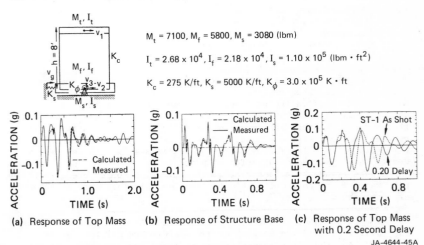

$M_t = 7100$, $M_f = 5800$, $M_s = 3080$ (lbm)

$I_t = 2.68 \times 10^4$, $I_f = 2.18 \times 10^4$, $I_s = 1.10 \times 10^5$ (lbm · ft^2)

$K_c = 275$ K/ft, $K_s = 5000$ K/ft, $K_\phi = 3.0 \times 10^5$ K · ft

(a) Response of Top Mass   (b) Response of Structure Base   (c) Response of Top Mass with 0.2 Second Delay

JA–4644–45A

Figure 8. Calculated structure response for Test ST-1.

As expected, to achieve maximum structure response, the pulse frequency should be chosen so that the structure frequency is a multiple of the pulse frequency. As seen in Figure 8(c), reducing the delay between pulses for test ST-1 from 0.25 s to 0.20 s increased the maximum peak-to-peak acceleration by about 15%. We believe that the small increase in response is due in part to the high value of damping for the structure caused by flexibility of the base-column connection (measured as 10% of critical). We expect that increasing the number of pulses, especially for structures with less damping (2% critical is typical for steel-framed structures) and timing the pulses to excite resonant behavior will significantly increase the response.

For our second analysis, we performed a parameter study using the nonlinear finite element analysis program NONSAP to investigate modifying the array design to increase the ground motion. The soil was modeled as elastic-perfectly plastic with a Mohr-Coulomb yield condition. Figure 9 shows calculated and measured curves of ground displacement as a function of source pressure obtained for an earlier study using a one-third size array [ten 4-in.-diameter (10-cm) sources]. The curve shows an abrupt increase in slope at a critical pressure of about 40 psi. The slope of the curve is a measure of the transmittance of the array. Although the measured curve is shifted to the right from the theoretical curve and the increase in slope is not as abrupt, the large increase in transmittance at high pressure is apparent.

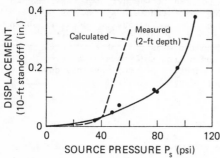

Figure 9. Relation between source pressure and soil displacement.

In our study, we varied the number of sources, source spacing, soil stiffness, soil cohesion, and soil friction angle. Figure 10 shows a typical mesh used in the analyses. For each configuration, we calculated the critical pressure and transmittance. We determined that the factor that has the greatest influence on the efficiency of the array is the source spacing. Figure 10 shows that decreasing the source spacing from 8 ft (2.4 m) to 4 ft (1.2 m), while otherwise keeping the same array design, would decrease the critical pressure by about 50% and increase the transmittance by a factor of about two. At a source pressure of 100 psi, we estimate that this would increase the displacements by a factor of about four.

The performance of the array could be enhanced even more by replacing the soil around the sources with a fluid. The critical pressure is then zero and the array transmittance is one. For the soil properties at our current site, we estimate that displacements of about 2 in. (5 cm) are obtainable for source pressures of 80 psi.

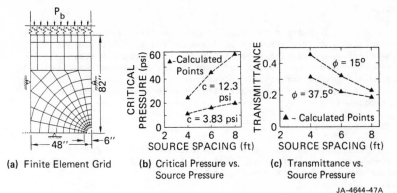

(a) Finite Element Grid

(b) Critical Pressure vs. Source Pressure

(c) Transmittance vs. Source Pressure

JA-4644-47A

Figure 10. Parameter study for array modification.

To replace the soil around the sources with a fluid, we would install the sources in rectangular steel boxes, as shown in Figure 11. The boxes would have dimensions of 8 ft (2.4 m) by 3 ft (0.91 m) by 39 ft (12 m). They would be placed side by side in a trench and filled with a slurry of fine sand, bentonite, and water that maintains a density near that of the soil. In addition to allowing greater motions, the boxes also have the following advantages: (1) individual sources can be removed or replaced easily, (2) the slurry would provide a less harsh environment for the rubber bladders, (3) the boxes would prevent the soil around the sources from blowing out, (4) the maximum pressure the soil experiences would be much less for a given displacement, and (5) the resulting motion will be more nearly planar.

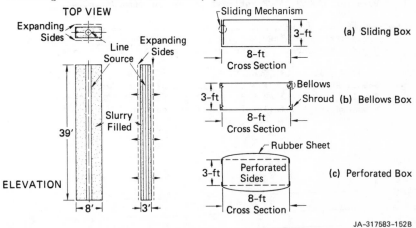

JA-317583-152B

Figure 11. Expandable boxes.

The key problem here will be to design a functional box that expands easily and is reusable. We are currently considering three designs for the boxes, shown in Figure 11: a sliding box, a bellows box, and a perforated box with an external rubber sheet.

Future Work

We are planning a two-year program to investigate methods of
maximizing the displacements produced by the RESCUE technique. During
the first year, we will develop the design concepts described above and
test them using small-scale models. During the second year, we will
build and test a large-scale prototype to demonstate the adequacy of the
design for simulating the soil-structure interactions that occur in
earthquakes.

Acknowledgments

This research was sponsored by National Science Foundation Grant No.
CEE-8200321. K. Thirumalai and A. J. Eggenberger were the NSF Program
Directors for the work reported here.

References

Abrahamson, G. R., H. E. Lindberg, and J. R. Bruce. 1977. "Simulation
of Strong Earthquake Motion with Explosive Line Source Array," Final
Report prepared for the National Science Foundation, SRI Project
6004.

Bruce, J. R., H. E. Lindberg, and G. R. Abrahamson. 1979. "Simulation
of Earthquake Motion with Contained-Explosive Line Source Arrays,"
Proceedings of the U.S. National Conference on Earthquake
Engineering, Stanford University, Palo Alto, CA.

Bruce, J. R., H. E. Lindberg, and L. E. Schwer. 1982. "Soil Motion from
Contained Explosions," Physical Modeling of Soil Dynamics Problems,
Preprint 82-063, ASCE National Convention, Las Vegas, NV.

Lindberg, H. E., R. Mak, and J. R. Bruce. 1984. "Calculation of Soil
Motion from Contained Explosive Arrays," Proceedings of the Eighth
World Conference on Earthquake Engineering, San Francisco, CA.

Simons, J. W., A. N. Lin, and H. E. Lindberg. 1985. "Dynamic Testing of
a Soil-Structure System Using the Technique of Repeatable Earth
Shaking by Controlled Underground Expansion (RESCUE)," Final Report
prepared for the National Science Foundation, SRI Project 4644.

Pseudo-Dynamic Testing of Structural Components

Haluk M. Aktan* and Ahmed A. Hashish**

The pseudo-dynamic testing procedure is a simultaneous simulation and control process in which inertia and damping properties are simulated and flexibility properties are acquired from the specimen. The procedure calculates a set of dynamic displacements based on active control theory, utilizing the simulated inertia and damping properties and acquired flexibility properties, under a hypothetical ground motion, and simulates the response of the structure in a quasi-static fashion. The application of the procedure for the testing of structural components is simulated on a seven-story reinforced concrete frame wall building.

Introduction

The principles of the psuedo-dynamic testing have been successfully used in Japan for seismic behavior evaluations of simple structures under the name "Computer Simulation on Line System" [8]. The pseudo-dynamic testing procedure was recently implemented in the testing of flexible framed structures [4] and the theoretical background of the application of the method to general structures was also recently developed [1].

The pseudo-dynamic procedure is based on classical modal control theory [2,5,6,7]. Selected vibration modes of the test structure is controlled by the lateral loads applied to the defined degrees of freedom while being hypothetically subjected to an earthquake motion. In implementing active control principles, the loads corresponding to the degrees of freedom are envisioned as eliminating, rather than causing, the deformations of the selected vibration modes generated by the input earthquake. Therefore, instead of subjecting the base of the structure to an earthquake motion, the calculated forces are considered to generate an opposite input earthquake motion, and the displacements generated by these active control forces will in fact represent the pseudo-dynamic displacement to be applied to the structure. A simplified block diagram of the procedure is shown in Figure 1.

*Associate Professor, Department of Civil Engineering, Wayne State University, Detroit, MI 48202.

**Graduate Student, Department of Civil Engineering, Wayne State University, Detroit, MI 48202.

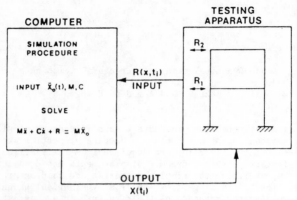

Figure 1. Simplified Block Diagram of the Pseudo-Dynamic Method

Pseudo-Dynamic Procedure

In implementing the procedure, the mass characteristics of the test structure are analytically generated in discrete form. The discretization is arbitrary, yet, the degrees of freedom should coincide with the points where the displacements are experimentally monitored. The flexibility properties are experimentally determined from the test structure at the points of discretization. The procedure requires the damping to be only in the viscous type and proportional to mass and stiffness in the form $\alpha M + \beta K$.

The differential equation of motion of the discretized test structure is transformed into canonical form by linear transformation, and for a specific mode the transformed equation is written as:

$$\begin{Bmatrix} \dot{y}_n \\ \dot{v}_n \end{Bmatrix} = \begin{bmatrix} 0 & 1 \\ -2\xi_n\omega_n & \omega_n^2 \end{bmatrix} \begin{Bmatrix} v_n \\ y_n \end{Bmatrix} + \begin{Bmatrix} 0 \\ 1 \end{Bmatrix} F_c - \begin{Bmatrix} 0 \\ \Gamma_n \end{Bmatrix} \ddot{z}_g \tag{1}$$

where v_n is velocity, y_n is displacement, ξ_n is damping ratio, ω_n is circular frequency, Γ_n is participation factor of the nth mode, F_c is external modal control force, and $\ddot{z}_g$ is ground acceleration.

The optimal modal control forces are calculated by the minimization of a functional, and determined as:

$$F_c = -(g_1 \cdot y_n + g_2 \cdot v_n)$$

and

$$g_1 = -\omega_n^2 + \sqrt{(\omega_n^2)^2 + q_1^2} \tag{2}$$

$$g_2 = -2\xi_n\omega_n + \sqrt{(2\xi_n\omega_n)^2 + 4\xi_n\omega_n \cdot g_1 + q_2^2}$$

where q_1 and q_2 are constants chosen by the analyst that specify the control accuracy. The selection procedure for q_1 and q_2 and the details of the derivation is presented in [1].

Substituting Eq. (2) into Eq. (1) and solving the differential equation, v_n and y_n are obtained. The modal pseudo-dynamic displacements are calculated by substituting the control forces in place of the earthquake forces as:

$$\ddot{a}_n + 2\omega_n \xi_n \dot{a}_n + \omega_n^2 a = g_1 v_n + g_2 y_n \tag{3}$$

Where a_n is the pseudo-dynamic displacement of the nth mode.

Once, a desired number of modal pseudo-dynamic solutions are obtained by using Eqs. (1), (2) and (3) the test structure is subjected to the following displacements calculated as follows:

$$\{x\} = [\phi]\{a\} \tag{4}$$

where $[\phi]$ is the orthogonal transformation matrix and $\{x\}$ is the pseudo-dynamic displacement vector.

Eqs. (2), (3) and (4) constitute the closed loop equations of the pseudo-dynamic testing method. The non-linear restoring force vector does not directly appear in the closed loop equations. The force feedback data are used to calculate the flexibility matrix of the structure for the time step in question. Updated values of modal frequencies (ω_n) and the modal transformation matrix $[\phi]$ are calculated from the flexibility matrix constructed using the restoring force feedback.

Implementation of the Procedure for Testing of Structural Components

The implementation of the procedure is simulated on a wall component of a seven-story frame, wall type R/C structure. The elevation of the test structure is shown in Figure 2 [3]. The isolated wall component, representing the specimen, in reference to the structural wall of the building is also shown in Figure 2. The objective of the simulation is to obtain the shear stiffness variation of the wall component during a prescribed service level earthquake load. The wall component will be tested in the apparatus shown in Figure 3 under quasi-statically applied axial loads and shear deformations, calculated at each time step, with the pseudo-dynamic procedure.

The analytical simulation is based on an assumed elastic behavior of the wall. The flexibility properties of the building and the properties of the wall are given in Table 1. The flexibility properties are updated at each time increment and new values for modal parameters are calculated. Utilizing Eqs. (1) through (4) first story pseudo-dynamic displacement is determined.

The first story displacement is assumed to be the summation of shear and flexural deformations in ratio with the respective flexibilities [3]. The axial load on the specimen is determined from the flexural

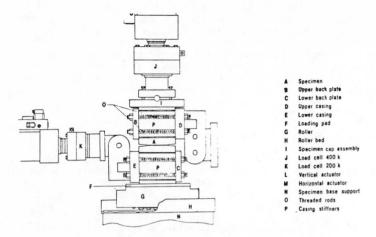

A	Specimen
B	Upper back plate
C	Lower back plate
D	Upper casing
E	Lower casing
F	Loading pad
G	Roller
H	Roller bed
I	Specimen cap assembly
J	Load cell 400 k
K	Load cell 200 k
L	Vertical actuator
M	Horizontal actuator
N	Specimen base support
O	Threaded rods
P	Casing stiffners

Figure 3.   Test Apparatus for the Pseudo-Dynamic Component Testing

theory of cracked concrete section. The shear deformations are assumed to be the sum of slippages along horizontal cracks. The slip
applied on the specimen is calculated for a fixed number of cracks
along the first story height.

The axial load and shear slip calculations are based on the contribution of only the first three modes of vibration. The results are
presented in Figure 4, and Figure 5 gives the axial load variation on
the boundary member with the sending moment. Figures 4(a) and (b)
show the earthquake and first floor displacement histories. Figure 5
shows the axial load on the specimen against the bending moment on the
cross section.

In reality, during an experiment the shear stiffness of the wall is
updated from the measured force corresponding to the imposed slip of
that time step and the updated sitffness are used in the pseudo-dynamic displacement calculations. For this example all other structural
components are assumed to remain elastic. During an actual experiment
the procedure should accommodate inclusion of higher modes as nonlinearity effects increase. In highly non-linear systems with degrading stiffness contribution of higher modes may significantly change
pseudo-dynamic displacements.

Table 1

Flexibility Properties of the Example Structure [2]

$(10^{-3}$ inch/kip)

columns rows	7	6	5	4	3	2	1
7	9.51	7.94	6.21	4.64	3.14	1.81	0.78
6	7.94	6.99	5.58	4.24	2.91	1.73	0.75
5	6.21	5.58	4.87	3.77	2.67	1.57	0.70
4	4.64	4.24	3.77	3.30	2.36	1.49	0.68
3	3.14	2.91	2.67	2.36	2.04	1.34	0.62
2	1.81	1.73	1.57	1.49	1.34	1.10	0.56
1	0.78	0.75	0.70	0.68	0.62	0.56	0.40

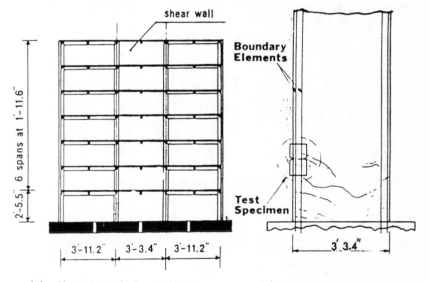

(a) Elevation of the Building

(b) Shear Wall with the Location of the Test Specimen

(c) Cross-Section of the Shear Wall

Figure 2. Configuration of the Component in Reference to the Test Structure

Conclusion

The application of the pseudo-dynamic procedure to structural components is presented here. The procedure utilizes the adaptive systems theory in calculating the realistic forces and deformations of the specimen. The changing flexibility properties of the specimen is fedback to the structural model for updating the dynamic properties continually.

The example given in this paper is an analytical simulation for presenting the applicability of the procedure. The pseudo-dynamic apparatus shown in figure 3 will be used in the near future for the testing of shear wall boundary members using the procedure described in this paper.

The pseudo-dynamic procedure is most appropriate for large structural systems, where the capacities of earthquake simulators are inadequate and for calculating accurate control variables to impose on the structural components.

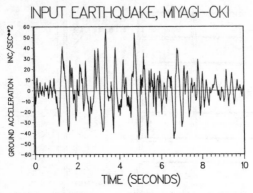

Figure 4 (a)   Scaled Miyagi-Oki Earthquake Record as Input Motion

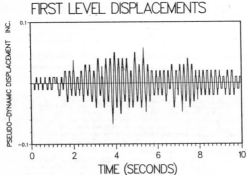

Figure 4 (b) Pseudo-Dynamic Displacements of the First Floor Level of the Test Structure

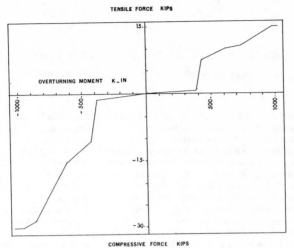

TENSILE FORCE KIPS

OVERTURNING MOMENT K_IN

COMPRESSIVE FORCE KIPS

Figure 5. Axial Load on the Wall Component vs Bending Moment at the Cross-Section, Calculated from Pseudo-Dynamic Story Displacements

References

1. Aktan, H.M., "Pseudo-Dynamic Testing of Structures," Journal of ASCE, EM Division, Vol. 112, No. 2, February 1986.
2. Athans, M and Falb, P.L., Optimal Control, McGraw-Hill Book Company, 1966.
3. Bertero, V.V., et al., "Earthquake Simulation Tests and Associated Experimental, Analytical and Correlation Studies of 1/5th Scale Model," ACI, SP-84, J.K. Wight, Ed., Earthquake Effects on R/C Structures, August 1985, pp. 375-424.
4. Popov, E.P., et al., "Inelastic Response of Tubular Steel Offshore Towers," Journal of ASCE, ST Division, Vol. 111, No. ST10, October 1985, pp. 2240-2258.
5. Martin, C.R. and Soong, T.T., "Modal Control of Multistory Structures," Journal of ASCE, EM Division, Vol. 102, No. 7, August 1976, pp. 613-623.
6. Meirovitch, L. and Silversberg, L.M., "Control of Structures Subjected to Seismic Excitations," Journal of ASCE, EM Division, Vol. 109, No. 2, April 1983, pp. 604-615.
7. Porter, B. and Crossley, R., "Modal Control, Theory and Applications," Taylor and Francis Ltd., 10-14 Macklin Street, London, 1972.
8. Takanashi, K., et al., "Nonlinear Earthquake Response Analysis of Structures by a Computer-Actuator On-Line System," Bull. of the Earthquake Resistant Structure Research Center, Inst. of Indust. Sci., Univ. of Tokyo, Nov. 8, 1974.

SEISMIC TESTS OF FRICTION DAMPED STEEL FRAMES

A. Filiatrault*   and   S. Cherry,** F.ASCE

ABSTRACT

This paper presents the results obtained from qualification tests of a new friction damping system, which has been proposed in order to improve the response of Moment Resisting Frames and Braced Moment Resisting Frames during severe earthquakes. The system basically consists of a special inexpensive mechanism containing friction brake lining pads introduced at the intersection of frame cross-braces. Seismic tests of a three storey Friction Damped Braced Frame model were performed on an earthquake simulator table. The experimental results are compared with the findings of an inelastic time-history dynamic analysis. Two different computer models are used for this purpose. The results, both analytical and experimental, clearly indicate the superior performance of the Friction Damped Braced Frame compared to conventional building systems.

INTRODUCTION

Recently a novel structural system for the aseismic design of steel framed buildings has been proposed and patented by Pall (1). The system basically consists of an inexpensive mechanism containing friction brake lining pads introduced at the intersection of frame cross-braces. During an earthquake, these pads develop additional energy dissipating sources which can be marshalled to protect the main members from structural damage. The device is designed not to slip under normal service loads and moderate earthquakes. During severe seismic excitations, the device slips at a predetermined optimum load, before any yielding of the main members has occurred. Slipping of a device changes the natural frequency of the structure and allows the structure to alter its fundamental mode shape during a severe earthquake. The phenomenon of quasi-resonance between the structure and the earthquake excitation is prevented because of this de-tuning capability of the structure. These devices also can be conveniently incorporated in existing framed buildings to upgrade their earthquake resistance.

The main objective of the investigation reported in this paper was to study experimentally the seismic performance of steel braced frame structures incorporating these friction damped devices. This was done by placing a 1/3 scale model of a 3 storey friction damped structure on a shaking table and subjecting it to representative ground motion time-histories. The experimental results are compared to the predictions of an inelastic time-history dynamic analysis using the computer program "DRAIN-2D".

---

* Graduate Research Assistant and **Professor, Department of Civil Engineering, University of British Columbia, Vancouver, B.C., Canada, V6T 1W5.

Two different computer models were used to predict analytically the response of the Friction Damped Braced Frame. The first one, originally proposed by Pall (1), is based on an equivalent hysteretic model and is only approximate, since it does not take into account the complete behaviour of the friction devices. As a result, a new, refined computer model was developed and the results obtained from analyses using both models are compared.

DESCRIPTION OF THE FRICTION DAMPING SYSTEM

The general arrangement of an actual friction device is presented in Figure 1. The friction device can be used in any configuration of the bracing system. If the diagonal braces of an ordinary braced frame structure were designed not to buckle in compression, a simple friction joint could be inserted in each diagonal. In this case each slip joint would act independently of the other. The slip load should be lower than the yield load of the braces, so that the joint can be activated before the member yields. However, it is not economical to design the braces in compression and, more often, since the braces are quite slender they are designed to be effective in tension only. In such cases, a simple friction joint would slip in tension but would not slip back during reversal of the tension load and in the compression (buckled) regime. The energy absorption would be relatively poor since the brace would not slip again until it was stretched beyond the previous elongated length.

The braces can be made to slip both in tension and compression by connecting the friction mechanism at the intersection of frame cross-braces as indicated in Figure 2, which illustrates five stages during a typical load cycle. When a seismic lateral load is induced in the frame, one of the braces goes into tension while the other brace buckles very early in compression. If the slip load of the friction joint is lower than the yield load of the brace, when the load in the tension brace reaches the slip load it forces the joint to slip and activates the four links. This in turn forces the joint in the other brace to slip simultaneously. In this manner, energy is dissipated in both braces in each half cycle. Moreover, in each half cycle, the mechanism straightens the buckled brace and makes it ready to absorb energy immediately when the load is reversed. In this way, the energy dissipation of this system is comparable with that of a simple friction joint when used with braces which are designed not to buckle in compression.

To be effective the slip joints must present very stable, non-deteriorating hysteresis loops. Several static and dynamic friction tests on slip joints having different surface treatments have been conducted under repeated reversals of loads and are reported in the literature (1). On the basis of these tests, heavy duty asbestos brake lining pads, inserted between the sliding steel surfaces, were chosen for the slip joint used in the present study.

During the study reported in this paper, cyclic load tests on the friction devices were carried out in a standard testing machine. One end of a diagonal link of the friction device was bolted to a rigid testing bench while the opposite end was attached to a vertical hydraulic actuator. Figure 3 shows a typical load-displacement curve obtained from this series of tests.

The performance of the brake lining pads is seen to be reliable and repeatable. The hysteresis loop is very nearly a perfect rectangle and exhibits negligible fade even after 50 cycles. The imperfections at the two opposite corners of the hysteresis loop are the result of fabrication tolerances in the friction devices. A perfectly rectangular hysteresis loop presumably could be obtained by completely eliminating the difference between the diameter of the bolts and of the bolt holes.

DESCRIPTION OF TEST FRAMES

A 1/3 scale model of a 3 storey Moment Resisting Frame was used in the shake table tests. The overall dimensions of the model frame are 2.05 × 1.4 m in plan and 3.53 m in height. All the main beams and columns are made of the smallest S shapes (S75 × 8) available. The dead load is simulated by concrete blocks at each floor. Two identical frames were fabricated for the experimental study. The general arrangement of the test frame is shown in Figure 4.

The beam-column connections were designed such that the Moment Resisting Frame could be transformed easily into a Braced Moment Resisting Frame or a Friction Damped Braced Frame as needed. The fundamental natural frequencies of the three model types were found to be: Moment Resisting Frame = 2.86 Hz; Braced Moment Resisting Frame = 5.29 Hz; and Friction Damped Braced Frame = 7.03 Hz. The fundamental frequency of the Friction Damped Braced Frame was measured under very low amplitude excitations when none of the devices slip and all the braces behave elastically in tension and compression.

ANALYTICAL STUDY

The energy dissipation of a friction device is equal to the product of slip load by the total slip travel. The optimum slip load is the load which leads to maximum energy dissipation in the friction device. To determine the optimum slip load of the model frame, inelastic time history dynamic analyses were performed for different values of the slip load using the computer program "DRAIN-2D". In the program, we define a fictitious yield stress in tension, which corresponds to the stress in the tension brace when the device slips. We also assign a very low fictitious compression yield stress to the material, corresponding to the buckling stress of the compression brace. This simple model was originally proposed and used by Pall (1). The optimum slip load study was carried out for 3 different earthquake records: El Centro 1940 N-S and Parkfield 1966 N65E, both scaled to 0.52 g; Newmark-Blume-Kapur artificial accelerogram scaled to 0.3 g. Typical results of the optimum slip load study are given in Figure 5, which shows envelopes of maximum bending moment developed in the beams of the model frame for global slip loads ranging from 0 to 10 kN, representing the elastic region of the cross-braces. Similar curves were obtained for response envelopes of lateral deflections and column moments and shears. Results for zero global slip load represent the response of a moment resisting frame. The global slip load is defined as the load in the tension brace when slippage occurs, while the local slip load represents the load in each friction pad at the same time.

The figure clearly shows the effectiveness of the friction devices in improving the seismic response of the frame. As the global slip load is increased, the bending moments decrease steadily up to a global slip load of 6 kN. For a global slip load between 6 kN and 10 kN, there is very little variation in the response.

This lower bound value of 6 kN for the global slip load is observed for the 3 different earthquakes studied. This suggests that the optimum value of the global slip load may be independent of the ground motion input. If this is found to hold true for all cases, the optimum slip load can be considered as a structural property. This observation may greatly simplify the development of a design procedure for the friction devices. On the basis of the results obtained, an optimum global slip load value of 7 kN was subsequently used for the study of the Friction Damped Braced Frame model. The optimum slip load is 20% of the weight of the model.

The Pall model is only approximate since it assumes that the device slips at every cycle and that the slippage of the device is large enough to straighten completely the buckled brace; this is not the case in an actual earthquake. Therefore, this simple model over-estimates the energy absorption of the friction device. For this reason a more accurate or refined model was developed to evaluate the significance of these simplifying assumptions on the overall response of the structure under a major earthquake. This refined model considers each element of a friction device as an individual member with its own material (stress-strain) properties. It can therefore represent the complete behaviour of the friction devices, but requires many more degrees of freedom so that the computer costs are much more expensive than the corresponding costs associated with the simplified model. For this reason, the results of analyses using the two models are compared for the El-Centro 1940 earthquake only and are presented in Figure 6. The response parameters obtained using the two models of the Friction Damped Braced Frame are also compared to the corresponding responses of the Moment Resisting Frame and the Braced Moment Resisting Frame. The refined model gives higher member forces and deflections than the simplified model since the refined model takes into account the real deformation pattern of the friction devices; however, the results are reasonably close and within the variations typically expected in earthquake analysis. Therefore the simplified model seems adequate for practical application.

SEISMIC TESTS ON SHAKING TABLE

The test frames were mounted on a shaking table and subjected to various earthquake accelerograms with different intensities expressed in terms of peak acceleration. A variety of sensors were used to measure displacement, acceleration, friction pad slippage and strain at critical locations in the frame.

The Moment Resisting Frame and the Braced Moment Resisting Frame did not perform well during the tests. Very large strains occurred in

the base column, and in the first and second floor beams of the Moment
Resisting Frame. Although the main structural members of the Braced
Moment Resisting Frame remained elastic, many cross-braces yielded in
tension. The elongation of the braces was very large and they buckled
significantly in the compression regime; this indicates that heavy
non-structural damage would have occurred in a real building (cracks in
walls, broken glass, etc.). However, the Friction Damped Braced Frame
performed very well; no damage occurred in any member and the deflections and accelerations were far less than the values measured in the
two other types of construction.

The envelopes of the measured horizontal accelerations for the
Newmark-Blume-Kapur artificial earthquake at 0.3 g peak acceleration
are shown in Figure 7. The influence of the new damping system in
reducing seismic response is apparent. The maximum acceleration ampli-
fications experienced by the three frames under this excitation are
1.98, 5.08 and 6.63 for the Friction Damped Braced Frame, Braced Moment
Resisting Frame and Moment Resisting Frame respectively.

The envelopes of the bending moments in the beams are presented in
Figure 8. Good agreement is observed between the measured and predicted values. The figure clearly shows the effectiveness of the friction
devices in improving the seismic response of the frame.

Figures 9 and 10 show typical response characteristics of the
three structural configurations when subjected to an extremely severe
earthquake: Taft 1952 scaled to a peak acceleration of 0.9 g. Figure 9
illustrates the superior performance of the Friction Damped Braced
Frame, expressed in terms of the measured horizontal accelerations. A
peak acceleration of 1.42 g was measured at the top of the Friction
Damped Braced Frame compared to peak acceleration values of 2.24 g and
2.67 g for the Braced Moment Resisting Frame and the Moment Resisting
Frame respectively. Since the excitations developed were extremely
severe, it was possible to measure significant slippage in the friction
devices of the Friction Damped Braced Frame. Figure 10 shows the slippage time-history of the second floor device. A peak slippage of 10.2
mm was recorded at a time which coincided with the time at which the
peak ground acceleration occurred. The flat plateaux indicate time
increments during which the device remains frozen: the load in the
tension brace is less than the local slip load and the structure
behaves as a normal braced frame.

CONCLUSIONS

1. A refined computer model of a Friction Damped Braced Frame was
    developed to eliminate the non-conservative assumptions used with
    the simplified model originally proposed by Pall (1). By compar-
    ing these two models it is concluded that the simplified model is
    simpler and cheaper to use than the refined model and yields
    results which satisfy the accuracy normally associated with earth-
    quake analysis.

2. An optimum slip load study was performed to determine the value of
    the slip load which optimized the energy dissipation of the fric-
    tion devices. The results seem to indicate that the optimum slip
    load may be independent of the ground motion time-history and is

rather a structural property.

3. The devices were first tested under cyclic loads in order to study the stability of the brake lining pads and to calibrate their slipping loads. The results of the tests clearly indicate that the behaviour of the pads is very stable even after 50 cycles. However, a purely rectangular load-deformation curve can only be obtained if the fabrication tolerances of the friction devices are minimized.

4. Seismic testing of the model frames on the shaking table under simulated earthquake loads confirmed the superior performance of the Friction Damped Braced Frame compared to conventional aseismic building systems. Even an earthquake record with a peak acceleration of 0.90 g did not cause any damage to the Friction Damped Braced Frame, while the Moment Resisting Frame and the Braced Moment Resisting Frame underwent large inelastic deformations.

The work reported was carried out under Contract No. 1ST84-00045 with the Department of Supplies and Services, Canada. The authors acknowledge the fruitful discussions which were held with Dr. A.S. Pall, and the help of the Canadian Institute of Steel Construction.

REFERENCES

(1) PALL, A.S. and MARSH, C., "Response of Friction Damped Braced Frames," ASCE, Journal of Structural Division, Vol. 108, ST6, June 1982, pp. 1313-1323.

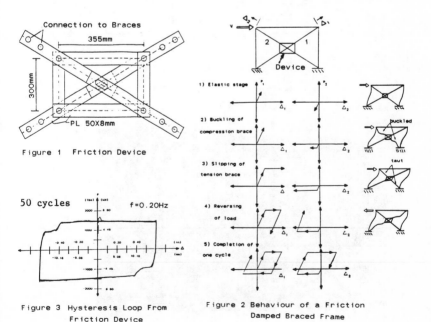

Figure 1 Friction Device

Figure 3 Hysteresis Loop From Friction Device

Figure 2 Behaviour of a Friction Damped Braced Frame

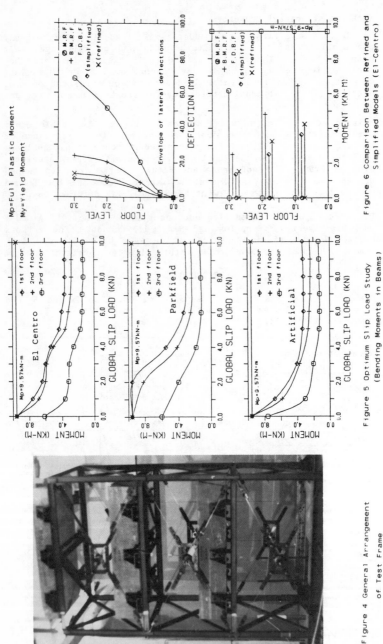

Figure 4 General Arrangement of Test Frame

Figure 5 Optimum Slip Load Study (Bending Moments in Beams)

Figure 6 Comparison Between Refined and Simplified Models (El-Centro)

Mp=Full Plastic Moment
My=Yield Moment

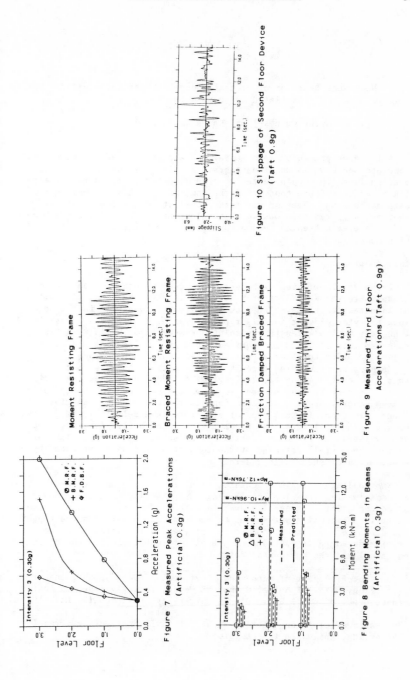

Figure 10 Slippage of Second Floor Device
(Taft 0.9g)

Figure 9 Measured Third Floor
Accelerations (Taft 0.9g)

Figure 7 Measured Peak Accelerations
(Artificial 0.3g)

Figure 8 Bending Moments in Beams
(Artificial 0.3g)

Nonlinear Behavior of Precast Concrete Shear Walls under
Simulated Earthquake Loading

Vincent Caccese[1] and Harry G. Harris[2]

This paper presents experimental and analytical results on precast
concrete shear walls under earthquake loading. It was observed in
both small scale models tested and the computer analysis that the nonlinear effects found consist of the combined action of rocking and
slip mechanisms. Rocking was more prevalent at the lower joints,
whereas, slip response was most prevalent at the upper levels. Damage
and effective softening tended to increase the amount of rocking and
slip observed.

Introduction

Large Panel building systems are composed of precast concrete components where the walls transfer load directly to the foundation without
an intermediate frame. The precast concrete shear wall is the major
load carrying element in this type of system. Most often, these buildings rely on the strength of isolated simple shear walls when resistance to lateral loads is required. Precast wall panels and floor
planks are tied together at well defined connection regions. Applied
loads must be transfered through the connections that, when compared
to the precast components, are regions of reduced strength and stiffness. Thus, a typical precast shear wall system can be thought of as
a deep concrete beam with known regions where nonlinear behavior will
occur. Inherent in this type of structure is a relatively large dead
weight which will cause the transmission of high inertia forces even
at low levels of ground vibration. When subjected to large lateral
forces the overall response of the connection region is affected by a
combination of shear slip and rocking action.

The nonlinear rocking mechanism is attributed to an overturning moment originating from inplane lateral forces acting on panels above.
Since the joint is generally weak in tension, uplift of the panel may
occur if the tensile stress due to bending is greater than the superimposed compressive stress due to the active gravity loads. Once the
tensile strength is exceeded, a gap will form between the end of the
end panel and the horizontal joint.

[1]Assistant Professor of Mechanical Engineering, Dept. of Mechanical Engineering, Univ. of Maine, Orono, ME 04473, formerly Research Assistant, Dept. of Civil Eng., Drexel University, Philadephia, PA 19106

[2]Professor of Civil Engineering, Dept. of Civil Engineering, Drexel
University, Philadelphia, PA 19106

In an analytical study, Becker and Llorente (1,2) quantified the inplane behavior of precast concrete shear walls and reported the shear slip and the rocking behavior to be the two significant mechanisms of nonlinearity. They showed that when shear slip occurs, seismic forces are limited as energy is dissipated. With rocking action, a reduced fundamental period exists as the joint stiffness decreases. The decrease in fundamental period is manifested by an increase in response for typical earthquake inputs. When rocking and slip occur together, and uplift exists at the tensile side, shear must be transferred through the unopened portion of the connection. The reduced contact area, along with the thrusting action of the rocking mechanism, causes high compressive stress to occur at panel corners; thus, damage to the panel corners was predicted.

Shricker and Powell (9) used a modified version of the Drain 2-D finite element analysis code to study the behavior of the precast concrete shear wall system subjected to inplane seismic loads. In their analysis they discovered that the computed resonse was very sensitive to the post yield or post slip behavior assumed for the connection. They reported that slip tended to concentrate in a joint if it was initially weaker than other joints. Also, when a joint lost strength, slip tended to concentrate in the joint that slipped first. Slip and gap openings were effective in reducing forces induced into the Large Panel structure.

The combination of rocking and shear slip creates a complicated mechanism of nonlinear, inelastic behavior in the seismic response of precast concrete simple shear walls. A series of models tested on a small shaking table facility were used ro reproduce some of these effects experimentally (Caccese and Harris (3-5)). According to Clough (6) the effect of nonlinearities due to joints that open and close during earthquake motion need further study.

Small Scale Model

Four small scale microconcrete models (1/32 scale) were constructed to simulate a five story stack of precast concrete walls and floors appropriately removed from the remaining structure. The models were tested on the Drexel University shaking table, described by Caccese and Harris (3), and were subjected to a uniaxial, simulated earthquake base vibration that limited the study to inplane inertia effects only (see Fig. 1). Base accelerations representing the El Centro 1940 N-S, Taft 1952 N21E, and Pacoima Dam 1971 S16E were used as input. To properly simulate the dynamic ineria effects, the addition of mass plates was necessary (Sabnis et al (8)). Lead mass plates were placed so that the inertia effect would be distributed to the model as it would be in the real structure without adding undue stiffness or strength to the joint.

Precast wall panels, hollow core floor plank, and footing components were designed according to PCA (7,10) recommendations and scaled down by a factor of 1/32. Since the small scale floor plank were very thin (1/4 inch thick) a hollow core was cast into the floor planks only at the connection end to insure simulation of the connection geometry. The footing component was heavily reinforced as it was designed

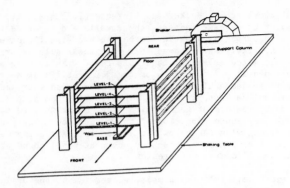

Fig. 1 - Schematic of Small Scale Model Showing Reference
          Levels.

to be rigid. Wall panels were reinforced for temperature and shrinkage
using a mesh of deformed wire placed at the center of the wall panel
crossection. A steel vertical tie, whose properties were varied, was
placed near the end of each wall panel. The first three Models (1-A,
1-B, and 1-C) contained a high strength, 117 ksi. ultimate, 4-40
threaded steel rod as the vertical tie. The vertical tie of the final
Model (2-A) was constructed with an annealed 2-56 threaded rod having
a yield strength of 42 ksi.

Experimental Study

    The initial elastic vibration was determined for each model by two
distinct methods. Firstly, an initial displacement test was performed
where the results were analyzed in the frequency domain. Secondly,
the initial flexibility matrix was determined by a static load test
and the frequencies were determined from flexibility coefficients
using an eigenvalue computer code. Results of these studies, reported
by Caccese and Harris (5) and given in Table 1, show a close correlation between the two independent methods.

    For the initial simulated earthquake test each small scale model
was subjected to a low level of base acceleration. For subsequent
tests, the base acceleration was increased in a stepwise manner so

Table 1 - Comparison of the Fundametal Frequencies Observed

Model	Initial Static Test (Hz.)	Dynamic Test (Hz.)	Percent Diff (%)
1-A	42.5	44	3.4
1-B	48.6	50	2.8
1-C	46.0	53	13.2
2-A	55.0	55	0.0

that increased levels of nonlinear action would occur. During the
testing, displacements were monitored in the horizontal direction at
the floor level. Figs.2 and 3 show two typical level 5 horizontal
displacement signatures. Fig. 2 shows results of the initial low le-

signature in Fig. 3, the signature in Fig. 2 is characterized by a
vel base motion (0.76 g max.) input to Model 2-A. When compared to the
waveform that has a higher frequency comosition. Also, an amplified
response not proportional to the increase in base motion is typical.

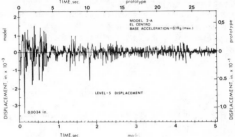

Fig. 2 - Typical Level -5 Horiz. Displ. Signature at 0.19g Max. Acc.

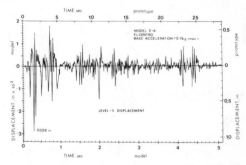

Fig. 3 - Typical Level-5 Horiz. Displ. Signature at 0.76g Max. Acc.

The base motion increased by a factor of 4.0 whereas the level 5 max-
mum displacement increased by a factor of 8.4. This amplification of
response is attributed to the increased shear slip and rocking and was
typical for all four models tested.

The slip phenomenon was monitored for Model 2-A using two displacement transducers oriented in the horizontal direction at the top and
the bottom of the joint being measured. Slip was observed to be amplified for an increase in ground motion (Table 2).

Table 2 - Maximum Slip at Level 3 for Model 2-A

Test No.	Max. Base Acceleration	Maximum Slip (in.)	
		Model	Prototype
1	0.19	**	**
2	0.38	0.0011*	0.035
3	0.76	0.0053	0.170
4	1.02	0.0092	0.294
5	1.41	0.0410	1.312
6	1.66	0.0680	2.176
7	1.64	0.0790	2.528

**not detectable *approximated

The rocking action was measured by locating displacement transducers in a vertical orientation across the front and back of the joint being observed, this was typically the level 1 joint. Rocking action, shown in Fig. 4, is characterized by a displacement signature that was biased toward the negative, uplift, direction. An increased joint stiffness, as compression of the joint microconcrete takes place, causes a reduction of displacement shown on the positive axis of the curve. This behavior is typical for panels where rocking occurs and the rocking action becomes more prevalent as the base motion is increased.

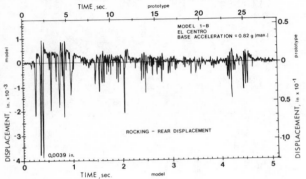

Fig. 4 - Typical Rocking Displacement Signature.

Finite Element Model for Earthquake Analysis

A nonlinear inelastic mathematical analysis was performed using a version of the Drain 2-D program modified by Schricker and Powell (3). The finite element mesh used in the investigation is shown in Fig. 5. The five wall panels of the simple shear wall system were idealized by twelve 8 DOF plane stress rectangles having a thickness of 8 inches. These wall panel elements were linear elastic with a modulus of elasticity and Piosson's ratio of 4500 ksi and 0.15, respectively, chosen to match the microconcrete wall panel. The vertical tie was modelled by a truss element with elasto-plastic behavior. The ties were connected at node lines 2 to 114 and 5 to 118 corresponding to the location of the vertical tie in the small scale model (Fig. 5).

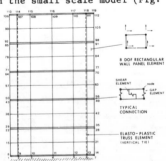

Fig. 5 - Finite Element Model.

Elasto-plastic shear connection elements of 2 different material types were used across each horizontal joint. The first element type modelled the frictional resistance of the grout-concrete interface. Its strength was given as the coefficient of friction (0.3 or 0.6) times the normal force. the shear stiffness for these elements (K_i) was related to the initial shear stiffnss (K_o) by a scalar multiplication factor (α). The proportionality factor was used as a parameter in this analysis and was varied with values of 0.1, 0.2 and 0.4. The second type of shear element was the steel vertical tie acting as a dowell. Its strength was taken as the shear strength of the tie used.

A gap element modelled the extensional properties of the horizontal joint. The compressive properties of microconcrete were input as the element properties in compression. Zero strength and stiffness were assumed in tension. Compressive stiffness was used as a parameter and taken to be either 100% or 50% of the full elastic stiffness.

Comparison of the Mathematical Model to the Small Scale Model Tests

Using Drain 2-D, a parametric study was performed for coefficients of friction of 0.3 and 0.6 and a shear stiffness reduction factor (α) varied from 0.1 to 0.4. The response was determined not to be sensitive to the extensional stiffness of the gap element and therefore, results presented are for the extensional stiffness equal to the full elastic stiffness of the joint material. Base motions input to the mathematical analysis were representative of the El Centro 1940 earthquake.

The maximum level 5 horizontal displacements are plotted versus the maximum base acceleration for the small scale models and the Drain 2-D analyses in Fig. 6. The lowest magnitude base acceleration input for

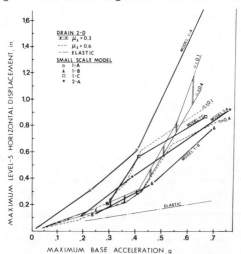

Fig. 6 - Comparison of the Max. Level-5 Horiz. Displ. of the Small
Scale Model to the Mathematical Analysis.

Models 1-B, 1-C, and 2-A tests were results of models that were not previously damaged. The inelastic Drain 2-D analysis gave a very good estimation of the level 5 displacement for these tests with all results falling in the predicted range. All model tests showed a large increase in response when compared to the results of the elastic analysis. The Drain 2-D analyses showed that regions of the model behaved in a nonlinear manner for inputs as low as 0.08g maximum base acceleration. The model behavior is characterized by an amplification of maximum response; as the ground motion increases so does the amount of nonlinear action exhibited.

For Model 1-A, that had an initially weak level 1 joint, the excessive maximum displacement shows how the response of the small scale model is more dependent on the initial condition of the structure. For lower values of base acceleration, sliding is not as prevalent, therefore, a decrease in shear stiffness imposed by initial damage will cause the response to attain a higher value. In the higher excitation levels, global slip is more prevalent and will be induced at an earlier stage in the loading history. Once sliding is induced, a previously uncracked model will behave more like a cracked model.

The maximum rocking displacements for Models 1-A, 1-B, and 1-C were scaled up to the prototype domain and plotted in Fig. 7 along with the

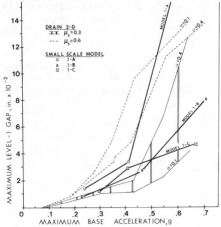

Fig. 7 - Comparison of Max. Level-1 Gap Opening of Small Scale Model to Math. Analysis.

Drain 2-D coefficient of friction study. The rocking comparison shows that the small scale model exhibits a lower response when compared to the 0.6 coefficient of friction model, as the 0.6 friction coefficient model tends to over predict the rocking response. The lower coefficient of friction of 0.3 however, did give a better prediction of the rocking with most of the small scale model tests falling within its bounds. The tests that did fall outside the 0.3 coefficient of friction study were somewhere in between the response predicted by the 0.3 and 0.6 coefficient studies.

The finite element model with the higher coefficient of friction tended to overpredict the rocking response of the small scale model. The rocking response observed in Models 1-B and 1-C was predicted by a 0.3 coefficicient of friction model for the first simulated earthquake input.   These two models had not undergone any excitation prior to this test and therefore, the initial properities were better matched by the finite element model. An accidental initial damage to Model 1-A greatly affected its behavior as its loss in stiffness increased the rocking.

Conclusions

    The results of the simulated earthquake study show that the shear wall structure behaves in a nonlinear inelastic manner at even very low base acceleration levels. Since the typical precast concrete Large Panel building is relatively massive, high inertia forces will exist even at low magnitudes of base acceleration. For low amounts of base acceleration (as low as 0.10g), an elastic analysis will underpredict the response of the simple shear wall model. The nonlinear mechanisms observed in both the small scale model and the computer analysis consist of the combined action of the rocking and slip mechanisms. Rocking was most prevalent at the lower joints; whereas, slip response was most prevalent at the level 5 and the level 4 joints. Damage and effective softening tended to increase the amount of rocking and slip observed.

    The response of the small scale model was sensitive to the magnitude of the input base acceleration. The displacement of the model was amplified for higher base acceleration levels. This same behavior was also observed in the finite element model. The El Centro input was composed mostly of low frequency components that tended to amplify model response as the model softened and its natural frequencies decreased. Other conclusions pertaining to the seismic behavior of the simple shear wall model are summarized as follows:

1) A linear elastic earthquake analysis will not correctly predict the seismic response of the five story simple shear wall system.
2) A nonlinear inelastic mathematical study can be used to predict model response; however, choice of stiffness and strength parameters is based on limited amounts of experimental testing. A parametric study is recommended if this type of model is used.
3) Further research is needed to determine the cyclic behavior of typical horizontal joints so that a more exact representation of the horizontal joint behavior can be input into a mathematical model.
4) There is much variability in the behavior of the simple shear wall system. Factors such as quality control of the concrete and grout mixes, and workmanship during construction of the structural system should be monitored closely to help decrease this variability.
5) When subjected to the simulated El Centro earthquake, the model response increased as the model softened.
6) Nonlinear action caused a decrease in the behavioral frequency of the model. This frequency decrease was shown by a decrease in frequency of the response waveform.

    The small scale mode withstood high levels of base acceleration without total collapse. In some instances, the model remained standing

even after base accelerations of more than 3 times the recorded El
Centro motion were input. Ductility was inherent in the system due to
the force limiting behavior of the slip mechanism that tended to decrease force levels as slip occurred.

Acknowledgements

    The work reported herein was conducted in the Dept. of Civil Eng.,
Drexel University, under Grants No. PFR-7924723 and CEE-8342561 from
the National Science Foundation. The cognizant NSF Program official
for these grants was Dr. John B. Scalzi.

Appendix A References

1. Becker, J.M. and Llorente, C., "Seismic Design of Precast Concrete
   Panel Buildings," paper presented at the Workshop on Earthquake
   Resistant Reinforced Concrete Building Construction, July 11-16,
   1977, University of California, Berkeley, sponsored by NSF.
2. Becker, J.M. and Llorente, C., "The Seismic Response of Simple Precast Concrete Panel Walls," Proc. of the U.S. National Conference
   on Earthquake Engineering, Stanford, California, August, 1979.
3. Caccese, V. and Harris, H.G., "Seismic Behavior of Precast Concrete
   Large Panel Buildings Using a Small Shaking Table," Report 1 – Description and Operation of the Drexel Univ. Structural Dynamics Laboratory, Structural Dynamics Laboratory Report No. D82-01, Dept.
   of Civil Eng., Drexel University, Philadelphia, PA, June 1982.
4. Caccese, V. and Harris, H.G., "Seismic Behavior of Precast Concrete
   Large Panel Buildings," Report 2 – Small-Scale Tests of Simple Precast Concrete Shear Wall Models Under Earthquake Loading, Dept. of
   Civil Engineering, Drexel University, Philadelphia, PA Dec. 1984
5. Caccese, V. and Harris, H.G., "Seismic Behavior of Precast Concrete
   Large Panel Buildings," Report 3 – Correlation of Experimental and
   Analytical Results, Dept. of Civil Eng., Drexel Univ., Philadelphia,
   PA, June 1985, Also, Doctoral Dissertation of First Author under Direction of the Second Author, Presented ot Drexel Univ. June 1985.
6. Clough, D.P. "Design of Connections for Precast Prestressed Concrete Buildings for the Effects of Earthquakes," ABAM Engineers
   Inc., Washington, NSF Grant No. CEE-8121733, March 1985.
7. Fintel, M., Schultz, D.M. and Iqbal, M. "Report 2: Philosophy of
   Structural Response to Normal and Abnormal Loads," Design and Construction of Large-Panel Concrete Structures, Office of Policy Development and Research, Dept. of Housing and Urban Development,
   Washingtin, DC, March 1976.
8. Sabnis, G.M.,Harris, H.G., White, R.N. and Mirza, S.M., Structural
   Modeling and Experimental Techniques, Prentice Hall Inc., Inglewood
   Cliffs, New Jersey, 1983.
9. Schricker, V. and Powell, G.H., "Inelastic Seismic Analysis of
   Large Panel Buildings," Report No. EERC 80-38, University of
   California, Berkeley, CA, Sept. 1980.
10. Schultz, D.M., Burnett, E.F.P., Fintel, M., "Report 4: A Design Ap-
    proach to General Structural Integrity," Design and Construction of
    Large Panel Concrete Structures, Office of Policy Development and
    Research, Dept. of Housing and Urban Development, Washington, DC,
    October 1977.

Hysteretic Behavior of Composite Diaphragms

By William S. Easterling,[1] A. M. ASCE and Max L. Porter,[2] M. ASCE

Abstract

Floor systems constructed with steel-deck-reinforced concrete
are described. When used to resist lateral loads these systems are
referred to as diaphragms. A description of a test arrangement,
along with test results for 14 diaphragms are presented. The relative
energy dissipation levels between the different tests are discussed.

Introduction

Floor systems constructed with steel-deck-reinforced concrete
are prevalent in steel frame structures. Many tests have been performed on elements of such floor systems to determine behavioral
and strength characteristics when subjected to gravity type loading.
This extensive testing has led to the recent publication of specifications and commentary (ASCE, 1985; Porter, 1985) pertaining to composite slabs. When structures that utilize composite floor systems
are subjected to lateral forces, such as those produced by wind or
earthquakes, the floor systems are required to transfer loads in-plane.
Floor systems performing in this manner are typically referred to
as diaphragms. Figure 1 shows a building schematic in which examples
of the diaphragm elements are shown.

In many instances the forces that the floor diaphragms are required to transfer, in the particular structural configuration, may
be relatively small and thus sophisticated analytical models of the
diaphragm are not required. There are cases, however, when the diaphragms may be subjected to large forces such that simple behavior
of the system may not be presumed. In these cases, more information
is needed regarding the behavior and capacity of the diaphragms.
To that end, research is currently underway at Iowa State University
(ISU) in which composite diaphragms are being studied. This study
is a continuation of previous work conducted at ISU (Porter and Greimann, 1980). To date, a total of 30 diaphragm specimens have been
tested in the research programs. The primary aim of this paper is
to describe the test arrangement and to present comparative results
describing relative energy dissipation between various diaphragms.

---

[1]Research Assistant, Dept. of Civ. Engr., Iowa State Univ., Ames, Iowa
[2]Prof. of Civ. Engr., Iowa State Univ., Ames, Iowa

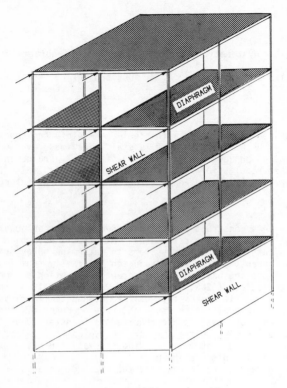

Fig. 1. Building schematic.

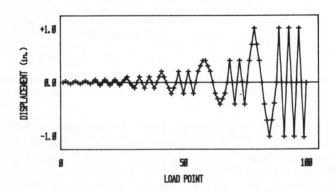

Fig. 2. Typical displacement history.

Test Arrangement

The testing was performed in the Structural Engineering Laboratory at ISU utilizing a previously designed test frame (Arnold, Greimann, Porter, 1978). Figure 3 shows the test frame configuration, which is a cantilever diaphragm frame with a fixed edge. The fixed edge consists of three large concrete blocks anchored to the structural tie-down floor. The remaining sides consist of W24x76 steel members fastened together with flexible tie connections.

The loading is applied via two hydraulic cylinders attached to the north frame member (see Figure 1). A quasi-static cyclic loading was used for all tests. Figure 2 shows a typical displacement history. This displacement history was selected, not with the intent of representing a particular natural event, but rather, with the intent of studying the basic behavior of composite diaphragms under cyclic loading.

Data collection devices included electrical resistance strain gages, direct current linear variable displacement transducers (DCDT), load cells, mechanical dial indicators and still photography. Continuous load versus displacement plots were made during testing. Examples are shown in Figures 4 and 5. A DCDT attached at the northeast corner (see Figure 2), in line with the load application, provided the displacement value. For discussion purposes in this paper the only data presented will be the area within the hysteresis loops, which can be thought of an energy dissipated by the diaphragms. These areas were calculated using a simple numerical integration scheme.

Test Results

Results from 14 diaphragm tests are presented in Table 1 for consideration in this paper. Each of the tests has been categorized according to the type of failure associated with the ultimate load. A complete listing and description of failure modes associated with composite diaphragms has been made previously (Porter and Greimann, 1980). Only the three considered in this paper are described here. They are diagonal tension (DT), shear transfer mechanism (STM) and stud zone failure (SZF).

Diagonal tension failure is characterized by the occurrence of cracks that form diagonally across the surface of the concrete. These cracks are typically at angles of approximately 45 degrees and extend over a large portion of the diaphragm. The ultimate load occurs just prior to the development of the diagonal tension crack.

The load transfer paths of these diaphragms is such that the load must pass from the loading frame through the fasteners (arc spot welds) and steel deck into the concrete. This transfer of load through the connectors and steel deck is primarily through shear transfer and has been shown to take place in a finite edge zone (Porter and Greimann, 1980). When degradation of the composite system occurs to a degree such that load cannot be transferred adequately to the concrete, failure of the shear transfer mechanisms is said to have occurred.

When headed shear studs are used as fasteners, failure of the concrete around the stud is referred to as a stud zone failure. If this occurs prior to the attainment of diagonal tension capacity of the diaphragm, the ultimate load is said to be associated with

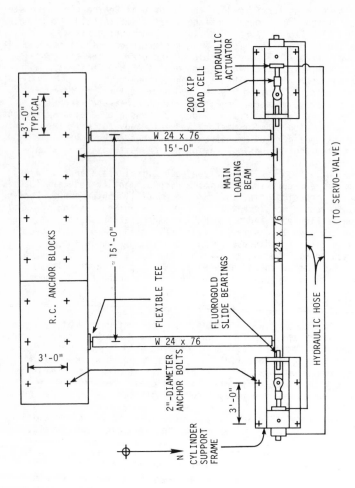

Fig. 3. Diaphragm test frame schematic.

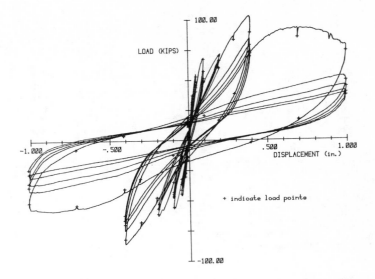

Fig. 4. Load-displacement diagram for Test A.

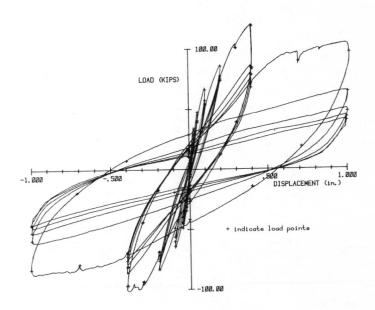

Fig. 5. Load-displacement diagram for Test B.

Table 1

Slab #	Failure Mode	Total Energy Dissipated (k-in.)
A	STM	411
B	DT	548
C	DT	678
D	DT	682
E	STM	304
F	STM	350
G	DT	688
H	STM	381
I	DT	825
J	DT	550
K	SZF	274
L	SZF	289
M	DT	341
N	SZF	271

Table 2

Diagonal Tension	Shear Transfer Mechanism	Stud Zone Failure
B-548	A-411	K-274
C-678	E-304	L-289
D-682	F-350	N-271
G-688	H-381	
I-825		
J-550		
M-341		

a stud failure. This is similar to the shear transfer mechanism failure, which is associated primarily with the use of arc spot welds.

Discussion of Test Results

The data presented in Table 1 has been arranged and listed in Table 2 according to the mode of failure. An important point to be emphasized is that the trend of the data is considered herein as opposed to specific values. Another point that warrants stating is that the failure mode classifications were made independent of any consideration of the energy dissipation values.

As can be seen in Table 2, distinct trends in the data are noteable. Specifically, the diaphragms exhibiting a diagonal tension failure dissipated more energy than the diaphragms in either of the two failure mode categories. Those failing by way of the shear transfer mechanism showed slightly more dissipative capacity than those where the stud failure occurred. Although not previously mentioned, the upper limit with respect to capacity, and therefore, the limiting failure mode, is associated with a diagonal tension failure of the diaphragm.

Summary and Conclusions

A test arrangement and methodology has been described for an investigation that is underway at ISU to determine behavioral characteristics of steel-deck-reinforced concrete floor diaphragms. Three types of failure modes have been described. Test results were presented for 14 diaphragms. These results consisted of failure mode classification and total energy dissipation. Diaphragms failing in the diagonal tension mode were shown to dissipate more energy than those failing in either the shear transfer mechanism or stud zone failure modes.

Acknowledgements

The authors are grateful to the National Science Foundation for the sponsorship of this research under contract no. CEE-8209104. Mssrs. S. M. Dodd, M. K. Neilsen, M. D. Prins and D. L. Wood were instrumental in the research effort and their assistance is gratefully acknowledged.

References

ASCE, "Specifications for the Design and Construction of Composite Slabs", ASCE Standard, Published by American Society of Civil Engineers, New York, New York, October, 1985.

Porter, M. L., "Commentary on Specifications for the Design and Construction of Composite Slabs", ASCE Standard, Published by American Society of Civil Engineers, New York, New York, October, 1985.

Porter, M. L. and Greimann, L. F., "Seismic Resistance of Composite Floor Diaphragms", Final Report ERI-80133. Engineering Research Institute, Iowa State University, Ames, Iowa, May, 1980.

PRECAST ROOFS FOR BLAST SHELTERS

S. B. Garner*

Abstract

A shallow-buried reinforced concrete blast shelter has been
constructed and dynamically tested by personnel at the U.S. Army
Engineer Waterways Experiment Station (WES). Design loading was
50 psi incident overpressure from a simulated 1-Mt weapon. To develop
design options and to determine the effects of soil-structure
interaction on the loading and response of a simply supported roof,
three thicknesses of precast roof panels were designed for the
shelter. One-quarter-scale models were constructed at WES and
dynamically tested at Fort Polk, LA, using a 16-kt, 50-psi
simulation. Panels were placed on a reaction structure designed to
model the blast shelter and were covered with 1 foot of compacted
sand. The test bed was instrumented to determine blast pressures and
free-field pressures and accelerations. The structure was
instrumented to determine pressures across the roof, roof deflection,
and roof and floor accelerations. Deflections in all tests were small
(less than L/40). Nonuniform loadings due to soil arching were
observed in all three tests with the greatest reduction in loading
occurring in the most flexible panels.

Introduction

Shallow-buried reinforced concrete blast shelters are usually
designed with cast-in-place roofs. Research at the U.S. Army Engineer
Waterways Experiment Station (WES) has shown that these structures
resist significantly greater loads than predicted by analysis
(References 1, 2, and 3), possibly due to soil-structure interaction
(soil arching).

Soil arching occurs when loads are transmitted from compressible
areas to stiffer areas by shear stresses in the soil. Active arching
occurs when a buried structure is compressed more than the surrounding
soil. Passive arching takes place when the soil is compressed more
than the structure. In this paper, arching due to a difference in
deflection between the structure and the surrounding soil will be
referred to as global arching and will be defined by an arching ratio,
C_a, the ratio of average pressure on the roof to the free-field
pressure in the soil at roof level.

* Research Structural Engineer, USAE Waterways Experiment Station,
P.O. Box 631, Vicksburg, MS 39180-0631.

Very rigid structures deflect uniformly, resulting in a uniform loading. Flexible structures such as a simply supported roof deflect nonuniformly, resulting in a redistribution of load from "soft" areas that move away from the soil to "hard" areas at supports. This pressure redistribution will be referred to as local arching.

A three-bay shelter of cast-in-place reinforced concrete has been dynamically tested by WES personnel. Design loading was a 50-psi incident overpressure from a simulated 1-Mt nuclear weapon. To develop design options for the shelter roof and to determine the effects of soil arching on the loading and response of a simply supported roof, three thicknesses of precast roof panels were selected for dynamic testing.

Structural response calculations were carried out using a computer code developed at WES (Reference 4) that includes dynamic soil-structure interaction effects. Based on these calculations, prototype roof thicknesses of 10, 8, and 6 inches were selected for testing and 1/4-scale models were constructed. Using cube root scaling, a 16-kt nuclear simulation was selected for the tests.

Testing

Effective depth, span-to-thickness ratio (L/t), and tensile and compressive reinforcing ratios for the 1/4-scale slabs are given in Table 1. Design concrete compressive strength was 4,000 psi for the roof panels and 3,000 psi for the reaction structure. Design principal reinforcing yield strength was 60,000 psi. Results of concrete cylinder tests for roof panels are given in Table 2. Average test yield strength for the principal reinforcing bars (D3 wire) was 78,000 psi.

Table 1. Experimental Parameters

Test No.	Slab Thickness (in.)	Effective Depth (in.)	L/t	Reinforcing Ratios		Support
				Midspan		
				ρ	ρ'	ρ & ρ'
1	2-1/2	2-1/4	14	0.0105	0.004	0.0105
2	2	1-3/4	17.6	0.0148	0.004	0.0148
3	1-1/2	1-1/4	23.5	0.0151	0.0076	0.0151

Table 2. Roof Panel Concrete Compressive Strength

Test Element	Batch	Compressive Strength (psi)		
		7-Day	28-Day	Day of Test
Roof sections, Tests 1 and 2	1	2,770	3,890 3,980 4,600	–
Roof sections, Test 3	2	2,930	4,350 4,170 4,280	4,244

The models were tested in sand backfill at a burial depth of approximately L/3. A single reaction structure designed to model two bays of the blast shelter was used in all three tests. The roof panel plan and sections are shown in Figure 1. The reaction structure plan and section are shown in Figure 2.

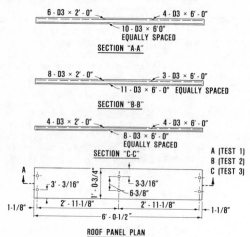

Figure 1.   Roof Panel Plan and Sections

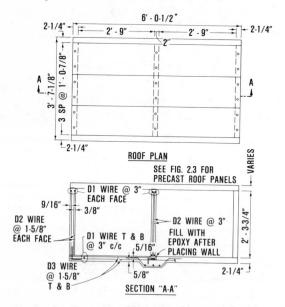

Figure 2.   Reaction Structure Plan and Section

Roof elements were instrumented to determine reinforcing strain, acceleration, and mid- and quarter-span deflections. Roof pressures were measured with soil-stress (SE) gages lockated 1/8 inch to 2 inches above the roof. Additional soil-stress gages were placed in the free field, and floor and free-field accelerations were also measured. Blast pressures were measured by gages located 2 inches below grade. Soil stress and blast pressure gage locations are shown in Figure 3.

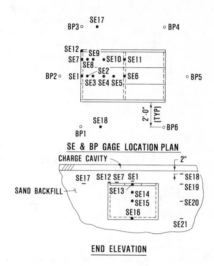

Figure 3. Soil-Stress and Blast Pressure Gage Locations

In each test, a 6-foot-deep hole was backfilled with compacted sand to the bottom-of-structure elevation. The structure was set in place and sand was placed and compacted in 6-inch lifts to provide 1 foot of cover over the roof. A 12- by 12- by 3-foot, wood-framed, plywood-covered HEST charge cavity was placed directly over the backfill. Each cavity contained 17 uniformly spaced strands of 50-grain/foot detonating cord. A 4-foot uncompacted sand overburden was used over the charge cavity to confine the blast pressure and produce the long duration characteristic of a 16-kt weapon. Weapon yield, overpressure, and maximum midspan response for each test are given in Table 3.

Table 3. Test Results and Weapon Simulations

Test No.	Weapon Yield (kt)	Approximate Overpressure (psi)	Midspan Deflection (in.)
1	16	56	0.19
2	16	60	0.43
3	16	62	0.84

Analysis

The nonuniform load distribution due to local arching is accounted for by the load factor, K_L, the ratio of equivalent to actual total load.

Average values for C_a and K_L for the period from 0 ms to the time of maximum midspan deflection are given for each test and for both free-field SE gages in Table 4. For a uniformly distributed load, the load factor is 0.5. In all three tests, load factors were less than 0.5, indicating local arching across the roof panels. Since maximum deflections were relatively small, arching ratios for all tests were close to 1.0, indicating that global passive arching was taking place at the same time.

Table 4. Average C_a and K_L Calculated from Test Data

Test	Time to Maximum Deflection (ms)	Using SE-17 Data		Using SE-19 Data	
		C_a(AVG)	K_L(AVG)	C_a(AVG)	K_L(AVG)
1	10	1.124	0.383	1.231	0.383
2	15	0.740	0.245	0.913	0.245
3	32	0.983	0.229	0.802	0.229

The load distribution for the precast roof tests can be best represented by a third-order polynomial, as shown in Figure 4. For comparison, pressure distributions for Tests 1 and 3 at 10 ms are given in Figures 5 and 6. Center pressure for the most flexible roof slab (Test 3) was 0 at 10 ms, with the load redistributed to supports as the center of the slab deflected away from the soil.

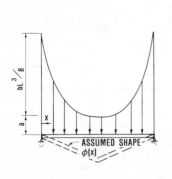

Figure 4. Roof Loading

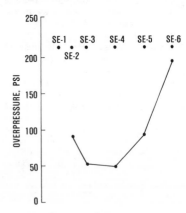

Figure 5. Test 1, 10 ms

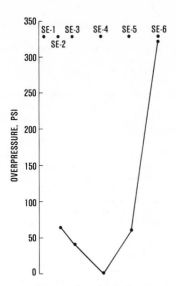

Figure 6.   Test 3, 10 ms

Conclusions and Recommendations

1. Precast panels are easily constructed and installed, and can be designed to resist a 50-psi overpressure from a 1-Mt weapon. However, a minimum design to resist 50 psi is not a practical alternative, since handling and installation can become a problem with very thin sections. It is recommended that 6-inch-thick precast roof panels be considered as an alternative to a poured-in-place roof.

2. Global soil arching can develop at shallow depths and is influenced by structural stiffness.

3. Roof capacity is enhanced by local soil arching, which results in a redistribution of the loads from the center of the roof panels to "hard" areas at the supports.

References

1. Getchell, J. V. and Kiger, S. A., "Vulnerability of Shallow-Buried Flat-Roof Structures, Foam HEST 4"; Technical Report SL-80-7, Report 2; U.S. Army Engineer Waterways Experiment Station, Vicksburg, Mississippi; October 1980.

2. Kiger, S. A., "Static Test of a Hardened Shallow-Buried Structure"; Technical Report N-78-7; U.S. Army Engineer Waterways Experiment Station, Vicksburg, Mississippi; October 1978.

3. Kiger, S. A.; Eagles, P. S.; and Baylot, J. T.; "Response of Earth-Covered Slabs in Clay and Sand Backfills"; Technical Report SL-84-18; U.S. Army Engineer Waterways Experiment Station, Vicksburg, Mississippi; October 1984.

4. Kiger, S. A.; Slawson, T. R.; and Hyde, D. W.; "Vulnerability of Shallow-Buried Flat-Roof Structures; a Computational Procedure"; Technical Report SL-80-7; U.S. Army Engineer Waterways Experiment Station, Vicksburg, Mississippi; September 1984.

Acknowledgement

The research discussed in this paper was sponsored by the Federal Emergency Management Agency, Washington, D.C., under the direction of Program Manager Tom Provenzano. Models were constructed at the Structures Laboratory, WES. Technician Billy Benson assisted with construction supervision. Field tests were supervised by Randy Holmes. Instrumentation was by William Strahan and Phil Parks of the Instrumentation Services Division, WES.

FIELD TESTS OF WIND-STRUCTURE INTERACTION

Theodor Krauthammer[1], M.ASCE, Roberto T. Leon[1], M.ASCE
and Paul A. Rowekamp[2]

ABSTRACT

The present study was initiated to investigate the possible causes for excessive lamp failure rates on High Mast illumination towers in the State of Minnesota. The study consisted of two methods of analysis that were combined for obtaining information on the motion of the towers and the luminaries. The first step was to analyze numerically the complete system by employing a finite element code and to compute the motions of the complete system under typical wind conditions at the site for wind velocities between 5 to 60 mph (8 to 96 kph). The second phase of the study consisted of an experimental effort during which one of the illumination towers was instrumented and motion data was collected for various wind conditions. The final phase of the study included the evaluation of the numerical as well as the experimental data that had been obtained from the preceding steps. This evaluation included the identification of modes of vibration in the frequency domain, filtering of data for accessment of modal effects, and comparisons between experimental and numerical results. The experimental work is presented and discussed in this paper.

INTRODUCTION AND BACKGROUND

This paper is based on the results from a recent study performed at the University of Minnesota for the Minnesota Department of Transportation (MN/DOT) concerning excessive failure rates of high pressure sodium bulbs installed in High-Mast Illumination towers (8). These towers were 120 ft. high, and at the top are the luminaries, as will be further described. Preliminary reviews of the issue by MN/DOT personnel determined that such failures were not caused by electrical or electromagnetic problems, and the issue of wind induced vibrations needed to be investigated. This study consisted of an analytical/numerical phase, an experimental phase, and a comparison and evaluation phase. The analytical/numerical part included implementation of the finite element method for the assessment of the tower behavior under various static and dynamic loading conditions in order to assess the expected experimental data. In the experimental phase one of the towers was instrumented with twelve single-degree-of-freedom (SDOF) accelerometers, and the recorded data was studied both in the time and frequency domains. The numerical and experimental results were compared at various stages of this study for obtaining

---

[1] Department of Civil and Mineral Engineering, University of Minnesota, Minneapolis, MN 55455.
[2] T.Y. Lin International, 8085 Wayzata Boulevard, Minneapolis, MN 55426.

an accurate description of the wind induced vibrations and the relationship with the observed bulb failures.

Structural response to dynamic loads can be evaluated by employing either classical methods in the linear domain, or when complex structures are considered, one may have to use numerical techniques for the analysis. For structural analysis in the nonlinear domain of behavior a reliable numerical approach is almost always required for obtaining acceptable solutions which can be interpreted for engineering purposes. A substantial amount of information on dynamic structural analysis in the linear and nonlinear domains of behavior is provided in Ref. (10), but it is primarily oriented to address the problem of earthquake effects. Nevertheless, the analytical approach presented does apply to a wide range of structural dynamics problems. Also, there are similarities between the structural response of systems under the effects of dynamic loads transmitted through the ground and the behavior of structures under wind induced loads as presented in (5). A fundamental treatment of wind engineering was presented in (15 and 16) and this information can be employed for the analysis of structural systems under the effects of wind loads. These texts also provide an abundance of information concerning wind forcing functions, drag coefficients and vortex shedding effects. The issue of loading conditions associated with air flow has been studied quite extensively, and the information available from the literature was used in this study. Those references also contain several typical wind records with durations of about 2 minutes, and such instantaneous records were an essential part of the numerical approach because actual wind data from the test site were not available. Cheung and Chiu (6), provided information concerning probabilistic determination of extreme winds from short duration records, and other publications (2,9,11,12,14, and 17) contain relevant information on wind forces and corresponding structural response for various systems, but primarily on tower-like structures (which are similar to the High-Mast system under consideration). There are many excellent references available concerning the finite element approach, for example.

DESCRIPTION OF TOWERS

The specific towers deal with on this project measure 120 ft. (36.6m) high and taper in diameter from the base to the top. The average base diameter is approximately 24 inches (61m) and the average top diameter is approximately 7 inches (17.8m), as illustrated in Figure 1. The tower is actually composed of three sections of cold formed steel plate. Each section is approximately 40 ft. (12.2m) high and is formed into a sixteen sided cylinder fitted together to form the mast. All three tower sections have different plate thicknesses. The lower section is 5/16 inch (0.8cm) thick, the middle section is 1/4 inch (0.6cm) thick, and the top section is 3/16 (0.5m) inch thick steel. The material is high strength, low alloy steel per ASTM A588 with a 50,000 psi (345 MPa) minimum yield strength, and 70,000 psi (483 MPa) minimum tensile strength. The approximate weight of the shaft is 7200 lbs. (327 Kg). The section modulus at the base is nearly 190 inches (3114 cu.cm.) cubed which results in a 100 percent yield moment of 785,00 ft.-lbs. (1,064 m KN). At the base of the tower there is a 36 inch (91 cm) high tube section made from folded plate that tapers from 39 inches (99 cm) in diameter at the bottom to 24 inches (61 cm) at the top. The top of this section is connected to the bottom of the lower section of the shaft with 100 percent penetration welds. This base section also has a door for access to the electrical connections and steel cable spools for lowering the

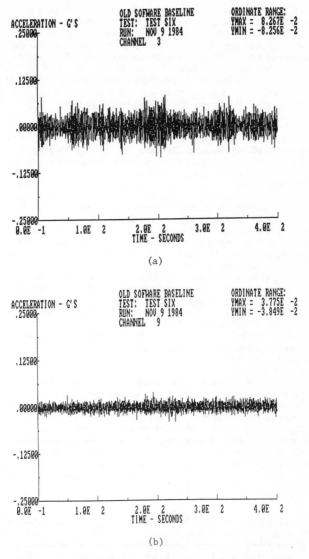

(a)

(b)

Figure 1 – Recorded Accelerations (a) Horizontal,
and (b) Vertical.

ring assembly, and luminaries. The largest ring of the assembly is a 2.5 inch (6.3 cm) steel pipe that is cold formed into a 47.25 (120 cm) inch diameter ring. Four smaller steel pipes extend from the ring and the luminaries are mounted on these smaller pipes. Each luminary weighs approximately 65 lbs. (289 N) and houses one 1000 watt high pressure sodium bulb. These bulbs are cylindrical in shape measuring about 13 inches (33 cm) long and 2.75 inches (7 cm) in diameter. At the base of the tower shaft there is a spool of steel cable that is fed through the inside of the shaft and is connected to the ring assembly at the top of the tower. When this spool is unwound the ring disengages from the mast head and is slowly lowered. The mast head assembly at the top of the tower houses several pulley mechanisms that thread the electrical and steel cables as the ring is raised or lowered. When the ring assembly is raised to its standard uppermost position it is locked into place by the mast head.

METHODOLOGY

As briefly discussed earlier, the analysis of the High Mast Tower system consisted of an analytical/numerical phase and an experimental phase. The analytical study was carried out first to gain a better understanding of the tower's basic properties including stiffness, damping, and natural frequencies, and to obtain preliminary assessments of the relationship between wind-induced effects, as recommended in the literature (15 and 16); and the corresponding levels of motion. This information was obtained by applying simulated wind loads to the model tower and observing its response. These wind loads could be varied by changing the frequency and magnitude of the applied forces. Different loading patterns including uniform loads, concentrated loads, and mass proportional loads were included. This type of analysis should provide an adequate spectrum of expected values for the displacements and accelerations. Since it is nearly impossible to exactly model the complicated head and ring assemblies the second stage of the analysis included placing accelerometers on the ring and luminary to observe the actual motion of the structure. The results of these separate studies could then be used for comparison and updating of the theoretical models. In the analytical evaluation of the problem use was made the finite element method (FEM) for the dynamic analysis of the structural system, and the program ADINA (1) was employed for this purpose. This is a commercially available multi-purpose finite element program for linear or nonlinear analysis of structural systems in static or dynamic domains of behavior, in one, two, or three dimensional space. The program can provide time histories of forces, strains, stresses, displacements, velocities, accelerations, and also natural frequencies and mode shapes of the structure.

EXPERIMENTAL APPROACH

The experimental phase of this study included mounting accelerometers on an existing High-Mast tower in Eagan, Minnesota. A total of 12 accelerometers were mounted on the ring assembly of the tower and the tower motions were monitored on 5 different days in late October and early November 1984. The average duration of each record was 7 minutes with a sampling rate of 10 readings per second. Several other records were also taken with sampling rates of 100 readings per second. After processing all the raw data into acceleration-time histories the next step in the analysis was to process the data through a double integration scheme in order to compute the corresponding deflections. For preventing the accumulation of errors that tend to occur during integration the data was baseline corrected before each integration, as suggested in (18), and the results from the experimental study are summarized next.

On November 9, 1984, afternoon wind gusts were measured in excess of 25 mph (40 Kph), and the resulting filtered tower accelerations show that the maximum horizontal acceleration was 0.08 g's, and the corresponding peak displacement for the lower end of the lamp was approximately 2 in. (5 cm). These accelerations and deflections compare well with the response from five other data sets recorded on that day. In Figures 1(a) and 1(b) are the horizontal and vertical acceleration-time records at the top of the tower, respectively. Comparing these plots shows that the vertical accelerations magnitude was approximately half the horizontal acceleration magnitude. From the comparison it is quite clear that the numerical prediction based on a "generic" wind force are acceptable, and that this approach can be employed for reliable pre-test predictions. In order to calculate the lowest natural frequencies of the tower the acceleration data was processed through a Fast Fourier Transform (FFT) program. The FFT program employs the principle of the Fourier integral in discrete form to transfer a function or data from the time domain to the frequency domain. After reviewing all the relevant FFT data from these tests the lowest 3 frequencies were found to be 0.4, 1.67, and 4.44 Hertz, comparing well with the numerical results. The final step of the experimental analysis was to isolate each mode and calculate the modal participation factors. This was carried out by processing the displacement data through the bandpass filtering scheme and filtering all frequencies except those surrounding each natural frequency. Upon computing of the displacements for the filtered frequency data it was found that no displacements greater than 0.15 in. was associated with the higher modes of vibration.

DISCUSSION

Comparing the results of the analytical/numerical study (8) with the experimental data shows good agreement on almost all counts. The values calculated for the lowest natural frequencies of the structure were identical in both studies, and the calculated values of maximum acceleration and displacement due to winds under 30 mph (48 kph) were also nearly identical. A very interesting point brought out in both phases of the study was the prevalence of first mode response. Data from numerical study indicated that the fundamental mode response governed the behavior, and similar results were obtained from the experimental data. Close examination of the finite element results show that even when the tower was subjected to very high winds the response was mainly at 0.4 Hertz. After reviewing the experimental displacement data it was obvious that almost all of the horizontal response was in the first mode. Both the analytical and experimental results clearly showed that the magnitude of acceleration was less than 0.20 g's, and the displacements less than 3 in. for wind speeds under 30 mph (48 kph).

CONCLUSIONS

The present study was initiated to study possible wind-induced vibrations on lamps that are mounted on the High-Mast Tower illumination systems. Initially, it was expected that vibrations were the principal source for the lamp excessive failure rates. However, after preliminary data became available during this study is seemed that mechanical vibrations cannot be the only cause for such failures. As a result of this study it was concluded that the peak acceleration would not be severe enough for causing the observed lamp damage. No problem was detected which concerns the tower design and its possible effect on the lamp performance. At this time it seems that the principal cause for sever stresses in the arc tube is associated with thermally-induced deformations in the two metal support wires that are located in parallel to the arc tube, as was discussed previously. Nevertheless, it is quite reasonable to expect that the low magnitude vibrations would further enhance the failure rate since the dynamic effects will contribute to the state of stresses in the arc tube.

ACKNOWLEDGMENTS

The authors wish to extend their thanks to the Minnesota Department of Transportation for funding this study, and for their technical support during the experimental phase. Also, thanks are due to ANCO Engineers, Inc. for their support with the VIPAC data acquisition software.

APPENDIX I - REFERENCES

1. ADINA Engineering, "ADINA User's Manual", Report AE-81-1, Sept. 1981.

2. Ahmad, M.B., Pande, P.K., and Krishna, P., "Self-Supporting Towers Under Wind Loads", Proc. Journal of The Structural Division, American Society of Civil Engineers, Feb. 1984, pp. 370-384.

3. Baker, D.G., "Climate of Minnesota - Wind Climatology and Wind Power", Tech. Bulletin, Univ. of Minn. Agricultural Station, AD-TB1955, 1983.

4. Bathe, K.J., "Finite Element Procedures in Engineering Analysis", Prentice-Hall, 1982.

5. Cevallos-Candau, P.J., and Hall, W.J., "The Commonality of Earthquake and Wind Analysis", SRS No. 472, Department of Civil Engineering, University of Illinois, January 1980.

6. Cheung, E., and Chiu, A.L., "Extreme Winds Simulated From Short-Period Records", Journal of The Structural Division, American Society of Civil Engineers, Jan. 1985, pp. 77-94.

7. Clough, R., and Penzien, J., "Dynamics of Structures", McGraw-Hill, 1975.

8. Krauthammer, T., and Rowekamp, P.A., "The Effect of Wind Induced Vibrations on Highway Lighting Bulbs", Final Report, Minnesota Department of Transportation, July 1985.

9. Kwok, K.C.S., Hancock, G.J., Bailey, P.A., and Haylen, P.T., "Dynamics of a Freestanding Steel Lighting Tower", Engineering Structures, Jan. 1985, pp. 46-50.

10. Newmark, N.M., and Rosenblueth, E., "Fundamentals of Earthquake Engineering", Prentice Hall, 1971.

11. Paquet, J., "Measurement and Interpretation of Wind Effects on a 50 Meter High Tower", Proceedings, Wind Effects on Buildings and Structures, Tokyo 1971.

12. Penzien, J., "Wind Induced Vibration of Cylindrical Structures", Proc. ASCE, Journal of The Engineering Mechanics Division, Jan. 1957.

13. Plate, E., "Engineering Meterology", Elsevier, 1982.

14. Ross, H.E., and Edwards, T.C., "Wind Induced Vibration in Light Poles", Proc. ASCE, Journal of the Structural Division, Vol. 96, June 1970, pp. 1221-1235.

15. Sachs, P., "Wind Forces in Engineering", Pergamon Press, 1978.

16. Simiu, E., and Scanlan, R.H., "Wind Effects on Structures: An Introduction to Wind Engineering", John Wiley and Sons, 1978.

17. Vickery, J., and Kao, K.H., "Drag or Along-Wind Response of Slender Structures", Proc. ASCE, Journal of the Structural Division, January 1972, pp. 21-36.

18. VIPAC, "Vibration Testing Work Station - Software Description and User's Manual", ANCO Engineers, July 1984.

TABLE 1 - PEAK ACCELERATIONS, NUMERICAL STUDY

Record	Peak Wind Velocity mph (kph)	Maximum Along Wind Displacement in inches (cm)	Maximum Along Wind Acceleration (g's)
Extreme	46 (74)	14 (35.6)	0.15
Moderate	25 (40)	3.8 (9.6)	0.07
Calm	<15 (24)	2.2 (5.6)	0.03

VIBRATION TESTS OF RESERVOIR STRUCTURES
OUTLET TOWER AND FOOTBRIDGE

William E. Gates, Member ASCE (1)
Lawrence G. Selna, Member ASCE (2)
Dennis K. Ostrom, Member ASCE (3)

SUMMARY

This paper presents a field test procedure used in determining the dynamic properties of an outlet tower and footbridge for a major reservoir. Such factors as virtual water mass, soil-structure interaction, and structure-bridge interaction were investigated in the test program. Two methods of vibratory excitation were used to produce varying levels of strain within the structures and foundation soils to study nonlinear variations in stiffness and damping characteristics. Results from the test program were used to verify mathematical three-dimensional finite element models for use in dynamic analyses of the structures under earthquake conditions.

INTRODUCTION

This report presents the highlights of a field test program to assess the dynamic properties of an outlet tower and footbridge. These structures were identified as critical to the seismic safety of the Garvey Reservoir Dam, owned and operated by the Metropolitan Water District of Southern California. Seismic review of the reservoir was initiated by the State of California, Resource Agency, Department of Water Resources, Division of Safety of Dams, under a national program for inspection of dams authorized by Public Law 92-367. The total program included review of the earth embankments of the reservoir, the inlet-outlet structures, and related underground structures and piping. The co-authors served as principal investigators and participants in the field vibration test program of the Inlet-Outlet Structures.

---

(1) Associate, Dames & Moore, Los Angeles, California.

(2) Professor, Department of Civil Engineering, University of California, Los Angeles, California.

(3) Senior Engineer, Southern California Edison, Rosemead, California.

The Garvey Reservoir was constructed in Monterey Park, California in the early 1950s. It was designed under standards for earthquake safety that have seen fairly extensive revision over the last 15 years. The reservoir embankments and pertinent structures were designed to static building code force levels. The dynamic effects of earthquake motions on the tall, slender outlet tower were not directly considered in the original static seismic design approach.

The objective of the dynamic test program was to provide data for verifying mathematical models of the structures to be used in the earthquake safety analysis. The dynamic properties of concern included: natural frequencies of vibration, mode shapes, and structural damping.

DESCRIPTION OF STRUCTURES

Figure 1 illustrates the outlet tower, footbridge, and connecting underground pipe gallery with inlet-outlet piping. The 16-foot-diameter tower stands 86.5 feet high and is founded on a 45-foot-diameter reinforced concrete mat that is 8 feet thick. The foundation mat, along with 20 feet of tower shaft, are embedded in compacted fill in the eastern embankment of the reservoir. A 40-foot segment of the tower shaft is submerged in the reservoir under high water conditions and the remaining 26 feet extend above the water surface. The tower wall cross section varies in thickness from 2'6" at the base to 16" at the top.

The access bridge spans from the embankment of the reservoir to the outlet tower. The bridge is anchored at the tower in all three principal directions. It is supported vertically and transversely on rocker bearings at the midpoint pier and the end abutment, at the top of the reservoir embankment. Thus, longitudinal motion of the bridge is transferred directly into the tower through the bridge connection at the tower corbel.

The outlet tower functions as a vertical cantilever supported on a semi-rigid base in resisting lateral motions induced under earthquake. The waters of the reservoir surrounding the tower adds virtual mass and a nominal degree of damping to the structure. The soil surrounding the tower shaft provides a varying degree of stiffness and damping depending on the level of strain in the soil under vibratory motion. The bridge interacts with the tower under dynamic loads due to its method of support and flexibility.

TEST PROGRAM

Key Test Considerations

The water level of the reservoir could fluctuate by as much as 30 feet under seasonal usage. Thus, two levels of reservoir depth were investigated, high water level, Elevation +570, and low water level, Elevation +540 to assess whether the influence of the variation in virtual water mass was significant upon the dynamic characteristics of the tower.

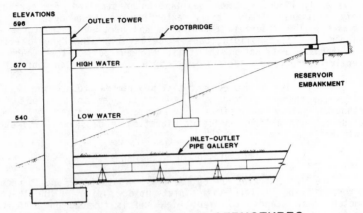

FIGURE 1 INLET-OUTLET STRUCTURES

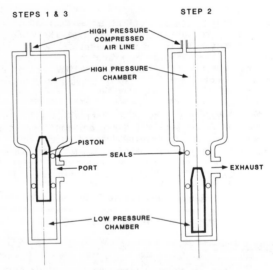

FIGURE 2 "POPPER" CROSS SECTION

The stiffening and restraining effects of the soil surrounding the tower shaft and its mat foundation are another potential variable under earthquake conditions. Monitoring the restraining effects of the soil at low strain levels may provide misleading information on the true stiffness at higher strain levels compatible with earthquake conditions. The test program included two levels of vibratory motion in order to determine if a basis could be established to assess the shift in dynamic properties associated with nonlinear behavior in the surrounding soils.

The mass and stiffness of the footbridge are less than the tower. However, it was unknown whether the bridge mass could produce a significant effect on the dynamic properties of the tower and its foundation system under dynamic earthquake response. The test program was designed to determine the natural frequencies of the tower and footbridge to assess whether these frequencies were in close resonance and whether they could introduce significant interaction forces upon the tower.

Vibratory Loading

Two forms of vibratory loading were selected for the test program. The first relied on ambient ground motion conditions as the exciting force. The second utilized an induced compression wave in the water generated by a pneumatic/mechanical device known as a "popper". This device was developed and documented by Ostrom (Ref. 4) for exciting dams and hydraulic structures for dynamic test purposes. The device is illustrated in Figure 2. The popper consists of an aluminum cylinder with a piston inside that is driven by compressed air. A quick release valve action on the part of the piston permits compressed air to be released suddenly into the water, producing a compression pulse with peak pressures on the order of 5 atmospheres. By placing this cylinder in the reservoir near the tower, pulses were induced in sequence of up to one per second. The force levels produced by the popper are considerably higher than those produced under ambient vibrations, yet well below the thresholds that could cause damage to the tower or footbridge.

Instrumentation

The motions of the tower under the ambient as well as low-level forced vibration tests were monitored with a pair of accelerometers selectively located at each of the points shown on the tower footbridge and inlet pipe in Figure 3. The equipment used in the test program consisted of the following:

1. Tow strong motion force balance accelerometers -- Kinemetrics Model SMA-3

2. Realtime Spectrum Analyzer -- Hewlett-Packard Model HP-3582A

3. Strip chart recorder -- Gould brush No. 220

4. X-Y Plotter -- Hewlett-Packard Model HP-7015B

5. Signal conditioner -- Kinemetrics Model SC-1

6. Magnetic tape recorder -- Hewlett-Packard Model 3960.

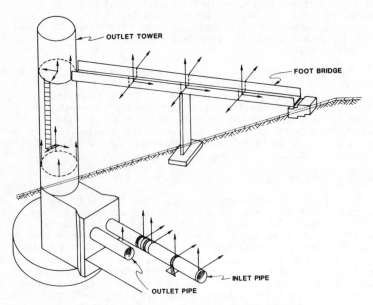

FIGURE 3 LOCATION AND ORIENTATION OF ACCELEROMETERS

Preliminary Analysis

A three-dimensional elastic finite element model (shown in Figure 4A) was formulated of the tower and footbridge prior to the test program to provide preliminary dynamic properties for the structure for use in comparison with those experimentally determined. This preliminary analysis also provided guidance in selecting the locations on the tower and footbridge where the accelerometers were placed to monitor the maximum modal response.

The model of the tower and footbridge was analyzed in the SAP IV and EASE 2 computer programs (Ref. 1 and 2). The tower shaft was treated as a three-dimensional beam column using the gross, uncracked concrete section properties for the stiffness. The supporting and surrounding soils were represented by a series of springs. The stiffness of these springs was based on low-strain passive soil stiffness characteristics. The rocking and vertical stiffness at the base of the foundation was modeled using elastic half-space theory.

Lumped mass properties of the tower included the concrete shaft, soil surrounding the tower shaft above the concrete base mat, water trapped in the tower shaft, and virtual water mass moving with the

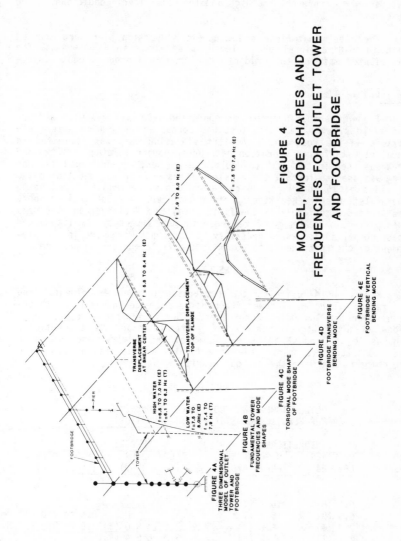

FIGURE 4

MODEL, MODE SHAPES AND
FREQUENCIES FOR OUTLET TOWER
AND FOOTBRIDGE

tower shaft around its exterior perimeter. This virtual water mass was varied to represent the high and low water level conditions in the reservoir.

The steel footbridge and concrete supporting pier were modeled with three-dimensional beam elements, as shown in Figure 4A. The distributed mass of the bridge and pier were lumped at equal intervals.

Test Sequence

The entire test program was conducted over two days with a four-man field crew. The first test was conducted under high water conditions in the reservoir. Ambient vibration data was collected in parallel with forced vibration data under popper loadings. A two-man crew was used in the boat to deploy the popper; an additional two-man crew was used to install the accelerometers and monitor the test data. A series of 15 to 30 pops were recorded and spectrum analyzed using the Hewlett-Packard Realtime Spectrum Analyzer. The spectral response was plotted immediately in the field using an X-Y plotter and the data interpreted to assess modal response through frequency, amplitude and phase comparisons. Where necessary, adjustments were made in the placement of the instruments to optimize the field monitoring program.

RESULTS

Figure 4B illustrates the comparison between experimental and theoretical fundamental tower frequencies and mode shapes for both the high water and low water conditions. The experimental results provided slightly higher periods of vibration than the theoretical. Figures 4C, 4D and 4E illustrate some of the primary modes of tower and footbridge response. The experimental frequencies of vibration are compared with the theoretical results in Table I.

TABLE I

OUTLET TOWER AND FOOTBRIDGE FREQUENCIES FOR ELASTIC LOW-STRAIN TEST CONDITIONS

MODE NO.	THEORETICAL		EXPERIMENTAL	
	LOW WATER (ELEV. 540)	HIGH WATER (ELEV. 570)	LOW WATER (ELEV. 540)	HIGH WATER (ELEV. 570)
1	6.5*(B)	6.1 (T)		
2	6.9 (B)	6.2 (T)		6.5 to 7.0 (T)
3	7.4 (T)	6.6 (B)	7.5 to 8.0 (T)	7.9 to 8.0 (B)
4	7.9 (T)	7.0 (B)	5.8 to 6.4 (B)	
5	8.1 (B)	8.1 (B)		7.5 to 7.6 (B)
6	10.0 (B)	9.9 (B)		

*Frequencies are in hertz and elevations are in feet.
T = Tower mode of vibration.
B = Footbridge mode of vibration.

The computed frequences of the tower were reduced by the presence of water surrounding the tower. This fact was confirmed directly by the test program. Close correlation between the theoretical and experimental models was obtained using a virtual water mass on the exterior of the tower that was 85 percent of the volume of entrained water within the exterior perimeter of the tower. This result agreed well with the theoretical values published by Liaw and Chopra (Ref. 3) in their study of axisymmetric towers in water.

There was little apparent shift in the natural period of tower vibration under ambient conditions versus low-level forced vibration. Thus, nonlinearity in the surrounding soils at the base of the tower at the lowest strain levels tested did not appear to be a significant factor. In fact, best correlation occurred between experimental and theoretical results when the base of the tower was fixed at ground surface. However, under large earthquake conditions, strain levels are several orders of magnitude greater than those monitored in the test program. Therefore, in the dynamic analysis a reduction in soil stiffness was used to account for softening of the surrounding soils under large earthquake strains. This reduction is soil stiffness was based upon laboratory tests of soils samples.

The fundamental frequencies of the tower and footbridge fall in about the same frequency range as shown in Table I. Thus, dynamic earthquake response of the footbridge could impose significant loads on the tower as a result of tower-bridge interaction. In the final dynamic analysis of the tower to assess its strengths, the footbridge was left in the model.

First mode damping in the outlet tower under the low-level forced vibration (popper) test was found to be on the order of 5 percent of critical. This was an order of magnitude greater than the damping measured under ambient vibration conditions. Under large earthquake strain levels, higher damping than 5 percent of critical would be experienced by the tower-bridge system.

DISCUSSION AND CONCLUSIONS

Low-level forced vibration tests using a compression wave in the reservoir surrounding the outlet tower proved to be a practical method for exciting and monitoring the dynamic properties of this class of structure. The exciting force produced strain levels in the structure that were sufficiently large to provide clear, clean motion records for spectral analysis. The strain levels in the structure and soil were sufficiently high to excite structural damping consistent with the levels anticipated under low level earthquake conditions. Soil stiffness and virtual water mass were readily determined through the program of testing outlined in this report.

The major weakness in the program was the number of instrument channels simultaneously used to recorded the vibration of the structure. The two-channel recording system worked effectively with the field monitoring and data reduction equipment available. However, a more accurate assessment to tower and bridge mode shapes could have been obtained through a multi-channel recording system.

ACKNOWLEDGEMENTS

    The authors wish to acknowledge the support and encouragement of the Metropolitan Water District of Southern California in sponsoring the Garvey Reservoir seismic stability investigation and the test program described in this paper. Kinemetrics of Pasadena, California provided the monitoring equipment and two-man crew to perform the recording and data reduction. The entire investigation was conducted under contract by Dames & Moore for the Metropolitan Water District of Southern California.

REFERENCES

1. Bathe, K.J., Wilson, E.L., Peterson, F.E., 1972, "SAP IV, A Structural Analysis Program for Static and Dynamic Response of Linear Systems," EERC No. 73-11, Earthquake Engineering Research Center, University of California, Berkeley, California.

2. Blissmer, F.P., Kirby, R.W., Peterson, F.E., 1979, "EASE2 Elastic Analysis for Structural Engineering," Engineering Analysis Corporation, Redondo Beach, California.

3. Liaw, C.Y. and Chopra, A.K., 1973, "Earthquake Response of Axisymmetric Tower Structure Surrounded by Water," Report No. EERC 73-25, Earthquake Engineering Research Center, University of California, Berkeley, California.

4. Ostrom, D.K. and Kelly, T.A., 1977, "Method for Dynamc Testing of Dams," ASCE Journal of the Power Division, 13044, July.

Some Aspects of Bridge Instrumentation

Emmanuel Maragakis* and Gary Norris**

Abstract

This paper provides some guidelines about the proper arrangement of strong motion instruments for the proper assessment of some particular phenomena that have been either observed in past earthquakes or revealed as the result of theoretical studies. These phenomena concern the interaction between the bridge deck and the abutments which is particularly important for short skew bridges and the change in the rotational stiffness of pile groups over several cycles of loading tending towards a stabilized value.

Introduction

The extensive damage of highway bridges which occurred in the 1971 San Fernando Earthquake demonstrated how little was known about the behavior of bridges during earthquakes. As a consequence, strong motion instrumentation programs have been initiated by state and federal government agencies in order to obtain additional information regarding the dynamic behavior of bridges in future earthquakes. As a result, a detailed report has already been published to provide guidelines about the proper instrumentation of certain types of highway bridges in order to obtain the most information relative to specific anticipated dynamic responses or failure mechanisms during earthquakes (Rojahn and Raggett, 1981). It is pointed out in this report that the selection of the strong motion instrumentation scheme depends primarily on three factors: the objective of the instrumentation program, the expected bridge behavior, and the quantities to be recorded.

The purpose of this paper is to provide some guidelines about the proper arrangement of strong motion instruments in order to assess the importance of particular phenomena that have either been observed during the 1971 San Fernando earthquake or revealed as the result of studies with analytical models. In summary, the phenomena of interest herein are the following two: the interaction between the bridge deck and the abutment occurring with short bridges (particularly skew bridges) and the change in the rotational stiffness of pile groups over several cycles of loading tending toward a stabilized value.

The presentation is divided into two parts. In each part, a brief theoretical description of a particular phenomenon is provided followed by the description of a suggested instrumentation scheme which could

---

* Assistant Professor of Civil Engineering, University of Nevada-Reno
** Associate Professor of Civil Engineering, University of Nevada-Reno

provide adequate information relative to this phenomenon in the event of a future earthquake.

Interaction Between the Bridge Deck and the Abutments

One of the observations from the damage to freeway structures caused by the 1971 San Fernando earthquake was that several moderate span bridges with relatively large skew angles showed a tendency to rotate in a horizontal plane in a direction that increased their skewness (Jennings, 1971). The same behavior was later observed during the recent Coalinga Earthquake of May 1983. In the San Fernando Earthquake, this susceptibility of skewed bridges to rotational displacements caused, in some cases, severe damage to columns and abutments. The damage to bridges was relatively minor during the Coalinga Earthquake.

It has been concluded (Wood, Jennings, 1971) that this rotation was a direct result of the interaction between the structure and the approach fill, and it was suggested that research on this phenomenon was required. In fact, this phenomenon was investigated through recent analytical studies (Maragakis, 1985) and was explained as a direct consequence of the impact between the bridge deck and the abutments. More specifically, in these studies, a theoretical model was developed in which the bridge deck was represented as a one-dimensional rigid bar having three degress of freedom (lateral translation, longitudinal translation, and in-plane rotation). The bridge piers and the elasto-meric pads were modeled by bilinear hysteretic translational springs resisting the translation of the bridge deck in both directions, by rotational springs resisting the in-plane rotations of the bridge deck, and by viscous dampers. Each abutment was represented by a gap in the longitudinal direction, usually used to allow thermal expansion of the bridge deck, and by translational elastic-bilinear hysteretic springs which account for the resistance of the abutment and the soil deposit behind it. The abutment springs are activated after the closure of the gaps (i.e., when an impact between the bridge deck and the abutment occurs). More details about the model, as well as about the evaluation of the parameters, are given in Ref. 2 (Maragakis, 1985). Based on several parametric studies of the above briefly described model, one can draw the following general remarks about the kinematic mechanism and the overall response of the model.

(i) The planar rigid body rotations of the deck are induced primarily as a result of the skewness of the deck and the impact between the deck and the abutment. Thus, after the closure of either of the gaps between the ends of the deck and the abutments, impact forces are created; the moment of these forces about the center of the mass of the deck induces rotational vibrations and couples the equations of motion. The rotation of the deck induces significant displacement of the deck in the transverse direction even when the excitation is in the longitudinal direction only. The results of the impact between the bridge deck and the abutments are shown in Fig. 1, which illustrates the phases of the motion of the bridge deck between the first two impacts between the deck and the abutments.

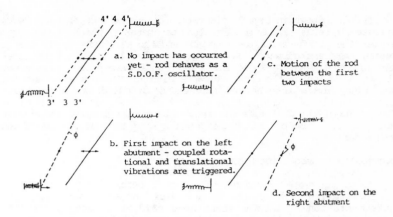

Figure 1. The Phases of the Motions of the Bridge Deck
Between the First Two Impacts

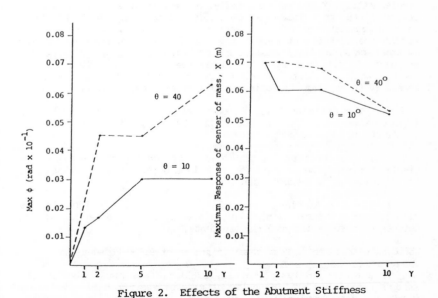

Figure 2. Effects of the Abutment Stiffness

(ii) As shown in Fig. 2, the variation of the abutment stiffness, expressed in terms of the dimensionless parameter γ, which is the ratio between the stiffness of the abutment and the stiffness of the bridge column, has the following effects: the maximum longitudinal displacement of the bridge deck decreases with an increase in γ, while the maximum rotation of the deck increases with γ. Thus, one can see the importance of the interaction between the bridge deck and the abutments: on the one hand, it is the source of significant in-plane vibrations of the bridge deck; and, on the other hand, it provides a restraint that would limit the extensive longitudinal motion of the bridge deck.

Suggested Instrumentation Scheme

To obtain proper field data for a more detailed investigation of the interaction between the bridge deck and the abutments, the instrumentation of simple short skew bridges is suggested. The selection of such bridges will make a comparison between the field data and the theoretical results easier. For this type of bridge, an appropriate instrumentation scheme should include the installation of at least three pairs of accelerometers on the deck and a pair at each abutment. One of the pairs of the accelerometers on the deck should be placed at the middle of the bridge, and the other two at the ends. The directions and the locations of the accelerometers are shown in Fig. 4 (accelerometers 1-6). As noted from the figure, the directions of the accelerometers, Y and X, are parallel and perpendicular to the axis of the piers and the abutments respectively. This is because it was found that, in the theoretical development of the model, the rotational vibrations of the deck were triggered primarily by movement of the bridge deck along the X direction (see Fig. 1). A comparison between the records obtained from the accelerometers on the deck and the records obtained from the accelerometers on the abutments will be necessary to confirm the following theoretical observations.

(i) The rotational vibrations of the deck are induced by the impact between the deck and the abutments, i.e., after the closure of either one of the two abutment gaps.

(ii) The rotational vibrations of the deck generate extensive displacements of the ends of the deck in a direction perpendicular to the direction of the excitation.

(iii) The displacements of the deck of the bridge in the direction of the excitation are significantly reduced after the impact between the bridge deck and the abutments.

The Cyclic Increase in Rotational Stiffness of Piled Foundations

In a recent analysis (Norris, 1985), it was shown that a pile group's rotational stiffness increases over its first few cycles of excitation corresponding to its movement toward a stabilized response. This phenomenon is shown schematically in Figs. 3a, 3b, and 3c relative

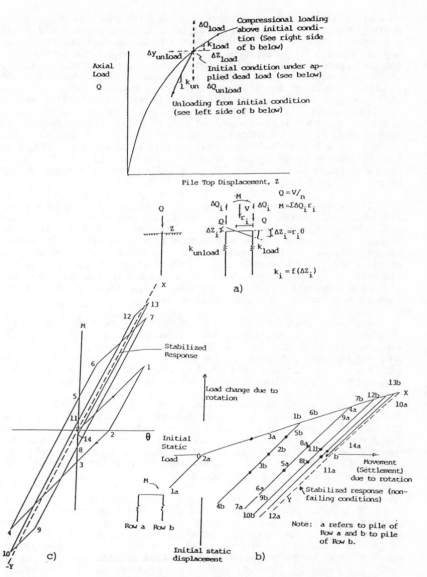

Figure 3. a) Axial load-unload response of pile relative to
               initial static condition
            b) Rotationally induced axial pile response relative
               to the initial static condition
            c) M-θ response of a two pile row system

to a symmetric two row pile group. Fig. 3a indicates that all piles in a group are under an initial axial load, Q. The pile tops are all deflected a distance, z, corresponding to this initial condition. An applied moment, M, is resisted by axial load changes, ΔQ, acting about the corresponding center of rotation. Such axial load changes, ΔQ, are of the same magnitude though opposite in sense in the unload versus the load piles on opposite sides of the axis of rotation. Note, however, that, because of the difference in stiffness in unloading versus further loading, the associated deflection, Δz, of the unload (row a) piles is less than that of the load (row b) piles. Corresponding to this situation, the instantaneous center of rotation is shifted toward the stiffer unload/reload piles. All of this is shown schematically in Fig. 3a along with the applied moment, M, and the angle of rotation, θ. Such a response is represented by path 0-1 in the ΔQ-Δz graph of Fig. 3b and the M-θ curve of Fig. 3c.

If one now considers a reversal of rotation from θ, then row a piles will reload (path 1a-2a in Fig. 3b) and row b piles will unload (path 1b-2b) corresponding to equal stiffnesses and displacement changes, Δz_{1-2}, about a centrally located axis of rotation. The moment change, ΔM_{1-2}, leading to moment $M_2 = 0$ is shown in Fig. 3c. At this point, row a piles have reloaded so that they will thereafter travel the backbone Q-z curve (path 2a-3a-4a) while row b piles will continue along the same unload path (2b-3b-4b). Again, the axis of rotation shifts toward the stiffer unload/reload piles.

Point 4 in both Figs. 3b and 3c represents a rotation θ_4 equal but opposite to θ_1. If rotation is again reversed at this point, then row a piles start to unload (path 4a-5a) while row b piles reload (path 4b-5b). Corresponding to point 5, one full cycle of rotation has occurred with the result that there are residual forces in both pile rows; and a rotationally induced settlement equal to Δz_5 has developed.

Further response evaluation yields the path 5-6-...-13-14 as shown in both Figs. 3b and 3c. Of particular note with respect to Fig. 3c is the fact that there are lines of only two distinct slopes comprising the M-θ curve. They are associated, respectively, with conditions where pile rows a and b are simultaneously on unload/reload paths (e.g., 4-6) or where one pile row is loading along the backbone curve while the other is unloading (e.g., 2-4).

In very few cycles, however, stabilized behavior (path X-Y in Figs. 3b and 3c) develops corresponding to purely unload/reload axial pile response and a centrally located axis of rotation. While there will be a certain amount of associated rotationally induced settlement, of greater interest is the large value of stabilized rotational stiffness of the group as compared to its initial stiffness. As shown in Fig. 3c, the stabilized stiffness can easily be twice its initial stiffness. Whether the equivalent effective stiffness in an earthquake is closer to the initial or stabilized stiffness depends upon the amplitude and number of cycles of near peak excitation.

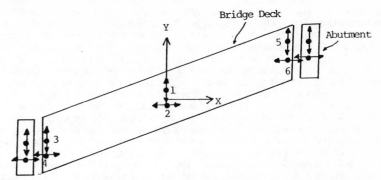

Figure 4. Suggested Instrumentation for the Interaction
Between the Bridge Deck and the Abutments

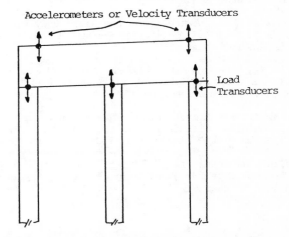

Figure 5. Suggested Instrumentation for the Cyclic
Increase in Rotational Stiffness of Piled
Foundation

Suggested Instrumentation Scheme

In order to establish the rotational behavior of a piled foundation about a longtudinal axis, it will be necessary to employ vertical
accelerometers (or, better yet, velocity transducers) at opposite sides
of the pile cap normal to the centerline of the bridge. By double
integration of the acceleration (or single integration of the velocity)
records over the first several cycles of significant excitation, one
should be able to assess movements and rotation and, in turn, the shift
in the axis of rotation. In order to assess the corresponding moment,
various piles will require instrumentation relative to axial load,
which can only be realistically achieved during the construction of the
bridge. Such axial load data would be of considerable importance since
geotechnical engineers are still unsure as to the load distribution in
the group under static conditions. Such instrumentation for load
transfer along the pile length relative to cast-in-place concrete piles
has been employed for many years. The problem will be to assure that
it will remain functional over a prolonged period of time. The suggested instrumentation scheme is shown in Fig. 5.

References

1. Jennings, P.C., editor, "Engineering Features of the San Fernando
   Earthquake," EERL 71-02, California Technical Institute, Pasadena,
   California, June 1971.

2. Maragakis, E.A., "A Model for the Rigid Body Motions of Skew
   Bridges," EERL 85-02, California Technical Institute, Pasadena,
   California, 1985.

3. Norris, G., "The Lateral and Rotational Stiffness of Pile Groups
   for the Seismic Analysis of Highway Bridges," CCEER Report,
   University of Nevada-Reno, 1985.

4. Rojahn, C. and Raggett, J.D., "Guidelines for Strong-Motion
   Instrumentation of Highway Bridges," Report No. FHWA/RD-82/016,
   U.S. Department of Transportation, Federal Highway Administration.

5. Wood, J.H. and Jennings, P.C. "Damage to Freeway Structures in the
   San Fernando Earthquake," Bulletin of the New Zealand Society for
   Earthquake Engineering, Vol. 4, No. 3, pp. 347-376, December 1971.

DYNAMIC TESTS OF LNG STORAGE TANKS
FOR EARTHQUAKE ANALYSIS OF LIQUID-TANK-PILE-SOIL INTERACTION

William E. Gates, Member ASCE (1)
David P. Hu, Member ASCE (2)
Lawrence G. Selna, Member ASCE (3)

SUMMARY

This paper presents a field test procedure for determining the dynamic properties of large liquid natural gas (LNG) storage tanks under operating conditions. Two methods of vibratory excitation were used to produce varying levels of strain within the vessels and foundation soils to assess strain dependent stiffness and damping characteristics of the liquid-tank-pile-soil system. Results from the test program were used to verify mathematical two-dimensional and three-dimensional finite element models of the tank foundation system for use in dynamic earthquake analysis.

INTRODUCTION

This report presents the highlights of the field test program to assess the dynamic properties of two large LNG storage tanks that were investigated for seismic safety. These tanks were owned and operated by San Diego Gas & Electric Company and are located along the southeastern edge of San Diego Bay in the community of Chula Vista, California.

The objective of the dynamic test program was to provide data for verifying dynamic mathematical models of the tank-foundation system for use in the earthquake safety review of the existing tanks.

DESCRIPTION OF LNG TANKS

The two LNG tanks are situated within an earth-containment dike in relative close proximity to the process plant used to convert natural gas to LNG. Figure 1 is a cross-section of LNG storage Tank T-1. The tank vessel consists of two concentric cylindrical flat-bottomed steel tank shells with the angular space between filled with perlite insulating material. The entire mat is supported on a closely

---

(1) Associate, Dames & Moore, Los Angeles, California.

(2) Civil-Structural Engineering Supervisor, San Diego Gas and Electric Company, San Diego, California.

(3) Professor, Department of Civil Engineering, University of California, Los Angeles, California.

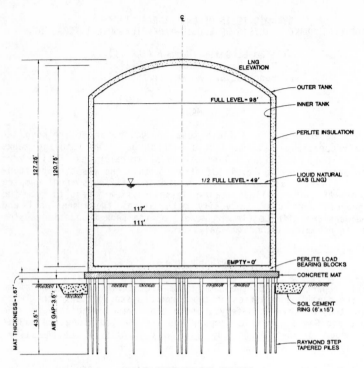

FIGURE 1 **LNG STORAGE TANK**

spaced gridwork of vertical piles. The tank is elevated above the ground surface about two-feet to provide air circulation under the tank mat to prevent the ground from freezing.

A soil cement ring 6 feet thick and 15 feet wide was placed around the perimeter of the pile foundation during construction. This ring acted to constrain the outer row of piles in a manner that was unusual for normal pile foundation construction. The presence of the ring complicated both the field test program and the analytical modeling of the tank-foundation system.

The site is situated along the San Diego Bay in an area of relatively deep silts and sands and clays with a high water table. Fill materials have been placed on the site in the upper ten to twenty feet as part of the site development process.

THEORETICAL DYNAMIC BEHAVIOR

Figure 2 illustrates some of the primary modes of vibration that may be excited in the liquid-tank-pile-foundation system under ground motion excitation. These modes include: liquid sloshing, horizontal rigid body tank translation; rigid body tank-fluid rocking; vertical dimpling of the tank dome; shell bending of the tank in the cantilevel bending mode; and flexing in the foundation mat under vertical excitation. Depending on the relative stiffnesses of the soil and tank system and their contributing masses, any combination of the preceding primary modes may be blended in an individual mode. Thus, rocking and translation are normally coupled along with cantilevel bending. Mat flexing and dome dimpling may show up in the same mode shape.

TEST PROGRAM

TEST CONSIDERATIONS AND CONSTRAINTS

The theory of dynamic behavior of flat-bottom tanks has been well developed in recent years (Ref. 3, and 4) However, the theoretical development for dynamic modeling of elevated tanks supported on closely spaced piles in soft soils is still a developing science. There are theoretical solutions for individual piles under lateral and vertical load, there are solutions for group action effects for piles constrained by a rigid pile cap, however, the presence of a constraining concrete ring around a cluster of closely spaced piles has not, to the knowledge of the authors, been developed. Other complicating factors besides the constraining soil cement ring where the presence of high water table and a large degree of variability in soil stiffness due to filling processes at the site as well as pile driving that further densified the soils directly under the tank. The flexible mat-pile cap for the tank was located above the ground surface and thus friction between the mat and the soil was eliminated. Instead, the piles functioned as short piers in transfering lateral and vertical loads into the soil. All of these factors added a significant degree of uncertainty to the analytical modeling process. Thus the primary emphasis of the test program was to verify those modes of tank-pile-foundation vibration associated with rigid body translation and rocking of the tank on its foundation system. The second objective of the field test program was to measure the fundamental period of vibration for the site.

To further complicate the field test program compressors and pumps used in the adjacent liquifaction plant had to continue to function through the test. The background vibrations from these mechanical equipment tended to dominate certain frequencies of the vibration spectrum and mask potential tank-foundation frequencies under investigation. Thus the test program had to provide a means for distinguishing between mechanical stationary source vibrations and those associated with the vibratory modes of the tanks under investigation.

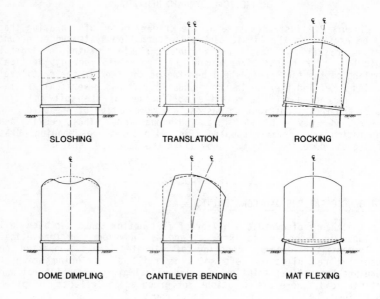

SLOSHING   TRANSLATION   ROCKING

DOME DIMPLING   CANTILEVER BENDING   MAT FLEXING

FIGURE 2   PRIMARY MODES OF LIQUID-TANK-PILE VIBRATION

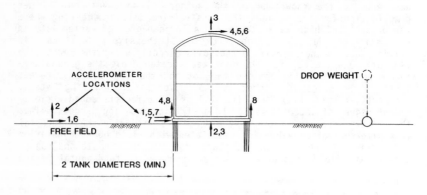

FIGURE 3   AMBIENT AND FORCED VIBRATION TESTS

VIBRATORY LOADING

To monitor the effects of strain dependent soil stiffness and damping as well as to distinguish between stationary and transient vibrations in the site soils, two forms of vibratory loading were selected for the test. The first relied on ambient ground motion conditions along with transient wind gusts as the exciting force for the tanks. This source of vibratory motion induced the lowest strain in the tank and its foundation system. Higher strain levels were produced through a forced vibration technique utilizing a 4,000 pound wrecking ball that was dropped from a height of 25 feet using a crane as its hoist. The impact of the ball against the soft sandy soils produced a crater that ultimately penetrated the ground to the water table. Penetration on the order of 10 to 12 feet was reached after approximately 100 drops. The impact of the wrecking ball against the ground imparted sufficient energy to excite many of the dominant modes in the tank-soil system including the first mode of the site soil profile.

INSTRUMENTATION

The motions of the tank and its supporting soil foundation were monitored during the vibration tests with a pair of accelerometers selectively located at each of the key points on the tanks, its foundation, and in the free field as shown in Figure 3. The equipment used in the test program consisted of the following:

1. Two strong motion accelerometers -- Kinemetrics Model SMA-3

2. Real time spectrum analyzer -- Hewlett-Packard Model HP3582A

3. Strip chart recorder -- Gould Brush Number 220

4. X-Y Plotter -- Hewlett-Packard Model HP7015B

5. Signal conditional -- Kinemetrics Model SC-1

TEST SEQUENCE

The entire test program was conducted over two days with a four-man field crew. The dynamic properties of the two tanks were monitored in two orthognal directions. A total of four test setups analogous to Figure 3 were performed. The pair of force balance accelerometers were located at each of the numbered points shown on Figure 3. Each number represents a specific test run. Once the instruments were in place, both an ambient and a drop weight or forced vibration test were conducted. In the forced vibration test the ball was dropped 15 to 20 times per test run. The resulting records were computer processed in the Hewlett-Packard Real Time Spectrum Analyzer. Spectral response was plotted immediately in the field using an X-Y Plotter and the data was interpreted. Where necessary, adjustments were made in the placement in the drop weight, or instruments to optimize the field monitoring program.

The recorded accelerations were also plotted in real time on the Gould Brush Strip Chart Recorder. These records provided insight into the vibrations sensitivity or responsiveness at various locations on the tank and the soil.

EXPERIMENTAL RESULTS

Figure 4 illustrates the type of spectral data obtained from the Hewlett-Packard Real Time Spectrum Analyzer. The upper half (Figure 4A) is a plot of two spectra from horizontal test runs 18 and 21 of Tank T-1. Run 18 was ambient and run 21 was force vibrated with the drop weight. The tall highly responsive and very narrow spikes at 8.8, 10, 17.6, and 19.8 Hertz represent stationary source vibrations from compressor equipment in the liquifaction plant. These stationary sources were verified through manufacturer's specifications on the equipment as well as by inspection of the comparative plots from ambient and forced vibration responses shown in Figure 4A. Under ambient conditions (run 18 monitored at transducer A and shown as a solid line) the dominant spectral values are those associated with the steady state vibration sources. Under the drop weight (run 21) the horizontal response of tank and soil field suddenly loom above the quite zones in the ambient spectrum. This is shown graphically by the dashed line. The amplitude of the stationary sources did not change between the ambient test case 18 and the forced vibration test case 21.

Figure 4B illustrates a similar comparison between ambient and forced vibration for vertical response of a tank. In this instance the A transducer has been located at the center of the tank foundation mat and has been oriented vertically to monitor vertical motions in the flexible mat and the soil system. Under ambient conditions (run 7) the four equipment frequencies show up very strongly as plotted by the solid line. For the forced vibration case (run 6) the vertical tank frequencies again project above the ambient spectrum at all the frequencies except those associated with the steady state equipment vibration sources. Through analysis of the correlation functions between the two transducers, A and B, modes of tank vibration were identified as indicated in Figure 4.

ANALYTICAL RESULTS

In the analysis of the LNG storage tanks, a axisymmetric finite element model of the tank, fluid, piles, soil and soil cement ring was formulated. A description of the model and its analysis is the topic of independent papers by Gates and Hu (References 1 and 2).

For purposes of direct comparison with the experimental test program, soil stiffness properties consistent with low strain ambient and forced vibration test conditions were introduced in the axisymmetric finite element model. The mathematical model was excited with a step pulse acceleration loading at its soil boundaries and the resulting acceleration response recorded and spectrum analyzed by

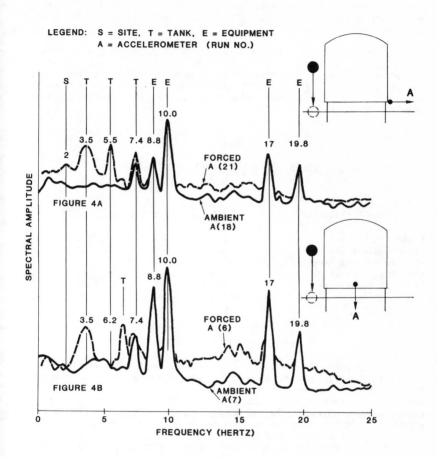

FIGURE 4 COMPARISON OF AMBIENT AND FORCED VIBRATION
SPECTRAL DATA FROM LNG TANK T-1

computer.   The acceleration responses were computed at various locations along the tank and in the soil field similar to those shown in
the test program for the tank, Figure 3.   The resulting spectral plots
were very analagous to those in Figure 4 without the steady state
equipment vibrations.

Table I presents a comparison between the computed frequencies
and field measured frequencies for various modes of tank and soil
vibration.   There is relatively good correlation between the field
results and those determined through the axisymmetric finite element
analysis.   This correlation substantiated the axisymmetric computer
model of the tank.

TABLE I

FREQUENCIES OF TANK-PILE-SOIL SYSTEM

MODE OF VIBRATION	FIELD MEASURED	COMPUTED FROM SSI/FEM
Fluid 1st Vertical	3.5	2.86 Hz
Fluid 2nd Verical	6.2	5.0 Hz
Site & Tank 1st Vertical	7.4 Hz	8.0 Hz
Site 2nd Vertical	11.0 Hz	12.0 Hz
Site 1st Horizontal	2.0 Hz	1.8 Hz
Tank Rocking	3.5 Hz	4.0 Hz
Tank Horizontal	5.5 Hz	

DISCUSSION AND CONCLUSIONS

The low force level dynamic test program described in this paper
provided valuable data on the dynamic properties of the LNG storage
tanks and their foundation systems.   This data proved useful in
verifying the mathematical axisymmetric finite element models used in
the earthquake safety review of the tanks as described in references
1 and 2.   The exciting force from the drop weight generated strain
levels in the tank and surrounding soils that were sufficiently large
to produce clear motion records for spectral analysis.   By using two
separate levels of exciting force in the structure it was possible to
distinguish between steady state motions from nearby mechanical equipment and the transient response of the tank and its supporting soil
field.   Natural frequencies of vibration and associated mode shapes
were affectively determined through the field test program.   The
correlation between field test data and theoretical models was also
good.

One of the major weaknesses in the test program was the number of
instrument channels used to simultaneously record the vibrations in
the tank and surrounding soils.   The two channel recording system
worked effectively with the field monitoring and data reduction equipment available.   However, extensive time was required in post-
processing these records to determine the mode shapes associated with

each key frequency. A multi-channel recording system could have significantly reduced the post-processing time and effort.

ACKNOWLEDGEMENTS

The authors wish to acknowledge the technical assistance from John Diehl and Kinemetrics of Pasadena, California that provided the monitoring equipment and two-man crew to perform the recording and field reductions of data.

REFERENCES

1.   Gates, W.E. and Hu, D.P., 1984, "Seismic Risk Assessment of LNG Storage Tanks", proceedings of the 8th World Conference on Earthquake Engineering, San Francisco, California, July.

2.   Gates, W.E. and Hu, D.P., "Seismic Risk Analysis of LNG Storage Tanks", in preparation for presentation at the 1986 ASCE Annual Spring Convention, Seattle, Washington.

3.   Haroun, M.A. and Housner, G.W., 1981, "Seismic Design of LNG Storage Tanks", preceedings of the Journal of the Technical Counsel of ASCE, Volume 107, Number TC1, Pages 191 through 207, April.

4.   Haroun, M.A. and Ellaithy, H.M., 1983, "A Model for Flexible Tanks Undergoing Rocking", report to the National Science Foundation.

Bruce L. Hutchison[1] and Daniel W. Symonds[2]

Abstract

Four large concrete floating bridges located in Washington State have been
the subject of extensive analytical modeling for dynamic response to storm
waves. The dynamic analysis of the bridges was performed using impulse
response methods in the frequency domain to account for the frequency
dependent hydrodynamic mass and damping properties. The short-crested
character of the directional storm wave spectra was incorporated in the
stochastic integral through the use of a coherency function. Hysteretic
structural damping was included in the model, and the responses proved to be
quite sensitive to the magnitude of the structural damping. Sensitivities to the
central heading angle for the incident directional wave spectra and sensitivities
to the magnitude of the wave energy spreading were also examined. Responses
were determined for both cracked and uncracked structural properties. Phase
plane diagrams for the joint vertical and lateral bending processes were
developed. When these demand diagrams are compared with the structural
capacity diagrams it becomes apparent that cracking limits structural demand.

Introduction

Three of the most substantial floating bridges in the world are located in
Washington State, and a fourth is under construction. In 1979 the western half of
the Hood Canal Bridge broke up and sank in a severe storm. The west half was
re-built to a new design and the bridge re-opened in 1982.

These bridges are of continuous reinforced concrete box beam construction,
and all excepting the oldest bridge are post-tensioned. The bridges are held in
place by a system of lateral moorings and gravity anchors. Principal dimensions
of these bridges are shown in figure 1. They have floating spans ranging from
5758 feet (1755 meters) to 7518 feet (2291 meters).

---

[1] ASCE Member; The Glosten Associates, Inc., 605 First Avenue, Seattle, WA
    98104-2224

[2] KPFF Consulting Engineers, 999 Third Avenue, Seattle, WA 98104

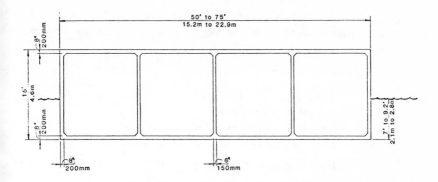

Figure 1: Cross Section of Typical Pontoon

Since 1983 all four of the bridges have been subject to extensive analysis for dynamic loads using the frequency dependent impulse response methods of Hutchison (1984). These methods provide an efficient means to evaluate the response processes for floating bridges and breakwaters subject to short-crested incident wave fields describable by directional variance spectra, $S_{\zeta\zeta}(\omega,\theta)$.

Dynamic Model

The bridges are modeled using a finite element dynamic structural model:

$$[M{+}A(\omega)]\,\ddot{\vec{x}} + [B_1(\omega)]\,\dot{\vec{x}} + [B_2]\frac{\dot{\vec{x}}}{|\dot{\vec{x}}|}|\dot{\vec{x}}| + [K{+}C]\,\vec{x} = \vec{f}(\omega)\,e^{i\omega t} \qquad (1)$$

where: M   is the structural mass-inertia matrix

$A(\omega)$ is the hydrodynamic added mass matrix

$B_1(\omega)$ is the hydrodynamic damping matrix

B_2 is the hysteretic structural damping

K   is the structural stiffness matrix

C   is the hydrostatic stiffness matrix

$\vec{f}(\omega)$ is the normalized (for unit wave amplitude) frequency dependent complex amplitude vector for wave forces and moments

$\vec{x}$ is the complex amplitude vector for nodal displacements

ω is the angular wave frequency

The structural stiffness matrix includes the effects of mooring lines.

This system of equations is solved in the complex phase plane using a LDL^T (Cholesky) decomposition to obtain the frequency dependent impulse response functions:

$$X_\alpha(\omega, \theta, j; k)$$

where: α is the mode of response, e.g., roll rotation, sway translation, or vertical bending moment

       θ is the wave heading angle (90° is beam seas)

       j is the node index where the response occurs

       k is the node index where the wave force impulse is applied

Responses to Directional Wave Spectra

The incident directional wave spectrum is decomposed into the product of a scalar wave spectral density function $S_{\zeta\zeta}(\omega)$ and a spreading function, $\Psi(\theta)$. Hutchison (1984) has shown that the response cross-spectral densities may be expressed as:

$$S_{\alpha\beta}(\omega, j; k) = X_\alpha(\omega, \tfrac{\pi}{2}, j; k) \; \overline{X_\beta(\omega, \tfrac{\pi}{2}, j; 1)} \; S_{\zeta\zeta}(\omega) \; \sqrt{\gamma_{k1}(\omega)} \qquad (2)$$

where: $\gamma_{kl}(\omega)$ is the complex valued scalar coherency function between the location of the nodal points k and l in the wave field.

The scalar coherency function between two points $\vec{A}$ and $\vec{B}$ is given by:

$$\gamma_{AB}(\omega) = \int_{-\pi}^{\pi} \Psi(\theta) \, e^{i(\vec{k} \cdot \vec{\xi})} \, d\theta \qquad \int_{-\pi}^{\pi} \Psi(\theta) \, e^{-i(\vec{k} \cdot \vec{\xi})} \, d\theta \qquad (3)$$

where: $\vec{\xi} = (\vec{A} - \vec{B})$ is the position vector from the point $\vec{B}$ to the point $\vec{A}$;

$$\vec{k} = \frac{\omega^2}{g} [\vec{u}_1 \cos(\theta) + \vec{u}_2 \sin(\theta)] \qquad \text{is the vector wave number.}$$

The spectral moments are obtained as the summation of complex integrals over the range of possible combinations of jointly excited nodes:

$$m_{\alpha\beta}^n(j) = \sum_k \sum_l \int_0^\infty X_\alpha(\omega, \tfrac{\pi}{2}, j; k) \; \overline{X_\beta(\omega, \tfrac{\pi}{2}, j; 1)} \; \omega^n \, S_{\zeta\zeta}(\omega) \; \sqrt{\gamma_{k1}(\omega)} \; d\omega \qquad (4)$$

where: $m_{\alpha\beta}^n(j)$ is the nth cross spectral moment between response modes α and β as seen at node j.

Response statistics are obtained using standard methods from the spectral moments m^0, m^2, and m^4.

Hydrodynamic Coefficients

The frequency dependent hydrodynamic coefficients for added mass, damping and the wave forcing functions in the equations of motion are obtained by solving the appropriate Cauchy problems for two-dimensional flow in the presence of the free surface. These problems were solved using a close-fit technique. A plot showing the frequency variation of the sway added mass for a typical pontoon is shown in figure 2.

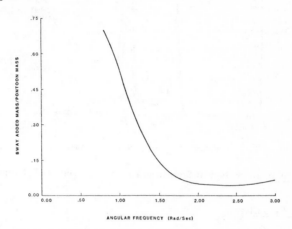

Figure 2: Variation of Hydrodynamic Added Mass With Frequency

Structural Damping

Structural damping was included in addition to the hydrodynamic damping. Equations of motion for this project were developed on an elemental stiffness basis but a nodal (lumped) mass formulation. This formulation is quite accurate for the small nodal spacings used. However, it introduces some difficulties in formulating viscous (Coulomb) damping. In a problem such as this, the structure not only experiences relative velocities due to structure vibrations, but it also undergoes rigid body motions similar to a ship on an ocean. For this reason we chose to use displacement dependent (mass independent) hysteretic damping. Hysteretic damping is in phase with the velocity but proportional to the displacement as indicated in equation (1).

The response proved to be quite sensitive to the structural damping, as hydro-dynamic damping had little impact on the structural vibrations. Response calculations were performed using three levels of hysteretic damping in the model of the bridge currently under construction. Responses were computed with non-dimensional hysteretic damping values of zero, 0.02π and 0.06π (roughly corresponding to zero, 3.4% and 10.2% of critical viscous damping). Figure 3 shows torsional moment response spectra for the three levels of hysteretic damping.

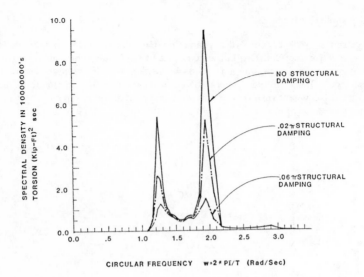

CIRCULAR FREQUENCY w=2*PI/T (Rad/Sec)

Figure 3: The Effect of Structural Damping on Torsional Response

Directional Wave Effects

The form of wave energy spreading function selected was that of an even cosine power function constructed as a unary operator.

$$\Psi(\theta) = C \cos^n(\theta-\theta_c) \qquad \text{on } -\frac{\pi}{2} < (\theta-\theta_c) < \frac{\pi}{2}$$

$$\Psi(\theta) = 0 \qquad \text{otherwise}$$

where: C   is a constant such that the unary operator condition,

$$\int_{-\pi}^{\pi} \Psi(\theta)d\theta = 1.0, \text{ is satisfied}$$

θ_c is the central heading angle

n   is an even integer, e.g., 2, 4, 6, ...

A limited quantity of scalar coherency measurements was available from the Evergreen Point floating bridge on Lake Washington (Seltzer, 1979). A re-analysis of those data indicated that a cosine to the twelfth power (n = 12) was a reasonable representation of the directional spreading on these narrow inland bodies of water. The sensitivity of responses to the power of the spreading function is indicated in figure 4 which shows vertical bending moments plotted over the length of the bridge.

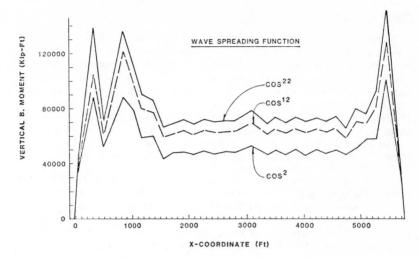

Figure 4: The Effect of Wave Spreading Function on Bending Response

The responses are even more sensitive to the central heading angle. Figure 5 shows the variation in lateral bending moment which results from central heading angle changes from 90 degrees (beam seas) to 45 degrees.

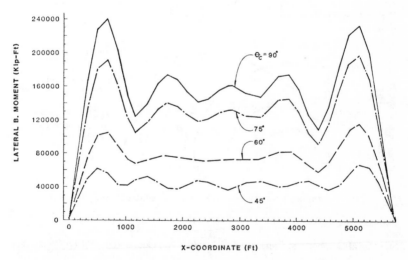

Figure 5: The Effect of Central Heading Angle on Bending Response

Joint Responses

Inspection of the east span of the Hood Canal Bridge indicates that there have been instances where loads in some areas significantly exceeded the cracking limit, but yield has not been reached. The analytical model also indicated that cracking should occur in severe storms. Accordingly dynamic responses were evaluated using both cracked and uncracked section properties. Structural damping was increased for those analyses with cracked sections. The results are shown in figure 6, which is a polar plot of structural demand and capacity for combined bending processes (combined vertical and lateral bending). The demand processes were combined using the cross co-spectral moment methods of Hutchison (1982). Capacity lines are drawn for both cracking and yielding.

In figure 6, the demand curve which is based on uncracked section properties exceeds the cracking limit and there is also an indication for yielding. However, as can be seen from the demand curve based on cracked section properties, the effect of the increased damping and cracked section properties is to limit demand to a value exceeding cracking but less than yield. Thus the structural mechanics provide a load limiting mechanism which protects the bridge.

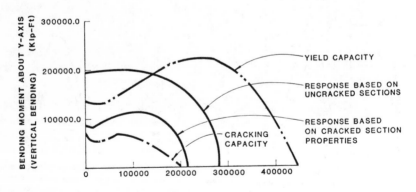

BENDING MOMENT ABOUT Z-AXIS (LATERAL BENDING) (Kip-Ft)

Figure 6
Structural Demand Vs. Capacity for Cracked and Uncracked Section Properties

Summary

Procedures have been developed for the dynamic analysis of large elastic floating structures such as bridges and breakwaters which are subject to short-crested seas. These procedures are flexible and computationally efficient when compared with the alternative analysis methods of time domain simulation or frequency domain superposition of long-crested wave responses.

Responses of floating bridges are shown to be sensitive to the directional distribution of wave energy. Demand processes are quite sensitive to the degree of structural damping. Cracking is shown to work to limit structural demand. Results obtained correlate well with actual observations of the major floating bridges investigated.

Acknowledgements

The engineering research summarized in this technical paper was supported by the Washington State Department of Transportation. The authors wish to express their particular appreciation to Mr. C. S. Gloyd, State Bridge and Structures Engineer, for his support and encouragement during the course of these projects.

References

Hutchison, Bruce L., "A Note on the Application of Response Cross Spectra," SNAME, Journal of Ship Research, Vol. 26, No. 2, June 1982.

Hutchison, Bruce L., "Impulse Response Techniques for Floating Bridges and Breakwaters Subject to Short-Crested Seas," SNAME, Marine Technology, Vol. 21, No. 3, July 1984.

Seltzer, George H., "Wave Crests - How Long?" University of Washington, Department of Civil Engineering, October 1979.

Cyclic Energy Absorption by Reinforced Concrete

Levon Minnetyan,[1] M. ASCE and Victor M. Ciancetta, Jr.[2]

Results are presented on the basis of experiments conducted on conventional and steel fibrous reinforced concrete beams. The effects of the amount and nature of steel reinforcement and presence and amount of axial compressive load are investigated with respect to durability, stiffness degradation and energy absorption capability of tested beams. The degradation behavior of conventional and steel fibrous reinforced concrete beams are compared.

Introduction

The scope of this paper is to examine the degradation characteristics of reinforced concrete structural members under cyclic loading. The results are based upon tests conducted with beams of relatively small cross section, measuring four inches (100 mm) wide and six inches (150 mm) high. All specimens presented here measure four feet (1.22 m) between the pinned-end support points. All specimens contain identical longitudinal reinforcement, having a #4 deformed bar at each corner. The conventional shear reinforcement, when used, is fabricated from #2 smooth bars and placed at two-and-a-half inch (63 mm) spacing throughout the length of a specimen. The steel fibers are of the usual commercial type with crimped ends, one inch (25 mm) in overall length, that are supplied in water-soluble glued bundles. When used, steel fiber content is constant at 1.25 percent by volume of the concrete mix. Maximum aggregate size is limited to 3/8 inch (9 mm) peastone to be consistent with the length of steel fibers. The test specimens are loaded by a cyclic load, acting in the stronger direction of the specimen at midspan and a constant axial compressive load. Test specimens that are loaded only by a transverse cyclic load are referred to as beams. Those specimens that are subjected to an axial compressive load as well as a transverse cyclic load are referred to as beam-columns. The effect of the loading rate is rather significant, especially for beams. Slower loading rates cause more deterioration per cycle compared to faster loading rates (2). All test data presented in this paper, however, is obtained at a constant transverse cyclic loading frequency of 1.0 Hz. Specific load amplitudes are selected with the aim of reaching failure after a

---

[1]Associate Professor, Department of Civil and Environmental Engineering, Clarkson University, Potsdam, NY 13676

[2]Graduate Student, Department of Civil and Environmental Engineering, Clarkson University, Potsdam, NY 13676

limited number of loading cycles. Failure is defined as the reaching of 1.0 in. (25.4 mm) midspan deflection amplitude during cyclic loading. End supports are designed to remain moment-free under dynamic loading.

Model Development

The overall objective of the present research is to develop an analytical model to describe the stiffness degradation of conventional and steel fibrous reinforced concrete structural members. In this paper a single degree of freedom model is used to summarize the overall response of a beam or beam-column under transverse loading. The equation of motion is written as:

$$M \ddot{u} + C \dot{u} + K(t) u = P(t) \tag{1}$$

where
- M = equivalent mass of the single degree of freedom system
- C = linear damping coefficient
- $P(t)$ = applied transverse load
- $K(t)$ = variable stiffness of the system to be described by a stiffness model.

Eq. (1) serves as a basis upon which a mathematical model can be constructed to predict the remaining strength and stiffness of the beam after a given loading history. An examination of the overall characteristics of the test data indicates that the stiffness of this system could be approximated by

$$K(t) = K_o (1 - D(i/N)^E) \tag{2}$$

where K_o = stiffness during the first loading cycle
- i = the load cycle number at time t
- N = the total number of loading cycles to failure
- D, E = parameters to be determined from test data

The stiffness model $K(t)$ is assumed to be effective during the loading phases of each cycle. The initial stiffness K_o is used during the unloading phases. The parameters K_o, D and E are computed by minimizing the time history error between the stiffness model and the actual test data (1). Table 1 summarizes some of the results for individual beams.

The parameter D represents the amount of stiffness degradation at the time of failure. For example, if $D = 0.75$, stiffness degradation is 75 percent of the initial stiffness K_o , and the remaining stiffness is 25 percent of Ko . The parameter E represents the changes in the rate at which the stiffness is reduced. If $E = 1$, the rate of stiffness degradation is constant at all stages of loading. If $E < 1$, stiffness degradation rate is greater at the beginning of

loading. On the other hand if E > 1 , stiffness degradation rate is
greater at the end of loading when the accumulated damage is larger.

A study of Table 1 reveals that for the behavior of beam-columns
under cyclic transverse loading, the parameter D is in the range of
0.65 to 0.99 with little variation due to the effect of reinforcement
type and axial loading. A typical value for the parameter D is 0.8 ,
indicating an eighty percent reduction from the original stiffness at
the time of failure. On the other hand the parameter E is rather
sensitive to the amount of axial compression applied to the beam at
the time of transverse loading. When there is no axial compression,
the Parameter E varies in the range of 0.09 to 0.74 , indicating a
higher rate of stiffness degradation at the earlier stages of these
tests. Type A reinforced beams with both conventional and steel fiber
reinforcement give the most consistent values for E due to the
stabilizing effect of the fibers. Type C beams with only conventional
shear reinforcement give the least consistent values for E because of
the greater vulnerability of these beams to normal fabrication defects
typical for reinforced concrete. The greater variation in the measured
values of the parameter E indicates the more unpredictable behavior of
the beam.

Table 1

Summary of Test Results and Corresponding Model Parameters

Spec-imen No.	Loading (Kips) Trans	Axial	Reinf. Type	Total Cycles N	Absorbed Energy (K-in)	Optimized Values Ko	D	E	Error per cycle
1	9	0	A	39	251	46	.84	0.14	.47
2	9	0	A	39	248	43	.80	0.15	.66
3	12	16	A	130	506	53	.80	1.3	.13
4	12	16	A	57	266	50	.82	1.7	.11
5	14	32	A	272	520	71	.72	1.8	.02
6	9	0	B	28	145	43	.86	0.46	.60
7	9	0	B	31	130	56	.87	0.11	.22
8	12	16	B	76	266	62	.89	2.2	.16
9	12	16	B	91	359	55	.83	1.9	.11
10	14	32	B	134	264	67	.70	3.5	.06
11	9	0	C	24	121	37	.80	0.09	.41
12	9	0	C	21	150	31	.99	0.74	4.63
13	12	16	C	91	358	48	.79	2.6	.29
14	12	16	C	35	209	54	.94	1.1	.62
15	14	32	C	270	242	69	.65	11.9	.04

TYPE	REINFORCEMENT
A	1.25% fiber, stirrups, and longitudinal reinforcement
B	1.25% fiber, and longitudinal reinforcement
C	stirrups and longitudinal reinforcement

For beam-columns subject to axial compression, the parameter E is always greater than 1.0 , indicating a slow rate of stiffness degradation at the beginning stages of the test and a higher rate of stiffness degradation at the later stages of loading. The larger the value of the parameter E the higher the rate of stiffness degradation near the actual failure of the beam-column, and consequently the more sudden and precipitous the failure. In general, the larger the axial compression, the larger is the value of the parameter E. Again, as it was the case at the absence of axial compression, for beam-columns also type C specimens without steel fibers show the larger variations in the parameter E under the same applied loading, indicating the sensitivity of these beam-columns to normal variations in the fabrication of reinforced concrete. In comparison, beam-columns containing steel fibers show less variation of the parameter E under the same load conditions, indicating less sensitivity to variations in fabrication. Nevertheless, at the presence of axial compression, the difference between fibrous and conventional beam-columns is not as pronounced as it is for the beams.

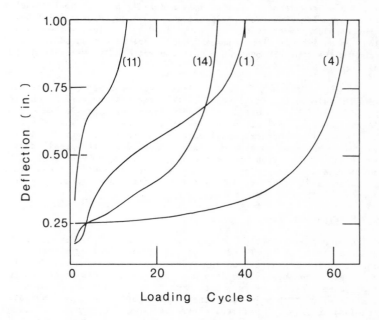

Fig. 1 - Midspan Deflections of Selected Specimens under Loading
        Specified in Table 1 (numbers in parantheses indicate
        specimen numbers from Table 1).

Effect of Axial Compression on Cyclic Deformations

The effect of axial compression may be examined by considering the interactions of the beneficial confining effect and the damaging secondary moment effect. At relatively small transverse deformations axial compression is beneficial for the durability of the structural member due to the predominance of the confining effect and the negligible secondary moment effect. However, as the transverse deflections are increased, the secondary moment developed becomes more and more predominant. The beneficial confining effect is neutralized, and the beam fails rather precipitously within a few loading cycles. Transversely loaded beams without axial load tend to deteriorate more uniformly over the test duration and the failure mode is usually gradual. On the other hand, beams subject to axial compression deteriorate slower at the beginning of testing, but the deterioration rate is accelerated as the beam sustains more damage and cyclic transverse deflection amplitudes are increased.

Fig. 1 shows typical plots of the midspan deflection amplitude for some of the specimens. Deflections of Specimen 11 are plotted as an example representing typical behavior of conventionally reinforced beams. This beam deteriorates rapidly and fails after a small number of cycles. In comparison, Specimen 1 that contains steel fibers, endures through almost twice as many loading cycles. The cyclic deflection amplitudes of Specimens 14 and 4 are also plotted in Fig. 1 to compare the typical endurance of conventionally reinforced and steel fibrous beam-columns.

The last column in Table 1 lists the error per cycle for each specimen. This is the final value of the time history error function that is minimized to obtain the model parameters. Smaller values of the error function indicate better matching of the model with the test data. Beam-columns under axial compression have less error than the beams. Under the same loading condition, specimens with steel fibers have less error than conventionally reinforced specimens. The unpredictable and irregular behavior of conventionally reinforced beams makes it difficult to define a model that can represent the beam response.

Energy Dissipation

The amount of energy dissipated by each test specimen is determined from the work done on the specimen during a test. The amount of energy dissipated during a loading cycle is equal to the area within the hysteresis loop of that cycle. Plots of the energy dissipated per cycle versus the number of cycles are shown in Fig. 2 . Also, Table 1 lists the total amount of energy dissipated during loading of each specimen. There is some correlation between the total number of load cycles and the total energy dissipated. However, since the energy dissipation depends on the deformation of the beam as well as the loading, there is a significant difference between the amount of energy dissipated and the number of cycles of load endured.

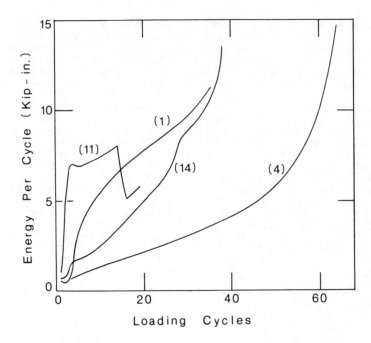

Fig. 2 - Energy Dissipation per Cycle for Selected   Specimens
         under Loading Specified in Table 1 (numbers in
         parantheses indicate specimen numbers from Table 1)

The variation in the rate of energy dissipation follows a similar
pattern as the stiffness degradation. As shown in Fig. 2 for
Specimen 4, most beam-columns exhibit an exponential growth in energy
dissipation per cycle, resulting in large amounts of energy dissipated
over the last few loading cycles before failure. Beam-column
Specimen 14 is the exception to this observation. This is a
conventionally reinforced specimen. It develops large deformations at
the beginning of the test and fails prematurely compared to the other
beam-columns; demonstrating again the difficulty of predicting the
response of conventionally reinforced structural members.

Exponential growth in energy dissipation per cycle is more
pronounced for those beam columns that are subject to the higher axial
compression. This is again due to the interactions of the confining
and secondary moment effects of the large axial compressive load. At
the initial stages of loading, the confining effect helps to limit the
dilatation of cracks, maintain stiffness, and impede transverse
deflections. Smaller transverse deflections result in less energy

dissipation. However, as transverse deflection amplitude grows energy dissipation is increased. But the secondary moment effect also becomes increasingly more significant, causing a rather precipitous stiffness degradation and failure at the end of the test. As can be observed from Table 1, the energy dissipated by steel fiber reinforced beam-columns is significantly higher.

In the case of beams without axial compression, the amount of energy dissipated per cycle is larger at the beginning of a test as compared with the energy dissipated by a beam-column at the same stage of loading. However, during the later stages of loading energy dissipated increases at a much slower rate compared to beam-columns. In some cases, cyclic energy dissipation is decreased because of the spalling of concrete and the resulting pinched hysteresis curve that is typical under high amplitude cyclic loading of reinforced concrete that has lost its confinement. Pinching of the hysteresis curves and the reduction in cyclic energy dissipation is most pronounced for conventionally reinforced beams. Steel fiber reinforced beams do not show a reduction in the energy dissipation capability at the later stages of loading. Also, the hysteresis curves are not pinched for fibrous beams. This is due to the fact that even after significant cracks are developed, steel fibers still have a beneficial effect in preventing the concrete from spalling out.

Conclusions

In general, it can be stated that steel fibrous reinforced concrete is most effective under flexural loads at high shear regions, at the absence of axial compression. If an axial compression is applied, the need for steel fibers in the concrete matrix is reduced. Due to the initial favorable confining pressure exerted by axial compression, crack expansion is effectively inhibited during cyclic loading. When there is a moderate axial compressive load, the transverse cyclic load carrying capacity of structural members is increased by seventy percent. There is a similar increase in the corresponding total energy absorption capability of the tested specimens. The beneficial effect of the addition of steel fibers to the concrete matrix also depends on presence and amount of the compressive load. For specimens tested without axial compression, the energy absorption capacity is increased by eighty percent due to steel fibers. However, the same steel fiber content improves the energy absorption capacity by only forty percent when there is a moderate axial compression.

Although not presented in this paper, slower loading rates cause more structural damage with fewer load cycles. The loading rate effect is most pronounced at the absence of axial compression. The failure mode is also modified depending on the presence of axial compressive loading. Beams that fail by the formation of shear cracks when there is no axial load, fail only by transverse flexural cracks when axial load is present (1).

In summary, transverse cyclic loading capacity, durability, and energy absorption capacity are increased with steel fibers and axial compressive loading. An accumulated damage model with bilinear load-displacement relationship and cyclic stiffness degradation, based on the amount of energy dissipated by a structural member, appears to be the most appropriate to represent steel fibrous and conventional reinforced concrete structural behavior under seismic loading.

The results indicate that steel fibers can contribute effectively to the durability of a reinforced concrete structure under an earthquake type loading. The conventional earthquake resistant design philosophy of providing stiff inertial force absorption elements such as shear walls and diagonal bracing is reinforced. If beams are designed as energy absorption elements, their deformation must be limited to assure that they do not prematurely deteriorate. Prestressing or post tensioning beams may also be useful in reducing the deterioration rate of a reinforced concrete structure under earthquake loading.

APPENDIX._ References

1. Ciancetta, V. M., Jr., "Seismic Behavior of Fiber Reinforced Concrete Beam Columns," M.S. Thesis, Department of Civil and Environmental Engineering, Clarkson University, August 1985

2. Minnetyan, L. and Batson, G. B., "Steel Fibrous Concrete under Seismic Loading," Proceedings of the Second International Conference on Recent Advances in Structural Dynamics, 9-13 April 1984, University of Southampton, England, Vol. 2, pp. 589-597

Significance of Nonlinear Modeling of R/C Columns to Seismic Response

C. A. Zeris † St.M . ASCE, and S. A. Mahin ‡ M . ASCE

Columns in reinforced concrete (R/C) structures exhibit complex hysteretic behavior during earthquakes. This can significantly influence predictions of overall response and damage distribution. Current analysis approaches often simplify this behavior, failing to properly model the influence of axial load on strength and stiffness as well as other important phenomenon. A finite element model is presented that overcomes some of these limitations. Emphasis is placed herein on comparisons of experimental data with analytical results obtained using the new element as well as with more conventional models. Topics requiring further research are identified, as are improvements required for the proposed scheme.

Introduction

Reinforced concrete members under load and deformation reversals exhibit complex global hysteretic behavior, depending heavily on load history, member detailing, constituent material characteristics and distribution of damage along and across the member. For flexurally dominated behavior, physical phenomena such as concrete cracking, spalling of the concrete cover, degradation of the concrete core, Bauschinger effects in the reinforcement, bond deterioration and so on contribute to this complexity. Analytical modeling simplifications often rely on phenomenologically based assumptions on the load-deformation relationship for the entire member. In the case of beams, guidelines have been proposed assuming that the member behaves in a stable hysteretic way for a given load pattern and extent of nonlinearity (Mahin 1975, Anagnostopoulos 1981). For columns, the coupling between the axial and flexural stiffnesses complicates such phenomenological approaches. Moreover, more theoretically based plasticity approaches, are not necessarily justified unless the member behaves in the implicitly assumed positive definite manner. This is rarely the case, unless the axial loads are low and the section confinement is adequate.

The importance of refined R/C beam-column modeling is demonstrated herein using a formulation where global phenomenological approximations are avoided. The reliability of the proposed model is demonstrated by analyzing two single members and a shake table specimen that have been previously tested under controlled conditions. Finally, the significance of the element is demonstrated by analyzing a frame where a large variation of axial forces is expected in the columns.

Description of the Formulation

Different approaches towards modeling of R/C single and multi degree of freedom systems are discussed in (Mahin 1975, Shiohara et al.,1983,Takizawa et al. 1976). The approaches can be generally classified as lumped or distributed nonlinearity models. In the former, series or parallel representation are possible, with the physical phenomena being modeled at the global level. For beams, phenomenological description of nonlinear spring behavior allows modeling of bond slip, stiffness degradation, etc. An extension of this approach to relate global lumped nonlinearity to section deformations has been proposed in (Lai et al.,1984). In distributed nonlinearity models, element flexibility is interpolated based on section behavior, obtained either from the constituent material description (Kaba et al.,1984,Mahasurevachai 1982,Menegotto et al. 1977) or phenomenologically (Brancaleoni et al.,1983). Alternative descriptions model the plastic hinge length with appropriate

† Research Assistant, Department of Civil Engineering, Berkeley, California 94720.
‡ Associate Professor, Department of Civil Engineering, Berkeley, California 94720.

constitutive relations assumed separately for the elastic and the yielded parts (Arzoumanidis et al.,1981), or assume that the element is subdivided into very small constant property segments. Standard beam interpolation functions and static condensation to the external degrees of freedom are often used to predict global nonlinear behavior of a member (Aktan et al.,1973, Kang et al.,1977).

The theory and use of the proposed formulation is only briefly described herein. The member is modeled as a distributed flexibility, fiber element. The element flexibility is monitored at selected interior sections as well as at the ends. Section flexibilities are assumed to vary linearly between these locations. This basic approach has been used for biaxial R/C column static analyses (Menegotto et al.,1977) and for steel tubular member (Mahasurevachai 1982) dynamic analysis; the latter formulation has been extended to R/C members by Kaba and Mahin (1984). While this latter formulation produced useful results, its applicability was limited. The standard displacement formulation and iterative strategy used in this element resulted in large internal equilibrium errors and numerical instabilities under certain conditions. The element has been rewritten resulting in a significant improvement in its applicability and reliability. An equilibrium constraint is enforced throughout the entire element and the search strategy has been improved. In addition, a more descriptive mild reinforcing steel model has been incorporated.

The element is implemented in a general purpose structural analysis program for simulating the nonlinear dynamic response of planar systems (Gollafshani et al.,1974). The program employs a constant acceleration time integration scheme, with event to event state determination to minimize equilibrium errors. Such a scheme will follow, during the state determination phase, each change in stiffness of the constituent materials (which is considered an event). Nonlinear stress-strain models currently implemented, however, are treated as continuous rather than discrete multilinear relations. Special provisions are incorporated so that the event to event scheme can be used in the case where negative definite stiffnesses occur at the element level. In the current program version, no events are permitted in the static load solution. Small displacements are assumed throughout. Geometric stiffness effects are accounted for at the element level based on gravity loads and assuming the geometric stiffness corresponding to a simple truss element.

Analytical Correlations with Single Member Experimental Results

Kent No.24 Beam. The first set of experimental data considered were obtained from a simply supported beam (Number 24), tested by Kent (1966). The rectangular beam is 4.84 in. (12.29 cm) wide by 8. in. (20.32 cm) deep, symmetrically reinforced. Top and bottom reinforcement to gross section area used is 1.1 % (Fig.1). Grade 40 steel and 6.95 ksi (48 MPa) compressive strength concrete were used. Due to longitudinal symmetry the beam is analyzed as a cantilever under an imposed tip vertical displacement. Six slices equally spaced along the beam are monitored, with concrete properties specified separately for the confined core and the unconfined cover.

The resistance vs. tip displacement history of the model is compared in Fig.(1) with the experimental results (in circles) and with the analytical results obtained by Kent (1966). Small amplitude cycles at the beginning of the test (which caused yielding of the bottom reinforcement) were omitted from the current analyses; hence the initial loading stiffness is overestimated. Analytical and theoretical hysteretic behavior compare well both in strength and in global stiffness. Following yielding of the steel in tension, the crack at the critical section remains open on reversal so that only the steel resists moment (Fig.1). The assumed steel characteristics obey an initial elasto-plastic hardening behavior, with a nonlinear explicit relation as proposed in (Menegotto et al.,1977) used to model the Bauschinger effect.

Ma, Bertero and Popov T1 Beam. A T-beam with a large span to depth ratio was tested under force reversals by Ma, Bertero and Popov (1976) and is analyzed here. The ratio of steel area to effective section area is 1.4% and 0.74% for the bottom and top reinforcement, respectively. Grade 60 steel is used, and the concrete has a maximum compressive strength of

4.8 ksi (33.1 MPa) (Fig.2). Due to the lack of confinement, the slab concrete was considered herein as being unconfined. Five unequally spaced monitoring slices were used.

The unequal amount of reinforcement in the top and bottom of the T beam causes the yield moments in the positive and negative bending directions to be different (Fig.2). As in the case of rectangular sections, as long as the flexural cracks are open the steel couple provides the only effective resistance. Upon crack closing, the positive moment flexural stiffness becomes considerably higher than the stiffness in the opposite direction due to the participation of the slab concrete. Stiffness degrading phenomenological type analytical models generally fail to account for this type of pinched behavior. The inherent modeling assumption that plane sections remain plane, however, causes a sudden transition in the flexural stiffness during crack closing. In reality, strain gauge readings across the slab show that edge stresses lag those at the centerline (until ultimate resistance is fully mobilized and a yield line forms across the entire slab). Bond deterioration along the beam and anchorage related fixed end rotations are neglected in this analysis. Depending on the geometry of the problem at hand, the current formulation will tend to be too stiff. Significant elongation of the beam at the reference axis can be noted (Fig.3). This is a result of the continuous opening of the cracks with cycling. This behavior is also ignored in phenomenological elements.

Correlation Study of a Shake Table Specimen

The shaking table specimen selected for study is a simple two story bare R/C frame. Its design and test sequence are described in Clough et al. (1976). The structure is a 1/1.4 scale model of a typical two story office building. Grade 40 reinforcement and an average concrete compressive strength of 4.4 ksi (30.3 MPa) were used. The specimen was tested under a unidirectional earthquake input. For test run W2, analytically reproduced using the current formulation, the 1952 Taft N69W record was used. The actual table response is used in the analysis, since the earthquake signal was somewhat modified by table-structure interaction. Due to this interaction the vertical acceleration component was found to be significant; however, it was ignored for simplicity in the following analyses. The peak table acceleration was recorded to be 0.52g. The analytical study spans the first seven seconds of the strong motion response.

The initial free vibration periods of the model were experimentally determined to be 0.32 sec. and 0.11 sec. for the two lateral modes. Prior testing of the specimen under a 0.07g maximum ground input (test W1) resulted in an increase of the periods by 20.%. Hence, the model was already cracked prior to the test W2. First mode damping was empirically estimated to be about 4.2% of critical. The period of the vertical flexural vibration mode of the beams was computed to be about 0.02 sec, depending on the axial stiffness assumed. Consequently, a time step of 0.005 sec was selected for the time history analysis. The theoretical model is initially excited by a long risetime steadily increasing vertical acceleration pulse that stabilizes to 1.0g at 0.7 seconds, in order to establish the dead load force distribution. The modal damping used (mass proportional) was similar to the measured value. For the purpose of comparison, the analysis is repeated with a standard bilinear, two-component, elastoplastic element with axial load-moment interaction. The average cracked section properties estimated by Clough and Gidwani (1976) were used. Post yielding member stiffnesses were assumed to be 1% of the elastic value.

The results of the time history analyses are shown in (Fig.4a), compared to the experimental and the two component model response (Fig.4b). The specimen initially responds with a period of about 0.35 seconds (first mode), with major inelastic excursions at the 4.5 and 6.0 seconds. The fiber model's waveform follows closely the true response until the first major displacement peak, which is underestimated by about 20%. The next large cycle in the opposite direction is at 6.2 seconds and is well simulated. The intermediate behavior is comparable. The slight period elongation exhibited by the real structure is not followed by the model. The analytical response is generally lower in amplitude, although of the right periodicity. Most possible explanation is that bond deterioration in the columns and joints allows for limited slip of the steel, resulting in larger amplitudes. Such a phenomenon cannot be modelled presently.

The response of the two component model is much stiffer right from the beginning, since this model cannot allow for change in the flexural stiffness of the members, prior to yield. Such cracking is evident in the base-moment curvature histories (Fig.5) of the first story columns, obtained with the fiber model. After the first inelastic excursion the phenomenological model attains a residual displacement which is not observed in the true response. Comparison of the predicted base moment history at the left column, obtained by the two models, reveals that the phenomenological model significantly underestimates the internal force demands, after the first peak at the 6.0 second (Fig.6). This moment (and shear) underestimation should be considered in design, when using such simplified models.

Analysis of a Simple Structure

The significance of refined modeling of axially loaded members to global and local response estimation is demonstrated by analyzing a simple structure. This structure is based on an idealized representation of the East transverse bay of the Imperial County Services Building (ICSB). Analyses of the six story building (Zeris et al.,1984) have demonstrated the significance of a) the local detailing of the ground story columns, b) the inadequacy of the concept used for the structural system and c) the bi-directional building response, in reducing the deformability of the structure at the ground story. In particular, the discontinuous shear wall in the transverse direction, which were supported by four columns at the ground story, resulted in a concentration of damage at the ground level. Overturning effects in this wall resulted in exceedence of the columns flexural capacities under tensile as well as compressive loads at the two edges of the building.

A simplified system is analyzed here, to assess the basic response characteristics of such a structure. In the actual building, shear walls are located at the ground level, though they are offset from the walls in the upper stories. The frame considered here neglects the contribution of these ground story walls. The presence of these wall results in less shear being taken by the four columns, but increases the tendency of the upper wall to rock thereby accentuating the axial effects on the supporting members. Shear deterioration and certain detailing weaknesses are also neglected in the model presented. To assess the lateral response characteristics of this system, an intense impulsive ground motion is used in the analyses. Again, both fiber and two-component models were used. The flexural stiffness of the two component model is taken equal to the initial stiffness of the fiber model. A ratio of post-yield hardening to elastic flexural stiffness of 10% was assumed, while the axial rigidity is based on the gross section area.

The dynamic response of the entire system is primarily in rocking; lateral translations being small. For the kind of ground acceleration input used, the global lateral displacement at the first floor is almost identically predicted by both models considered. The overturning rotation of the wall imposes opposite axial force increments to the extreme columns, while causing first yielding of the structure to occur at the base of these members.

The shear history of the individual fiber columns is compared in Fig.(8) to that obtained by the simpler model. The shear resisted by each member depends strongly on the imposed axial load and rotational deformations. Progressive flexural softening of the two columns at the tensile side of the structure transfers shears to the opposite (compressive) end of the building, with a considerable increase in the demanded curvature at this side of the structure (Fig.7). Such behavior is not evident in the two component model, where the demands at the extreme edges of the building are equal in magnitude.

The fiber model is generally stiffer axially than the two component model. As a result, axial forces taken by the columns are higher, resulting in an increased base shear capacity of the frame, with increasing first floor lateral deformation (Fig.9). However, the non-ductile behavior of the columns limits the energy absorption capacity of the structure. In the simpler formulation, the strength and stiffness are underestimated and the lack of deformability is not predicted.

Conclusions and Recommendations for Further Research

The practicability of modeling R/C beam-column members using a refined finite element approach is examined. In spite of its complexity, the current formulation is unable to account for several factors, such as bond deterioration and shear lag in slabs. However, this modeling approach will have significant advantages in certain situations. For example, it is inherently capable of accounting for local member behavior thereby providing insights into ductility demands and the adequacy of details. Moreover, the formulation is capable of simulating the complex hysteretic characteristics exhibited by R/C members with nonsymmetric cross sections or varying axial loads. Conventional models may not be able to adequately simulate these types of behavior and provide the types of information needed by the designer. Additional research is needed to account for phenomenon not currently considered, to explore methods for simplifying the computational requirements, and to extend it to considerations of full biaxial load-moment interaction.

Acknowledgements

The authors would like to acknowledge the helpful suggestions of V. Bertero and A. Scordelis. The financial support of the National Science Foundation is gratefully appreciated.

Appendix I - References

[1] Aktan A., Pecknold D. and Sozen M., Effects of Two Dimensional Earthquake Motion on a Reinforced Concrete Column Univ. of Illinois, Rep. No UILU-ENG 73-2009, Urbana, May 1973.

[2] Anagnostopoulos S., Inelastic Beams for Seismic Analysis of Structures , *Jnl. Am. Soc. of Civil Engrs.*, **107**, ST 7, July 1981 , p. 1297-1311.

[3] Arzoumanidis S. and Meyer C., Modeling Reinforced Concrete Beams Subjected to Cyclic Loads, *Dept. of Civil Eng. Techn. Rep.*, Columbia University, New York, March 1981.

[4] Brancaleoni F., Ciampi V. and Di Antonio R. Rate-Type Models for Non Linear Hysteretic Structural Behavior *EUROMECH Coll..*, Inelastic Structures Under Repeated Loads, Palermo, Oct 1983.

[5] Clough R.W. and Gidwani J., Reinforced Concrete Frame 2:, Seismic Testing and Analytical Correlation, *Earthqu. Eng. Res. Center.*, Rep. No EERC-76/15 , June 1976.

[6] Golafshani A. and Powell G., DRAIN - 2D 2 A Computer Program for Inelastic Seismic Response of Structures, PhD Disrtn, Univ. of California, Berkeley, 1974.

[7] Kaba S. and Mahin S., Refined Modelling of Reinforced Concrete Columns for Seismic Analysis, *Earthqu. Eng. Res. Center.*, Rep. No EERC-84/03 , April 1984.

[8] Kang Y. and Scordelis A., Nonlinear Geometric, Material and Time Dependent Analysis of Reinforced and Prestressed Concrete Frames, Univ. of California , Rep. UC-SESM No. 77-1, Berkeley, Jan. 1977 .

[9] Kent D., Inelastic Behavior of Reinforced Concrete Members with Cyclic Loading, Ph D Disrtn, Department of Civil Engineering, Univ. of Canterbury, Christchurch, 1969.

[10] Lai S., Will G. and Otani S., Model for Inelastic Biaxial Bending of Concrete Members, *Jnl. Am. Soc. of Civil Engrs.*, **110**, ST 11, Nov. 1984 , p. 2563-2584.

[11] Ma S., Bertero V. and Popov E., Experimental and Analytical Studies of the Hysteretic Behavior of Reinforced Concrete Rectangular and T-Beams, *Earthqu. Eng. Res. Center* , Rep. No EERC 76-2, Berkeley, May 1976 .

[12] Mahasurevachai M., Inelastic Analysis of Piping and Tubular Structures, Ph D Disrtn, Department of Civil Engineering, Univ. of Calif. , Berkeley, Nov. 1982.

[13] Mahin S., An Evaluation of the Seismic Response of Reinforced Concrete Buildings, PhD Disrtn, Univ. of California, Berkeley, 1974.

[14] Menegotto M. and Pinto P., Slender RC Compressed Members in Biaxial Bending, *Jnl of*

Am. Soc. of Civil Eng., **103**, ST 3, March 1977, 587-605.

[15] Shiohara H., Otani S. and Aoyama H. Comparison of Various Member Models for Reinforced Concrete Earthquake Response Analysis, *Trans. of Japan Conc. Inst..*, **5**, 1983 , p. 269-276.

[16] Takizawa H., Notes on Some Basic Problems in Inelastic Analysis of Planar R/C Structures, Parts I and II, *Trans of Arch Inst of Japan*, **240**, Part I in Feb. 1976, p.51-62, Part II in Mar. 1976, p.65-77.

[17] Zeris C. and Altmann R., Implications of Damages to the Imperial County Services Building for Earthquake Resistant Design, *Proc., VIII World Conf. of Earthqu. Eng.*, **IV**, San Fransisco, 1984.

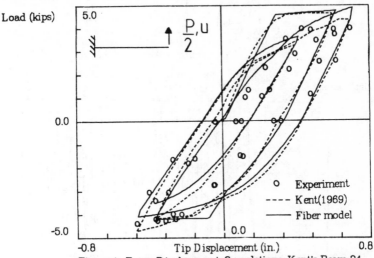

Figure 1. Force Displacement Correlations, Kent's Beam 24

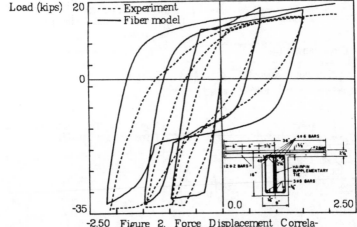

Figure 2. Force Displacement Correlations, Ma Bertero and Popov (1976) Beam T1

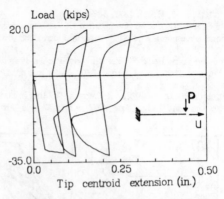

Figure 3. Theoretical Load Extension Behavior, M a et al. Beam T 1

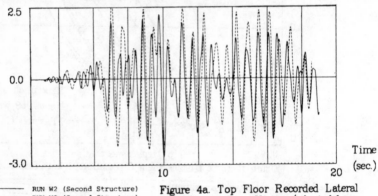

——— RUN W2 (Second Structure)
------ RUN W3 (Second Structure)

Figure 4a. Top Floor Recorded Lateral Displacement Time History (adapted from Clough et al.,1976).

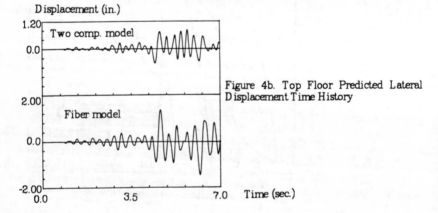

Figure 4b. Top Floor Predicted Lateral Displacement Time History

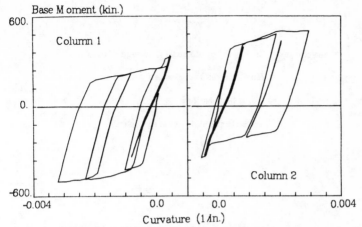

Figure 5. Predicted Moment Curvature
Behavior, First Storey Columns, Base.

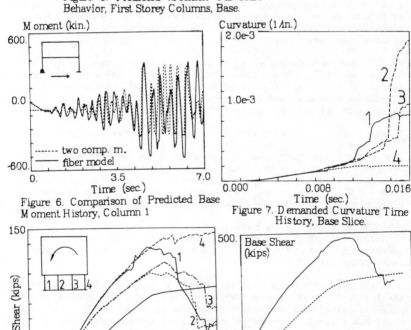

Figure 6. Comparison of Predicted Base
Moment History, Column 1

Figure 7. Demanded Curvature Time
History, Base Slice.

Figure 8. Comparison of Column Shear
Force Histories, Fiber and 2-component
Models.

Figure 9. Total Base Shear First Floor
Lateral Response Behavior.

Earthquake Analysis of Liquid Storage Tanks

Wing Kam Liu[*], Member ASCE

Dennis Lam

Ted Belytschko, Member ASCE

and

Bak Wong

Abstract

In the design of modern liquid storage tanks, it is of technical importance to examine their safety against earthquake damage. The methodologies used in design do not account for the major collapse mechanisms and reflect inadequate understanding of this complex fluid-structure system.

To provide a better understanding of storage-tank-collapse, a cylindrical shell subject to vertical compression and a cylindrical tank subject to horizontal fluid load has been studied. Nonlinear finite element methods are used for the model, both large deflections and material nonlinearities are included.

Buckling Analysis of a Cylindrical Shell under Axial Compression

A nonlinear finite element procedure which accounts for finite deformation is used in this analysis. The fundamental aspects of the 3D nonlinear finite element shell analysis can be found in the article by Hughes and Liu (1981); and the development of the degenerated shell element with stabilization is described by Liu et al. (1985).

The dimension of the cylindrical shell is depicted in Fig. 1. It is set up in such a way that both edges of the shell are constrained except in the axial direction; the end displacements are prescribed incrementally.

By taking advantage of three planes of symmetry, one eighth of the shell is modeled with 420 quadrilateral elements. The assumption of symmetry rules out the possibility of odd number circumferencial wave modes and anti-symmetric axial modes in the deformed geometry.

[*]Dept. of Mechanical and Nuclear Engineering, Northwestern University, Evanston, IL 60201.

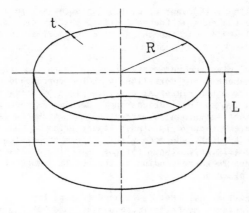

$R = 100mm$ (3.94in)
$L = 71.9mm$ (2.83in)
$t = 0.247mm$ (0.0097in)
$E = 567$ kg/mm^2 (0.805×10^{6}psi)
$\nu = 0.3$

Fig. 1

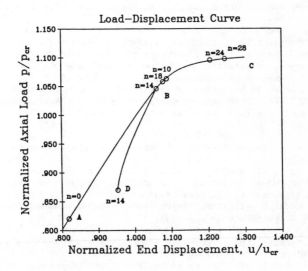

Load–Displacement Curve

Fig. 2

As the cylinder is compressed, bending stresses are induced in the region close to the ends of the cylindrical shell. Away from the edges, the cylindrical shell sustains the applied load primarily by the membrane stress.

In the load displacement curve (Fig. 2), both the axial load p and the end displacement u are normalized with the corresponding critical values derived from the classical analysis given by Timoshenko and Gere (1961). Initially, the cylindrical shell deforms asymmetrically with an axial displacement less than 20 percent of the thickness (see Fig. 3). The load-displacement curve is almost linear prior to the bifurcation point B. At that point, the mode with a circumferencial wave number of 14 is developed (Fig. 4). Also, from the same bifurcation point, a postbuckling path BD corresponding to the buckling mode of n=14 is traced out by unloading.

When the cylindrical shell is loaded further beyond the point B, the shell deforms axisymmetrically though a number of buckling modes appear. This is also shown in Fig. 2. The curve ABC represents the load path of the axisymmetric deformation. On the branch BC, several buckling modes are identified. The shapes and axial stress contours of these modes are shown in Fig. 5 to Fig. 8. All these modes appear to be unstable, and the only stable bifurcation path that can be traced out is at the lowest buckling load with a buckling mode of n=14.

Dynamic Analysis of a Liquid Storage Tank

The dimensions of the tank model used in this analysis are shown in Fig. 9. The tank is partially filled with water and is excited by the 1940 N-S El Centro earthquake time history. The analysis is divided into two parts.

In the first part, a linear fluid-structure interaction analysis is performed to obtain the pressure time history on the tank wall. The fluid-structure interaction algorithm is due to Liu and Chang (1985). From the pressure time history, the pressure load which delivers the maximum base overturning moment is estimated. This pressure load is then applied as a step load in the second part of the analysis, which is used to study the buckling modes by a nonlinear finite element procedure.

Since the interaction effect between the tank wall and the fluid is accounted for in obtaining the pressure time history, the pressure load used in the buckling analysis is a realistic approximation. With this analysis procedure, the fluid-tank buckling problem is reduced to a tank buckling problem with the fluid removed. This reduces the size of the computer model drastically. An explicit program WHAMS was used for this analysis, Belytschko et al. (1984).

The results from the dynamic simulations are shown in Fig. 10, which depicts the deformation of the tank at different times. Fig. 11 shows a magnified view of the geometry of the tank base. The buckling appears to be triggered at the base where the stress is maximum. The deformation then propagates upward along the axial direction.

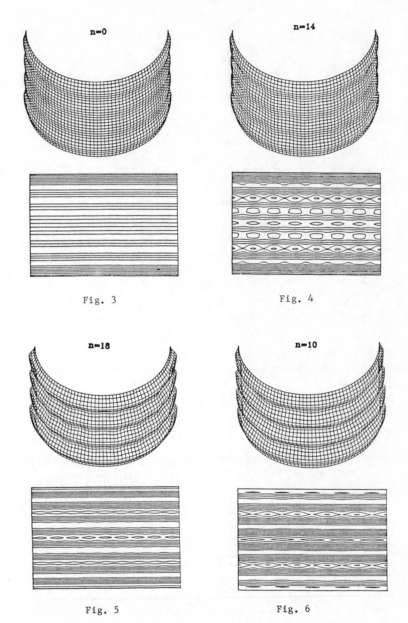

Fig. 3

Fig. 4

Fig. 5

Fig. 6

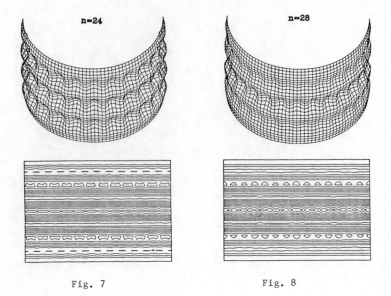

Fig. 7   Fig. 8

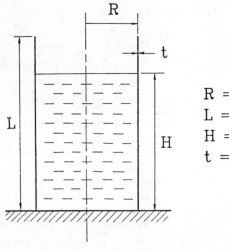

R = 61.5in (1.55m)
L = 240in (6.10m)
H = 200in (5.08m)
t = 0.078in (2.0mm)

Fig. 9

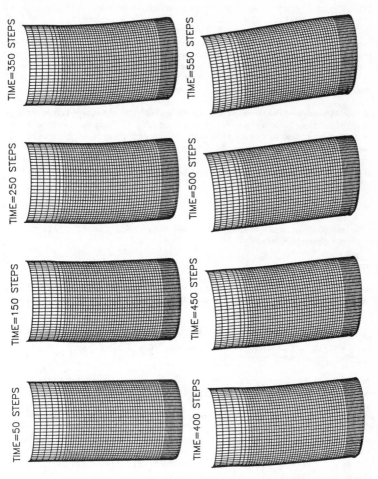

TIME=350 STEPS
TIME=550 STEPS
TIME=250 STEPS
TIME=500 STEPS
TIME=150 STEPS
TIME=450 STEPS
TIME=50 STEPS
TIME=400 STEPS

Fig. 10

Summary

    The cylindrical shell under axial compression is studied via
incremental displacement formulation. Cos nθ diamond modes (n=0, 14,
18, 10, 24 and 28) are found to be the possible postbuckling shapes of
the cylindrical shell. The transition from the axisymmetric mode (n=0)
to a diamond mode of collapse (n=14) is studied in detail. This
postbuckling behavior agrees well with the experimental results obtained
by Yamaki et al. (1975). From the analysis, it is also found that the
cylindrical shell could buckle in a number of different modes. The
actual buckling mode depends very much on other factors, such as
geometric imperfections and material properties.

    A nonlinear analysis was also made using the dynamic loads from a
linear fluid structure system subjected to the El Centro earthquake
motion. The maximum horizontal fluid load acting on the tank wall was
then applied to a nonlinear model. It is found that the buckling mode
is triggered at the base. This mode of deformation does not appear to
exhibit any diamond buckles and is concentrated at the bottom where the
compressive stress is maximum.

Acknowledgment

    The support of this research by NSF Grant No. CEE 82-13739 is
gratefully acknowledged.

References

Belytschko, T., Lin, J. I. and Tsay, C. S., "Explicit Algorithms for the
Nonlinear Dynamics of Shells," Comp. Meth. Appl. Mech. Engrg., Vol. 42,
1984, pp. 225-251.

Hughes, T. J. R. and Liu, W. K., "Nonlinear Finite Element Analysis of
Shells: Part I, Three-dimensional Shells," Comp. Meth. Appl. Mech.
Engrg., Vol. 26, 1981, pp. 331-362.

Liu, W. K. and Chang, H., "A Method of Computation for Fluid Structure
Interaction," Comp. Struct., Vol. 20, 1985, pp. 311-320.

Liu, W. K., Law, E., Lam, D. and Belytschko, T., "Resultant Stress
Degenerated Shell Element," to appear in Comp. Meth. Appl. Mech. Engrg.,
1985.

Timoshenko, S. P. and Gere, J. M., Theory of Elastic Stability, McGraw
Hill Book Co., Inc., 1961, pp. 457-459.

Yamaki, N., Otomo, K. and Matsuda, K., "Experiments on the Postbuckling
Behavior of Circular Cylindrical Shells Under Compression," Experimental
Mechanics, Vol. 15, Jan. 1975, pp. 23-28.

REFINEMENTS IN PSEUDO-DYNAMIC TEST PROCEDURES

By C.R. Thewalt[1], S.M. ASCE, S.A. Mahin[2], M. ASCE, S.N. Dermitzakis[3], A.M ASCE

One of the major concerns in seismic testing is whether the loading conditions imposed on a structure adequately represent those that might occur during a major earthquake. Response simulation using shaking tables can resolve many of these concerns for some types of structures. The recently developed pseudo-dynamic test method may enable these advantages to be extended to a broader range of test specimens. This paper outlines some recent advances in this method, discusses some of its limitations and identifies future research areas.

INTRODUCTION

Considerable research has been devoted recently to developing an on-line computer controlled test method for seismic performance evaluation. This hybrid analysis procedure is commonly known as the pseudo-dynamic method. Using this relatively new method, the seismic response of a test specimen can be simulated by slowly imposing displacements using generally available laboratory equipment. Software in the controlling computer is used to account for dynamic effects. This approach appears to combine the simplicity and economy of conventional quasi-static tests, the versatility and convenience of analytical studies, and the realism of shaking table testing. However, until recently only very limited experience has been gained with the method.

A coordinated series of investigations has been undertaken as part of the U.S. - Japan Cooperative Earthquake Research Program to assess the reliability of the pseudo-dynamic method. These investigations have been performed in Japan at the Building Research Institute, Ministry of Construction, Tsukuba, and in the U.S. at the University of Michigan, Ann Arbor, and at the University of California, Berkeley.

Several studies at Berkeley on the pseudo-dynamic method have been reported previously by Shing and Mahin (1983, 1984, 1985). These relate to the stability and accuracy of the underlying numerical procedures, the effect of experimental errors on the accuracy of test results, and verification of the procedure by means of simple experiments and analyses. In an effort to expand the applicability of the pseudo-dynamic method or to solve some of the problems encountered with its use, several extensions to these studies have been made recently. In this paper, the recent work at Berkeley will be summarized.

NUMERICAL FORMULATIONS

Multiple Component Base Excitation

The pseudo-dynamic test method has been primarily used to test complete planar structures subjected to a single lateral base excitation. This corresponds to the way a structure might be tested on a shaking table. Although this mode of testing is quite useful, the pseudo-dynamic method can easily be extended to permit testing of complete structures subjected to multiple component base excitation. Using this extension, it would be possible to obtain

---

[1]Research Assistant, University of California, Berkeley, CA.
[2]Associate Professor of Civil Engineering., University of California, Berkeley, CA.
[3]Senior Engineer, Impel Corp., Walnut Creek, CA.

information about the three dimensional response of structures that could not be easily acquired using conventional shaking tables. The feasibility of three dimensional testing is currently being studied at Berkeley (Thewalt 1984), and a three degree of freedom (DOF) specimen is being used to compare the results of pseudo-dynamic tests, shaking table tests and analytical simulations.

Equations of Motion.-- The time discretized equations of motion, as used in pseudo-dynamic testing, are given by:

$$\mathbf{m}\,\mathbf{a}_i + \mathbf{c}\,\mathbf{v}_i + \mathbf{r}_i = \mathbf{m}\,\mathbf{b}\,\mathbf{a}_{g_i} \tag{1}$$

where $\mathbf{m}, \mathbf{c}$ = mass and damping matrices;
         $\mathbf{a}_i, \mathbf{v}_i$ = acceleration and velocity vectors;
         $\mathbf{r}_i$ = structural restoring force vector;
         $\mathbf{b}$ = ground acceleration transformation matrix;
         $\mathbf{a}_{g_i}$ = ground acceleration vector.

The component $\mathbf{b}_{ij}$ is the acceleration at structural DOF i when the structure acts as a rigid body under a unit acceleration for ground component j.

In performing three dimensional tests, it has been found convenient to allow the user to supply a linear transformation matrix so that the equations of motions are solved a coordinate system other than that determined by the arrangement of the hydraulic actuators. This new coordinate system simplifies calculation of the mass and damping matrices, often allowing the use of a diagonal mass matrix.

Geometric Corrections.-- In three dimensional tests, specimen displacements may be large enough to cause angular changes in the actuators that will in turn induce errors in both the force and global displacement vectors. The example shown in Fig. 1 shows that movement of a single actuator will give an incorrect global specimen position, if the displacements are measured by monitoring the change in actuator length. Moreover, the measured actuator forces may need to be corrected for actuator angle. After measuring the force vector, a user supplied routine is used to correct the force vector based on the current deformed position. Using the corrected forces, the next displacement vector can be calculated using standard pseudo-dynamic algorithms. A second user specified routine can then calculate the actuator movement necessary to correctly impose the computed global displacements. These user specified routines may be nonlinear and would depend on actuator length and the specimen geometry.

Geometric Stiffness.-- In some specimens it may not be necessary to include the full design weight during a test. For example, added masses for scaling effects may not be needed in certain pseudo-dynamic tests (Shing and Mahin 1984). However, it may still be desirable to account for these masses when considering the effects of geometric stiffness. In this case, a user supplied geometric stiffness matrix may be specified to modify the measured restoring force vector in terms of the current displacement pattern.

Substructuring Concepts in Pseudodynamic Testing

Thus far, applications of the pseudo-dynamic method have been limited to tests of complete systems. Tests of complete structural models are not only expensive, but require special test facilities as well. Where detailed information on the local behavior of critical regions is required, it is more efficient and economical to test structural subassemblages. In addition, one may be interested in the dynamic response of components or equipment mounted in structures which are subjected to ground excitation. However, since the ground motion is not directly applied to the base of the equipment, the supporting structure has to be accounted for in such tests. This leads to a costly and inefficient test setup or significant simplifications that may reduce the accuracy of the results.

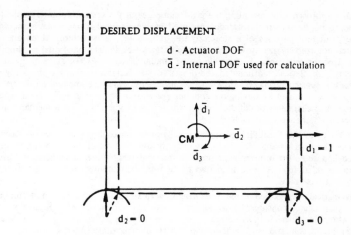

Fig. 1   Position Error Due to Geometry for Displacements
Measured Across the Actuators

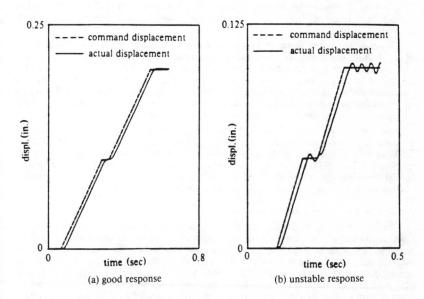

Fig. 2 Electrohydraulic Control Loop Response

One approach to overcoming these difficulties with the pseudo-dynamic method is by application of the substructuring concepts used in conventional dynamic analyses. In such analyses, different portions of a structure are grouped into substructures which are treated separately for convenience in formulating the data as well as for computational economy. In a pseudo-dynamic test it may be possible to use similar methods, except that certain substructures may be analytically formulated and others are subassemblages that are physically tested. By means of substructuring techniques the displacements which are imposed on the test structure would be obtained by solving the equations of motion of the complete system, where the restoring force characteristics of the portion which is not subjected to experimental testing are provided by mathematical models.

A study of the feasibility of applying substructuring concepts has recently been completed by Dermitzakis and Mahin (1985). The theoretical basis of the method has been formulated and some preliminary experimental and analytical studies have been performed to evaluate its reliability. A brief discussion of the underlying numerical formulations for various applications follow.

Numerical Formulation of Substructuring Concepts.-- Using substructuring techniques , a test structure can be considered as an assembly of two distinct parts : (i) a physical subassemblage which is experimentally tested using load applying actuators ; and (ii) analytical subassemblages consisting of mathematical models of structural elements. The equations of motion of a complete system now takes the following form :

$$\mathbf{M} \, \mathbf{a} + \mathbf{C} \, \mathbf{v} + \mathbf{R}' + \mathbf{R}^* = \mathbf{F} \tag{2}$$

In which $\mathbf{R}^*$ contains the restoring force vector of the physical specimen and $\mathbf{R}'$ contains the restoring force vector of the analytical subassemblage. For linear elastic systems, $\mathbf{R}'_{i+1} = \mathbf{K} \cdot \mathbf{d}_{i+1}$, whereas for nonlinear systems, $\mathbf{R}'_{i+1} = \mathbf{R}'_i + \mathbf{K}^*_{i+1} \cdot (\mathbf{d}_{i+1} - \mathbf{d}_i)$. For nonlinear systems, the tangent stiffness matrix $\mathbf{K}^*_{i+1}$ for the analytical subassemblage is continually updated.

Solution by Explicit Newmark Methods.-- So long as a finite mass can be associated with all active DOF, the solution of Eq. 2 can be carried out conventionally using explicit integration methods, such as the Newmark method (Shing and Mahin, 1984, 1985). This approach is identical to the general solution of the equations considered in a pseudo-dynamic test of a complete structure, except for the addition of the restoring forces resulting from the analytically modeled components. To obtain bounded solutions, the stability criteria for the particular explicit integration algorithm used must be satisfied for the complete system.

Solution by an Implicit-Explicit Integration Algorithm.-- When massless degrees-of-freedom, such as rotations, must be included in the equations of motion, the explicit Newmark algorithms fail to provide a stable solution. A mixed implicit-explicit algorithm (Hughes and Liu, 1978) can be applied to pseudo-dynamic systems having such properties. Special considerations are necessary when massless DOF lie on the boundary between the physical and analytical substructures (Dermitzakis and Mahin, 1985).

Pseudo-Dynamic Testing Under Force Control

In extremely stiff structures, such as those containing shear walls, the use of displacement control in pseudodynamic testing may lead to poor results. Actuator forces in such systems will change rapidly with very small displacement variations. Verification tests indicate that very accurate displacement control may be possible. However, this may require the use of special high performance equipment and instrumentation, and error propagation considerations may still become critical. An alternate test method has been devised to help overcome these problems. This is based on force control. The proposed method removes problems associated with discrete time integration and essentially solves the equations of motion in analog form using transducer outputs. The rate at which an experiment is performed may be varied continuously up to the electro-hydraulic system capacity. Real inertial and viscous damping forces are explicitly accounted for in this method. This approach also provides for the possibility of performing tests in real time, which would be necessary for testing systems that exhibit rate sensitive behavior.

To implement this procedure, the test specimen must be constructed so as to satisfy all standard similitude relationships, as required for a shaking table specimen. Thus, reduced scale models would be constructed with mass added so that the modal frequencies change by a factor of the square root of the physical model scale. The method then applies forces to the structure that are based on the ground motion inputs and on the real acceleration and velocity of the structure during testing. These measured quantities are used to modify the force signal so that actual inertial and damping forces in the test specimen are accounted for in the equations of motion. A complete description of how these signals are combined in analog form is given by Thewalt (1984).

VERIFICATION STUDIES

Electro-hydraulic Control of Test Specimens

In view of the importance of accurate control of specimen's position during pseudo-dynamic testing, a series of simple tests has been performed. To study the behavior of the electro-hydraulic actuators under displacement control both single and multiple actuator systems have been investigated (Thewalt, 1984). The single actuator test, consisting of an actuator mounted on a wide flange steel cantilevered beam, was used to compare the performance of three different size servovalves (5, 10 and 25 gallon/minute rated flows), and also to study the effect of loop gain, force level, ramp shape and specimen stiffness on the accuracy of the control loop.

These tests demonstrated that increasing loop gains resulted in more accurate response, as long as the loop remained stable. Examples of stable and unstable response are shown in Fig's. 2(a) and 2(b). The unstable response was obtained by introducing very high loop gains. It was also found to be beneficial to use more smoothly varying ramp shapes, such as haversine curves, since this reduced the amount of overshoot at the end of the ramp. Also, as valve size increased, the ability to accurately track the desired signal decreased.

The experiments suggest the use of low flow capacity, high performance servovalves with high control loop gains. The use of haversine or other ramps without discontinuities at the ends also appear to be preferable to linear ramps. Additional experiments and analytical studies are needed to characterize the performance of these systems. In particular, efforts to establish guidelines for selecting equipment and control system parameters must be further developed.

Multiple Components of Excitation

A simple three degree of freedom structure shown in Fig. 1 is being used to test the reliability of performing pseudo-dynamic tests with multiple components of base excitation. The structure has been tested on a shaking table under a single component of lateral excitation. This component, however, was skewed at forty-five degrees to the principal axis of the structure. The measured lateral table accelerations as well as the three components of rotational base acceleration, (due to imperfect table control), are used as input to the pseudo-dynamic test. Once this phase of verification testing is completed, further comparisons between pseudo-dynamic tests and analytical results will be made using two independent components of lateral base excitation.

The pseudo-dynamic tests are currently in the system adjustment phase, where low level elastic runs are performed to study the nature and magnitude of experimental errors. These tests have emphasized the difficulty in performing multiple degree of freedom displacement controlled tests, since the actuators interact and it has been found to be difficult to properly adjust the electro-hydraulic control loops. However, the use of high performance displacement transducers and servovalves has lead to good results, even for the difficult low level tests, as shown for one of the DOF of this system in Fig. 3.

Analytical Substructuring

A variety of tests have been performed to evaluate the reliability of substructuring methods as applied to pseudo-dynamic testing. These relate to structural subassemblage and equipment tests using the implicit-explicit formulation mentioned previously as well as to nonlinear substructure tests based on an explicit formulation. Additional information on these tests is given by Dermitzakis and Mahin (1985).

Nonlinear Substructure specimen.-- A single-bay, three-story steel frame building was considered to demonstrate the application of substructuring techniques to specimens with nonlinear subassemblages. For the dynamic analysis, horizontal ground excitation was considered. The prototype structure was idealized as having one lateral DOF at each floor and masses were concentrated at these levels. To facilitate this verification test, beams were assumed rigid.

Analytical results for the complete structure are compared with results from a pseudo-dynamic test using substructuring. During the pseudo-dynamic test, the top two stories were modeled analytically and the bottom story was tested physically. To simplify the experimental setup, inflection points were assumed at mid-height of the first story, and overturning effects were disregarded. With these assumptions, the experimental specimen was reduced to a single degree-of-freedom cantilever column. This specimen was a 4 ft. (1.22 m) long W6×20 wide flange steel section.

Since there are no massless degrees-of-freedom in this structure, the Newmark explicit algorithm can be used as the integration method. Restoring force values in Eq. 2 are obtained from the experimental specimen and from the inelastic analytical models of the substructured stories. The inelastic behavior of the top two (substructured) stories was represented by means of a Menegotto-Pinto model. The inelastic behavior of the frame was examined considering 10 seconds of the 1940 El Centro (NS) ground motion scaled arbitrarily to 1.0g peak acceleration.

The displacement time histories for the three degrees-of-freedom are shown in Figs. 4(a) to 4(c). The experimental specimen experienced significant inelastic deformations, and its response compared well with analytical simulations. The results of this and other pseudo-dynamic tests with substructuring indicate that substructuring techniques may be successfully applied.

However, it must be recognized that the reliability of the results is limited by the accuracy of the analytical models used. Thus, care must be used in selecting the types of structures and subassemblages to be tested in this fashion. Additional research is needed to extend the applicability of the method and to more completely characterize error propagation effects.

CONCLUSIONS

The pseudo-dynamic test method has been demonstrated to be a powerful and versatile procedure for assessing seismic performance. In particular, it may provide insight on the seismic response of structures that are too large, massive or strong for testing using available shaking tables. In addition, extension of the basic procedures will permit tests considering complex support excitations not possible with other means. The use of analytical substructuring techniques will allow for more economical and realistic seismic performance evaluations of large structural subassemblages. However, like all experimental procedures, the pseudo-dynamic method has inherent limitations and errors. These must be fully recognized in developing pseudo-dynamic test programs. In particular, the need for high performance control equipment and instrumentation must be taken into account.

A variety of future developments and investigations should be undertaken to make the pseudo-dynamic method more convenient and reliable. This is especially true for methods

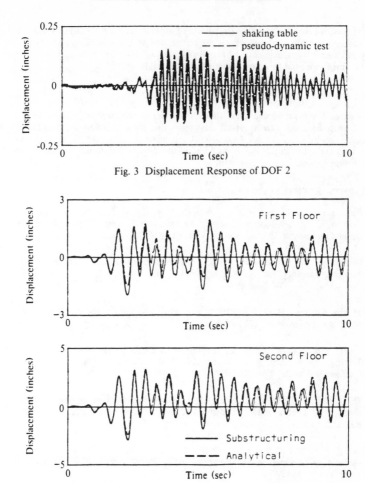

Fig. 3 Displacement Response of DOF 2

Fig. 4 Displacement Time Histories

associated with controlling the adverse effects of experimental errors and with substructuring methodologies. Additional research is needed related to understanding and improving our ability to control test specimens. Not until reliable guidelines for the use of pseudo-dynamic test equipment are established can the method be used dependably.

ACKNOWLEDGEMENTS

This work was performed under the U.S. - Japan Cooperative Earthquake Research Program. The authors are grateful for the advice of other participants in the program as well as the financial support of the National Science Foundation. However, opinions, findings and conclusions expressed in this report are those of the authors and do not necessarily reflect the views of the National Science Foundation.

APPENDIX.-- REFERENCES

1   Dermitzakis, S.N., and Mahin, S.A., "Development of Substructuring Techniques for On-Line Computer Controlled Seismic Performance Testing," *UCB/EERC-85/04,* Earthquake Engineering Research Center, University of California, Feb. 1985.

2   Hughes, T.G.R., and Liu, W.K., "Implicit-Explicit Finite Elements in Transient Analysis: Stability Theory," ASME, Journal of *ASME, Journal of Applied Mechanics,* Vol. 45, pp. 365-368, June 1978.

3   Mahin, S.A., and Shing, P.B., "Pseudo-Dynamic Method for Seismic Testing," *Journal of Structural Engineering,* ASCE, July 1985.

4   Shing, P.B., and Mahin, S.A., "Experimental Error Propagation in Pseudo-Dynamic Testing," *UCB/EERC-83/12,* Earthquake Engineering Research Center, University of California, Berkeley, June 1983.

5   Shing, P.B., and Mahin, S.A., "Pseudo-Dynamic Test Method for Seismic Performance Evaluation: Theory and Implementation," *UCB/EERC-84/01,* Earthquake Engineering Research Center, University of California, Jan. 1984.

6   Shing, P.B., and Mahin, S.A., "Computational Aspects of a Seismic Performance Test Method using On-Line Computer Control," *Earthquake Engineering and Structural Dynamics,* Aug. 1985.

7   Thewalt, C.R., "Practical Implementation of the Pseudo-Dynamic Test Method," *CE 299 Report* University of California, Berkeley, Apr. 1984.

Building Records from 1984 Morgan Hill Earthquake

M.J. Huang, A.M. ASCE, A.F. Shakal, R.W. Sherburne
and R.V. Nutt, M. ASCE *

Abstract

The dynamic response of 23 structures instrumented by the California
Strong Motion Instrumentation Program was recorded during the magnitude
6.2 (ML) Morgan Hill earthquake of 24 April 1984. The structures range
in complexity from a single-story warehouse to a 13-story office
building. The number of acceleration sensors in the structures varied
from 6 to 22, depending on structural complexity and specific
measurement objectives. As an example, the Santa Clara County Office
Building in San Jose was instrumented with a total of 22 sensors
configured to record foundation rocking and the translational as well
as torsional motion on five of the 13 stories.

The strong motion records from seven buildings and one bridge have
been digitized and processed. Maximum horizontal ground accelerations
at the buildings were in the range 0.03 - 0.11 g; maximum roof
accelerations were 0.17 - 0.41 g. Processing of the digitized data
indicates that the maximum horizontal displacements at roof level
ranged from 1 to 19 centimeters. Significant structural response
lasted for less than 35 seconds for most buildings. However, the
steel-frame Santa Clara County Office Building had significant response
for about 80 seconds. The records from this structure and from a
single-story warehouse of tilt-up construction are considered in some
detail in this paper.

Introduction

The Morgan Hill earthquake occurred on April 24, 1984 on the
Calaveras fault southeast of San Jose, California with local magnitude
(ML) of 6.2 (BRK). It triggered the strong-motion accelerographs at
nearly 50 stations instrumented by the California Strong Motion
Instrumentation Program (CSMIP). Twenty-three of those stations are
extensively-instrumented structures. The records obtained at these
structures as well as the instrumentation descriptions are given in
Shakal et al. (1984). Additional information about the Morgan Hill
earthquake and the associated damage is available in the compilation by
Bennett and Sherburne (1984) and a special issue of Earthquake Spectra
(1985).

The records obtained at 7 buildings have been digitized and
processed. A listing of the buildings and the maximum amplitudes of
motion at the base and the top of each is given in Table 1. The

* Office of Strong Motion Studies, California Division of Mines and
Geology, 630 Bercut Drive, Sacramento, CA 95814

241

complete results of processing the records from these buildings are available in Huang et al. (1985). In the interests of brevity, this paper presents only the records obtained at the Santa Clara County Office Building in San Jose and the Glorietta Warehouse in Hollister. The records from the other buildings are only briefly discussed here.

Santa Clara County Office Building in San Jose

The 13-story Santa Clara County Office Bldg. in San Jose is about 21 km from the earthquake epicenter. The structure is 167 by 167 feet in plan and 187 feet in height. The exterior cores at the west and south ends extend from the lower level to 24 feet above the roof. The lateral load-resisting system of the building is composed entirely of moment-resisting steel frames. The vertical load-carrying system consists of concrete floor slabs overlying metal decks supported by the steel framing. The foundation is a concrete mat.

Twenty-two accelerometers were installed at locations throughout the building. The signals from the accelerometers were recorded by two central recorders on the lower level. The locations of the accelerometers are shown in Fig. 1. Sensors 1 - 3, on the lower level, are mounted vertically to record the vertical motion at the base as well as to indicate the presence of any foundation rocking. Channels 20 and 21 record the east-west translational motion of the foundation at the north and south ends. This east-west pair of sensors is repeated on the 2nd, 7th and 12th floors and on the roof. These sensor pairs allow analysis of translational as well as torsional motions at these levels. These pairs are complemented by pairs sensing north-south motion on the same levels.

The digitized records obtained during the 1984 Morgan Hill earthquake are plotted in Fig. 2. The records show that the building vibration lasted about 80 seconds with the last 50 seconds in almost purely fundamental mode motion. The period of this fundamental mode motion is about 2 seconds in both directions, which is 50% longer than the period given by T=0.1N for a steel frame building. The processing results included in Huang et al. (1985) indicate that the maximum relative displacement between the roof and lower level is about 16 centimeters. This is 5 to 6 times that at the other two 10-story reinforced concrete buildings in the same area (see Table 1). In addition, the duration of vibration of this building is about twice as long as that of the other two buildings.

The extensive instrumentation of this building warrants thorough and detailed studies of the records. Studies could include estimation of modal periods and damping, investigations of the torsional response, of the relationship between design loads and structural response, and of the performance of the building if subjected to stronger shaking than that during the Morgan Hill earthquake. Preliminary analysis of the data indicates the presence of torsion in the building response. Further, the foundation torsional motion appears to be relatively small compared to that on the higher floors. Thus, most of the torsional motion at the upper levels of the structure represents torsional response to translational input motion, or alternatively, amplification of the torsional input motion.

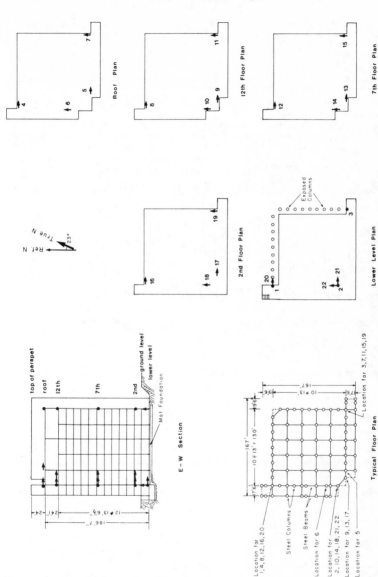

Figure 1. Location of Acceleration Sensors in the Santa Clara County Office Bldg. in San Jose during the 1984 Morgan Hill Earthquake.

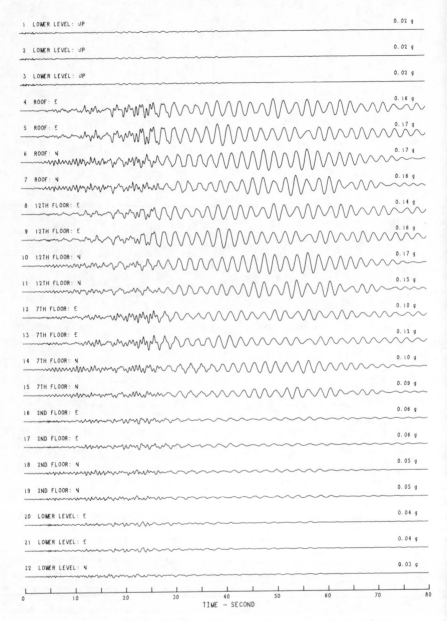

Figure 2. Acceleration Records from the Santa Clara County Office Bldg. in San Jose during the 1984 Morgan Hill Earthquake.

Glorietta Warehouse in Hollister

The Glorietta (now Tri-Valley Growers) warehouse in Hollister is a single-story structure approximately 300 by 100 feet in plan and 30 feet in height, and is approximately 37 km from the Morgan Hill earthquake fault. The lateral force-resisting system consists of concrete shear walls on the perimeter and a plywood roof diaphragm. The glu-lam roof beams, interior steel pipe columns, and concrete perimeter walls carry vertical loads. The structure is of tilt-up construction.

Thirteen accelerometers were installed on the walls and roof of the structure, as shown in Fig. 3. The record obtained during the 1984 Morgan Hill earthquake is also shown in Fig. 3. Some important observations and preliminary interpretations can be made through direct inspection of this record. The transverse in-plane vibration of the roof diaphragm is readily apparent by comparing the record from sensor 4, at the middle of the roof, with those from sensors 2 and 3, at the end walls. Similarly, comparison of the record from sensor 11 with those from 10 and 12 indicates the in-plane longitudinal vibration of the roof diaphragm. The side-wall out-of-plane motion can be noted by comparing the record from sensors 5 and 6 with the base motion indicated by sensor 7.

The in-plane vibration of the roof diaphragm in the transverse direction, indicated by sensor 4, suggests that the principal mode of vibration of the structure is that shown in Fig. 4a. From the response spectrum it is estimated that this mode has a frequency of about 1.7 Hz. Comparison of the records from sensors 5, 6 and 7, on the sidewall, indicates that much of the high-frequency motion present at the mid-height (sensor 6) did not occur at the top of the wall (sensor 5). This may reflect structural response in the vibration mode shown in Fig. 4b, in which the center line of the roof is a node. Analysis of the spectrum indicates this mode has a frequency of about 4.2 Hz.

The record obtained at this warehouse is the first strong motion record obtained from a tilt-up building. Many tilt-up structures were heavily damaged in the 1971 San Fernando earthquake. Most failures arose from a weak connection between the roof diaphragm and tilt-up wall. The building code was revised after that earthquake. This warehouse was designed and constructed in 1979 with the roof diaphragm built according to the new code. It performed well during the Morgan Hill earthquake. Detailed study of this record will provide guidance in code revision.

Features of Records From Other Buildings

The Town Park Tower Apartment building and the Great Western S & L building in San Jose are both 10-story reinforced concrete buildings. The records from these buildings, quite close to the Santa Clara County Bldg., have been digitized and are included in Huang et al. (1985). The records from the gymnasium at West Valley College in Saratoga indicate large in-plane vibrations of the roof diaphragm. Another building, the 4-story Watsonville Telephone building, is important because, constructed in 1948, it predates most of the modern code

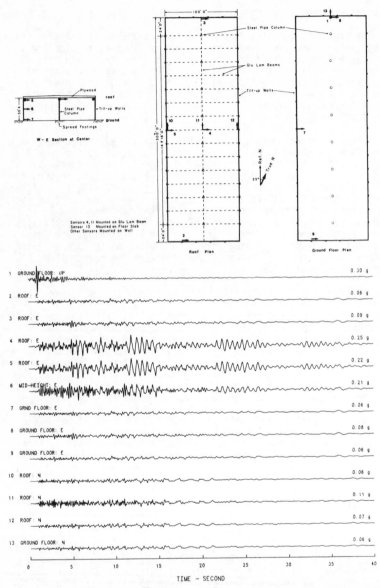

Figure 3. (Upper) Location of Acceleration Sensors in the Glorietta
Warehouse in Hollister. (Lower) Record Obtained during the
1984 Morgan Hill Earthquake.

provisions. Study of that record can lead to a better understanding of
ways of retrofitting old buildings. Finally, the record from the
Kaiser Medical Center in South San Francisco is of interest because of
the large response amplitudes of this 4-story steel-frame hospital,
nearly 80 km from the earthquake source.

Data Availability

The 1984 Morgan Hill earthquake generated an important set of
records for the analyses of structural response. Although the
instrumented buildings did not suffer structural damage, the extensive
instrumentation provided detailed data on the response of several types
of structures. For each structure, the raw and processed data, copies
of the structural drawings, and details on the locations of each sensor
can be obtained upon request to: Office of Strong Motion Studies,
Division of Mines and Geology, California Department of Conservation,
630 Bercut Drive, Sacramento, California 95814.

Table 1 - Summary of Processed Building-Response Records from
the 1984 Morgan Hill Earthquake

Structure	No. Stories (above/below ground)	No. of Sensors	Type of Bldg.	Epicentral Distance (km)*	Peak Acceleration**		Peak Velocity (cm/sec)		Peak Displacement (cm)	
					Ground	Structure	Ground	Structure	Ground	Structure
San Jose - Town Park Tower Apartment Bldg.	10/0	13	RC Shear Walls	19 [19]	0.06g H 0.04g V	0.21g H	11.5 H 3.5 V	23.6 H	3.0 H 1.1 V	5.2 H
San Jose - Great Western S & L	10/1	13	RC Frame & Shear Walls	19 [19]	0.06g H 0.03g V	0.22g H	11.9 H 3.9 V	27.4 H	3.6 H 0.9 V	6.4 H
San Jose - Santa Clara County Bldg.	12/1	22	Steel Frame	21 [21]	0.04g H 0.02g V	0.17g H	9.2 H 3.4 V	57.4 H	3.2 H 1.0 V	19.2 H
Saratoga - West Valley College Gym	1/0	11	RC Columns & Shear Walls	30 [30]	0.10g H 0.03g V	0.41g H	5.8 H 2.5 V	18.2 H	1.0 H 0.6 V	1.1 H
Watsonville - Telephone Bldg.	4/0	13	RC Shear Walls	45 [30]	0.11g H 0.08g V	0.33g H	9.1 H 3.9 V	16.2 H	1.8 H 0.7 H	2.2 H
Hollister - Glorietta Warehouse	1/0	13	Perimeter Shear Walls	57 [37]	0.08g H 0.30g V	0.25g V	7.3 H 11.7 V	22.4 H	3.1 H 0.8 V	3.5 H
So. San Francisco - Kaiser Medical Center	4/0	11	Steel Frame	78 [78]	0.03g H 0.02g V	0.26g H	3.9 H 1.3 V	19.5 H	0.7 H 0.3 V	2.4 H

* - Distance given relative to the epicenter, 37.309 N, 121.678 W. Bracketed
number is the distance to the nearest point on the fault inferred from
the aftershock distribution.
** - Corrected acceleration, H - horizontal component, V - vertical.

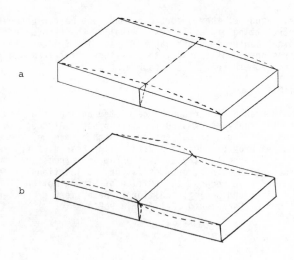

Figure 4. Inferred Mode Shapes of Vibration of the Glorietta Warehouse
during the 1984 Morgan Hill Earthquake.

References

Bennett, J.H. and R.W. Sherburne (1985) (eds.) The Morgan Hill,
    California earthquake, Calif. Div. Mines and Geol., Report SP-68,
    271 pp.

Earthquake Spectra, issue devoted to Morgan Hill earthquake, Vol. 1,
    No. 3, 1985, Earthq. Eng. Res. Inst., Berkeley, California.

Huang, M.J., A.F. Shakal, D.L. Parke, R.W. Sherburne and R.V. Nutt
    (1985), Processed Data from the Strong-Motion Records of the Morgan
    Hill Earthquake of 24 April 1984, Part II. Structural-Response
    Records, Calif. Div. Mines and Geol., Office of Strong Motion
    Studies, Report OSMS 85-05.

Shakal, A.F., R.W. Sherburne and D.L. Parke (1984) CDMG strong-motion
    records from the Morgan Hill, California earthquake of 24 April
    1984, Calif. Div. Mines and Geol., Office of Strong Motion Studies,
    Report OSMS 84-7.

Use of Vibration Tests to Determine Lateral
Load Paths in Medium Rise Buildings

Peter R. Sparks* M.ASCE

Abstract

A structural system identification procedure is described in which an initial analytical model of a building is modified until it predicts accurately the mode shapes and natural frequencies observed in a forced vibration test on the structure. The nature of such tests and possible experimental errors are discussed together with the implications of using an inaccurate structural model when checking the seismic capacity of an existing builidng. The procedure is illustrated by its application to an eleven story reinforced concrete building.

Introduction

When eccentric-mass vibrators were first used to test buildings more than 20 years ago, the primary aim was to determine the natural frequencies, mode shapes and damping of buildings likely to be subjected to earthquakes. Although it was recognized that the results were unlikely to be representative of behavior in a severe earthquake, they were considered a valuable contribution to the body of knowledge of structural behavior and a guide to the accuracy of the dynamic analysis techniques developing at the time.

As buildings became lighter and more slender, their response to the wind became an important serviceability design criterion. Testing programs were undertaken to determine the dynamic characteristics of typical buildings so that realistic values of natural frequencies, mode shapes and damping could be included in design procedures (Jeary and Sparks, 1979, Jeary and Ellis, 1981).

Although tall, slender buildings are the most likely to exhibit an unacceptable response to the wind, the majority of tests using eccentric-mass vibrators were actually conducted on low to medium-rise buildings. This body of data proved particularly valuable later when it was recognized that such buildings could have extremely complex lateral load carrying systems and that the identification of these was an essential step in checking the capacity of the structure.

* Associate Professor, Department of Civil Engineering, Clemson University, Clemson, SC 29634-0911

Structural Systems in Medium-Rise Buildings

In high rise buildings the resistance to lateral loads, wind or earthquake, is of such prime concern that it often dictates the structural form of the building. In addition, such buildings are normally designed by specialists who take steps to separate the structural system from cladding and partitions, both of which are usually kept as light as possible. As buildings become shorter the gravity loads increasingly dictate the structural form, the lateral load carrying system sometimes being incidental to the main structural system. Additionally, cladding and partitions are often substantial and connected rigidly to the main structural system. Initially this was thought to be beneficial, adding stiffness and strength to the structure, but there is growing evidence that in some cases the effect may be detrimental, causing premature failure of members due to unanticipated loads.

This uncertainty about the nature of the actual structural system is of particular concern when the safety of an existing structure is considered in an area where, although there is a moderate risk of earthquake damage, buildings have not traditionally been designed for seismic effects.

Forced-Vibration Tests

In a typical forced vibration test, the vibrator is attached to the building at a suitably high position and a search is made for natural frequencies. The vibrator is started and set to some low frequency, below the lowest expected natural frequency, but high enough for the vibrator to generate sufficient force for the response to be monitored. The generated frequency is increased incrementally and accelerometers, placed strategically around the structure, are used to note the maximum responses which occur when the vibrator and structure are in resonance at a natural frequency.

In the next phase of the investigation a detailed examination is made of each mode. This involves setting the frequency of the vibrator to a previously noted resonance, then placing an accelerometer at various locations in turn to allow a comparison to be made of its response with that of a reference accelerometer placed near the top of the building. At the end of the investigation of each mode the vibrator is suddenly stopped and a measurement of damping is made from the ensuing decay of oscillations.

While the structural stiffness can be computed directly from these measurements, such computations rely on accurate measurements of force, acceleration and damping. Unfortunately these measurements can be subject to considerable experimental error for the following reasons. Firstly, the force causing the response may, in fact, be larger than that deduced from the arrangement of masses on the eccentric-mass vibrator. Particularly when a small vibrator is used, considerable excitation may come from other sources such as the wind. Secondly, when acceleration levels are small, any tilting of the surface upon which an accelerometer is mounted will cause a component of the acceleration of gravity to be included in the accelerometer response, normally assumed to be only the result of translation. Thirdly, measurements of damping are notoriously erratic and have been shown to be dependent on the amplitude of excitation (Jeary and Ellis, 1981). Fourthly, modes may be coupled and this may seriously affect the interpretation of the response.

However, identification of natural frequencies and mode shapes is relatively straightforward and independent of accelerometer calibration and floor tilt. In addition, these measurements are virtually independent of amplitude, for, although Jeary and Ellis have noted that a fifteen-fold increase in amplitude of response can produce a 30% change in damping, the corresponding frequency change was found to be less than 3%.

For these reasons only the observed mode shapes and natural frequencies are used in a system identification procedure rather than a direct use of a derived stiffness.

System Identification Techniques

In recent years a number of system identification techniques have been proposed which attempt to provide a direct mathematical approach to the determination of the physical characteristics of a structure, based on observed dynamic characteristics or reponse to a given dynamic loading. Whilst theoretically attractive, these procedures have not generally been found suitable for buildings, since they require the identification of large numbers of modes and natural frequencies which are not normally available from typical vibration tests.

The approach that has been adopted in this case is to develop as good an analytical model as possible based on the apparent structural system. This is then modified in an iterative manner, to include the effects of "non-structural" partitions, shear lag and stiff cladding, until the observed natural frequencies and mode shapes are predicted by the analytical model. Particular care is taken to match the mode shapes since experience has shown that matching natural frequencies without regard to mode shapes can lead to serious errors in the prediction of load paths (Mirtaheri, 1982, Sparks and Mirtaheri, 1983).

Once an accurate analytical model has been obtained any type of lateral load can be applied and the distribution of forces between members can be determined. Parametric studies can also be conducted to determine the sensitivity of the lateral load carrying system to the effects of partitions and cladding and the inefficiency of members due to shear lag.

Illustrative Example

The building shown in Figure 1 is typical of many designed solely for wind loads. It is a reinforced-concrete office building rising eleven stories above the ground. There are three elevator cores,

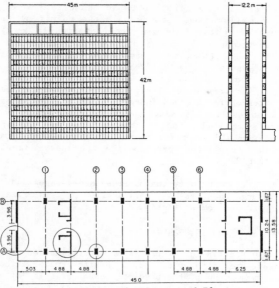

Figure 1.   Building Elevations and Plan

three pairs of flat shear walls and two rows of columns.   Beams
connect the columns in the longitudinal direction and the floors are
ribbed in the transverse direction, except in the elevator lobbies
where they are solid.   The central section of the building is open
plan but there are some concrete block partitions in the end areas.
The arrangement of shear walls is clearly designed to resist the
large transverse wind loads with little attention being paid to the
longitudinal load bearing characteristics.   Unfortunately in an earthquake significant lateral loads are likely to occur in the longitudinal direction.   A building such as this would therefore warrant
special attention in any seismic safety check.

No design calculations could be traced for the building but, considering its age, it is unlikely to have been designed using a computer and the lateral loads in the longitudinal direction were
probably assumed to have been carried by the elevator cores.   In a
seismic safety check a three-dimensional dynamic analysis might be
made using the apparent structural system in a frame and shear wall
program such as TABS77.   A typical analytical model for the longitudinal direction might be to assume two frame lines, each containing
the exterior columns, three shear walls and 1-1/2 elevator cores. Use
of this arrangement actually predicted a natural frequency of 1.00Hz
and a mode shape shown as CX in Figure 2.   In a forced vibration
test, using a small eccentric-mass vibrator, the measured natural
frequency was 1.25Hz with the other mode shape shown in Figure 2
(Jeary and Sparks, 1979).   The correct frequency and mode shape were

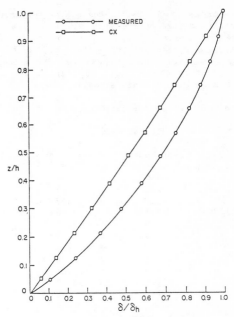

Figure 2. Mode Shapes

eventually obtained from an analytical model in which the end floor
slabs had been stiffened to account for the effect of the block
partitions and the stiffness of the elevator shafts had been reduced
to account for shear lag. Details of the sensitivity of natural
frequencies and mode shapes to these actions can be found in
Mirtaheri (1982), but basically floor stiffening corrected the
frequency and adjustments to the elevator cores corrected the mode
shape.

In the most sophisticated of seismic safety checks a representative ground acceleration record would be applied to the analytical
model and the maximum forces in each member determined. In order to
judge the effect of incorrect modelling on these peak forces, a
ground acceleration proportional to the 1940 El Centro earthquake was
applied in the longitudinal direction, both to the original analytical model and that which predicted the most accurate fundamental
natural frequency and mode shape. In doing this it was assumed that
the structure would remain in the elastic range during excitation and
that the level of damping would be as measured in the forced vibration test. As such, the results reflect the effects of errors in
earthquake loads associated with incorrect natural frequencies and
mode shapes as well as errors associated with the incorrect distribution of loads due to an inappropriate structural model.

Table 1 shows the extent of the errors in load specification for
the three structural elements ringed in Figure 1. They are expressed

as ratios of the forces predicted by the use of the accurate structural model to those predicted by the initial structural model.   It
is clear that the end shear walls, by virtue of their stiff connection to the elevator cores, are attempting to carry much of the load
shed by the elevator cores due to their reduced moment-carrying
efficiency brought about by shear lag. Apparently the columns are
being relieved of some of their loads as well.

Table 1 - Ratios of Earthquake Forces

End Shear Wall

Level	Bottom Moment	Top Moment	Axial Force	Shear Force
11	3.28	2.01	1.86	2.38
10	4.83	3.33	2.23	3.82
9	4.71	3.06	2.42	3.58
8	4.79	3.52	2.56	3.93
7	4.77	3.98	2.69	4.20
6	4.83	4.69	2.81	4.62
5	5.09	5.73	2.93	5.42
4	5.45	6.67	3.06	6.04
3	5.64	7.93	3.21	6.77
2	6.27	11.82	3.37	8.15
1	2.31	2.18	3.48	4.02

Elevator Core

Level	Bottom Moment	Top Moment	Axial Force	Shear Force
11	0.19	1.27	3.10	0.49
10	0.16	0.54	2.97	1.38
9	0.23	0.51	2.55	1.07
8	0.32	0.57	1.92	0.90
7	0.48	0.67	1.53	0.85
6	0.78	0.69	1.25	0.86
5	0.56	0.76	1.10	0.96
4	0.49	0.81	1.00	1.07
3	0.44	1.03	0.94	0.97
2	0.40	0.74	0.88	0.86
1	0.74	0.34	0.83	1.08

Column A2

Level	Bottom Moment	Top Moment	Axial Force	Shear Force
11	0.70	0.70	0.66	0.70
10	0.74	0.74	0.70	0.74
9	0.74	0.74	0.58	0.75
8	0.74	0.74	0.50	0.75
7	0.75	0.75	0.67	0.76
6	0.79	0.79	0.53	0.80
5	0.86	0.86	0.50	0.87
4	0.96	0.96	0.51	0.97
3	1.06	1.05	0.55	1.06
2	1.18	1.16	0.63	1.19
1	1.70	1.57	0.76	1.63

Discussion

Earlier it was observed that the elevator cores were probably assumed to be carrying all of the lateral loads applied in the longitudinal direction and thus the shear walls were probably not designed for bending about their weaker axes. It is interesting to note that the exterior shear walls of this building are, in fact, badly cracked in a manner consistent with the bending about those axes, presumably caused by the action of wind loads.

It might be argued that in the event of a severe earthaquake in the longitudinal direction the shear walls would fail and the load would then be carried as originally envisaged by the elevator cores and columns. However, an earthquake might excite the structure in both the longitudinal and transverse directions. This building relies heavily on the shear walls for its resistance to transverse loads and these might not be in a condition to carry such loads if the building were to be subjected to simultanerous excitation in both directions.

As it happens, this building is located in a part of the world where the seismic risk is low, however it is constructed in a similar manner to many in the eastern United States which could be subjected to moderate earthquakes. It has been shown that long, thin buildings of this type which have been designed only for wind loads are often seriously underdesigned even for moderate earthquakes in the longitudinal direction(Sparks, 1984, Schatte, 1985). Any seismic checking of these buildings would therefore probably concentrate on the longitudinal direction. Such errors in the identification of load paths as indicated in Table 1 could have a serious effect on any decisions concerning remedial action.

Conclusions

This paper has pointed out some of the difficulties in using eccentric-mass vibrators to conduct forced vibration tests on structures. It has, however, also been pointed out that certain parameters, namely mode shapes and natural frequencies can be obtained easily and with a high degree of accuracy. When this information is used to correct analytical models of the structure, a better estimation of the lateral load paths can be made and the effects of partitions, cladding and shear lag can be quantified. The importance of a good analytical model in assessing the capacity of an existing building has been illustrated by showing the errors in assessing the loads carried by individual members when an inappropriate model is used.

Since forced vibration tests can be conducted relatively easily and their use can result in a much improved understanding of the structural system, it would seem wise to conduct such tests routinely in any situation where the ability of a structure to carry either wind or earthquake loads is being checked.

Acknowledgements

The experimental part of this work was carried out in the United Kingdom while the author was a member of the Structural Design Division of the Building Research Establishment. The assistance of A. P. Jeary and B.R. Ellis of that establishment and C. Williams of Plymouth Polytechnic is gratefully acknowledged.

The analytical work was carried out at Virginia Polytechnic Institute and State University and supported by the National Science Foundation under grant CEE 88105784, M. P. Gaus, Monitor. This support and the assistance of M. Mirtaheri, a graduate student at V.P.I. and SU at the time, is also gratefully acknowledged.

Appendix - References

1. Jeary, A.P. and Sparks, P.R., "Some Observations on the Dynamic Sway Characteristics of Concrete Structures", part of "Vibrations of Concrete Structures", Publication SP-60, American Concrete Institute, Detroit, 1979.

2. Jeary, A.P. and Ellis, B.R., "Vibration Tests on Structures at Varied Amplitudes", Proceedings of the Second Speciality Conference on Dynamic Response of Structures: Experimentation, Observation, Prediction and Control, Atlanta, GA, 1981, American Society of Civil Engineers, New York, NY, 1981.

3. Mirtaheri, M. "Assessment of Lateral and Torsional Stiffness Characteristics of Medium Rise Concrete Buildings", Ph.D. Thesis, Virginia Polytechnic Institute and State University, Blacksburg, VA, 1982.

4. Schatte, D.E., "A Comparison of Wind and Earthquake Design Requirements for a Concrete Shear-Wall Building', M. Eng. Report 10S-85, Department of Civil Engineering, Clemson University, Clemson, SC, 1985.

5. Sparks, P.R. and Mirtaheri, M., "The Influence of Structural Performance on the Distribution Wind Loads on Buildings", Journal of Wind Engineering and Industrial Aerodynamics, Vol. 13, No. 1-3, December, 1983.

6. Sparks, P.R., "Earthquake - How Will South Carolina's Buildings Fare?", The South Carolina Engineer, Vol. 34, No. 2, Spring, 1984.

Explosion Hazards to Urban Structures

A. Longinow,* A. Wiedermann,** H. S. Napadensky,*** T. Eichler**

Manufacture, storage and transportation of chemicals capable of exploding at energy levels comparable to TNT has increased in recent years. Land areas adjoining such facilities and transportation routes are now more densely populated and people in these areas are at a higher risk than previously. This paper focuses on these problems and suggests solutions so that the potential for damaging effects can be considered and mitigated by planners in the vicinity of hazardous facilities and transportation routes.

Introduction

There is a need to re-evaluate current regulations governing the siting of "inhabited buildings" relative to stores of hazardous substances capable of exploding and relative to transportation routes along which such substances may be moved.

This paper reviews the general problem of explosion hazards to inhabited buildings due to both military explosives and commercial products capable of exploding. The following areas are covered.

1) Existing DOD regulations are compared with results of full-scale explosions obtained from accident and field tests; 2) The mechanics of a vapor phase explosion model are discussed; 3) The problem of risk to inhabited buildings from accidental explosions is briefly addressed, and 4) Current state-of-the-art of predicting the response of buildings when subjected to blast effects produced by such explosions is briefly discussed.

Existing Quantity - Distance Criteria

Existing DOD quantity-distance regulations (4) used for the specification of minimum separation distances between various types and quantities of stored explosives and surrounding communities have evolved primarily from studies of actual blast damage results (1,2). Each separate explosion was analyzed with respect to quantity of explosive, nature and extent of damage produced. Each result was plotted on a quantity-distance chart, and a curve was finally drawn to

*   Consultant, Wiss, Janney, Elstner Associates, Inc., 330 Pfingsten Rd., Northbrook, Illinois 60062.
**   Principal, AT/Research, 94 North Main Street, Glen Ellyn, Illinois 60137.
*** Director of Research, Explosion Science and Engineering, IIT Research Institute, 10 West 35th Street, Chicago, Illinois 60616.

form an envelope on the safe side of the data points. Separation distances for quantities of explosives for which no experimental (or accident) data existed were determined by scaling existing data. In these charts damage is expressed in terms of quantity of explosive and distance to the target and is primarily related to the overpressure at the site of the target.

Current regulations (4) governing storage of explosives permit conventional, inhabited buildings to be located at distances of from $40W^{1/3}$ to $50W^{1/3}$ ft ($16Q^{1/3}$ to $20Q^{1/3}$ m). Where W is the weight of explosive in pounds of TNT and Q is in kg. The larger number applies to quantities greater than 250,000 lbs of TNT. The value of $40W^{1/3}$ represents approximately the range to the 1.2 psi contour. The increase from 40 to 50 attempts to consider the effect of longer pressure durations of higher yields on damage production. How this regulation reflects the actual response of buildings subjected to blast loads in theory and practice is discussed next.

TABLE 1 - SAFE SEPARATION DISTANCES, PEAK OVERPRESSURES AND
IMPULSES FOR INDICATED QUANTITIES OF EXPLOSIVE

Target	Quantity of Explosive (lbs)											
	1,000			10,000			100,000			1,000,000		
	R	P	I	R	P	I	R	P	I	R	P	I
House (split, level) frame-brick)	290	1.79	30	776	1.36	54	2089*	1.0	92	4500	1.0	190
Church (A-frame construction)	250	2.15	35	1185*	0.77	35	3481*	0.52	54	8000*	0.48	106
School (one-story, masonry)	180	3.64	48	496	2.49	84	1671	1.36	116	5500*	0.77	155
Office Bldg. (multistory, R/C frame, block walls)	140	5.68	61	323	4.60	128	928	3.0	205	2500	2.15	340
Mobile Home	340	1.45	26	790	1.23	50	1900	1.08	102	4950	0.82	172
Bus (Passenger)	230	2.45	38	550	2.10	72	1280	1.90	151	3100	1.59	272
Camper Pick-up Unit	250	2.20	35	620	1.75	64	1520	1.45	128	4000	1.10	212
Commercial Jet Aircraft	200	3.00	43	539	2.15	73	1253	1.97	155	2700	1.97	315
Person (Male, 168 lbs)	230	2.5	38	496	2.50	80	1068	2.50	180	2300	2.5	370
DOD minimum distance requirements (ft):												
(a) blast hazard	400			862			1855			5000		
(b) blast and fragment hazard	1250			1250			1855			5000		

R = Separation Distance, ft
P = Peak Overpressure, psi
I = Impulse, psi ms
1 ft = 0.3048 m, 1 psi = 6.89 kPa
* "safe" distance is greater than that required by regulations.

A study reported in (3) was concerned with establishing safe separation distances for a number of potential civilian targets with respect to five quantities of explosives. Results for nine of the targets are given in Table 1. Included are safe separation distances, peak overpressures and impulses.

Results were generated, based on experimental results and analysis, by first determining what constitutes damage resulting in injury to occupants. The greatest separation distance was then determined which for the given quantity of explosive would just preclude the occurrence of this level of damage.

Separation distances based on current regulations (1) are included for comparison in the lower part of Table 1. It will be noted that in five of the cases (see asterisks in Table 1), the separation distances stipulated by the current regulation are inadequate. This occurs mostly for buildings with large span roofs. It is noted that potentially lethal effects of debris from failing neighboring buildings were not considered in the analysis described.

Figure 1 is a compilation of data (2) from accidental explosions, categorized by types of damage. Many different weight quantities are involved. It is seen that even at a scaled distance of 50 ft/lb$^{1/3}$ (20 m/kg$^{1/3}$) all levels of damage except "demolished" are found. Although most of such exceedences are found in the lesser damage categories, it is important to emphasize that low blast intensities cover larger land areas than do high blast intensities. The potential for damage and injuries in the larger areas of low-level blast effects can, therefore, be very significant.

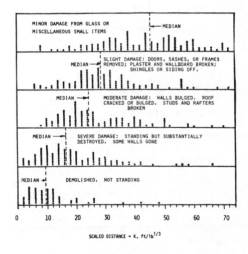

Fig. 1 - Catagories of Damage

The only other agency which stipulates separation distances for explosives is the NRC (U.S. Nuclear Regulatory Commission). NRC has issued a regulatory guide (9) which describes a method for determining distances from critical plant structures to a railway, highway, or navigable waterway beyond which any explosion that might occur on these transportation routes is not likely to have an adverse effect on plant operation or to prevent a safe shutdown. The safe distance is defined by $R > KW^{1/3}$. Here, R is the distance (range) and W is the weight of the charge as previously defined. The factor K = 45 when W is in pounds and K = 18 when W is in kg. For vapor clouds, NRC suggests a mass equivalence of 240 percent. This guide is limited to solid explosives and hydrocarbons liquified under pressure.

Quantification of the Free-field Explosion Environments

The necessary conditions for a vapor cloud explosion to occur are the formation of a mixture of vaporized fuel (generally hydrocarbons), and air within specific concentration limits, and the presence of a suitable ignition source. The mobility characteristics of vapor clouds, and the spreading and transport of liquid pools which generate some of the vapor, can move the explosion source significant distances from the release site, and with the addition of the far reaching effects of the explosion environment itself can result in extending the hazard to significant distances from the accident site.

The generation of most vapor cloud explosion environments require, as a minimum, a specific sequence of events. These events are:

o   Release dynamics
o   Pool dynamics and/or vaporization
o   Vapor cloud formation and dispersion

These phenomena are discussed in the following paragraphs.

Release Dynamics - The physical state of the hazardous material at the time of release can be varied, however, with the exception of process systems where specific material states must be met, the preferred storage and transport state is the liquid state since this optimizes the quantity per unit volume requirement. To achieve this storage state for most materials, the material must be cooled (at essentially atmospheric pressure) - the cryogenic state, or pressurized (at essentially ambient temperature) - the pressurized state; some, but few, materials will be in a liquid state at ambient pressure and ambient temperature conditions - the naturally liquid state.

The accidental failure of the containment system (which may involve a shell puncture or tear, pipe connection break, etc.) will result in the release of the material to the atmosphere over some period of time. The release of liquids is readily determined in most cases since these containment systems are usually vented and, thus gravity is the primary driving force.

The release of fluids from pressurized containment system is complex since this process is dependent upon a variety of

thermodynamic properties and usually results in the development of
two-phase flow. A simple, fluid behavioral model (10) was developed
by means of which the transient mass flow rate efflux can be evaluated
for virtually all fluids of this class, and the influence of other
parameters can also be determined (11). An important characteristic
of the release flows from pressurized containment systems is that, in
many instances, the two-phase exit state is a sonic state, and that
the additional expansion of the fluid from this state to the ambient
pressure level results in a significant mass fraction of the efflux
being in the vapor phase (approximately 10 to 30 percent quality).
This vapor, referred to as direct vapor production, is immediately
available for the formation of a vapor cloud. Additionally, this
vapor will generally be heavier than the surrounding ambient air since
the molecular weight of these materials is usually greater than that
of air and the temperature of the vapor will also generally be colder
than the ambient air due to the expansion (i.e., the decompression)
process. The remainder of the efflux will appear as a cold liquid and
spill on the ground.

Pool Dynamics and Vaporization - The liquid efflux which spills on the
ground will spread and vaporize due to both thermal and wind effects.
The spreading will be influenced and/or controlled by topographical
features such as shorelines for water spills and dikes and ditches for
ground spills. Additionally, thermal effects, which can be quite
complex and stochastic in nature, will play a significant role in the
spreading and vaporization of the liquid hazardous material. Finally,
a variety of other effects, such as surface roughness (puddling
effects) for ground spills and pool breakup effects for spills on
water, as well as many site specific parameters are important.

Vapor Cloud Formation and Dispersion - The formation and initial
characteristics of vapor clouds will depend upon the two basic vapor
production processes, i.e., direct vapor production from the release
point and pool vaporization which is generally of significant areal
extent and sometimes quite mobile. Clearly wind, both local and
general, as well as the atmospheric stability are important
parameters. The nature of the cloud and its dispersion will also
depend upon the relative buoyancy of the vapor and the ambient air.
Most clouds which result from the release of hazardous fluids will be
negatively buoyant and initially move along the ground in a somewhat
global manner until air entrainment dillutes the cloud sufficiently
such that it becomes neutrally buoyant. Neutrally buoyant clouds are
controlled by rather well understood turbulent diffusion mechanisms
and are commonly treated by rather conventional atmospheric dispersion
models. Vapor clouds which are positively buoyant due either to
elevated temperature effects and/or low molecular weight will rise,
primarily vertically, until a stable elevation is reached, after which
an essentially neutrally buoyant behavior will be exhibited.

Explosion Environments - The evaluation of the formation, growth, and
transport details of potentially detonable vapor clouds provides the
shape, size, location, and concentration distributions (both spatially
for quasi-steady cloud formations and time-wise for transient cloud
formations) such that in-cloud, near and far field explosion

environments, can be ascertained. Of particular importance is the observation that negatively buoyant vapor clouds are frequently very thin and flat such that the explosion environment is significantly influenced by the geometry of the cloud. The explosion environment resulting from such cloud geometries was evaluated (12) and somewhat reduced far field explosion environments resulted, both with respect to the overpressure and the impulse as compared to spherical source configurations. For hemispherical vapor cloud geometries the conventional TNT equivalency approach or spherical detonable cloud methodologies are available. Clearly, the point of ignition is important since these locations are frequently along the outer edges of the cloud. Thus, strong directional effects can be expected to occur in many applications.

    In order to obtain some measure of validation, the above solution methodologies were applied to a major industrial accident event which occurred at Flixborough, England in 1974 (5). The sudden double ended rupture of a 27-in. diameter pipe filled with pressurized cyclohexane rapidly released a large quantity of liquid and vapor. The initial negatively bouyant vapor cloud was ignited 54 seconds after the start of the release and about 50 tons of cyclohexane vapor was involved in the resulting explosion event. Since many of the parameters of this accident were unknown or unavailable, nominal values were assumed when needed, and their use may have some impact on the validation attempt. Results were compared with corresponding estimates made from damage assessment data. Comparisons, both with respect to detonable cloud size and location and the overpressure details are very good (Fig. 2). These comparisons as well as results of sensitivity studies allow two conclusions to be reached, namely, 1) the above sequence of solution methodologies is <u>adequate</u> and, 2) the explosion environment is somewhat <u>insensitive</u> to some of the parameters and phenomenological events which are involved. Finally, it should be noted that the geometry of the detonable vapor cloud is important in defining the explosion hazard.

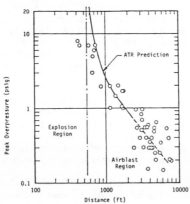

Fig. 2 - Peak Overpressure - Distance Data for Flixborough Event

The Problem of Risk

Explosive incidents can result in or near urban areas during normal transportation of hazardous materials, or during storage of such materials. The extent of the problem is illustrated by an accident which occurred in East St. Louis, Missouri on January 22, 1972 (6). A railroad car filled with propylene was punctured by another car during normal operation in a railroad yard. The released propylene formed an unconfined vapor cloud which subsequently exploded. In addition to damage in the railroad yard, some 650 homes were damaged, and an estimated $6.5 million of damage resulted. Strehlow reviewed 108 such explosions over a 42-year period. Aside from (7) and individual accident reports, no attempt appears to have been made to establish the risk of such accidents to inhabited buildings and other structures.

Loading and Response of Structures

The problem of loading and response of structures when subjected to blast loads can be very complex depending on the size of the structure, its makeup, response range and the character of the load. The need for simplified but accurate methods of analysis is acute.

During the past several decades a fair amount of attention has been focused on the design and analysis of protective structures against blast effects of conventional explosives, primarily TNT. Methods for predicting loads exist (8). Simplified structural analysis methods, including failure criteria also exist, but mostly for simple geometries and only for reinforced concrete and steel. There is a need to extend this methodology to more complex geometries and materials found in conventional buildings.

When it concerns vapor cloud explosions, very little information exists that may be directly used for predicting a reasonable load-time-geometry history given a potential threat. A great deal remains to be done to develop a load prediction methodology at least at the level of (8).

Conclusions and Recommendations

There is increasing pressure for land adjacent to hazardous facilities, previously used for agricultural purposes, to be used for commercial development. Land developers should be aware of the potential for damage near chemical and explosive manufacturing and storage facilities and along transportation routes where hazardous material may be carried.

An effort should be undertaken to develop quantity-distance charts on a rational probabilistic basis, taking into account the differences in response from types of structural systems, components, connections, materials, building orientation, etc. In those cases where buildings cannot meet stipulated separation distances, guidelines for the design and retrofitting of buildings to resist accidental explosions should be developed.

Appendix: References

1. Army-Navy Explosives Safety Board, Igloo Tests, Naval Proving Ground, Arco, Idaho, Technical Paper 3, 1945.
2. "Barricade Effectiveness Evaluated from Records of Accidental Explosion," Armed Services Explosives Safety Board, Department of Defense, AD 487554, July 1966.
3. Custard, G. H., et al, "Evaluation of Explosives Storage Safety Criteria," Falcon Research and Development Company, March 1970.
4. "DOD Ammunition and Explosives Safety Standards," DOD Explosives Safety Board, DOD 6055.9-STD, July 31, 1984.
5. Sadee, C., Samuels, D. E., and O'Brien, T. P., "The Characteristics of the Explosion of Cyclohexane at the Nypro (UK) Flixborough Plant on 1st June 1974," Journal of Occupational Accidents, 1 (1976/1977) 203-235.
6. Somes, N. F., "Abnormal Loading on Buildings and Progressive Collapse," National Bureau of Standards, Washington, D.C., NBSIR 73-211, May, 1973.
7. Strehlow, R. A., "Unconfined Vapor Cloud Explosions - An Overview," Report AAE TR 72-1, UILU-Eng. 72, 0501, University of Illinois, Urbana, Illinois, February 1972.
8. "Structures to Resist the Effects of Accidental Explosions," Department of the Army Technical Manual, TM5-1300, June, 1969.
9. U. S. Nuclear Regulatory Commission, Office of Standards Development, Regulatory Guide 1.91, "Evaluation of Explosions Postulated to Occur on Transportation Routes Near Nuclear Power Plants," Rev. 1, February 1978.
10. Wiedermann, A. H., "An Expansion Isentrope Formulation for Two-phase Fluids," 1982 Annual Meeting of the AIChE, Symposium on Basic Aspects of Multiphase Flow Systems, Paper No. 39a, Los Angeles, California, November 1982.
11. Wiedermann, A. H., "The Decompression of Pressurized Vessels Filled with Two-phase Fluids," 1984 ASME Pressure Vessels and Piping Conference, ASME Symposium Publication of Topics in Fluid Structure Interaction, pp 49-56, San Antonio, Texas, June 1984.
12. Wiedermann, A. H., Eichler, T. V., and Kot, C. A., "Air Blast Effects on Nuclear Power Plants from Vapor Cloud Explosions," 6th International Conference on Structural Mechanics in Reactor Technology (SMiRT), Paper No. j10/8, Paris, France, August 1981.

Some experimental results on the dynamic behaviour
of a 40m guyed mast

Giuliano Augusti (') and Vittorio Gusella (")

Abstract

This paper presents the first results and elaborations obtained
from the tests of an experimental 40m tall guyed mast. The instruments
installed to date are a cup anemometer on the mast top and two
accelerometers at an intermediate level, whose readings are recorded on
magnetic tape and elaborated on an HP computer. So far, records due to
impulsive and wind loads have been obtained.

Introduction

In the frame of a programme of theoretical and experimental
studies on wind effects on deformable structures (Augusti, Borri and
Spinelli, 1984; Augusti et al., 1985a, 1985b), it has been possible to
erect on the grounds of the School of Engineering of the University of
Florence, specifically for structural dynamics experiments, a 40m tall
prototype of many guyed mast antennas currently built in Italy.

This paper, after a short description of the experimental
structure and instrumentation, presents the first test results and
their elaborations. Further experiments are in progress at the time of
this writing.

Experimental mast and instrumentation

The mast, 40 m tall, is constituted by a steel truss of triangular
section with 70 cm sides, supported by a spherical hinge at the base
and restrained at two levels (16.50 and 36.00 m) by three guys anchored
to concrete blocks, as diagrammatically indicated in Fig.1. The
vertical elements of the truss are L-profiles 65x65x5 mm; the diagonals
are L-profiles 35x35x5 mm and are connected to the vertical elements at
74 cm intervals. The guys are steel 7-strand ropes of 14 mm external
diameter.

On top of the mast, a wind recording station Windmaster MK II,
constituted by a flag direction detector and a cup anemometer of the
Robinson type, has been installed; the eccitation threshold of the
anemometer is 0.3 m/sec, the sensitivity 0.5 m/sec; the recording range
so far used in the experiments has been 0.7-20 m/sec, with an exit
potential of 50 mV per m/sec (the instrument would allow to double the
recording range, at the expenses of precision).

(') Professor, Universita' di Roma "La Sapienza", Dept. Structural and
Geotechnical Engrg., Via Eudossiana 18, 00184 Rome, Italy;
(") Graduate Student, Universita' di Firenze, Dept. Civil Engrg., Via
di S.Marta 3, 50139 Florence, Italy.

A preliminary dynamical analysis of the mast, by means of the well known SAP computer code, has yielded the mode shapes and corresponding own frequencies shown in Fig.2.

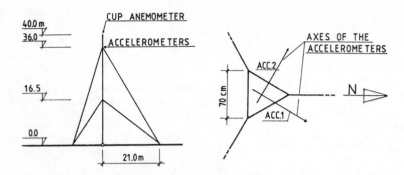

Fig.1: Diagram of experimental mast

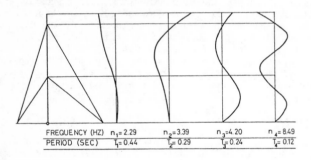

FREQUENCY (HZ)	$n_1 = 2.29$	$n_2 = 3.39$	$n_3 = 4.20$	$n_4 = 8.49$
PERIOD (SEC)	$T_1 = 0.44$	$T_2 = 0.29$	$T_3 = 0.24$	$T_4 = 0.12$

Fig.2: Modes and frequencies of free oscillation of the mast

Only two accelerometers have been available so far for the experiments, and have been placed horizontally at 90 degrees to each other at the level of the upper guy connection (Fig.1). The accelometers are Hottinger-Baldwin (H.B.M.) B12/200, with own frequency 200 Hz and reliable recording range 0-100 Hz; they are connected to an H.B.M. KWS 6-AS amplifier. The readings are stored by an 8-bit 16-channels digital data logger (BAS 161), which can also perform some real time elaborations and in particular draw the power spectrum of the record segment shown on the display (Fig.3). For more accurate elaborations, the data are trasnferred from the data logger to a personal computer HP 150, hence to the Departmental HP 1000 computer.

Procedure for data elaboration

As well known, the wind velocity is a random process, that for each record has been assumed as stationary and ergodic; the same assumption has been introduced for the acceleration process (Augusti, Baratta and Casciati, 1984). The elaboration of the recorded data

follows standard procedures, and in particular those suggested by
Newland (1980). Thus, the linear trend has been eliminated from the
accelometer records with a least square approach, and the spectral
density has been obtained - for both wind and acceleration records - by
means of the so-called "radix-2 FTT" algorithm, introducing a
triangular window on the delay time and an appropriate spectral window
(Newland, cit., p.135); the accuracy of the experimental spectrum has
been improved by averaging 3 or 5 adjacent estimates (ibid., p.138).

Experimental results and discussion

Preliminary experiments were aimed at the determination of the
first (fundamental) frequency of free oscillation of the guyed mast. To
this aim, a cable tied to the mast top was put in tension and suddenly
released. The first segment of the ensuing records of the two
accelerometers are shown in Fig.3, together with the spectral analysis
perfomed in real time by the data logger. It can be seen that the
experimental frequency of 2.3 Hz compares very well with the first
calculated frequency (Fig.2).

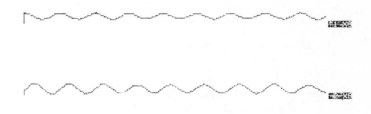

UNIVERSITA' DI FIRENZE-FAC. DI INGEGNERIA-DIPARTIMENTO DI ING.CIVILE
CHANNEL ON: 2*SAMPLING PERIOD:0.01SEC*DISK:Y*THRESHOLD: 1*SIZE: 1*

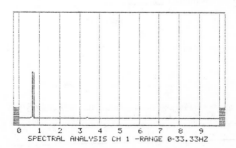

SPECTRAL ANALYSIS CH 1 -RANGE 0-33.33HZ

Fig.3: First interval of accelerometer records and corresponding
        spectral density after impulsive load

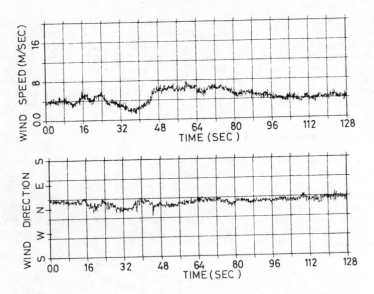

Fig.4: First interval of wind velocity and wind direction histories, recorded on August 2, 1985.

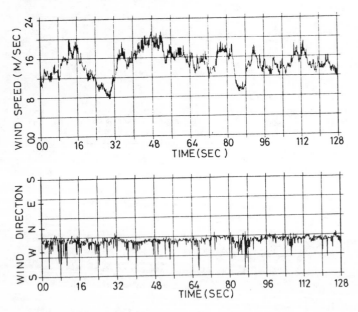

Fig.5: Wind velocity and direction histories, recorded on Aug.6, 1985.

Figs. 4 and 5 show the records of wind speed and direction, obtained respectively on August 2 (average wind speed 4.35 m/sec) and August 6, 1985 (average wind speed 14.7 sec): note that the former recording lasted 410 seconds, of which only the first 128 sec are shown in the Figure, while the whole latter record (of 128 sec duration) is reported. Fig.6 shows the accelerometer records obtained at the same time of Fig.5. For either wind and acceleration records, the sampling interval was 0.1 sec and the corresponding Nyquist frequency 5 Hz.

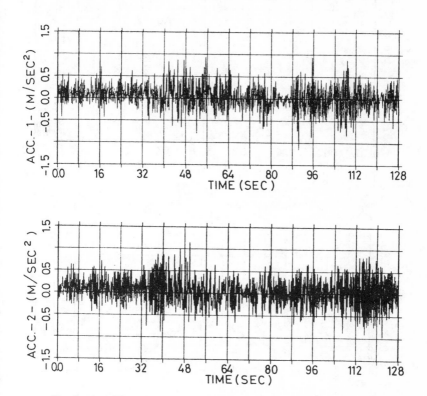

Fig.6: Accelerometer records of August 6, 1985 (cf. Fig.5).

The spectral densities of the wind velocity records reported respectively in Fig.4 and Fig.5, normalized assuming a rugosity length 30 cm, are shown in Fig.7 in comparison with two of the most usual spectral densities suggested by the specialist literature, namely the spectrum valid for the inertial subrange and the Kaimal spectrum (Simiu and Scanlan, 1978, pp.53-54). It can be seen that the agreement is rather bad in the low frequency range: the possible reasons for this disagreement are being investigated. Also, the spectrum of both records shows significant irregularities on the right-hand end of the range: these may be due in part to movements of the mast top, with frequencies of the same order of the fundamental frequency of the mast (2.3 Hz), in part to an aliasing phenomenon, due to the vicinity of the Nyquist frequency (5 Hz).

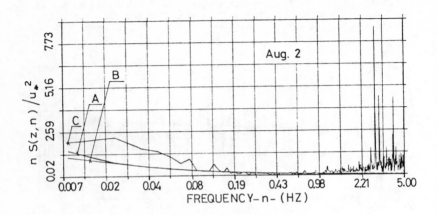

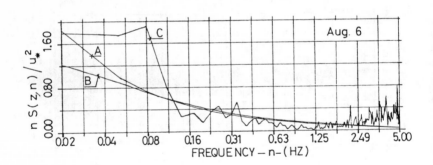

Fig.7: Spectral densities of wind velocity, respectively for average speed 4.35 m/sec (August 2: cf.Fig.4) and 14.7 (August 6: cf.Fig.5):

      (A) Spectral density of inertial subrange;
      (B) Spectrum suggested by Kaimal;
      (C) Spectral density of record.

Finally, Fig.8 shows the spectral densities of the accelerometer records reported in Fig.6. It can be seen that a very concentrated peak exists at a frequency practically coincident with the fundamental calculated frequency.

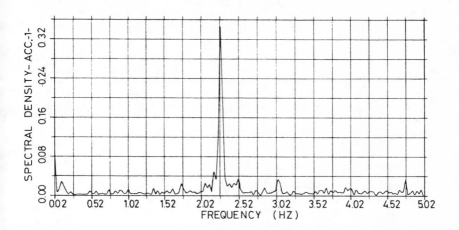

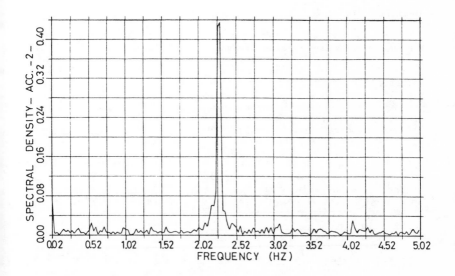

Fig.8: Spectral density of accelerometer records of Fig.6.

Some concluding remarks

Only preliminary results of a current experimental programme have been presented in this paper: further measurements and elaborations are in progress, while instrumentation and data acquisition system are being improved. However, the data appear already of some interest, and an appreciable agreement has been reached between experimental and calculated results.

It is worth noting that our results compare well also with the first results obtained at the Polytechnic of Central London, applying to an analytical model of the same guyed mast a specially developed computer programme which includes numerical simulation of the input wind loading histories and non linear analysis of the mast response, by means of an algorithm based on minimization of the total potential energy of the structural system (Iannuzzi, 1985).

Acknowledgements

The work reported in this paper is financed by a "National Interest Research Programme" of the (Italian) Ministry for Public Education. The support of ELCA Elettro Carpenterie S.p.A. of Milan, who has provided the experimental mast, is gratefully acknowledged. Thanks are also due to Prof.P.Spinelli and Dr.C.Borri for their help during the preparation of the paper.

Appendix: References

G.Augusti, A.Baratta, F.Casciati: Probabilistic Methods in Structural Engineering, Chapman & Hall, London, 1984, pp.xxix+556.

G.Augusti, C.Borri, P.Spinelli: "Interazione dinamica vento-strutture: impostazione del problema ed alcuni risultati"; 12.th Nat. Conf. Ital. Ass. Stress Analysis (AIAS), General Lecture; Naples, Sept.1984.

G.Augusti, C.Borri, L.Marradi, P.Spinelli: "On the time-domain analysis of wind response of structures; 6.th Colloquium on Industrial Aero-dynamics; Aachen, BRD, June 1985; Part 2, pp.191-232.

G.Augusti, C.Borri, A.Iannuzzi, P.Spinelli: "Three-dimensional analysis of wind response of suspension and cable-stayed bridges"; Int. Congress on Theory and Experimental Investigation of Spatial Structures; Int. Ass. Shell & Spatial Structures; Moscow, USSR, Sept. 1985; Vol.I-3, pp.653-667.

A.Iannuzzi: Personal communication, July 7, 1985.

D.E.Newland: Random Vibrations and Spectral Analysis, 3rd ed., Longman, London and New York, 1980; Chapters 10-12.

E.Simiu, R.H.Scanlan: Wind Effects on Structures: an Introduction to Wind Engineering, J.Wiley, New York etc., 1978; Chap.2.

RESPONSE OF THE SAN JUAN BAUTISTA 156/101 BRIDGE TO THE 1979 COYOTE LAKE EARTHQUAKE

John C. Wilson*

ABSTRACT

Strong-motion accelerograms are used to examine the response of a multiple-span highway bridge to earthquake loading. Estimates of modal frequencies and damping values for the first two horizontal modes of the bridge are obtained from the recorded structural responses using a technique of system identification. These values are compared with the results of a finite element dynamic analysis of the bridge. Throughout the study, it has been desirable to extract from the data set as much information as possible, because of the limited earthquake response data available from such structures.

INTRODUCTION

For engineering purposes, the basic source of data on the earthquake response of structures is strong-motion accelerograms. Although many buildings are instrumented with strong-motion accelerographs, and many excellent records have been obtained from these installations, it was not until the mid-1970's that a program of strong-motion instrumentation of bridges and other transportation structures was initiated in California. At the present time five California bridges are instrumented to record earthquake shaking. It is fortuitous that, since the beginning of the instrumentation program, three of these five have yielded significant data, so that now there exists a limited supply of the accelerograms needed to examine the actual seismic response of various types of highway bridges. The work described in this paper is the first investigation of the strong-motion records from one of these structures, the San Juan Bautista 156/101 Separation Bridge. The overall objective of this study is to understand the seismic response of the bridge using the strong-motion data recorded during the 1979 Coyote Lake earthquake.

THE SAN JUAN BAUTISTA 156/101 BRIDGE

The San Juan Bautista 156/101 Separation Bridge is located approximately 3 km (1.8 mi) north-west of the town of San Juan Bautista in San Benito County, California. This two-lane bridge was constructed in 1959 and is typical of the late 1950's - early 1960's style of highway bridge design in the United States. The San Juan Bautista bridge consists of six simple spans of steel girders composite with a reinforced concrete deck, as shown in Figure 1. The spans are simply-supported on two-column, reinforced concrete bents with a fixed bearing at one end of each span (the left-hand end of each span in Figure 1) and an expansion bearing at the other end. The design and orientation of the bearings is such as to allow for longitudinal movement (in a direction parallel to the centerline of the roadway) across the expansion bearings. Foundation support consists of a 7x12x2.5-foot (2.13x3.66x0.76 m) spread footing at the base of each column. Soil tests at the bridge site prior to construction gave Standard Penetration Test values of N of approximately 50, indicating a very dense soil. For later discussions, a global X-Y-Z coordinate system is defined as shown on Figure 1.

---

* Assistant Professor, Dept. of Civil Engineering, McMaster University, Hamilton, Ontario Canada L8S 4L7

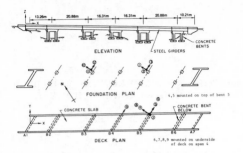

Figure 1.   Strong-Motion Instrumentation on the San Juan Bautista 156/101 Bridge

In May 1977 the San Juan Bautista bridge was instrumented by the Office of Strong Motion Studies of the California Division of Mines and Geology with twelve channels of strong-motion instrumentation, all linked to a central recording system having a common trigger and time signal. Six transducers were placed at ground level to measure the input motions to the structure, three at bent 3 (B3 in Figure 1) and three at bent 5 (B5). The remaining six transducers were placed at various locations on the superstructure as shown by the numbered instrumentation plan of Figure 1. The main shock of the August 6, 1979 Coyote Lake earthquake ($M_L = 5.9$) triggered the system and resulted in the recording of approximately 27 seconds of acceleration on each of the twelve channels. The peak recorded ground acceleration (channel 1) was 0.12g and the peak recorded structural response (on channel 8) was 0.27g (corrected absolute values) with the duration of strong motion lasting about 10 seconds. There was no structural damage to the bridge.

STRUCTURAL IDENTIFICATION

An output-error system identification technique, developed by Beck (1978), is utilized in this study to identify linear models of the form

$$M \ddot{\underline{x}} + C \dot{\underline{x}} + K \underline{x} = \underline{0} \tag{1}$$

which are capable of closely reproducing the observed time-domain earthquake response of the San Juan Bautista bridge when the models are subjected to the same input motion as the real structure.

Beck's technique of structural identification is based upon using a single recorded input and a single recorded structural response to obtain reliable estimates of the parameters which appear in the uncoupled modal equations of motion for planar, linear, structural models. The parameters of interest in this study are frequency f, and damping ratio ζ, for each identifiable mode. An optimal time-domain match between the model output and the recorded response of the real system is achieved by minimizing a normalized integral mean square output error called J, where

$$J = \int_{t_i}^{t_f} (y_0 - y)^2 \, dt \Big/ \int_{t_i}^{t_f} y_0^2 \, dt \tag{2}$$

In Equation (2), γ represents the computed relative displacement, velocity or acceleration of the model in Equation (1), and γ_{ϱ} represents the corresponding response quantity recorded on the structure over the time interval $[t_i, t_f]$. The parameter estimates which minimize J are called optimal estimates and will be denoted as $\hat{f}$ and $\hat{\zeta}$. In practice Beck found that values of J less than 0.1 represented excellent time-domain matches, whereas values of J as high as 0.5 indicated relatively poor matching.

Time-invariant models

One-mode optimal models were determined by matches of displacement, velocity and acceleration over the time interval 0 to 20 seconds. The separate identifications made for motions perpendicular to the bents (N23W, using data of channels 1 and 4), and in the plane of the bents (N67E, using data of channels 3 and 5) are summarized in Table 1. Optimal estimates of modal frequencies are consistent for matches of all three response quantities, in each direction, thereby providing a strong measure of confidence that they are reliable optimal values describing the first two dominant modes of bridge response. In this case, the motion associated with the first mode was polarized in the N23W direction and the motion in the second mode was polarized in the N67E direction. This considerably eased the task of structural identification for this bridge.

TABLE 1. Optimal Time-Invariant One-Mode Models

Direction	Match	$\hat{f}$ (Hz)	$\hat{\zeta}$ (%)	J
N23W	Displ.	3.50	11.0	0.13
	Velocity	3.47	10.3	0.21
	Accel.	3.46	8.7	0.40
N67E	Displ.	6.33	10.2	0.37
	Velocity	6.33	10.0	0.29
	Accel.	6.21	7.5	0.40

It is interesting to note that the best fit in the N23W direction was obtained using displacements, while velocity matching worked best in the N67E direction. Although there is a difference of a factor of approximately two in the minimum J values for the two directions, a visual comparison of actual and computed responses, shown in Figures 2a and 2b, indicates that one-mode time-invariant models performed very well indeed over the 20 second interval. For many of the cycles, the model response (dashed curve) is indistinguishable from the measured structural response (solid curve).

The higher J values for N67E data as compared to N23W data are rather difficult to explain. One possible reason is that the dynamic response of the bridge in the N67E direction was not described as well by models of the class given by Equation (1) as was the response in the N23W direction. It is also possible that a higher mode in the N67E direction contributed to the larger J values. While these factors contribute to a somewhat less precise match for the N67E direction, the results depicted in Figure 2 are, nonetheless, very encouraging.

Time-varying models

To investigate the possibility of changes occurring in the stiffness of the San Juan Bautista bridge during the Coyote Lake earthquake, optimal linear models were determined for five successive time segments, each of four seconds duration. To compare the time variation of modal parameters with results of the time-invariant models, the type of matching was kept the same in both cases. For the N23W components this was displacement matching; for the N67E data velocity matching was used. The results presented in Table 2 indicate that the

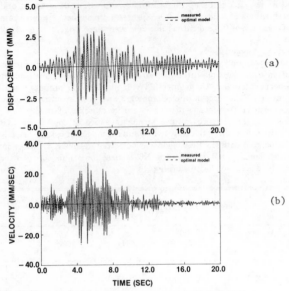

Figure 2. Optimal One-Mode Models (a) N23W, (b) N67E.

frequencies of the two identifiable modes experienced a gradual decrease during the first twenty seconds of the earthquake. In interpreting these results, it should be recalled that the time from 0 to ~12 seconds is of greatest engineering significance since it covers the interval of strongest response. During the first 12 seconds the change in the two modal frequencies amounts to a 2.3% decrease in fundamental frequency (N23W) and a 9.3% decrease in frequency of the second mode (N67E). These percentage changes are similar to those found

TABLE 2. Optimal Time-Varying One-Mode Models						
Time Interval (sec)	$\hat{f}$ (Hz)	N23W $\hat{\zeta}$ (%)	J	$\hat{f}$ (Hz)	N67E $\hat{\zeta}$ (%)	J
0-4	3.53	5.4	0.068	6.85	13.4	0.41
4-8	3.46	12.0	0.088	6.21	7.3	0.22
8-12	3.45	7.4	0.124	6.21	11.0	0.27
12-16	3.62	3.5	0.053	6.76	12.6	0.33
16-20	3.39	3.1	0.035	6.13	8.1	0.40

for time-varying models of the Union Bank building and JPL Building 180 during the 1971 San Fernando earthquake (Beck, 1978). Both buildings suffered only minor damage to nonstructural components.

Damping for the N23W motion shows a large increase during the segment of strong response from 4 to 8 seconds. Over this time, the damping approximately doubled from the initial value of 5.4%, indicating that certain energy dissipation mechanisms in the bridge became activated at the higher levels of response, or alternatively, these mechanisms have a nonlinear response with respect to amplitude. One possible mechanism is an increased energy loss with amplitude through soil-structure interaction. The low-to-moderate levels of damping observed at the beginning and end of the N23W record when excitations were fairly low are probably indicative of the damping that would be observed in the fundamental mode of response during ambient or forced vibration testing. Thus, at low levels of dynamic response one might reasonably expect the bridge to have 3% to 6% damping in the fundamental mode.

Optimal estimates of damping for the one-mode model in the N67E direction tend to maintain a consistently high level throughout the 20 seconds. However, the 13.4% damping in the first time segment seems excessively high and is thought to be a result of a rapid change in frequency over the first few seconds of response. Since the system identification procedure attempts to find a "best-fit" to the changing frequency, the damping will be adjusted to try to make up for deficiencies in the frequency match. The overall effect is to produce a rather poor match over the interval 0 to 4 seconds. This is reflected in the high J value of 0.41.

DYNAMIC ANALYSIS OF THE BRIDGE

A three-dimensional linear elastic finite element beam model of the San Juan Bautista bridge was developed as shown in Figure 3. The left end of each span (orientations as in Figure 1) was modeled as a fixed bearing, having only a rotational degree-of-freedom about the Y axis; the right end was modeled as an expansion bearing having degrees-of-freedom for X translation and rotations about both Y and Z axes. End nodes on adjacent spans were rigidly linked together to provide continuity across the joint for translations in the Y and Z directions and rotations about the X axis. Rotational foundation springs were placed at the base of each column on all five bents, allowing rotation of each column footing about the X and Y axes of the bridge.

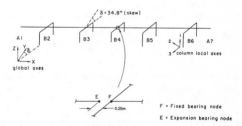

Figure 3. Finite Element Model of the Bridge

Despite the details introduced into the finite element model, the computed response was found not to correlate well with the observed response of the bridge during the Coyote Lake earthquake. In particular, the problem appeared to lie in providing the finite element model with the capability of correctly reproducing the behavior of the expansion joints. The results of the analysis indicated that consideration must be given to a locking effect of the expansion joints, wherein opening and closing of the joints in the X-direction is prevented. The reason that such locking may occur is possibly a result of a certain amount of corrosion at the bearing interfaces, and the accumulation of windblown debris over a period of years. Such locking behavior has been noticed by Douglas and Reid (1982) in full-scale field tests of a bridge. Smith (1983) observed a significant amount of debris in the bearings of the San Juan Bautista bridge during a 1981 field inspection and questioned their capability to move freely (in the intended, longitudinal direction). In a revised finite element model, each span is assumed to have pinned-pinned connections for longitudinal motions to simulate the locking effect. To include an allowance for abutment stiffness in the revised finite element model, a linear translational spring was placed in the X direction at both ends of the deck. Using the results of the system identification analysis as a guide, it was found that a spring stiffness of 3×10^6 lbs/ft (4.38×10^7 N/m) was required in order that the fundamental model frequency match the optimal fundamental frequency of 3.50 Hz. This value of stiffness is somewhat low compared to recommended design values and possibly indicates that the abutment stiffness during the Coyote Lake earthquake resulted from the mobilization of only a small amount of soil resistance due to the low displacement amplitudes.

Mode shapes calculated for the first two horizontal modes are shown in Figure 4 with the corresponding frequencies being: $f_1 = 3.50$ Hz and $f_2 = 6.27$ Hz. These results indicate that, once the fundamental frequency of the model is forced to match the results from system identification, the model also has a second mode at a frequency of 6.27 Hz, very close to the 6.33 Hz frequency found in the system identification. Verification of more than two horizontal modes in the finite element model is not possible with the present data set, as it appears that longitudinal and transverse response in higher modes are indistinguishable from high frequency recording noise.

Figure 4.   First Two Modes of the Bridge Response in the Horizontal Plane

By comparing Fourier amplitudes from the earthquake records with the modal amplitudes at nodes of the finite element model which correspond to instrument locations, it is possible to obtain an indication of how well the model simulates the actual seismic response of the bridge. In Table 3, values of fundamental modal amplitudes computed from the finite element model are compared to Fourier spectral amplitudes of the fundamental mode and are found to be in reasonably good agreement. Each amplitude has been normalized with respect to the amplitude in the X direction at the top of bent 5 and the motions at XB5 and XD5 and the motions at YB5 and YD5 are both in-phase, respectively. In this respect, the model, which assumes deck joints locked in the X direction, provides a reasonable representation of the fundamental mode of response of the bridge.

Dynamic response in the vertical direction

The deployment of strong-motion instruments on the bridge, shown in Figure 1, makes it clear that very little experimental information can be gained concerning the vertical

TABLE 3. Comparison of Fundamental Modal Amplitudes
with Fourier Spectral Data

Component	Normalized Fourier Amplitude	Normalized Modal Amplitude
XB5	1.00	1.00
YB5	0.80	0.75
XD5	0.92	1.08
YD5	1.09	0.80

XB5	= X-motion at top of bent 5; similarly for Y-direction in YB5.
XD5	= X-motion on deck at bent 5; similarly for Y-direction in YD5.

modes of response because there is only one vertically-oriented transducer located close to the end of span 4. However, a Fourier spectrum of the motion at this location did suggest a structural resonance of span 4 at a frequency of 7.13 Hz. In Table 4, values of the fundamental frequency of vertical vibration computed for each span are compared to data obtained during an ambient vibration survey (AVS) of the bridge conducted by Caltrans in April 1981.

TABLE 4. Comparisons of Fundamental Vertical Mode Frequencies

Data Source	Vertical Frequencies (Hz)					
	Span 1	Span 2	Span 3	Span 4	Span 5	Span 6
Finite Element Model	10.59	5.26	7.34	7.32	5.26	not calculated
Caltrans AVS	10.01	5.57	7.81	7.91	5.62	18.65

The vertical frequency of 7.32 Hz computed by the finite element model for span 4 is in good agreement with the 7.13 Hz frequency observed during the earthquake. The ambient vibration frequencies for spans 2 through 5 are from 6% to 8% greater than those predicted by the finite element analysis, possibly as a consequence of some minor amount of rotational restraint existing at the bearings during the low levels of ambient excitation.

Although the vertical seismic responses of bridges such as the San Juan Bautista bridge are not of as great concern to engineers as are the longitudinal and horizontal motions, the close agreement between results from the analysis and from experiments, including both ambient tests and the limited earthquake data, helps provide confidence in the structural idealizations and model synthesis.

CONCLUSIONS

The earthquake response of a six-span highway bridge has been studied using strong-motion records obtained on the bridge during the 1979 Coyote Lake earthquake. The moderate levels of shaking and the undamaged condition of the bridge after the earthquake provided reasonable grounds for assuming linear elastic behavior of structural components. Using time-invariant models, best estimates of frequencies of the first two horizontal modes of the bridge were found to be 3.50 Hz and 6.33 Hz, with associated damping values of approximately 10% of critical in each mode. A study of the time variation of frequencies and damping values, indicated a general trend towards a decrease in the frequency of each mode as the intensity of shaking increased. In the fundamental mode, at low levels of excitation, the damping was found to be in the range of 3% to 6%, but increased to 12% during the time of strongest response. A three-dimensional finite element model of the bridge was able to predict horizontal responses in reasonable agreement with observed values only when the expansion joints were assumed to be locked, thereby preventing relative motion between adjacent spans in the longitudinal direction. Together, the results for horizontal and vertical responses suggested that the bearings were essentially free to work in rotation but not in longitudinal translation. Similar analyses, applied to the strong-motion records from other bridges, should lead to a substantially better understanding of the earthquake responses of these structures.

ACKNOWLEDGEMENTS

The author would like to thank Professors P.C. Jennings and J.L. Beck for their contributions and valuable suggestions in connection with this research. Mr. James Gates of the California Department of Transportation and Mr. John Ragsdale, formerly of the California Division of Mines and Geology, were also very helpful. The financial support of the Natural Sciences and Engineering Research Council of Canada, the National Science Foundation and the California Institute of Technology is gratefully acknowledged.

REFERENCES

Beck, J.L. (1978) "Determining Models of Structures from Earthquake Records", Earthquake Engineering Research Laboratory, EERL 78-01, California Institute of Technology, Pasadena, California.

Douglas, B.M. and Reid, W.H. (1982) "Dynamic Tests and System Identification of Bridges", Journal of the Structural Division, ASCE, Vol. 108, No. ST10, October, pp. 2295-2312.

Smith, M.J. (1983) California Dept. of Transportation, Sacramento, California, personal communication.

Wilson, J.C. (1984) "Analysis of the Observed Earthquake Response of a Multiple Span Bridge", Earthquake Engineering Research Laboratory, EERL 84-01, California Institute of Technology, Pasadena, California.

SEISMIC BEHAVIOR OF LONG-SPAN CABLE-STAYED BRIDGES

Ahmed M. Abdel-Ghaffar[1], M.ASCE and Aly S. Nazmy[2]

ABSTRACT

Special features of the earthquake-response analysis of cable-stayed bridges, subjected to multiple-support excitation, are presented including a basis for calculation of seismic forces and deformations of such long-span, three-dimensional and flexible structures. Significant comments and recommendations for the earthquake-resistant design of such bridges are also presented.

INTRODUCTION

Despite the fact that several cable-stayed bridges with long spans have been built and are planned for construction in seismically active zones, very few, if any, have yet experienced strong earthquake shaking. Accordingly, experience with the earthquake-response analysis of such new, flexible, structural systems is quite limited at present. In addition, the absence of large magnitude and frequent earthquakes in the eastern and central U.S., where several of such long structures are planned, has resulted in complacency, or perhaps unawareness on the part of the general population of the existance of any earthquake threat. Technically speaking, even in geographical areas where seismic risk is assumed to be low as a result of infrequent earthquake activity, the consequences of a large earthquake need careful engineering consideration. Furthermore, it is difficult to estimate or assess the reliability of using the existing specifications for seismic design of highway bridges which are directed toward standard short-to-medium span highway bridges and are not tailored to special bridges such as cable-stayed and suspension bridges. Hence, special detailed analysis and investigation are needed for the earthquake safety and design of cable-stayed bridges. This paper is intended to provide a summary of the earthquake-response analysis of cable-stayed bridges; in addition, significant comments and recommendations for the earthquake-resistant design of such bridges are presented.

SPECIAL FEATURES OF THE EARTHQUAKE-RESPONSE ANALYSIS OF CABLE-STAYED BRIDGES

The relatively flexible and extended-in-plane configuration as well as the three-dimensionality of cable-stayed bridges make them susceptible to a unique class of vibration problems. Unlike classic suspension bridges, vibrations of cable-stayed bridges cannot be categorized as vertical (or bending), lateral (or sway), and torsional; instead, three-dimensional motion is associated with almost every mode of vibration. Figure 2 shows some of the three-dimensional vibrational mode shapes of a typical steel alternate design of a cable-stayed bridge (shown in Fig.

(1)Associate Professor, Civil Engrg. Dept., Princeton University, Princeton, NJ 08544
(2)Research Assistant, Civil Engrg. Dept., Princeton University, Princeton, NJ 08544

2) to be constructed in the southeastern region of the United States.

Furthermore, earthquake excitations introduce special features into the response analysis because of their complicated interaction with the whole structure. For example, vertical vibrations of these structures are introduced to a great extent by both the longitudinal and the lateral shaking in addition to the vertical excitation. In addition, for these flexible structures, understanding of the multimodal contribution to the final response is of extreme importance to provide representative values of the response quantities. Finally, the long-span character of such bridges necessitates the formulation of a dynamic response analysis to deal with multiple-support excitations. Existing strong motion records can be used to define representative and appropriately correlated multiple-support seismic inputs. Some of the ground motion records taken from the Imperial Valley (El Centro), California earthquake (M_L=6.6) of October 15, 1979 [2] are employed in this analysis to define the multiple-input support motion as well as the uniform motion.

BASIS FOR CALCULATION OF SEISMIC FORCES AND DEFORMATIONS

As only oscillations with relatively small amplitudes are of interest, the earthquake-response analysis can be based on a linearization of the force-displacement relation corresponding to the tangent stiffness derived from the dead load condition [3]. The nodal dynamic displacements of the bridge may be decomposed into quasi- (or pseudo) static displacements and relative (or vibrational) displacements [1]. Quasi-static displacements are those resulting from the static application of support displacements at any time (see Fig. 3).

METHOD OF ANALYSIS

1. Equation of Motion

The equations of motion of the three-dimensional vibration of the bridge (with N degrees of freedom) when subjected to multiple-support seismic excitations at the two anchor piers and the two tower bases can be expressed in matrix form as:

$$[M]\{\ddot{u}\} + [C]\{\dot{u}\} + [K]\{u\} = 0 \tag{1}$$

where [M] is the mass matrix of the bridge, [K] is the stiffness matrix and $\{u\}$ is the total vibrational displacement vector. Equation (1) may be written in partitioned form as

$$\begin{bmatrix} M_{ss} & M_{sg} \\ M_{gs} & M_{gg} \end{bmatrix} \begin{Bmatrix} \ddot{u}_s \\ \ddot{u}_g \end{Bmatrix} + \begin{bmatrix} C_{ss} & C_{sg} \\ C_{gs} & C_{gg} \end{bmatrix} \begin{Bmatrix} \dot{u}_s \\ \dot{u}_g \end{Bmatrix} + \begin{bmatrix} K_{ss} & K_{sg} \\ K_{gs} & K_{gg} \end{bmatrix} \begin{Bmatrix} u_s \\ u_g \end{Bmatrix} = \begin{Bmatrix} 0 \\ 0 \end{Bmatrix} \tag{2}$$

where the subscript "g" designates the degrees of freedom corresponding to the points of application of ground motions and the subscript "s" corresponds to all other (fixed-base) structural degrees of freedom of the bridge.

2. General Solution

The nodal displacements may be decomposed into quasi- (or pseudo) static displacements and relative (or vibrational) displacements as follows:

$$\begin{Bmatrix} u_s \\ u_g \end{Bmatrix} = \begin{Bmatrix} u_{ps} \\ u_{pg} \end{Bmatrix} + \begin{Bmatrix} u_{vs} \\ 0 \end{Bmatrix} \tag{3}$$

where the subscript "p" denotes the pseudo-static displacements and the subscript "v" denotes the vibrational displacements.

The pseudo-static displacement vector can be expressed as

$$\begin{Bmatrix} u_{ps} \\ u_{pg} \end{Bmatrix} = \sum_{i=1}^{G} \begin{Bmatrix} g_{psi} \\ g_{pgi} \end{Bmatrix} f_i(t) \tag{4}$$

where g_{psi} is the ith quasi-static functions that result from unit displacement at the ith supporting point; $f_i(t)$, $i = 1,2,...,G$ are the input displacement motions to the supporting points of the bridge; and g_{pgi} is an Gx1 vector whose ith element is equal to unity, with all its other elements being zero.

In this case, the three-dimensional ground motions at the two anchor piers and the two tower bases are taken to be different.

Substituting Eqs. (3) and (4) into Eq. (2) gives:

$$[M_{ss}]\{\ddot{u}_{vs}\} + [C_{ss}]\{\dot{u}_{vs}\} + [K_{ss}]\{u_{vs}\} = -\sum_{i=1}^{G} \Big([M_{ss}M_{sg}]\ddot{f}_i(t)$$

$$+ [C_{ss}C_{sg}]\dot{f}_i(t) + [K_{ss}K_{sg}]f_i(t)\Big)\begin{Bmatrix} g_{psi} \\ g_{pgi} \end{Bmatrix} \tag{5}$$

The quasi-static vectors can be found (or defined) in the following manner [1]:

$$\{g_{psi}\} = -[K_{ss}]^{-1}[K_{sg}]\{g_{pgi}\} \quad , \quad i = 1,2,...,G \tag{6}$$

and thus the term containing the displacement input in the right-hand-side of Eq. (5) vanishes. Note that Eq. (5) is excited by the three-dimensional ground acceleration and velocity terms.

3. Modal Solutions - Earthquake Response

The solution to Eq. (5) is obtained by modal superposition; that is, the vibrational displacement is taken to be

$$\{u_{vs}\} = \sum_{n=1}^{N} \{\phi_n\}q_n(t) \tag{7}$$

where $\{\phi_n\}$ is the nth vibration mode shape; $q_n(t)$ is the nth generalized coordinate; and N corresponds to the total number of degrees of freedom in the finite element model. Substituting Eq. (7) into Eq. (5) and utilizing the orthogonality of the modes yields the governing equation for the nth generalized coordinate

$$\ddot{q}_n(t) + 2\xi_n\omega_n\dot{q}_n(t) + \omega_n^2 q_n(t) = \sum_{i=1}^{G} \Big(\alpha_{ni}\ddot{f}_i(t) + \beta_{ni}\dot{f}_i(t)\Big), \quad n=1,2,...,N \tag{8}$$

where ξ_n is the damping ratio of the nth vibration mode, and the modal participation coefficients α_{ni} and β_{ni} are given by

$$\alpha_{ni} = \frac{-\{\phi_n\}^T[M_{ss}M_{sg}]\{g_{psi}g_{pgi}\}^T}{\{\phi_n\}^T[M_{ss}]\{\phi_n\}} , \quad \begin{array}{l} i = 1,2,...,G \\ n = 1,2,3,...,N \end{array} \tag{9}$$

and

$$\beta_{ni} = \frac{-\{\phi_n\}^T [C_{ss} C_{sg}] \{g_{psi} g_{pgi}\}^T}{\{\phi_n\}^T [M_{ss}] \{\phi_n\}} , \quad \begin{array}{l} i = 1, 2, \ldots, G \\ n = 1, 2, 3, \ldots, N \end{array} \qquad (10)$$

It is important to note that for the purpose of calculating the modal participation factors, β_{ni}, involving the damping matrix $[C]$, the following approach is used. For each mode, the damping matrix is assumed diagonal. Those degrees of freedom corresponding to the translation of the bridge are entered as $[(2\xi_n \omega_n m) \cdot L]$ where m is the mass of the element per unit length and L is the element length. The matrices $[C_{ss}]$ and $[C_{sg}]$ are isolated from the diagonal $[C]$ matrix, and the modal participation coefficients β_{ni} can be calculated using Eq. (10).

The solution to Eq. (8), assuming quiescent initial conditions is given by the convolution integral. In general, the three-dimensional ground motions at the two anchor piers and the two tower bases are taken to be different; thus there are twelve different components of input ground motion for the multiple-support excitation case while there are only three components of ground motion for the uniform case.

4. Sectional Forces and Deformations

The dynamically-induced bending moments, shear forces and normal forces can be calculated from the nodal displacements and the stiffness properties of each bridge element. Figures 4 through 8 display some of these response quantities. It is important to note that the multiple-support seismic excitation can have a significant effect, particularly on the displacement of both the bridge deck (Fig. 5) and the towers (Figs. 7 and 8).

Finally, it should be noted that because relative displacements arise from out-of-phase motion of different parts of the bridge, the design displacement and forces are considered equally important in the design methodology. Furthermore, it was found that the use of energy absorption devices, at the connections of deck-tower, anchor-pier-deck, and the bridge-approach spans, can provide a viable solution for earthquake resistant design.

FINAL COMMENTS AND ANALYSIS-DESIGN RECOMMENDATIONS

1. To make a realistic earthquake-response analysis of cable-stayed bridges, at least three different time histories should be used to assess the seismic safety.
2. There is strong coupling in the three orthogonal directions within each mode of vibration, and, in addition, several modes of vibration will in general contribute to the total response of the structure (Fig. 6). Thus, the three basic features of the earthquake response of cable-stayed bridges are: (i) modal coupling of effects, (ii) multi-modal contribution to the final response, and (iii) multiple-support seismic excitions.
3. Because of the ductility of the structure, the earthquake-induced sectional forces can be carefully reduced.
4. All earthquake-induced sectional forces and deformations are absolute (maximum) spectral values and are completely reversible in sign with the exception of the cable forces which are only considered tensile.
5. The P-Δ moment magnification of the towers should be considered for the dynamic earthquake loading (see Figs. 7 and 8).

6. In order to reduce the differential longitudinal displacement between the bridge deck and the tower bearing at the strut, it is possible to employ "cable restrainers" either vertically between the deck and the strut, or diagonally, between the tower legs and the bridge deck.
7. The characteristics of the soil conditions at the site of the bridge can have a significant effect on the earthquake response and accordingly they should be considered in the analysis.
8. For the approach spans, because free horizontal movement of adjacent spans toward each other is allowed, a bumper device set in the joint would be appropriate to assure the serviceability of the bridge following a large seismic event.

ACKNOWLEDGMENT

This research is supported by the National Science Foundation through Grant No. ECE-8501067. This support is gratefully acknowledged. The data and permission provided by Greiner Engineering Science, Inc. of Tampa, FL, regarding the steel cable-stayed bridge is greatly appreciated.

REFERENCES

1. Abdel-Ghaffar, A. M., Scanlan, R. H., and Rubin, L. I., "Earthquake Response of Long-Span Suspension Bridges," Report No. 83-SM-13, Department of Civil Engineering, Princeton University, Princeton, NJ 08544, May 1983.
2. Brady, A. G., Perez, V., and Mork, P. N., "The Imperial Valley Earthquake, October 15, 1979: Digitization and Processing of Accelerograph Records," U.S. Geological Survey, Seismic Engineering Branch, Open-File Report 80-703, April 1980, Menlo Park, CA.
3. Fleming, J. G., and Egeseli, E. A., "Dynamic Response of Cable-Stayed Bridge Structures," Engineering Symposium on Cable-Stayed Bridges, Federal Highway Administration, Pasco, WA, December 1977.

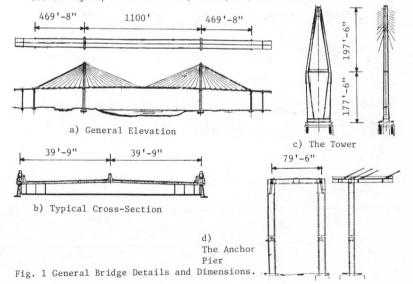

469'-8"   1100'   469'-8"

a) General Elevation

197'-6"

177'-6"

c) The Tower

39'-9"   39'-9"

b) Typical Cross-Section

79'-6"

d)
The Anchor
Pier
Fig. 1 General Bridge Details and Dimensions.

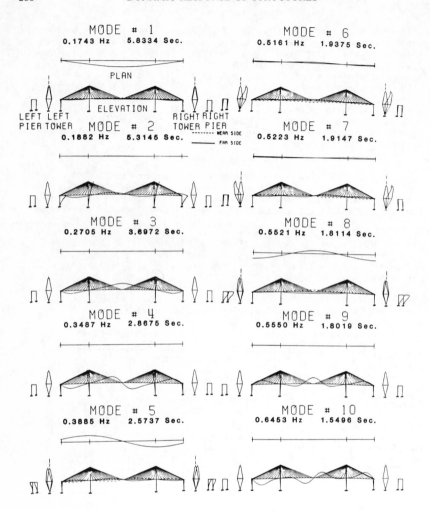

Fig. 2 Some of Computed Three-Dimensional modes of vibration.

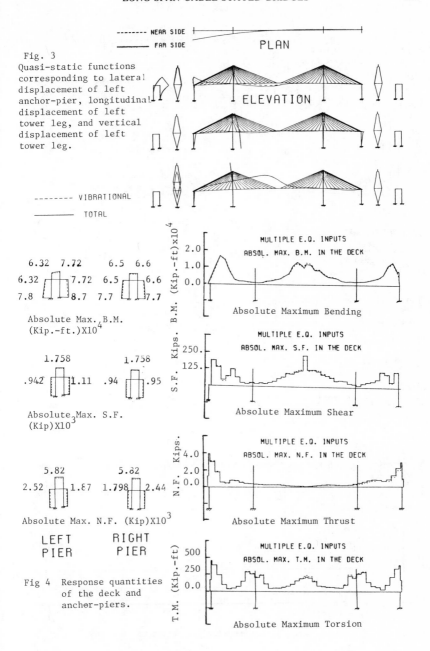

Fig. 3 Quasi-static functions corresponding to lateral displacement of left anchor-pier, longitudinal displacement of left tower leg, and vertical displacement of left tower leg.

-------- NEAR SIDE
———— FAR SIDE
PLAN

ELEVATION

-------- VIBRATIONAL
———— TOTAL

6.32 7.72   6.5   6.6
6.32 ⌐──┐7.72 6.5 ⌐──┐6.6
7.8 ⌊──⌋8.7   7.7 ⌊──⌋7.7

Absolute Max. B.M.
(Kip.-ft.)X10⁴

1.758   1.758
.942 ⌐─┐1.11 .94 ⌐─┐.95

Absolute Max. S.F.
(Kip)X10³

5.82   5.82
2.52 ⌐─┐1.87 1.798⌐─┐2.44

Absolute Max. N.F. (Kip)X10³

LEFT   RIGHT
PIER   PIER

Fig 4 Response quantities of the deck and anchor-piers.

MULTIPLE E.Q. INPUTS
ABSOL. MAX. B.M. IN THE DECK
B.M. (Kip.-ft.)x10⁴
2.0
1.0
0.0
Absolute Maximum Bending

MULTIPLE E.Q. INPUTS
ABSOL. MAX. S.F. IN THE DECK
S.F. Kips.
250.
125.
Absolute Maximum Shear

MULTIPLE E.Q. INPUTS
ABSOL. MAX. N.F. IN THE DECK
N.F. Kips.
4.0
2.0
0.0
Absolute Maximum Thrust

MULTIPLE E.Q. INPUTS
ABSOL. MAX. T.M. IN THE DECK
T.M. (Kip.-ft)
500
250
0.0
Absolute Maximum Torsion

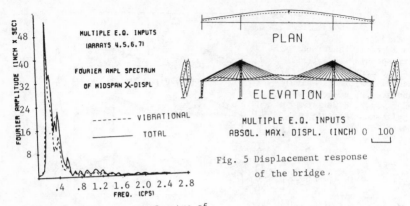

Fig. 6 Fourier Amplitude Spectra of
         mid-span displacement.

Fig. 5 Displacement response
       of the bridge.

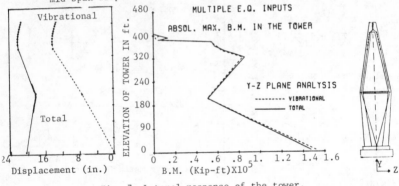

Fig. 7   Lateral response of the tower.

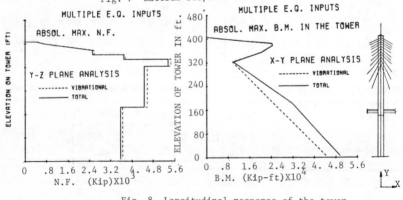

Fig. 8 Longitudinal response of the tower.

Seismic Response Analysis of Two Major Bridge Structures

W.D. Liu*, J. Lea* and R.A. Imbsen**

ABSTRACT

The seismic response analyses of two major bridge structures have been conducted. In these two bridges, portions of the piles are either submerged in the water (Bridge No. 1) or free-standing within the isolation sleeves (Bridge No. 2). Major seismic responses occurred below the footing level. The accuracy of the computed pile forces is essential to the foundation design.

INTRODUCTION

Seismic response analyses of two major bridge structures have been conducted to evaluate the adequacy of their respective designs, in particular, foundation pilings.

Bridge No. 1 is a seven-span approach structure to the Third Lake Washington Floating Bridge (Highway I-90). Five of the seven bents are supported on lake footings with pile groups partially submerged in the water.

Bridge No. 2 is a double-leaf concrete swing bridge located in the Port of Seattle. By providing wider channel clearance (250 feet of channel width and 55 feet vertical clearance above Mean High Water), the bridge will allow 85% of the maritime traffic without opening. The remaining 15% of the shipping traffic will result in an average of seven openings per day.

The common feature of these two bridges is that portions of the pile group foundations extend above the mudline, either in the water (Bridge No. 1) or within isolation sleeves (Bridge No. 2). These piles serve not only as foundation elements but also as structural columns. The force and deformation evaluations of these members should be consistent with other structure members and the modern ductile design concept.

A pile group superelement procedure is employed in the computer program SEISAB [1] to satisfy the dual role of these piles.

* Research Engineer, Engineering Computer Corporation, 3217 Ramos Circle, Sacramento, California.
** President, Engineering Computer Corporation, 3217 Ramos Circle, Sacramento, California

ANALYTICAL PROCEDURE

The Substructure Method is a simple but rigorous approach to the complicated foundation-structure interaction problem. Engineers of different expertises may work on different substructures. Varying degrees of sophistication may be employed for different portions of the total structure-foundation system.

In the seismic response analysis of the soil-structure interaction using the substructure method, the system is divided into the soil-foundation subsystem and the structure subsystem. Appropriate boundary conditions must be imposed to assure continuity and equilibrium between the two subsystems. Three steps are usually required to account for the complexity of the nonlinear soil behavior:

1. Based on assumed soil response, the best estimate of foundation stiffness is computed and used as boundary conditions in Step 2.

2. Conduct the structural response analysis to obtain the maximum force and deformation estimates.

3. Evaluate the forces and deformations in the soil-foundation subsystem.

Iterations may be necessary if the computed soil response in Step 3 deviates significantly from the assumed soil condition in Step 1. For typical highway bridges, this may not be needed if sound engineering judgments and experiences are exercised in Step 1 [2].

Typically, a member is either in the structure subsystem (considered in Step 2) or in the soil-foundation subsystem (considered in Steps 1 and 3). Since foundation failures are not allowed in a design, the emphasis in Step 3 is usually placed on the soil responses.

In the pile group foundations, the stiffness matrix at the pile cap, which includes the effect of all piles in the group and their surrounding soil, is evaluated first. Structural response analysis is then conducted. The maximum displacements or forces obtained at the pile cap are used to analyze the soil-pile-group subsystem. If the analysis is conducted by the Response Spectrum Method, the signs of maximum response quantities are lost in the modal combination process and maxima of different components do not usually occur at the same time. In this method, additional uncertainties are introduced in the computation of pile forces.

If a pile acts also as a structural column, it is important that this pile be designed in accordance with the ductile design concept. To do so, estimations of the pile forces consistent with other structural members are necessary. In other words, the pile forces should be computed directly from the modal responses. This may be accomplished by implementing a pile group superelement in which all individual piles are included. Each pile is represented by a soil stiffness matrix at mudline and beam-column element for the free-standing portion (if any). The pile heads are connected to the centroid of the rigid pile cap through rigid body transformation. Special allowance can be made for pinned connections. To keep the size of the matrix equation at a minimum, all degrees of freedom in the pile group which are not connected to the superstructure may be consistently eliminated by the substructure condensation technique [3]. During the condensation process, a displacement transformation matrix for each pile is formed and saved which may be used later on to compute modal internal pile forces. At this stage, the total pile group foundation is represented by the generalized stiffness and mass matrices at the pile cap.

For the two bridges described later, this expanded feature of SEISAB is particularly appealing. Maximum pile forces and maximum pile displacements at mudline, accounting for the modal correlation and directional uncertainty of earthquake excitations, are obtained directly from the computer output in a single run.

BRIDGE NO. 1

This bridge is the East Approach structure to the Third Lake Washington Floating Bridge. Figure 1 is the schematical idealization of the bridge. Abutment 1 is the east pontoon of the floating bridge, and span 1 is the transition span which is hinge-connected at Pier 9 to the rest of the bridge. The only other hinge is located at Pier 15 between the bent and the continuous superstructure.

Piers 9 thru 13 are lake footings supported on 48 inch diameter steel pipe piles. A typical bent is shown in Figure 2. Column heights vary from 9 feet at Pier 9 to 33 feet at Pier 13. Significant portions of the piles are submerged in the water. The submerged depths at Pier 9 thru 13 are 85, 87.5, 79.5, 75.5, and 35.5 feet, respectively. The number of piles used under Piers 9 thru 13 are 6, 10, 8, 8 and 14, respectively. At these locations, the pile caps (7 feet in thickness), flared columns and the continuous superstructure form a relatively rigid system above the pile caps. It is expected that the major contribution to the seismic response of this bridge will occur below the pile cap. Piers 14 and 15 are supported on the more conventional land-based, embedded spread footings.

As shown in Figure 2, the deck superstructure is actually a twin box with continuous overhang slab. To verify that a single unit representation of the deck (Figure 1) is appropriate, an independent 3-D space frame model was established which models the deck as two parallel units with cross beam. Because of the rigidity above the footing, the difference between the two idealizations were negligible.

The subsurface material under the lake consists of 10 to 28 feet of loose material overlying on very dense silt, sand, gravel with cobbles and boulders. The shear wave velocity of these dense material is about 3000 ft/sec. Neglecting the loose surface layer, this site condition fits into the general description of Soil Type I in the AASHTO design guidelines, and the corresponding elastic response spectrum with maximum ground acceleration of 0.32g was used in the analysis.

The previously described methodology is used to idealize the structure including all piles. The hydrodynamic inertia effect of the water is considered by adding appropriate mass density to the piles. [4]

To predict the pile forces correctly, it is essential that higher mode contributions be considered, particularly when there are apparent differences in the horizontal stiffnesses of the piers. A 15 mode response spectrum analysis is conducted. The natural frequencies and modal participation factors are listed in Table 1. The first five horizontal modes are plotted in Figure 3 which shows significant translational and rotational displacements at lake footings. The maximum displacements at deck and footing levels, based on CQC rule [5], are listed in Table 2. Maximum shear force distributions at pile caps are listed in Table 3. Table 4 lists maximum pile axial forces and the corresponding biaxial bending moments. Note that the pile forces are obtained directly from modal responses for the portion above mudline. These pile force estimates may be checked immediately against:

1. pile capacity as a foundation element, and
2. ultimate capacity as a structural column.

BRIDGE NO. 2

This bridge, as shown schematically in Figure 4, is composed of two 3-span approach structures, one on each side of the channel, and a double-leaf concrete swing bridge (with two identical movable leaves) across the channel. The approach structures are conventional highway bridges. The main focus of interest is on the swing bridge foundation piling.

The swing bridge includes two identical movable leaves and two pivot piers. Each movable leaf is comprised of a 173.5 feet tail span cantilever, a 240 feet channel span cantilever and a 12 feet diameter pivot shaft extending down into the pivot pier. The connections between the approach structures and the movable leaves are two tail locks and a central lock which allow free longitudinal movement and rotations but prevent vertical and transverse relative movement. Figure 5 shows a section of the pivot pier. In the closed position, the movable leaf is supported by service bearings on top of the machinery housing. When opened, the leaf is supported by the lift piston vertically and the guided bearings horizontally.

The pivot piers, each located on one side of the channel, are supported on pile group foundations. The piles are 36" diameter steel pipe piles filled with concrete. There are 32 piles under each pier. Due to the sloping bank topography, piles within a group would have varying effective length and, therefore, varying stiffness. This results in an undesirable situation under earthquake excitation. Piles with deeper embedment carry heavier loads until the soil capacity is reached. When this happens, loads will be shifted to the rest piles and a series of domino-type failures may follow.

The direct cause for this to occur is the unequal effective lengths of piles within a group. This may be easily alleviated by using isolation sleeves in order for the embedded lengths of all piles to be equal. Another assumption in the aforementioned scenario is that soil failure occurred before any ductile yielding in the pile. This is consistent with current foundation design practice which does not allow yielding to occur below the ground surface. By using isolation sleeves, pile segments within the sleeve act more like a structural column and the combined axial and biaxial bending response, which is ductile, may be developed in the free-standing portion within the sleeve. An ideal situation would be that the ductile yielding in the piles occurs first and therefore limits the forces being transmitted to the soil. To take advantage of this, designers must ascertain that there is sufficient redundancy in the system. In the swing bridge case, piles are the only ductile elements and one has to be sure that yielding, under the most severe loading, will not penetrate the whole pile group. Another possible advantage by using isolation sleeves would be the elongation of natural periods.

The analytical procedure previously described was used to model the structure and pile group foundations. The bridge was modeled in both the closed and opened positions. The dynamic characteristics of the two cases

are listed in Table 5 and 6, respectively. Fundamental
modes in both cases involve the torsional vibration of the
pivot pier (bearing + pile group). Except for vertical
bending modes of leaf's cantilevers, flexibility of the
pile group is the major contributor factor in the
structural response.

Maximum pile forces are computed directly from combination
(CQC rule) of the modal responses. The maximum static
plus seismic pile forces in eight outside corner piles,
which are most severely loaded, are shown in Figure 6.
Also shown in the Figure are the yield surface according
to Bresler's interaction relationship [6] and the pile
compressive and uplifting capacities. It is noted that,
under the design earthquake, a few piles may yield in
combined tension and flexure, but well within the pile
uplifting capacity.

Since the pile group is the only lateral force resisting
elements, it would be catastrophic if yielding penetrates
the entire pile group. A nonlinear incremental analysis
was conducted to estimate the maximum lateral load
capacity of the Pivot Pier pile group. The simplified
idealization is shown in Figure 7. Piles are idealized as
3-D elasto-plastic column elements. At the end of the
sleeve, a soil stiffness matrix is attached to each pile.
The pseudo static lateral load distribution is assumed to
be proportional to the distribution of maximum
acceleration. The incremental nonlinear analysis was done
by the Program NEABS [7]. Figure 8 shows the lateral
load-displacement relationship. The ultimate shear
capacity at the footing level is 8925 kips which is well
above the maximum shear of 4500 kips predicted by the
Response Spectrum Method. The load paths of the two most
severely loaded piles are shown in Figure 6. It is
interesting to note that, only after yielding occurred,
the force point barely reached the compressive limit.

CONCLUSIONS

A simple and efficient substructure technique has been
implemented in the computer program SEISAB to include all
piles in a pile group explicitly in the analytical modal.
Two major bridges have been conveniently analyzed using
this computational tool. Most of the design variables may
be obtained from a single computer run.

Also discussed in the paper is the concept of sleeved pile
foundations. This concept has been studied in the past
[8] with emphasis focused mainly on the elongation of
natural periods and therefore reduction of the earthquake
loading. In the case considered here, however, the period
elongation is not a major factor. Most benefit is gained
from shifting of the brittle soil failure to the ductile

flexural yielding.

REFERENCES

1. "SEISAB-I, Seismic Analysis of Bridges", Workshop Manual, Engineering Computer Corporation, Sacramento, California, October, 1982.

2. Lam, I. Martin, G.R., "Seismic Design for Highway Bridge Foundations", Lifeline Earthquake Engineering: Performance Design and Construction, ASCE Convention, San Francisco, California, October, 1984.

3. Wilson, E.L., "The Static Condensation Algorithm", International Journal of Numerical Methods in Engineering, 1974, pp. 199-203.

4. Liaw, C.Y. and Chopra, A.K., "Earthquake Response of Axisymmetric Tower Structures Surrounded by Water", Earthquake Engineering Research Center, Report No. UCB/EERC-73/25, University of California, Berkeley, October, 1973.

5. Wilson, E.L., Der Kiureghian, A., and Bayo, E.P., "A Replacement for the SRSS Method in Seismic Analysis", Earthquake Engineering and Structural Dynamics, Vol. 9, 1981, pp. 187-194.

6. Tseng, W.S., and Penzien, J., "Analytical Investigations of the Seismic Response of Long Multiple Span Highway Bridges", Earthquake Engineering Research Center, Report No. UCB/EERC 73-12, University of California, Berkeley, June, 1973.

7. Penzien, J., Imbsen, R.A., and Liu, W.D., "NEABS - Nonlinear Earthquake Analysis of Bridge Systems", Earthquake Engineering Research Center, University of California, Berkeley, May, 1981.

8. Biggs, J.M., "Flexible Sleeved-Pile Foundations for Aseismic Design", Publication No. R82-04, Department of Civil Engineering, Massachusetts Institute of Technology, March, 1982.

TABLE 1: VIBRATION CHARACTERISTICS (BRIDGE NO. 1)

Mode	Period (Sec)	Frequency (Hz)	SA	Participation Factors X	Y	Z
1	1.492	0.670	0.29	-1.274	0.039	24.953
2	1.062	0.942	0.37	0.282	0.039	20.097
3	0.931	1.074	0.40	-37.567	-0.323	-1.003
4	0.719	1.391	0.48	-0.036	-8.420	-0.001
5	0.670	1.493	0.50	-1.018	-1.952	-0.058
6	0.577	1.734	0.55	-1.119	-0.053	13.205
7	0.509	1.964	0.60	-2.168	2.597	-0.144
8	0.428	2.333	0.68	-2.766	2.183	-0.214
9	0.367	2.722	0.75	-0.519	-0.136	-9.816
10	0.352	2.838	0.77	-3.703	10.111	-0.185
11	0.330	3.029	0.80	0.611	20.605	0.082
12	0.289	3.456	0.80	0.398	0.446	-3.261
13	0.275	3.636	0.80	3.969	4.565	-0.671
14	0.247	4.049	0.80	0.286	-0.553	7.133
15	0.245	4.078	0.80	0.458	1.980	4.666

SA = Spectral Acceleration

TABLE 2: MAXIMUM (CQC) DISPLACEMENTS (BRIDGE 1) (FEET)

Pier	Deck Transv.	Deck Longit.	Footing Transv.	Footing Longit.
9 (85.0 ft*)	0.784	0.294	0.784 (0.026)++	0.295 (0.010)
10 (87.5 ft.)	0.559	0.291	0.686 (0.043)	0.321 (0.026)
11 (79.5 ft.)	0.351	0.284	0.487 (0.037)	0.309 (0.013)
12 (75.5 ft.)	0.309	0.273	0.327 (0.018)	0.306 (0.028)
13 (35.5 ft.)	0.191	0.260	0.141 (0.024)	0.175 (0.040)
14+	0.070	0.251	0.004	0.012
15+	0.025	0.253	0.005	0

* Depth from lake footing to mudline
+ Land-based spread footings
++ Numbers in ( ) are pile displacements at mudline

TABLE 3: MAXIMUM (CQC) FORCES AT PILE CAPS (BRIDGE 1)

Pier	Longitudinal EQ Shear (kip)	Moment* (kip-ft)	Axial* (kip)	Transverse EQ Shear (kip)	Moment* (kip-ft)	Axial* (kip)
9	112	1009	176	214	18133	464
10	709	28097	179	1077	17872	4032
11	82	5430	291	1040	29764	3523
12	979	26759	309	432	1205	831
13	2039	33797	651	921	14187	2265
14	3888	75458	266	1444	25100	1771
15	0	0	669	732	15471	146

* Moment and axial force correspond to the maximum shear force.

TABLE 4: MAXIMUM PILE FORCES DUE TO COMBINED LONGITUDINAL AND TRANSVERSE EARTHQUAKE EXCITATION IN ACCORDANCE WITH AASHTO (BRIDGE 1)

Pier	Axial Force (kip) Dead Load	Seismic*	Maximum Tension	Maximum Compression	Corresponding Biaxial Moments* (kip-ft)
9	565	±534	-	1099	1920
10	691	±1959	-1268	2650	4072
11	882	±1647	-765	2529	2599
12	992	±1904	-912	2896	2804
13	591	±1976	-1385	2567	1226

* Seismic forces are obtained by either (L+0.3T) or (0.3L+T) where;
L = forces due to longitudinal earthquake excitation alone
T = forces due to transverse earthquake excitation alone

TABLE 5: VIBRATION CHARACTERISTICS (BRIDGE NO. 2, CLOSED POSITION)

Mode	Period	Frequency	SA	Participation Factors X	Y	Z
1	1.861	0.537	0.21	0.161	0.006	1.487
2	1.546	0.647	0.24	-0.891	10.704	0.024
3	1.185	0.844	0.32	-2.080	-0.087	19.354
4	1.089	0.919	0.34	-17.119	-0.601	-1.383
5	1.039	0.962	0.36	0.378	-0.003	25.962
6	0.989	1.011	0.37	-22.131	-1.618	0.064
7	0.920	1.087	0.40	-11.756	2.621	-0.001
8	0.719	1.391	0.50	-14.649	0.246	-0.000
9	0.681	1.468	0.53	0.888	0.008	4.600
10	0.604	1.654	0.60	-0.573	0.007	-5.684
11	0.566	1.767	0.64	-1.541	0.054	0.650
12	0.557	1.797	0.65	5.518	-0.531	0.174
13	0.496	2.015	0.70	-0.155	-0.009	0.722
14	0.467	2.143	0.70	0.657	2.716	0.021
15	0.451	2.216	0.70	-0.233	1.729	0.012
16	0.421	2.373	0.70	1.574	0.292	0.030
17	0.412	2.426	0.70	-0.044	0.048	-8.034
18	0.393	2.544	0.70	0.577	-0.898	-0.001
19	0.348	2.871	0.70	0.118	-10.518	0.005
20	0.338	2.962	0.70	-0.353	0.524	-0.000

SA = Spectral Acceleration

TABLE 6: VIBRATION CHARACTERISTICS (BRIDGE NO. 2, OPENED POSITION)

Mode	Period	Frequency	SA	Participation Factors X	Y	Z
1	3.541	0.282	0.06	-0.000	0.000	0.462
2	2.324	0.430	0.09	6.826	-1.760	0.000
3	1.217	0.822	0.16	-0.000	0.000	-23.710
4	1.089	0.918	0.17	-12.418	-10.431	-0.000
5	0.903	1.108	0.21	-20.374	6.315	0.000
6	0.461	2.170	0.35	0.000	0.000	-7.277
7	0.455	2.196	0.35	-2.031	-7.627	-0.000
8	0.258	3.883	0.35	-1.564	6.948	0.000
9	0.224	4.457	0.35	0.000	0.000	3.484
10	0.188	5.325	0.35	1.079	11.617	-0.000
11	0.166	6.016	0.35	-2.077	3.131	0.000
12	0.158	6.327	0.35	0.000	0.000	0.667
13	0.123	8.112	0.32	0.049	14.556	0.000
14	0.091	10.931	0.27	-0.136	1.767	0.000
15	0.089	11.221	0.26	0.000	0.000	-0.194

SA = Spectral Acceleration

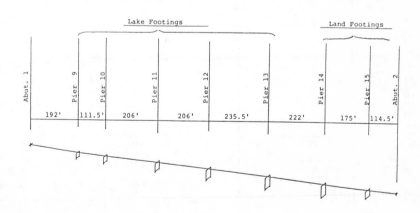

Figure 1: East Approach Structure to the Third Lake Washington
Floating Bridge (Bridge No. 1)

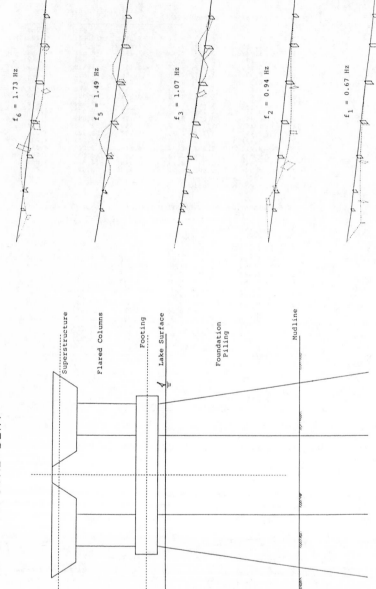

Figure 3: Mode Shapes (Bridge No. 1)

$f_6 = 1.73$ Hz

$f_5 = 1.49$ Hz

$f_3 = 1.07$ Hz

$f_2 = 0.94$ Hz

$f_1 = 0.67$ Hz

TYPICAL BENT

Superstructure

Flared Columns

Footing

Lake Surface

Foundation Piling

Mudline

Figure 2: Typical Bent (Bridge No. 1)

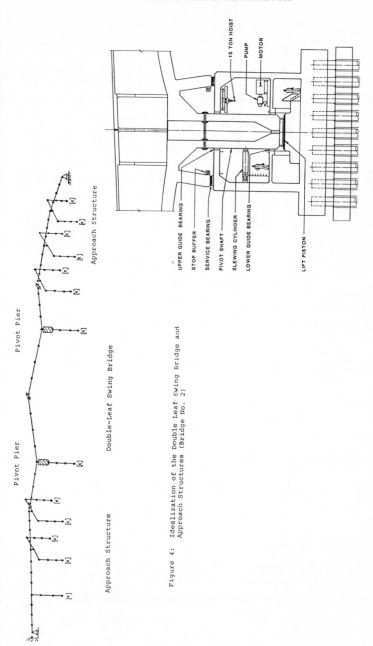

UPPER GUIDE BEARING
STOP BUFFER
SERVICE BEARING
PIVOT SHAFT
SLEWING CYLINDER
LOWER GUIDE BEARING
LIFT PISTON

15 TON HOIST
PUMP
MOTOR

SECTION THROUGH PIVOT PIER

0   5   10
FEET

Figure 5: Swing Bridge Pivot Pier

Approach Structure

Pivot Pier

Pivot Pier

Double-Leaf Swing Bridge

Approach Structure

Figure 4: Idealization of the Double Leaf Swing Bridge and
Approach Structures (Bridge No. 2)

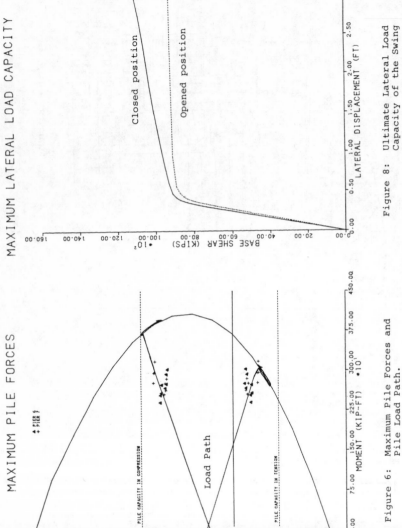

MAXIMUM LATERAL LOAD CAPACITY

Figure 8: Ultimate Lateral Load
           Capacity of the Swing
           Bridge Pile Group

MAXIMUM PILE FORCES

Figure 6: Maximum Pile Forces and
           Pile Load Path.

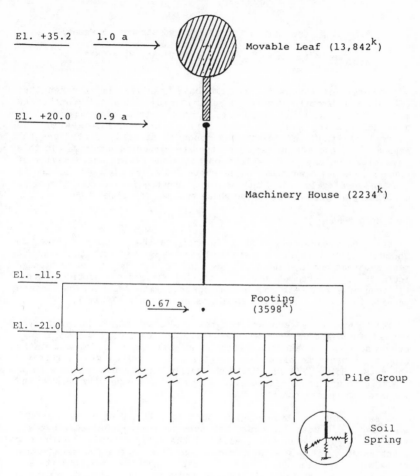

Lateral Force = Weight x Coefficient (a)

El. +35.2   1.0 a   Movable Leaf (13,842^k)

El. +20.0   0.9 a

Machinery House (2234^k)

El. -11.5

0.67 a   Footing (3598^k)

El. -21.0

Pile Group

Soil Spring

Figure 7:   Equivalent Lateral Load Distribution for Estimating the Ultimate Lateral Load Capacity of the Pivot Pier

Attaching Steel Members to Reinforced Concrete Elements

David F. Wiener* and James O. Jirsa, M. ASCE**

A variety of structural systems have been devised to strengthen existing buildings for improved seismic resistance. Strengthening methods currently in use for reinforced concrete structures include diagonal bracing systems. The connection detail between the structural steel and reinforced concrete strongly influences the response and the effectiveness of the strengthening program. In this investigation the load-deflection characteristics and behavior of connections between existing concrete and steel elements loaded in tension were evaluated. Anchors epoxy grouted into a concrete block simulating a column transferred shear between the elements.

Test Specimens

Six specimens were tested to investigate the capacity of steel to concrete connections. The steel element was loaded in tension. The test specimen configuration is shown in Fig. 1. The variables studied include: (1) number of anchor bolts, (2) steel cross-sectional area, (3) type of steel section, and (4) concrete-steel interface surface preparation. Details of the specimens are given in Table 1.

The anchor bolts were cut from the same stock of 3/4 in. mild steel threaded rod and cleaned with a wire brush to ensure a good steel-epoxy bond. The anchors extended a maximum of 2 in. above the concrete surface to allow the use of a socket wrench when tightening the nuts. Holes (7/8 in. dia.) were drilled in the base block with an electric rotary percussion drill. All holes were carefully brushed and vacuumed to remove concrete dust.

A two component liquid epoxy bonding agent, Concresive 1001-LPL, was used. One batch of epoxy was used to install bolts in all six specimens. To assure the quality of the epoxy adhesive resin, two slant shear specimens were prepared according to the recommendations of AASHTO T237. The shear tests indicated that the epoxy quality was satisfactory. Additionally, three threaded rod pullout tests were run to ensure that the grouting was done properly. All three rods yielded in the pullout test.

*Graduate Research Assistant, Ferguson Structural Engineering Laboratory, University of Texas at Austin.

**Ferguson Professor of Civil Engineering, University of Texas at Austin, Austin, Texas 78712

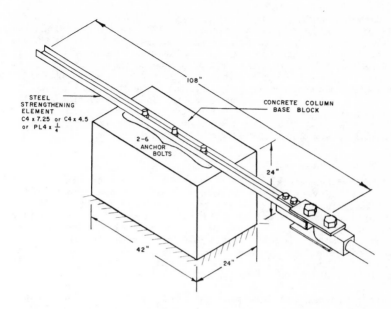

Fig. 1   Test Specimen

No.	Base Block Details		Steel Section			Interface Details				Maximum load (kips)
	f'_c (psi) @ 28 days @ test date	Reinforcement Longitudinal stirrups	Type (A36)	Net Area (in.)	Web Thickness (in.)	Surface Preparation	No. of 3/4" Ø A325 bolts	Distance between adjacent bolts (in.)	Embedment (in.)	
2C	2650 3100	2 Ø11's Ø3's @ 12"	C-4x7.25	2.13	0.32	smooth surface untreated	2	24	6	29
3C	2650 3100	2 Ø11's Ø3's @ 12"	C-4x5.45	1.59	0.18	smooth surface untreated	3	12	6	43
3P	2400 2800	2 Ø11's Ø3's @ 12"	PL-4x1/4	1.00	0.25	smooth surface untreated	3	12	6	42
6C	2650 3100	2 Ø11's Ø3's @ 12"	C-4x7.25	2.13	0.32	smooth surface untreated	6	6	6	84
3CE	2650 3100	2 Ø11's Ø3's @ 12"	C-4x7.25	2.13	0.32	smooth surface gel epoxy at interface	3	12	6	51
3CER	2400 2800	2 Ø11's Ø3's @ 12"	C-4x7.25	2.13	0.32	roughened surface gel epoxy at interface	3	12	12	45

Table 1   Description of Test Specimens

Before a steel element was set into place, the concrete surface was cleared of dust and loose material using an air hose. For two tests, 3CE and 3CER, Sikadur 31 Hi-Mod Gel epoxy was applied at the steel-concrete interface to improve shear capacity and stiffness of the connection. It is a high solid, high modulus, moisture insensitive,

two-component epoxy resin. Prior to epoxy gel application, base block
3CER was lightly chipped to a depth of 1/16 in. to 3/16 in.   The
channel sections were cleaned with acetone, placed over the bolts, and
the nuts were tightened in stages. Mild steel washers, 2 in. in
diameter and 1/8 in. thick, were placed between the nut and the steel
section.   Each anchor was tightened to 200 ft/lb in increments of 50
ft/lb using a torque wrench.   All nuts on a specimen were torqued to
the same load before proceeding to the next higher load stage.   Bolts
were tightened 24 to 48 hours before testing, except in the case of
specimens 3CE and 3CER which were tightened at the time of gel epoxy
application, approximately 10 days prior to testing.

Test Procedure

Each specimen was subjected to repeated cycling.   The direction of
loading was reversed when interface slip or load exceeded limits
established for each stage of the test.   A total of eight load cycles
were applied to all specimens.   First, one elastic cycle at a load
level of about 5 kips per anchor bolt was applied.   Second three
cycles displacing the steel section 0.1 in. in each direction at load
levels not exceeding the steel section design capacity were applied,
followed by three cycles at displacements of approximately 0.25 in. in
each loading direction.   Finally, the section was loaded to failure.

Displacements and deformations were measured using linear
potentiometers.   The placement of transducers was the same for all
specimens and is illustrated in Fig. 2.   Potentiometers 0 and 2 at one

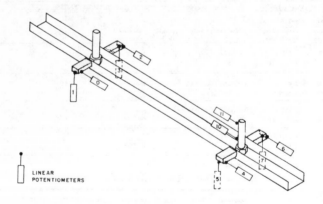

LINEAR
POTENTIOMETERS

Fig. 2   Slip Instrumentation

end of the bolt, and 4 and 6 at the other end were used to measure
lateral displacement at the interface parallel to the direction of
loading. Devices 1, 3, 5, and 7 on the strengthening elements were
oriented normal to the interface plane to detect uplift at the end
bolt locations. Displacement transducers 10 and 11 recorded rotation
of one bolt in each test.

Each strengthening element was instrumented with strain gages to evaluate the mechanism of load resistance. Approximately ten strain gages were used in each test to determine the forces in the steel between anchor bolts, and to determine the load resisted by friction and by surface epoxies.

Load vs. Interface Slip

A continuous plot of the load-slip relationship was obtained for each test, as shown in Fig. 3. The value of slip plotted was at the end of the connection closest to the active ram. In the first

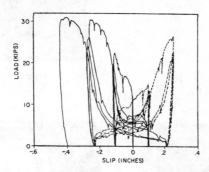

Fig. 3   Measured Load-Slip Curves--Test 2C

(elastic) loading stage, the curves were nearly vertical. Just before first slip, the connection fully developed the static frictional force along its surface. In Table 2 the loads resisted by the specimens at first slip (taken as 0.01 in.) and at ultimate are presented. The values at first slip for tests in which epoxy was used between the steel element and the concrete were considerably greater (about 15 kips/bolt) than in those without interface epoxy (about 6 kips/bolt) but were nearly the same for all tests at ultimate.

In tests 3CE and 3CER, the cracks propagated through from the end of the connection through the concrete or between the mill scale and steel section until the steel element completely debonded from the concrete. Higher loads may have been resisted by the connections with surface gel epoxy had the concrete surfaces been prepared by mechanical abrasion and the steel contact surfaces sandblasted.

After the steel-epoxy interface failed, specimens 3CE and 3CER responded better than the other specimens in all stages of testing. The improved stiffness was probably due to epoxy seated between the anchor and the oversized hole. Figure 4 shows the gap between the bolt and steel section. The bearing of the bolt against the steel caused considerable yielding and elongation of the hole. Epoxy effectively reduced the gap as shown in Fig. 5. The epoxy acted to distribute bearing stresses uniformly, thereby eliminating deformations where the steel element and anchor were in contact. The epoxy in the gap suffered little damage under bearing loads because it

Test	Maximum Load Resisted Before First Slip (kips)		Avg. of tests	Peak Load (kips)		Avg. of tests
	Total	Per Bolt		Total	Per Bolt	
2C	12.7	6.35		29.1	14.6	
3C	12.5	4.17	5.75	42.5	14.2	14.2
3P	18.7	6.23		42.0	14.0	
6C	37.4	6.23		84.0	14.0	
3CE	51.2	17.1	14.6	51.2	17.1	16.1
3CER	36.2	12.1		45.4	15.1	

Table 2   Loads at first slip
and ultimate

Fig. 4   Bearing of bolt
against steel section

Fig. 5   Epoxy-filled gap
between bolt and
section

Fig. 6   Ring of epoxy
adhering to belt
after failure of
connection

was confined in all directions. The epoxy ring could be removed
intact, as shown in Fig. 6.

Load-Slip Envelopes

For each test a load-slip envelope was constructed. The envelope
was a composite of load-slip curves in both loading directions. The
envelopes are plotted in Fig. 7. The envelopes can be used to compare
stiffness and energy absorption characteristics of the various

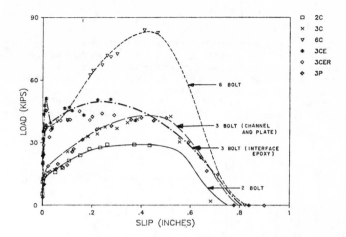

Fig. 7 Load-slip envelopes

connections. In Fig. 8 the initial regions (up to 0.04 in. slip) of
the envelopes are shown. The stiffness of all the specimens in the
elastic range can be seen clearly. After slip occurred, stiffness
decreased causing large deformations upon application of significant
loads. Connections 3CER and 3CE show a 25% reduction in the load
resisted after slip reached 0.015 in. due to failure of the epoxy
interface. Interface epoxy increased maximum load per bolt in the
elastic range more than three times and a higher load was maintained
up to nearly 0.5 in. slip. Overall, connections 3CE and 3CER clearly
performed better.

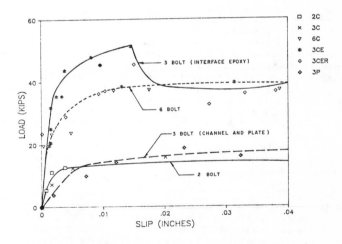

Fig. 8 Initial portion of load-slip envelopes

Estimated Bolt Loads

All steel strengthening elements were instrumented with strain gages to determine the distribution of load to anchor bolts. Local stress concentration and bending made conversion of strains to stresses difficult in some cases. However, the strain data yield valuable insight as to connection behavior in resisting applied loads. The percentage of total applied load resisted by each bolt are shown in the bar graphs in Fig. 9. Values were obtained by averaging data

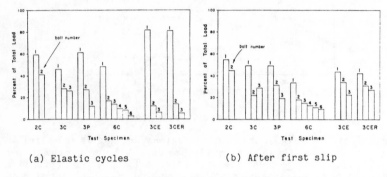

(a) Elastic cycles   (b) After first slip

Fig. 9   Estimated load carried by bolts

in both loading directions at the maximum slip value reached in a given cycle. During non-slip (elastic) loading, frictional and adhesive forces resisted shear loads along interface surfaces; the bolts did not directly resist shear. Once relative movement between steel and concrete began, most load was resisted in bearing.

In every test the lead bolt resisted the greatest load. Remaining load was divided between consecutive bolts based on their distance from loaded end. During non-slip loading in the elastic range, the lead bolt resisted about 50% of the load when the interface was not prepared, and over 80% when surface epoxy used. When no interface epoxy was used the lead bolt continued to resist more load than the other bolts in all cycles. In tests 3CE and 2CER interface epoxy at the lead bolt resisted most of the load until the second or third cycle. Then load was distributed to all bolts fairly evenly. Surface epoxy improved connection behavior even after the interface fractured. By filling gaps between the anchors and bolt hole walls, it served to distribute loads between bolts. Once slip began, the bolts in tests 3CE and 3CER shared load more equally than in the other tests.

Conclusions

The following conclusions can be drawn from the test data: The bolt closest to the applied force resists the most load and fails first. Connections with epoxy applied at the steel-concrete interface were stiffer and had higher elastic and ultimate capacities. The epoxy distributed loads more equally to all bolts than in connections without surface epoxy.

Acknowledgments

This study was part of a broad research project to investigate the repair and strengthening of reinforced concrete structures. It was conducted at the Phil M. Ferguson Structural Engineering Laboratory of The University of Texas at Austin in a cooperative program with H. J. Degenkolb Associates, San Francisco. The work was supported under Grant No. CEE-8201205 from the National Science Foundation. The results reported here were used in the design of a steel bracing system for strengthening a 2/3 scale reinforced concrete frame tested at The University of Texas.

References

1. Chon, C. "Epoxies for Anchoring Dowels in Concrete," Thesis submitted for the M.S. degree, University of Texas at Austin, Aug. 1984.

2. Luke, P. C., "Strength and Behavior of Rebar Dowels Epoxy-Bonded in Hardened Concrete," Thesis submitted for M.S. degree, University of Texas at Austin, May 1984.

3. Wiener, D. F., "Behavior of Steel to Concrete Connections Used to Strengthen Existing Structures," Thesis submitted for M.S. degree, University of Texas at Austin, Aug. 1985.

PERFORMANCE OF EXISTING BUILDINGS: RESULTS OF BENCHMARK
STRUCTURE ANALYSES PERFORMED IN U.S.-JAPAN WORKSHOPS

Richard N. White[1], F. ASCE and Peter Gergely[2]. F. ASCE

Introduction

In response to the need for better understanding of current
safety evaluation methodologies, three U.S.-Japan workshops have been
conducted on the evaluation of the performance of existing buildings
for resistance to earthquakes. The workshops, which were held in
1983, 1984, and 1985, were organized through Task Committee D,
Evaluation of the Performance of Structures, under the auspices of the
U.S.-Japan Cooperative Program in Natural Resources (UJNR). The
workshop series was supported by the National Science Foundation
(Grant CEE-8217190 to Cornell University).

The first workshop, held in Japan, treated five theme areas: (1)
Overview of evaluation of buildings for seismic resistance, (2)
Practical methodology, (3) Computer programming, (4) Evaluation of
structural performance through full scale tests, and (5) Post-
earthquake evaluations. The second workshop was held in the U.S. on
(1) Safety evaluation methods, (2) Evaluation methods applied to
benchmark structures, and (3) Damage and analysis of other
structures. The third workshop was held in Japan and concentrated on
additional comparisons of evaluations of benchmark structures and on
updating recent work in methodology. The details of the workshops are
covered in three volumes of papers White and Gergely (1985), (1986).

Evaluation methodologies discussed at the workshops may be
classified into the following categories:

1. Screening methods for quick evaluation of large groups of
    buildings.

2. Rapid evaluation procedures for individual buildings.

3. Multi-Level procedures for progressively more accurate evaluation
    of individual structures, including the 1976 Japanese Standard for

1. Professor, Department of Structural Engineering, Cornell
    University, Ithaca, N.Y. 14853.

2. Professor and Chairman, Department of Structural
    Engineering, and Director, School of Civil and
    Environmental Engineering, Cornell University, Ithaca,
    N.Y. 14853.

Seismic Capacity Evaluation of Existing Reinforced Concrete
Buildings, and the U.S. Navy Procedure.

4. Detailed analytical approaches, usually involving dynamic
   analysis, and computer-based methods such as WOODAM (for
   evaluation of timber structures).

5. Special approaches still under active development, including rule-based inference procedures for formal damage
   assessment, and the use of expert systems.

The present paper concentrates on the results of comparative
analysis of benchmark structures.

Comments on Evaluation Methodologies

Brief comments on several methods of evaluation are in order
prior to discussing results of the comparative benchmark structures
analysis. A more complete discussion is provided by Gergely, White,
and Fuller (1986).

The capacity spectrum method, which was adopted for the U.S. Navy
Procedure for building evaluation, is described by Freeman in Paper
II-1 of the 1983 Workshop Proceedings.* In this rapid evaluation
method the inelastic load-deflection curve is first estimated for roof
displacement versus the base shear for a lateral load representing the
inertia force vector, either by hand calculations or with a computer.
In the next step the roof displacement is related to the corresponding
spectral displacement S_d, and the base shear is related to the
spectral acceleration S_a through the well-known modal equations.
These transformations are ordinarily based on the first mode only.
The periods corresponding to S_d and S_a values are calculated.

The plot of S_a versus the period T is called the capacity
spectrum. It is next superimposed on the design spectrum curve. The
relative positions of the capacity and response spectra determines the
safety of the structure. An interpolation is necessary because the
damping changes as inelasticity develops.

The capacity spectrum method is a relatively simple and
attractive method when the first mode dominates, although the contribution of higher modes to the roof deflection can be incorporated
in a approximate manner. Often good accuracy can be obtained using an
estimated bilinear stiffness of the structure.

The Japanese Standard for Seismic Capacity Evaluation of Existing
Reinforced Concrete Buildings was compiled in 1977 under the Ministry
of Construction. It is described in some detail by Aoyama (1981)**

*Hereafter, for convenience, individual papers in the Workshop
 Proceedings volumes will be referred to by citing the paper number
 and the workshop year--1983, 1984, or 1985.

**This paper is reprinted as Paper I-2 in the 1983 Workshop
 Proceedings.

and is intended to be applied primarily to low rise buildings up to 6 or 7 stories.

This three-level procedure involves calculation of Seismic Index $I_s = E_0 G S_D T$ where $E_0 = \phi (C \times F)$ is the basic seismic index. Within the term E_0, C is a strength matrix, F is a ductility matrix, and ϕ is a story index that relates single DOF system response to that of the multi-story system. Thus it is apparent that this method evaluates both strength and ductility capacities and expresses the result E_0 as a product of these capacities.

The other necessary definitions are G = geological index, S_D = structural design index ranging from 0.4 to 1.2, and T = time index ranging from 0.4 to 1.0 to cover such items as previous damage, shrinkage and settlement effects, and deterioration of the structure from loadings and environmental factors.

The first level procedure is very easy to apply -- it is based on the average stress in the column and wall areas. It is well-suited for wall-dominated structures and is quite conservative for ductile frames. The second level procedure concentrates on potential problems in vertical members by identifying five types of vertical elements: extremely brittle short columns, shear-critical columns, shear walls, flexure-critical columns, and flexural walls. Simple equations are given for treating each type of element.

The third level procedure recognizes flexural hysteresis by considering the response of three additional types of vertical members: columns governed by shear beams, columns governed by flexural beams, and rotating walls with uplift effects from over-turning. Equations are again provided for these calculations.

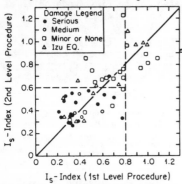

a. Seismic index and damage in 1978 earthquakes in Japan

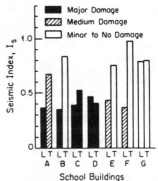

b. Seismic index and damage of school buildings in 1968 Tokachi-Oki EQ (2nd level procedure)

Fig. 1   Seismic indices and damage, Japanese Standard (1977)

Results of the application of these procedures to damaged structures are given in Fig. 1, where the correlation between first and second level procedures applied to buildings from 1978 earthquakes in Japan is given in Fig. 1a. A comparison of damage levels and seismic index I_S values calculated for both the longitudinal (L) and transverse (T) motions of a group of school buildings is shown in Fig. 1b.

While seismic capacity is best judged by comparing I_S to an E_t index that is tied to specific ground acceleration levels and other critical factors, general guidelines on performance may be stated. Superior performance is indicated by I_S values of 0.9, 0.7, and 0.6 for levels 1, 2, and 3 evaluation procedures, respectively. I_S values less than 0.4 at the 2nd and 3rd levels indicate questionable performance, and the borderline of damage for level 1 is often taken as 0.8 (Paper II-2, 1983).

Applications to Benchmark Structures

Six benchmark structures were selected for evaluation by the various methods; the framing details are described in Appendix A of the 1984 Workshop Proceedings. The buildings are: 1) 7 story full-scale reinforced concrete frame tested in Japan under the U.S.-Japan Cooperative Research Program, 2) Imperial County Services Building, California, 3) 9 story ATC benchmark structure, reinforced concrete, 1964, 4) 7 story reinforced concrete building (Holiday Inn), instrumented, 5) Reinforced concrete building, Japan, and 6) 6 story full-scale steel frame tested in Japan under U.S.-Japan Cooperative Research Program.

1. 7 Story U.S.-Japan Reinforced Concrete Building:

This combination frame-shear wall structure had three bays of 19.7, 16.4, and 19.7 ft (6, 5, and 6 m) in the longitudinal direction and two spans of 19.7 ft each (6 m) in the transverse direction. A shear wall was placed parallel to the direction of loading in the middle bay of the center frame, continuous from the first to the seventh story.

Several papers (Yoshimura, IV-2 and Kabeyasawa, IV-3, 1983) provide detailed experimental and analytical results.

Okamoto (Paper IV-1, 1983) evaluated the seven story reinforced concrete building using the Japanese Standard described above. The evaluation was made only for the direction in which the seismic forces were applied during the loading in the laboratory, and the weight of each story was taken as the dead weight of the floor level plus the weight of the loading frames. The Ground Motion Index G, the Shape Index S, and the Time Deterioration Index T were all taken as unity.

Results are quoted directly from Okamoto: "The Basic Structure Performance Index, E_0, was computed as 0.36 (1st Screening), 0.68 (2nd Screening), and 0.59 (3rd Screening). The index given by the first screening is found to be significantly lower than those given by the

second and third screenings. This low value of index as obtained
primarily because of small areas of the columns and shearwalls. In
the second and third screenings, on the other hand, E_0 was much larger
because the structure was estimated to exhibit flexural failure.
Nevertheless, if the weight is increased to 1 ton/m^2 by taking live
load into account, E_0 will decrease significantly to a level at which
the structure is considered to be unsafe".

2. Imperial County Services Building:

The Imperial County Services Building is well-known for its
unsatisfactory performance in the 1979 Imperial Valley earthquake, and
has received considerable attention in various post-mortem analyses.
It was designed in 1968 in accordance with the 1967 UBC and
construction was completed in 1971.

The poor performance of this building under seismic loads has
been attributed partially to the poor shear transfer properties at one
end of the building.

Aoyama (Paper II-8, 1984) applied the Japanese Standard
procedures (levels 1 and 2 only) to the structure, using S_D = 0.95, G
= 1.0, and T = 1.0, and obtained the following results.

	Story	E-W I_s	N-S I_s
	6	0.65	1.06
	5	0.31	0.52
(First Level Procedure)	4	0.23	0.37
	3	0.18	0.31
	2	0.17	0.29
	1	0.11	0.48
(Second Level Procedure)	1	0.22	0.48

In comparison with accepted Japanese standards, it was found that the
Imperial County Services Building fell in a region regarded to be
dangerous under earthquake loads.

By Japanese standards, the upper level walls would be regarded as
"miscellaneous walls" with only 1/3 the effectiveness of regular shear
walls. The average axial stress level in the vertical elements was
about twice that of typical Japanese building designs. While there is
a question on the validity of using the Japanese approach on the N-S
framing, the E-W direction framing does fit within assumptions.

The analysis by Hart (Paper II-10, 1984) showed that the normal
assumption of fixed base shear walls was not valid and that the
soil-building system must be modeled. He also showed that the floor
diaphragms were flexible, not rigid, and that using a rigid diaphragm
analysis is misleading in terms of the distribution of loads at the
ground floor column level.

Sozen (Paper II-9, 1984) summarizes the key findings from the PhD dissertation of M. Kreger (U. Illinois, August 1983) who analyzed the Imperial County Services Building with particular emphasis on the causes of the column failures   It is shown that it is difficult to explain the failures by considering forces in only on direction at a time. Sozen concludes his discussion as follows: "It is very unlikely that calculation alone would have identified the problem and it is very likely that a structural engineer who is informed of similar events would have avoided the conditions at the east end of the structure."

The shear resistance at the base, calculated by standard U.S approaches, is about 1/3 the building weight in the N-S direction and about 1/4 the weight in the E-W direction. Such values are acceptable for U.S. design practice (and in fact are even on the high side).

3. 9 Story Reinforced Concrete ATC Benchmark Building:

This structure was designed with the 1964 UBC provisions, and has been the subject of re-design in the ATC project. The building framing is concrete pan-joist on 20 by 24 ft (6.1 by 7.3 m) bays framing into girders and columns, with shear walls along both axes of the building to resist lateral loads. The plan dimensions of the main portion of the building are 141.5 by 80.67 ft (43.1 m by 24.6 m).

Lew (Paper II-7, 1984) applied the rapid, Level 1 Field Evaluation Method of the National Bureau of Standards, and concluded that for an earthquake intensity of 9 MMI, the building rated <u>fair</u>. The unsymmetrical distribution of shear walls in the transverse direction controlled the rating of the structure.

Okamoto and Watanabe (Paper II-2, 1984) and Okamoto and Teshigawara (Paper I-3, 1985) applied the Japanese Standard 1st level procedures. In the longitudinal direction, I_s values ranged from 1.37 at the top floor to 0.34 at the base; the fifth floor had I_s = 0.37. In the transverse direction, I_s ranged from 2.30 to 0.67, with 0.68 at the 5th floor. These I_s values are low by Japanese standards. However, it must be noted that the procedure is very conservative for ductile frames and that the result for this particular building cannot be regarded as conclusive.

Freeman (Paper II-3, 1984) used the capacity spectrum method for the ATC building, in a very quick one page calculation. He estimated the static capacity for the uncoupled walls and the spandrel frames and plotted the capacity spectrum curve as two straight lines. The intersection with a design spectrum is at about 0.19g.

4. 7 Story Holiday Inn

The Orion Holiday Inn building in Los Angeles is a reinforced concrete flat plate structure with 8 bays (18.75 ft or 5.71 m) in the E-W direction and 3 bays (about 20 ft or 6.1 m) in the N-S direction, with spandrel beams in the periphery. The building is instrumented and measured response during the 1972 San Fernando earthquake is available.

This structure was studied by Freeman, Shimizu (Paper II-1, 1984), and Moehle (Paper II-4, 1984). Shimizu applied the Japanese Standard and obtained E_0 values between 0.20 and 0.36 by rapid evaluation, and 0.24 to 0.45 by detailed evaluation. By Japanese standards, E_0 should be at least 0.6. The validity of the procedure for this flat plate building is somewhat questionable, however.

Moehle made a relatively simple analysis plus an assessment of connections and details. In the N-S direction, T_0 as calculated was 0.8 sec, and the measured value was 0.5 sec. He concluded that (a) the structure is fairly flexible and excessive nonstructural damage could be expected in a moderate earthquake, (b) inadequate ductility may be a problem during a more intense earthquake, (c) analysis underpredicts initial stiffness by nearly two to one, and (d) the first floor slab to column connection details are considered inadequate.

The Capacity Spectrum method results in spectral accelerations in the 0.15g - 0.21g range, depending on the contribution of non-structural elements, in the transverse direction. The range is 0.20g to 0.26g in the longitudiual direction.

5. 6 Story Steel Frame (U.S. - Japan Program)

The six story steel frame building tested in Tsukuba, Japan consisted of three two-bay frames of 24.6 ft (7.5 m) spans with a first story height of 14.8 ft (4.5 m) and upper story heights of 11.1 ft (3.4 m). The two exterior frames were rigid and the interior frame had K-bracing in one bay. The structure was designed to satisfy the requirements of both the 1976 UBC and the 1980 Japanese Aseismic Design Code, with a design base shear coefficient of 0.197.

Lew (Paper II-7, 1984) rated this building using the NBS rapid FEM, and concluded that the structure rated fair for a MMI 9 earthquake. The semi-rigid nature of the steel deck with concrete topping was a critical factor in reaching this rating.

Hart (Paper II-6, 1984) did a detailed structural analysis of the system, based on a perspective which recognizes the use of site dependent earthquake response spectra. The analysis required about 100 engineering labor hours. It provides a rather detailed prediction of the development of plastic hinges in the structure, the lateral displacement (nonlinear), and the variation of period with base shear. Typical U.S. steel frames have a much flatter curve for change in period with change in base shear.

Paper II-5, 1984 (by Yamanouchi et al) provides detailed test results for this structure.

Summary and Conclusions

Methods of seismic safety evaluation of existing buildings were studied in three U.S.-Japan workshops. The methods range from rapid classifications of groups of buildings to detailed dynamic analysis of individual structures.

Methods developed in Japan and in the U.S. are "tuned" to the characteristics of typical buildings in each country. Comparative analyses of benchmark structures reflect the relatively more conservative Japanese seismic design approaches.

The comparisons reveal that the various methods are suitable for the expected level of evaluation accuracy for buildings with good arrangement of lateral strength elements, and that a general idea of expected damage level can be estimated. However, additional developments are required to validate and improve the methods to make them more useful, especially for buildings of unusual geometry or with special secondary elements, and to include effects of soil-structure interaction.

Acknowledgements

The financial support of the National Science Foundation is greatly appreciated. The efforts of Dr. M. Watabe, Dr. S Okamoto, and G.R. Fuller in helping to organize the workshops are also much appreciated, as is the enthusiastic involvement of the many dozens of individuals from Japan and the United States who participated in the workshops.

References

1. Aoyama, H., "A Method for the Evaluation of the Seismic Capacity of Existing Reinforced Concrete Buildings in Japan", Bulletin of the New Zealand National Society for Earthquake Engineering, Vol. 14, No. 3, September 1981.

2. Gergely, P., White, R.N., and Fuller, G. R., "Seismic Performance of Existing Buildings", Proc. of Third U.S. National Conference on Earthquake Engneering, Charleston, S.C., August 1986.

3. White, R.N., and Gergely, P., (Editors), Proceedings of the Workshops on Seismic Performance of Existing Buildings, (3 Volumes - May 1983, July 1984, and May 1985), Department of Structural Engineering, Cornell University, Ithaca, N.Y., 14853, published April 1985 (Vols. 1 & 2) and February 1986 (Vol. 3).

The opinions, findings, conclusions, and recommendations expressed in this paper are those of the individual contributors and do not necessarily reflect the views of the NSF and other private or governmental organizations.

STRENGTHENING OF A MULTI-STORY
BUILDING WITH DEFLECTION CONSTRAINTS

by Gary C. Hart*, M. A. Basharkhah**, J. Elmlinger***

INTRODUCTION

The strengthening of buildings in seismic zones often
requires that additional structural members be added to
absorb the seismic force. Occasionally the existing
structural or nonstructural systems of the building
require that we impose deflection constraints or limits
in order to avoid an undesirable behavior of existing
(old) components. This usually takes the form of
restricting the interstory displacement estimated to
exist due to a design earthquake input. This paper
discusses one method of providing deflection constraints
without compromising window openings or the operation of
the structure.

THE STRUCTURAL STRENGTHENING

The stiffness of the beams in a building structure often
provide a major contribution to total frame stiffness.
For the structure under consideration our desire was to
provide very stiff beams to limit deflection. This was
not accomplished by adding one large beam but two smaller
beams as part of a new external frame.

Figure 1 shows a computer model of the east and west
perimeter reinforced concrete frames. These frames
needed to be strengthened without a significant loss of
window openings which are desirable for ventilation and
light. We strengthened the building by adding one frame
to the exterior of the building on both the east and west
faces. Figure 2 shows the form of the added external
frame with its "pyramid" shape. The frame was added to
the outside of the building in order to minimize
inconvenience to the existing building occupants.

---

  * Professor, Department of Civil Engineering,
    University of California, Los Angeles and Principal,
    Englekirk and Hart, Inc.
 ** Engineer, Englekirk and Hart, Inc.
*** Corporate Associate, Englekirk and Hart, Inc.

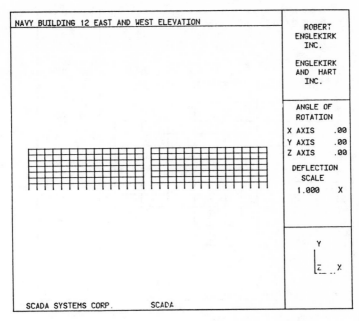

Figure 1: Computer Model of the East and West Elevation

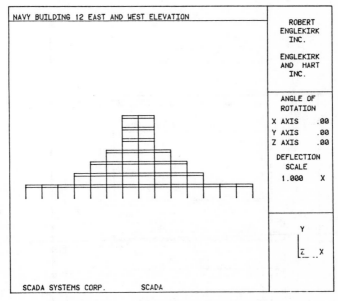

Figure 2: New Added Ductile External Frame

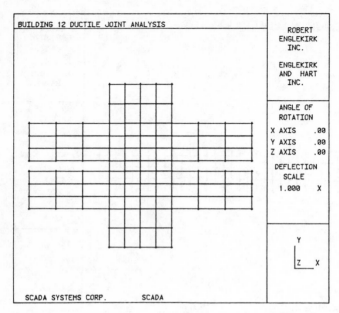

Figure 3:   Finite Element Model of the Ductile Joint

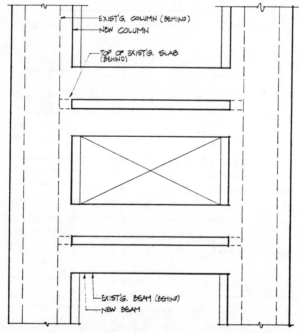

Figure 4:   Elevation of the Double Beam Scheme

Double beams were selected for the design because they provided the added frame stiffness necessary to limit the lateral frame deflections and also because they controlled the level of joint shear force. Stated differently, we have two hinges at the end of a bay, each hinge providing a plastic moment that transfers a shear to the joint approximately one-half that required for one large beam.

Figure 3 shows a finite element representative of the model. Figures 4 and 5 show the designed double beam deflection constraint control system.

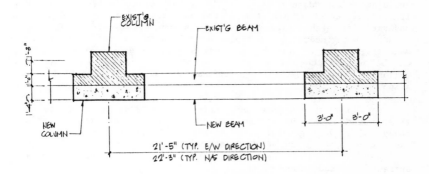

Figure 5:   Plan View of the New Added Concrete Frame

REGIONAL RISK ASSESSMENT OF EXISTING BUILDINGS

Howard M. Thurston,[1] A.M. ASCE
Haresh C. Shah,[2] M.ASCE

Abstract

A problem facing city building officials, engineers, architects, and other persons involved in implementing existing building strengthening programs for seismic safety is determining which buildings in a region pose the greatest risk. A building must be viewed in the context of its total risk, taking into account such attributes as regional seismic hazard and building vulnerability, importance, and occupancy. Obviously, there is not a single "correct" evaluation of a building's risk based on a particular combination of these attributes. It is believed, however, that if the attributes are described as low, moderate, or high, a general consensus can be reached among expert structural engineers in describing a building's risk as very low, low, moderate, high, or very high. In this way, an expert knowledge base can be built which can be used to screen the existing buildings in a region in order to determine which ones warrant a more detailed evaluation.

For most regions, however, there are too many buildings to consider, even using the simple preliminary screening method described above. Given very general data on a region's geological hazard, land use and occupancy, and predominant construction type, it is possible to identify high risk subareas within the region. The preliminary screening process could then be applied to those subareas only. Of course, this method may miss some high risk buildings which are not within the high risk subareas. An important part of the process, then, would be to have local authorities identify "special" buildings, important buildings in the region which must be considered, irrespective of their location in a high risk subarea.

The next step would be to perform a detailed inspection of the "special" buildings plus those identified as high risk by the preliminary screening process. Based on detailed data obtained from these inspections, a final list of high risk buildings could be obtained for the region.

---

[1] Graduate Research Assistant, The John A. Blume Earthquake Engineering Center, Department of Civil Engineering, Stanford University, Stanford, CA 94305.
[2] Professor and Chairman, Department of Civil Engineering, Stanford University, Stanford, CA 94305.

This paper describes current research underway at Stanford University designed to develop a methodology for determining high risk buildings in a region. The work involves developing a method to identify the high risk subareas. Once done, the buildings within those areas would be evaluated using an expert knowledge base to identify a preliminary list of high risk structures. Stanford researchers have developed two knowledge bases, along with a survey questionnaire designed to extract knowledge from experts to aid in creating a working knowledge base synthesized from the opinions of multiple experts. Two proprietary expert system "shell" programs have been used to implement such knowledge bases on a personal microcomputer. The buildings identified as high risk by the expert system plus the "special" buildings would be evaluated in detail using a building inspection questionnaire developed by Stanford researchers. The data obtained from the inspection would be used to update the risk factors for each building, which would then be input into an expert system to obtain a final list of high risk structures for the region.

The research described herein represents a new method of identifying high risk buildings in a region using an expert knowledge base. It is being validated in part with the cooperation of a local building department to ensure its practical relevance.

Introduction

During the past year, a group of researchers at Stanford University has been engaged in the task of developing a methodology to identify buildings in a region which are high risk due to potential seismic activity. The underlying philosophy of the research team has been to view a building in the context of its total risk, taking into account attributes such as regional seismic hazard, building function, building vulnerability, and building importance. Although intuitively appealing, it is obvious that there is not a single "correct" evaluation of a building's risk based on a combination of attributes. Rather, the evaluation is highly subjective. It is believed, however, that a general consensus exists among most earthquake engineering experts concerning the evaluation of a building in the high risk range. It is for this reason that the research team has adopted the idea of an expert system approach to combining risk attributes in order to assess a building's total risk. The problem then becomes one of being able to accurately assess a building's risk attributes. However, it is here that an expert system methodology works to its best advantage. While it is impossible to assess a building's risk attributes with perfect accuracy, an expert earthquake engineer should be able to assess a building's total risk given general descriptions of those attributes. For example, if a building is in a highly seismic region, of high importance and high vulnerability, the building is clearly high risk. Also, if a building in the same region is of very high vulnerability but of very low importance (perhaps a rarely used storage building), then the risk is certainly less than the first building. Although this may seem somewhat trivial, it is believed that an expert knowledge base can be built and generally agreed upon which can consistently assess the total risk of buildings in a region based on general descriptions of risk attributes, thereby making it possible to identify a preliminary list of buildings which should be evaluated in detail. This has the advantage of greatly simplifying the problem facing persons in knowing where to begin the process of addressing a region's existing building situation.

An expert system could then be used again with information obtained from a detailed evaluation of each of the preliminary buildings to obtain an updated assessment of each building's total risk. In this way a final list of high risk buildings can be obtained.

For a region with many buildings, the procedure described above is not enough to simplify the problem. Stanford researchers are working on a method of identifying high risk subareas within a region based upon very general descriptions of geological hazard, land use and occupancy, and predominant construction type. The expert system evaluation procedure would then be applied only to the buildings within those high risk subareas. Of course, certain buildings in a region must be classified as "special" buildings. These buildings would be a part of the preliminary list to be evaluated in detail described above, irrespective of their location in a high risk subarea.

Discussion of Methodology

The first step in simplifying the assessment of the existing building situation in a region is to reduce the number of buildings which must be evaluated. The reason for this step is obvious. Even in a moderate sized city such as Palo Alto, California (approx. population 50,000), the number of existing nonresidential buildings is approximately 2000. The manpower and budget available in most cities simply do not allow a detailed inspection of every building. In order to address this problem, Stanford researchers have developed a procedure known as the "Bird's Eye View" approach. For the region under consideration, the following data are required.

First, the geological hazard is considered. Geological hazards such as liquefaction potential, landslide potential, ground rupture potential, and strong ground shaking must be considered. The methodology developed so far calls for initially defining geological hazard in terms of strong ground shaking. The other hazards are considered later. So, the strong ground shaking is mapped. In Palo Alto, for example, the variation of strong ground shaking is due to either proximity to the San Andreas fault or the presence of so-called Bay Mud, a soil type which amplifies ground motion in certain frequency ranges.

Second, from zoning information, land use and occupancy are mapped. This information will indicate areas of commercial buildings and high occupancy.

Third, it is usually possible to identify, in very general terms, the distribution of predominant building types throughout the region. For example, again for the City of Palo Alto, the "downtown" area consists of buildings of which approximately 15% are unreinforced masonry, 60% are reinforced concrete, 15% are steel frame, and 10% are wood frame. Another section of the city consists of industrial buildings of which approximately 80% are pre-1973 tilt-up, 5% are post-1973 tilt-up, 5% are steel frame, and 10% are wood frame. Although this data is admittedly very coarse, it will tend to point out the areas of the city which contain the buildings of most concern, based on general performance data of broad building classes which have been collected by various researchers. Of course, there may be buildings of a type for which poor seismic performance is usually expected, located in an area containing generally better buildings. Thus, for the sake of simplicity, the process may tend to overlook some buildings. However, as will be discussed later, the inclusion of "special" buildings in the preliminary list will preclude overlooking important buildings.

Once these three pieces of data are collected, they must be combined in some way. The Stanford researchers are currently working on this step. As a simple first try, this information was gathered for Palo Alto and drawn on separate maps. These maps were then simply overlayed. This process indicated very distinct subareas of concern as follows: the downtown area, where most of the city's unreinforced masonry buildings are located and an area of high density during business hours, two industrial areas consisting of high percentages of pre-1973 tilt-up buildings located in areas of very strong ground motion potential (Bay Mud), and another industrial area with a high percentage of pre-1973 tilt-up buildings housing industries which are important to Palo Alto's economy. The Stanford team is currently working on a more refined method to combine these data. The methods under study represent theoretical approaches such as "fuzzy" techniques, or more intuitive ones such as an expert knowledge base.

Once the high risk subareas are identified, the proposed methodology will obtain a "first cut" list of buildings selected from all buildings within those subareas. The research team believes that an expert system approach is the simplest and most consistent way of identifying those buildings for which a detailed evaluation is warranted. During the course of the research, two expert systems have been developed, both utilizing available proprietary expert system "shell" programs designed for use on personal microcomputers. One, described schematically in Figure (1), is designed for the purpose of identifying the "first cut" list of buildings to be examined in detail. As such, it assigns a level of risk to a building based on risk attributes obtained from a few very general pieces of data. The other expert system, described in Reference (1), takes into account more detailed information and is more appropriate at a later stage in the methodology. In general, both systems involve creating a knowledge base consisting of rules which indicate the conclusions an expert might reach concerning a building's level of risk given information on geological hazard, importance, vulnerability, and occupancy (risk attributes). The geological hazards of liquefaction, landslide, and ground rupture potential and strong ground shaking are considered along with the seismic hazard to define the geological hazard for input to the expert system. The rules are written in linguistic form, using words such as low, moderate, or high to describe the various risk attributes, and very low, low, moderate, high, and very high to describe the building's total risk. The rules describe the relationship between a particular combination of risk attributes and the corresponding total risk. Obviously, such rules are very subjective,

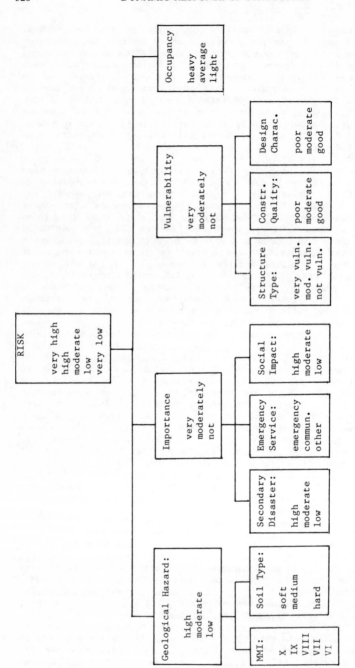

FIGURE 1

RISKFILE: A BUILDING RISK ASSESSMENT KNOWLEDGE BASE
Knowledge Structure of Risk Attributes

so the knowledge base will have to evolve until general consensus is reached by the experts involved in creating it. Stanford researchers have developed a questionnaire to be filled out by experts in order to build this knowledge base. However, it is important to understand that it is not necessary for the assigned risk levels to be absolutely accurate. Rather, the expert system will indicate relative risk levels, indicating which buildings in a subarea are at higher risk than the others. These, then, are the buildings which should be evaluated in detail, in combination with any especially important buildings in a region. The Stanford research team believes that identifying high risk buildings by the expert system approach is very important, because it allows any building in a high risk subarea to emerge as a candidate for strengthening, not just those buildings preselected for strengthening on the basis of their importance. The team believes that this aspect of the methodology is important in optimizing the benefits of a regional strengthening strategy in terms of reducing life, property, and economic losses. It is for this reason that a major portion of the research team's efforts have been directed toward the development of a methodology to identify which buildings in a region should be strengthened. It should be noted that the methodology explicitly does not consider historical buildings, but focuses rather on ordinary buildings.

The expert system to determine a building's overall risk consists of a hierarchical structure of smaller expert systems. For example, the geological hazards of liquefaction, landslide, and ground rupture potential, strong ground shaking, and seismic hazard must be combined to express an overall level of geological hazard. This geological hazard must then be combined with the other risk attributes, each of which consists of lower level combinations of data, to obtain an overall risk level. The Stanford researchers are considering also "fuzzy" techniques to be used either in lieu of, or to supplement, the expert systems at each hierarchical level.

Every region has important buildings which should always be considered in a strengthening program. Such buildings might include police and fire stations, hospitals, communication centers, electrical power substations, etc. The definition of these "special" buildings may vary from region to region. The proposed methodology would call upon regional authorities to identify such buildings in their region. These buildings, together with those identified by the expert system as high risk, would be evaluated in detail as described in the next step.

The Stanford research team has developed a building inspection questionnaire designed to be used by a competent structural engineer in order to obtain detailed information about a particular building. A detailed inspection will be performed on all "first cut" and "special" buildings described earlier. With the information so obtained, more accurate descriptions of the risk attributes can be used by an expert system to obtain a revised assessment of a building's risk. In this way, a final list of high risk buildings may be obtained.

In order to carry out this process, Stanford researchers have created a Building Classification Scheme based on structural system and material. Then, based on available data obtained from various researchers, it was possible to plot Mean Damage Ratio (DR) versus MMI for each of the building classifications, where DR is defined as the ratio of repair cost to replacement cost. Since these plots describe mean DR for broad building classes, there may be considerable difference between the predicted DR for a specific building of a particular class and that building's actual DR. However, this information can be used to describe the "first cut" vulnerability of a building for use in the expert system. Once again, of course, the actual vulnerability is not as important as the relative vulnerability between the building categories. The information obtained from the inspection questionnaire would then be used to assess a specific building's vulnerability. The research team is currently investigating the possibility of using a technique in which an expected DR at a specific intensity level ("equivalent" intensity) can be convolved with the probability of occurrence of that intensity level to determine an expected DR over a prescribed time period. The advantage of this technique is that using the usual convolution equation to obtain expected DR requires evaluating an expected DR for each intensity level. It is more feasible to inspect a building with only one particular intensity level in mind, namely the "equivalent" intensity, then determine the expected DR given the "equivalent" intensity, $E(DR/I^*)$, and the probability of occurrence of the "equivalent" intensity, $P(I^*)$. Convolving these two quantities would then give an expected DR for the particular building under consideration, $E(DR)$. Two problems arise, however. First, how is it possible to obtain $E(DR/I^*)$ for a particular building based on an inspection? Second, assuming the first problem can be solved, how will the expected DR so obtained be used to describe a building's vulnerability? This is an area of active work by the Stanford team.

The final step is to use an expert system again, using updated information on the risk attributes to determine a better informed assessment of a building's total risk. It is here that the expert system described in Reference (1), utilizing more building specific information, could be used to obtain a final list of high risk buildings. It should be noted that the final output of the expert systems developed to date is a list of buildings grouped according to risk level. The next step will be to determine which of these buildings should be strengthened in order to optimize the money spent in strengthening in terms of life safety and other socio-economic considerations. This process may involve the development of another expert system to prioritize the targeted buildings using information on collapse hazard, optimum strengthening level and option, building and market life, and strengthening versus replacement cost.

Two important points should be mentioned. First, the methodology has been designed to identify high risk buildings independently of any strengthening requirements which may be in force in a given city or region. Second, the methodology addresses the issue of buildings which, though not necessarily collapse hazards, may be at high risk for other reasons. (Of course, the methodology also considers buildings which pose a hazard due to collapse).

Conclusion

The work described in this paper represents the development of a framework methodology designed to identify high risk buildings in a region. Considerable work remains to be done to obtain a final, workable methodology which can be applied to cities and regions.

Also it is important to insure a consistent and compatible approach to the problem of first identifying the high risk buildings in a region and second, determining the strengthening priority and optimum methods. It is believed that the complex problem of identifying high risk buildings in a region is an appropriate arena for considering the application of an expert system based methodology.

Reference

(1) Miyasato, G.H., Dong, W., Levitt, R.E., Boissonnade, A.C., "Implementation of a Knowledge Based Seismic Risk Evaluation System on Microcomputers," August 1985.

Acknowledgement

The research work described herein was supported under National Science Foundation Grant No. CEE 84-03516.

ON RELIABILITY–BASED STRUCTURAL OPTIMIZATION FOR EARTHQUAKES

Mohammad Khalessi[1] and Lewis P. Felton[2]

ABSTRACT

Previous studies of reliability–based structural optimization have been extended in order to (1) investigate the applicability of the linear statistical model in cases where random parameters have relatively large coefficients of variation such as occur in earthquake related problems, and (2) compare results from the linear statistical model with those of a nonlinear statistical model. The validity of these formulations has been assessed by Monte Carlo Simulation. Applications to earthquake resistant designs are presented which include statistical characterizations of earthquake response spectrum shape, structural stiffness characteristics and peak ground acceleration. Examples include elastic designs and design of a structure which may yield significantly but which must avoid collapse within its lifetime.

INTRODUCTION

Problems of structural design must generally be resolved in the face of various uncertainties. Uncertainties arise not only in the assessment of loads which the structure has to sustain and from the occasional lack of control during the production processes of the materials and components, but also from idealizations and possible inadequacies inherent in the analytic models describing the response of the structure and it capacity to sustain the loads. As a consequence, the evaluation of "reliability", or the related "probability of failure", during the structure's lifetime is highly desirable when searching for an optimal balance between the technical, economic and social consequences of failure and initial construction expenditure.

Much of the literature in the area of structural reliability has been devoted to procedures for the analysis of probabilities of failure of structures. A significant portion of the literature has dealt with design of structures with random parameters. In recent years

---

1. Civil Engineering Department, University of California, Los Angeles, California 90024. Presently with Airesearch Manufacturing, Company, Torrance, California.

2. Civil Engineering Department, University of California, Los Angeles, California 90024.

optimization techniques have been applied to a wide range of structural design problems, including problems involving non-deterministic parameters which must be treated in a probabilistic format (3-9). Optimum design of earthquake-resistant structures, however, is a subject which has received relatively little attention. In Reference 3 a general formulation of the non-deterministic optimization problem based on a linear statistical model was presented for the minimum-weight design of structures subjected to an earthquake ground motion and constrained by a specified upper bound on overall probability of failure. The formulation included a probabilistic representation of earthquake response spectrum shape and structural stiffness characteristics, but peak ground acceleration was treated as a deterministic quantity.

The main objectives of the present appear are extensions of Reference 3 aimed at (1) an investigation of the applicability of the linear statistical model in cases where the random parameters have relatively large coefficients of variations, such as occur in earthquake related problems, and (2) comparison of results from the linear statistical model with those of a suitable non-linear statistical model. The validity of these formulations is assessed by Monte Carlo Simulation. Additional objectives of the present work, for particular application to earthquake resistant design are (3) the extension of the data in Reference 3 to include a statistical characterization of peak ground acceleration, and (4) an exploration of the optimum design of a structure which may yield significantly but must avoid collapse within its lifetime.

GENERAL FORMULATION

For a structure with random parameters, the optimization problem may be stated in the following form:

$$\text{Minimize} \quad W(\underline{X}) \tag{1}$$

$$\text{subject to} \quad P_o(\underline{X}) \leq p_o \tag{2a}$$

$$P_i(\underline{X}) = \Pr[Z_i(\underline{X}) \leq 0] \leq p_i, i=1,2,\ldots M \tag{2b}$$

where $\underline{X}^T = [X_1, X_2, \ldots X_N]$ is a vector of N random parameters in which it is assumed $X_k, k=1, \ldots n, n \leq N$, are design variables. $Z_i(\underline{X})$ in (2b) relates the ith response quantity of the system to its allowable limit; p_o is the specified probability of failure for the whole structure and $P_o(\underline{X})$ is the actual probability of failure for the entire system; $P_i(\underline{X})$ is the probability of the ith response exceeding its allowable limit and p_i is the specified allowable probability for this occurrence.

Assuming the random variables to be normally distributed, the final general form of the inequality-constrained optimization problem is (5)

$$\text{Minimize} \quad W(\bar{\underline{X}}) \tag{3}$$

$$\text{subject to} \quad g_i(\underline{X}) = \sum_{i=1}^{M} P_i(\underline{X}) \leq p_o, i=1,2,\ldots M \tag{4a}$$

$$g_{i+1} = \beta_i S_{z_i} - \bar{Z}_i \leq 0, i=1,2,\ldots,M \tag{4b}$$

where $(\bar{\underline{X}})$ is the vector of the mean random variables. $\bar{Z}_i$ and S_{z_i} are the mean and standard deviation of z_i, respectively, calculated at the mean values of the design variables, and $\beta_i = \bar{Z}_i / S_{z_i}$.

The solution of this optimization problem will provide the mean values of the design variables. These values will also be used to compute $\bar{Z}_i$ and S_{z_i}, $i=1,\ldots M$, by Monte Carlo simulation. The validity of the present formulation will be assessed by comparing the obtained results.

EXAMPLES BASED ON LINEAR STATISTICAL MODEL

Details of procedures for computations of means values and standard deviations of response quantities based on a linear statistical model have been presented in Ref. 5 and will not be repeated here. Illustrative optimum designs for several examples are presented directly for comparison with results of Monte Carlo simulations.

The structures considered are the uniformly loaded, simply supported beam shown in Figure 1, and the two-story single-bay frame shown in Figure 2. Geometries of these structures are assumed to be specified and deterministic. In all cases applied loads, Young's moduli and the design variables (cross-sectional dimensions in Figure 1, moments of inertia in Figure 2) are assumed to be random parameters. Each of these random parameters is characterized by its mean value and coefficient of variation, with the coefficient of variation taken as a prespecified quantity. In all examples, constraints are placed on overall probability of failure, (4a), on the probabilities of design variables falling below minimum allowable values, and on the probabilities of member stresses and system displacements exceeding allowable values, Eq. (4b). The objective function is taken as mean weight.

Since response quantities for these examples are system displacements and element stresses, it is necessary to compute partial derivatives of these displacements and stresses with respect to the random parameters in order to obtain the standard deviations for use in the constraint equations. These partial derivatives are computed by finite differences.

All inequality-constrained minimization problems of the form of Eq.

(3) were solved herein by using the COPES/CONMIN computer program.

Beam Example

The beam in Figure 1 is designed to transmit a unformly distributed random load with mean amplitude per unit length q as shown in figure 1. Denoting the cross-sectional dimensions (design variables) by h and b, and allowable normal and shear stresses by σ_a and τ_a, respectively, the vector of independent random parameters, $\underline{X}$, may be written as

$$\underline{X}^T = \left\{X_1, \ldots X_N\right\} = \left\{b, h, \sigma_a, \tau_a, q\right\}$$

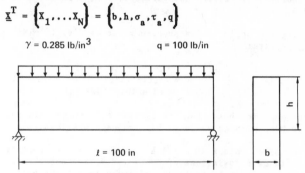

$\gamma = 0.285 \text{ lb/in}^3$ \qquad $q = 100 \text{ lb/in}$

$\ell = 100 \text{ in}$

Figure 1. Uniformly loaded beam.

A general statement of the design problem for this example will contain five constraint equations of the form of Eq. (4b), namely two stress constraints, one displacement constraint, and two minimum cross-sectional dimension constraints.

In the present example, mean values of maximum allowable normal and shear stresses are taken as $\sigma_a = 47$ ksi (324,065 kPa) and $\tau_a = 20$ ksi (137,900 kPa) with standard deviations $S_\sigma = 3.0$ ksi (20,685 kPa) and $S_\tau = 1.2$ ksi, (8270 kPa), respectively, and the mean value of the applied load as $\bar{q} = 100$ lb/in (17.5 kN/m) with standard deviation $S_q = 80$ lb/in (14.0 kn/m). Note that this implies a coefficient of variation for load of 0.80. Maximum allowable beam deformation, U_a, is taken as $U_a = 0.075$ in (1.91 mm, deterministic value), and the coefficients of variation for h and b are assumed to be equal to 0.1. The density is assigned the hypothetical value 0.285 lb/in^3 (7889 kg/m^3), and Young's modulus $E = 10 \times 10^{10}$ psi (6.895×10^{10} kPa). The minimum allowable value for h and b is 0.01 in (0.25 mm).

Note that Z_i is defined as allowable response minus actual response; therefore

$$Z_1 = U_a - \frac{60 q l^4}{384 E b h^3} \qquad\qquad (5a)$$

$$Z_2 = \sigma_a - \frac{3q l^2}{4bh^2} \tag{5b}$$

$$Z_3 = \tau_a - \frac{3a l}{4bh} \tag{5c}$$

The overall probability of failure, p_o, is taken to be 0.001, and the probability of failure associated with Z_1 through Z_3 is taken as P_i = 0.000233.

Solving the optimization problem given by Eqs. (3)-(4) yields the following results:

$$\overline{b} = 0.0336 in. (0.85mm)$$
$$\overline{h} = 44.2 in. (1123mm)$$
$$W = 42.4 \ lb. \ (19.2 \ kg)$$

Since buckling has been neglected and no constraint has been imposed on h/b, a large value of h/b has been obtained.

The obtained values of $\overline{b}$ and $\overline{h}$ have been used to compute the mean value and standard deviations of Z_i, i=1,2,3, by Monte Carlo simulation. The results are shown in Table 1, and indicate good agreement with the results of the linear statistical model. The maximum difference is 10.4% for standard deviations and 6.4% for mean values.

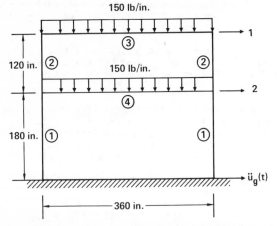

Figure 2. Two-story single-bay frame

Frame Examples

Two illustrative designs are presented for the frame shown in Figure 2. The objective of each example is to obtain the minimum-weight structure for which the overall probability of failure does not exceed a specified allowable value when the frame is subjected to a ground motion which is characterized by a response spectrum of the type shown in Figure 3 (1). This spectrum is based on a study of thirty-three historical earthquakes. The values in this figure correspond to a peak ground acceleration normalized to 1.0g. The corresponding coefficients of variation of spectral displacements are also shown in the figure. It may be noted that these coefficients are generally large, particularly at low values of natural frequency.

The use of this spectrum for the analysis of structural response implies that the behavior is linear; consequently, the designs under consideration will be taken as limit designs for which $P_i(\underline{X})$, i=1,...,12, are probabilities of stresses at the ends of each beam and column member exceeding material yield strength. $P_{13}(\underline{X})$ and $P_{14}(\underline{X})$ are taken as probabilities of displacements at coordinates 1 and 2 (Figure 2) exceeding allowable values, Eq. (4b). In addition to the stress and displacement constraints, constraints are placed on overall probability of failure, Eq. (4a), and on design variables, $X_i = I_i$, i=1,...,4, falling below minimum specified values, where I_i is the moment of inertia of member i.

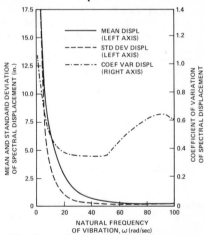

Figure 3. Generalized response spectrum shape parameters for 0.02 damping ratio, ref. 1 (Normalized with respect to a peak ground acceleration of 1.0g; 1 in. = 2.54 cm)

Example 1: The frame in Figure 2 is designed to resist the ground motion which is characterized by the response spectrum shown in Figure 3, such that the overall probability of failure does not exceed 0.001. The allowable probability of individual constraints being violated, P_i, is taken to be 0.5. The vector of random parameters, $\underline{X}$, may be written as

$$\underline{x}^T = \left\{x_1, \ldots, x_8\right\} = \left\{I_1, I_2, I_3, I_4, \rho_1, \rho_2, E, \ddot{u}_g\right\}$$

where ρ_1 and ρ_2 are the displacement response spectrum values for the first two natural frequencies, E is Young's modulus, and $\ddot{u}_g$ is the ground acceleration. Mean modulus of elasticity, $\bar{E}$, is taken as 30×10^6 psi (20.7×10^{10} kPa) and the corresponding coefficient of variation as 0.15. Mean ground acceleration $\ddot{u}_g = 0.1g$ and standard deviation for ground acceleration $S_{\ddot{u}_g} = 0.05g$ (10). Coefficients of variation of all I_i are taken as 0.05, and the minimum allowable value for moments of inertia is taken to be 50 in^4 (20.8×10^6 mm^4). All cross-correlation coefficients are assumed zero. All random variables are assumed to be normally distributed.

The yield strength of steel and allowable displacements at coordinates 1 and 2 (Figure 2) are assumed to be deterministic, and are taken as 36,000 psi, (248,220 kPa), 3.0 in, (76.2 mm) and 1.8 in, (45.7 mm) respectively. Also, damping is assumed to be a deterministic value equal to 0.02.

Mean stresses and weights are obtained from the approximate relations between section modulus, weight per unit length, and moment of inertia for wide-flange economy sections given in Ref. 2.

$$S_e = (60.6I + 84100)^{1/2} - 290 \quad 0 \le I \le 9000 \text{ in}^4 \tag{6a}$$

$$S_e = (I - 80563)/1.876 \quad 9000 \le I \le 20,000 \text{ in}^4 \tag{6b}$$

$$w = \frac{\sqrt{I}}{7.516} \quad 0 \le I \le 9000 \text{ in}^4 \tag{6c}$$

$$w = (I + 2300)/903.63 \quad 9000 \le I \le 20000 \text{ in}^4 \tag{6d}$$

Element weights, plus an additional static load of 150 lb/in (26.25 kn/m) on the beam, are assumed to contribute to lumped masses at coordinates 1 and 2.

A general statement of the design problem contains eighteen constraint equations of the from of Eq. (4b), namely, twelve stress constraints (one constraints at each end of each member), two displacement constraints and four minimum- moment-of-interia constraints. Typical stress constraints have the form

$$g_k = |\bar{\sigma}_{ia}| - \beta_k S_{\sigma_{ia}} - \sigma_y \le 0$$

$$\beta_k = \frac{\bar{Z}_k}{S_{Z_k}}$$

(7)

where σ_y is yield stress which is assumed constant for both tension and compression, σ_{ia} and $S_{\sigma_{ia}}$ are the mean and standard deviation of stress in each member i at end a, respectively. A similar inequality constraint can be written for displacement.

As may be seen from Table 2, a minimum-weight design based on the probabilistic response spectrum defined by the parameters in Figure 3, with magnitudes modified to reflect the peak acceleration, leads to I_i, i=1,...,4, of 2812.6 in^4 (11.7x10^8 mm^4), 628.7 in^4 (2.6x10^8 mm^4), 579.7 in^4 (2.4x10^8 mm^4), and 974.3 in^4 (4.1x10^8 mm^4), respectively, with the corresponding weight equal to 5926.0 lb (2688 kg).

It may also be seen from Table 2 that the design is actively constrained by the prescribed overall probability of system failure (0.001).

The obtained values of I_i were used to calculate mean and standard deviation of Z_i i=1,...14 using Monte Carlo simulation. These results (Ref. 5) show maximum differences of 1.26% for mean values and 2.05% for standard deviations, and indicate very good agreement between the linear statistical model and the Monte Carlo simulation.

Example 2: An alternative to the preceding elastic design approach may be states as follows (10):

   (i) the structure should respond elastically to a "moderate" earthquake of an intensity reasonably anticipated within its lifetime and,

   (ii) the structure may yield significantly but must avoid collapse during a maximum credible earthquake.

It is important to note here that this approach to design seeks to avoid collapse. There are several procedures which are purported to lead to designs which will yield without collapsing. The most popular of these is the so-called strong column-weak girder design philosophy (10). In this approach a structure is designed so that during a strong earthquake inelastic activity is confined as much as possible to the beams rather than the columns. The basic belief is that if the columns remain elastic, or nearly so, then collapse is very unlikely.

In order to obtain a reasonable measure of the non-linear behavior of a structure without carrying out a true nonlinear analysis, the ductility-factor method has been used. The basic assumption of this method is that the deflections produced by a given earthquake are essentially the same, whether the structure responds elastically or yields significantly.

In order to conduct a detailed examination of the design problem at hand an explicit formulation of the system constraints must be presented. For the purpose of this exposition it is not necessary that these constraints be formulation exactly as they appear in standard practice but rather that they represent only good approximations. To accomplish this, all of the internal force constraints may be expressed in terms of moments rather than stresses as is typically done in practice. Thus, internal force limits are placed on member plastic moment rather than yield stress. In the case of beam members the constraint functions which result from the moment and stress approaches are virtually identical; differences occur only when dealing with columns.

Normally, in the specification of dynamic system constraints, dead/live load effects on the beams are accommodated in addition to the earthquake loading results. This is the format followed here.

As noted, the strong column-weak girder philosophy attempts to confine inelastic activity to the beams as much as possible. In terms of ductility ratio this means that the ductility demands of each member must be less than some specified allowable which, for the columns, is 1.0 or close to 1.0. For this discussion let M_T represent the total maximum moment in a particular member. Then by the ductility factor method, the general form of the moment constraints is $M_T/M_p \leq \mu_a$ where μ_a is the allowable ductility, and M_p is defined as follows:

$$M_p = \begin{cases} \sigma_y S_p & \text{for } P/P_y \leq 0.15 \\ 1.18 \, \sigma_y S_p (1-P/\sigma_y) & \text{for } P/P_y > 0.15 \end{cases}$$

where σ_y is yield stress, P is the axial load, S_p is the plastic section modulus which, for standard structural shapes, is given by (10)

$$S_p = 1.13 \, S_e \tag{9}$$

where S_e is the elastic section modulus. $P_y = A\sigma_y$ is the yield load where A is area, and is defined as (10)

$$A = 1.1 \, \frac{S_p^2}{I} \tag{10}$$

where I is moment of inertia.

The frame in Example 1 has been redesigned based on ductility allowables of 6.0 for the beams, 2.0 for the first-story columns, and 1.0 for the second-story columns (10). All other parameters are considered to be the same as in Frame Example 1.

As has been indicated in Table 2, a minimum-weight design for normally distributed random variables leads to mean moments of inertia of members 1-4 (Figure 2), equal to 2180.3 in^4 (9.1x10^8 mm^4), 724.4 in^4 (3.0x10^8 mm^4), 24.1 in^4 (0.1x10^8 mm^4), and 209.4 in^4, (0.9x10^8 mm4), respectively, with the corresponding weight equal to 3981.8 lb (1806 kg). This is 33% lighter than the design in example 1.

The obtained values of I_i were used to calculate mean and standard deviation of Z_i, i=1,...14, using Monte Carlo simulation. Comparison of the results again shows good agreement between the two methods (5).

EXAMPLES BASED ON NONLINEAR STATISTICAL MODEL

Consider a single function, $Z(\underline{x})$, which is dependent upon n independent random variables, $X_1, X_2, ... X_n$. If second order terms of Taylor series expansions about mean values of independent random variable, $X_1, X_2, ... X_n$, are used to derive the mean value and standard deviation of functions $Z, \bar{Z}$, and S_z, it can be shown that (5)

$$\bar{Z} = E[Z(\underline{X})] = Z(\underline{\bar{X}}) + \frac{1}{2} \sum_{i=1}^{N} \frac{\partial^2 Z(\underline{\bar{X}})}{\partial X_i^2} S_{X_i}^2 \qquad (11)$$

$$S_Z^2 = \sum_{i=1}^{N} \left(\frac{\partial Z(\underline{\bar{X}})}{\partial X_i} S_{X_i} \right)^2 + \frac{3}{4} \sum_{i=1}^{N} \left(\frac{\partial^2 Z(\underline{\bar{X}})}{\partial X_i^2} S_{X_i}^2 \right)^2$$

$$+ \sum_{i=1}^{N-1} \sum_{j=i+1}^{N} \left(\frac{\partial^2 Z(\underline{\bar{X}})}{\partial X_i \partial X_j} S_{x_i} S_{x_j} \right)^2 + \frac{1}{2} \sum_{i=1}^{N-1} \sum_{j=i+1}^{N} \left(\frac{\partial^2 Z(\underline{\bar{X}})}{\partial X_i^2} \frac{\partial^2 Z(\underline{\bar{X}})}{\partial X_j^2} S_{X_i}^2 S_{X_i}^2 \right)$$

$$- \frac{1}{4} \left(\sum_{i=1}^{N} \frac{\partial^2 Z(\underline{\bar{X}})}{\partial X_i^2} S_{X_i}^2 \right)^2 \qquad (12)$$

Using Eqs. (11) and (12) in the optimization problem given by Eqs. (3) and (4) for the beam example examined previously yields the following results:

$$\underline{\bar{b}} = .034 \text{ in } (.086 \text{ mm})$$
$$\underline{h} = 63.4 \text{ in } (1610.4 \text{ mm})$$
$$w = 60.7 \text{ lb } (27.5 \text{ kg})$$

Comparing these results with those for the preceding beam example, it is seen that the present design is 23% heavier.

A comparison of the present differences between Monte Carlo simulation and the nonlinear statistical model results with the differences obtained previously (i.e., differences between Monte Carlo simulation and the linear statistical model results) indicates that better results are obtained in the present case for the mean values of Z_2 and Z_3 (2.2% and 2.7% in comparison with 3.2% and 5.0% for Z_2 and Z_3, respectively) and the standard deviation of Z_3 (10.0% vs. 10.4%); however, better results are obtained for the mean value of Z_1 and standard deviation of Z_1 and Z_2, using the linear approximation. This clearly shows that the nonlinear statistical model may not necessarily be superior be the linear statistical model. It is worthwhile mentioning that for problems with a large number of random variables the nonlinear approach could be more costly than Monte Carlo simulation. For example, for a problem with 30 random variables, 30 first derivatives and 900 second derivatives should be calculated in order to formulate the problem. The same problem can be formulated using Monte Carlo simulation with 500 samples for each constraint. Therefore the use of a nonlinear statistical model cannot be recommended for problems with a large number of random variables.

CONCLUSIONS

The present study has demonstrated the feasibility of using linear statistical model and reliability–based optimization techniques to obtain earthquake–resistant structural designs. This approach is particularly attractive for earthquake engineering applications which involve parameters most logically characterized in probabilistic form. The relatively direct problem statement formulated herein enables the designer to specify maximum allowable probabilities of occurrence of individual failure modes and of overall system failure. Results are primarily dependent on the reasonable expectation that the designer should be able to provide realistic estimates of coefficients of variation for the various random parameters in the system.

The present study, which has been performed up to the maximum value of coefficient of variation of 0.8, shows that a linear statistical model may be used for reliability–based optimization of earthquake resistant structural designs. However, it is worthwhile mentioning that even though the obtained results are very satisfactory, they are dependent upon the values of the coefficients of variation.

For the beam example the statistics of structural response were also formulated based upon a nonlinear statistical model. The results clearly showed that a nonlinear statistical model may not necessarily lead to better results than does the linear statistical model. It is also worthwhile repeating that for problems with a large number of random variables the nonlinear approach could be more costly than Monte Carlo simulation.

This study has also shown that lighter designs can be achieved by allowing the structure to yield but not collapse.

ACKNOWLEDGEMENTS

This research was partially supported by National Science Foundation Grant PFR78–23093.

REFERENCES

1. Blume, J.A. & Associates, "Recommendation for Shape of Earthquake Response Spectra", WASH–1254, UC–11, February 1973.

2. D.M. Brown and A.H. Ang, "Structural Optimization by Nonlinear Programming", J. Struct. Div., ASCE 92(ST6) pp. 319-340 (Dec. 1966).

3. W. Davidson, L.P. Felton and G.C. Hart, "On Reliability-Based Optimization for Earthquakes", Journal of Computers and Structures, Vol. 12, 1979, pp. 99-105.

4. Hilton and M. Feigen, "Minimum Weight Analysis Based on Structural Reliability", J. Aerospace Sci. 27(9) pp. 641-652 (Sept. 1974).

5. Khalessi, M.R., Reliability Based Optimization for Design, Ph.D. Dissertation, University of California, Los Angeles, 1983.

6. F. Moses and D.E. Kinser, "Optimum Structural Design with Failure Probability Constraints", AIAA J. 5(6) pp. 1152-1158 (June 1967).

7. P.N. Murty and G. Subramanina, "Minimum Weight Analysis Based on Structural Reliability", AIAA Jour. No. 6(10), pp. 2037-2039 (Oct. 1968).

8. M.K. Ravindra, N.C. Lind and W. Sin, "Illustrations of Reliability-Based Design", J. Struct. Div., ASCE 100 (ST9), pp. 1789-1811 (Sept. 1974).

9. M. Shinozuka and T.N. Yang, "Optimum Structural Design Based on Reliability and Proof-load Testing", NASA Tec. Rept. 32-1042 (June 15, 1969).

10. N.D. Walker, Jr., "Automated Design of Earthquake Resistant Multistory Steel Building Frames", Earthquake Engineering Research Center, University of California, Berkeley.

Linear Stat. Model	Monte Carlo simulation	% Difference
Z_1 (in)	0.0560	0.0524	6.4
Z_2 (psi)	40.31×10^3	38.99×10^3	3.2
Z_3 (psi)	15.02×10^3	14.27×10^3	5.0
S_{Z_1} (in)	0.0160	0.0157	1.9
S_{Z_2} (psi)	6.31×10^3	5.7×10^3	9.3
S_{Z_3} (psi)	4.22×10^3	3.78×10^3	10.4

Table 1   Beam problem: means and standard deviation of Z_i, i = 1,2,3 (1 m = 25.4 in; 1 psi = 6.895 kPa)

Example 1		Example 2	
W lb	5926.0	W lb	3981.8
P_0	0.001003	P_0	0.0009995
I_1, in^4	2812.6	I_1, in^4	2180.3
I_2, in^4	628.7	I_2, in^4	724.4
I_3, in^4	579.7	I_3, in^4	24.1
I_4, in^4	974.3	I_4, in^4	209.4
ω_1 rad/sec	15.401	ω_1 rad/sec	10.03
ω_2 rad/sec	56.698	ω_2 rad/sec	48.37
ρ_{1G} ,in	4.43	ρ_{1G} ,in	0.682
$s_{\rho_{1G}}$,in	1.62	$s_{\rho_{1G}}$,in	0.373
ρ_{2G} ,in	0.29	ρ_{2G} ,in	0.434
$s_{\rho_{2G}}$,in	0.153	$s_{\rho_{2G}}$,in	0.182

Table 2. Frame problem: design results
(1 in. = 25.4 mm; 1 in^4 = 4.16 $\times 10^5$ mm^4 ; 1 lb = 0.4536 kg)

SEISMIC RESTRAINT OF HAZARDOUS PROCESS PIPING

V.N. Vagliente,[1] Bahman Lashkari[2] and C.A. Kircher[3]

Abstract

The lack of appropriate guidelines for the design and installation of hazardous process piping in industrial facilities has created
a potentially dangerous situation. In the electronics industry, for
example, corrosive and flammable liquids are routinely used in the
manufacture of semiconductors and electronic circuitry. Process piping
carrying these liquids is usually supported only in the vertical
direction. Seismically generated forces require the piping to be
restrained horizontally. At the present time no building code addresses the design and installation of flexible piping systems in industrial facilities. This paper proposes a methodology from which piping
system design guidelines can be developed and in so doing, attempts to
fill a gap in the building codes.

I. Introduction

The installation of process piping systems carrying hazardous
materials, such as flammables or caustics in high technology industrial facilities, presents an added seismic hazard. Overhead process
piping is usually supported vertically for gravity loads only.
Horizontal seismic restraints are usually ignored. In fact, the
seismic loading of these piping systems is not routinely included in
the design process. The failure of a piping system carrying hazardous
materials could exacerbate the situation caused by an earthquake by
adding fire or a burn hazard to the occupants of a building.

This situation will become more severe as plastic piping become
more readily available. The lack of seismic restraint is critical to
the survivability of plastic piping during an earthquake. Plastic
piping does not possess the strength and ductility of the metal piping
which it is replacing. Hence, the failure of its support system or
the lack of an adequate support system will decrease its chance of
surviving an earthquake.

---

[1]Assoc. Professor, Department of Civil Engineering
University of Nevada, Reno, NV   89557
[2]Senior Engineer, Jack R. Benjamin & Assoc., Inc.
Mountain View, CA   94041
[3]President, Jack R. Benjamin & Assoc., Inc.
Mountain View, CA   94041

The lack of appropriate guidelines for the design and installation of hazardous process piping in conventional industrial facilities should be of concern to all design engineers. The purpose of this paper is to focus attention on the principal contributors to the added seismic risk and to make recommendations for their mitigation.

II. Identification of Present Code Requirements

The requirements of present day building codes are too vague for piping design. For example, Chapter 23 of the Uniform Building Code (UBC) (1) specifies lateral loads for rigid piping only without providing design procedures or loads for flexible or flexibly-supported piping. Further, the UBC does not address the question of possible dynamic amplification caused by building vibration or by the flexibility of the piping systems. There are discussions of pipe bracing systems in industrial literature. Unfortunately, these presentations are not adequate since they focus on the adequacy of the bracing rather than on the response of the piping system and neglect the dynamic characteristics of the piping and the difference in the properties of the various pipe materials. Furthermore, they do not address the design of piping systems containing hazardous materials.

The American Society of Civil Engineers (ASCE) has published a guideline for the design of oil and gas pipelines (12). However, the publication presents a theoretical approach to the design and installation of piping systems and not a ready-to-use design basis procedure. The same comment can be applied to a recent American Society of Mechanical Engineers (ASME) (13) publication. These publications, and others like them, do not present a design procedure for use by engineers who do not have the comprehensive knowledge required to apply the theoretical aspects of the problem.

The nuclear industry has developed a sound approach for the seismic analysis and design of potentially hazardous power plant piping systems. However, their approach would be extremely costly to apply to conventional industrial piping systems. The reason for this is that, in the nuclear industry, each piping system is individually analyzed and designed using finite element computer programs or through a combination of analysis and test.

There is, at present, no code which directly or adequately addresses the design and installation of piping systems in industrial facilities.

III. Categorization of Hazardous Materials

The degree of the hazard presented by the transported material will determine the design basis for the piping system. There are two sets of criteria which determine the categorization and, thereby, the importance of the piping to the seismic hazard.

The criteria defined by the Environmental Protection Agency (EPA) must be met if there is a chance that the contaminants could get into the environment. If the working area could be contaminated, the requirements of the Occupational Safety and Hazards Administration (OSHA) have to be met. These requirements often overlap and are made more stringent by additional state requirements.

Briefly stated, these criteria classify flammables, caustics, and pyrophorics by one of five categories:

0   No Hazard,

1   Slight Hazard,

2   Moderate Hazard,

3   High Hazard, and

4   Extreme Hazard.

In providing seismic restraint to the process piping, the level of the material hazard must be considered.

IV. Background

It is the motion of the building itself which transfers loads and/or imposes deflections on piping systems which are supported by structures which, in turn, are attached to the buildings. In the case of an earthquake, the motion of the soil beneath the building is transmitted throughout the structure of the building and, consequently, to all the structure attached to the building. A piping system will respond to the motion of the building. Its response can be quite different from the motion of the building because of its inertia and the stiffness of the structure by which it is attached to the building (4). In any case, the dynamic characteristics of the piping system must be considered in addition to those of the building.

The motion of the building which we consider is initiated by an earthquake, but the excitation could be caused by a hurricane, internal or external explosion, or any other event of such force to set the building in motion. When the building responds to any event by moving, the piping system will also be exposed to this motion. Thus, besides the vertically directed weight of the pipes, contents, and attachments, there will be additional vertical and horizontally imposed loads or deflections. One horizontal component of the inertia load can be taken perpendicular to the axis of the piping and the other parallel to this direction. To these loads must be added the internal pressure of the fluids or gases being transported by the piping and any loads associated with the motion of these fluids or gases. These loads constitute a load combination which must be reacted by the piping system and its supporting structure. Where deflections rather than loads are more easily recognizeable, they can be reduced to equivalent loads and included in the load combination.

Thus, the response of the piping system is due to a multiplicity of loads which could all be acting simultaneously. Due to the diverse geometry of a piping system, different segments can be loaded quite differently. One segment might be loaded primarily in bending; while, in another segment, the primary load might be applied axially.

Since different types and sizes of buildings will respond diversely to a seismic or other external excitation, a properly designed piping system should reflect the characteristic dynamic behavior of the building. A flexible building might transmit larger loads to a piping system than a relatively stiff building.

However, it would not be economical to design a piping support system solely on the dynamic characteristics of the building. The type of fluid or gas carried by the piping system should also be considered. If a relatively inert substance which poses no danger should spill from the piping system while being transported, design requirements should be less stringent than if a toxic substance were being transported. Consequently, there are two significant criteria for establishing the design basis of a piping system in a building; the first is the dynamic characteristics of the building and the second is the danger or risk which failure of the piping system would create. The danger, of course, is due to the hazardous material transported by the piping. This latter consideration is designated the "importance" of the piping system. These two considerations constitute the design basis for the piping system.

In References 1 and 5, very general guidelines are presented for the analysis of rigid piping systems and for the definition of loads resulting from an earthquake. Unfortunately, this latter document does not distinguish between the diverse dynamic response of building types or take into account the flexibility of the piping; nor does it treat the importance of the piping system. This document presents loads which are probably too conservative in some cases and unconservative in others. In order to make the installation of a piping system cost-effective, more precise guidelines reflecting the dynamic characteristics of the building and piping and the importance of the piping must be considered.

In References 6 through 8, additional information is available on the analysis and design of piping systems. Much of this information is not applicable to the installation of piping in industrial buildings. Furthermore, what is applicable is not in a form which is suitable for the needs of those engineers installing piping systems in industrial buildings.

V. Recommended Approach to Analysis

Our approach, takes into account all the information presently scattered in diverse building codes. Utilizing previous experience with the analysis of piping systems, this information is integrated into a rational procedure for the design and installation of piping systems in buildings. Finally, using this approach a streamlined design guideline can be developed for metal and plastic piping which provides practical spacing limits and bracing details as a function of the piping contents and the seismic environment.

Furthermore this approach is applicable to all piping systems from those carrying hazardous materials to those carrying nonhazardous materials. But the level of pipe support and restraint will, of course, increase with the "importance" of the piping and will, perhaps, consist only of vertical supports for nonhazardous piping.

An analytical approach which will be simple to use can be developed based on the established Response Spectrum Method. This approach has been shown to be applicable to piping systems by virtue of its many years of use in the nuclear power industry for piping analysis. Furthermore, computer programs have been developed to calculate building response at various elevations. This routine analysis includes seismic zonation, the soil properties underlying the building, the diverse systems of resistance of buildings to lateral forces. Using this approach the floor response of buildings can be calculated efficiently and economically. The floor response spectrum defines the loads on the piping system. Consequently, a second analysis can be performed again using the Response Spectrum Method. This analysis will provide the stresses in the piping system. The stresses in the system can be calculated as a function of the support spacing. This could be an iterative process where the support system is acceptable when all the stresses in the system are within established allowables. A second approach would be to introduce probabilistic system analysis. The main advantage to this approach is that a probability of failure could be selected for the design of the piping system which is consistent with the probability of failure for the building itself. Using this overall criteria, generic support designs for piping of given diameter, material, and wall thickness could be developed. The results of the analysis will permit the calculation of the spacing interval for the gravity and seismic restraints. It is this final analytical procedure which can be reduced to a design procedure using tables and charts.

Since plastic piping is seeing more application, the methodology developed above will have to include plastic piping. This will require that performance criteria be established for polyvinyl chloride (PVC), fiber reinforced (FRP), and other polymer piping used for industrial piping. Not only will the definition of allowable stresses be required but also the failure strengths of glued and threaded pipe joints are necessary. To develop this information may require both analytical and test investigations. For PVC piping, a handbook has been developed by Uni-Bell (14). However, this handbook does not address hazardous piping; nor does it take into account the seismic environment of the piping.

CONCLUSION

It can be seen that, at the present time, there is no set of guidelines suitable for the design of piping systems transporting hazardous materials in industrial facilities. This paper presents the significant aspects of a methodology from which piping system design guidelines can be developed. The significant aspect of this methodology is that it will explicitly consider the importance of the piping (i.e., hazardous vs. nonhazardous contents), the dynamic environment (i.e., regional seismicity and building dynamics), and the piping system's dynamic characteristics.

The need for plastic piping guidelines is particularly important since plastic materials are being used to replace ductile steel and copper as a basic piping material. Plastic piping offers significant cost savings since it is much cheaper to purchase and since it is easier to assemble and install than steel piping. The resistance of plastic piping to corrosion complements the lower cost associated with its use. Because plastic piping is resistant to corrosion, the maintenance, inspection, and routine replacement of degraded pipes can be significantly reduced. Corrosion resistance also makes plastic piping desirable for use with hazardous acids and other fluids which could degrade metal piping.

However, plastic piping systems, if not supported in a fashion which takes into account their reduced strength and ductility, present potentially increased risk to the employees of facilities using plastic pipes, increased economic risk, and liability to operators of these facilities. Consequently, this paper addresses a current problem which will only become more serious as the use of plastic piping increases.

REFERENCES

1. Uniform Building Code, Whittier, California: International Conference of Building Officials, 1985.

2. Minimum Design Loads for Buildings and Other Structures, American National Standard, ANSI A58.1-1982.

3. Recommended Lateral Force Requirements and Commentary, Seismology Committee Structural Engineers Association of California, 1980.

4. Department of the Army, the Navy, and the Air Force, "Seismic Design of Buildings," TM 5-809-10, NAVFAC P-355, AFM 88-3, February 1982.

5. M.W. Kellogg Company, Design of Piping Systems, New York: John Wiley & Sons, 1956.

6. ITT Grinnel Industrial Piping, Inc., Piping Design and Engineering, ITT Grinnel Corporation, USA, 1981.

7. King, Reno C., Piping Handbook, Fifth Edition, New York: McGraw-Hill Book Company, 1973.

8. National Fire Protection Association, Inc., "Installation of Sprinkler Systems 1980," ANSI/NFPA 13, November 20, 1980.

9. Tentative Provisions for the Development of Seismic Regulations for Buildings, Applied Technology Council (ATC), Publication ATC 3-06, NSF Publication 78-8, U.S. Department of Commerce, National Bureau of Standards, National Science Foundation, June 1978.

10. Lashkari, B., "Seismic Analysis of Multi-Span Piping," 4th International Modal Analysis Conference, Los Angeles, Feb. 3-6, 1986.

11. Lashkari, B., "Seismic Analysis of Multi-Diameter Piping," ASME Pressure Vessels and Piping Conference, June 23-27, 1985.

12. Guidelines for the Seismic Design of Oil and Gas Pipeline Systems, Committee on Gas and Liquid Fuel Lifelines of the ASCE Technical Council on Lifeline Earthquake Engineering, ASCE 1984.

13. "Effects of Piping Restraints on Piping Integrity," ASME PVP-40, 1980.

14. Handbook of PVC Pipe, Design and Construction, Dallas, Texas: Uni-Bell Plastic Pipe Association, 1982.

THE ATLANTIC PALACE: PART I
STRUCTURAL SYSTEM SELECTION BASED ON WIND TUNNEL TESTING

Socrates A. Ioannides[1] M. ASCE

James H. Parker[2] A.M. ASCE

Introduction

The Atlantic Palace, a 34-story high-rise condominium in Atlantic City, New Jersey, was used as an example to demonstrate the inter-action process between the structural engineer and the wind tunnel laboratory to achieve an optimum structural system. Part II of this paper deals with the wind tunnel techniques. This part demonstrates the structural engineer's part.

The location of the structure in a coastal area susceptible to high winds, the irregular shape and height of the building and the desire for more accurate wind loading and acceleration information contributed to the decision to perform a wind tunnel test in conjunction with the design.

The design process consisted of several different computer models using the program COMBAT [1] incorporating special modeling techniques for shear walls to achieve optimal results with the least computer effort. Different structural features were included in separate models to evaluate their effects. The effective inertias of beam and column members in the structural system were approximated at 75% of I-gross for beams and 90% of I-gross for columns. These approximations were checked before the final design was completed. Rigid end zones were specified on the beams but not for the columns to approximately model the beam column joint and approximately account for panel zone deformations occurring within the joint.

Preliminary Design

The preliminary structural system is shown in Figure 1. The beginning model, Model 1, consisted of 12 inch shear walls along with 12 x 18 inch perimeter beams and 18 x 36 inch perimeter columns only. A 100 mph wind velocity was used to calculate the preliminary wind loads because no wind tunnel loading information was available at that time. The resulting top level displacement for this initial computer run was much too large, 2.72 ft. In Model 2, the wall thickness was doubled to 24 inches and the deflection was reduced to a more accept-

[1]Vice President, Stanley D. Lindsey & Assoc., Ltd., 1906 West End Avenue, Nashville, Tennessee 37203; Adjunct Professor at Vanderbilt University, Nashville, Tennessee.

[2]Structural Engineer, Stanley D. Lindsey & Assoc., Ltd., 1906 West End Avenue, Nashville, Tennessee 37203.

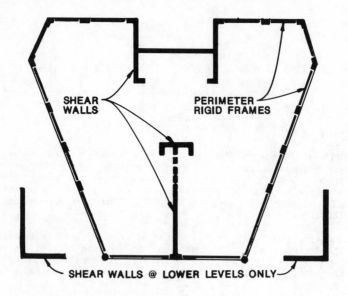

FIG. #1. PRELIMINARY STRUCTURAL SYSTEM.

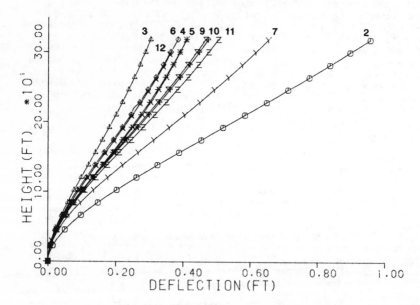

FIG. #2. DEFLECTED SHAPES.

able level. The displacement at the top of 0.95 feet produced a Drift Index of 0.003. This might have been an acceptable limit if drift were the only serviceability criterion. Another important serviceability criterion is acceleration.

Comfort of the occupants is evaluated on the basis of the accelerations of the structure under turbulent wind [2]. Using the procedure found in the National Building Code of Canada (NBCC) for calculating approximate accelerations, the accelerations for the structure were estimated to be 33 mg's in the X direction and 31 mg's in the Y direction, not in the desired range of 16 to 25 mg's. The mass of the structure was well defined; therefore, the stiffness of the structure was increased consequently in order to reduce the periods which in turn reduce the accelerations. A 3.0 sec period was selected as the goal by substitution of the desired acceleration level into the acceleration equations.

In Model 3, the beams were enlarged to 12 x 48 inches. Notched beams found in corners of the structure were also introduced by the use of the appropriate stiffness coefficients for the beam stiffness matrices. Beams at the lower four levels were reduced to 12 x 24 to improve clearance requirements in the parking levels.

Observing the deflected shape of the structure (Fig. 2), particularly at the lower levels, it was decided the additional stiffness was required at the base. Lower walls were introduced in Model 3; the walls were 12 inches thick and only spanned the first four floors. Because of their location they also increased the torsional stiffness of the building. The above-mentioned changes in the structural system reduced the first natural period of vibration to 3.0 sec.

Additional model refinements were introduced and the system described as Model 5 in Table 1 was selected as the best preliminary design. The results of these preliminary analyses are presented in Table 1.

It is important to note that reducing the shearwall thickness in the upper floors did not significantly change the stiffness of the building, proving the well known fact that in shearwall/frame interaction the frames resist most of the load at the upper levels.

At this point, the dynamic properties (periods of vibration, mode shapes, masses, deflection, etc.) for the structural system preliminary selected were sent to the wind tunnel facilities for use in the wind tunnel study.

After Model 5, a change of layout occurred due to value engineering cost estimates. These changes included a different wall arrangement (Fig. 3) as well as a requirement that the exterior beams be up-turned on the 10 inch slab to minimize formwork cost and construction time. The use of up-turned beams limited the beam depth to 2'-6" above the floor slab. Extrapolating from the preliminary design, an approximation of the amount of walls required was presented to the architect.

TABLE 1
COMPARISON OF VARIOUS MODELS

Model #	Features	Periods	Top Level Deflection
1	12" x 18" beams 1'-0" shearwall properties	8.3 sec Y 7.5 sec X 4.1 sec Rot	x = 2.72 ft. y = 1.41 ft.
2	12" x 18" beams 2'-0" shearwall properties	4.4 sec X 3.8 sec Y 2.5 sec Rot	x = 0.95 ft. y = 0.41 ft.
3	12" x 48" beams 12" x 48" notched beams 24" walls and 12" lower walls	3.0 sec Y 2.6 sec X 1.9 sec Rot	x = 0.30 ft. y = 0.32 ft.
4	12" x 48" beams/notched beams 24" walls and 12" lower walls No coupling beam	3.0 sec Y 3.0 sec X 1.9 sec Rot	x = 0.41 ft. y = 0.32 ft.
5	12" x 48" beams/notched beams 24" walls to 17/12" walls to top 12" lower walls No coupling beam	3.0 sec Y 3.0 sec X 1.9 sec Rot	x = 0.41 ft. y = 0.32 ft.
6	12" x 48" beams 18" x 36" columns	2.9 sec Y 2.8 sec X 1.7 sec Rot	x = 0.38 ft. y = 0.29 ft.
7	12" x 24" beams 18" x 36" columns	3.9 sec Y 3.6 sec X 2.2 sec Rot	x = 0.63 ft. y = 0.46 ft.
8	No perimeter frame Only shear walls	5.6 sec Y 5.0 sec X 2.8 sec Rot	x = 1.25 ft. y = 0.81 ft.
9	12" x 36" beams 18" x 36" columns	3.4 sec Y 3.2 sec X 1.9 sec Rot	x = 0.47 ft. y = 0.37 ft.
10	12" x 36" beams 14" x 48" columns	3.4 sec Y 3.1 sec X 1.9 sec Rot	x = 0.47 ft. y = 0.38 ft.
11	12" x 33" beams 18" x 36" columns	3.6 sec Y 3.2 sec X 2.0 sec Rot	x = 0.50 ft. y = 0.40 ft.
Final	12" x 34" beams/12" x 48" col. Wind tunnel loads Revised story heights & masses	3.4 sec Y 3.1 sec X 2.0 sec Rot	x = 0.35 ft. Y = 0.36 ft.

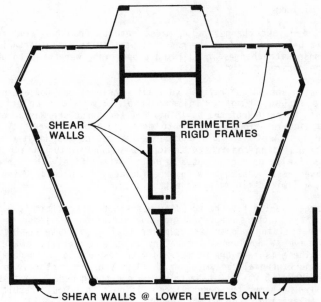

FIG. #3. MODIFIED STRUCTURAL SYSTEM.

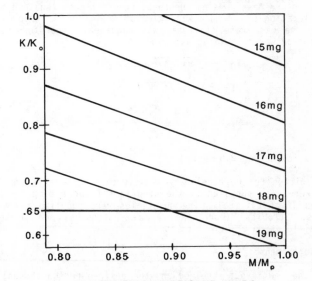

1.5% DAMPING-HURRICANE WINDS

FIG. #4 CORNER ACCELERATIONS.

Wind Tunnel Input

   With the layout changes being made, results from the wind tunnel
tests were being received. The results, particularly the variation of
the response with changes in building properties, were used to better
select the desired period.

   The design period was refined to limit accelerations for a ten-year
storm to around 18 mg's. This limit of 18 mg's was selected by
looking at the nature of the building. If the structure was an office
building, accelerations of up to 25 mg's would be allowed. A hotel
would be limited to 20 mg's and permanent residences limited to 16
mg's. The building consists of resort area condominiums which will,
more than likely, be rented out a majority of the time. The structure
will be used as something between a residence and a hotel. Using this
criteria, a limit of 18 mg's was established.

   The target period for the building was set to approximately 3.5 sec
by using the isobars representing the variation of response with
changes in building properties, similar to Fig. 4, developed at the
Boundary Layer Wind Tunnel Laboratory of the University of Western
Ontario. These isobars are normalized with respect to the generalized
stiffness and generalized mass of the base case structure; in this
instance, Model 5 was the base case. Since the generalized mass was
essentially the same, the generalized stiffness that would result in a
corner acceleration of 18 mg was sought.

   Assuming the mass remains constant, the stiffness required to yield
an acceleration of 18 mg's was selected from Fig. 4. Approximately
0.65 of the base case stiffness was required. Therefore, the new
target period was calculated as:

$$T/T_o = \sqrt{(M/K)/(M_o/K_o)}$$

$$T = T_o \sqrt{K_o/K}$$

$$= 3.0 \sqrt{1/0.65}$$

$$= 3.7 \text{ sec}$$

which was rounded to 3.5 sec.

   In addition to the accelerations, overturning moments and lateral
drifts were also supplied by the wind tunnel for a damping ratio of
1.5% and 2% of critical. The quantities were supplied for two wind
climates: tropical (hurricane) and extratropical due to the geographical location of Atlantic City.

Design Refinement

   With the new layout received and the desired period chosen, design
was continued.

Model 6 was established utilizing the new wall arrangement along with perimeter rigid frames with 18 x 36 inch columns and 12 x 48 inch beams. The period for this model was 2.95 sec., well below the goal of 3.5 sec. In Model 7 beams 24 inches deep were used instead of 48 inches to further reduce the stiffness and increase the period. This beam size resulted in a period of 3.86 sec. higher than the target. A beam between 24 inches and 48 inches was needed.

Due to continuous pressure from the contractor to remove the perimeter beams, Model 8 was established with the shearwalls only as lateral load resisting elements. The period jumped to 5.6 sec and the deflection to 1.25 ft., demonstrating the assistance the frame action between the beams and columns provide to the structural stiffness.

Various other beam and column size combinations were investigated and are shown in Table 1 as Models 9, 10 and 11. As may be seen, the beam sizes affect the performance proportionality more than the columns.

Final Design

With the results of the different analyses available the architect and developer were consulted in finalizing member sizes and configurations before proceeding with the final design. It was decided to use 12 x 34 inch beams and 12 x 48 inch columns. The beams on the back of the building were removed from the structure, thus leaving frames only on the front and two sloping sides. The height was also reduced to 32 stories.

These changes were incorporated into the model for the final analysis. The story heights for the model were changed to reflect changes in the structure's story heights that occurred during design development. The masses were also revised to reflect the structure's actual mass. Before an analysis was performed with the final model, the wind loads were changed to the wind loads obtained from the wind tunnel study.

The wind tunnel results were in the form of base shears and overturning moments in each direction plus the resulting torsional moments. Since a pressure model was not tested, the exact distribution of the applied forces was not known. An approximate pressure distribution, based on experience with other tests, was supplied by the wind tunnel and was integrated over the tributory areas to produce the applied story forces. The pressure distribution was such that the resulting base moment and shears were the same as predicted by the test.

The base forces obtained are the maximums produced with the wind acting from all different directions and integration with the wind climate. Therefore, they do not necessarily act together. For that reason, design load cases should include combination of all the forces. It was recommended by the wind tunnel to use a combination factor of 0.9 for X or Y moments plus torsion or a 0.8 factor for a combination of all three to account for the fact that the maximums do

not all occur together. The following load combinations were therefore used:

$$
\begin{array}{c}
X \\
Y \\
0.9(X+T) \\
0.9(X-T) \\
0.9(Y+T) \\
0.9(Y-T) \\
0.8(X+Y+T) \\
0.8(X+Y-T) \\
0.8(X-Y+T) \\
0.8(X-Y-T) \\
0.8(-X+Y+T) \\
0.8(-X+Y-T)
\end{array}
$$

The combination producing the largest forces in each member is used
as the design case for the member.

The final model was analyzed, including the wind tunnel loads and
load combinations, and gave an acceptable period of 3.4 sec.

The dynamic properties of the final model were also forwarded to
the wind tunnel facility for incorporation in their final report. The
new properties were used to predict loads and responses by extrapolation of the model test results. The predicted responses for this
model were within the established limits and the applied loads were
shown to be conservative. Therefore, the structural design produces
an acceptable building.

Conclusion

Wind tunnel testing offers distinct advantages in the design of
high-rise building. Wind tunnel testing and structural design interaction enables the designer to conduct a more efficient and accurate
analysis and design by considering the dynamic action of the building
under turbulent wind. Testing provides more accurate loading information for designing the structural system along with providing dynamic
response characteristics. In some cases, wind tunnel testing may give
rise to a more efficient design than existing code procedures, whereas
in other cases it provides useful information on occupant comfort
which cannot be otherwise estimated. As shown with this example, this
serviceability criterion controlled the design and not strength or
drift.

References

1. Computech Engineering Services, Inc., COMBAT-Comprehensive Building
   Analysis Tool, NISSEE/Computer Applications, University of
   California, Berkley, December, 1983.

2. Chen, P.W. and Robertson, L.E., "Human Perception Thresholds of
   Horizontal Motion", Journal of the Structural Division, ASCE, Vol.
   98, pp. 1681-1695 No. St8, August, 1972.

THE ATLANTIC PALACE - PART II

THE APPLICATION OF WIND TUNNEL/FORCE BALANCE TECHNIQUES

David Surry[1] and Gary R. Lythe[2]

ABSTRACT

A relatively new approach for determining overall wind-induced loads on a structure is described. This force balance technique is economical and flexible, and able to provide predictions of loads and responses for varying building dynamic properties without the need for repeated testing. The Atlantic Palace Building is used as an example to show the power of the technique.

1.0 INTRODUCTION

The design of the Atlantic Palace building provides a useful example of the application of state-of-the-art wind tunnel techniques to structural design. Part I of this paper has detailed the structural design considerations and the role played by the wind tunnel test results in structural optimization. Part II provides a more detailed background to the wind tunnel test procedures themselves.

The wind tunnel test procedures centre on the use of a lightweight rigid model mounted on a sensitive but stiff balance system that allows direct measurement of the unsteady generalized wind forces acting in the lowest three modes of vibration of the structure. The spectra of these forces, together with estimates of the structural damping, can be used to analytically derive the resonant contributions to the loading and resulting motion for all wind directions tested. The total structural responses to wind can then be integrated with the statistics of the wind climate to provide statistical response predictions. The power of the force-balance technique lies in its capability of providing parameterized response data for differing structural properties (i.e. stiffness, mass and damping), all based on the same basic wind tunnel experiments.

The Atlantic Palace project presents some interesting nuances on these techniques. First, the effective centre of rotation is substantially above ground level, requiring on-line correction to the ground-level generalized forces sensed by the balance. The paper also briefly discusses other corrections to the data required by the force-balance technique, associated with differences between model/balance mode shapes and those of the actual building, and illustrates the strengths and weaknesses of the technique. Suggestions are made as to when more elaborate aeroelastic model studies might be necessary. Second, the site has a mixed climate. That is, both extratropical and cyclonic storms are

---

[1]Associate Research Director, [2]Research Engineer: Boundary Layer Wind Tunnel Laboratory, The University of Western Ontario, London, Ontario, Canada, N6A 5B9

important. The latter, in their most severe form as hurricanes, dominate the overall structural loading; however, extratropical storms govern for more common events such as the serviceability criteria associated with accelerations. The paper illustrates this and discusses the importance of considering the acceleration results from the two climates both independently and in combination.

2.0   THE FORCE BALANCE TECHNIQUE

Multi-degree-of-freedom aeroelastic models have traditionally been used to study wind sensitive structures. Although these provide the most reliable estimates of overall loads and responses, they are expensive and time consuming. Two-degree-of-freedom aeroelastic models simulate the wind-induced response in the two fundamental sway modes and are less expensive, but do not provide torsional information which may be significant for buildings of unusual shape and structural properties.

A recently-developed alternative (Tschanz, 1982) utilizes a sensitive force balance, capable of measuring the aerodynamic modal wind forces (in both fundamental sway modes and in torsion) acting on a rigid, light, geometric scale model of the building attached to it. Provided that aerodynamic damping effects are negligible, which is usually the case, the balance/model system can measure the modal loads directly. These modal loads are a function only of the wind structure, the mode shapes and the building's aerodynamic shape. Thus, simple acrylic foam models can provide an economical and rapid description of the modal forces needing correction only for deviations in mode shapes between actuality and those inherent in the balance technique. This technique has already been used for numerous buildings, a few of which have also had aeroelastic model tests performed. Comparisons of the two techniques are quite favourable (Tschanz, 1982; Thomas and Isyumov, 1985; Steckley et al, 1985).

The technique assumes a linear response of an elastic system which can be expressed in the general form, $R = \bar{R} + g\sigma_R$, where $\bar{R}$ is the mean response, σ_R the rms response and g a dimensionless time varying factor. A particular response of interest is the peak response $\hat{R}$, in which case the corresponding value of g is roughly 3 to 4.

The deflection response can be broken down into two components: a quasi-static (or background) component σ_{back}; and a resonant component σ_{res} which can be written

$$\sigma_{back} = \frac{\sigma_F}{K}; \qquad \sigma_{res} = \sqrt{\frac{\pi}{4}\frac{1}{\zeta}\frac{f_oS_F(f_o)}{K^2}} \qquad (1)$$

In these, σ_F is the rms generalized force; $K = \{2\pi f_o\}^2 M$ is the generalized stiffness in the fundamental mode of vibration; M are the generalized mass defined as $M = \int_o^H m(z)\mu^2(z)dz$; $m(z)$, $\mu(z)$ are the mass per unit height and normalized modal deflection at height z; ζ is the total damping as a ratio of the critical damping; and $S_F(f_o)$ is the power spectrum of the generalized force at the natural frequency f_o.

Equation (1) is the well known result for the response of a lightly damped system (the structure), with a second order mechanical admittance function, to a random force whose spectrum is fairly smooth, such as is the case for the aerodynamic forces. Using mean square addition of the components above gives:

$$\sigma_R = \frac{\sigma_F}{K} \sqrt{1 + \frac{\pi}{4}\frac{1}{\zeta}\frac{f_o S_F(f_o)}{\sigma_F^2}} \qquad (2)$$

This representation has been given in various references (Davenport, 1966, for example) and is also the basis of gust factor approaches. The difficulty in evaluating this equation has always been the determination of the rms generalized force, σ_F and its normalized spectrum $f_o S_F(f_o)/\sigma_F^2$. The balance can now be used to measure these data directly. The peak response can then be calculated using a suitable value of g (see Davenport, 1966).

The acceleration response can be estimated in a straightforward manner from the above by noting that acceleration = $\{2\pi f_o\}^2$ (deflection) and that the background response contributes almost nothing. Thus, the root-mean-square acceleration is simply given by $\{2\pi f_o\}^2 \sigma_{res}$. Peak accelerations are estimated as for peak deflections.

The aerodynamic damping is body-motion dependent, and can only be measured with conventional aeroelastic models. There are, however, indications (Davenport and Tschanz, 1981) that the value is small and positive for typical buildings in the drag direction, and is also positive in the lift direction up to the velocity where vortex shedding is a major contribution to the response. This velocity is usually higher than typical design wind speeds, and is indicated in the force spectra by a narrow-band peak. The result is that equation (2) gives a slightly conservative value, except for lift forces at high velocities.

The calculation of responses for a building with 3-dimensional modes is somewhat more complex but follows the same general procedure. Applications to date have ignored cross-coupling of the aerodynamic forces, simply because measuring all of the cross-spectra is tedious and, where examined, these cross-spectra have proven to be small. Further, Tallin and Ellingwood (1985a), have found that "statistical correlations among forces do not have a pronounced effect on the total building acceleration . . . ".

Inherent in the force balance measurement technique is the assumption that the mode shapes in the sway direction are linear and that the mode shape in torsion is constant. Since actual mode shapes deviate from these, particularly in torsion, some correction becomes necessary. A detailed analysis is given in Vickery et al (1985) where it is shown that corrections derived from application of quasi-steady theory for line-like structures (Davenport 1967, 1977) agree very well with experiments in both the lift and drag directions. Vickery et al found that mode shape effects are not pronounced for the resonant component of the bending moments and are of even less importance for peak moments which include the mean and the quasi-static components; however, the accelerations can be significantly overestimated if the generalized mass is based on the actual mode shape and no corrections of the force balance model data are made, see Figure 1. Deflections can be in even greater error. While not examined in detail in the paper, the base torque measured with the force balance is a significant overestimate of the generalized torque and must be corrected for mode shape. The correction to adjust from a constant to a linearly varying mode shape has been determined analytically at the Laboratory to be approximately 0.70 which is in agreement with Tallin and Ellingwood (1985b).

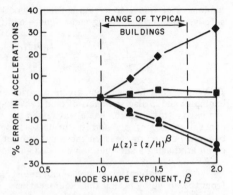

- ◆ TRUE DYNAMIC PROPERTIES, GENERALIZED FORCE BASED ON $\mu(z) = (z/H)$

- ■ TRUE DYNAMIC PROPERTIES, GENERALIZED FORCE ADJUSTED

- ● DYNAMIC PROPERTIES AND GENERALIZED FORCE BASED ON $\mu(z) = (z/H)$ MASS DISTRIBUTION OF $\rho_t/\rho_b = 1.0$

- ▲ DYNAMIC PROPERTIES AND GENERALIZED FORCE BASED ON $\mu(z) = (z/H)$ MASS DISTRIBUTION OF $\rho_t/\rho_b = 0.75$

ρ_t = BUILDING DENSITY AT TOP
ρ_b = BUILDING DENSITY AT BASE

FIG. 1   EFFECT OF MODE SHAPE ON ESTIMATES OF TIP ACCELERATIONS
(after Vickery et al, 1985)

In practice, it has been found that the corrected force balance results are in excellent agreement with aeroelastic results even for complex building shapes. Nevertheless, aeroelastic models are still the most reliable approach and are still used when the force balance analysis indicates results which are of particular concern in the final design, particularly if the mode shapes are very unusual, or when aerodynamic cross-coupling between modes may be important, or when mode shape corrections become large.

3.0   THE ATLANTIC PALACE BUILDING

The basic tool used in the study was the Laboratory's Boundary Layer Wind Tunnel. The 1:500 scale test model was placed on the force balance on a turntable surrounded by a 2000 ft. radius model of the nearby city, built in reasonable detail (see Figure 2), all of which could be rotated to simulate different wind directions. Upstream of this, various appropriate roughnesses were used, over fetches equivalent to about 6 miles, in order to develop boundary layer characteristics which are similar to the wind over the terrain approaching the actual site. This methodology has been highly developed and is detailed elsewhere (Davenport and Isyumov, 1968; Whitbread, 1963; Surry and Isyumov, 1975).

Two separate and self-contained test procedures were used. The first measured the generalized forces in terms of the mean and rms base bending moments along orthogonal x and y building axes as shown in Figure 3, as well as mean and rms base torque. The second procedure recorded measurements of spectra of the base loads as required to develop the resonant component of response. All of these measurements were made over the complete range of wind direction at 10° increments.

For this building, the effective point of rotation of the building is 45.5 ft. above ground level. In order to determine the moments about this height and still allow the balance to be mounted conveniently at local "ground level" in the wind tunnel, an on-line computer program was used to compute the instantaneous moment at the 45.5 ft. height using the instantaneous ground level moments and shears. The results are shown in Figures 4 and 5. Figure 4 shows the mean and rms moments and torques, all in the form of coefficients. Figure 5 shows an example of the spectra measured at one wind angle.

FIG. 2: PHOTO OF MODEL
IN WIND TUNNEL

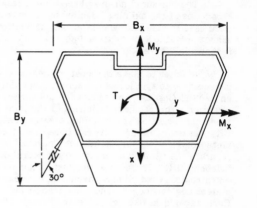

FIG. 3: SIGN CONVENTION USED

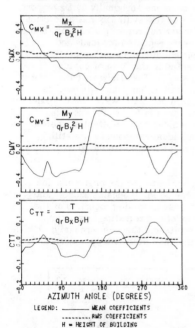

FIG. 4: MOMENT AND TORQUE
COEFFICIENTS

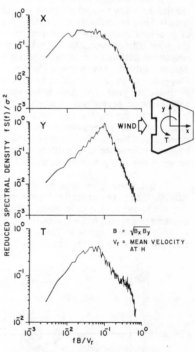

FIG. 5: EXAMPLE SPECTRA

The measured mean and rms quantities, the spectra and the dynamic properties of the building (supplied by the engineers), corrected where necessary, were used to calculate accelerations, deflections and moments for the structure using the methodology outlined above. An example of results (for accelerations) is shown in Figure 6.

In order to make the most rational use of the wind tunnel information, it is necessary to synthesize the aerodynamic data with the wind climate characteristics, i.e., to account for the variation of likely wind speeds with direction. This requires definition of the joint probability distribution of wind speed and direction at a suitable reference height. This definition for Atlantic City was complicated by its exposure to hurricanes, which arise from different meteorological phenomena than do "normal" extratropical winds. It is safest to separate the two and examine the structural performance under each independently. Therefore, two wind climate models were developed; an extratropical wind climate, excluding hurricanes, was developed using historical meteorological records, and a hurricane wind climate was developed using Monte Carlo techniques (Georgiou, 1985). These models are described in detail in Lythe et al (1985).

The detailed methodology used to integrate the aerodynamic wind tunnel results with the two wind climate models, as developed originally by Davenport, is described in Lythe et al (1985). The product of this synthesis is predictions of accelerations, deflections and moments for various return periods for each of the wind climate models. These predictions were repeated for loads and responses that were calculated using different building dynamic properties as the design evolved. These predictions were presented in the form of contour diagrams, examples of which are shown in Figure 7. They show that the once-in-one hundred year structural loads were dominated by hurricane winds. The prediction of accelerations at the 10-year return period, however, were found to be very similar for the two wind climates, as shown in Figure 8. Assuming independence, the responses can be simply combined to a resultant as shown; however, in the case of accelerations, the individual components are still important because of the different psychological factors associated with human acceptance of the two types of storm.

Subsequent to these predictions being presented to the engineers, the building geometry underwent some changes; the building became shorter and slightly wider in one dimension. Consequently the dynamic properties of the building changed. In lieu of building a new model and retesting, approximate corrections were made to the original test results and they were reanalysed with the new dynamic properties. This provided the engineers with approximate

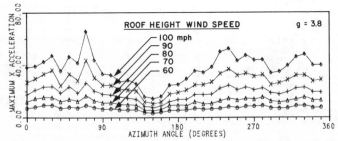

FIG. 6: ACCELERATIONS (in milli-g's) FOR VARIOUS WIND SPEEDS

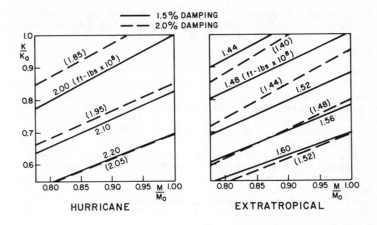

FIG. 7   CONTOURS OF PREDICTED MOMENTS
(45 ft. above grade)

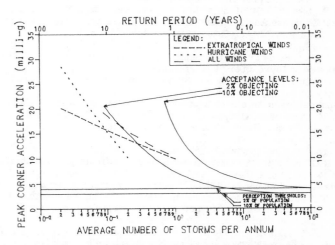

FIG. 8   PREDICTED 10-YEAR ACCELERATIONS

predictions of the revised building's loads and responses without the expense of another wind tunnel test, albeit with more uncertainty.

4.0   SUMMARY

The force balance technique provides an economical and flexible approach to determining the overall loads and responses of a building due to wind action. Predictions of responses for varying building dynamic properties can be derived from one set of wind tunnel measurements in order to help the structural engineer optimize the structural system. In the case of the Atlantic Palace building, loads were measured about a point of rotation substantially above ground level and consideration was given to hurricane, as well as "normal" winds experienced at the building site.

REFERENCES

Davenport, A.G. (1966), "The Treatment of Wind Loading on Tall Bldgs.," Symposium on Tall Buildings at the University of Southampton, April 13-15, 1966, Pergamon Press, pp. 3-44.

Davenport, A.G. (1967), "Gust Loading Factors," ASCE, J. Struct. Div., Vol. 93, No. ST3.

Davenport, A.G., Isyumov, N. (1968), "The Application of the Boundary Layer Wind Tunnel to the Prediction of Wind Loading," Int. Research Seminar on Wind Effects on Bldgs. and Structures, Ottawa, Canada, September 1967, University of Toronto Press.

Davenport, A.G. (1977), "The Prediction of the Response of Structures to Gusty Wind," Int. Res. Seminar on Safety of Structures Under Dynamic Loading, Trondheim.

Davenport, A.G., Tschanz, T. (1981), "The Response of Tall Buildings to Wind: Effects of Wind Direction and the Direct Measurement of Force," Proc. 4th U.S. National Conf. on Wind Engineering Research, July 27-29, Seattle, WA.

Georgiou, P.N. (1985), "Design Wind Speeds in Tropical Cyclone-Prone Regions," Ph.D. Thesis, University of Western Ontario.

Lythe, G.R., Surry, D., Mikitiuk, M.J., Georgiou, P.N. (1985), "Wind-Induced Overall Loads and Accelerations on the Atlantic Palace Building, Atlantic City, New Jersey," BLWT-SS10-1985, The University of Western Ontario.

Steckley, A., Lythe, G., Isyumov, N., Davenport, A.G. (1985), "Aeroelastic Wind Tunnel Model Study of the New Bank of China Building, Hong Kong," BLWT-SS2-1985, The University of Western Ontario.

Surry, D., Isyumov, N. (1975), "Model Studies of Wind Effects - A Perspective on the Problems of Experimental Technique and Instrumentation," Int. Cong. on Instrumentation in Aerospace Simulation Facilities, 1975 Record, pp. 76-90.

Tallin, A., Ellingwood, B. (1985a), "Wind-Induced Lateral-Torsional Motion of Buildings," ASCE J. Struct. Div. Vol. 111, No. 10, pp. 2197-2213.

Tallin, A., Ellingwood, B., (1985b) "Analysis of Torsional Moments on Tall Buildings," J. Wind Eng. and Ind. Aero., 18, pp. 191-195.

Thomas, R.G., Isyumov, N. (1985), "A Study of Wind Effects on the Fountain Place Project, Dallas Texas, Part 3 - Overall Wind-Induced Structural Loads and Responses," BLWT-SS1-1985, University of Western Ontario.

Tschanz, T. (1982), "The Base Balance Measurement Technique and Applications to Dynamic Wind Loading of Structures," Ph.D. Thesis, The University of Western Ontario, Faculty of Engineering Science.

Vickery, P.J., Steckley, A., Isyumov, N., Vickery, B.J. (1985), "The Effect of Mode Shape on the Wind-Induced Response of Tall Buildings.," 5th U.S. National Conf. on Wind Engineering, Lubbock, Texas, pp. 1B-41/1B-48.

Whitbread, R.E. (1963), "Model Simulation of Wind Effects on Structures," NPL Int. Conf. on Wind Effects on Bldgs. and Structures, Teddington, England.

THREE PHASE DESIGN FOR
WIND LOADS ON STRUCTURES

by Suzanne M. Dow*, Gary C. Hart**, Robert E. Englekirk***, Jon D. Raggett****

Introduction

The successful design of a structure requires three phases of work. These phases are: conceptual design, design development and design verification. The conceptual design is undertaken during the first stages of the product's development. Often, the details of the structure aren't well defined and the contract may not even have been signed. The design development phase is when the majority of the design work is carried out. During design development, fairly accurate design loads are determined, and the structure required to resist these loads is determined, based upon some analytical model and assumed design strength of the material. The last phase, design verification, is used to verify that the assumptions used in determining the design loads, resistances of materials, and analytical models were correct, and that the structure, as designed, possesses the required level of safety.

This paper describes the three phases as they apply to wind loading on structures. The flowchart shown in Figure 1 shows the various steps in the procedure. These steps are, simply:

_____

  * Project Engineer, Englekirk and Hart, Inc., Los Angeles, California
 ** Professor of Engineering, University of California, Los Angeles and Principal, Englekirk and Hart, Inc., Los Angeles California
*** Adjunct Professor of Engineering, University of California, Los Angeles and Principal, Englekirk and Hart, Inc., Los Angeles, California
**** President, J. D. Raggett and Associates, Inc., Carmel, California

1. Conceptual Design - using code determined loads and material strengths with simplified analytical models.

2. Design Development - using separately determined loads and resistances based upon the statistics of the variables involved, and a specified reliability index, β_S.

3. Design Verification - a combined analysis considering load and resistance variables simultaneously.

An example is given describing how these three phases would apply in the design of a cladding component. This example was chosen for its simplicity, however, the procedure is just as applicable to the design of the main lateral force resisting system of a building, or any other structure subject to wind loading.

Conceptual Design

The conceptual design phase is used to help determine an overall budget for a project. As is implied in its name, this phase is used to determine concepts, not details. Therefore, in determining design wind loads, code determined loads are adequate for this phase. Alternatively, an estimate could be made, based on past experience with similar structures in the area, of the design wind loads that may be determined in the design development phase. Either way, these loads would then be used, with an approximate structural model and code determined material strengths, to determine typical member sizes.

Design Development

The design development phase contains the most effort expended on a project. During this phase, the details of the project are determined. In the context of this paper, design wind loads are determined for a project, using a more accurate method than is presented in the code, namely a wind tunnel test. A procedure to determine design wind loads for a building or structure is presented in Dow[1]. This procedure uses reliability analysis and a wind tunnel test to determine design wind loads for a given reliability index. The determination of an appropriate reliability index is beyond the scope of this work, but is discussed in Ellingwood, et.al.[5] and Davenport[6].Using this procedure, and a compatible method of determining member capacities (i.e., ultimate strengths defined by the same reliability index), the design is completed, including the design of all elements and details of the system. Often, this is the end of a design, however, in a complete design there actually exists one additional step.

Design Verification

The design verification phase is performed to assure that
the completed structure, if built as designed, possesses
the required reliability. In other words, the true
loading and the true capacity of the structure are
compared, using reliability analysis, to determine the
safety of the structure in terms of the reliability
index. This analysis is performed using the Hasofer-Lind
method of reliability analysis as described in Thoft-
Christensen and Baker[2]. This method is invariant with
respect to the choice of failure function, even for
nonlinear problems. Therefore it is ideal for a problem
such as wind loading, where even the simplest form of the
problem is nonlinear (since the load is proportional to
the square of the velocity).

Example

In the following analysis, the method described above is
applied to the design of a cladding element. Figure 2
shows the project concept for the structural support of
the cladding elements on a twenty story office building
located in Los Angeles. The terrain surrounding the site
is suburban in all directions.

The conceptual design phase of the cladding design
requires the engineer to determine sizes of typical
cladding elements. The sixteenth story was chosen as
representative of an average load on the component (a
high average, to be somewhat conservative at this stage).
The design wind pressure was determined using the 1982
Uniform Building Code[3], Section 2311. The interpretation of the code is beyond the scope of this work,
however, it suffices to say that the design wind pressure
for the design of a cladding element at this location on
this building is 25 psf (maximum of positive or negative
load). The design was continued and it was determined
that the vertical element maximum spacing is 3'3" o.c. as
shown in Figure 3. This design is then used for
preliminary pricing of the cladding system.

During the design development stage of a major project, a
wind tunnel study is performed. In this example, the
design wind pressures for the cladding were determined
using the procedure presented in Dow[1]. Once again, the
details of the load definition are beyond the scope of
this work, but are described in Dow, Hart and Raggett[4]
and in detail in Dow[1]. However, the cladding loads are
presented in Figure 4, which gives the maximum negative
design pressure as a function of location on the
building. Similar figures show the maximum positive
design pressures, and both positive and negative
pressures are presented for all faces of the building.
These pressures were determined using a wind tunnel test

and a specified reliability index of 3.0. Winds from eight principle wind directions were considered. Using these design loads, the cladding was then designed using load and resistance factor designs with ultimate member capacities determined using a reliability index equal to 3.0 (same as that used for the load determination.) Figure 5 shows the resulting design. Note that the vertical member spacing varies over the face of the building. Since this is the design development phase, it is important to realize cost savings due to reduced wind loading at some locations on the building. This is very simply done by changing the spacing of the vertical elements of this truss system, thus saving steel, with minimal additional planning work, and basically no additional labor costs.

The final phase of this design is the design verification phase. During this phase, the safety of the design resulting from the design development phase is checked. The Hasofer-Lind method was used, incorporating all the random variables in both the load and resistance parts of the failure equation. Figure 6 shows the failure surface and design Point A for a two dimensional problem. In this example, the failure surface has many dimensions, one for each random variable in the failure equation (a function of all load and resistance variables). However, the design point is still that point on the failure surface whch is closest to the origin. Then the distance from the design point to the origin is equal to the reliability index, β. This β is then compared to the required β_S (3.0 as chosen in the design development phase above). If $\beta \geq \beta_S$ then the design is acceptable, i.e. provides the required safety. Otherwise, the design is unacceptable and a redesign and reanalysis is required. Similarly, if the actual β is much higher than β_S, the design capacity may be reduced and reanalyzed to achieve a more economical structure.

References

[1]   Dow, S.M., Design Wind Load Formulation Using a Wind Tunnel and Consistent Risk, Thesis submitted in partial satisfaction of the requirements for Master of Science, UCLA, 1986.

[2]   Thoft-Christiansen, P. and Baker, M.J., Structural Reliability Theory and Its Applications, Springer-Verlog Berlin, Herdelberg, 1982.

[3]   Uniform Building Code, ICBO, Whittier, 1982.

[4] Dow, S.M., Hart, G.C., and Raggett, J.D., "Design
     Wind Load Formulation Using a Wind Tunnel and
     Consistent Risk," Proceedings, Fifth U.S. National
     Conference on Wind Engineering, Nov. 6-8, 1985,
     Lubbock, Texas.

[5] Ellingwood, B., Galambos, T.V., MacGregor, J.G., and
     Cornell, C.A., Development of a Probability Based
     Load Criterion for American National Standard A58,
     National Bureau of Standards Special Publication
     577, June 1980.

[6] Davenport, A.G., "On the Assessment of the
     Reliability of Wind Loading on Low Buildings," 5th
     Colloquium on Industrial Aerodynamics, Aachen,
     Germany, June 14-16, 1982.

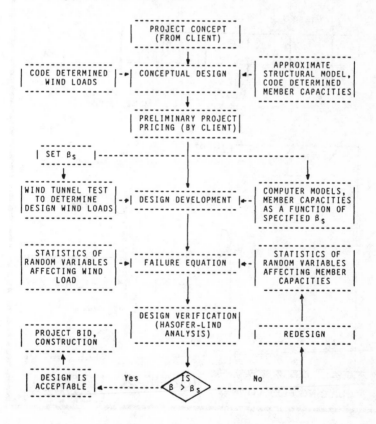

FIGURE 1:   DESIGN PROCEDURE

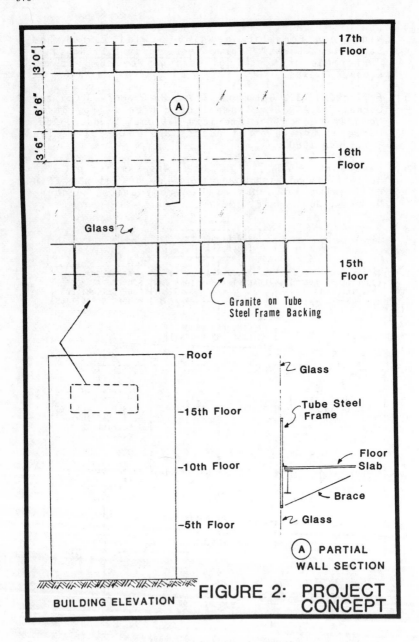

FIGURE 2: PROJECT CONCEPT

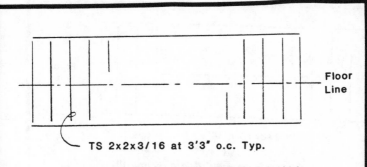

TS 2x2x3/16 at 3'3" o.c. Typ.

**FIGURE 3: TUBE STEEL CURTAINWALL
SUPPORT FRAME (Conceptual Design)**

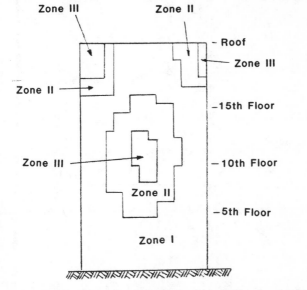

Zone III

Zone II

Roof

Zone III

Zone II

–15th Floor

Zone III

–10th Floor

Zone II

–5th Floor

Zone I

Zone I: 17 psf
Zone II: 23 psf
Zone III: 39 psf

**FIGURE 4: MAXIMUM NEGATIVE
DESIGN PRESSURES $\left(\beta=3\right)$**

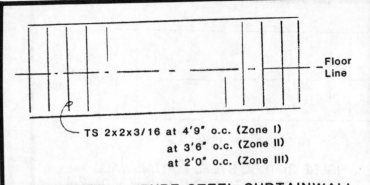

TS 2x2x3/16 at 4'9" o.c. (Zone I)
at 3'6" o.c. (Zone II)
at 2'0" o.c. (Zone III)

FIGURE 5: TUBE STEEL CURTAINWALL SUPPORT FRAME (Design Development)

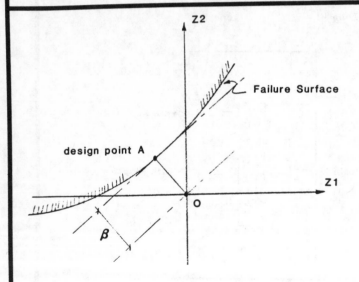

FIGURE 6: GRAPHICAL REPRESENTATION OF HASOFER–LIND ANALYSIS

Nonlinear Dynamic Response of Window Glass Plates Using Finite Difference Method

C. V. Girija Vallabhan, M. ASCE[1] and R. Panneer Selvam[2]

Abstract

Von Karman nonlinear plate equations have been found to be very useful to represent static responses of thin window glass plates subjected to uniform lateral pressure. Vallabhan (1983) has established an efficient iterative procedure to solve the above nonlinear equations using finite difference method for a rectangular plate. Here the method is extended for dynamic analysis of window glass subjected to transient pressure loading. In this model, it is assumed that in-plane displacements are small compared to lateral displacements; hence the corresponding in-plane inertia forces are neglected. The dynamic nonlinear von Karman equations are solved using finite difference method in space and Newmark β method in time. An iterative procedure is used to achieve convergence of the solution of the nonlinear equations. For a model square glass plate subjected to triangular pulse loading, displacements, velocities, and accelerations are calculated and presented. Global maximum principal tensile stresses are computed and discussed.

Introduction

The behavior of thin window glass plates subjected to lateral forces becomes nonlinear as the value of the maximum lateral displacement becomes larger than the thickness of the plate. Membrane or in-plane stresses are produced in addition to the conventional bending stresses, and these membrane stresses stiffen the plate, making it behave nonlinearly. This behavior is completely elastic and considered as geometrically nonlinear. Glass researchers in the past (Tsai and Stewart, 1976; Moore, 1979; Beason, 1980, to mention a few) have confirmed this nonlinear behavior through theoretical and experimental research. While theoretical solutions representing this nonlinear behavior are possible through numerical methods such as finite element methods, Vallabhan and Wang (1981) have shown that the von Karman nonlinear plate equations are indeed very useful to represent the static response of thin rectangular plates subjected to uniform lateral pressure.

Vallabhan and Wang have demonstrated that their model is computationally very efficient and requires very little input data preparation. Recent experiments on thin glass plates at Texas Tech University (Behr et al., 1983; Vallabhan and Minor, 1984) have indicated excellent

[1]Professor, Department of Civil Engineering, Texas Tech University, P.O. Box 4089, Lubbock, Texas 79409.
[2]Lecturer, Department of Civil Engineering, Texas Tech University, P.O. Box 4089, Lubbock, Texas 79409.

correlation between experimental and theoretical finite difference model results. The main objective in this paper is to extend the finite difference model for the nonlinear dynamic analysis of thin window glass plates subjected to transient dynamic pressure loads. The dynamic analysis is accomplished by a step-by-step integration procedure using the Newmark beta method. An iterative procedure similar to that used in static analysis is incorporated here to achieve convergence of the solution at the end of every time step. Example problems of square plates subjected to a triangular pulse load are solved for displacements, velocities, and accelerations. Stresses due to bending and membrane forces are computed at every time step. These stresses are combined algebraically and the corresponding principal stresses at the top and bottom faces of the corresponding plates are computed. A brief discussion on the discretizations in space and time used in the analysis is made at the end of the paper.

Theoretical Formulation of Dynamic Glass Plate Model

The assumptions made for the development of the equations used in this study are the same as those made by von Karman for developing his static plate equations. The full list of assumptions used for the von Karman theory are presented in Timoshenko and Woinowsky-Krieger (1959) and Szilard (1974). The extension of these equations to accommodate dynamic response conditions is accomplished by introducing inertia forces into the equations using D'Alembert's principle (Langhaar, 1962). In this study, it is assumed that in-plane and rotational accelerations at the middle surface are small compared to the accelerations in lateral directions. The governing equations in the domain of the plate are:

$$D \nabla^4 w = p + h L (w,\phi) - \rho \ddot{w} \tag{1}$$

and

$$\nabla^4 \phi = - \frac{E}{2} L (w,w) \tag{2}$$

where
  D is the flexural rigidity of the plate
  w is the lateral deflection of the middle surface
  p is the lateral pressure on the plate which is a function of time
  ϕ is the Airy stress function
  ρ is the mass of the plate per unit area
  E is the Young's modulus of elasticity of the plate
  ∇^4 is the biharmonic operator
  $L(w,\phi) = (w_{,xx}\phi_{,yy} - 2w_{,xy}\phi_{,xy} + w_{,yy}\phi_{,xx})$

The subscript comma notation represents differentiation with respect to the variables following it. The Airy stress function   represents the membrane action in the plate induced by large displacements. From inspection, it can be seen that the functions $L(w, \phi)$ and $L(w,w)$ are nonlinear. The above nonlinear equations have to be solved by considering appropriate boundary conditions.

Following the previous investigations (Moore, 1979; Vallabhan, 1983), simply supported boundary conditions are used here. There are two sets of boundary conditions to be satisfied: flexural boundary conditions and membrane restraints on the boundary. The flexural boundary conditions are

$w = 0$, and $M_n = 0$ on the edges $\qquad\qquad\qquad\qquad\qquad\qquad$ (3)

Since $M_n = -D (w_{,nn} + \nu\, w_{,ss})$ and $w_{,ss} = 0$, then $w_{,nn} = 0$ $\qquad$ (4)

Here the subscripts n and s represent the normal and tangential coordinates along the edges. The second set of boundary conditions involve membrane restraints on the boundary. It is assumed that the in-plane restraints of the edges are negligible; i.e., the surface tractions on the edges are assumed to be zero. Thus, on the edges we have

$$\text{Normal membrane stress} = \sigma_n^m = \phi_{,ss} = 0 \qquad\qquad\qquad (5)$$

$$\text{Membrane shear stress} = \tau_{ns}^m = -\phi_{,ns} = 0 \qquad\qquad\qquad (6)$$

Equations 3, 4, 5, and 6 give the complete set of boundary conditions to be satisfied for the plate at all times during the dynamic response.

For a complete dynamic analysis of the plate, initial conditions of the plate have to be established. In this study, it was assumed that the plate is flat, stress free, and stationary at time $t = 0$.

Numerical Model

Employing the same technique used in the finite difference model for solving the static von Karman equations, researchers can transform Equations 1 and 2 into algebraic equations of the type

$$[A] \{w\} = \{Q\} + \{f_1(w,F)\} - [M] \{\ddot{w}\} \qquad\qquad (7)$$

and

$$[B] \{F\} = \{f_2(w)\} \qquad\qquad\qquad\qquad\qquad (8)$$

Here matrices A and B are square matrices representing the corresponding biharmonic operators; w, F, and Q are vectors representing discrete nodal displacement, Airy stress function, and load vectors, respectively. The vectors f_1 and f_2 are appropriate vectors representing $L(w,\phi)$ and $L(w,w)$ in Equations 1 and 2. The matrix M represents discrete lumped masses at the nodes and is a diagonal matrix, and $\ddot{w}$ is the discrete nodal acceleration vector. Also, the matrices A and B are symmetric and banded; therefore, only the half banded portion of the matrix was used on the computer. The details of the matrices A and B and the vectors f_1 and f_2 are available in Vallabhan and Wang (1981). Since the plate equations are nonlinear, the solution can be achieved only by an iterative procedure.

The dynamic analysis was also accomplished using a similar technique. The load being a transient loading, a step-by-step integration procedure with iteration was employed here. This step-by-step integration technique is a marching procedure, where discrete values of displacement, velocity, and acceleration at a particular time, t, are used to calculate similar values at a time, t + dt, where dt is a small increment of time, called a time step. The solution marches in time where Equations 1 and 2 are satisfied at every discrete time, t. There are several methods available in the literature for step-by-step integration procedures. They can be broadly classified as explicit methods, implicit methods, or combined methods (Vallabhan et al., 1973).

When a step-by-step integration procedure is used to solve a set of dynamic equations, many factors must be studied such as accuracy, stability of the solution, and above all its computational efficiency. These factors are linked to the magnitude of the time step dt. These methods can be unstable, conditionally stable, or stable for a given time step, depending on the methodology used and the system characteristics.

After a detailed study of all these methods, the Newmark-β method was selected for the present analysis. In this method, the velocity and displacement at time t + dt are computed from the values of displacement, velocity, and acceleration at times t and t + dt. The equations, respectively, are

$$\dot{w}_{t+dt} = \dot{w}_t + \left[(\tfrac{1}{2}-\gamma)\ddot{w}_t + \gamma\ddot{w}_{t+dt}\right](dt) \tag{9}$$

and

$$w_{t+dt} = w_t + \dot{w}_t dt + \left[(\tfrac{1}{2}-\beta)\ddot{w}_t + \beta\ddot{w}_{t+dt}\right](dt)^2 \tag{10}$$

Here γ and β are interpolation parameters. Newmark used $\beta = 1/2$ and studied the suitability of the method for different values of β. The details of this method, its system properties, and convergence characteristics are described in Vallabhan et al. (1973). For linear dynamic problems with $\beta = 1/6$, the acceleration between time steps is considered as linearly varying, but the method is only conditionally stable. On the other hand, for $\beta = 1/4$ the method is stable for all discrete time steps, and the acceleration between time intervals is a constant and equal to the average of the values of accelerations at the beginning and end of the time step dt. In this study, the value of β was kept as 1/4.

Equations of the Dynamic Plate Model

It should be noted that in Equations 7 and 8 the unknown vectors at any particular time t are the displacement vector w_t, the acceleration vector $\ddot{w}_t$, and the Airy stress function vector F_t, which are interconnected nonlinearly. A recursive set of equations suitable for iteration was developed, where the acceleration and Airy stress function vectors were calculated at time t + dt. Considering Equations 7 and 8 at time t + dt, substituting for w_{t+dt} using Equations 9 and 10, and rearranging, we get

$$[\tilde{M}]\{\ddot{w}_{t+dt}\} = \{Q_{t+dt}\} + \{f_1(w_{t+dt}, F_{t+dt})\} - \{\tilde{w}_t\} \tag{11}$$

and

$$[B]\{F_{t+dt}\} = \{f_2(w_{t+dt})\} \tag{12}$$

where

$$\{\tilde{w}_t\} = [A]\{w_t + \dot{w}_t dt + (\tfrac{1}{2}-\beta)\ddot{w}_t(dt)^2\} \tag{13}$$

and

$$[\tilde{M}] = [\beta(dt)^2 A + M] \tag{14}$$

The vector $\tilde{w}_t$ is a constant vector at time t containing displacement, velocity, and accelerations at time t only and premultiplied by the A matrix; therefore, this vector is computed once in the beginning of the iterative procedure made at time t + dt.

Computational Procedure

To achieve maximum efficiency in the computations, the following procedures are taken initially.

1) Establish a suitable time step dt to yield a satisfactory solution.

2) Develop matrices A, B, and $\tilde{M}$. Since these matrices are symmetric and positive definite, they are formed in half band width to save computer storage. Matrices B and $\tilde{M}$ are decomposed using Cholesky-type decomposition to minimize the number of computations for simultaneously solving Equations 11 and 12.

3) At time t = 0, $w_t = 0$, $\dot{w}_t = 0$, $\ddot{w}_t = 0$, $F_t = 0$.

4) Start the step-by-step integration procedure, t=n dt, for n=1,N, where N is the desired total number of time steps.

5) At the beginning of time t + dt, first calculate $\tilde{w}_t$ and assume $w_{t+dt} = w_t$ and $F_{t+dt} = F_t$ as first estimates for use in computing the vector f_1.

6) Start the iteration at time t + dt, for i = 1,NITER where NITER = maximum number of iterations specified.

7) Calculate $\ddot{w}_{t+dt}$ using Equation 11.

8) Calculate $w_{t+dt}^{(i)}$ using Equation 10. The superscript i corresponds to the i-th iteration.

9) Calculate F_{t+dt} using Equation 12.

10) Check convergence using the following as an acceptable tolerance:

$$\frac{\sum \left\| w_{t+dt}^{(i+1)} - w_{t+dt}^{(i)} \right\|}{\text{Number of components in the vector}} \leq \varepsilon \tag{15}$$

Repeat steps 7 through 10 until convergence is achieved.

11) Repeat step 4 until the desired number of time steps are completed.

The selection of the proper time steps makes it possible to achieve convergence in less than three iterations for all the cases reported in this study, except when the nonlinearity as denoted by (w_{max}/h) reaches larger than 10. From the displacement response, the bending and membrane stresses are computed at each nodal point within the plate at any time t.

Verification of the Model for Linear Response

Since the model developed here is a numerical one, it is necessary to verify the model with some known theoretical solutions in the linear case. A simply supported square glass plate 66 x 66 x 0.222 in. subjected to a constant pulse of magnitude 0.01 psi was taken for the verification. The Young's modulus of elasticity E, Poisson's ratio υ, and mass density of the glass material were taken as 10^7 psi, 0.22, and 4.85 slugs/ft^3, respectively. Only one quarter of the plate was used for the analysis. Even for a 5 x 5 grid and time step = 1 ms, the displacement response matched very well with the theoretical response. The velocity and acceleration responses show slight variations. For

obtaining better correlation for velocity and acceleration responses
the grid size has to be increased to 10 x 10 or more.

Nonlinear Dynamic Analysis

For this case, the same size plate was used as an example. The
pressure pulse used here was a triangular pulse with a peak pressure
equal to 0.3 psi and time duration equal to 100 ms, as shown in Figure
1. Both space and time discretization studies were made in order to
establish the most optimum grid size and time step in the analysis.
The results presented here are for a grid 10 x 10 with a time step
equal to 1 ms.

The displacement, velocity, and acceleration responses at the center
of the plate were computed for a total time of 100 ms. and these are
shown in Figure 1 (a, b, and c). The maximum displacement calculated
is 1.2 in. and this occurs in 43 ms. The values of maximum velocity
and acceleration at the center of the plate were calculated as 140
in./sec and 63000 in.^{2}/sec, respectively. It is seen that the acceleration response contains several high frequency responses (modes)
which is to be expected in the vibration of plates subjected to this
type of loading. The numerical values of the acceleration response
depend on the space and time discretizations. Since the velocity and
displacement responses are integrated quantities of the accelerations,
a considerable amount of smoothing of data occurs during the integration procedure. This smoothing of data yields fairly consistent
displacement behavior for all grid sizes such as 5 x 5, 10 x 10, and 15
x 15.

From the displacement and Airy stress function vectors, the bending
and membrane stresses are calculated using the equations of the nonlinear von Karman theory and the finite difference method. Components of
bending and membrane stresses are added together algebraically at
discrete nodel points separately for the outward (loaded side) and
inward faces of the plate. Of course, maximum tensile stresses occur
on the inward face of the plate as the plate begins to deform to its
first node. These stress components are transformed to obtain their
corresponding principal stresses. The following table illustrates the
maximum displacement at the center, principal stress at the center and
the global maximum principal tensile stress in the plate.

Table
Displacements and Stresses on the Unloaded Face

Time (ms)	Central Displacement (in.)	Principal Stress at Center (Ksi)	Global Maximum Principal Tensile Stress (Ksi)
5	0.057	-0.004	0.912
10	0.275	-0.575	1.877
15	0.787	3.130	3.130
20	1.099	3.603	5.574
25	1.172	2.419	6.961
30	1.153	4.661	5.821
35	0.666	1.276	4.254

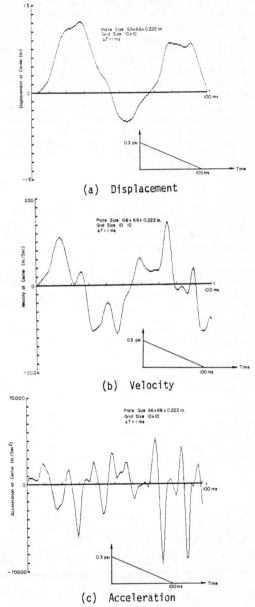

(a) Displacement

(b) Velocity

(c) Acceleration

Figure 1. Displacement, Velocity, and Acceleration at the Center of the Plate, 10 x 10 Grid, $\Delta T = 1$ ms

It is to be noted that during the initial phase of the loading, due to the sudden application of the load, the displacement pattern contains some higher modes, which result in a negative curvature at the center of the plate. This negative curvature produces compressive stresses at the center of the plate on the unloaded face for time equal to 5 and 10 ms respectively. At these instances, the global maximum principal stresses occur in a region very close to the corner. For later time, as the response contains the first mode predominantly the plate has positive curvature, which brings tensile stresses at the center of the plate on the unloaded face. It is not necessary that the global maximum principal stresses occur at the center of the plate at any time. In the table given above, it can be seen that the global maximum principal tensile stresses which occur in a region close to the corners are substantially larger than those occurring at the center of the plate.

Conclusions

A discrete numerical model is developed here using the finite difference for space and Newmark β method for time discretization, to solve the nonlinear dynamic von Karman equations. The value of beta is kept as 0.25. The following conclusions are mode based on the studies reported here:

1) Because of symmetry, only a quarter of the plate is used for the discretization. A grid size 10 x 10 for a quarter of a square plate yields satisfactory results for displacement and velocity responses. For getting accurate acceleration response, the grid size has to be increased.

2) For the triangular pressure loading, lasting up to 100 ms, results obtained using time steps 0.5, 1, 2, and 5 ms indicate that a time step equal to 2 ms yields displacement response comparable to using 0.5 and 1 ms.

3) For other sizes of plates and loading, it is found convenient to use a time step at least equal to 1/20-th of the period of the first oscillation.

4) The method requires less than 3 to 4 iterations at amplitudes illustrated here. For larger nonlinearities, more iterations may be necessary.

5) The CPU time required for the analysis shown here is less than 10 seconds for the IBM3033, which indicates that the procedure is very efficient.

Acknowledgements

This paper is based on research accomplished in the Civil Engineering Department at Texas Tech University, Lubbock, Texas.

The authors wish to thank Dr. Joseph E. Minor, Director of the Glass Research and Testing Laboratory, Texas Tech University, Lubbock, for his encouragement and support during the course of this research. Assistance from Mr. Richard Dillingham for proofreading the manuscript is also acknowledged.

References

Beason, W. L., 1980: "A Failure Prediction Model for Window Glass," Institute for Disaster Research, Texas Tech University, Lubbock, TX (NTIS Accession No. PB81-148421), May.

Behr, R. A., Minor, J. E., Linden, M. P., and Vallabhan, C. V. G., 1985: "Laminated Glass Units Under Uniform Lateral Load," Journal of Structural Engineering, ASCE, Vol. 111, No. 5, May.

Langhaar, H. L., 1962: Energy Methods in Applied Mechanics, John Wiley and Sons, Inc., New York, NY.

Moore, D., 1979: "Design Curves for Non-Linear Analysis of Simply Supported, Uniformly-Loaded Rectangular Plates," Preprint of paper presented at AIAA/ASMA/ASCE/AHS 20th Structure, Structural Dynamics and Materials Conference, St. Louis, MS, April 4-6.

Szilard, R., 1974: Theory and Analysis of Plates--Classical and Numerical Method, Prentice-Hall, Inc., Englewood Cliffs, NJ.

Timoshenko, S. and Woinowsky-Krieger, S., 1959: Theory of Plates and Shells, McGraw-Hill Book Company, Inc., New York, NY.

Tsai, C. R. and Stewart, R. A., 1976: "Stress Analysis of Large Deflection of Glass Plates by Finite Element Method," Journal of American Ceramic Society, Vol. 59, Nos. 9-10, pp. 445-448.

Vallabhan, C. V. G., 1983: "Iterative Analysis of Nonlinear Glass Plates," Journal of Structural Engineering, ASCE, Vol. 109. No. 2, February.

Vallabhan, C. V. G. and Minor, J. E., 1984: "Experimentally Verified Theoretical Analysis of Thin Glass Plates," Proceedings, Second International Conference on Computational Methods and Experimental Measurements, Springer-Verlag, June/July.

Vallabhan, C. V. G., Vann, W. P., and Iyer, S. M., 1973: "Step-by-Step Integration of the Dynamic Response of Large Structural Systems," Department of Civil Engineering, Texas Tech University, December.

Vallabhan, C. V. G. and Wang, B. Y-T., 1981: "Nonlinear Analysis of Rectangular Glass Plates by Finite Difference Method," (NTIS Accession No. PB82-172552) Institute for Disaster Research, Texas Tech University, Lubbock, TX.

CHANGES IN BRIDGE DECK FLUTTER DERIVATIVES CAUSED BY TURBULENCE

Robert H. Scanlan*, Mem. A.S.C.E.

and

Dryver R. Huston*

ABSTRACT: There has been much speculation and a certain amount of experimental evidence to suggest that the aeroelastic stability of suspended span bridges may be highly dependent upon the size and intensity of turbulence present in the oncoming wind. Field studies have shown that the average wind gust size is several times larger than the width of a bridge deck.   This paper will discuss some of the results of a wind tunnel study that tested the aeroelastic stability of section models in smooth and gusting flow.   Wind gusts several times longer than the section model deck widths were created by the method of active gust generation.   The   stability characteristics of the section models was ascertained by using two different methods to measure the flutter derivatives under smooth and gusting flow regimes.

DISCUSSION: A suitable simulation of the aeroelastic behavior of a suspended span bridge that is dominated by aerodynamic loading on a flexible deck is created in a wind tunnel by placing a geometrically faithful slice of the deck in properly scaled flow.   This bridge deck slice, called a section model, is placed on well-tuned elastic springs that permit vertical displacements and rotation about the deck axis.   The section models used in this study were 5 ft. (152.4 cm.) long sections of the Golden Gate and Deer Isle Bridges.   The Golden Gate model has a deck width of 14.75 in.   (37.5 cm.), corresponding to a model-prototype length scale ratio of 1:80.   The Deer Isle section model has a deck width of 12 in. (30.5 cm.), corresponding to a model-prototype length scale ratio of 1:25.   Both models were mounted on elastic springs such that the natural torsional and vertical frequencies of oscillation ranged from 2.0 Hz. to 5.0 Hz.

* Department of Civil Engineering
  The Johns Hopkins University
  Baltimore, MD 21218

   The stability of suspended span bridges, the corresponding
section models and other dynamical systems about equilibrium points
is assessed from the linearized self-excited forces acting about
equilibrium.   The equations of motion , (1.) and (2.), of a section
model contain self-excited aerodynamic lift and moment forces.   The
coefficients appearing in the linearized self-excited aerodynamic
lift and moment forces are known as flutter derivatives (Scanlan and
Tomko 1971). Positive trends in the A*(K) derivative are indicative
of torsional instabilities. Positive trends in the H*(K) derivative
are indicative of vertical instabilities.   The A*(K) flutter
derivative is a measure of the effective torsional aerodynamic
stiffness.   A possible mechanism of turbulence altering the
stability characteristics of a bridge is to change the shape of the
flutter derivative curves.

$$m\ddot{h} + S_\alpha\ddot{\alpha} + c_h\dot{h} + K_h h = \tfrac{1}{2}\rho U^2(2B)\left[KH_1^*(K)\frac{\dot{h}}{U} + KH_2^*(K)\frac{B\dot{\alpha}}{U} + K^2 H_3^*(K)\alpha\right] \quad (1.)$$

$$S_\alpha\ddot{h} + I\ddot{\alpha} + c_\alpha\dot{\alpha} + K_\alpha\alpha = \tfrac{1}{2}\rho U^2(2B^2)\left[KA_1^*(K)\frac{\dot{h}}{U} + KA_2^*(K)\frac{B\dot{\alpha}}{U} + K^2 A_3^*(K)\alpha\right] \quad (2.)$$

h,α = vertical and torsional degrees of freedom, respectively

 m = bridge deck mass per unit length

 I = bridge deck torsional mass moment of inertia per unit length

S_a = bridge deck mass product of inertia per unit length

ρ = density of air

 B = bridge deck width

 U = mean wind velocity

 K = reduced frequency

   Large scale two dimensional gusts were developed in the wind
tunnel by an Active Gust Generator that consisted of two arrays of
symmetric airfoils positioned upstream of the section model (Huston,
1984 and Cermak et al. 1985). In this first use of the Active Gust
Generators two principal flow patterns were created. The first flow
pattern was created by holding the airfoils in a stationary neutral
position. This flow, hereinafter called Neutral flow, proved to be
remarkably smooth. The turbulence intensity was less than 2%. The
second flow pattern was created by flapping the set of airfoils
nearest to the section model with a large amplitude 2.0 Hz. low-pass
white noise driving signal. The second flow pattern, hereinafter
referred to as Low-Pass flow, was a highly turbulent flow with
intensities in excess of 20%. Both flow patterns were created at
mean wind speeds in the range of 5-35 ft./sec. (1.5-10.5 m/sec.).
Figs. 1.a and 1.b contain plots of the vertical turbulence spectra,
normalized by the bridge deck width and mean velocity, for the
Neutral and Low-Pass flows, respectively. The Deer Isle section

model was also tested in two additional flow patterns that were
developed with the Active Turbulence Generators removed from the
wind tunnel. The first of these was an undisturbed laminar flow
with turbulence intensities less than 0.5%. The second additional
flow pattern was created by placing a wood lath grid upstream of the
model. The grid formed a small scale turbulent flow with
intensities less than 12%.

The flutter derivative coefficients were measured by noting the
change in the natural frequency and damping ratio of the aeroelastic
section model system as it was subjected to a wide range of mean
wind velocities. The system properties of damping and natural
frequency were identified by a modified least squares logarithmic
decrement procedure where the section model was given an initial
vertical or torsional displacement and then released. The presence
of random buffeting forces acting on the model after the initial
conditions raised questions about the validity of the log-decrement
procedure. A second method of extracting the flutter derivatives
was developed to augment and verify the modified log-decrement
technique. This method is a least squares spectral curve fitting
procedure. The aeroelastic natural frequency and the damping ratio
are extracted from the spectra of the stationary section model
buffeting response caused by gusts and signature turbulence.

The experimental results demonstrate that there exists a
measurable difference in the flutter derivative curves for the
different flow regimes. Fig. 2. is a plot of the A*(K) curves for
the Golden Gate section model obtained by the modified log-decrement
method. Fig. 3. is a plot of the A*(K) curves for the Golden Gate
Section model obtained by the spectral curve fitting technique. The
two methods show similar, although not identical results. The large
scale gusting of the Low-Pass flow appears to cause the Golden Gate
model to loose torsional stability at a lower wind speeds than the
smooth Neutral flow. The Golden Gate A*(K) and H*(K) coefficients,
figures 4. and 5. show measurable although not extremely
significant changes in behavior at higher wind velocities. The Deer
Isle A*(K) coefficient curves also showed a torsional destabilizing
effect caused by various levels of turbulence. The Deer Isle H*(K)
coefficient appeared to indicate an increased vertical stability
caused by the turbulence. However, the flutter derivative
coefficient curves extracted from the Deer Isle section model should
be viewed with a certain amount of caution since the H-shaped cross-
section is highly prone to nonlinear vortex shedding action.

CONCLUSION: A preliminary study of the effects that large-scale
gusts have upon the aeroelastic stability of suspended span bridges
has been conducted in a wind tunnel using the method of active gust
generation. The stability was assessed by noting the change in the
flutter derivatives caused by different amounts of turbulence in the
flow. The large scale turbulent flow, while not actually
corresponding to field conditions, did indeed induce noticeable
changes in the stability of the bridge deck section model. The
turbulence appeared to exert a large destabilizing effect on the
Golden Gate section model. The Deer Isle model appeared to exhibit
mixed changes in flutter stability.

ACKNOWLEDGMENT: The authors are greatly indebted to C. Galambos and H. Bosch of the Turner-Fairbank Research Center, McLean, VA for their permission and assistance in the use of the Federal Highway Administration, U.S. Department of Transportation, wind tunnel and gust generator facilities at which all of the experiments of the present study were carried out. This study was supported, in part, by a National Highway Institute research fellowship grant.

REFERENCES:

1. Cermak, J. E., Bienkiewicz, B. and Peterka, J. A., 1985: "Active Modeling of Turbulence." Federal Highway Administration No. FHWA/RD-82/148, Washington, D.C.

2. Huston, D. R., 1984: "Active Turbulence Generation for Section Model Studies" presented at the 14th Joint U.S. Japan Conference on Wind and Seismic Effects, Gaithersburg. MD

3. Scanlan, R. H. and Tomko, J. J., 1971: "Airfoil and Bridge Deck Flutter Derivatives." Journal of the Engineering Mechanics Division, ASCE, 97, 1717-1737

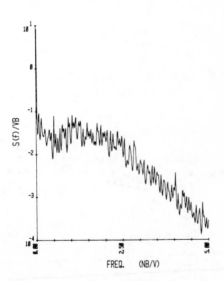

Fig. 1.a. Neutral flow normalized vertical turbulence spectrum, B = 1 Ft. (30.5 cm.), V = 7 Ft./sec. (2.13 m/sec.)

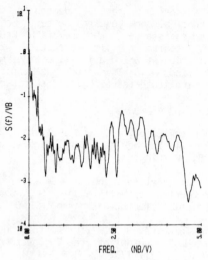

Fig. 1.b Low-Pass flow normalized vertical turbulence spectrum,
B = 1 Ft. (30.5 cm.), V = 7 Ft./sec. (2.13 m/sec.)

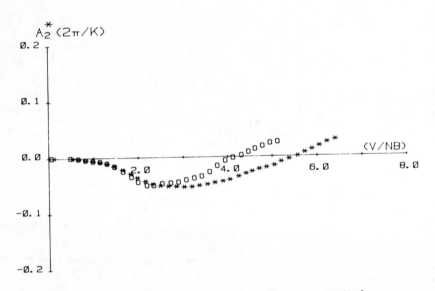

Fig. 2. Golden Gate Flutter Derivative, flow ∗ = Neutral,

o = Low-Pass

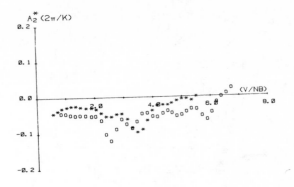

G. G. SPECTRAL WIDTH

Fig. 3. Golden Gate Flutter Derivative, extracted by spectral curve fitting, flow ₀ = Neutral, * = Low-Pass

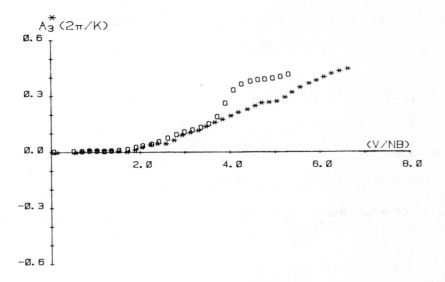

Fig. 4. Golden Gate Flutter Derivative, flow * = Neutral
            ▫ = Low-Pass

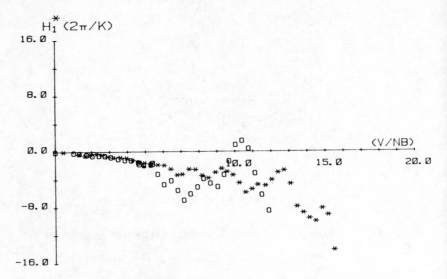

Fig. 5. Golden Gate Flutter Derivative, flow ＊ = Neutral,

□ = Low-Pass

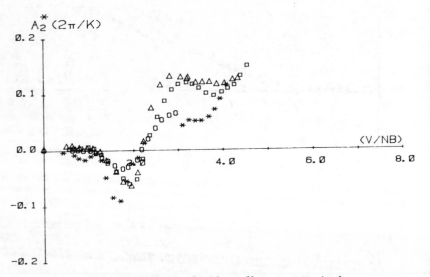

Fig. 6. Deer Isle Flutter Derivative, flow □ = Neutral,

△ = Low-Pass, ＊= laminar and ○ = grid turbulence

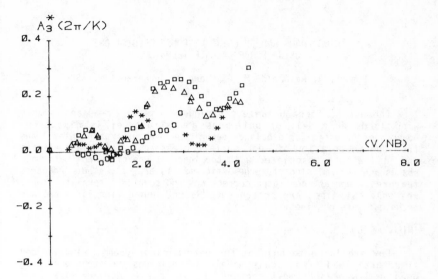

Fig. 7. Deer Isle Flutter Derivative, flow □ = Neutral,

△ = Low-Pass, ∗ = laminar and ◇ = grid turbulence

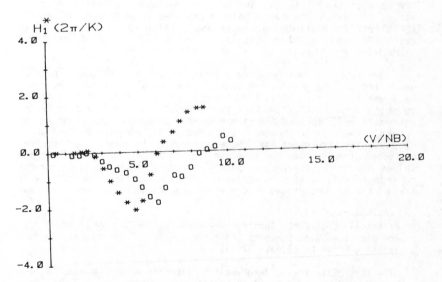

Fig. 8. Deer Isle Flutter Derivative, flow ∗ = Neutral,

□ = Low-Pass

WIND LOADS AND BUILDING RESPONSE PREDICTIONS
USING FORCE BALANCE TECHNIQUES

Timothy A. Reinhold[1] M. ASCE and Ahsan Kareem[2] M. ASCE

The use of force balance techniques to measure mean and dynamic wind loads on models of buildings and structures is discussed. Advantages of using force balance techniques over estimating wind loads from the integration of mean surface pressures or aeroelastic tests and limitations associated with the inherent use of approximate mode shapes and the omission of aeroelastic effects are discussed. Advanced features, such as mode shape corrections and coupled lateral-torsional response estimates are presented. Second generation force balance techniques are outlined.

INTRODUCTION

Force balance techniques for determining dynamic wind induced structural loads from scale models of buildings and structures have been implemented at most centers of commercial boundary layer wind tunnel testing in North America. These techniques have dramatically reduced both the time and cost required to obtain estimates of wind loads and building response levels. From a practical standpoint, the models can be built so easily that it is literally possible to build the model, conduct tests, and produce preliminary estimates within a matter of days. The force balance provides dynamic load information for a specific building geometry and setting which can be used to calculate loads and response levels for a wide range of structural characteristics, damping values, and building masses.

The basis of the techniques is the measurement of power spectral densities of mode generalized (modal) wind loads on the building model as a whole. This requires that the balance has sufficient sensitivity to measure small fluctuations in the modal loads while either having extremely well defined and stable dynamic properties or a high enough natural frequency to insure that the mechanical admittance function is nearly unity throughout the frequency range of interest. Furthermore, estimated dynamic loads and response levels can only be reliable if the shapes of normalized modal spectra and magnitudes of loads are not dependent on Reynold's number effects or aeroelastic effects. Use of force balance techniques has now matured to the point where various

---

[1] Principal Engineer, Applied Research Engineering Services, Inc., and Director, Wind Dynamics Laboratory, 4917 Professional Court, Raleigh, North Carolina, 27609.

[2] Director, Structural Aerodynamics Laboratory and Ocean Systems Modeling, Department of Civil Engineering, University of Houston, Houston, Texas 77004.

levels of sophistication can be incorporated in the analysis of test results. The level of sophistication usually depends on the complexity of the building and the level of analysis employed by the structural engineer.

Force balance tests offer significant advantages in accuracy and reliability of gust loading estimates over loads obtained from code procedures or the integration of surface pressure which are multiplied by a gust factor. However, there are definite limitations to the technique and there are instances where aeroelastic model tests should be undertaken. This paper describes the role of force balance tests on a continuum of techniques for determining wind induced structural loads which extends from code procedures to multi-degree-of-freedom aeroelastic model tests. Analytical refinements which can be applied to enhance or improve force balance predictions are discussed. Second generation force-balance techniques are described which directly allow measurement of loads for more complex mode shapes and, thus, avoid the need for analytical corrections.

THE ROLE OF FORCE BALANCE TESTS

There are basically four approaches which can be used to estimate wind induced loads for design of structural frames. These approaches are:

1. Code and analytical methods for calculating loads acting in the direction of the mean wind.

2. The integration of mean pressure to produce mean loads which are multiplied by a gust factor chosen to account for both the gustiness of the wind and inertial loads associated with the expected response of the building or structure.

3. Force balance tests which involve direct measurement of wind loads acting on the building as a whole and utilize basic principles of structural dynamics to obtain dynamic loads from the calculated response of the building or structure.

4. Aeroelastic model tests where building properties (mass, stiffness, and damping) are simulated in the model and wind induced loads and response can be measured directly.

Code and analytical methods have been developed for estimating mean and peak wind loads acting in the direction of the mean wind (along-wind)(ANSI, 1982; NRCC, 1980; Solari, 1984). However, these methods are most appropriate for isolated rectangular buildings when the wind is blowing perpendicular to one of the building faces. The across-wind and torsional loads have been much more difficult to treat analytically since no generalized theory has been developed which adequately describes load fluctuations in separated flow regions. The procedures which have been developed (Kareem, 1984; Solari, 1985) rely on the use of experimentally determined modal force spectra for across-wind and torsional loads on isolated buildings with simple shapes. Site specific aspects such as the directional dependence of the wind climate or the influence of specific surrounding buildings or terrain features are difficult or impossible to include. Finally, only rudimentary procedures are available for estimating building response levels used in assessing occupant comfort.

While it is adequate to use code specified structural loads for many buildings, including most mid-rise buildings (10-25 stories), there are many instances where the use of more accurate methods which account for site and building specific aspects, would produce designs which are more economical and have a more uniform level of reliability. There are also instances where, due to complex shape, slenderness or unusual structural systems, the code specified loads are not adequate to provide suitable factors of safety or to insure occupant comfort. Thus, in many instances, it is worthwhile either from a cost or reliability/serviceability standpoint to obtain load information from a wind tunnel test.

Due to the large investment associated with modern curtain wall systems and their relative flexibility, an increasing number of buildings are being wind tunnel tested to determine cladding loads. Once a cladding load study has been conducted, it is a relatively inexpensive add-on to integrate the mean pressures and determine mean loads on the building as a whole. It is common practice to then multiply the mean loads by a gust factor obtained from a analytical procedure such as those outlined in References 1, 2, and 3 to produce estimates of peak loads. However, it is important to be conservative in the selection of the gust factors because this procedure shares many of the limitations of the analytical procedures (i.e., as the mean load becomes small for the across-wind component the dynamic loads may remain large and the variation in magnitude of the across-wind or torsional load is not well correlated with the mean load). The practical implication of this for most buildings is that combined loads may be underestimated for critical wind directions. For slender or extremely flexible structures the peak across-wind load may be significantly larger than the peak along-wind load and the strength requirements may be seriously underestimated if the design is based on along-wind loads (Melbourne, 1975; Reinhold, 1979). A typical comparison of overturning moments obtained from integrating mean pressures multiplied by a gust factor and from a force balance test is illustrated in Figure 1 to demonstrate the limitations associated with estimating peak loads from mean values.

With these considerations in mind, adequate loads for many buildings can still be estimated based on integration of mean pressures provided appropriately conservative gust factors are selected. The gust factors should be sufficiently increased to compensate for expected increases in loads on critical members due to possible underestimation of combined loads. A more appropriate approach which eliminates the uncertainty is to use the peak loads determined from the pressure study as a quick look at possible magnitudes of the structural loads and follow up with a force balance or aeroelastic model test.

Force balance tests are the least expensive and least time consuming test method available for obtaining realistic estimates of mean and dynamic wind induced structural loads and response levels. The primary limitations are that aeroelastic effects are not included

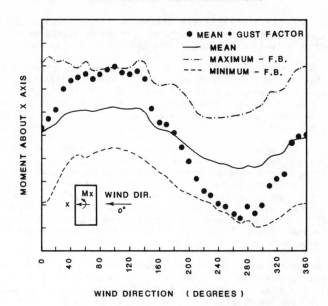

FIGURE 1. COMPARISON OF OVERTURNING MOMENTS
MEAN LOAD STUDY VS. FORCE BALANCE

and that for current balance forces and moments are measured at the
base using a rigid model of the building or structures. For most
buildings and typical design wind velocities, aeroelastic effects
appear to have little influence or will tend to reduce the amplitude of
motion (Reinhold, 1979; Davenport and Tschanz, 1981). The use of a
rigid model on a balance where forces and moments are measured at the
base implies a linear mode shape for lateral forces and a uniform mode
shape for torsional moments. These mode shapes arise from the use of
the base overturning moment and base torque to determine the mode
generalized forces and torques. Thus, torsional forces must always be
adjusted and results of conventional force balance tests may under-
estimate or overestimate the appropriate modal forces if sway mode
shapes differ significantly from a linear mode shape (Vickery,
Steckley, Isyumov, and Vickery, 1985). Mode shape corrections are
discussed in more detail in the following section.

The dynamic loads measured by the force balance are applied to a
mathematical model of the building or structure to compute response
levels. Total loads are determined which include both the dynamic wind
loads and inertial loads associated with the motion of the building or
structure (dynamic amplification). Building properties can be readily
changed in the mathematical model and hence results can be obtained for
a range of building properties.

Aeroelastic models have constituted the most recognized type of
model for use in determining wind induced motion of a building and in
determining resultant fluctuating loads acting on the building as a
whole (base bending moments and base shear). However, force balance

tests have recently replaced the aeroelastic model test as the most commonly used technique. The range of complexity available in aeroelastic models extends from semi-rigid models (linear sway mode shapes) which incorporate a gimbal and a set of springs to a leg type model which may have four or more lumped masses connected by flexible legs. Installation of torsional flexures, which are rigid in sway, in a semi-rigid model provides a means for approximating the lowest torsional model shape in a stepwise fashion. The lowest sway mode shapes and the lowest torsional model shapes are all modeled in a stepwise fashion by the leg type model.

Aeroelastic model tests are conducted for a number of different wind speeds at each wind direction considered and the structural properties may be varied to bracket possible values which may ultimately exist in the prototype. Alternatively, effects due to changes in properties can be assessed by calculation. The primary limitation to the use of calculation in assessing effects of changes in structural properties is the case where changes in frequency result in strong coupling between torsional and sway vibration that did not exist in the model as tested.

The primary advantage of force balance tests over aeroelastic model tests is the timing. Aeroelastic tests are also significantly more expensive than force balance tests; but, the greatest problem is that by the time the developer and design team decide that wind tunnel tests are required, there may not be sufficient time for aeroelastic models to be constructed and tested. On the other hand, multisegment aeroelastic models will provide more reliable information on torsional effects than conventional force-balance tests and stick type aeroelastic models can approximate more complex mode shapes.

ANALYTICAL REFINEMENTS

In a basic force balance test, mean and dynamic overturning and torsional moments are measured at the base of the model. Several balances also allow measurement of base shear forces. The dynamic moments are analyzed to obtain root-mean-square values and power spectral density functions. In the basic application, these moments are taken as representative of the mode generalized forces (with a nominal correction for the torsion) and the response in each of the fundamental modes is calculated assuming that there is no mechanical coupling and no correlation between the forces. The balance is usually located such that its center coincides with the expected center of rigidity of the building so that torques are measured about that point.

A more general approach is to measure not only the spectra of the overturning moments and torques but also the cross-spectra between the components. This formulation allows a more complete treatment of the response and estimation of loads for different orientations of the axes and about various centers of rotation.

Differences between mode shapes associated with the mode generalized forces measured with the model and the actual mode shapes of the building or structure will introduce uncertainty in the computed loads and response levels. An analysis of torsional moments measured at various levels throughout the height of a tall building (Tallin and Ellingwood, 1985) indicated that the mode generalized torsional moment for a linear mode shape could be approximated as two thirds of the base torque measured using a uniform mode shape. A reduction factor of 0.70 is commonly used.

Approximate bounds on the errors introduced by differences between the model mode shapes and actual mode shapes have been suggested based on an application of quasi-static theory (Vickery, et al., 1985). However, as they demonstrated by comparison of the predicted corrections with actual corrections, implied by the comparison of base shears (uniform mode shape) and base overturning moments (linear mode shape), the corrections tend to break down in a complex real environment where wakes of surrounding buildings have a strong influence on the distributions of loads. Fortunately, many buildings have a near linear mode shape and the errors associated with force balance test results for sway are relatively small. Since it is common to find that response levels for winds from six to eight directions contribute about ninety percent or more of the total probability of exceeding various acceleration levels, analytically derived corrections to force levels and accelerations may have a high level of uncertainty for buildings in a real setting.

Analytically derived corrections are useful for estimating the order of magnitude of possible corrections. However, when mode shape corrections based on approximate analytical techniques indicate that significant corrections should be applied to conventional force balance test results or when torsion is important, consideration should be given to using a more advanced model technique unless conservatively biased force balance test results are acceptable. Stick built aeroelastic models or an advanced force balance, as described in the following section, would provide the more accurate information directly without requiring analytically derived corrections.

SECOND GENERATION TECHNIQUES

Force balance techniques have become widely accepted and used as an inexpensive yet accurate means for determining wind induced loads and responses. However, current techniques provide only an approximate means for determining mode generalized torsional moments and loads and response for sway may be inaccurate if the sway mode shapes of the building or structure differ significantly from a linear mode shape.

In order to overcome these limitations, Applied Research Engineering Services is developing a second generation force balance which includes several torsional flexures mounted on a stiff spine. Figure 2 illustrates the balance concept. Segments of the shell of the building are constructed and attached to the torsion flexures. Lateral loads are transmitted through the torsion flexure to the central spine. The torsional flexures are sized such that for nominal model weights, the torsional frequency of vibration of each segment is about 100 Hertz. Similarly the spine is designed to insure that the overall sway frequencies are about 100 Hertz. Each segment of the spine between attachment points of the torsion flexures is sized to provide maximum sensitivity to moments while maintaining the sway frequencies. Actual

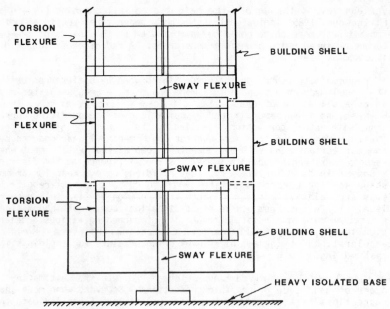

FIGURE 2.   SCHEMATIC OF ADVANCED FORCE BALANCE

frequencies vary from model to model depending on the weight and flexibility of the shell segments. However, for most cases the final frequencies are expected to exceed 70 Hertz. Test velocities are selected such at for most buildings studied and model scales used, the maximum load fluctuation frequency of interest will be less than 70 percent of the lowest model frequency. Within this frequency range, corrections applied to the loads for model resonance are insensitive (less than 2 percent difference) to model damping between zero and 4 percent of critical.

Initially, the emphasis is being placed on obtaining torsional moments where the fundamental mode shape is approximated in the same way that multisegment aeroelastic models simulate the mode shapes of a building. The mode generalized torsional moments are obtained by weighting the torques measured with each torsional flexure according to the mode shape and adding the torques either digitally or by means of a simple analog summing circuit. The second step is to instrument the spine so that lateral forces can be measured for each segment. These lateral forces can in turn be combined to produce mode generalized lateral loads which more closely correspond to the expected mode shape of the prototype building or structure.

SUMMARY

Force balance tests provide an inexpensive yet relatively accurate means for determining the wind induced dynamic response of buildings and structures. Equivalent static loads can be readily calculated which represent the gust loading on the building and dynamic amplification due to the building response.

Primary limitations of the technique are that aeroelastic effects are not included and that corrections must be applied to the torsional moments. These corrections for torsion arise from the fact that base torques correspond to a uniform mode shape rather than the linear mode shape which is common for many buildings. Other limitations are that mode generalized lateral forces obtained by current balances correspond to a linear mode shape. Magnitudes of possible errors can be estimated analytically but actual errors depend on the influence of surrounding buildings. An advanced balance is described which will allow a direct measurement of mode generalized torsional moments and lateral forces for more complex mode shapes.

APPENDIX A

American National Standards Institute, 1982: Minimum Design Loads for Buildings and Other Structures, ANSI-A58.1-1982, New York.

Davenport, A.G., and Tschanz, T., 1981: "The Response of Tall Buildings to Winds: Effects of Wind Direction and the Direct Measurement of Dynamic Force," Proc. of the 4th U.S. National Conference on Wind Engineering Research, University of Washington, Seattle, Washington.

Kareem, A., 1984: "A Model for Prediction of the Acrosswind Response of Buildings," Engineering Structures, Vol. 6, No. 2.

Melbourne, W.H., 1975: "Cross-Wind Response of Structures to Wind Action," Proc. of the 4th International Conference Wind Effects on Buildings and Structures, Heathrow, Published by Cambridge University Press.

National Research Council of Canada, 1980: The Supplement to the National Building Code of Canada, Associate Committee on the National Building Code, NRCC No. 17724, Ottawa.

Reinhold, T. A., 1979: "The Influence of Wind Direction on the Response of a Square-Section Tall Building," Proc. of the 5th International Conference on Wind Engineering, Fort Collins, Colorado, Published by Pergamon Press.

Solari, G., 1984: "Analytical Estimation of the Alongwind Response of Structures," Journal of Wind Engineering and Industrial Aerodynamics, Vol. 14.

Solari, G., 1985: "A Mathematical Model to Predict 3-D Wind Loading on Buildings," Journal of Engineering Mechanics, ASCE, Vol. 111, No. EM2.

Tallin, A., and Ellingwood, B., 1985: "Analysis of Torsional Moments on Tall Buildings," Journal of Wind Engineering and Industrial Aerodynamics, Vol. 18.

Vickery, P.J., Steckley, A., Isyumov, N., and Vickery, B.J., 1985: "The Effect of Mode Shape on the Wind-Induced Response of Tall Buildings," Proc. of the 5th U.S. National Conference on Wind Engineering, Texas Tech University, Lubbock, Texas.

Behaviour of Full-Scale Reinforced Concrete Structure
on Realistic Foundation

Shunsuke OTANI[*], Masaru EMURA[**] and Hiroyuki AOYAMA[*]

A full-scale seven-story reinforced concrete building, tested as a part of U.S.-Japan Cooperative Research Program, performed well even under a severe earthquake loading condition. The structure was tested under an idealized situation. It was anticipated, therefore, that the structure might not behave sufficiently if the structure had been tested in more realistic conditions such as on deformable foundation. Consequently, nonlinear earthquake response analyses were carried out taking into account the base uplifting of a shear wall on the flexible foundation. It was noted that the overall behaviour of the structure, with or without uplifting at the base of the shear wall, was quite similar except for the ductility demand in the foundation girders. The influence of soil flexibility was found remarkable under medium intensity earthquake excitation.

Introduction

A full-scale seven-story reinforced concrete frame-wall building was tested in 1981 as a part of U.S-Japan Cooperative Research Program at Building Research Institute [Refs.1-6]. The test structure performed well to a story drift angle of 1/64 rad. under a series of simulated earthquake excitations with a ductile energy dissipation at the beam ends and at the base of the shear wall without brittle shear failure. However, the test structure did not represent a typical construction in Japan; i.e., a) The structure did not satisfy Japanese Building Code due to the limitation in load capacity of the testing facilities, the member dimensions and reinforcement being chosen smaller than those normally used in Japan. b) The test structure was fixed rigid on the test floor during the test. c) A single-degree-of-freedom pseudo-dynamic testing procedure was used to apply lateral load history to the structure, in which an earthquake motion was distorted to excite the fundamental mode and the lateral load distribution along the height of the structure was predetermined to a calculated fundamental mode.

This paper studies the nonlinear earthquake response of the building under more realistic environment; i.e., the structural was supported on deformable foundation. The edge of the wall base was allowed to uplift under a net tension. The structural members were designed with Japanese earthquake loads. Natural earthquake motions were used to excite the building in nonlinear earthquake response analyses.

----

* Dept. of Architecture, Faculty of Engineering, University of Tokyo
** Ohbayashi-gumi, Ltd.

Analytical Models

The full-scale seven-story building [Ref.5] had three bays in the longitudinal direction, and two spans in the transverse direction (Fig.1). A shear wall was placed, parallel to the direction of loading, in the center frame. Column dimensions were 500x500 mm with 8-D22 (No.7) bars, and girder size was 300x500 mm with 3-D19 (No.6) bars at the top and 2-D19 bars at the bottom, whereas that of transverse beams was 300x450 mm. The thickness of the shear wall was 200 mm with 2-D10 bars at 200 mm on centers. Foundation girders were 500x1,310 mm with 5-D25 (No.8) bars at the top and bottom. The compressive strength of concrete was assumed 270 kg/cm^2 (3,900 psi), and yield strength of reinforcement 3,500 kg/cm^2 (50,000 psi). The weight of the structure was evaluated as the dead weight including the weight of loading apparatus, but without the live load.

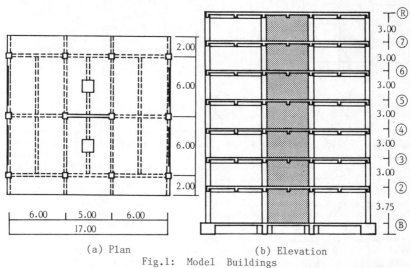

(a) Plan   (b) Elevation
Fig.1: Model Buildings

Two structural models were constructed; i.e., Model F1 simulated the test structure. In Model F2, the flexural reinforcement in the wall boundary columns was increased to 2.5 times that in the test structure, and the longitudinal reinforcement in the girders (boundary beams) connected to the wall and the wall reinforcement were increased to approximately 1.5 times to meet Japanese lateral load requirements. In other words, 20-D22 bars were placed in a column section, and 4-D22 bars at the top and 3-D22 bars at the bottom of a beam section. The wall was reinforced with 2-D10 bars at 120 mm on centers. The reinforcement ratio of Model F2 became excessive because the dimensions were kept the same.

The model buildings were supported by 400 mm-diameter piles of 12 m long; four piles under an independent column and five piles under a wall boundary column. The vertical stiffness of the pile under compression was estimated from the pile section and the dynamic soil coefficients of 0.32 and 8.0 $tonf/m^3$, corresponding to soil shear wave velocities Vs of 100 and 500 m/s, respectively. The edge of the shear wall base was allowed to uplift from the pile support when the tension force due to overturning moment cancelled the vertical gravity loads (Fig.2). As a reference, the structures were also analyzed on the rigid fixed base.

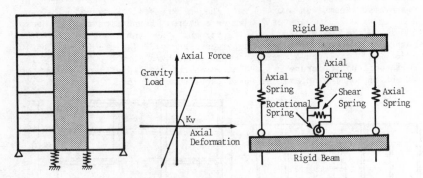

(a) Base Modelling   (b) Base Hysteresis   (c) Wall Model
Fig.2: Modelling of Foundation and Wall

The structural models were analyzed by a nonlinear dynamic response analysis program DANDY [Ref.3], in which the columns and girders were represented by line elements with two nonlinear rotational springs at the ends. Axial deformation was considered in columns. The beam-column connection was assumed rigid. The load-deflection relationship of beams and columns under monotonically increasing load was determined on the basis of section analysis and material stress-strain relations. The member end rotation was assumed proportional to the curvature at the critical section with a trilinear relation. The shear wall was idealized as three vertical line elements (Fig.2). The vertical truss element, the deformation of which represented the elongation at the tensile extreme fibre of the wall under bending, was assumed to remain elastic under compression, and to yield in tension when the column reinforcement yielded. The transverse beams connected to the shear wall were included in the analysis to take into account the vertical actions of the wall transmitted from the beams. The hysteresis model of member models were Takeda-Slip model for beams, Takeda model for columns, the origin-oriented model for shear and rotational springs of the shear wall. The analysis method including the choice of various constants was proven to simulate the overall and local behaviour of the test structure [Ref.3]. The stiffness of the foundation was assumed to be elastic when a net compression acted in the base vertical spring (Fig.2), while the wall edge was allowed to uplift under net tension.

Static Analysis

The models were analyzed under monotonically increasing static lateral loads of triangular distribution. The total base shear and wall base shear are plotted with respect to the roof-level displacement relation in Fig.3. The drift angles, the roof-level displacement divided by the total height, are given at the flexural yielding (YW), at the uplifting of the wall (UW), and at the flexural yielding of the foundation girders (YB) and transverse foundation girder (YT).

In Model F1, foundation deformation contributed to the overall deformation of the structure. However, the effect of the foundation flexibility diminished after the wall yielding. The flexural yielding of the wall on the rigid foundation was calculated at a story drift of 1/540 rad., while the flexural yielding of the wall on the deformable foundation was calculated at 1/207 rad. after the wall uplifting at a drift angle of 1/500 and the yielding of the foundation girders at 1/265 rad. The Model F1 formed the collapse mechanism at 1/44 story drift, when the load was approximately equal to the lateral load resisting capacity of the test structure on the fixed base. The loads at the wall flexural yielding on the rigid foundation and at the wall uplifting happened to be comparable.

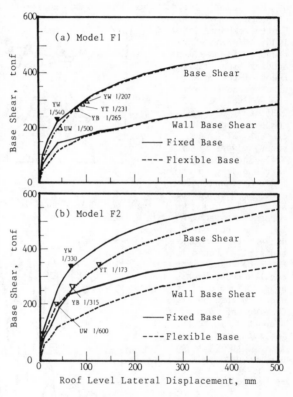

Fig.3: Load-Displacement Relation under Monotonic Load

In Model F2 designed with larger earthquake loads, the lateral load was higher at the wall flexural yielding than at the uplifting. Consequently, the collapse mechanism was formed by uplifting of the shear wall on the deformable foundation rather than wall uplifting.

In the elastic stage, the shear wall carried 70 to 85 % of the total base shear. The share of the wall on the flexible foundation was less than that on the rigid foundation by approximately 10 %. The ratioes at the wall uplifting and flexural yielding were comparable for the walls on the two foundation conditions.

Earthquake Response Analysis

The models were subjected to first 20 seconds of the EW component of Hachinohe record, the 1968 Tokachi-oki earthquake, and the NS component of El Centro record, the 1940 Imperial Earthquake. The maximum acceleration amplitudes of the Hachinohe record were chosen to be 175, 263 and 350 gals (cm/sec^2), and those of the El Centro record were 316, 474, and 632 gals corresponding to the maximum ground velocity amplitudes of 30, 45 and 60 cm/sec, respectively. The two earthquake motion presented similar trend in response. Hence, the results from Hachinohe record are shown in this paper. Damping was assumed proportional to the instantaneous stiffness, the initial elastic damping factor being 3 % of the critical.

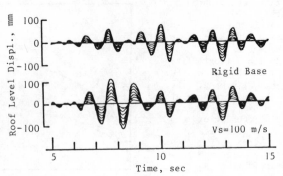

(a) Maximum Ground Velocity of 30 cm/sec

The response displacement waveforms are shown in Fig.4 for Model F1 under the Hachinohe record at 30 and 45 cm/sec intensities. Note that the structure oscillated in the fundamental mode with uniform interstory deflection.

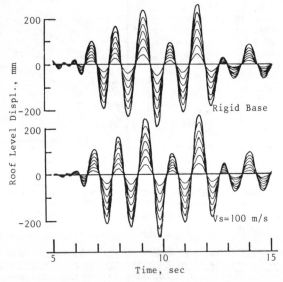

(b) Maximum Ground Velocity of 45 cm/sec
Fig.4: Displacement Waveforms
(Model F1, Hachinohe Record)

Table 1: Maximum Response Values

Model	F1			F2		
Ground Motion,cm/sec	30	45	60	30	45	60
(a) Structure on Fixed Rigid Base						
Base Shear, ton	302	446	526	348	547	600
Lateral Drift	1/256	1/79	1/60	1/357	1/136	1/61
Ductility Factor						
Boundary Beam	1.2	3.5	3.9	–	1.2	2.6
Column Base	–	3.3	4.7	–	1.9	4.8
Beams	1.2	3.5	3.8	–	2.1	4.3
(b) Structure on Flexible Foundation (Vs=100m/sec)						
Base Shear, ton	347	447	513	340	478	533
Lateral Drift	1/189	1/79	1/61	1/201	1/88	1/61
Ductility Factor						
Boundary Beam	1.7	3.6	4.2	1.9	5.4	8.4
Column Base	2.0	4.6	6.0	2.1	5.7	4.8
beams	1.6	3.4	3.8	1.5	3.1	3.9
(c) Structure on Flexible Foundation (Vs=500m/sec)						
Base Shear, ton	281	445	510	322	481	530
Lateral Drift	1/272	1/81	1/61	1/252	1/94	1/62
Ductility Factor						
Boundary Beam	1.2	3.5	4.5	1.5	5.3	8.4
Column Base	1.4	4.7	6.0	1.6	5.4	8.4
Beams	1.1	3.5	4.5	1.5	2.9	4.7

Maximum response values are summarized in Table 1. The maximum ductility factor of beams and columns was defined as the maximum response rotation divided by the yield rotation. The natural period of Model F1 was 0.43 sec on the fixed base base, and 0.51 sec on the ground of Vs=100m/s. The maximum story drift of a structure on the soil of Vs=100 m/s was calculated greater than that on the soil of Vs=500 m/s). The difference was remarkable when the structures were subjected to low to medium intensity ground motion and when the structures responded in the elastic range. The foundation stiffness and the natural period were important in the maximum response values under medium intensity base motion, in which the flexural yielding was not calculated at the wall base. However, the difference became smaller for a large intensity base motion.

The maximum vertical displacement of the wall at the base of the uplifting wall and maximum elongation deflection of the boundary column in the first story shear wall are shown in Fig.5. The former indicates the amount of deflection due to uplifting at the wall base. while the latter deflection corresponds to the flexural deformation at the base of the shear wall.

When the building with an uplifting wall was subjected to the Hachinohe record at 45 and 60 cm/sec intensities, the flexural yielding occurred after the wall uplifting. The maximum drift angle of the building and base shear were as large as those calculated for the model with fixed base. The influence of supporting conditions was observed less after the formation of a mechanism in Model F1.

In Model F2, the drift angle of the building was kept small under the Hachinohe record at 30 and 45 cm/sec intensities. The maximum drift angle was calculated comparable for the models on the fixed base and on the deformable foundation. This implies that a major deflection of the model on the fixed base resulted from the elongation of the tensile boundary column, whereas that with a uplifting wall resulted from the vertical uplifting displacement. However, the uplifting deflection

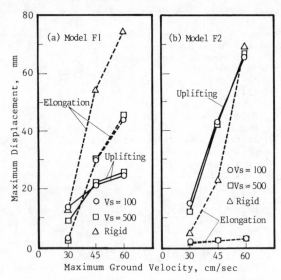

and the elongation became comparable for the ground motion at 60 cm/sec intensity.

Fig.5: Elongation of Boundary Column and Uplifting Displacement

Concluding Remarks

The flexural yielding at the wall base must have occurred in the test structure with and without uplifting at the wall base. The extent of damage and the deflection of the structures were almost identical in the two structures. Under intense ground motion, a large difference was noted in the flexural deformation at the base of the wall and the deflection in the foundation girders.

When the design earthquake loads were increased from the test structure level, the collapse mechanism of a building was controlled by the uplifting of the wall, but the building showed ductile deformation efficiency was almost the same as the test building on the fixed base. Therefore, it is possible to design a structure with the uplifting shear wall as ductile as the moment resisting frame.

The influence of the soil stiffness was significant on the response amplitudes under small to medium intensity earthquakes, but became less with the intensity of base motion.

Acknowledgement

    The writers wish to express a sincere gratitude to Mr. Kabeyasawa for providing the witers with earthquake response analysis program DANDY.

Appendix 1: References

1. Kaminosono, T., S. Okamoto, Y. Kitagawa, and M. Yoshimura, "Testing Procedure and Preliminary Test Results of a Full Scale Seven Story Reinforced Concrete Building", Earthquake Effects on Reinforced Concrete Structures, U.S.-Japan Research, SP-84, American Concrete Institute, 1985, pp.105-132.

2. Okamoto, S., J. Wight, S. Nakata, M. Yoshimura, and T. Kaminosono, "Testing, Repair and Strengthening, and Retesting of a Full Scale Seven Story Reinforced Concrete Building", Earthquake Effects on Reinforced Concrete Structures, U.S.-Japan Research, SP-84, American Concrete Institute, pp.133-162.

3. Otani, S., T. Kabeyasawa, H. Shiohara, and H. Aoyama, "Analysis of the Full Scale Seven Story Reinforced Concrete Test Structure", Earthquake Effects on Reinforced Concrete Structures, U.S.-Japan Research, SP-84, American Concrete Institute, 1985, pp.203-240.

4. Wight, J., S. Nakata, and T. Kaminosono, "Construction and Instrumentation of the Full-Scale Specimen", Earthquake Effects on Reinforced Concrete Structures, U.S.-Japan Research, SP-84, American Concrete Institute, 1985, pp. 49-72.

5. Wight, J., V.V. Bertero, and H. Aoyama, "Comparison between the Reinforced Concrete Test Structure and Design Requirements from U.S. and Japanese Building Codes", Earthquake Effects on Reinforced Concrete Structures, U.S.-Japan Research, SP-84, American Concrete Institute, 1985, pp.73-104.

6. Yoshimura, M., and Y. Kurose, "Inelastic Behavior of the Building", Earthquake Effects on Reinforced Concrete Structures, U.S.-Japan Research, SP-84, American Concrete Institute, 1985, pp.163-202.

Modeling 3-D Response in R/C Frame-Wall Structures

Finley A. Charney*

Introduction

As part of the recent U-S Japan Cooperative Research Program on the seismic performance of buildings, a 1/5-scale model of a 7-story reinforced concrete frame-wall structure, Fig. 1, was constructed and subsequently tested on the earthquake simulator at the University of California at Berkeley [3]. The response of this structure was three dimensional in nature and was dominated by the inelastic rocking (neutral axis migration) and axial growth of the centrally located shearwall and by the outriggering restraint provided by the longitudinal and transverse girders which framed into the wall. In order to analytically trace the response of this structure using a computer program restricted to nonlinear dynamic planar behavior, special modeling techniques had to be developed which would capture the important three-dimensional aspects of the response.

Description of Model and Experimental Program

The 1/5-scale, 7-story test structure consisted of three parallel frames in the direction of horizontal table shaking: two exterior moment resisting frames (Frames A and C) and one interior frame-wall system (Frame B) incorporating a central barbell type wall. In the orthogonal direction there were four parallel moment resisting frames (Frames 1 through 4) and four end-walls. These walls were included within the end frames to reduce accidental torsional response. The floor system consisted of a solid one-way slab which spanned beams and girders running between frames A, B and C. All of the principal structural elements were very lightly reinforced as described in detail in reference 3. After initial testing, the model was subjected to the series of simulated earthquake accelerations as described in Table 1.

Description of Experimental Response

Prior to the MO14.7 test, the model responded in a near linear fashion, with no cracking damage being visually observed after the test. During the MO14.7 test, a horizontal flexural crack was observed to have formed at the wall-foundation interface, with this

*Technical Analyst, KKBNA, Inc., Consulting Engineers, Denver, CO

crack spanning the full length of the shearwall. The crack grew in width during the M024.7 test and by the end of the M028.3 test the crack had opened to a maximum width of approximately 0.156 inches (3.96 mm). As a result of this crack, the shearwall rocked back and forth during simulated earthquake excitation, with the center of rotation (at maximum displacement) being the compression-side boundary element of the wall. At the time of maximum displacement, the tension side of the wall was displaced upwards, with this movement being resisted by the longitudinal and transverse beams which framed into the tension side of the wall, as shown in Fig. 2. Since relatively little downward deformation was simultaneously occurring in the compression side boundary element, the net reactions from the restraining beams caused the compressive force in the shearwall to increase significantly. The peak compressive force increases, as well as other pertinent data obtained from the several tests, are summarized in Table 1.

Since no vertical accelerations were applied during these tests, the increases in shearwall axial compression were equilibrated by decreased compression in the exterior columns of frames 1, 4, A, and C.

Modeling Techniques for 3-Dimensional Response

In order to analytically predict the response of the test structure, the following components of behavior must be included in the mathematical model:

1) Inelastic rocking and axial growth of the shearwall and associated outriggering restraint provided by longitudinal and transverse beams

2) Increase or decrease in column and wall flexural stiffness and strength and in shear stiffness and strength with variations in axial force.

The only computer program available at Berkeley for the analysis of the test structure was the program DRAIN-2D [2], which, unfortunately, is limited by design to predicting the nonlinear dynamic response of planar structures. Furthermore, none of the elements provided with the program allow for the explicit modeling of the conditions stated in 1) and 2) above. Since the scope of the project was limited to the use of currently existing computer programs, it was decided to make do with DRAIN by implementing modeling techniques which might emulate the response characteristics observed during the testing of the model. An overview of these techniques used is now presented.

Modeling the Shearwall

It is clear that the phenomenological point hinge idealizations used by the one-dimensional, two-component beam-column element of DRAIN will not display the rocking behavior observed during the experimental response. However, through the use of an assemblage of simple nonlinear truss elements, it was possible to create a physical model of the base of the wall which displayed response characteristics similar to those observed.

Consider the simple model of Fig. 3 which consists of a rigid horizontal bar attached to a foundation by a series of vertical nonlinear springs. This simple modeling concept, which consists of superimposing the nonlinear response of several constrained elements (in this case extensional springs) is referred to as the LINKS model in the following discussion. Each spring in the LINKS model has the capability to yield in tension or compression, or to buckle in tension or compression, as shown in Fig. 4. The yield-yield springs may be used to represent reinforcement, and the yield-buckle springs may be used to represent concrete. By judiciously selecting the placement, stiffness, and yielding or buckling strength of the individual springs, the simple model, as subjected to monotonically increasing axial load and moment, very closely emulates the response of a fiber model which divides the section into more than one hundred segments. For example, in Fig. 5 the response of an eight-spring (four concrete and four steel springs) LINKS model of the shearwall of the test structure is compared to the response of a fiber model of the same section.

It was noticed after reducing the data from the experiments, that the hysteretic moment-rotation response of the base of the wall was significantly pinched. By using springs which buckle in tension and yield in compression (Fig. 4), a pinched moment-rotation response as shown in Fig. 6 is produced. If necessary, a degrading stiffness response may be modeled by replacing the steel fiber springs with a horizontal cantilever beam whose hysteretic response degrades under cyclic deformation reversal.

The final DRAIN model of the base of the shearwall is shown in Fig. 7, where it can be seen that it was necessary to place the concrete springs above the rigid bar in order to cause them to "buckle in tension".

Modeling the Longitudinal Girders

Considerable uncertainties existed with regard to computing the strength and stiffness of the principal girders. As shown in Fig. 8, the computed monotonic strength and ductility of a typical girder of Frame B varies considerably with the effective width of the reinforced slab. For negative moment with fully effective slab (center-to-center dimensions), the flexural strength is more than twice that computed on the basis of the ACI [1] flange width. For positive moment, the ratio is approximately 1.4 for wide slab strength to ACI slab strength. These uncertainties in girder strength relate directly to uncertainties in the increase in wall compressive force and in the decrease in column compressive force, and, therefore, to the three-dimensional behavior of the entire structure.

It must be noted that the above strength ratios are upper bounds since they are based on the assumption that plane sections remain plane during deformation. In reality, this is not going to be the case, particularly for the wider slab. Negative moment analysis performed on the basis of stress and strain attenuation across the widest possible slab width (plane sections not remaining plane) produced the moment-curvature response shown by the dashed line of Fig. 8. The maximum negative moment strength shown for this assumption correlated very well with tests performed on beam-column-slab subassemblage tests.

Modeling the Transverse Girders

In order to model the affect of the transverse girders framing orthogonally into the plane of the wall, a fictitious transverse beam element was created which displays behavior similar to that observed during the testing of the model. Consider the isolated portion of the structure which consists of real girders G4 spanning Frame A and B, or B and C, as shown in Fig. 9. These girders, when attached to the rigid arm (representing the shearwall) will deform as shown when the wall is subjected to a unit rotation θ and to a unit vertical displacement Δ. The stiffness matrix for this subassemblage is shown at the bottom of Fig. 9.

Now consider the fictitious subassemblage shown in Fig. 10 which consists of two simple beams, each cantilevered out from the centerline of the wall and pinned to the appropriate column of Frame A or C. In order for this model to work, it was necessary to lump Frames A and C into one Frame, A', and to superimpose Frame B and Frame A' onto a single plane. For a unit rotation θ and vertical displacement Δ, the stiffness matrix for the fictitious subassemblage is shown in at the bottom of Fig. 10.

By equating the matrices of Figs. 9 and 10, it can be seen that the following relationship holds:

$$\frac{(I_{TL}+I_{TR})}{(I_{TL}^{*}+I_{TR}^{*})} = \frac{aL_T^3}{12L_W^3} \quad \text{where } a = \left[1 - \frac{3}{4+\dfrac{K_S L_T}{EI_{TR}}} \right] \quad (1)$$

These equations are uniquely satisfied when $I_{TL} = I_{TR}$ (system is elastic) or when either I_{TL} or I_{TR} are zero (system is plastic). In either case, the fictitious transverse beams stiffness may be found from equation (1). If there is some post-yielding stiffness in the tension-side beam, the equations are not exactly satisfied. However, the error introduced is insignificant in comparison with the uncertainties in establishing the stiffness and strength of the real transverse beam.

The flexural strength of the fictitious beam was determined in such a manner that the shear force at flexural yield would be the same for the real and fictitious element.

Results of Monotonically Increasing Lateral Load

The response of the DRAIN model as subjected to three patterns of monotonically increasing lateral load is shown in Fig. 11, together with the envelope of peak base shear as measured from the shaking table tests. As can be seen from the figure, the analysis consistently underestimates the base shear corresponding to a particular displacement, with the error being the greatest for the triangular load. For the MO28.3 load pattern, the computed shears are about 87 percent of those measured experimentally.

While the increase in analytical wall axial force (at a roof displacement of 2.0 inches (50.8mm) was a significant 14 kips (62kN), this is considerably less than the corresponding experimental value of 29 kips (129kN). The low analytical wall axial force increase results in a lower moment and in a lower base shear, and this is the primary reason for the differences in response as shown in Fig. 11. It is interesting to note, however, that the 14 kip analytical compressive force increase in the wall is more than 85 percent of the maximum increase available if the maximum available increase is computed on the basis of the predicted transverse beam strength alone. The higher force increase as measured during the experimental response may be due to several effects, including slab membrane action, accidental vertical accelerations, vertical accelerations due to wall rocking, interaction with mass tie-down rigging, or error in measurement.

Summary and Conclusions

The modeling of three dimensional response in frame-wall structures is extremely important, since the mechanisms predicted on the basis of planar analysis may be unconservative from the point of view of both strength and ductility supply and demand. Methods have been presented for modeling three-dimensional structural response with computer programs which are designed for planar response analysis. The methods, as applied to the US-Japan R/C test structure provided a fair degree of correlation with the experimental response.

References

1) American Concrete Institute Committee 318, "Building Code Requirements for Reinforced Concrete (ACI 318-77)", American Concrete Institute, Detroit, Michigan, 1977.

2) Kaanan, A.E., and Powell, G.H., "DRAIN-2D, A general Purpose Computer Program for Dynamic Analysis of Planar Structures", Report #UCB/EERC 73-6, EERC, University of California, Berkeley.

3) Wight, J.K. (Editor), "Earthquake Effects on Reinforced Concrete Structures", ACI SP-S4, American Concrete Institute, Detroit, Michigan, 1985.

TABLE 1

Summary of Peak Response Quantities Measured During Testing

Test	Table Accl (% g)	Roof Disp (in)	Base Shear (kips)	Wall Comp. (kips)	Increase
MO 5.0	5.0	0.093	10.5	0.8	
MO 9.7	9.7	0.158	18.4	1.9	
MO14.7	14.7	0.520	28.9	10.0	
MO24.7	24.7	1.040	44.2	18.2	
MO28.3	28.3	1.600	49.6	27.1	
T40.3	40.3	2.520	54.0	34.0	

Notes: Total weight of model = 105 kips
        Total height of model = 171 inches
        Initial wall axial force = 26 kips

1 kip=4.45 kN, 1 inch=25.4 mm

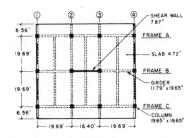

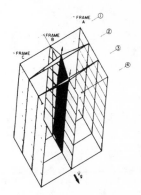

Fig. 2   3-Dimensional Behavior

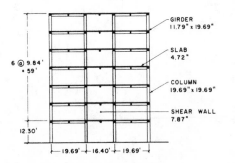

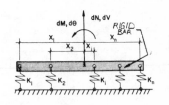

Fig. 1   Plan and Elevation
         of Prototype

Fig. 3   LINKS Modeling Concept

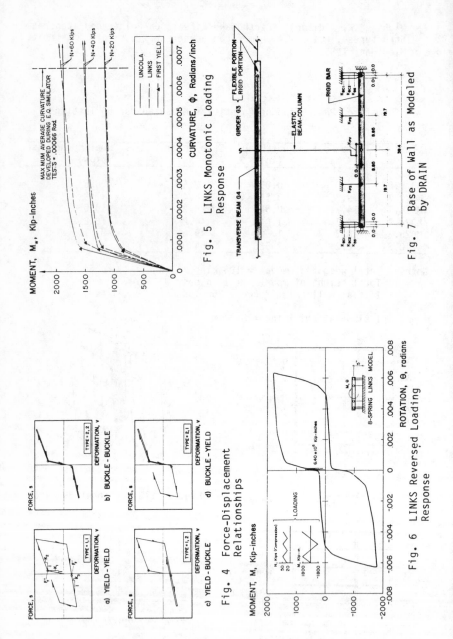

Fig. 5 LINKS Monotonic Loading Response

Fig. 7 Base of Wall as Modeled by DRAIN

Fig. 4 Force-Displacement Relationships

Fig. 6 LINKS Reversed Loading Response

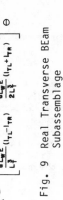

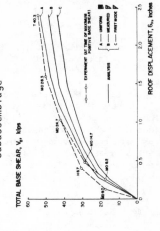

Fig. 9 Real Transverse BEam
        Subassemblage

Fig. 11 Global Monotonic Loading
         Response

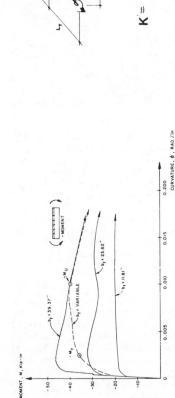

Fig. 8 M-ϕ Response of Typical
        Girder

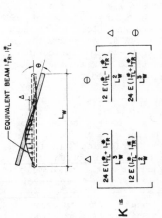

Fig. 10 Fictitious Transverse Beam
         Subassemblage

Seismic Behavior of a Full-Scale K-Braced
Steel Frame
- Part I Gross Behavior -

Hiroyuki YAMANOUCHI (I), Mitsumasa MIDORIKAWA (II)
Isao NISHIYAMA (II) and Akio HORI (III)

As a part of the U.S.-Japan Cooperative Research Program Utilizing Large-Scale Testing Facilities (Ref. 3), a six-story, 2 by 2 bay, concentrically K-braced steel building was constructed in full-scale, and subjected to Miyagi-Ken-Oki simulated earthquake of different intensities by the pseudo-dynamic (PSD) test facilities (on-line computer-actuator test). To simulate working load conditions, the earthquake motion was scaled to 65 gal; for a moderate earthquake the peak intensity was set at 250 gal; the maximum earthquake was run at 500 gal.

Full-Scale Test Building

The test building was designed to satisfy the requirements of both the 1976 Uniform Building Code and the 1981 Seismic Design Code of Japan. The plan and elevation are shown in Fig. 1. This building structure consisted of two unbraced moment-resisting frames and a concentrically K-braced frame. The actual dead load of the whole building was 523.6 tons (1154 kips) while the design dead load was 634.7 tons (1399 kips).

Member size of the test structure is in Table 1. Principal design criteria and details were: 1) The design base shear coefficient should be 0.197. 2) Live load and exterior wall weight should not be included in the design earthquake lateral forces. 3) Columns and girders should be wide flange shapes of ASTM A36 structural steel, and braces be square tubing section of ASTM A500 Grade B structural steel. 4) Girders and floor beams should act compositely with the floor using shear studs. 5) Braces should resist both tension and compression. 6) Girder-to-column connections should be moment connections in loading direction. 7) Column bases should be fixed. 8) Braces should be directly welded to surrounding frames without gusset plates.

Test Program and Procedure

All of the tests, exclusive of free and forced vibration tests, were run as a six degree of freedom pseudo-dynamic (PSD) system. The input excitation was the NS component of the 1978 Miyagi-Ken-Oki Earthquake (Tohoku University). To maximize the amount of knowledge gained from the program, the test sequence in Table 2 was adopted, involving the three level major seismic tests (the elastic PSD,

---

(I) Head of Structure Division, Struct. Engrg. Dept.,
    Building Research Institute, Tsukuba, Japan
(II) Research Engineer, Struct. Engrg. Dept. BRI.
(III) Research Engineer, Hazama-gumi, Co. Ltd.

inelastic PSD Moderate and inelastic PSD Final tests). At the head and
end of the test sequence, the free and forced vibration tests were
performed to obtain modal parameters and damping characteristics of the
test structure and the change of them between before and after the
maximum intensity earthquake input.

The test structure was loaded by eight servo-controlled actuators
through rigid loading beams installed at the edge of each floor. The
roof floor had two actuators, with a capacity of $\pm$ 100 ton (220 kips)
load and $\pm$ 100 cm (39.4 in.) stroke, to prevent the test structure from
overall torsion by same displacements given to the two jacks. Also at
the 6th floor, two jacks were arranged with a capacity of $\pm$ 100 ton and
$\pm$ 50 cm (19.7 in). Only one of these two jacks, however, took part in
the PSD test control. The other actuator was controlled to generate
the same load as the PSD-controlled actuator. Excluding the upper two
floor levels, one actuator with the same capacity as the actuators at
the 6th floor was set up for each floor. The PSD tests were run with
time intervals of 0.01 sec for the elastic level and 0.005 sec for the
Moderate and Final tests, respectively.

The basic instrumentation plan consisted of the following two
phases. The first one was to measure floor-level displacements and
actuator forces for the PSD tests at each time step, by using
digital-type displacement transducers and load cells. The second is to
detect deformations of members of the steel skeleton, such as rotation
angles of members, panel zone distortions of girder-to-column
connections, as well as strains on structure components such as
girders, columns, braces and reinforcement bars in slabs, to estimate
the load-deformation characteristics of the structural members of the
building. Approximately 1,000 channels of data were collected for this
purpose; the data were sampled at every fourth time step during the
major three PSD tests.

Test Results and Examinations on Gross Behavior

1. Overall Response of Test Structure
    Response waveforms of floor level displacements as a function of
time, from the Moderate and Final tests are shown as solid lines in
Figs. 2 and 3.

In the Moderate test, the girder panel-zone at the brace junction
in the middle of the span failed at the 2nd floor. Also the braces at
the 2nd, 3rd and 5th stories buckled slightly. The girder panel-zone
was repaired by replacing it with a new plate before the Final test.

In the Final test, at the 1st through 5th stories, the braces
buckled in-plane and/or out-of-plane of the frame. In particular, the
braces at the 2nd and 3rd stories buckled laterally 21 to 24 cm (8.27
to 9.54 in.) with local buckling and tearing at the mid-span and both
ends. Many panel-zones of girder-to-column connections and member ends
also suffered, mainly in the lower three stories. It should be noted
that the maximum roof displacement reached was 22 cm (8.66 in.) and the
maximum base shear was 331 tons (729 kips) at a displacement of 2.3 cm
(0.91 in.) in the 1st story. Fig. 4 shows the relationship of story

shear force vs. interstory displacement in the Final test (solid lines). In this figure, it can be seen that the lower stories reached their load carrying capacity. On reaching this force level, several braces degraded much their capacity. The vertical displacement at the brace junction reached 3.5 cm (1.38 in.) at the 3rd floor by a significant loss in capacity of the braces. Finally the test structure sustained an interstory drift angle of 1/57 at the 2nd story, and the north side brace at the 3rd story ruptured completely .

In Figs. 2, 3 and 4, analytical results are also shown in dotted lines. The DRAIN-2D computer program developed by Kanaan and Powell (Ref. 2) was used for the analysis. In this analysis, a degrading hysteresis model (Fig. 5) was applied to the post-buckling and cyclic behavior of braces. This model was developed by modifying the Jain-Goel-Hanson hysteresis rule of braces (Ref. 1). The comparison indicates that the analytical model is very good in predicting the overall behavior of the test building.

2. Lateral Elastic Stiffness of Each Story
Lateral elastic stiffness of each story is defined as a value of story shear divided by interstory drift. By the Elastic test, the stiffness of each story was obtained as listed in Table 3. On the contrary, an analytical stiffness of each story was obtained by using a computer program for static analysis on the basis of the DRAIN-2D. However, the results of this analysis (Analysis 0) had significant deviations from the test results as shown in Fig. 6. This discrepancy would be due to 1) three dimensional(3-D) effects of the framing of the test structure and 2) inaccurate estimation of each member's stiffness such as the stiffness of composite girders.

To study the effects of the above factors on the story stiffness in the Elastic test, further analysis were performed by taking the following items into account:

1) The nodal horizontal displacements at the nodes of B1 and the mid-span of the braced girders must be independent of those at the other nodes. Also reasonable shear stiffness of slabs between interior and exterior frames as a diaphragm effect may be effective; Analysis 1.
2) Axial deformations of B1-columns would be restrained by X-braces installed in the transverse direction; Analysis 2.
3) The stiffness of concrete should be taken into account to calculate the "negative" bending stiffness of composite girders in the Elastic test; Analysis 3.
4) Rigid zones in girder-to-column connection panels must be considered, which shorten the effective length of columns. Also shear deformations of columns may be considered; Analysis 4.

Each analysis, herein, was done containing successively the results of the preceding analyses. Fig. 6 shows how the analytical results were developed from Analysis 0 to Analysis 4. From this figure, it can be seen that the 3-D effects and rigid zones estimated for columns are considerably significant to evaluate the actual story stiffness.

3. Story Shear Force Carried by Braces and by Moment Frames

The ratio of story shear forces carried by seismic resistant braces and by moment frames is a basic parameter not only to design a steel sturcture with such braces but also to evaluate the actual seismic performance, in particular, after buckling of the braces.

The ratios, in this sense, were calculated for each story. Here, properties of the braces at each story are listed in Table 4. Effective slenderness ratio of the braces varied from 51.7 to 78.5 assuming an effective length coefficient K=0.7.

Fig. 7 gives time histories on ratios of story shear forces carried by the braces and by the moment frames to those by the actuator forces respectively, for the 2nd story in the Final test (solid lines). It can be seen in this figure that the line on the braces turns down, and that on the moment frames turns up, having an intersection at the response time of 7.25 sec. Around this time the north side brace began to develop severe buckling with the compressive strength much decreased. It means that the deterioration of the brace capacity forced the moment frames into inelastic activity. On the contrary in the Elastic test, even in the Moderate test, the ratios were very stable. Table 5 gives the ratios on the braces (in average) for the two tests. These values were close to those expected in the design process (Table 5). For the Final test, the ratios on the braces were also fairly predicted by the analysis as shown in Fig. 7 (dashed lines). However, this prediction was considerably disturbed by local failure such as local buckling, kinks and cracking of the braces.

Conclusions

1) The overall behavior of a full-scale K-braced steel structure was obtained by a lot of capacity of the pseudo-dynamic testing technique that simulates real earthquake responses.

2) Three dimensional effects of framing and more accurate estimations of members' stiffness than usual ways made the analytical story stiffnesses close to the experimental ones.

3) The ratio between the story shear forces carried by the K-braces and by the moment frames were stable and predictable before the local failure of the braces. However, after such failure of the braces, the ratios changed so largely that further studies are needed to estimate the behavior of such failure or to prevent the braces from it.

Acknowledgements

The authors wish to express their gratitude to the members of the JTCC (the Joint Technical Coordinating Committee of the U.S./Japan Cooperative Research Program: Co-chairman; H. Umemura and J. Penzien) who encouraged the authors and cordially gave advice.

References

1. Jain, A. K., Goel, S. C. and Hanson, R. D., "Hysteresis Behavior of Bracing Members and Seismic Response of Braced Frames with

Different Proportions," Report No. UMEE 78R3, University of Michigan, July, 1978

2. Kanaan, A. E. and Powell, G. H., "General Purpose Computer Program for Inelastic Dynamic Response of Plane Structure," Report No. EERC 73-6, University of California, Berkeley, April, 1973

3. U.S.-Japan Planning Group, "Recommendations for a U.S.-Japan Cooperative Research Program Utilizing Large-Scale Testing Facilities," Report No. UCB/EERC-79/26, Sept., 1979

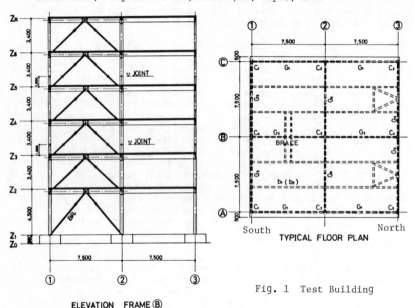

Fig. 1 Test Building

Table 1 Member Size and Proportion

COLUMN SCHEDULE

MARK STORY	C1	C2	C3	C4	C5
6-5	W 10 x 33 H-247.1x202.2 x7.37x11.05	W 10 x 33	W 10 x 49 H-253.5x254.0 x8.64x14.22	W 10 x 33	W 12 x 40 H-303.3x203.3 x7.49x13.08
4-3	W 10 x 39 H-252.0x202.8 x8.00x13.46	W 12 x 53 H-306.3x253.9 x8.76x14.60	W 12 x 65 H-307.8x304.8 x9.91x15.37	W 10 x 60 H-259.6x256.0 x10.67x17.27	W 12 x 72 H-311.2x305.8 x10.92x17.02
2	W 12 x 50 H-309.6x205.2 x9.40x16.26	W 12 x 65	W 12 x 79 H-314.5x306.8 x11.94x18.67	W 12 x 79	W 12 x 106 H-327.4x310.4 x15.49x25.15
1	W 12 x 65	W 12 x 87 H-318.3x308.0 x13.08x20.57	W 12 x 87	W 12 x 106	W 12 x 136 H-340.6x315.0 x20.07x31.75

GIRDER SCHEDULE

MARK FLOOR	G1	G2	G3	G4
RFL-6FL	W 16 x 31 H-403.4x140.3 x6.98x11.18	W 16 x 31 H-403.4x140.3 x6.98x11.18	W 18 x 35 H-449.6x152.4 x7.62x10.80	W 21 x 50 H-529.1x165.9 x9.65x13.59
5FL	W 16 x 31	W 18 x 35	W 18 x 35	W 21 x 50
4FL	W 18 x 35	W 18 x 35	W 18 x 35	W 21 x 50
3FL	W 18 x 35	W 18 x 40 H-454.7x152.8 x8.00x13.34	W 18 x 35	W 21 x 50
2FL	W 18 x 40	W 18 x 40	W 18 x 35	W 21 x 50

BRACE SCHEDULE

MARK STORY	BR1
6	Tube 4x4x1/5.56 Box-101.6x101.6x4.57x4.57
5	Tube 5x5x1/5.56 Box-127.0x127.0x4.57x4.57
4	Tube 5x5x1/4 Box-127.0x.127.0x6.35x6.35
3-2	Tube 6x6x1/4 Box-152.4x152.4x6.35x6.35
1	Tube 6x6x1/2 Box-152.4x152.4x12.7x12.7

Table 2   Test Sequence

1. Free and Forced Vibration Test (VT) #1
2. Each Floor Level Loading Test (FLL) #1
3. Trial Elastic Pseudo-Dynamic Test (PSD) #1
4. Elastic Pseudo-Dynamic Test #1&2
   (Maximum Input Acceleration 65 gal)
5. Pseudo-Dynamic Free Vibration Test #1
6. Trial Elastic PSD Test #2
7. Elastic PSD Test #3
   (Maximum Input Acceleration 65 gal)
8. Inelastic PSD Moderate Test
   (Maximum Input Acceleration 250 gal)
9. PSD Free Vibration (PSD-F) Test #2
10. Repair Works of Girder Panel
11. Each Floor Level Loading (FLL) Test #2
12. PSD Free Vibration (PSD-F) Test #3
13. Inelastic PSD Final Test
    (Maximum Input Acceleration 500 gal)
14. Each Floor Level Loading (FLL) Test #3
15. Free and Forced Vibration (VT) Test #2

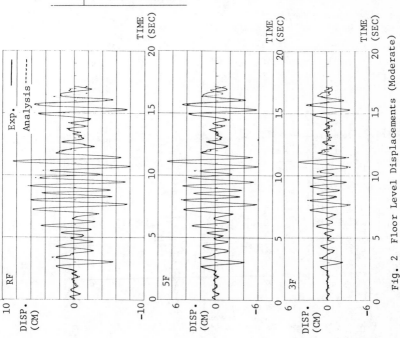

Fig. 2 Floor Level Displacements (Moderate)

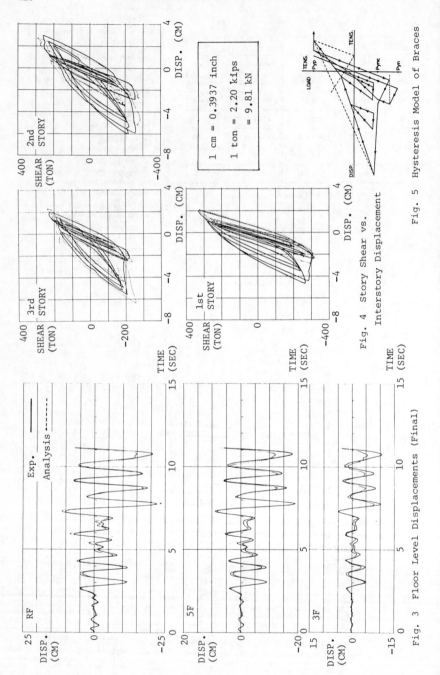

1 cm = 0.3937 inch
1 ton = 2.20 kips
       = 9.81 kN

Fig. 3 Floor Level Displacements (Final)

Fig. 4 Story Shear vs. Interstory Displacement

Fig. 5 Hysteresis Model of Braces

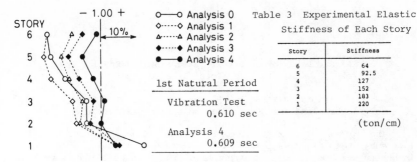

- 1.00 +

STORY

Vibration Test 0.610 sec

Analysis 4 0.609 sec

○——○ Analysis 0
◇·····◇ Analysis 1
△·····△ Analysis 2
◆·····◆ Analysis 3
●——● Analysis 4

1st Natural Period

Table 3 Experimental Elastic Stiffness of Each Story

Story	Stiffness
6	64
5	92.5
4	127
3	152
2	183
1	220

(ton/cm)

Fig. 6 Deviations of Analytical Stiffness from the Test Result

Table 4 Brace Properties

Item Story		A (cm²)	r (cm)	L (cm)	L/r	KL/r (K=0.7)	P_{yn} (ton)	P_{yp} (A·F) (ton)
6	S	17.21	3.94	442.3	112.3	78.6	40.6	74.5
	N							
5	S	21.85	4.98	441.0	88.6	62.0	59.5	84.8
	N							
4	S	29.61	4.88	435.7	89.3	62.5	81.1	117.0
	N							
3	S	36.06	5.92	434.0	73.3	51.3	109.5	140.3
	N						113.8	147.8
2	S	36.06	5.92	432.9	73.1	51.2	109.6	140.3
	N						111.0	142.8
1	S	66.84	5.61	513.4	91.5	64.1	192.1	292.8
	N			512.4	91.3	63.9	191.3	290.1

1 cm = 0.3937 inch

1 ton = 2.20 kips
      = 9.81 kN

Table 5 Share Ratio of Braces in Story Shear Forces

STORY	ELASTIC	MODERATE	FINAL	ELASTIC ANALYSIS	DESIGN PROCESS ELASTIC	ULTIMATE
6	0.59	0.59	0.59	0.67	0.74	0.76
5	0.67	0.67	0.67	0.72	0.78	0.49
4	0.71	0.71	0.71	0.73	0.78	0.51
3	0.75	0.75	0.75-0.51	0.76	0.82	0.56
2	0.72	0.72-0.67	0.72-0.26	0.72	0.78	0.48
1	0.83	0.83-0.65	0.83-0.64	0.81	0.85	0.58

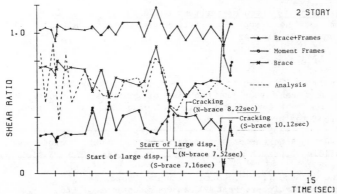

Fig. 7 Ratios of Story Shear Forces Carried by Braces and by Moment Frames to Those by Actuator Forces

Seismic Behavior of a Full-Scale K-Braced Steel Frame
- Part II Local Behavior -

Isao NISHIYAMA (I), Hiroyuki YAMANOUCHI (II)
Shinnichi IGARASHI (III) and Takahiro MIKAWA (IV)

The full-scale six-story concentrically K-braced steel building was subjected to Miyagi-Ken-Oki simulated earthquake whose peak intensity was 500gal. Local behavior of the test building as well as overall behavior were obtained and considerations focused on the local behavior were carried out.

Introduction

In a series of seismic tests on a full-scale six-story concentrically K-braced steel building (Ref.5), the overall inelastic behavior, for instance story shear vs. story drift relationships, of the three dimensional full-scale structure was obtained, and considerations were carried out from various viewpoints (Ref.6).

Elastic or inelastic behavior of many members (local behavior) was also obtained as well as overall behavior. To examine local behavior in the full-scale structure is very useful for the verification of the propriety of common knowledge derived from tests and/or analysis on isolated members or sections, for the confirmation of the existence of the redistribution of the stress in members and for the clarification of the three dimensional action.

From these points of view, the following members were selected to be considered; i) braces, ii) girder-to-column panels (joint panels), iii) columns, iv) girders and v) transverse braces.

Plan and elevation of the test building are shown in Fig. 1. The locations of the members or sections considered their local behavior are shown in the figure.

Behavior of Braces

A pair of K-braces of each story carried about half of the story shear force even after the buckling of compression brace. Therefore the overall behavior of the test building depended so much on the behavior of the braces.

I   Research Engineer of Structure Division,Struct. Engrg. Dept.,
     Building Research Institute, 1-Tatehara, Oho, Tsukuba, Ibaraki,
     Japan
II   Head of Struct. Div., Struct. Engrg. Dept., BRI
III Research Engineer, Penta-Ocean Construction Co.,Ltd.
IV   Research Engineer, Kounoikegumi Co.,Ltd.

The axial force vs. axial displacement relationships for a pair of the braces in the 2nd story are shown in Fig. 2. As seen from the figure, both braces showed rather large inelastic deformation under the axial thrust after buckling but they did not show tensile yielding at all.

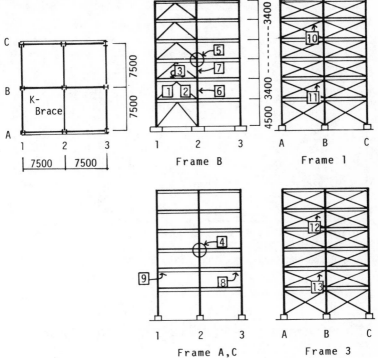

Fig. 1. Plan And Elevation Of Test Structure

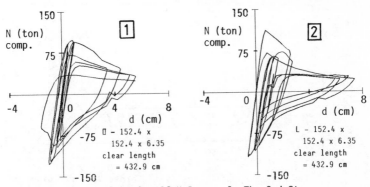

Fig. 2. Behavior Of K-Braces In The 2nd Story

The buckling strength of each brace of the test building is plotted and compared with European Column Curve (Ref. 3) in Fig. 3, where column strength curve 2 for rectangular tubes with residual stresses is shown. Herein, the nondimensional slenderness ratio λ for each brace is so estimated that the effective buckling length is taken to be 70% of the clear length of each brace (K=0.7). The test results agree well with the column strength curve 2.

As for no tensile yielding in the 2nd story braces, it is because the girder yielded at the intersection of the braces under the vertical

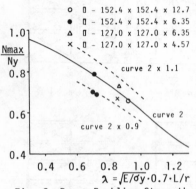

Fig. 3. Brace Buckling Strength

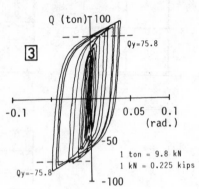

Fig. 5. Shear Panel Deformation

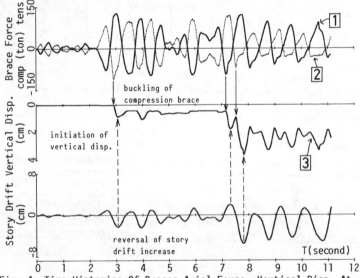

Fig. 4. Time-Histories Of Braces Axial Force, Vertical Disp. At Braces-To-Girder Joint And Story Drift

force produced by the difference of the axial force of the braces and
the increase of the tensile force in a brace was interrupted by the
reduction of the compression force by buckling of the other brace.   In
Fig. 4, shown are the time-history of the axial force of the 2nd story
braces, that of the vertical displacement at the intersection of the
girder and the braces of the 2nd floor level and that of the story
drift in the 2nd story of the test building.   As seen from these
time-histories, the increase of the vertical displacement at the
braces-to-girder joint took place when the initiation of the decrease
of the compressive strength of the buckled brace starts and it
continued till the reversal of the increase of the story drift.

The unbalance of the axial force of the braces effected not only
on the flexural yielding of the girder but also on the shear yielding
of the shear panel formed in the girder.   The shear panel showed rather
large shear deformation but the hysteresis loop is quite stable as
shown in Fig. 5.   The maximum shear deformation amplitude reached 0.05
rad. but no cracks were observed at all in and around the shear panel.

Behavior of Girder-to-Column Panels (Joint Panels)

The strength of girder-to-column panels (joint panels) was
designed to be much less than that of the adjacent members in this test
building.   The interior joint panels in the frames parallel to the
loading direction were especially weaker in strength than exterior
ones, relative to the strength of the adjacent members.

The shear deformation vs. panel moment relationships of the
interior joint panels of 4th floor level are shown in Fig. 6.   The
panel moment was estimated as the sum of the face moment of the upper
and lower columns (the moment diagram was evaluated from the readings
of the strain gauges attached to the intermediate sections of the
columns).   The calculated yield panel moment according to the AIJ
(architectural Institute of Japan) steel design standard (Ref. 1) is
shown by broken line in the figure.   As seen in the figure, the joint
panels showed large inelastic shear deformation and the maximum
amplitude of the shear deformation reached about 0.02 rad..   The

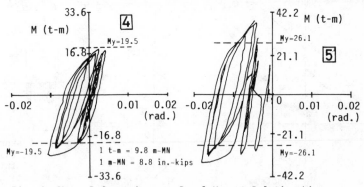

Fig. 6. Shear Deformation vs. Panel Moment Relationship

plastic shear deformation in the joint panel of the braced-bay column concentrates in one side, because the axial thrust in the column affected the increase of plastic shear deformation.

Behavior of Columns

Axial forces of the braced-bay columns became very large when the structure was subjected to large horizontal force. Axial force vs. axial deformation relationships are shown in Figs. 7 and 8 for 2nd and 3rd story interior braced-bay columns. Obviously they showed likely behavior of yielding. However, the calculated simple axial yield strength is about 2 times as high as the force level corresponding to the apparent yielding. The moment-axial force diagrams of both ends of each column are shown in Figs. 9 and 10, respectively. Fig. 9 shows

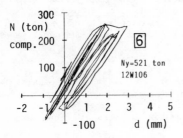

Fig. 7. Axial Force vs. Axial
        Deformation Relation-
        ship (2nd story)

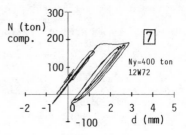

Fig. 8. Axial Force vs. Axial
        Deformation Relation-
        ship (3rd story)

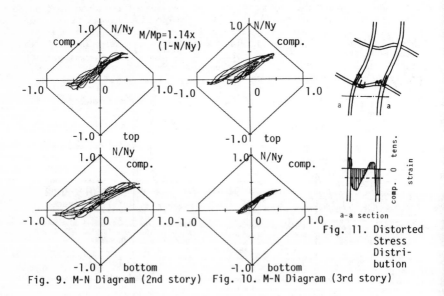

Fig. 9. M-N Diagram (2nd story) Fig. 10. M-N Diagram (3rd story)

Fig. 11. Distorted
         Stress
         Distri-
         bution

that the bottom end of the 2nd story column should have yielded under the combination of flexural moment and axial force. As for 3rd story column, the combination of flexural moment and axial force is a bit smaller than the yield criterion as seen in Fig. 10. As mentioned in the behavior of joint panels, the joint panel adjacent to the top of the 3rd story column discussed herein showed a large amount of yielding (Fig. 6). According to the investigation by Nakao et al. (Ref. 4), the member adjacent to the yielded joint panel yields under a relatively small amount of flexural moment because of the stress distribution distorted as schematically shown in Fig. 11. This effect should have made the early initiation of the yielding of column top of the 3rd story column (Figs. 8 and 10). As a matter of fact, results of the strain measurement at the location close to the top end of the column showed much yielding.

Behavior of Girders

All the girders were so designed to act compositely with the concrete slab using shear studs and their strength is much higher than that of the joint panels. Therefore the remarkable yielding was not observed in the girder. Here presented are the moment-curvature relationships measured at two sections of the 2nd floor girder in Frame-A (Fig. 12). Possitive and negative stiffnesses calculated according to the AIJ recommendations (Ref. 2) and the stiffness of the bare steel girder are shown in the figure. Test results show fairly good agreement with calculated stiffnesses.

As shown in this figure, apparent yielding under the possitive bending and stiffness reduction under the negative bending were observed at section- ⑧ . The theoretical full plastic moment of the composite girder is far larger. Slippage between the steel girder and the concrete slab can explain the apparent yielding in the moment-curvature relationship. Namely, it is supposed that the composite girder comes not to be able to sustain the increase of the flexural moment when the slippage occur, and then the moment-curvature

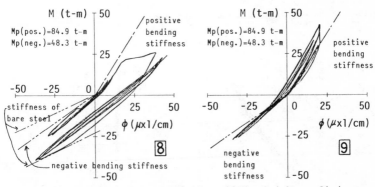

Fig. 12. Moment-Curvature Relations Of The 2nd Story Girders

relationship shows apparent yielding.

As for the stiffness reduction under the negative bending, the stiffness after the reduction is well agree with the stiffness estimated for bare steel (Fig. 12), so that the slippage between the steel girder and concrete slab or between the concrete slab and the reinforcing bars could be the reason for the stiffness reduction.

The stiffness at the location ⑨ in the figure increases in the positive bending. The reason of this phenomenon is not explained yet.

Behavior of Transverse Braces

So as to restrain the torsional deformation of the test building under horizontal loading, transverse braces made of angle sections were installed in the exterior frame 1 and 3 perpendicular to the loading direction. These transverse braces restrained also the vertical movements of the column ends which arose from the flexural deformation of the braced-bay frame.

The axial strains measured in the transverse braces are presented in Fig. 13. From the figure, it can be seen that transverse braces connected to the braced-bay columns (members ⑩ and ⑪) were subjected to considerable strains, but those not connected to the brace-bay columns (members ⑫ and ⑬) were subjected to almost zero axial strains.

It is because the difference between the vertical displacements of the column in braced-bay and of the column in Frame A or C are relatively large. Yet, the difference between the vertical displacements of the column not in the braced-bay and of the column in

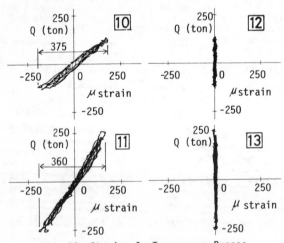

Fig. 13. Strains In Transverse Braces

Frame A or C is quite small.

Moreover, the axial strains of the transverse braces were larger in higher stories. It is because the flexural deformation of the braced-bay frame produced large vertical displacements in higher stories.

From the results of the strain measurement of the transverse braces, it can be said that three dimensional action (transverse action) should be considered to understand the overall behavior of the test building.

Conclusions

1. Very effective data were obtained by the seismic test of a full-scale six-story concentrically K-braced steel building not only on the overall behavior but also on the local behavior.
2. Obtained local behavior was considered from various viewpoints and the behavior of many members and sections was made clear. Furthermore, data on the members and/or sections obtained in the test were verified to be reliable and useful for attaining knowledge on the behavior of individual members in actual structures.
3. It is clearly shown that the three dimensional action by the transverse braces is important to understand the overall behavior.

Acknowledgements

The authors wish to express their gratitude to the members of the JTCC (the Joint Technical Coordinating Committee of the U.S./Japan Cooperative Research Program: Co-chairman; H. Umemura and J. Penzien) who encouraged the authors and cordially gave advice. The authors are also thankful to the visiting researchers at BRI from the general contractor companies, for their excellent help in the analysis, tests and data processing.

References

1.   Design Standard for Steel Structures, AIJ, Tokyo, 1970
2.   Design Recommendations for Composite Constructions, AIJ, 1985
3.   Johnston, B.G., "Guide to Stability Design Criteria for Metal Structure," SSRC, John Wiley & Sons, New York, 1976
4.   Nakao, M et al., "Strength and deformation capacity of joints and members in steel structures Part 3: Girder-to-Column Joints," Quarterly CULUMN 79, Nippon Steel Corporation, January, 1981 (in Japanese)
5.   Yamanouchi, H., Midorikawa, M., Nishiyama, I. and Kato, B., "Full-Scale Seismic Tests on a Six-Story Concentrically K-Braced Steel Building - U.S.-Japan Cooperative Research Program -," Proceedings of Special Conference of EM Division/ASCE, Laramie, Wyoming, August 1-3, 1984
6.   Yamanouchi, H., Midorikawa, M., Nishiyama, I. and Hori, A., "Seismic Behavior of a Full-Scale K-Braced Steel Frame - Part I Gross Behavior -," ASCE Conference, UCLA, April, 1986

BUILDING IDENTIFICATION: NEEDS AND MEANS

A. E. Aktan*, M.ASCE

Abstract

The need to conduct "system identification" of existing buildings by nondestructive testing is discussed.   A recommended flow-
chart is exemplified.   Existing dynamic test procedures and associated algorithms permitted reliable identification of only several
frequencies and mode shapes of a 7-story test building.   A direct
measurement of the flexibility coefficients was possible by static
tests.   This permitted a reliable and comprehensive identification.

Introduction

The author advocates "building identification" in the context of
"system identification" of structures [6], i.e., quantifying selected
parameters of an analytical model of the building by subjecting it to
tests and processing its measured responses.

The following comprises a flow-chart of operations for building
identification: (a) Establishing the need(s) for identification of
the structure; (b) Constructing an analytical model of the building,
compatible with the needs for identification; (c) Determining how
comprehensive an identification, in conjunction with the conceived
analytical model, is required.   An assessment of the required vs
available experimental capability; (d) Selection of the suitable
algorithm(s), i.e., theoretical basis of the system identification
procedure to be applied in conjunction with the available experimental capability; (e) Designing and conducting the experiments,
acquisition of data and its on-line and/or subsequent processing.
Correlation of the identified parameters with predicted ones;
(f) Several iterations between steps (b) and (e) until it is verified
that the analytical model is sufficient in reflecting the stiffness,
mass and energy dissipation characteristics of the structure; and
(g) An assessment of the reliability in the identified parameters of
the analytical model.

Objectives

The objectives of this paper are to discuss the steps in
identification, exemplifying them by an application carried out on a
1/5-scale 7-story building model shown in Fig. 1.   This model was
fabricated and tested on the earthquake simulator at U.C. Berkeley
[1,3], as part of a joint U.S.-Japan research program.

---

*Associate Professor of Civil Engineering, Louisiana State Univer-
sity, Baton Rouge, LA 70803.

FIG. 1 7-STORY, 1/5-SCALE BUILDING

LUMPED TRANSLATIONAL MASSES
RIGID END ZONES
SHEAR WALL
FRAME A & C
FRAME B

FIG. 2 ANALYTICAL MODEL OF BUILDING

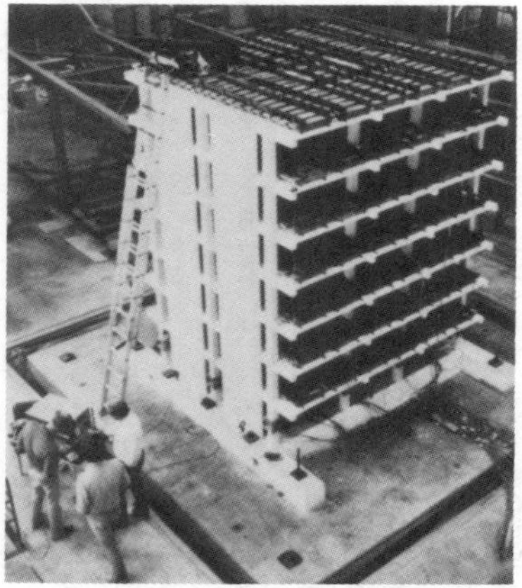

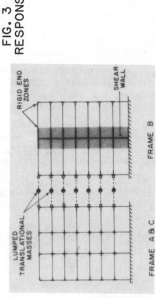

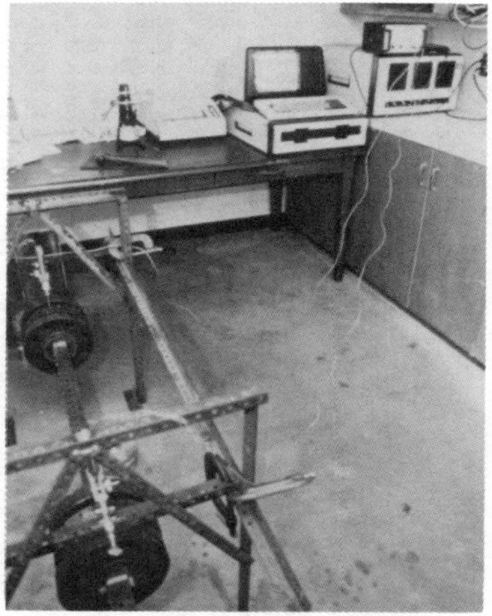

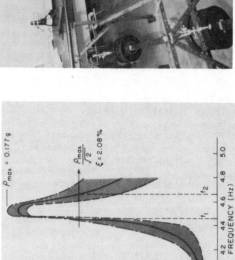

ρ_{max} = 0.177 g
$\frac{\rho_{max}}{\sqrt{2}}$
ξ = 2.08%
AMPLITUDE (10^{-3} g)
180
160
140
120
100
80
60
40
20
0
f_1
f_2
4.0
4.2
4.4
4.6
4.8
5.0
FREQUENCY (Hz)

FIG. 3 FORCING FREQUENCY-RESPONSE AMPLITUDE RELATION

FIG. 4 IDENTIFICATION OF IDEALIZED PHYSICAL MODEL

Table 1. Identified and Predicted Parameters: Lateral Frequencies (Hz)/Damping (% of critical).

Method	1st Freq.	2nd Freq.	3rd Freq.
Analytical Prediction	5.09	18.39	36.61
Ambient Vibrations	4.75	---	---
Free Vibrations	4.78/1.94	---	---
Harmonic Excitation	4.55/2.08	---	---
Random Excitation	4.78/2.20	17.88/2.52	---
Computed from Measured Flexibility	4.79	18.10	35.40

Needs for Identification

One compelling reason to identify buildings is in order to improve the state-of-the-art in their analytical modeling. This would comprise a fundamental need and the primary motivation of many engineers in academia who are engaged in related research.

Identification of existing full scale structures presents a unique opportunity to understand some of the mechanisms which are seldom correctly incorporated in modeling buildings for design or response prediction, but which may significantly influence their responses. Such mechanisms would include the effects of soil-foundation flexibilities; nonstructural and secondary components; stresses and strains which arise due to construction, shrinkage, creep and temperature; multi-dimensional nature of the environmental effects and the response, as opposed to the planar idealization; other possible mechanisms which engineers have not yet conceived and simulated in the laboratory.

Unfortunately, existing buildings normally cannot be tested to damageability limit states. It is therefore possible to point out that identification based on experiments conducted at the service-ability limit state may not be useful to predict response at the damageability limit states. Indeed, the mechanisms of stiffness and resistance of a structure, and their contributions to the global stiffness and strength would change as different levels of damage occur. On the other hand, the serviceability limit state response characteristics of a structure would constitute the initial conditions for more advanced limit state responses, and their accurate modeling would be prerequisite for reliable prediction of response at the subsequent limit states.

Furthermore, the present emphasis on damage control under minor and moderate seismic events necessitates accurate analyses for the serviceability limit states. Reliable prediction of particularly the local force and distortion responses of elements at the service-ability limit states has emerged as an important need for rational and optimum seismic design.

In addition to the need for building identification to improve analytical modeling, more practical applications are emerging. For example, building identification comprises an essential component in active-control of seismic or wind responses of critical buildings.

Integrity monitoring of: (a) Important buildings; (b) Buildings which are suspected of being subjected to degradation due to fatigue or overloads; (c) Buildings which experience damage in earthquakes and the extent of which needs to be assessed; may be achieved by experimental identification.

Finally, effectiveness of repair and retrofit may be reliably verified by the identification of structures before and after such work.

Analytical Modeling

The analytical model, parameters of which are to be identified, should at least represent the main mechanisms which contribute to global stiffness (or flexibility) and the distribution of reactive translational and rotatory inertia of the structure.

Discrete modeling approaches to structures include the lumped mass-spring representation, finite element representation in conjunction with lumped or consistent mass and the generalized coordinates representation [4]. Selection of one of these approaches should be carried out in conjunction with the stiffness and inertia characteristics of the structure as well as the reason for identification.

If the purpose of identification warrants, local response may be explicitly included in the model. In general, however, any mechanism or feature of the structure which cannot be properly conceptually described would only increase the complexity of the model and cannot be reliably identified.

A planar finite element model of the 7-story building, as shown in Fig. 2, was constructed. All the elements were idealized as 1-D beam-column finite elements. The axial and in-plane shear stiffness of diaphragms were assumed to be infinitely rigid. The lateral displacements at the floor levels along the response direction were selected as the dynamic degrees of freedom. Translational mass at each floor level was assumed lumped at that level and rotatory inertia of the element were neglected.

Since almost 90 percent of the mass at each floor level was due to a compact layer of lead ballast, lumping translational mass at floor levels was quite justified. In full-scale buildings, particularly with wall systems, however, the rotatory inertia moments may comprise an important component of the mass forces.

The mass and stiffness characteristics of the model were analytically predicted and expressed by a 7 by 7 stiffness matrix and 7 diagonal terms of the mass matrix. Frequency, mode shapes and participation factors could then be computed.

Parameters of Analytical Model Which May Be Identified

The 7 diagonal terms of the mass matrix and 28 terms of the symmetric stiffness or flexibility matrix numerically define the global characteristics of the model. If proportional viscous damping is assumed, 7 damping coefficients corresponding to the modes of the structure would be sufficient to describe the energy dissipation characteristics as well.

Several levels of identification of this analytical model would be possible. The global characteristics of the structure would have been completely quantified by individually identifying all the 42 parameters comprising the independent mass, stiffness and damping coefficients, if this was possible.

Identification of the stiffness characteristics in even greater detail, such as the contribution of different element stiffnesses to the global stiffness coefficients of the structure may also be conceived. This was in fact partially accomplished in the research by utilizing special internal force transducers and conducting static tests in addition to dynamic tests [1,2,3].

In the case of field testing existing full-scale buildings, how-ever, such a comprehensive level of identification may not be pos-sible to achieve. For example, static testing to explicitly measure flexibility coefficients of the model, or, directly measuring inter-nal forces may not be practical even if possible.

Considering that only dynamic tests may be practical, all the seven frequencies, the associated mode shapes and damping coeffi-cients could theoretically have been obtained by rigorous modal testing [5]. Such modal tests have been successfully conducted to identify many different types of structures, as documented in IMAC proceedings [8].

Experimental Techniques and Algorithms

Experimental procedures and the associated algorithms which were applied aimed at identifying the lateral response characteristics only. It was not possible, however, to excite the symmetric building without torsional response. Torsion had to be included in the iden-tification to understand and subtract its contributions from lateral response. The test procedures and algorithms were:

(a) Ambient vibration testing. Two seismometers, located on the exterior frames at the roof level were used to measure ambient velocity responses of the building. Analog signals were added or subtracted to yield lateral and torsional components of response, and were amplified and filtered prior to feeding to a real-time digital spectrum analyzer [1]. The fundamental frequency along the lateral response direction could be identified from the Fourier spectra of the responses (Table 1).

Several algorithms to extract more information from ambient vibration measurements have been proposed based on random vibrations theory, in connection with integrity monitoring of offshore platforms [9]. These concepts may be applicable to certain types of tall buildings and bridge superstructures, if the exceptional demands of this technique on transducer sensitivity, accurate calibration and signal conditioning are satisfied.

(b) Free vibration testing. The building was pulled from the roof level laterally along the symmetry axis and released suddenly by cutting the line. The lateral and torsional responses at each floor level were obtained by proper combination of measured signals. Spec-tral analysis of the free vibration decay responses at the roof level yielded fundamental frequency and logarithmic decrement was used to estimate damping of the first mode, as shown in Table 1.

Initial conditions activating higher mode responses may be devised to extract further information from free vibration tests. Several algorithms which may be used to identify not one, but a number of frequencies, mode shapes and viscous as well as non-viscous damping from free vibration decay responses of a structure are avail-able [5]. The author is particularly impressed with the Ibrahim time domain algorithm [7] which may prove promising in terms of practical applicability to civil engineering structures.

(c) Forced vibration testing. Harmonic excitation, which may be induced by rotating mass vibration generators, or, more general excitation, induced by electro-magnetic or electro-hydraulic inertial mass excitation generators may be utilized for this type of testing.

With the availability of inertial mass excitation generators which may be used as a component of a closed-loop system, any type of controlled excitation and particularly broad-band "white-noise" type of forcing functions may be generated. The maximum force amplitudes which may be generated by such devices, however, are considerably less than the corresponding capability of rotating mass vibration generators at particularly lower frequencies.

The 7-story building was tested under harmonic force, induced by a rotating weight vibration generator as well as under random broad-band excitation applied by an electro-magnetic shaker. The latter test was conducted by Measurement and Analysis   Division of URS/John A. Blume and Associates, now operating as Qest Consultants at Berkeley.

The rotating weight vibration generator was located along the symmetry axis coinciding with the main response direction, but eccentrically with respect to the transverse direction. By different phasing of the two rotating weights, the building could be excited without torsion along the main response direction, or coupled with torsion along the transverse direction. The forcing frequency vs steady-state response amplitude relations were generated for selected responses. The frequency range of the shaker was limited to only some of the lower frequencies of the building.

Since steady-state response exhibited beating, a unique amplitude could not be measured at different forcing frequencies. This is illustrated by the bounds of amplitude indicated in Fig. 3 in the frequency-response function for the main response direction. The average amplitude was used to extract damping. Beating in the steady-state responses was assessed due to changes in the characteristics of the building and building-shaker interaction as resonant response was approached.

The coupled torsion-transverse lateral responses did not lead to reliable identification of damping or mode shape. The frequencies were approximately 6 Hz and 8 Hz, respectively. These were too close and did not permit to generate uncoupled frequency-amplitude relations for these two modes. Only the fundamental frequency along the main response direction could therefore be identified reliably (Table 1).

The second type of forced-vibration test was conducted by exciting the building in lateral and torsional directions simultaneously, using random pulses induced by an electro-magnetic shaker. Time-histories of the excitation as well as lateral and torsional responses of the floors were measured, the corresponding transfer functions were generated and recorded. Polynomial least-square curve-fitting was applied to windows of transfer functions which contained modal frequencies. The fitted polynomials were used to extract frequency and damping of the corresponding modes. A synthesis of the relative amplitudes from transfer functions of different floors' responses, yielded mode shapes.

Only the lowest two frequencies, damping coefficients and associated mode shapes in the lateral response direction were reliably identified. The frequencies and damping coefficients are included in Table 1.

(d) **Static tests**. The building was laterally loaded at each floor level and the coefficients of the flexibility matrix were measured. The matrix was not symmetric due to nonlinearities in response, axial distortions of the diaphragms and any measurement errors. Maximum differences between measured and computed coefficients ranged between 25 percent at the main diagonal to 40 percent at off-diagonal. Errors were accentuated when the main wall was loaded with small moment to shear ratio. The relative contributions of different mechanisms to base shear and base overturning moment resistances were identified by measuring column internal forces at the first story utilizing special internal force transducers [2].

The measured flexibility matrix was used in conjunction with the computed translational mass of floors to generate the frequencies and mode shapes of the structure (Table 1). The off-diagonal elements were averaged to render the matrix symmetric. Good correlation was attained between the lowest three measured and calculated frequencies as shown in Table 1, indicating that mass modeling was justified, and the discrepancy between analytical and experimental frequencies was due to errors in modeling the stiffness.

Conclusions and Research Needs

Although the fundamental frequency and damping coefficient of the building could be reliably identified by several procedures (mean frequency was 4.73 Hz with a standard deviation of 0.1), only experimental modal analysis was successful in identifying both first and second frequency, damping and mode shapes. It appears that the practical limits in experimental system identification would permit reliable identification of only some of the lower frequencies, mode shapes and damping coefficients of actual buildings.

It would therefore be rational to generate as simple an analytical model as would suffice for the purpose of identification unless static tests and special experimental procedures are applied to complement dynamic tests. Even a limited amount of information which may be obtained by practical experimentation, however, may comprise an adequate data base to improve analytical modeling, and would be quite sufficient to serve many of the practical needs for identification discussed in this paper. The analytically predicted fundamental frequency of the bare symmetric model, geometry, material and support characteristics of which were controlled and documented, was still 8 percent higher than the measured value.

Research is needed to investigate the applicability of practical experimental procedures and associated algorithms for more comprehensive identification of civil engineering structures. For this reason, a 4-channel digital dynamic analyzer, capable of real-time data analysis as well as storage, was acquired at Louisiana State University.

Identification of simple physical models, parameters of which may be analytically generated with a high level of confidence, are in progress, as illustrated in Fig. 4. The objective of this research is to verify the applicability of several algorithms to the identification of different analytical models. Although all proposed algorithms would be naturally theoretically sound, their practical applicability require verification.

After proofing the algorithms and test procedures on idealized physical models as shown in Fig. 4, applications to actual structures will begin. Effects of experimental and computational error sources on the reliability of identified parameters appear as an important problem area.

Acknowledgements

The author gratefully acknowledges Prof. V. V. Bertero's guidance and contributions to the described research as principal investigator. Support was provided by the National Science Foundation, Grant CEE80-09478. Dr. Y. Ghanaat of Qest Consultants provided valuable assistance. The author is grateful to the support provided to him by Louisiana State University in his current research activities.

Appendix--References

1. Aktan, A. E., Bertero, V. V., Chowdhury, A. A. and Nagashima, T., "Experimental and Analytical Predictions of the Mechanical Characteristics of a 1/5-Scale Model of a 7-Story RC Frame-Wall Building Structure," Report No. UCB/EERC-83/13, Earthquake Engineering research Center, College of Engineering, University of California, Berkeley, August 1983.

2. Aktan, A. E. and Bertero, V. V., "Measuring Internal Forces of Redundant Structures," Experimental Mechanics, Dec. 1985.

3. Bertero, V. V., Aktan, A. E., Charney, F. and Sause, R., "Earthquake Simulator Tests and Associated Experimental, Analytical, and Correlation Studies on One-Fifth Scale Model," Earthquake Effects on Reinforced Concrete Structures, ACI SP-84, 1985.

4. Clough, R. W. and Penzien, J., "Dynamics of Structures," McGraw-Hill, Inc., 1975.

5. Ewins, D. J., "Modal Testing: Theory and Practice," John Wiley and Sons, Inc., 1984.

6. Hart, G. C. and Yao, J. P. T., "System Identification in Structural Dynamics," Journal of the Engineering Mechanics Division, Dec. 1977.

7. Ibrahim, S. R., "Time-Domain Modal Parameter Identification and Modeling of Structures," Proceedings of the American Control Conference, Paper FA4, 1983.

8. International Modal Analysis Conference Proceedings, 1982, 1984 and 1985, Union College, Graduate and Continuing Studies, New York.

9. Rubin, S. and Coppolino, R., "Flexibility Monitoring Evaluation Study," Minerals Management Service, U.S. Department of the Interior, Sept. 1983.

Slab Participation in Frames under Lateral Loads

Jorge N. Bastos,*   James O. Jirsa, M. ASCE,** and
R. E. Klingner, M. ASCE***

In earthquake-resistant reinforced concrete ductile moment
resistant frames (DMRF), the joint region is a critical link in the
lateral load resisting mechanism. Design requirements usually are
based on a mechanism involving beam hinging rather than column
hinging. Consequently, the structural engineer is faced with the
problem of evaluating the slab contribution to floor system strength
to ensure that the column strength is equal to or greater than that of
the floor system.

A unique opportunity to study slab participation was afforded by
the U.S.-Japan Cooperative project on large scale testing. The data
obtained from the (a) full-scale 7-story building (2) tested in Japan
at the Building Research Institute (BRI); (b) full-scale exterior and
interior subassemblages tested at The University of Texas at Austin
(3); and (c) 1/12.5 scale tests at Stanford University (4), provide a
vehicle for making an evaluation of the role of the slab in complete
structures as well as in subassemblages. The objective of this paper
is to provide an indication of slab participation in the lateral load
resistance of moment resisting frames.

Specimen Details and Material Properties

The full-scale building tested in Japan is shown in Fig. 1. The
exterior and interior subassemblages were portions of the full-scale
building shown by the shaded areas in Fig. 1. In determining
subassemblage geommetry, it was assumed that inflection points were
located at beam midspan and column midheight (Fig. 2). The
reinforcement characteristics were: (1) longitudinal beams - G1, G2:
0.30m x 0.50m, 3 - 19mm (top), 2 - 19mm (bottom) with stirrups at
0.10mm; (2) transverse beam - B1, B2: 0.30m x 0.45m, 3 - 19mm (top), 2
- 19mm (bottom), with stirrups 10mm at 0.10m; (3) Column - 0.50 x
0.50m; 8 - 22mm with 10mm ties at 0.10m; (4) slab - 10mm bars top and
bottom at 0.20m in the middle strip and 0.30m in the column strip.
The material properties are shown in Table 1. The concrete strength

*Graduate Research Assistant, Ferguson Structural Engieering
Laboratory, University of Texas at Austin.

**Ferguson Professor of Civil Engineering, University of Texas at
Austin, Austin, Texas 78712.

***Associate Professor of Civil Engineering, University of Texas at
Austin, Austin, Texas 78712.

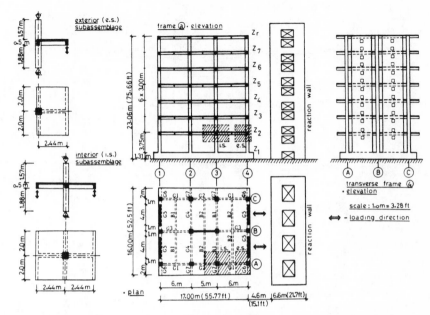

Fig. 1   Full-scale 7-story and isolated joint test specimens

is slightly higher for the small-scale Stanford University specimens because of difficulties in controlling microconcrete strength.

SPECIMEN		Japan – BRI Full-scale R/C Building	U. Texas @ Austin (1:1) Prototype Subassemblage		Stanford U. (1:12.5) Subassemblage w/ Slab	
			Exterior	Interior	Exterior	Interior
Concrete	f'_c (28 d.)					
	MPa	28.5	32.3* 16.1⁴	33.5* 23.7⁴	39.8	39.4
	ksi	4.1	4.7 2.3	4.9 3.4	5.8	5.7
Steel Reinforcement	1. Beam	f^s_y f^s_u	f^s_y	f^s_u	f^s_y	f^s_u
	Mpa	359. 537.	421.	662.	389.	561.
	ksi	52. 78.	61.	96.	56.	81.
	2. Col. MPa	371. 585.	545.	800.	427.	612.
	ksi	54. 85.	79.	116.	62.	89.
	3. Slab MPa	365. 544.	400.	565.	512.	779.
	ksi	53. 79.	58.	82.	75.	112.

Note : * - lower column and slab ; ⁴ - upper column .

Table 1   Material Strengths

Instrumentation and Load Histories

The University of Texas specimens were extensively instrumented with strain gages. The Stanford University interior subassemblages had only a top and bottom strain gage in the beam and two top bar strain gages in the slab. The Stanford University exterior joint

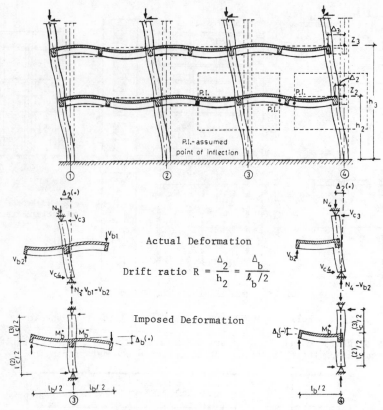

Fig. 2   Applied deformations on full-scale structure and isolated
         joint subassemblages

specimen was not instrumented with strain gages.   The BRI structure
unfortunately only had the bottom slab bars instrumented at the
interior joint.   The exterior joint was instrumented with gages on the
top and bottom slab bars.

The BRI structure was loaded laterally using hydraulic actuators
controlled by an on-line computer system.   The applied load history
consisted of a series of four modified earthquake ground motion
records.   Only the last two, PSD-3 and PSD-4 (Fig. 3) were considered
in this paper.   The subassemblages were loaded as shown in Fig. 1 by
vertical actuators that applied racking loads to the ends of the
beams.   Although a more realistic testing setup including P-Δ effects
could have been considered, the specimens were tested with the columns
in the vertical position (Fig. 2).   The applied load histories are
shown in Fig. 3.

In this study, strains in the slab reinforcement in the building
and in the isolated subassemblages are compared at selected drift

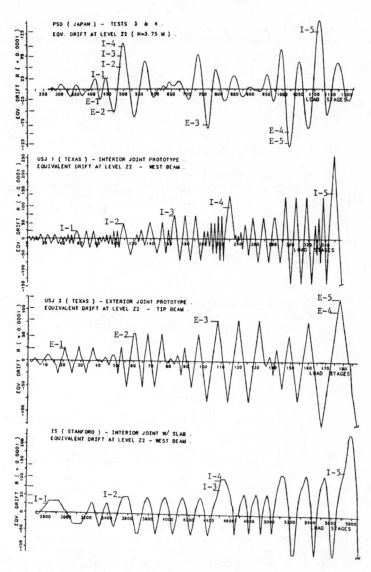

Fig. 3   Load histories

levels. The calculation for equivalent drift in the subassemblages
and the drift at the second level of the full scale structure are
shown in Fig. 2.   Strains were compared at different equivalent drift
values.   The drift levels were chosen to represent a point in the load
history in which the drift level was reached for the first time.   For
example, in Fig. 3, the slab reinforcement strains at the interior

joint at drift level 3 (I3) were plotted at load stage 490 in the BRI
structure, 165 in the Texas specimen, and 4490 in the Stanford test.

Test Results

Strain distributions were plotted across the floor section to allow
better insight into the response under various displacement levels.
The observed strain distributions in the exterior joints (Fig. 4) show

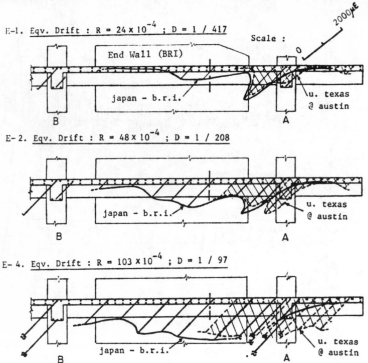

Fig. 4   Top slab reinforcement strains, exterior joints

that the end walls imposed a restraint on the transverse beam that
resulted in large fairly uniform strains across the entire slab at
early loading stages.   Near ultimate, yielding occurs over a large
portion of the cross section in both building and subassemblage
specimens (Fig. 5).   It also can be seen that the top slab bar strains
in the building overhang section remain low even at large
displacements.   This is likely the result of different imposed loading
or deformation, and different boundary conditions between the isolated
test specimens and the BRC structure.

The response of the interior joint located along transverse frame 3
in the BRI structure is shown in Fig. 6.   It can be observed that the
strains increase in the bottom slab rebars as lateral loads increase.

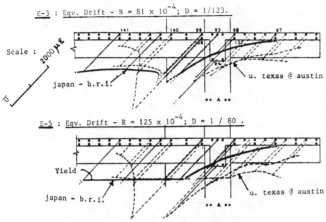

Fig. 5 Top and bottom slab reinforcement strains, exterior joints

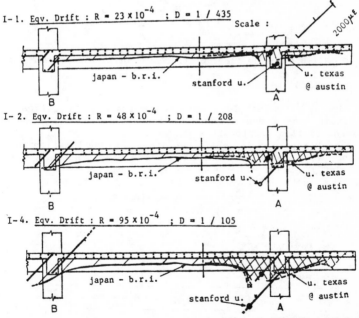

Fig. 6 Bottom slab reinforcement strains, interior joints

At large lateral displacements, full yielding occurred at considerable distances from the joint region, implying that the top slab steel also yielded under negative moment. As in the exterior joint, the observed strains in the overhanging section of the slab show that this region is not as heavily stressed as the "continuous" section between columns. The strain levels in the inner region compare well with the observed subassemblage values. Figure 7 shows the bottom strains in the BRI structure compared with the top and bottom strains in the subassemblage tests. Note that the top bar strains compare reasonably well in the Texas and Stanford tests.

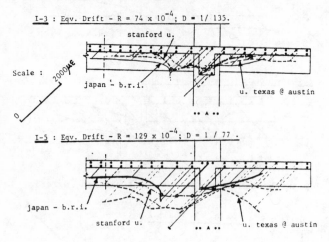

Fig. 7 Top and bottom slab reinforcement strains, interior joints

Slab Contribution to Floor Moments

The design of a ductile moment resisting frame requires that a predictable beam sidesway mechanism be considered. However, an assessment of floor member strength based on an incorrect assumption of slab participation may shift the plastic hinge location into the columns. With the actual material properties, the beam flexural capacity at yield and at a curvature of 10^{-3} rad/in. in the University of Texas structure (3) is given in Table 2. Current design code

	M_y		M_u @ 10^{-3} rad/in.[2]	
	ACI 318	Full Width	ACI 318	Full Width
1. Slab in Tension				
Interior Prototype	2000	4200	2485	4380
Exterior Prototype	2000	4150	2460	4320
2. Slab in Compression				
Interior Prototype	910	975	1240	1560
Exterior Prototype	915	975	1235	1555
3. Column	M_y = 3300 k-in. (either direction)			

Moment Ratios, $\Sigma M_{col}/\Sigma M_{floor}$

	Exterior Joint (M⁻only)		Interior Joint	
	@ First Yield	@ 0.001 rad/in.	@ First Yield	@ 0.001 rad/in.
ACI	3.30	2.68	2.27	1.77
Full Width	1.59	1.53	1.27	1.11

Table 2 Moment Capacities - University of Texas Tests

provisions, ACI 318-83 (Sec. A.4.2.2), require that the moment ratio be at least greater than 1.20.

Conclusions

1. It is clear that slab participation is highly dependent on the lateral deformation level. Considering the beam section only, severely underestimates the moments which are likely to develop in the floor system under negative loading (slab in tension). Considering the entire slab to the panel centerline as a flange acting monolithically with the beam overestimates the floor moments. It appears that at drift ratios of around 1/100, considering the slab steel in a segment 8 times the slab thickness on each side of the beam section provides a realistic estimate of the slab contribution. Within this section, the slab reinforcement (top and continuous bottom reinforcement) can be considered to act monolithically with the beam in calculating negative (slab in tension) floor flexural moments.

2. The overhanging portion of the BRI building developed lower slab strains in comparison with the component tests and the slab section between columns. This may be due to different boundary conditions and the way in which load is introduced into the specimen-- in the BRI building deformations were applied at floor levels with columns introducing moments to the floor system while in the tests, loads (or moments) were transferred from the floor slab and beams to the columns. Such differences are very difficult to quantify in a comparison such as this but illustrate the problems in extrapolating data on strains from an isolated test to the behavior of a continuous structure.

Acknowledgments

The support of the National Science Foundation under Grants PFR-8009039 and CEE-8320586 is gratefully acknowledged. The project was part of the U.S.-Japan Cooperative Program on Large-Scale Testing.

References

1. ACI Committee 318, "Building Code Requirements for Reinforced Concrete," American Concrete Institute, Detroit, 1982.

2. ACI SP-84, "Earthquake Effects on Reinforced Concrete Structures: U.S.-Japan Research," American Concrete Institute, Detroit, 1985.

3. Joglekar, M. R., "Behavior of Reinforced Concrete Floor Systems under Lateral Loads," Ph.D. dissertation, University of Texas at Austin, December 1984.

4. Wallace, B. J., and Krawinkler, H., "Small-Scale Model Experimentation in R/C Assemblages, U.S.-Japan Research Program Report No. 74, Blume Earthquake Engineering Center, Stanford University, June 1985.

EARTHQUAKE RESPONSE SIMULATION CAPACITY OF PSEUDO DYNAMIC TESTING
- EXPERIMENTAL DEMONSTRATION AND ANALYTIC EVALUATION -

by
Yutaka Yamazaki[1], Masayoshi Nakashima[1], and Takashi Kaminosono[1]

ABSTRACT

The response accuracy of the pseudo dynamic test was evaluated from the shake table and pseudo dynamic tests conducted for a two story steel braced frame model. The obtained responses were compared, and their correlation was estimated. It was found from the comparison as well as post numerical analysis that the accuracy of the pseudo dynamic response is very satisfactory.

1. INTRODUCTION

It was 1974 when Takanashi et al. of the Institute of Industrial Science of the University of Tokyo first developed a new method to directly simulate the earthquake response behavior of structural systems (Ref. 1). This method, referred to as the pseudo dynamic (PSD) test, can be classified as a combined experiment and numerical analysis, and, in this test, the actuators to apply forces to the test specimen are controlled by a computer program so that the specimen would take motion as if it were subjected to ground motion. A flow diagram of the PSD test is illustrated in Fig. 1. The PSD test possesses many advantages over the shake table test, known as the most direct method for earthquake response simulation. As the major advantages, we can find 1) that, as the loading can be quasi static, the less actuator capacity is required in the PSD test than in the shake table test if the same specimen is to be tested, 2) that, since the loading can be a repeated process of loading and pausing, conventional measuring devices used for quasi static test are sufficient in the PSD test, whereas in the shake table test, the measuring should be simultaneous and continuous, and 3) that, because of the discrete loading, the loading can be stopped any time upon request, and this enables us to make close observation of the local behavior of individual structural elements.

On the other hand, the PSD test includes various approximations in its algorithm. Some of those are 1) that the test structure, although it is basically a continuum, is assumed as a spring-mass discrete system (the space discretization), 2) that the governing equations of motion are treated as difference equations and solved discretely (the time discretization), 3) that the damping characteristics of the system need be presumed in advance to the test operation (except for the hysteretic damping), and 4) that, as the loading is quasi static, potential velocity effect on the hysteresis of the structure cannot be counted. In this condition, we should recognize that the response obtained from the PSD test is by no means identical with the true response. It is then imperative to estimate how much the obtained response is distorted in the PSD test.

1 Production Department, Building Research Institute, Ministry of Construction, 1 Tatehara, Oho-machi, Tsukuba, Ibaraki 305 JAPAN

For the purpose of obtaining basic information on the accuracy of the results obtained from the PSD test, PSD and shake table tests were conducted using a small scale steel braced frame model. This paper reports on the outline of the test including the test procedures and comparison of the results, and the post numerical analysis to further examine the possible differences between the two responses.

2. TEST PROGRAM

2.1 Specimen

The model prepared for this study, designated as BB model, was an approximate one-third scale model of a two story steel braced frame. A pair of specimens were fabricated for the test; one for the shake table test, and the other for the PSD test. The specimen was one meter in story hight and two meters in span length, and had two identical planar frames connected one meter apart by transverse beams. A H-shaped element was used for both columns and beams of the specimen frame, with the arrangement such that the beam bending would occur with respect to the strong axis of the element, whereas the column bending with respect to the weak axis. Then, the specimen frame would behave approximately as a shear type structure. Further, braces made of steel plate were arranged diagonally. The figure and basic dimensions of the specimen are illustrated in Fig. 2.

2.2 Shake Table Test

In order to preserve the prototype relationship, ballast was added to each floor level of the specimen. The total mass thus assigned was

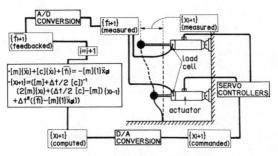

Fig. 1 Flow Diagram of PSD Test

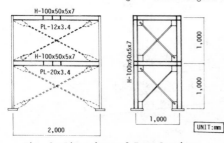

Fig. 2 Dimensions of Test Specimen

Table 1 Vibrational Properties of Specimen (1 kg = 9.8 N)

Mass $(kg \cdot sec^2/cm)$	1F:2.310	2F:2.280
	1st Mode	2nd Mode
Natural Period (sec)	0.126	0.052
Vibration Mode	1F: 0.45 2F: 1.00	1F: 1.00 2F:-0.51
Damping Ratio	0.012	0.015

2.310 (kg.sec^2/cm) and 2.228 for the first and second stories. To obtain the natural periods and vibration modes, resonance test was conducted prior to the main response test. Table 1 shows the results obtained from the resonance test. Two main tests were conducted. The first test (designated as Run-R1) was made with small input accleration, expecting the elastic response of the specimen, and the second test (Run-R2) with large input acceleration so that the specimen would sustain significant inelastic action. The input acceleration was a synthesized acceleration with its basis on the N-S component of the acceleration record obtained at Tohoku University during the 1978 Miyagiken-oki earthquake. The maximum accelerations for Run-R1 and Run-R2 were 491 and 2376 gals, respectively. The acceleration time history recorded on the shake table during the Run-R2 test and its Fourier spectrum are shown in Fig. 3.

2.3 PSD Test

The PSD test system installed at the Building Research Institute, Ministry of Construction (Ref. 2), was employed for the PSD test. To make possible the direct comparison between the shake table and PSD test results, the acceleration recorded on the shake table during the shake table test was used as the input acceleration of the corresponding PSD test. In this course, the acceleration record was digitized with the time interval of 0.01 sec. In the implementation of the PSD test, the weight equal to that added in the corresponding shake table test was attached to the specimen. The integration time interval used in the computation was 0.005 sec. For the viscous damping, the Rayleigh damping was assigned with the first and second mode damping ratios of 0.012 and 0.015, respectiely. Here, those values were obtained from the resonance tests. The test setup of the PSD test is shown in Fig. 4.

3. TEST RESULTS

Figures 5 and 6 show the Run-R2 story displacement and shear force time histories obtained from the shake table and PSD tests, and Fig. 7 the corresponding first story shear vs. displacement relationship. The correlation coefficients of the response histories between the two tests are also listed in Table 2. It is apparent from those figures that, in

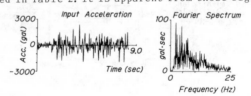

Fig. 3 Input Acceleration History and Fourier Spectrum

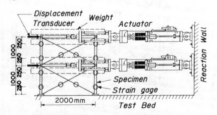

Fig. 4 Test Setup in PSD Test

general, the correlation between the two responses is satisfactory. Looking into Fig. 6, however, the first story displacement time history obtained from the PSD test is slightly biased on the positive side relative to that obtained from the shake table test. A closer look of the first story shear force vs. displacement relatioship (Fig. 7) exhibits that the shear force is nearly the same in the positive direction, while, in the negative direction, it is slightly greater in the PSD test than in the shake table test. This slight difference in the shear force was considered to be the source that had caused difference in the displacement time history. Discrepancy between the two shear force time histories is found occasionally such as at 5.4 sec and 6.1 sec. At 5.4 sec, the shear force is resisted only by the frame in the PSD test, whereas the force is resisted by both the frame and braces in the shake table test. At 6.1 sec, the resisting pattern is reversed. These are the points where the braces started carrying the force, and, because of the high stiffness in this range, only a slight difference in displacement seems to have lead significant change in the shear force.

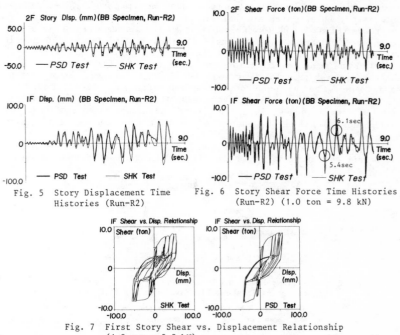

Fig. 5   Story Displacement Time
          Histories (Run-R2)

Fig. 6   Story Shear Force Time Histories
          (Run-R2) (1.0 ton = 9.8 kN)

Fig. 7   First Story Shear vs. Displacement Relationship
          (1.0 ton = 9.8 kN)

Table 2   Correlation Coefficients (Run-R2)

Entry	Correlation Coefficient		
	SHK Test-PSD Test	SHK Test-SHK Analysis	PSD Test-PSD Analysis
2F Story Disp.	0.8752	0.8544	0.8908
1F Disp.	0.9442	0.9508	0.9639
2F Shear Force	0.8459	0.7806	0.8639
1F Shear Force	0.8865	0.8580	0.9325

Figures 8 and 9 indicate the first story frame and brace shear force vs. displacement relationships of the Run-R2 shake table and PSD tests. To obtain those relationships, the shear resisted by the frame was estimated based upon the strain gage measurement, and the shear carried by the braces was taken to be the total shear force subtracted by thus obtained frame shear. These figures apparently indicate that the frame hysteresis is typical Masing type and the brace hysteresis typical bilinar slip type.

4. MODELLING OF SPECIMEN

Study was extended to investigate the sources of the differences between the two test results. For this purpose, analytical models that would represent the specimens used in the shake table and PSD tests were established. Examining the frame and brace hystereses (Figs. 7 to 9), the Ramberg-Osgood hysteresis model (Ref. 3, designated as the R-O Model hereinafter) was used to represent the frame behavior of the specimen, whereas the bilinear slip model for the brace behavior. In order to determine the values of the parameters included in those models, the nonlinear system identification techniques was employed. The frame and brace story shear force vs. displacement relationships obtained from the tests (Figs. 8 and 9) were utilized as the fundamental data, and the Levenberg-Marquard method was applied for parameter identification. The parameters included in the R-O model were the yield force in the positive and negative directions, and the two coefficients specifying the envelope shape. In the common R-O Model, the yield force is taken to be the same in both positive and negative directions. The experimental observation, however, exhibited that the shear resistance level differed between in the positive and negative directions, and, furthermore, this difference was speculated to be one of the major reasons that had generated the overall difference between the two tests. Considering this observation, in the modelling, the yield force was specified individually in accordance with the shear force between in

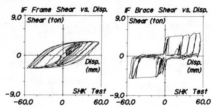

Fig. 8 First Story Frame and Brace Shear vs. Displacement
Relationships (Shake Table Test, Run-R2) (1.0 ton = 9.8 kN)

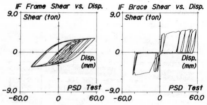

Fig. 9 First Story Frame and Brace Shear vs. Displacement
Relationships (PSD Test, Run-R2) (1.0 ton = 9.8 kN)

the positive and negative directions. Values of parameters thus derived are listed in Table 3.

5. CORRELATION OF BB MODEL

5.1 Effect of Approximations on Response
As discussed earlier, the PSD test contains various approximations. Using the analytical model defined to represent the specimen used in the PSD test, a series of numerical runs were carried out to examine the possible effect of the space and time discretizations and the selection of the visous damping on the response. As for the space discretization, comparison was made for the elastic response of the analytical model with the response obtained from the finite element analysis having distributed mass. Computation was also made for the elastic and inelastic responses of the model with various integration time intervals (0.0025 sec and 0.001 sec). Further, inelastic responses of the model having damping ratios different from those assigned in the test were examined. All results showed that the approximations nearly had no effect on the response, indicating that the difference in hysteresis between the two tests is the major source to have caused the difference in the obtained response.

5.2 Evaluation of Correlation
Figures 10 through 13 show the response obtained from the numerical models together with the corresponding experimental response, and Table 2 lists the correlation coefficients. These figures indicate that the analytical models can reproduce the experimental responses with reasonable accuracy. With respect to the displacement time histories, discrepancy was observed in the small deflection range after experiencing large deflection. In addition, similarly to the test, difference in the shear force time history appeared occasionally in the range in which the braces started carrying the shear force and, resultantly, the story shear force increased abruptly. In this range, only a slight change in the displacement caused great difference in the shear force, and, therefore, the numerical model had difficulty in duplicating the force at such instances.

Looking into Table 2, one can find that the correlation between the shake table test and model is no better than the correlation between the

Table 3 Parameters of Combined R-0 and Bilinear Slip Model

Entry	SHK		PSD	
	1F	2F	1F	2F
BRACES (Bilinear Model)				
Elastic Stiffness (ton/mm)	1.154	0.696	1.219	0.740
Yield Force (+) (ton)	4.334	2.610	4.277	2.252
Yield Force (-) (ton)	-4.540	-2.746	-4.104	2.074
Second Stiffness Relative to Elastic Stiffness (+)	0.0157	0.073	0.0233	0.0364
Second Stiffness Relative to Elastic Stiffness (-)	0.0087	0.041	0.0148	0.0522
FRAMES (R-0 Model)				
Elastic Stiffness (ton/mm)	0.1536	0.1434	0.1409	0.1519
Yield Force (+) (ton)	3.0615	3.1587	3.0502	4.6510
Yield Force (-) (ton)	-2.9949	-3.1937	-3.3011	-4.9811
α	0.7849	0.1147	0.8702	1.0672
γ	9.6437	10.8804	10.2328	10.9189

shake table and PSD tests. This observation infers that the shake model
may have its limit to represent the real behavior. As shown in Fig. 7,
the experimental story shear vs. displacement relationship in the shake
table test by no means exhibited smooth hysteresis. In fact, the
hysteresis had many zigzags. Such zigzags are speculated to have
occurred because of the slight phase lag between the force
(acceleration) and displacement measurements. Such unsmooth behavior
apparently contradict the R-O model, resulting in the aggravation of
correlation between the test and numerical responses.

Figure 14 shows the skeleton and hysteresis curves obtained from

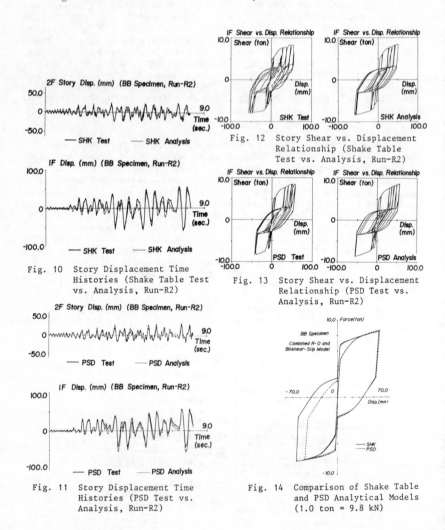

Fig. 10   Story Displacement Time
          Histories (Shake Table Test
          vs. Analysis, Run-R2)

Fig. 11   Story Displacement Time
          Histories (PSD Test vs.
          Analysis, Run-R2)

Fig. 12   Story Shear vs. Displacement
          Relationship (Shake Table
          Test vs. Analysis, Run-R2)

Fig. 13   Story Shear vs. Displacement
          Relationship (PSD Test vs.
          Analysis, Run-R2)

Fig. 14   Comparison of Shake Table
          and PSD Analytical Models
          (1.0 ton = 9.8 kN)

the shake table and PSD models. As evidenced from the figure, the correlation between the two models is very satisfactory except for some minor differences in the early inelastic range on the negative side. This finding clearly supports the observation that a pair of the shake table and PSD tests, a pair of the shake table test and model, and a pair of the PSD test and model, are correlated reasonably. This good correlation implies that there is no significant effect of the loading rate on the response, although the maximum velocity of 100 kine was induced during the test.

Considering other factors that may have caused the difference between the shake table and PSD test results, such as the possible difference in the clamping condition at the footing of the specimen, the possible rocking and torsion of the shake table, or the finite accuracies of the displacement and force sensors, the difference observed between the two tests should be considered to be very small.

6. COCLUSION

To study the accuracy of the response obtained from the PSD test, shake table and PSD tests were conducted using a two story steel braced frame model. Through comparison between the test results as well as comparison between the test results and the results of the analytical models that represented the test specimens, it was found that the response of the PSD test can duplicate satisfactorily the response of the shake table test. Various approximations included in the PSD test were estimated to have had minimal effect on the response, and the slight difference in response between the two results was believed to have caused by the difference in hysteresis between the two specimens. The effect of the loading rate on the response, however, was found to be insignificant.

ACKNOWLEDGEMENTS

This study was conducted as part of the U.S.-Japan Cooperative Research Program Utilizing Large Scale Testing Facilities. The writers wish to express their gratitude to all members in the Joint Technical Coordinating Committee (JTCC) of this Program for their support to this study. Opinions stated in this paper, however, do not necessarily reflect the views of the JTCC members.

REFERENCES
1. Takanashi, K., et al.,"Seismic Failure Analysis of Structures by Computer Pulsator On-Line System" Journal of Institute of Industrial Science, the University of Tokyo, Vol.26, No.11, 1974, pp.13-25 (in Japanese).
2. Okamoto, S., Kaminosono, T., Nakashima, M., and Kato, H.,"Techniques for Large Scale Testing at BRI Large Scale Structure Test Laboratory," Rersearch Paper, Building Research Institute, Ministry of Construction, No.101, May 1983, 38p.
3. Jennings, P. C.,"Periodic Response of General Yielding Structures," Journal of Structural Division, ASCE, Vol.90, No.EM2, April 1964, pp.131-166.

EVALUATION OF SEISMIC STRENGTHENING
OF AN EXISTING BUILDING USING BASE ISOLATION

by Gary C. Hart*, Ali Basharkhah**, Sunil Gupta**

INTRODUCTION

This paper summarizes a study performed by Englekirk and Hart, Inc. to evaluate the feasibility of using base isolation to seismically upgrade an existing highrise building. The specific objectives of this case study are:

(a) A rapid evaluation of the capacity of the existing building for the design response spectra.

(b) To evaluate the technical feasibility of using base isolation for the rehabilitation of the building.

(c) To assess some of the practical problems associated with incorporating base isolation.

(d) To establish the benefits of using base isolation.

(e) To evaluate the relative cost of using base isolation in comparison with a conventional scheme.

To achieve these objectives, we have:

(a) Determined the desired characteristics of the base isolation system required for the building.

(b) Developed mathematical models for both a conventional and a base isolated structure.

(c) Performed linear structural dynamic analyses of both mathematical models using the design earthquake response spectra.

(d) Calculated the values of the building's displacements, accelerations, forces and moments using the mathematical models.

---

 * Professor, Department of Civil Engineering, University of California, Los Angeles and Principal, Englekirk and Hart, Inc.
** Engineer, Englekirk and Hart, Inc.

(e) Developed the design of a base isolation scheme.

(f) Redesigned representative structural elements for seismic forces of the base isolated system.

(g) Developed a structural, seismic, upgrade construction cost comparison between a conventional structural system upgrade and a base isolated system upgrade.

THE BUILDING

The building was built in 1922. It is a seven-story building consisting of a complete vertical load carrying system. The building's primary lateral load resisting system is a concrete, moment resisting, perimeter frame. The frame consists of large rectangular columns with spandrel beams. The openings between the columns are infilled with 4 feet high concrete or brick walls. The building's floor slabs are of a flat plate design. The interior columns are not considered to be part of the lateral load resisting system because there are no girders to transfer positive or negative moments to the columns. The slabs only contain one layer of steel.

The building is approximately 24,400 feet square in plan. From a structural seismic design perspective, it is essentially symmetrical. The perimeter frames can be expected to carry approximately 100 percent of the seismic load. The 6-inch concrete slab on grade is essentially nonstructural except for the integral tie beams between piles. The typical floor system is a flat plate on column capitals and drop panels. The floor slab thickness varies from 9 inches at the second floor to 8 inches at the roof. Interior columns vary in size from 14 inches to 36 inches in diameter.

The building foundation system is comprised of concrete piles and pile caps. The foundation is believed to consist mostly of cast-in-place, drilled, concrete piers.

THE SEISMIC DESIGN CRITERIA

The building is located in San Diego. The Rose Canyon, La Nacion, Coronado Banks, San Diego Trough, Elsinore, San Clemente, San Jacinto and the San Andreas faults affect the potential seismic hazard at the site.

The design earthquake response spectrum has a 10% probability of being exceeded in 100 years. This level of earthquake corresponds to a 1,000-year return period. A 10% damping value is appropriate for reinforced concrete buildings.

THE BASE ISOLATORS FOR THIS BUILDING

The total weight of the structure, excluding the first floor, is about 26,000 kips. Assuming that the super-structure vibrates almost as a rigid body, the total horizontal stiffness of the base isolation system is selected to provide a period of 2.0 seconds. Therefore, for this building, the total stiffness of all base isolators is approximately 9,800 kips/feet. Due to this building's lateral displacement limitation because of an adjacent building, it is not possible to elongate the predominant period more than 2.0 seconds. This total horizontal stiffness would result from several reinforced elastomer bearings distributed under each column and providing support to the structure. The horizontal stiffness of each bearing is related to its physical dimensions, namely, the bearing area and total elastomer thickness. To limit the large displacements expected to occur due to induced flexibility at the base, energy absorption devices in the form of lead plugs are provided in the core of the rubber bearings.

The following guidelines are used to design the bearings and their layout:

(1) The bearing layout depends on the configuration of the structure immediately above the bearings.

(2) The vertical load carrying capacity of the bearings is a function of the plan dimension and internal layer thickness.

(3) The degree of isolation provided to the structure is defined by the total rubber thickness.

(4) Additional vertical loads induced by the earthquake must be added to the dead load plus any live load.

(5) Once the total rubber bearing thickness is deter-mined the bearing construction may be ascertained.

(6) To determine lead plug size, the following constraints must be kept in mind:

   (a) Plug diameter must be greater than one-sixth the plan dimension of the bearing.

   (b) Plug diameter must be less than one-third the plan dimension of the bearing.

   (c) Plug diameter must not be greater than overall height divided by 1.5.

REHABILITATION OF EXISTING BULDNG WITH A BASE ISOLATION SYSTEM

The first base isolation option is with the rubber bearings are located at the top of the slab in the first floor. The system has the following advantages:

(1) Minimal added structural costs.

(2) Separation at the base isolation level is simple to incorporate.

(3) Economical, if the first level is used for parking.

The disadvantages may be:

(1) May require cantilever elevator shaft below the first floor.

(2) No diaphragm is provided at the base isolation level.

In the second option, the rubber bearings are located at the top of the pilasters and below the first floor slab. The advantages of this model are:

(1) The base of the columns is connected by a diaphragm at the base isolation level.

(2) A back up system for vertical loads can be provided.

This system has the following disadvantages:

(1) The additional cost for excavation and demolition and building a new structural slab.

Option 1 was selected for this study because we believe that its construction costs will be less than Option 2. However, within the degree of accuracy intended for this study, the two options produced the same dynamic response. Therefore, no strong recommendation as to the most desirable option is made as part of this study. Both options are considered viable.

Let us first consider the existing frames with base isolation. This model is a combination of existing frames and lead rubber bearings. This combined system is analyzed for the response spectra with the total horizontal stiffness of the lead rubber bearings is computed to yield an effective fundamental period of 2 seconds. Due

to a constraint on the maximum horizontal displacement of
6 inches, because of an adjacent building to the south,
it is not practical to choose an effective period larger
than 2 seconds. It is assumed that this system has no
ductility and 10% damping.

The base shear is substantially reduced with the base
isolation. However, this reduced base shear of 930 kips
still exceeds the damage threshold capacity of 300 kips
which indicated the onset of damage to the beam. The
base isolation system by itself is not sufficient to
strengthen the building to meet mission essential
criteria.

REHABILITATION WITH NEW SHEAR WALLS AND A BASE ISOLATION
SYSTEM

Due to the significant damage to the beam in the existing
frames, we now propose new concrete shear walls supported
on a base isolation system will carry all of the lateral
seismic load. The rehabilitation systsem for the model
described herein is a combination of existing frames,
added shear walls and lead rubber bearings. The total
added length of the shear wall is approximately 100 feet.
All walls have an 18 inch thickness and a 4,000 psi
concrete strength. For the design earthquake motion, the
damping and ductility of the building increases to 10%
and 1.5, respectively.

The following results may be outlined for an anticipated
earthquake response using the proposed design system:

(1)   The base shear induced in the frames is about 1,000
       kips. Since the induced base shear in the frames is
       larger than the 300 kips base shear limit, plastic
       hinges will develop in most beams. However, the
       approximate component ductility demand of 3 is
       acceptable.

(2)   The roof displacement relative to rubber bearings
       will be 2.2 inches

(3)   Horizontal displacement at the top of rubber bear-
       ings will be 3.5 inches which is less than the 6.0
       in seismic gap.

(4)   The roof acceleration is 13%g.

(5)   The ground floor acceleration is 8%g.

Dynamics of 3-D Base Isolated Structures

Michalakis C. Constantinou*

Introduction

Base isolation is an accepted aseismic design technique in which a building is mounted on elastomeric bearings which increase the translational natural period of the structure to values beyond the "predominant" period of the earthquake excitation. While base isolation effectively reduces the level of forces imparted into the structure, it also creates a problem of very large relative displacements. The prediction and control of these displacements is critical because excessive displacements in combination with large vertical loads may cause instability to the bearings.

Base isolated structures respond in coupled lateral torsional motions because:
1. They have identical translational stiffnesses in the two horizontal directions and thus exhibit closely spaced natural frequencies of the lower modes.
2. The centers of mass (C.M.) and resistance (C.R.) of the isolation system do not coincide owing to the uneven distribution of the vertical load and the dependence of the stiffness of the bearings on the load they carry.

Pan and Kelly (4) have shown that linear base isolation systems with large eccentricities ($\approx 5\%$ of largest plan dimension) exhibit significant torsional response. This response, however, may considerably reduce if high damping bearings are used (damping 8 to 10%).

Lee (3) studied the coupled lateral torsional response of a single story structure mounted on four lead/rubber bearings and concluded that hysteretic base isolation systems are very effective in reducing the shears and torques generated in the structure. Both studies (3, 4) conclude that the high energy absorption capacity of the isolation system provides strong modal decoupling which results in a time lag between maximum translational and torsional responses.

The work described in this paper concerns the dynamic behavior of asymmetric base isolation systems consisting of laminated rubber bearings and hysteretic dampers. It represents an extension of the work described in Ref. 2 to torsionally coupled isolated structures.

It is shown in this analysis that if the stiffness of bearings and the number and orientation of hysteretic dampers are properly selected

*Assistant Professor, Dept. of Civil Engineering, Drexel University, Philadelphia, PA 19104.

(optimization) torsional coupling can be neglected even at very large eccentricities.

System Considered

The structural model considered is shown in Figure 1. It consists of N rigid floor decks and a rigid base which is mounted on axially inextensible laminated rubber bearings. Torsion bar hysteretic dampers, acting along x and y axes, provide high energy absorption capacity to the system. The configuration of Figure 1 which provides damping to torsional motions of the base will be refered to as system A. System B has the dampers rotated by 90°, thus provides damping only to translational motions of the base.

The equations of motion of the elastic superstructure may be written in the familiar matrix form with degrees of freedom the vector $\{U\}$ of the displacements and rotations of the C.M. of each floor relative to the base. The result is a system of 6N (N=number of stories) first order differential equations. In order to reduce the number of d.o.f., vibration in the first three modes is assumed

$$\{U\} = \sum_{i=1}^{3} \{\phi_i\} w_i \tag{1}$$

where w_i are the modal displacements and $\{\phi_i\}$ the mode shapes defined by

$$([K] - \omega_i^2 [M]) \{\phi_i\} = \{0\} \tag{2}$$

where $[K]$, $[M]$ = stiffness and mass matrices of the superstructure and ω_i = frequencies of free vibration.

The equations of motion of the base are derived from the equilibrium of the inertia forces and the forces from superstructure, bearings and hysteretic dampers which are shown in Figure 1 for the case of system A. The forces Q_i are provided by four groups of dampers which have identical properties and are equally distributed in opposite sides of the base. The restoring force Q_i, i = 1 to 4 is described by the following equations

$$Q_i = \frac{M_i}{2} [\alpha \frac{F_y}{Y}(U + \lambda V) + (1 - \alpha)F_y Z_i] \tag{3}$$

$$Y\dot{Z}_i + \gamma |\dot{U} + \lambda \dot{V}|Z_i + \beta(\dot{U} + \lambda \dot{V})|Z_i| - A(\dot{U} + \lambda \dot{V}) = 0 \tag{4}$$

where for i = 1 $U = U_{yb}$, $\lambda = \ell_x/r_b$, $M_i = M_y$
        i = 2 $U = U_{xb}$, $\lambda = -\ell_y/r_b$, $M_i = M_x$
        i = 3 $U = U_{yb}$, $\lambda = -\ell_x/r_b$, $M_i = M_y$ (5)
        i = 4 $U = U_{xb}$, $\lambda = \ell_y/r_b$, $M_i = M_x$

and $V = r_b U_{tb}$

As shown in Refs. 1 and 2 Eqs. 3 and 4 describe a hysteretic force-displacement relation which adequately represents the actual behavior of hysteretic dampers including torsion bar dampers and lead/rubber bearings. In these equations Z_i = hysteretic component of force Q_i, F_y = yield force of damper, Y = yield displacement, r_b = radius of gyration of base, α = post to preyielding stiffness ratio and β, γ, A = dimensionless parameters. For a particular torsion bar damper tested at U.C., Berkeley these parameters were determined to be F_y = 5 Kips, Y = 0.11 inches, α = 0.052, β = -0.01, γ = 1.05 and A = 1.

The reduced equations of motion of the superstructure, the equations of motion of the base and Eq. 4 form a system of 16 first order nonlinear differential equations which can be easily solved for the deterministic response of the system. More details are presented in Ref. 1.

Random Vibration

The probabilistic response when the earthquake excitation is described by a stochastic process is very important for the optimum design of the isolation system (1, 2). Furthermore, with probabilistic methods one may obtain reliable estimates of the extreme displacements which are important for the design of bearings for stability and for providing sufficient clearance between adjacent structures.

For the nonstationary probabilistic response the state-space approach is employed. To model the nonwhiteness of the excitation, a filter is included in the system (2). Furthermore, Eq. 4 is replaced by the linearized equation

$$Y\dot{Z}_i + C_{ei}(\dot{U} + \lambda\dot{V}) + K_{ei}Z_i = 0, \quad i = 1,\ldots, 4 \tag{6}$$

in which the coefficient C_{ei} and K_{ei}, determined by methods of equivalent linearization, are

$$C_{ei} = (\tfrac{2}{\pi})^{\frac{1}{2}}\left\{\gamma(E[\dot{U}Z_i] + \lambda E[\dot{V}Z_i])(\sigma_{\dot{U}}^2 + 2\lambda E[\dot{U}\dot{V}] + \lambda^2\sigma_{\dot{V}}^2)^{-\frac{1}{2}} + \beta\sigma_{zi}\right\} - A \tag{7}$$

$$K_{ei} = (\tfrac{2}{\pi})^{\frac{1}{2}}\left\{\gamma(\sigma_{\dot{U}}^2 + 2\lambda E[\dot{U}\dot{V}] + \lambda^2\sigma_{\dot{V}}^2)^{\frac{1}{2}} + \beta\frac{E[\dot{U}Z_i] + \lambda E[\dot{V}Z_i]}{\sigma_{zi}}\right\}$$

In these equations σ_x stands for the standard deviation of x and $E[\cdot]$ = expected value.

The evolutionary stochastic response is governed by

$$\frac{d[S]}{dt} = [P][S] + [S][P]^T + [Q] \tag{8}$$

in which [S] = zero time lag covariance matrix and [Q] = matrix of stochastic excitation (1, 2). Matrix [P] includes the coefficients C_{ei} and K_{ei} which are nonlinear functions of elements of the matrix [S]. This implies that Eq. 8 is nonlinear.

Numerical solutions were obtained by use of subroutine DGEAR of IMSL after reducing Eq. 8 to a system of 171 first order equations by utilizing the symmetry of matrix [S].

Optimization of Design

The optimum design is defined as
$$\text{minimize } f = \max_{\psi}(E[F^2])^{\frac{1}{2}} \tag{9}$$

$$\text{subject to } \max(U_{xb}, U_{yb}) \leq U_{cr} \tag{10}$$

with control variables the number of dampers M_x, M_y and the quantity ω_b:

$$\omega_b = (\frac{K_b}{m_b})^{\frac{1}{2}} \tag{11}$$

where K_b = total horizontal stiffness of bearings and m_b = mass of base.

The objective function f is the maximum over ψ (= direction of ground acceleration) of the square root of F^2:

$$F^2 = (\frac{V_x}{m_b})^2 + (\frac{V_y}{m_b})^2 + (\frac{T}{m_b r_b})^2 \qquad (12)$$

where V_x, V_y, T are the base shears and torque. Furthermore, U_{cr} is the critical displacement of the base the engineer may want to tolerate and $\max(U_{xb}, U_{yb})$ is the extreme base displacement relative to the ground.

Example Analysis

In order to illustrate the developed procedure, a 2-story building on base isolation is analyzed. The properties of the superstructure (with reference to Figure 1) are given in Table 1. In this table $r\ell$ = radius of gyration

Floor/ Story ℓ	Mass m_ℓ	r_ℓ	$K_{x\ell}/m$ (s^{-2})	$K_{y\ell}/m$ (s^{-2})	$K_{t\ell}/mr^2$ (s^{-2})	ey_ℓ/r	ex_ℓ/r
1	m	1.176r	1000	1000	1152	0.20	0.20
2	m	1.176r	1000	1000	1152	0.20	0.20
m = 1.4 Kip s²/in.		r = 250 in.		$\xi_1 = \xi_2 = \xi_3 = 0.02$			

Table 1. Properties of Example Building

of floor ℓ, $K_{x\ell}$, $K_{y\ell}$ = translational stiffnesses of story ℓ and $K_{t\ell}$ = torsional stiffness of story ℓ. The static eccentricities ex_ℓ, ey_ℓ correspond to 7% of the largest plan dimension. Furthermore, the mass of base $m_b = m$, $r_b = 294$ in., $\ell_x = \ell_y = 360$ in., and the eccentricities of the isolation system e_{xb}, e_{yb} are assumed to be 5% of the maximum base dimension.

The optimization scheme of Eqs. 9 and 10 was employed with constraint $U_{cr} = 4$ in., and the results are shown in Table 2. The excitation is a process with intensity, frequency content and duration of the

System	ξ %	Number of Dampers $(M_x = M_y)$	Optimum ω_b r/s	f in/s²	$\max(r_b U_{tb})$ in
Linear	5.8	0	6.90	37.706	2.087
Linear	10	0	6.00	27.273	1.694
B	5.8	2	5.55	26.866	2.054
B	5.8	4	4.13	19.362	1.570
B	5.8	6	3	17.796	1.079
B	5.8	8	3	20.774	0.734
A	5.8	2	5.85	27.272	0.372
A	5.8	4	4.20	19.256	0.098
A	5.8	6	3	17.732	0.038
A	5.8	8	3	20.774	0.024

Table 2. Results of Optimization

1952 Taft earthquake. From this table the optimum design for both systems A and B is $\omega_b = 3$ r/s and $M_x = M_y = 6$. This corresponds to a fun-

damental period of 0.8 - 2.2 sec. depending on the amplitude of excitation. In the case of linear systems (without dampers) the fundamental
period is 1.6 and 1.8 sec. depending on the damping factor of the bearings, ξ, which is 5.8% and 10% respectively. Of interest is the extreme rotation of the base which is substantially lower for system A
than for the other systems. For example, the corner point displacement
due to rotation alone is 0.05 in. for system A, 1.32 in. for system B
and 2.07 in. for the linear system with ξ = 10%. This illustrates the
significance of hysteretic dampers for the reduction of torsional response.

Table 3 shows the effect of eccentricity of the isolation system on
the extreme displacements of the optimized systems A, B and linear. In
this case the excitation was taken along the x-axis (ψ = 0). The eccentricities of the isolation system e_{xb}, e_{yb} were varied (for set values of structural eccentricity) from 1% to 10% of the maximum plan dimension. The superiority of system A over the other systems is apparent. Even at 10% eccentricity the extreme corner displacement, U, is
virtually the same as the motion of the C.M. In contrast, the contribution of rotation to the corner displacement of the linear system with
ξ = 0.10 is more than 12% even at 5% eccentricity and amounts to more
than 30% for 10% eccentricity.

Finally, Figure 2 presents the evolutionary response statistics of
the optimized isolated structure with system A for three different intensity envelopes of the stochastic excitation. It is interesting to
note that for slow built up intensity envelopes the maximum over time
of the statistical response coincides with the stationary response.
The time history of the base displacement when the structure is subjected to the actual record of the Taft earthquake is shown in Figure 3.

System	Eccentricity %	$\max(U_{xb})$ in.	$\max(r_b\ U_{tb})$ in.	max U in.
Linear, ξ=0.1	1	4.257	0.472	4.260
"	3	4.140	1.179	4.362
"	5	3.977	1.694	4.537
"	7	3.814	2.067	4.760
"	10	3.641	2.550	5.259
B	1	3.547	0.240	3.551
B	3	3.541	0.660	3.639
B	5	3.532	1.075	3.804
B	7	3.519	1.488	4.044
B	10	3.499	2.081	4.512
A	1	3.548	0.015	3.555
A	3	3.545	0.020	3.561
A	5	3.548	0.027	3.569
A	7	3.549	0.035	3.578
A	10	3.550	0.048	3.591

Table 3.   Effect of Eccentricity on Maximum Response

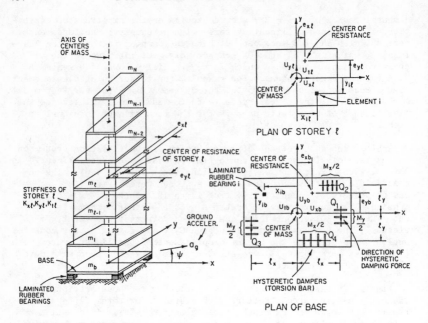

Fig. 1. Idealized Multistory Base Isolated Structure (system A)

Conclusions

A simplified method of analysis and optimization of design of torsionally coupled base isolated structures with hysteretic dampers has
been presented. As a result of this study the following conslusions
can be drawn:

1. Coupling induces torsional response and reduces the translational
   motion of the C.M. of the base in the direction of excitation for
   linear and B systems. The translational motion of system A, however, is effectively independent of the eccentricities of the isolation system.

2. The torsional response of the C.M. of the base increases with increases in the isolation system eccentricity.

3. For linear and B systems the corner displacement is virtually the
   same as the C.M. displacement for small eccentricities (<2%) but
   different at large eccentricities (>5%).

4. The corner displacement of system A is virtually the same as the
   C.M. displacement at all eccentricities (0 - 10%).

5. Hysteretic dampers if properly used (system A) are very effective
   in reducing the generated base shears and torque and in reducing

the torsional response even at extreme eccentricities. Base isolated structures with system A may be analyzed as plane structures
for all practical purposes.

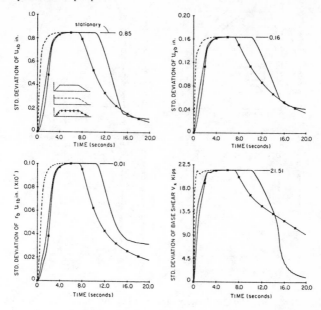

Fig. 2a.   Evolutionary Statistics of Optimized Structure

Appendix I - References

1.   Constantinou, M. C. (1984). "Random Vibration and Optimization of
     Design of Aseismic Base Isolation Systems," thesis presented to the
     Rensselaer Polytechnic Institute, in partial fulfillment of the re-
     quirements for the degree of Doctor of Philosophy.

2.   Constantinou, M. C. and Tadjbakhsh, I. G. (1985). "Hysteretic Damp-
     ers in Base Isolation: Random Approach." J. Struc. Engrg, ASCE,
     111(4), 705-721.

3.   Lee, D. M. (1980). "Base Isolation for Torsion Reduction in Asymme-
     tric Structures under Earthquake Loading." J. of Earthquake Engi-
     neering and Structural Dynamics, Vol. 8, 349-359.

4.   Pan, T. C. and Kelly, J. M. (1983). "Seismic Response of Torsional-
     ly Coupled Base Isolated Structures." J. of Earthquake Engineering
     and Structural Dynamics, Vol. 11, 749-770.

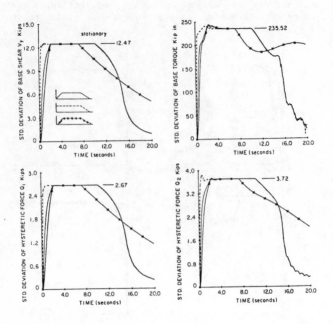

Fig. 2b. Evolutionary Statistics of Optimized Structure

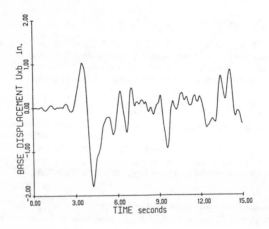

Fig. 3. Time History of Base Displacement (system A)

THE EARTHQUAKE BARRIER

Marc S. Caspe, P.E.*

The earthquake barrier protects buildings from the horizontal ground vibrations of an earthquake by installing a "force barrier and an energy barrier" at the underside of the building. This barrier limits these horizontal forces to a value well below the elastic strength of the building. The concept is well-suited to new construction as well as to the retrofit of both existing buildings and equipment that is found vulnerable to seismic damage.

By limiting the base shear in the fundamental mode to less than the building's elastic strength, the building can be maintained undamaged during the most catastrophic of strong motion earthquakes. In addition to reducing the seismic risk to life-safety and property damage, earthquake barrier protected buildings should cost less to construct than conventional earthquake resistant structures because of cost savings in structural, architectural and mechanical systems. The simplicity of the isolation devices, see Figure 1, make them inexpensive to install on both new construction and the retrofitting of older buildings.

The earthquake barrier establishes a barrier to the horizontal forces than a earthquake can exert when it tries to "pull the rug out" from under a building. No matter what Mother Nature throws at the building in the way of peak accelerations or the duration of ground motions, the earthquake barrier establishes a positive limit to the maximum horizontal force that the foundation can exert at the base of the building's superstructure. This earthquake barrier force is pre-engineered by the structural engineer to a level that is significantly higher than the design wind force but less than the elastic strength of the building. Since buildings are always designed to withstand these wind forces with an adequate safety factor, buildings having an earthquake barrier can be kept from being damaged during a major earthquake.

The earthquake barrier gives the structural engineer control over the forces of nature. To accomplish this, low friction teflon slide plates -- having a breakaway coefficient of friction of less than 0.02 -- are first inserted under each <u>column</u> and wall, see Figure 1, to destroy the grip that the

*President, M.S. Caspe Co., Inc., San Mateo, California 94403

foundation of a conventional building has on the superstructure. The slide plates have self-leveling elastomeric plates underneath to relieve construction tolerances for setting them level.

Having destroyed the shear transfer at the foundation, it can then be replaced by inserting high-friction prestressed slide plates -- having a breakaway coefficient of friction of 0.12 -- that firmly tie the building back to its foundation, as shown in Figure 1. The slide plates can be visualized as holding the building rigidly in place during all windstorms and minor earthquakes and then simply sliding -- while exerting a constant horizontal force on the building -- whenever a modest earthquake or a catastrophic earthquake were to cause the force to reach the earthquake barrier's preengineered limit. This constant earthquake barrier force pushes and pulls on the building so that the sliding motion is limited to only minor slippage.

The earthquake barrier system is a combination of these two simple and reliable devices. The low-friction telfon slide plates first destroy the foundation's grip on the building, then the high-friction prestressed slide plates are used to replace this grip with engineered materials. The prestress rods serve to control the earthquake barrier force level at which slippage occurs and to resist uplift (tension in the column) if this is induced by a combination of overturning forces and vertical accelerations. Before the earthquake barrier force is reached the slide plates remain rigid and will not slide during a major windstorm. However, during even a modest earthquake the slide plates will slip and exert a constant accelerating force on the building, until the direction of the ground motion (i.e. its relative velocity) reverses itself.

Upon reversal of the relative velocity the slide plates will automatically become stiff and rigid until the predetermined force is acting in the opposite direction; at which instant the slide plates will exert a constant decelerating force on the building. This is not unlike what the 1st story columns do in a frame building, except that instead of having the horizontal forces and deflections imposed imposed on the columns -- with their reduced ductility due to the weight they support -- the same magnitude of force, deflection and energy dissipation will be imposed on the slide plates using a much more stable configuration. In its simplest context, the earthquake barrier can be thought of as a separation of the vertical weight supporting system from the lateral force resisting system at the first story; each system being free to do what it does best.

It is the simplicity of the earthquake barrier devices that imparts such a high degree of reliability. The physical principles of sliding friction on teflon plates are well-tested in decades of use on bridges, nuclear power plants and buildings. In addition, prior to construction, the slide plates designed for each building will be dynamically tested -- in prototype -- under actual earthquake ground motions. This shake table testing of full-sized plates, which are the operative members during an earthquake, will give both the owner and the building official a high-level of confidence in the performance of the earthquake barrier before authorizing the start of construction. This level of confidence is particularly important when retrofitting older masonry buildings, where the plates serve both to limit the lateral forces below the elastic strength of the building and to provide an energy dissipating first story for an otherwise brittle building.

An understanding of the mode of operation of the earthquake barrier requires a thorough visualization of the relationships between the time-dependent parameters shown typically in Figure 2 for the 1940 El Centro earthquake. The variation of accelerations, velocities and displacements in an earthquake is not harmonic, hence the variation in amplitude of the acceleration curve is not symmetrical about the base axis of time. Since the velocity at any instant of time is the cumulative area under the acceleration curve, isolated spikes of acceleration do not necessarily cause a reversal in the direction of the ground velocity, even though they do cross the time axis. It is for this basic reason that the period of the velocity component is much longer than that of the individual acceleration spikes, as shown on the Figure 2.

Similarly, the displacement is an accumulation of the area under the velocity curve and the same logic applies. The correct picture of the earthquake ground motion is therefore that of a short period (0.2 seconds) inertial excitation or acceleration (i.e. F = Ma) that causes long period (2 to 6 second) displacements. That is, the ground vibrates rapidly but it moves relatively slowly off in one direction and then back and into the other direction.

Comprehension of this ground motion and its interaction with a building that is protected by an earthquake barrier is critical to understanding the force balance that exists within the slide plates during an earthquake. The slide plates cause the superstructure to follow the ground motion very closely. Since the ground velocity has a much longer period than does the acceleration, these viscous dampers have an extended interval of time during which to force the superstructure back toward (but not necessarily to) its original position with respect to the foundation. This distance ($s = vt + 0.5at^2$) would be transversed with a lesser acceleration of the superstructure than that experienced by the rapidly vibrating ground in displacing to this same point. This is because the distance transversed by the building under the influence of the earthquake barrier force is a function of the time interval squared.

The direction in which the continuously applied earthquake barrier force acts on the underside of the superstructure is dependent on the relative velocity, rather than the relative displacement, at every instant of time (i.e. the forcing function is velocity dependent). Since the velocity parameter is a longer period phenomenon than acceleration, see Figure 2, less frequent reversals can be expected during the duration of an earthquake. The relatively constant magnitude of the predetermined earthquake barrier force causes the superstructure to follow the ground motions at a relatively constant value of both building acceleration and base shear ($F = Ma$). Because the velocity is still a longer period phenomenon than is the acceleration, this constant force acts on the base of the building's superstructure for an extended interval of time, taking advantage of the (t^2) term in the above equation.

In essence, the slide plates act to smooth out the sharp vibratory accelerations of the ground. This results in a comparatively gentle motion for the superstructure. The motion is only a function of the the predetermined earthquake barrier force selected by the designer, giving the structural engineer total control of the building's behavior regardless of how severely the earthquake is or how long it lasts. It can therefore be anticipated that earthquake barrier protected buildings will not only be far safer during a catastrophic earthquake but will also be undamaged, saving the owner both the cost of repair and loss-of-use (i.e. revenue loss) after an earthquake.

Because the earthquake barrier protects the building from force levels that could cause damage, it accepts relative displacement in very much the same manner that a boxer does when he backs away from a blow to the chin. This relative displacement on the slide plates is limited to a few inches, not only by the high force level of the earthquake barrier but also because it is "biased" back toward the center of the earthquake barrier's force-deflection characteristic. This biasing prevents slippage from accumulating during an earthquake. When high safety factors against relative displacements are coupled with this biasing characteristic, safety levels for the earthquake barrier — against both forces and displacements — can be shown to be far in excess of conventional earthquake-resistant technology.

A concept similar to the earthquake barrier was recently announced by Taisei Corporation, a large engineering and construction company in Tokyo. Taisei announced that it had developed a barrier concept and had test it successfully on a shaker table. Taisei has also built the first commercial application in a control tower at Tokyo's Narita International Airport.

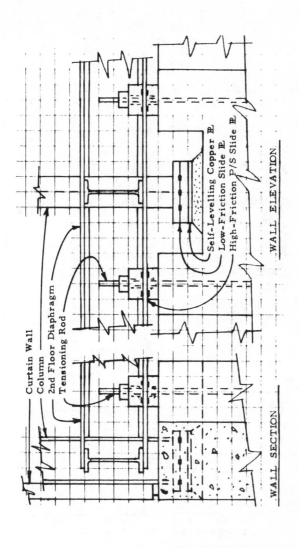

Figure 1 Earthquake Barrier

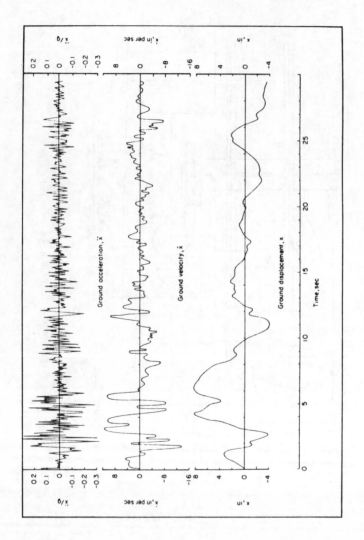

Figure 2 Recorded Ground Motions of the El Centro Earthquake

Earthquake Response of Five Story
Reinforced Concrete Masonry Test Building

Yutaka Yamazaki*
and
Masa-omi Teshigawara**

ABSTRACT
In 1986, a full-scale five story reinforced masonry test building
is to be constructed in BRI Large-Scale Structure Test Laboratory, and
through 1987 to 1988, the test will be carried out in order to evaluate
it's seismic performance.

At present, preliminary design and analyses of the test building
are being carried out. In this paper, results of ineleastic dynamic
response analyses of the test building are introduced.

1. Test Building

Figure 1 shows plan of a five story reinforced masonry building
specimen to be tested in BRI. The test is carried out in the x-
direction (actually, longitudinal direction) and, therefore, analyses
also are made in the same direction. There are four frames in this
direction ; those are frames-Y_1, -Y_2, -Y_3 and Y_4. The frames-Y_2 and -Y_3
have very wide shear walls in the center. Wall length rate at every
story of this test building is 15 cm/m^2 in the x-direction. All
structural elements, such as walls and beams except cast-in-place
reinforced concrete floor slabs and foundation beams, are made of
concrete masonry. An open-end type concrete block unit with two web
shells (390mm × 190mm × 190mm) is to be used for the test building.

Figure 2 shows arrangement of reinforcing bars in the frame-Y_1.

2. Inelastic Static Frame Analysis

In order to verify the seismic performance of the test building, a
precise static analysis on the ultimate strength and the failure
mechanism of the building subjected to external lateral forces is
carried out.

2.1 Model
1) Model

Walls and beams are idealized as beam models having rigid zones at
their both ends. The inside of all joint parts surrounded by walls and
beams is considered as rigid zone.

In order to represent the non-linearlity of members, a one-
component model consisting of series combination of flexural springs at
both ends and a shear spring at the center of the member is considered
as shown in Fig.3.
2) Concrete Block and Steel Strength Assumed

---

* Head, ** Research Engineer, Building Research Institute, Ministry of
Construction, Tatehara 1, Oho-machi, Tsukuba-gun, Ibaraki-ken, Japan

Concrete block
  prism strength f_m = 240 kg/cm^2 (23.5 MPa)
  Young's modulus   E = 1.8 × 10^5 kg/cm^2 (17.6 GPa)
  shear modulus   G = 0.9 × 10^5 kg/cm^2 (8.8 GPa)
Steel bar
  yield strength σ_y = 3800 kg/cm^2 (372 MPa)

3) Ultimate Strength of Walls and Beams
    The following experimental formulas are used to evaluate ultimate strength of walls and beams.[1]

Wall
  i. Ultimate Shear Strength

$$Q_{su} = \left\{ \frac{0.053\, P_t^{0.23}\,(f_m+180)}{M/(Qd) + 0.12} + 2.7\sqrt{P_w \cdot \sigma_{wy}} + 0.1\sigma_0 \right\} b \cdot j$$

where
    $M/(Qd)$: shear span ratio, $1 \leq M/(Qd) \leq 3$
       P_t : flexural reinforcement ratio (%)
      f_m : prism strength (kg/cm^2)

Reinforcement (If not Specified)
1. Beam STP.   D13@200
2. Wall Vertical   D10@200
       Horizontal D10@200
3. Foundation Beam STP. D10@200

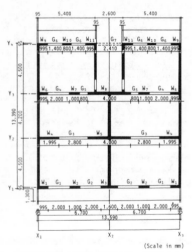

Fig.1 Plan of Test Building

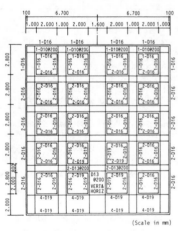

Fig.2 Reinforcing Bars (Frame-Y1)

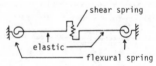

Fig.3 Column and Beam Model

P_w : horizontal reinforcement ratio
σ_{wy} : yield strength of horizontal reinforcement (kg/cm^2)
σ_0 : axial stress (kg/cm^2)
b : effective wall width (cm)
j : (7/8)d, d : 0.9Lw (Lw : wall length)

ⅱ. Ultimate Flexural Strength

$$M_u = a_t\, \sigma_y\, d + 0.5\, a_w\, \sigma_{wy}\, d + 0.5\, N\, d$$

where

a_t : cross section of flexural reinforcement (cm^2)
σ_y : yield strength of flexural reinforcement (kg/cm^2)
a_w : cross section of vertical reinforcement (cm^2)
σ_{wy} : yield strength of vertical reinforcement (kg/cm^2)
N : axial force (ton)

<u>Beam</u>
ⅰ. Ultimate Shear Strength

$$Q_{su} = \left\{ \frac{0.053\, P_t^{0.23}\, (f_m+180)}{M/(Qd) + 0.12} + 2.7\sqrt{P_w \cdot \sigma_{wy}} \right\} b \cdot j$$

ⅱ. Ultimate Flexural Strength

$$M_u = 0.9\, a_t\, \sigma_y\, d$$

4) Assumption of Restoring Force Characteristics of Members

Restoring force characteristics of flexural and shear springs in each member are, as shown in Figs.4 and 5, given by a degrading tri-linear and an origin-oriented hysteresis rules, respectively. In these figures, values of deflection angle of elements, θ_y and r_y, corresponding to their ultimate strengths were determined based upon results obtained from wall and beam tests which had been carried out in 1984 under this project.

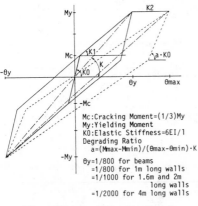

Mc:Cracking Moment=(1/3)My
My:Yielding Moment
K0:Elastic Stiffness=6EI/l
Degrading Ratio
a=(Mmax-Mmin)/(θmax-θmin)·K

θy=1/800 for beams
=1/800 for 1m long walls
=1/1000 for 1.6m and 2m
long walls
=1/2000 for 4m long walls

Fig.4 Flexural Spring Model

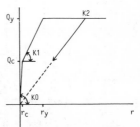

Qc : Cracking Shear Strength = (1/2)Qy
Qy : Ultimate Shear Strength
K0 : Elastic Stiffness = GA/1.2
ry = 1/800

Fig.5 Shear Spring Model

As an example, restoring force vs. deflection relationship of the beam G_1 at the first story in frame-Y_1, which is represented by the series combination of those of flexural springs at both ends and a shear spring at the center of the member in the analysis, is shown in Fig.6. The corresponding force vs. angle of deflection curve which was experimentally obtained from the test of the beam specimen GF3 is also shown in this figure. It is recognized that the analytical model of the element gives quite good agreement compared with the corresponding test result. It is also recognized from this figure that the angle of deflection of the beam G_1 is almost twice as much as the angle of story deflection at the first story.

5) Lateral Force Distribution Profile

The external lateral forces are distributed to each floor level of the test building in accordance with the A_j seismic force distribution profile over the height which is stipulated in the Japan Seismic Code. Those are 1.0P, 1.1P, 1.2P, 1.4P and 1.6P from lower to higher stories in order.

2.2 Results

Ultimate shear strength and story deflection of the building at the first story under the failure mechanism are as follows ;

Ultimate strength : Q_1 = 520.5 ton (5.1 MN)

(Base shear coefficient C_B = 0.47)

Angle of deflection : R_1 = 1/300

Shear force ratios of frames-Y_1 to -Y_4 to total base shear force, at which angle of deflection at the first story reaches R_1=1/100, are 19.8%, 28.5%, 36.6% and 15.1%, respectively.

Most of all plastic hinges are made at beam ends until an angle, R, reaches 1/400, where R means the angle of total deflection at the roof of the building. Shear failure occurrs at all short span beams except the second story ones in the frame-Y_3 until the angle, R, reaches 1/200. In the final stage, plastic hinges are made at all of the wall bottom ends except the first story center wall located in the frame-Y_2 (Wall-W_5).

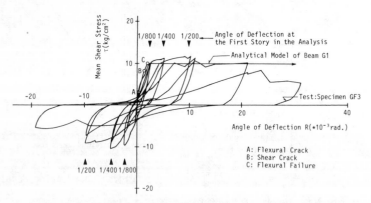

Fig.6 Analytical Model and Test Result
(Element:Beam G1, Test:GF3) (1 kg/cm^2= 98 kPa)

3. Dynamic Response Analysis
3.1 Analytical Model (Lumped-Mass-System)
1) Upper Structure Model

Restoring force characteristics of the shear spring for each story are given by a degrading tri-linear rule. Values of those parameters to represent the tri-linear restoring force characteristics are, as shown in Fig.7, determined based on the shear force vs. story deflection relationship which was obtained through the static frame analysis in the previous section.

Those values related to the lumped mass system are listed in Table 1. The value of viscous damping constant, h, taken through the analysis is 5 percent as a fraction of the critical damping.

2) Lower Structure Model (Soil-Structure Interaction Effect)

Most of all buildings in Japan are constructed on soft soil ground. It is believed that the effect of the soil-structure interaction is considerably large if a building has high rigidity. In case of masonry buildings, they are expected to have relatively very high rigidity. Therefore, three models including soil-structure interactive models are considered in the following analysis. They are :

Model-A --------- a model without soil-structure interaction effect (fixed condition)

Table 1 Dynamic Constants of the Test Building

Story	Weight W(ton)	Initial Stiffness K_0(ton/cm)	Stiffness Ratio		Shear Force	
			K_1/K_0	K_2/K_0	Q_1(ton)	Q_2(ton)
5F	218	2885	0.111	0.0121	45	115
4F	218	2811	0.180	0.0231	79	203
3F	218	2842	0.208	0.0359	108	275
2F	218	2935	0.240	0.0443	135	344
1F	218	4359	0.228	0.0381	170	424
Foundation	294	2000,Model-B 1000,Model-C	Note: 1 ton=9.8 kN			

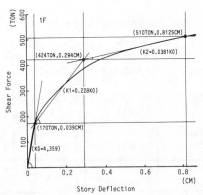

Fig.7 Tri-Linear Model (1 TON=9.8 kN)

Model-B and -C --- models in which values of sway spring constants
              are 2000 ton/cm (1.96 GN/m) and 1000 ton/cm (0.98
              GN/m), respectively

Values of sway spring constants, 2000 ton/cm (1.96 GN/m) and 1000
ton/cm (0.98 GN/m), are used for the Models-B and -C respectively. The
following equation was used to evaluate the values of these spring
constants.

$$_F K_s = \frac{8}{2-\nu_e} \cdot \frac{\gamma_e}{g} \cdot V_e{}^2 \cdot r_s \left\{ 1 + \frac{H_E}{r_s} \right\}$$

where

  $_F K_S$: sway spring constant of a foundation which is embedded on
       half space infinite ground
  ν_e : poison's ratio ($\rightarrow 0.48$)
  γ_e : soil density ($\rightarrow 1.80$ ton/m³ = 17.6 kN/m³)
  g : gravity constant (= 9.8 m/s²)
  V_e : shear wave velocity ($\rightarrow 100$ m/s and 141.4 m/s)
  r_s : equivalent radius of foundation = $\sqrt{(\text{foundation area})/\pi}$
                                          ($\rightarrow 7.7$ m)
  H_E : depth of embedded foundation ($\rightarrow 2.5$ m)

Linear restoring force vs. deflection relationship is assumed for
these sway springs through the analysis. This assumption generally
gives larger response to the upper structure compared with a model in
which inelastic restoring force characteristics are taken into account.
3) Input Ground Motions

As input ground motions to the models, three earthquake records are
used. They are :
  El Centro Record (NS)   1940 Imperial Valley Earthquake (341.7 gals)
  Taft Record (EW)   1952 (175.9 gals)
      and
  Hachinohe Record (EW)   1968 Tokachi-Oki Earthquake (182.9 gals)

The acceleration amplitude of these three earthquakes are linearly
enlarged or reduced so that the maximum values coincide with 200 gals
(2 m/s²) and 400 gals (4 m/s²).

3.2 Results

Natural period of three models are listed in Table 2. The first

Table 2   Natural Period of the Models

Model Mode	Model-A	Model-B	Model-C
First Mode	0.182(sec.) (1.0)	0.236(sec.) (1.30)	0.286(sec.) (1.57)
Second Mode	0.063 (1.0)	0.085 (1.35)	0.095 (1.51)
Third Mode	0.040 (1.0)	0.052 (1.30)	0.053 (1.33)

Values in the parentheses denote period ratios
to those of the Fixed Model.

natural periods of the models are 0.182, 0.236 and 0.286 seconds for the Model-A (Fixed Model), Model-B and Model-C, respectively.

According to the results for 200 gals input motions, the maximum angles of story deflection are very small and do not exceed the value, 1/1000 for any case.

The maximum responses when subjected to 400 gals input ground motion are shown in Figs.8a, 8b and 8c. It is recognized from these figures that story deflection responses generally increase with decreasing soil stiffness. The maximum values of the angles of story deflection (and ductility factors) are produced at the second story and are 1/400 (1.66), 1/360 (1.82) and 1/300 (2.20) for the Model-A (Fixed Model), Model-B and Model-C, respectively.

Sway ratio which is defined by d_G / d_{5F}, where the variables d_G and d_{5F} denote displacements at the foundation and the fifth story, respectively, are approximately $11 \sim 15\%$ and $20 \sim 27\%$ for the Model-B and Model-C, respectively.

4. Conclusions

Major findings obtained through a static and a dynamic response analysis of the test building are summarized as follows:

i) Under static loading, the failure mechanism of the building is made by flexural failure hinges which are produced at most of all beams.

ii) The ultimate strength and the corresponding base shear coefficient at the first story of the designed test building are 520.5 ton (5.1 MN) and 0.47, respectively. The angle of story deflection at the first story under the failure mechanism is 1/300.

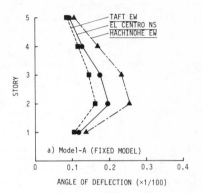

a) Model-A (FIXED MODEL)

ANGLE OF DEFLECTION (×1/100)

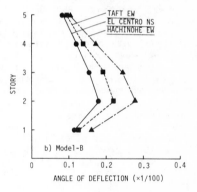

b) Model-B

ANGLE OF DEFLECTION (×1/100)

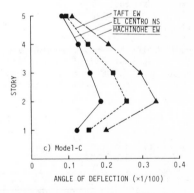

c) Model-C

ANGLE OF DEFLECTION (×1/100)

Fig.8 Maximum Angles of Story Deflection: The Max.Ground Acceleration=400gal

iii) Shear forces which are distributed to frames-Y_1 to $-Y_4$ at the first story are 19.8%, 28.5%, 36.6% and 15.1%, respectively, of the total story shear force under ultimate stage.

iv) The angle of deflection of the most standard beam G_1 of the test building is almost twice as much as the one at the first story.

v) Story deflection responses generally increase with decreasing stiffness of soil on which a building is constructed.

vi) Under 200 gals input motions, the maximum angles of story deflection are very small and do not exceed the value, 1/1000.

vii) Under 400 gals input motions, the maximum values of the angle of story deflection (and ductility factors) are produced at the second story and do not exceed the value, 1/300 (2.20).

According to items iv) and vii), the angle of deflection of all beams (and also walls) except very short span beams do not exceed the value, 1/150. Based upon the structural tests concerning walls and beams, those elements are expected to be distorted without rapid strength degradation under 400 gals earthquake ground motions.

In this paper, the test building was structurally idealized without considering effect of resisting elements which are located in the orthogonal direction. If these elements are taken into account, the base shear coefficient corresponding to the ultimate strength at the first story greatly increases in approximately 0.8 from 0.47. In this case, dynamic response is expected to be considerably decreased in than the model without considering these orthogonal elements.

References
1) "Design Guideline of Wall-type Reinforced Concrete Structures", The Building Center of Japan, 1983
2) "Tentative Provisions for the Development of Seismic Regulations for Buildings", Applied Technology Council (ATC) Associated with the Engineers Association of California, 1978

Acknowledgement
The authors wish to express their gratitude for Professor Tsuneo Okada, chairman of the Technical Coordinating Committee on Masonry Research (TECCMAR) under the U.S.-Japan Coordinated Earthquake Research Program, for his fruitful advise to this study.

The authors also express their sincere thanks to Messrs. Toshihiko Terada and Akio Naka-oka, visiting research engineers of BRI, for their technical coordination throughout the study.

Flexural Behavior of
Reinforced Masonry Walls and Beams

Tuneo   Okada *
Hisahiro Hiraishi **
Takashi Kaminosono ***
and
Masaomi Teshigawara ***

ABSTRACT
In order to resist severe earthquake, deformation capacity is one of the most important factors for medium-rise masonry building structures. Many reinforced masonry walls and beams were tested as a part of the U.S.-Japan cooperative research program on masonry building structures. Six concrete block and two clay brick masonry walls, one reinforced concrete wall, and four concrete block masonry beams were tested in order to evaluate their deformation capacity after yielding by flexure.
Major findings are that; 1) increases in amount of shear reinforcement and spiral reinforcement were effective on deformation capacity, 2) existence of splices did not affect badly on deformation capacity, and 3) ultimate shear stress increased, but deterioration of strength at large deformation became large by derease in shear span ratio of beam.

1. Introduction

Structural tests of 41 reinforced masonry walls, 3 reinforced concrete walls, and 21 reinforced masonry beams were conducted as a part of the U.S.-Japan cooperative research program on masonry building structures. This paper describes experimental results and discussions on flexural behavior of eight reinforced masonry walls, one reinforced concrete wall, and four reinforced masonry beams expected to be failed in flexure. The specimens were subjected to constant vertical load as well as cyclic lateral load.

Tests of the walls were conducted by focusing on the effects of the following four factors on the deformation capacity.
1) amount of shear reinforcement, and ratio of ultimate shear strength to ultimate flexural strength
2) spiral reinforcement at the critical compression zone
3) splices of main longitudinal bars
4) materials used for wall

Tests of the beams were conducted by focusing on the effects of the following two factors on the deformation capacity.
1) amount of shear reinforcement
2) shear span ratio (half value of beam length devided by beam depth)

----

* Professor, Institute of Industrial Science, University of Tokyo,
    Tokyo, JAPAN
** Senior Research Engineer,   *** Research Engineer,
    Building Research Institute, Ministry of Construction, JAPAN

2. Wall

2.1 Specimens

Nine walls were tested. The shape and dimensions of the Specimen WFJ1 is shown in Fig.1. The other specimens had almost same shape and dimensions. Outlines of the specimens are summarized in Table 1. The Specimens WF1, WF2, WFJ1, WFJ2, WFL1, and WFLM were reinforced concrete block masonry walls, and the Specimens WFR1 and WFR2 were reinforced clay brick masonry walls. The Specimen WFRC was a reinforced concrete wall. Splices of main longitudinal reinforcing bars (two deformed bars with 19mm diameter) were placed at the bottom of walls. The length of lap joint of the Specimen WFL1 was 40d (=76cm), and screw type mechanical splices of the Specimen WFLM were placed at 10cm above the bottom critical section of the wall. Mechanical properties of materials used in these specimens are tabulated in Table 2.

2.2 Loading Method

In order to obtain basic informations of the structural performance of masonry walls, a simple loading method which is usually used in Japan was selected for the wall tests. A testing facility named BRI/TTF, which was used in this test series, consists of one steel loading beam, three servo-actuators, one micro-conputer for vertical loading control, two mini-conputer for lateral loading control and data measuring, as showen in Fig.2. The loading conditions were as follows.

1) Axial stress of 5kg/cm^2 was kept to be constant.
2) The bottom end of the specimen was fixed to the testing floor, and rotation of the top end of the specimen is restrained.
3) Horizontal deformation varied with specified history.

Table 1 Outlines of the Specimens

Test Specimen	Tensile Longitudinal Reinf. (Pt %)	Vertical Shear Reinf. (Pwv %)	Lateral Shear Reinf. (Pwh %)	Spiral Reinf. 6∅, 120mm dia 40mm pitch h=600mm	Splice
WF1			D13 @400(0.17)		
WF2			2-D13 @200(0.67)	-------	
WFRC	2-D19	D16	D13 @400(0.17)		-------
WFJ1	(0.28)	@400mm		Top and	
WFJ2		(0.26)	2-D13 @200(0.67)	Bottom	
WFL1				Bottom	Lap
WFLM			D13 @400(0.17)		Mechanical
WFR1	2-D19	D16 @400		-------	-------
WFR2	(0.33)	(0.35)	2-D13 @200(0.67)		-------

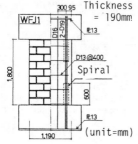

Fig. 1 Test Specimen

WFJ1 Thickness = 190mm (unit=mm)

Table 2 Mechanical Properties of Material

Test Specimen	Compression Strength of Concrete (kg/cm^2)	Compression Strength of Mortar (kg/cm^2)	Prism Strength (kg/cm^2)	Yield Stress of Reinf. Bar (kg/cm^2) D19	D16	D13
WF1	273	328	233			
WF2	252	243	220			
WFRC	254	-----	-----	3542		
WFJ1		177				
WFJ2	252	252	220		3289	3620
WFL1		243				
WFLM		324		3438		
WFR1	274	298	257	3542		
WFR2		324				

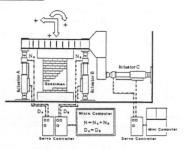

Fig. 2 Loading System

2.3 Crack and Failure Patterns

Figure 3 shows crack patterns of each specimen after test. All specimens show almost same crack patterns before a drift angle R, which is a value of deformation divided by clear height of wall, reached to 1/200. At first, flexural cracks appeared at both top and bottom parts of wall. Thereafter, Crack width became wider, and flexural shear and shear cracks at both top and bottom parts appeared and developed with increase in the drift angle. After a drift angle of 1/200, diagonal shear cracks appeared in seven specimens except the Specimen WF2 and WFJ2, and this diagonal shear cracks remarkably became wider with increase in the drift angle. The Specimen WFR2 finally failed at only top part of wall. This failure was caused by the incomplete filling of grout concrete at the top part. As for the Specimen WF2 and WFJ2, compression failure and slippage became remarkable with the development of cracks which appeared before a drift angle of 1/200. The face shell of concrete block units at both compression edges of the top part of the Specimen WF2 dropped out.

The strains of main longitudinal and vertical shear reinforcing bars of each specimen reached to 0.2% at a drift angle of 1/400. Therefore, all specimens were supposed to be yielded in flexure before a drift angle of 1/400.

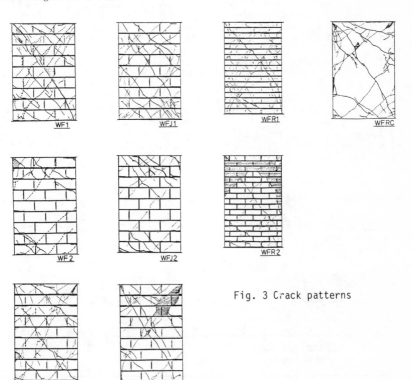

Fig. 3 Crack patterns

2.4 Comparison of Deformation Capacity

In Figures 4 through 7, skeleton curves of mean shear stress versus drift angle relationships of each specimen are shown by focusing on comparison of the effect of four factors on the deformation capacity.

Figure 4 shows the comparison between the Specimens WF1 and WF2, and the Specimens WFR1 and WFR2 whose amount of shear reinforcement differs each other. From this figure, the difference in the deformation capacity depending on the amount of shear reinforcement is obviously observed after a angle of 1/200. Concretely speaking, the Specimen WF2, in comparison with the Specimen WF1, showed stable deformation behavior without remarkable deterioration in strength at even a drift angle of 1/50, although the Specimen WFR2 did not show better deformation behavior at a drift angle of 1/50 than the Specimen WFR1 because of the partial failure mentioned before. This facts means that increase in the amount of shear reinforcement improves the deformation capacity. However, as ratios of ultimate shear strength to ultimate flexural strength are also increased by increase in the amount of shear reinforcement in these specimens, it is not obvious which increase is more effective to improve the deformation capacity.

Figure 5 shows the effect of spiral reinforcement at the critical compression zone. From the comparison of deformation behavior after a drift angle of 1/100, it is obviously observed that the Specimen WFJ1 showed a stable behavior while the Specimen WF1 showed negative slope in positive loading. The load carrying capacity of each specimen at a drift angle of 1/50 is in order of WFJ2>WF2, and WFJ1>WF1. This phenomena are explained by the reason why the developement of cracks and compression failure at edge zone might be prevented by the spiral reinforcement, so it is concluded that the spiral reinforcement at critical compression zone is effective on improvement of the deformation capacity under repeating cyclic loading.

Figure 6 shows the comparison regarding splices in a wall. There is no remarkable difference in the deformation capacity among three Specimens WF1, WFL1, and WFLM in the figure, so it may be concluded that an existence of splices and a kind of splices (lap splice, mechanical splice) scarcely affect on the deformation capacity of walls.

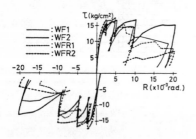

Fig. 4 Skeleton Curves of Walls
(Amount of Shear
Reinforcement)

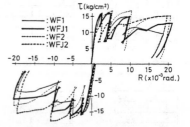

Fig. 5 Skeleton Curves of Walls
(Spiral Reinforcement)

Figure 7 shows the comparison among three Specimens WF1, WFRC, and WFR1 whose materials of wall are different. The Specimen WFRC of reinforced concrete wall showed a stable deformation behavior without a deterioration in strength before a drift angle reached to 1/100. Moreover, the deterioration in strength under repeating cyclic loading at the same deformation amplitude is little. Therefore, it is concluded that reinforced concrete wall has comparatively excellent deformation performance before large deterioration in strength will occur.

3. Beams
3.1 Test Specimen
Four beams were tested. The shape and dimensions of the Specimen GF1 is shown in Fig.8. Outlines of the specimens are summarized in Table 3. The parameters of the specimens are amount of shear reinforcement and shear span ratio. The amount of shear reinforcement are smaller in order, the Specimens GF3, GF4, and GF5. The Specimen GF6 has the same amount of shear reinforcement as the Specimen GF3. The shear span ratios are a value of 1.25 for the Specimens GF3, GF4, and GF5, and a value of 0.75 for the Specimen GF6. Tables 4 tabulates mechanical properties of materials used in the specimens of beam.
3.2 Loading Method
Loading facility used in this test series is illustrated in Fig.9. Alternating and antisymmetric bending moment were applied to the both ends of the specimen.

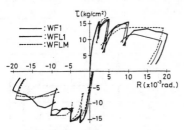

Fig. 6 Skeleron Curves of Walls
(Splice)

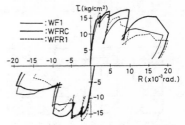

Fig. 7 Skeleton Curves of Walls
(Material)

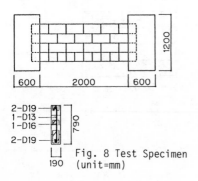

Fig. 8 Test Specimen
(unit=mm)

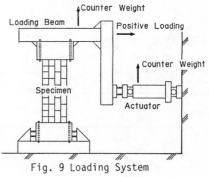

Fig. 9 Loading System

3.3 Crack and Failure Patterns

Figure 10 shows crack patterns of each specimen after the test. In the case of the Specimens GF3, GF4, and GF5 which shear span ratio is a value of 1.25, the first flexural cracks were observed along vertical mortar joints at the both ends of beam when the mean shear stress reached to a value from 3.3 to 4.0 kg/cm^2. Flexural shear cracks also occured, when the mean shear stress was about 8kg/cm^2 and a drift angle reached to 1/800. Tensile reinforcing bars yielded when the mean shear sterss was about 11kg/cm^2 and a drift angle reached to 1/400. Shear cracks were developed around the both ends of beam after the yielding of the tensile reinforcing bars. Thereafter, they failed in flexural with shear finally. In the case of the Specimen GF6 which shear span ratio is a value of 0.75, shear cracks occured when the mean shear stress was 6.7kg/cm^2. This specimen also yielded in flexure. However, it failed in shear finally when the mean shear stress was 19.0kg/cm^2.

Table 3 Outlines of the Specimen

Test Specimen	Clear Span Length (mm)	Tensile Reinf. Bar (Pt %)	Shear Reinf. Bar (Pw %)
GF3			D13 @400(0.17)
GF4	2000	2–D19	D13 @200(0.34)
GF5		(0.42)	2–D13 @200(0.67)
GF6	1200		D13 @400(0.17)

Table 4 Mechanical Properties of Material

		Stress (kg/cm^2)
Yield Strength of Reinf. Bar	D13	3620
	D16	3500
	D19	3542
Prism Strength		217

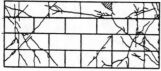

GF3 lo=2000mm D13@400mm

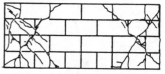

GF4 lo=2000mm D13@200mm

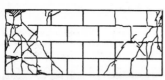

GF5 lo=2000mm 2-D13@200mm

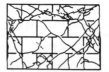

GF6 lo=1200mm D13@400mm

Fig.10 Crack Patterns

3.4 Comparison of Deformation Capacity

Table 5 tabulates the ultimate shear stress of the beam specimens obtained from the tests. In Figures 11 and 12, skeleton curves of mean shear stress versus drift angle relationships are shown by focusing on the effect of two parameters on the deformation capacity.

Figure 11 shows the relationships of the Specimens GF3, GF4, and GF5 in order to compare the effect of amount of shear reinforcement. The ultimate shear stresses of these specimens indicated almost same value of $12 kg/cm^2$ each other. After a drift angle of 1/100, there are discrepancies among these specimens in the deterioration in strength. The Specimen GF4 which has larger amount of shear reinforcement than the Specimen GF3 indicated a smaller deterioration in strength than that of the Specimen GF3. But the Specimen GF5 which has the largest amount of shear reinforcement indicated almost same deterioration of strength as the Specimen GF3. It is considerable that increase in the amount of shear reinforcement improves the deformation capacity. But according to the test results, it is not clear that amount of shear reinforcement is effective on the deformation capacity.

Figure 12 shows the relationships of the Specimen GF3 and GF6 in order to compare the effect of shear span ratio on the deformation capacity. The Specimen GF6 which shear span ratio is smaller than that of the Specimen GF3 indicated a value of $19.2 kg/cm^2$ in ultimate shear stress. This value is 1.6 times larger than that of the specimen GF3. The deterioration in strength started at a drift angle of 1/100 as for both Specimen GF3 and GF6, but the deterioration of the Specimen GF6 was larger than that of the Specimen GF3. It is concluded that decrease in the shear span ratio improves the ultimate shear stress and also makes the deterioration larger at a large deformation range.

Table 5 Shear Stresses obteained from the Test

Test Specimen	GF3	GF4	GF5	GF6
Flexural Crack	4.0	4.0	3.3	6.7
Shear Crack	11.1	10.4	11.3	13.3
Ultimate Strength	11.6	12.2	12.0	19.2

note:
values are
mean shear stress (kg/cm^2)
when the item occured.

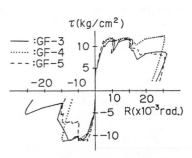

Fig.11 Skeleton Curves of Beams
(Shear Reinforcement)

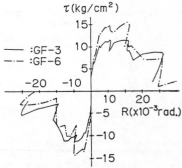

Fig.12 Skeleton Curves of Beams
(Shear Span Ratio)

4. Conclusions

Lateral loading tests of flexural type masonry wall and beam were performed. The followings are concluded regerding effects of some factors on the deformation capacity of wall and beam based on the test results.

1) Increase in the amount of shear reinforcement (consequently increase in the ratio of maximum shear strength to maximum flexural strength) improves the deformation capacity.

2) Spiral reinforcement at the critical compression zone increase the deformation capacity at a large amplitude of deformation and under repeating cyclic loading, because development of cracks and compression failure at the edge zone was prevented.

3) Splices of main longitudinal reinforcing bars scarcely affected on deformation capacity.

4) Diagonal shear cracks controlled badly a large extent deformation performance after yielding by flexure for the specimens which have small shear reinforcement and have not special reinforcement at compression zone.

5) Decrease in the shear span ratio of beam increase the ultimate shear stress, but it makes the deterioration in strength large after the ultimate shear stress appeared.

REFERENCES

(1) S.Okamoto et al., "U.S.-Japan Joint Earthquake Research Program Involving Large-Scale Experiments on Masonry Building Structures, Japanese Side Research Plan," The 17th Joint Meeting, U.S.-Japan Panel on Wind and Seismic Effects, UJNR, Tukuba, Japan, 1985

(2) "-U.S.-JAPAN COORDINATED EARTHQUAKE RESEARCH PROGRAM ON REINFORCED MASONRY BUILDINGS-, Seismic Capacity of Reinforced Masonry Walls snd Beams," Architectural Institute of Japan Annual Report, 1985 (in Japanese).

Isoishi et al., "Part-2 Loading System," pp.1085-1086
Kawashima et al., "Part-3 Flexural Behavior of Walls," pp.1089-1090
W.Y.Chung et al., "Part-11 Flexural Behavior of Reinforced Concrete Block Beam," pp.1103-1104

Acknowledgement

The authers wish to express their gratitude for Mr. Fumitosi Kumazawa, research associate, Institute of Industrial Science, Unversity of Tokyo and Messrs Hiroshi Isoishi and Toshikazu Kawashima, assistant research engineers of Building Research Institute.

Unit Conversion $1kg/cm^2 = 98kPa = 0.098MPa$

Stress-Strain Behavior of Grouted Hollow Unit Masonry

Gregory R. Kingsley, R. H. Atkinson[1], M.ASCE

ABSTRACT: Experimental results from axial stress-strain tests including post-peak response on hollow unit masonry prisms are compared to those of previous researchers.

INTRODUCTION

In the United States, many reinforced masonry structures in seismic zones are constructed from grouted hollow-unit masonry. Current design codes (Uniform Building Code 1985) incorporate working-stress design techniques for the design of these structures. Eventually, ultimate strength design techniques, like those used in concrete design, will be considered for use in U.S. building codes. This will require a thorough understanding of the complete stress-strain behavior of masonry up to and beyond peak stress.

Unlike concrete, however, the stress-strain behavior of grouted hollow masonry has not been well established, particularly in the post-peak or residual region. The collection or compilation of this data is difficult, as the results are highly sensitive to experimental technique, specimen configuration, and workmanship. Furthermore, the term "hollow-unit masonry" may include two distinct material types; clay or concrete. Present design codes make no distinction between the two types of masonry, thus implicitly assuming that they have identical properties; this assumption has not been verified for the purposes of ultimate strength or working stress design.

This study, which comprised the first task of the U.S.-Japan Coordinated Program for Masonry Building Research, had as its primary objectives; 1) to obtain the complete stress-strain curves for masonry in uniaxial compression, and 2) to conduct parallel compression tests on clay and concrete masonry prisms under identical conditions of testing to determine the degree to which these two materials exhibit common engineering behavior characteristics. Secondary objectives of this study were: 1) to provide data on materials considered for use in the U.S.-Japan Program, and 2) to establish standardized test procedures to be followed by all investigators in the U.S.-Japan Program.

The results of an experimental study designed to achieve these objectives are presented here, along with pertinent results from the work of other researchers.

EXPERIMENTAL PROGRAM

The experimental program consisted of ten series of tests, each

[1]Atkinson-Noland & Associates, Inc., 2619 Spruce St., Boulder, CO 80302

including five clay and five concrete prisms, in which one parameter was changed for each test series (Kingsley, Atkinson 1985). The prisms were tested in uniaxial compression to strains well beyond the strain at ultimate strength.

The "baseline" or reference masonry type consisted of grouted stack-bond concrete and clay prisms, nominally 6"(152mm) wide, and constructed with Type N mortar. The variables investigated in the other nine tests were:
  —Unit width— 4" (102mm) and 8" (204mm) wide units
  —Ungrouted prisms— constructed using 6"(152mm) and 8"(204mm) wide units
  —Mortar strength— one series constructed with type S mortar
  —Grout strength— one series with high strength grout
  —Bond pattern— one series with running bond
  —Load direction— one series was loaded parallel to the bed joint
  —Platen restraint— one series was tested using a teflon-grease-tef-
   lon interface between the prism and platen to reduce end shear
   tractions on the prism

The basic properties of the units, mortar, and grout are tabulated in Tables 1-3. Prisms were constructed using a special alignment jig to produce constant mortar bed thickness, plumbness of sides, and parallelism of the ends (Kingsley, Atkinson 1985). All prisms were instrumented with four LVDT's and four strain gages as shown in Figure 1.

Prisms were tested using a 1,000,000 lb (4.45 MN) servo-controlled hydraulic testing machine. This machine had a measured flexibility of 5.37×10^{-5} in/kip (0.0307 cm/kN). Prisms were loaded at a constant displacement, and LVDT transducer readings were recorded using a micro computer/data acquisition system.

The four LVDT's mounted to the sides of the prism provided an accurate measure of prism deformation up to near ultimate load strain. Cracking and face shell spalling rendered LVDT data useless beyond this point however. For post-peak response, deformations were obtained

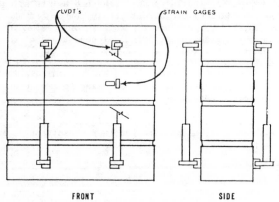

FIGURE 1
PRISM INSTRUMENTATION

TABLE 1

UNIT PROPERTIES

UNIT NOMINAL WIDTH	MATERIAL	fb^1 psi*	ft^2 psi*	IRA^3 gm/30in^2	$ABSORPTION^4$ % of UNIT WEIGHT	CORE AREA GROSS AREA
4"	CLAY	8974	–	15	9	0.28
6"	CLAY	12936	585	26	10	0.32
8"	CLAY	11344	–	25	10	0.44
4"	CONCRETE	3904	–	61	9	0.30
6"	CONCRETE	3583	300	55	10	0.37
8"	CONCRETE	3125	–	59	12	0.45

[1] Uniaxial Compressive Strength, measured according to ASTM C140-75

[2] Splitting Tensile Strength, measured according to ASTM C1006-84

[3] Initial Rate of Absorption, measured according to ASTM C67-78

[4] Absorption (24 hour immersion), measured according to ASTM C67-78

TABLE 2

GROUT PROPERTIES

GROUT	UNIT MATERIAL	MEAN COMPRESSIVE STRENGTH, psi*	NUMBER OF SPECIMENS	C.O.V.
NORMAL	CLAY	3205	3	0.02
NORMAL	CONCRETE	4226	4	0.08
STRONG	CLAY	3724	4	0.16
STRONG	CONCRETE	4437	4	0.18

TABLE 3

MORTAR PROPERTIES

MORTAR TYPE	MIX PROPORTIONS C:L:S	MEAN COMP. STRENGTH 2" CUBES, psi*	NO. SPEC.	C.O.V.
NORMAL	1:1:6	1717	162	.048
STRONG	1:1/2:4-1/2	3525	18	0.84

* 1 psi - 6.9 kPa

using the internal hydraulic ram LVDT. This transducer has included in its output the effects of the testing machine flexibility and initial machine take-up. The complete stress-strain curves were calculated using the ram LVDT output which was adjusted to account for machine flexibility and take-up effects following the method used by Priestley and Elder (Priestley, Elder 1983).

RESULTS

A typical set of stress-strain curves are presented in Figure 2 for the 6 inch (152 mm) grouted clay unit and concrete block unit prisms.

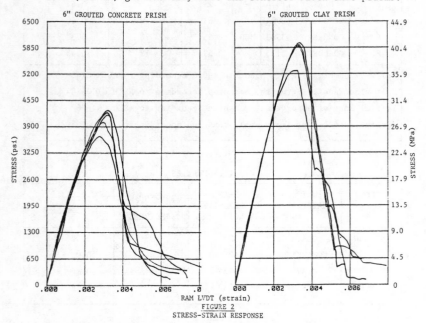

FIGURE 2
STRESS-STRAIN RESPONSE

Table 4 presents average strength and stiffness properties obtained from the first eight test series. The ultimate strength, f'm, was calculated using the net loaded area of the prism. The values of the strain at ultimate strength and secant modulus were calculated from the average of the four LVDT's mounted on the prisms because of the greater accuracy of these transducers compared to the ram LVDT for strains up to peak load. The secant modulus was computed at 50% of ultimate strength. The strain, ϵ_{50}, is the strain on the descending portion of the stress-strain curve at a stress level that is 50% of the ultimate strength.

EVALUATION OF RESULTS AND COMPARISON TO PREVIOUS RESEARCH

Table 5 shows a summary of mean strength and strain parameters for clay and concrete, grouted and ungrouted masonry. Included with the results from this project are summary results from the work of Hegemier at UCSD (Hegemier et.al. 1978) Brown at Clemson (Brown, Whitlock 1982),

TABLE 4
TEST RESULTS

NO.	DESCRIPTION	MATERIAL	ULTIMATE STRENGTH $f'm$ psi [††]	SECANT MODULUS E_s ksi[†††]	STRAIN AT ULTIMATE ϵ_p in/in	STRAIN AT 50% OF f'_m ON DESCENDING BRANCH ϵ_{50}
1	6" GROUTED	CLAY	5765	2317	.0030	.0040
		CONC.	4084	2340	.0024	.0032
2	4" GROUTED	CLAY	3946	2365	.0021	.0032
		CONC.	3037	1523	.0025	.0034
3	8" GROUTED	CLAY	4523	2627	.0022	.0028
		CONC.	4040	2609	.0023	.0032
4	6" UNGROUTED	CLAY	5704	1837	.0032	.0043
		CONC.	2602	1429	.0027	.0033
5	8" UNGROUTED	CLAY	5373	2167	.0028	.0038
		CONC.	2133	1234	.0028	.0033
6	6" RUNNING BOND	CLAY	4745	2151	.0024	.0032
		CONC.	3622	1918	.0016	.0037
7	6" TYPE S MORTAR	CLAY	5595	2342	.0024	.0032
		CONC.	4172	2334	.0025	.0034
8	6" STRONG GROUT	CLAY	5993	2249	.0030	.0040
		CONC.	4201	2270	.0025	.0033

[†] includes only prisms with h/t /2   [†††] 1 ksi - 6.9 MPa
[††] 1 psi = 6.9 kPa

and Priestley at the University of Canterbury in New Zealand (Priestley, Elder 1983). Brown studied grouted and ungrouted hollow clay masonry, and Hegemier and Priestley studied grouted hollow concrete masonry. While compiling the summary data in Table 5, no distinction was made between prisms of varying height, unit size, mortar type, grout type, or bond pattern. While these variables may have influenced prism strength to some degree, it was observed in this study that strain parameters were unaffected by them, and thus lumping together the mass of data was deemed justified.

Prism strength is very sensitive to materials, test methods, and prism configuration. This is illustrated by the variety of prism strengths represented in Table 5. Secant modulus values also vary rather considerably. The values of peak strain are more consistent; these are shown plotted in Figure 3. Mean values of peak strain are 0.0028 for clay, 0.0026 for concrete, and 0.0027 for both clay and concrete combined. Values of unloading strain, shown in Figure 4 are also fairly consistant, with mean values of 0.0036 for clay, 0.0040 for concrete, and 0.0038 for the combined data.

TABLE 5
COLLECTED RESULTS FROM PREVIOUS RESEARCH

RESEARCHER & MASONRY TYPE	f'm (psi)[††]			E_3 (ksi)[†††]			ϵ_p			ϵ_{50}		
	MEAN	COV	N	MEAN	COV	N	MEAN	COV	N	MEAN	COV	N
KINGSLEY GROUTED CLAY	5104	.16	27	2340	.10	26	.0025	.18	27	.0034	.17	27
BROWN GROUTED CLAY	2663	.26	18	1498	.23	17	.0030	.28	18	-		-
KINGSLEY UNGROUTED CLAY	5538	.08	10	2002	.16	10	.0030	.16	9	.0041	.15	9
BROWN UNGROUTED CLAY	3729	.34	20	1802	.35	20	.0030	.37	19	-		-
KINGSLEY GROUTED CONC.	3890	.12	27	2192	.18	26	.0025	.10	27	.0034	.09	27
PRIESTLY GROUTED CONC.	3840	-	3	1450	-	3	.0015		3	.0031		3
HEGEMIER [†] GROUTED CONC.	1737	.13	21	957	.12	21	.0027	.17	21	.0053	.20	20
KINGSLEY UNGROUTED CONC.	2360	.13	10	1331	.23	10	.0028	.15	10	.0033	.14	10

† includes only prisms with h/t /2   ††† 1 ksi - 6.9 MPa
†† 1 psi = 6.9 kPa

It is evident that the value of peak strain reported by Priestley
(0.0015)(Priestley,Elder 1983) is considerably less than mean values reported by other researchers (Brown, Whitlock 1982; Hegemier et.al. 1978,
Kingsley, Atkinson 1985.   It is suggested that the reason for this
disparity is the single prism configuration studied, (five course,
grouted running-bond), and the small number of specimens tested (three).
A total of 21 prisms were tested in his study, but only three were apparently included in the calculation for the mean for unconfined, unreinforced, slow-strain-rate tests. It may be noted that Hegemier also
reported a relatively low peak strain (0.0018) for prisms of similar
configuration. In the context of the large population of data studied
here, these slender, running-bond prisms appear to define the lower
bound for peak strain.

CONCLUSIONS

    The following conclusions are based on the experimental work perform-
ed in this study along with the published results of previous researchers.
    1. Hollow clay and hollow concrete masonry prisms considered in this
study had similar stress-strain characteristics, and thus may be considered to behave identically under uniaxial compressive loads.
    2. Hollow masonry prism strength varies considerably with prism
configuration and constitative material properties. Like concrete,
masonry strength must be determined individually for specific combinations of materials.
    3. A sample of 134 hollow masonry prisms compiled from four seperate

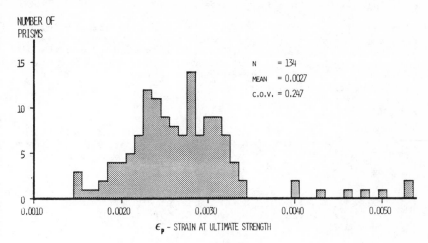

FIGURE 3
DISTRIBUTION OF PEAK-STRAIN DATA FROM STUDIES IN TABLE 5

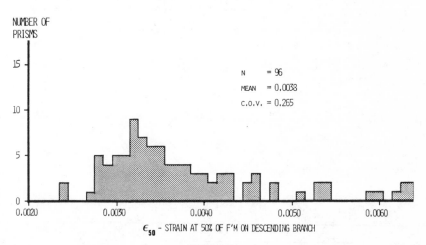

FIGURE 4
DISTRIBUTION OF STRAIN-ON-DESCENDING-BRANCH DATA FROM
STUDIES CITED IN TABLE FIVE

studies showed a mean value of peak strain $\epsilon p = 0.0027$ and a mean value of unloading strain (at 50% of ultimate strength) $\epsilon_{50} = 0.0038$.

The data presented here represents a wide range of prism configurations and materials for clay and concrete hollow unit masonry prisms under uniaxial compressive loads. In order to completely quantify the stress-strain behavior of masonry, however, further study is necessary to investigate the effect of confinement, strain gradients, and strain rate on critical strain parameters.

APPENDIX

Brown, R. H. and Whitlock, A. R., "Compressive Strength of Grouted Hollow Brick Prisms", Masonry: Materials, Properties, and Performance, ASTM STP 778, J. G. Borchelt, Ed., American Society for Testing and Materials, 1982, pp. 99-117.

Hegemier, G. A., Krishnamoorthy, G., Nunn, R. O., and Moorthy, T. V., "Prism Tests for the Compressive Strength of Concrete Masonry", Proc. First North American Masonry Conference, Boulder, Colorado, August 1978.

Kingsley, G. R., and Atkinson, R. H., " A Comparison of the Behavior of Clay and Concrete Masonry in Compression", Report No. 1.1-1, U.S. - Japan Coordinated Program for Masonry Building Research, Sept. 1985.

Priestley, M. J. N. and Elder, D. M., "Stress-Strain Curves for Unconfined and Confined Concrete Masonry", ACI Journal, No. 80-19, May-June 1983.

Uniform Building Code, 1985.

Full Scale Tests of Cladding Components

Marcy Li Wang*

A full scale, six story, steel frame at the Building Research Institute, Tsukuba, Japan was constructed for a series of projects in the U.S.-Japan Cooperative Research Program. This structure provided a unique opportunity to test nonstructural elements subjected to story drifts up to U.S. and Japanese code levels, in a realistic model. The U.S. side focused on precast concrete (PC) and glass fiber reinforced concrete (GFRC) cladding elements; the Japanese side tested internal nonstructural elements as well as cladding. This paper discusses several aspects of U.S. cladding performance revealed by the test, particularly those related to panel stresses induced by drift and affected by connection type.

Introduction

As part of the U.S.-Japan Cooperative Research Program, both Japanese and U.S. nonstructural components were installed onto a full scale, moment resistant steel frame. Static testing of the frame with external (cladding) and internal elements took place during three weeks of July, 1984 at the Building Research Institute in Tsukuba, Japan. During the static testing, strain gage and displacement transducer readings were automatically recorded at several steps per cycle, and technical staff noted joint changes and concrete cracking.

Typical floor height for the six story structural steel frame was 3.4 m. (11 feet); the plan was 15.24 m. (50 feet) square with two equal bays of 7.62 m. (25 feet) in each direction. U.S. cladding was installed on levels 2 and 4.

Objectives

The overall objective of this phase of the U.S.-Japan project is to investigate seismic performance of nonstructural elements as a result of story drift. On the U.S. side, PC and GFRC panels were tested with a variety of panel to frame connections that are commonly used in California.

* Assistant Professor, Department of Architecture
  University of California, Berkeley, CA 94720

Loading of the Steel Frame with Nonstructural Elements

    Loading jacks applied one direction of horizontal displacement on the frame at each story level. The static loading sequence (Figure 1) culminated in a 1/40 story drift ratio which nearly reached the jacks' capacity. This level closely corresponds to U.B.C. design drift requirements for nonstructural, external wall panels. The 1982 edition of the U.B.C. states that, "Connections and panel joints shall allow for relative movement between stories of not less than...(3.0/K) times the calculated elastic story displacement caused by required seismic forces..." Assuming a "K" value of about 0.75, the design drift is 3.0/0.75 x 0.005 = 0.02 or 1/50 story drift ratio. Since story height is typically 3,400 mm., this design drift is 68 mm. or about 2.75 inches. Joints most likely to be subject to such extreme changes in width are located where column covers attached to the column are adjacent to wall panels which are attached to the girders, or where panels attached to orthogonal frames meet at a corner. A range of joint widths were designed for the test, varying from about 20 mm. (3/4 inch) to huge gaps of 70 mm. (2-3/4 inches).

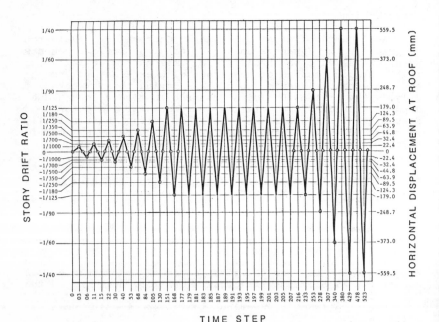

TIME STEP

Figure 1. – Loading Sequence

The Sway Mechanism of Cladding

All the U.S. panels were designed to incorporate the "sway" mechanism. This mechanism, shown in Figure 2, is commonly used as the basis of the design of cladding connections in California, and was adopted in the test design. The intention of the mechanism is to isolate the cladding elements from the steel frame in order to minimize interaction between structure and panels. By accommodating the frame motion, the cladding is less subject to stresses that would lead to failure of the panel or connections.

The sway mechanism requires bearing connections which transfer the panel's gravity load to the frame and lateral connections which have lateral degrees of freedom to accommodate interstory drift.

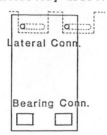

Figure 2. - Sway Mechanism

U.B.C. Seismic Force Design Requirements

The seismic force design values were determined from the 1982 U.B.C. formula (12-8), Fp = Z I Cp Wp. The requirements are summarized as follows:

SEISMIC FORCE DESIGN COEFFICIENTS OF CLADDING & CONNECTIONS

ITEM	(12-8) Cp VALUE	MULTIPLIER	FORCE
cladding panel	.3	x 1	= .3 Wp
connector body	.3	x 4/3	= .4 Wp
bolts, inserts, welds, dowels, etc.	.3	x 4	= 1.2 Wp

Panel and Connection Types

Three frame sides ("A" and "C" were parallel to load, and "1" was perpendicular to load) had U.S. cladding. The second floor wall panels had "clip angle" bearing connections and "long rod" lateral connections while the fourth floor panels had "tube" bearing connections and "slotted hole" lateral connections. (Figure 3) All column covers had clip angle bearing connections; the fourth floor column covers had short rod lateral connections and the second floor column covers had long rod lateral connections.

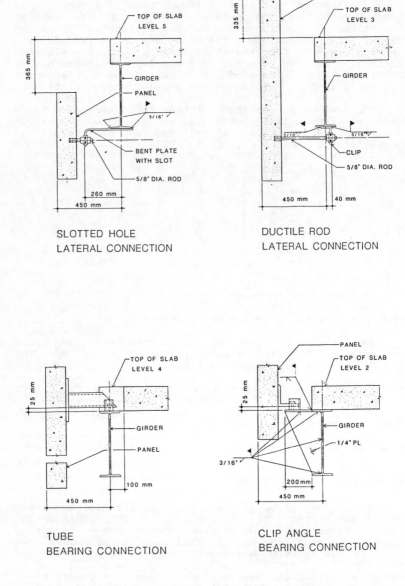

Figure 3. - Lateral and Bearing Connections

Lateral Connection Behavior

The long rod connections were predicted to perform better than the short rod/slotted type. As expected, the short rod/slotted hole connection suffered early problems. It was, in fact, the first to exhibit visible deterioration which quickly accelerated into total failure of some short rods on the frames loaded in the direction in-plane with the panels. (One reason for this degree of failure is due to excessive tightening of the connection by Japanese workmen not familiar with this U.S. detail.) Although some of these connections failed, others worked in sliding to various degrees. The detail is sensitive to installation, adding to its unreliability.

The long rod is much less subject to erection and weathering problems. Even in the most extreme cycles of the static test, this connection had sufficient ductility to accommodate large story drift without indication of impending failure. Drift in itself seems to pose no problem in this connection, however, the static test cannot address the question of whether the rod would have fractured at the panel interface after repeated dynamic load reversals.

Whether or not the lateral connection worked properly or not, it had little if any perceptible effect on the panel stresses. If the detail worked, the lateral motion precluded development of resisting forces at the connection; if the detail did not work, the connection ultimately failed rather than the panel cracked.

Bearing Connection Behavior

The cracking sequence and pattern of the concrete panels revealed the impact of various connections on the panel stresses. Generally, the cracking behavior of panels and performance of connections was not surprising. The greatest amount of cracking occurred near bearing connections, particularly the tube connections, while none was propagated at the lateral connections. From the cracking patterns of the concrete panels, the tube appeared to be less ductile than the angle, especially where new concrete filled the tube connection pockets of panel P-1, 4th Floor "1" frame. The difference in cracking on levels 2 and 4 (the wall panels of which respectively used clip angle and tube bearing connections) is radical as shown on the cracked elevations of Frame "1" in Figure 4. No bearing tubes or angles showed visible distortion of connection body, pullout of studs or cracking of welds.

GFRC panels and connections had virtually no damage at all. The two panels were installed on level 2 of "1" frame with long rod, lateral and clip angle, bearing connections. Connection types and loading direction had much to do with the excellent performance of these elements.

Comparison of Model Analysis with Test Results

Strain gages were located on concrete panel surfaces, on rebars, and on the bodies of bearing connections. The data from these gages are being examined to find aspects of behavior not discernable from cracking patterns and other visible evidence. They also allow us to quantify the effects of connections on panels subject to story drift.

Using a Sap 80 finite element program, models of the panels and connections are being generated, then analyzed to find theoretical stresses. The theoretical stresses resulting from analysis of models are compared to actual stresses indicated by the gages. The modeling of the bearing connections is the most crucial in arriving at a successful result (within 10% of actual results); the bearing connection, lateral connection and panel models are all adjusted until a sufficiently close match is achieved. In this way, each type of bearing connection (tube, tube with filled pocket, and angle clip) and lateral connection (long rod, short rod) can be compared to other connections in terms of stiffness and fixity. Thus far, theoretical stresses calculated thus far for connections are in the range of 4% to 25% of actual test results. The models are still being refined in order to decrease the discrepency.

The strain histories at gages located at bearing connections were examined to determine if cracking occurred prior to the first visible cracks. This determined up to what time step in the loading sequence, linear elastic behavior can safely be assumed. (P-1 panel, level 4, "1" frame, cracking was earliest and most severe and governed which step was examined.) Step 84 (-1/250), just prior to the 10 iterations of 1/125 cycles, was selected as the time step at which to compare theoretical with actual strains.

Panels on the "1" frame are loaded perpendicular to plane; joint widths didn't change on this side, thus, additional stresses from panel banging are never a factor. Panels on "1" frame were instrumented on both surfaces with vertically oriented strain gages. On each wall panel, a set of gages are vertically aligned at the bearing, mid height and lateral connection points. Panel P-1, has the most restraining bearing connections on the frame - tubes in concrete filled construction pockets. Panel, P-4, has the same bearing detail except that the construction pocket is left void of concrete. Panels on level 2, P-2 and P-6, both have angle clip bearing connections. The stress histories derived from readings on strain gages (vertically oriented on both sides of panels, can be compared for panels P-1 and P-2 in Figure 5. These plots verify that highest stesses occurred at bearing points, and decreased along the panel height toward the lateral connection points. All gages on panel P-1, (most restrained bearing connection) show the greatest stresses at all time steps.

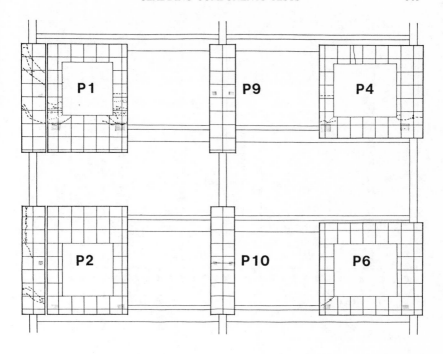

Figure 4. - Cracking Pattern on "l" Frame Panels, Levels 2 & 4

Preliminary Conclusions

The superior performance of long rod lateral connections compared to short rod/slotted connections occurred as expected. The degree of problem with the slotted connection, and the relatively problem free performance of the long rod connection, was in higher contrast than anticipated.

Type of bearing connection used had a great impact on panel concrete stresses and subsequent cracking. It was clear from the cracking pattern that induced concrete stresses were greatest in the case of tubes restrained by concrete, followed by tubes without concrete, and were least in the case of angle clips. The stress histories from the strain gages confirm this observation. The test results indicate that bearing connections which are relatively ductile and less stiff result in better cladding performance.

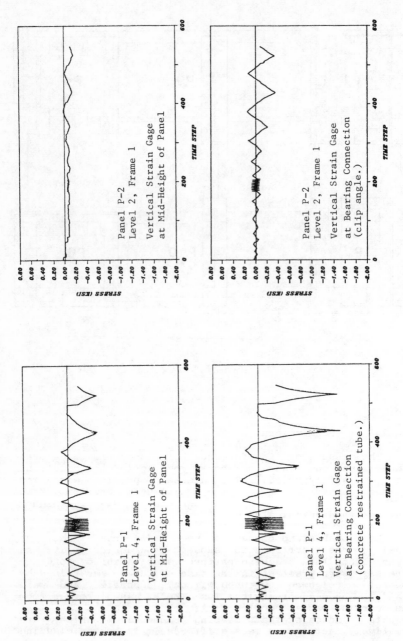

Figure 5.0 — Stress Histories from Strain Gages on Panels P-1 & P-2

System Identification Applied to Pseudo-Dynamic Test Data:
A Treatment of Experimental Errors

James L. Beck,* A.M. ASCE and Paramsothy Jayakumar,** S.M. ASCE

Abstract

A single-input single-output system identification technique, used to estimate the modal properties of a full-scale six-story steel structure tested within its elastic range by the pseudo-dynamic testing method, revealed negative damping for the third mode. This was attributed to the cumulative effect of feedback of experimental errors. In this paper, it is shown that the effect of feedback errors may be accounted for when estimating the modal parameters. By using the control errors, in addition to the ground motion, as inputs to a multiple-input multiple-output system identification technique, the equivalent viscous dampings for energy dissipated by the structure and for the apparent damping from the feedback errors are estimated. The identification results show that the feedback errors add energy to the test structure in its third mode while removing it from the lower modes. The estimates of the modal parameters of the full-scale structure are in very good agreement with those from a 0.3 scale model tested on a shaking table.

Introduction

Under the U.S.-Japan Cooperative Earthquake Research Program Utilizing Large Scale Testing Facilities, a full-scale two-bay six-story steel structure with eccentric bracing, no non-structural components and detailing according to U.S. practice (Fig. 1) was tested by the pseudo-dynamic testing method at the Building Research Institute, Tsukuba, Japan in 1984 during Phase II of the steel test program. The north bay of the interior frame was eccentrically K-braced while all three frames in the direction of loading were moment resisting. The floor system consisted of a formed metal decking with cast-in-place lightweight concrete acting compositely with the girders and floor beams. The Taft S21W accelerogram of the 1952 Kern County, California, earthquake with peak scaled to 6.5%g was used to excite the structure.

The purpose of this paper is to account for the cumulative effect of the feedback of the experimental errors during the estimation of the modal properties from the test data, and, in particular, to estimate the structural damping and also the apparent equivalent viscous damping produced by the feedback errors. The identification results are compared

---

* Assistant Professor **Graduate Student
  Earthquake Engineering Research Laboratory, California Institute of Technology, Pasadena, CA 91125.

with test results of a 0.3 scale model of the full-scale structure
(Bertero, et al., 1985).

Pseudo-Dynamic Testing Method

The pseudo-dynamic testing method is a relatively new technique
developed by Takanashi, et al. (1975) in which full-scale structures are
subjected to earthquake forces by means of on-line computer control of
hydraulic actuators. This method departs from regular quasi-static test
procedures by utilizing the actual restoring force characteristics of
the structure rather than postulating them a priori.

Consider a multi-degree-of-freedom system excited by ground accelerations $\ddot{z}(t)$. The equation of motion is

$$M\ddot{x} + C\dot{x} + R(x) = F(t) = -M\ddot{z}(t)\mathbf{1} \tag{1}$$

If Eq. (1) is discretized in time using the central-difference scheme,
it becomes

$$[M+C\Delta t/2]x_{i+1} = (\Delta t)^2[F_i-R_i] + 2Mx_i + [C\Delta t/2-M]x_{i-1} \tag{2}$$

Once the mass and damping matrices are prescribed, then the displacement
on the left of Eq. (2) can be calculated from the measured restoring
forces and calculated displacements at previous time steps. This displacement is imposed on the structure using hydraulic actuators and the
resulting restoring forces are measured using load transducers. This
information is fed back to an on-line computer to calculate the displacements to be imposed on the structure at next time step. The test
set-up available in the Building Research Institute, Tsukuba, Japan to
implement this method is shown in Fig. 2 (Okamoto, et al., 1983).

Experimental Errors in Pseudo-Dynamic Testing

The displacements calculated from Eq. (2) cannot be imposed exactly
on the test structure. Therefore, in practice an error tolerance bound
is defined within which the structure can either overshoot or undershoot
the calculated displacements at each time step. This may result in restoring forces being fed back to the computer for further computations
which do not correspond to the desired displacements. Although these
control errors, together with measurement errors, may be small at each
time step, the cumulative effect of feedback of these errors may
severely affect the response of the higher modes.

Let the displacement control errors and force measurement errors at
time step i be x_i^{ec} and R_i^{em} respectively, then the restoring forces measured will be

$$R_i = K(x_i+x_i^{ec}) + R_i^{em} \tag{3}$$

where K is the stiffness matrix for the elastic behavior of the test
structure. Substitution of Eq. (3) in Eq. (2) gives

$$[M+C\Delta t/2]x_{i+1} = (\Delta t)^2 E_i + [2M-(\Delta t)^2 K]x_i + [C\Delta t/2-M]x_{i-1}$$
$$-(\Delta t)^2[Kx_i^{ec} + R_i^{em}] \tag{4}$$

If the cumulative error at time step i is x_i^e, then

$$x_i = x_i^* + x_i^e \tag{5}$$

where x_i^* is the ideal (error-free) displacement, which satisfies the following equation:

$$[M+C\Delta t/2]x_{i+1}^* = (\Delta t)^2 E_i + [2M-(\Delta t)^2 K]x_i^* + [C\Delta t/2-M]x_{i-1}^* \tag{6}$$

Subtraction of Eq. (6) from Eq. (4) and the use of Eq. (5) leads to:

$$[M+C\Delta t/2]x_{i+1}^e = (\Delta t)^2[-Kx_i^{ec}-R_i^{em}] + [2M-(\Delta t)^2 K]x_i^e + [C\Delta t/2-M]x_{i-1}^e \tag{7}$$

Comparing Eq. (7) with Eq. (6), it is evident that the cumulative displacement error is the response of the structure to an excitation $[-Kx_i^{ec}-R_i^{em}]$.

System Identification Methods and the Identification Results

Single-Input Single-Output (SI-SO) Technique

The output-error approach for system identification developed by Beck and Jennings (1980) was used to study the elastic test data and to obtain the natural periods, dampings and modeshapes of the first three modes. The details of this study are given in Beck and Jayakumar (1986). The modal parameters estimated for a three-mode model using pseudo-velocity records from the test are given in Table 1. The estimated periods and dampings from each floor record are in excellent agreement, consistent with dynamics theory for linear systems.

The damping C in Eq. (1) represents the numerical damping used in the computer algorithm which was used to calculate the displacements during the test. But the SI-SO system identification method estimates the total damping D in the system according to the equation of motion:

$$M\ddot{x} + D\dot{x} + Kx = -M\ddot{z}(t)\mathbb{1} \tag{8}$$

Hence the damping D represents the sum of the numerical damping C and the equivalent viscous dampings modelling the damping due to some hysteretic action in the structure and the apparent energy effects of the feedback errors. It is seen from Table 1 that the latter effect led to negative total damping for the third mode in the elastic pseudo-dynamic test.

Multiple-Input Multiple-Output (MI-MO) Technique

The SI-SO technique was extended by Beck (Werner, et al., 1985) to the case where simultaneous excitations and responses were measured at several degrees of freedom. Assuming that force measurement errors are negligible in comparison with the force effects of displacement control errors, i.e., $||R_i^{em}|| \ll ||Kx_i^{ec}||$ in Eq. (7) where $||\cdot||$ is the usual

vector norm, the difference equation (7) suggests using the following
differential equation as a structural model:

$$M\ddot{x} + E\dot{x} + Kx = -M\ddot{z}(t)\mathbf{1} - Kx^{ec} \tag{9}$$

Let x_i^{em} be the displacement measurement error at the ith time step, then
the measured displacement x_i^m will be

$$x_i^m = x_i + x_i^{ec} + x_i^{em} \tag{10}$$

where x_i is the displacement calculated from Eq. (2). If it is assumed
that the measurement errors x_i^{em} are much smaller than the control errors
x_i^{ec} in the sense that $||x_i^{em}|| \ll ||x_i^{ec}||$, then x_i^{ec} can be calculated
from known quantities by using:

$$x_i^{ec} = x_i^m - x_i \tag{11}$$

In addition, x in Eq. (9) can be assumed to have significant contributions from only a small number, NM, of classical modes, so that:

$$x = \sum_{r=1}^{NM} x_r(t) \tag{12}$$

where the modal response x_r satisfies

$$\ddot{x}_r + 2\zeta_r\omega_r\dot{x}_r + \omega_r^2 x_r = -P_r f(t) \tag{13}$$

where P_r is the effective participation factor matrix for the rth mode
corresponding to the multiple inputs:

$$f(t) = \begin{bmatrix} \ddot{z}(t) \\ x^{ec}(t) \end{bmatrix} \tag{14}$$

The modal parameters, denoted by θ, are estimated by minimizing a
measure-of-fit $J(\theta)$ defined as:

$$J(\theta) = \frac{1}{V} \sum_{n=1}^{NR} \sum_{i=0}^{T/\Delta t} [a_n(i\Delta t) - \ddot{x}_n(i\Delta t;\theta)]^2 \tag{15}$$

where

$$V = \sum_{n=1}^{NR} \sum_{i=0}^{T/\Delta t} [a_n(i\Delta t)]^2 \tag{16}$$

$a_n, \ddot{x}_n$ = Pseudo-acceleration from the test, and model
acceleration at degree-of-freedom n

T = Length of the record considered

NR = No. of degrees of freedom at which
the response was measured.

As shown in Eqs. (13) and (14), the displacement control errors at all six floors and the specified ground motion record serve as inputs, and the outputs to be matched are the pseudo-acceleration from all six floors (i.e. NR=6). A three-mode model is considered (i.e. NM=3).

The damping E in Eq. (9) as determined from this method represents the sum of the numerical damping C and the structural damping, since the effects of the experimental errors are now explicitly taken into account. The estimated modal parameters are given in Table 2. The pseudo-acceleration of the test structure at the roof and the model accelerations corresponding to the optimal parameters agree very well, as seen in Fig. 3 where a representative six-second segment is shown. The identified modeshapes are also in good agreement with those of a 0.3 scale model of the test structure tested on a shaking table by Bertero, et al. (1985), as shown in Fig. 4. The plotted modeshape vectors for the full-scale structure are just the effective participation factors in the columns of Table 2, which in turn are the first columns of each modal matrix P_r, r=1,2,3, appearing in Eq. (13).

The structural damping is the difference between MI-MO damping values and the numerical damping values used in the test. The numerical damping for each mode is determined from the damping matrix C used in the test and the identified modeshapes and periods. The equivalent viscous dampings corresponding to the feedback errors are estimated by taking the difference between the damping values from the SI-SO and MI-MO system identification results. These various damping values along with the 0.3 scale model test values are shown in Table 3. The feedback errors contribute 0.3%, 0.9% and -2.4% damping to the first three modes, respectively. The structural dampings are 0.6%, 0.6% and 0.3% in the first three modes, which are similar for the first two modes to the scale model test values of 0.7% and 0.9% reported by Bertero, et al. (1985). These low values of structural damping are probably because of the absence of energy dissipation by nonstructural components and by radiation into the foundation.

Conclusions

It is apparent that the higher modes can be affected significantly by the feedback of experimental errors in pseudo-dynamic testing. However, the modal properties of the structure may be identified reliably for small-amplitude tests using system identification techniques, and the effective viscous damping values corresponding to the feedback errors may also be calculated. The feedback errors added energy to the third mode, but damped the contributions of the first two modes of the test structure during the elastic test. The identification results compare very well with those from the 0.3 scale model tested on the shaking table at the University of California, Berkeley.

Acknowledgment

This work was partly supported by the National Science Foundation under Grant No. CEE-8119962.

Appendix: References

(1) Beck, J.L. and Jennings, P.C., "Structural Identification Using Linear Models and Earthquake Records," Earthq. Engrg. and Struct. Dyn., $\underline{8}$, 2, pp. 145-160, 1980.

(2) Beck, J.L. and Jayakumar, P., "Application of System Identification to Pseudo-Dynamic Test Data from a Full-Scale Six-Story Steel Structure," to be published in Proc. Int. Conf. on Vibration Problems in Engrg., Xian, China, June 1986.

(3) Bertero, V., et al., "Progress Report on the Earthquake Simulation Tests and Associated Studies of 0.3 Scale Models of the 6-Story Steel Test Structure," Proc. 6th JTCC Meeting, U.S.-Japan Cooperative Research Program Utilizing Large-Scale Testing Facilities, Maui, Hawaii, June 1985.

(4) Okamoto, S., et al., "Techniques for Large Scale Testing at BRI Large Scale Structure Test Laboratory," Research Paper No. 101, Building Research Institute, Ministry of Construction, May 1983.

(5) Takanashi, K., et al., "Non-Linear Earthquake Response Analysis of Structures by a Computer-Actuator On-Line System," Bull. of Earthquake Resistant Structure Research Center, No. 8, Institute of Industrial Science, University of Tokyo, Japan, 1975.

(6) Werner, S.D., Levine, M.B. and Beck, J.L., "Seismic Response Characteristics of Meloland Road Overpass During 1979 Imperial Valley Earthquake," Agbabian Associates, March 1985.

Floor	Mode 1			Mode 2			Mode 3			J
	$\hat{T}$ (sec)	$\hat{\zeta}$ (%)	$\hat{p}$	$\hat{T}$ (sec)	$\hat{\zeta}$ (%)	$\hat{p}$	$\hat{T}$ (sec)	$\hat{\zeta}$ (%)	$\hat{p}$	(%)
Roof	0.553	1.2	1.39	0.191	2.2	-0.53	0.106	-0.05	0.03	0.75
6	0.553	1.2	1.22	0.191	2.2	-0.21	0.106	-0.04	-0.01	0.24
5	0.553	1.2	1.00	0.191	2.0	0.16	0.106	-0.05	-0.05	1.4
4	0.553	1.2	0.77	0.191	2.1	0.42	0.106	-0.06	-0.02	0.62
3	0.553	1.2	0.53	0.191	2.1	0.47	0.106	-0.05	0.02	2.4
2	0.553	1.2	0.30	0.191	2.1	0.36	0.106	-0.05	0.04	9.5

Table 1: Modal Parameters Estimated from Pseudo-Velocity Records Using a Three-Mode Model and SI-SO Technique

Mode	Numerical	Damping (%)				Scale Model
		SI-SO	MI-MO	Feedback Errors	Structural	
1	0.35	1.23	0.90	0.33	0.55	0.67
2	0.58	2.11	1.19	0.92	0.61	0.90
3	2.04	-0.05	2.32	-2.37	0.28	—

Table 3: Estimated Damping Values and Comparison with 0.3 Scale Model Test Values

Mode	1	2	3
Period (sec)	0.553	0.191	0.104
Damping (%)	0.90	1.19	2.32
Floor	Effective Participation Factors		
Roof	1.40	-0.53	0.15
6	1.22	-0.19	-0.05
5	1.01	0.16	-0.17
4	0.78	0.42	-0.05
3	0.53	0.48	0.11
2	0.30	0.37	0.15

Table 2: Modal Parameters Estimated from Pseudo-Acceleration Records Using a Three-Mode Model and MI-MO Technique (J=2.2%)

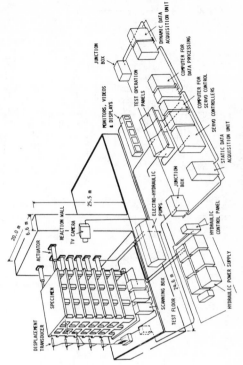

Fig. 2: Pseudo-Dynamic Testing System at the Building Research Institute, Japan

Fig. 1: Elevation of the Interior Frame of Test Structure (All dimensions in mm)

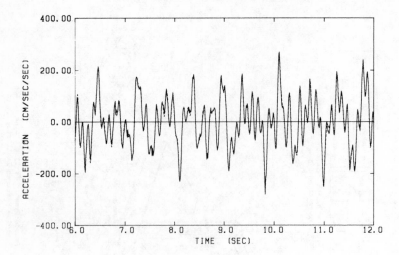

Fig. 3: Pseudo-Acceleration (——) at Roof and
         Calculated Acceleration (---)
         Using a Three-Mode Model Identified
         from Pseudo-Acceleration Records.

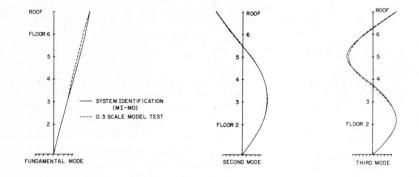

Fig. 4: Comparisons of Three Modes of Test
         Structure

Adaptation of Modal Testing Procedure to Shaker Tables

J-G Beliveau, ASCE Member[1]
F. R. Vigneron, Y. Soucy[2],
and S. Draisey[3]

This article presents a technique for adapting current modal testing procedures to shaker table tests commonly used for qualification tests of nuclear and space related equipment. It involves a modification of the measured frequency response function of the total acceleration, in order to make the assumptions of the modal testing systems applicable to this type of tests. Experimental results on a simple structure are given and demonstrate the simplicity and applicability of this method.

Introduction

Recent general purpose modal parameter estimation algorithms and software are designed to operate with frequency response measurements from dynamic tests, either in a random or sine sweep mode (SDRC, 1984). The structure is either fixed at several points, or alternately, supported on very low frequency mounts, for free-free tests. The measured input is a single force and the measured outputs are accelerations at several points on the flexible structure. Current widely-used algorithms to extract modal information are based on the complex exponential technique. The modal information consists of undamped resonant frequencies, modal damping ratios and elements of the complex or real mode shapes.

In many applications, particularly in the space vehicle and the nuclear industry, structural qualification is often done with dynamic test of a different configuration than that just described. The structure under test is attached to a rigid base that is driven uniaxially by a powerful electrodynamic or hydraulic exciter system. The input measurement is made with an accelerometer mounted to the base and the measured outputs are accelerations at several points on the flexible structure. The estimation algorithms and software referred to above are not directly applicable, because the configurations of measured input, measured output, and the structural dynamics model, are diffent for the two cases.

[1]University of Vermont, Burlington, Vermont 05405
[2]Department of Communications, Ottawa, Ontario, Canada K2H 8S2
[3]Spar Aerospace, Toronto, Ontario, Canada

A simple procedure to obtain the frequency response function for the relative acceleration from the experimentally obtained frequency response function of the total acceleration is developed. Once this is done, current modal testing techniques are applicable. The procedure is applied to a structure undergoing vertical vibration on a shaker table.

Measurements

The frame undergoing dynamic qualification tests while mounted to a movable table as shown in Figure 1 is used as an example. The total displacement at a point on the flexible structure, y, is the sum of the base motion, x_0, and the relative motion with respect to the base, x, which in vector form may be represented as:

$$\{y\} = \{x\} + x_0 \{g\} \tag{1}$$

in which the vector $\{g\}$ is function of the geometry. For the case shown this vector consists of 1's or 0's for accelerometers oriented along the excitation direction and perpendicular to it, respectively. The total velocity and acceleration are also related to relative values

$$\{\dot{y}\} = \{\dot{x}\} + v_0 \{g\} \tag{2}$$

$$\{\ddot{y}\} = \{\ddot{x}\} + a_0 \{g\} \tag{3}$$

with v_0 and a_0 the base velocity and acceleration respectively. In dynamic tests with base excitation, the total acceleration, $\ddot{y}$ and base acceleration, a_0 are measurable variables. $\ddot{x}$, the relative acceleration, is not direclty measurable.

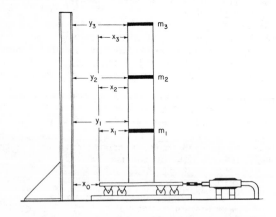

Fig. 1.   Base excitation by Shaker Table

Equations of Structural Dynamics

The kinetic energy of the structure, excluding the base, may be represented by the equation

$$T = \{\dot{y}\}' [M] \{\dot{y}\}/2 \qquad (4)$$

[M] is the mass matrix, and ' represents the transpose. The mechanical energy, U, and energy associated with viscous damping, F, are functions of relative displacements and velocities respectively. Lagrange's equations are then

$$U = \{x\}' [K] \{x\}/2 \qquad (5)$$

$$F = \{\dot{x}\}' [C] \{\dot{x}\}/2 \qquad (6)$$

$$[M] \{\ddot{x}\} + [C] \{\dot{x}\} + [K] \{x\} = -a_0 [M] \{g\} \qquad (7)$$

in which the square matrices [M], [C] and [K] represent the mass, viscous damping and stiffness, respectively, of the structure, excluding the base.

Frequency Response Function

In the case of sinusoidal base acceleration at a radial frequency, ω,

$$a_0 = A_0 e^{i\omega t} + A_0^* e^{-i\omega t} \qquad (8)$$

in which * represents the complex conjugate and i is the imaginary symbol, the steady state total acceleration is given by

$$\{a\} = \{\ddot{y}\} = \{A\} e^{i\omega t} + \{A^*\} e^{-i\omega t} \qquad (9)$$

in which from Eqns. 3-9,

$$\{A\} = A_0\big[[I] + \omega^2 [H] [M]\big] \{g\} \qquad (10)$$

For unsymmmetric matrices or for general viscous damping in which normal modes do not diagonalize the damping matrix, the frequency response, [H], is given by [Beliveau, 1979].

$$[H] = [U] [i\omega I - \lambda]^{-1} [U]^{-1} [G]^{-1} [M]^{-1}$$
$$+ [U^*] [i\omega I - \lambda^*]^{-1} [U^*]^{-1} [G^*]^{-1} [M]^{-1} \qquad (11)$$

in which $^{-1}$ represents the inverse λ and [U] are the eigenvalues with positive imaginary parts and corresponding matrix of eigenvectors of a quadratic eigenvalue problem, Eqn. 13. [G] is purely imaginary.

$$[G] = [U] [\lambda] [U]^{-1} - [U^*] [\lambda^*] [U^*]^{-1} \qquad (12)$$

$$\left[\lambda^2 [M] + \lambda [C] + [K]\right] \{u\} = \{0\} \tag{13}$$

For real normal modes [H] may be represented in a somewhat simpler form

$$[H] = [\phi] [D]^{-1} [\phi]^1 \tag{14}$$

in which the complex diagonal matrix [D] has elements given by

$$D_{jj} = \Omega_j^2 - \omega^2 + 2 i \zeta_j \Omega_j \omega \qquad j = 1,2 \ldots n \tag{15}$$

"n" is the number of degrees of freedom ϕ the matrix of real normal modes which diagonalize the mass and stiffness matrices.

$$[\phi]^1 [M][\phi] = \{I\} \tag{16}$$

$$[\phi]^1 [K] [\phi] = [\Omega^2] \tag{17}$$

$$[\phi]^1 [C] [\phi] = 2[\zeta] \{\Omega\} \tag{18}$$

$[\Omega]$ and $[\zeta]$ are diagonal matrices of the undamped natural frequencies and associated modal damping ratios. When the linear transformation of real normal modes $[\phi]$ does not diagonize the viscous damping matrix, then complex damped modes, Eqns. 11-13, should be used in formulating the frequency response function [H].

Complex Expontial

For random excitation with zero initial conditions, the Fourier Transform relationship between total acceleration and the base acceleration is (Beliveau, 1986)

$$F\{a\} = F\{a_0\} \left[[I] + \omega^2 [H] [M]\right] \{g\} \tag{19}$$

in which F is the Fourier transform.

$$F(a_0) = \int_\infty^\infty a_0(t) e^{-i\omega t} dt \tag{20}$$

The vector [Q] defined to be

$$\{Q\} = [H] [M] \{g\} \tag{21}$$

may then be obtained from the following relationship

$$\{Q\} = \frac{F\{a\} - F(a_0) \{g\}}{F(a_0) \omega^2} \tag{22}$$

For a damped structure, the inverse Fourier Transform of [H], $F^{-1}[H]$, Eqn. 11 is

$$[h] = F^{-1}[H(\omega)] = (1/2\pi) \int_{\infty}^{\infty} [H(\omega)]e^{i\omega t} d\omega \qquad (23)$$

$$= [U]\{e^{\lambda t}\}[U]^{-1}[G]^{-1}[M]^{-1} + [U^*][e^{\lambda^* t}][U^*]^{-1}[G^*]^{-1}[M]^{-1}$$

h is the impulse response matrix. The inverse Fourier Transform of $\{Q\}$, Eqn. 21, is then given by a sum of complex exponentials

$$\{q\} = F^{-1}\{Q\} = \sum_{j=1}^{n} w_j \{u_j\} e^{\lambda_j t} + w_j^* \{u_j^*\} e^{\lambda_j^* t} \qquad (24)$$

which are damped exponential-sinesoidal expressions in time. For base excitation, the vector $\{w\}$ is given by

$$\{w\} = [U]^{-1} [G]^{-1} \{g\} \qquad (25)$$

For real normal modes u_i is replaced by ϕ_i in Eqns. 11, 12, 24 and 25 and the eigenvalue with positive imaginary root is given by

$$\lambda_j = -2\zeta_j \Omega_j + i\Omega_j \sqrt{1-\zeta_j^2} \qquad (26)$$

The complex exponential algorithm may then be used to determine the corresponding undamped natural frequency, Ω, modal damping ratio, ζ, and elements of the complex, U, or real, ϕ, mode shapes. This is done by finding appropriate parameters to the time series obtained by taking the inverse Fourier Transform of elements of $\{Q\}$.

Simulated Example

The Fourier Transform of the relative accelerations for base excitation is

$$F\{\ddot{x}\} = F(a_0)\omega^2 [H] [M] \{g\} \qquad (27)$$

The amplitude and phase of this complex function is compared to that of the Fourier Transform of the total acceleration, Eqn. 19, for a simple five degree of freedom shear building model in Fig. 2. $F(a_0)$ is equal to a constant in the frequency range shown, and parameters for the structure are given in Table 1, including the five calculated undamped nature frequencies, Ω, and modal damping ratios, ζ. Near the resonants frequencies the amplitude of total and relative motion are close to each other as would be expected for a lightly damped structure. This is not true, however, away from the resonances. Thus a modal testing procedure based on frequency response functions of the total accelerations, those that are measured, is not appropriate. They should be modified, as expressed in Eqn. 22, before the inverse Fourier Transform is performed. For structures with light damping, this modification is of lesser importance.

Table 1 – Values of Parameters for Simulated Example
(Five story shear building)

PARAMETERS	VALUES	UNITS	Ω (RAD/SEC)	ζ(%)
$k_1 - k_5$	2 000	kN/m		
$m_1 - m_5$	1	kNs²/m		
c_1	25	kNs/m	12.8	3.7
c_2	5	kNs/m	37.7	7.8
c_3	5	kNs/m	59.1	11.3
c_4	1	kNs/m	75.6	9.6
c_5	0.1	kNs/m	83.3	7.5

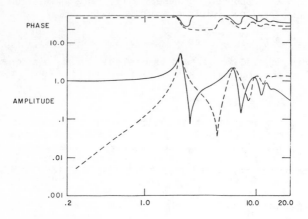

Fig. 2.   Frequency Response Function of Acceleration of Mass No. 1
Simulated Example (_____ total; - - - - - relative).

Experimental Verification

The structure tested, referred to as T-SAT (Fig. 3) was designed
to model the primary structure of a communications satellite.   It
represents a very simplified scaled version of a thrust tube and
equipment shelf configuration.   The thrust tube was manufactured from
carbon composite.   The equipment shelf is of balsa core with aluminum
face sheets.   The test series for the T-SAT exercise included single
point random input at four distinct points and vertical base
acceleration input, both in a random, and a sine sweep mode.

The testing was part of an exercise which involved modal testing,
finite element verification, and substructure coupling.   Further
description of the testing procedure is given elsewhere (Draisey,
1985).

Fig. 3.   T-SAT Structure
          on Shaker Table

Table 2.   T-SAT Modal Testing Parameters

Total   Relative

FREQUENCY (hz)	DAMPING ζ	AMP.	PHASE (rad)	FREQUENCY (hz)	DAMPING ζ	AMP.	PHASE (rad)
23.292	.00981	.4468	-.016	23.291	.01011	.4225	1.140
26.766	.02023	8.452	2.645	26.783	.01929	12.61	2.127
27.038	.02150	5.743	-.018	27.047	.02237	6.873	-1.375
30.258	.02092	15.54	1.168	30.264	.02051	17.43	1.410
30.776	.01775	51.94	-1.397	30.781	.01774	58.31	-1.475
31.199	.01433	16.44	2.933	31.199	.01437	12.92	2.829
75.009	.00513	6.271	2.545	75.011	.00512	6.105	2.669

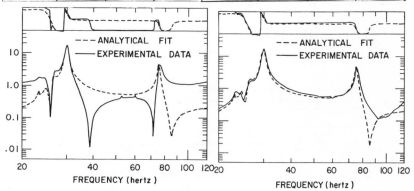

Fig. 4.   F.R.F. Total Acc.   Fig. 5. F.R.F. Relative Acc.

The frequency response functions of accelerations at a number of points was used in the modal testing and the polyreference method was used for modal parameter extraction (SDRC, 1984). Results of the measured and best fit to the frequency response parameter of the total acceleration for vertical motion at one point on the equipment shelf are shown in Fig. 4. Results and best fit for the frequency response functions of the relative acceleration are shown in Figure 5. Modal parameters for the first seven modes of both tests are given in Table 2. Obviously, the fit is much better for relative motion. The modal damping ratios and natural frequencies are, however, very close for both methods, as would be expected in a lightly damped structure. But, the mode shape information is quite different for the two approaches.

Conclusion

The use of uniaxial base excitation for qualification testing in the aerospace and nuclear fields is widespread. Dynamic testing has traditionally been done on fixed or free structures loaded by one or many exciters operating either in a random or sine sweep mode for a given frequency range. Modal extraction techniques and software have concentrated on the latter type of test. The complex exponential technique has been used for parameter estimation. The method proposed in this paper is, however, applicable to other parameter estimation techniques. We have shown that a simple algebraic pre-processing of the frequency response function obtained from base excitation renders this type of test amenable to modern modal testing procedures.

References

Beliveau, Jean-Guy, "First-Order Formulation of Resonance Testing", Journal of Sound and Vibration, Vol. 65, No. 3, 1979, pp. 319-327.

Beliveau, J-G, Vigneron, F.R., Soucy, Y., and Draisey, S., "Modal Parameters Estimation from Base Excitation", scheduled to appear in Journal of Sound and Vibration, Vol. 107, No. 1, July 1986.

Draisey, S., Vigneron, F.R., Soucy, Y. and Beliveau, J-G, "Modal Parameter Estimation from Driven-Base Tests", Paper No. 85-34, Proceedings Second Internation Symposium on Aeroelasticity and Structural Dynamics, Aachen West Germany, April, 1985.

SDRC, Structural Dynamics Research Corporation, "Modal-Plus User's Manual - Section 4, Modal Analysis Theory, 1984.

Problems of Spectral Analysis in Dispersion Tests

R. A. Douglas[1], Member ASCE, and G. L. Eller[2], Assoc. Member ASCE

The use of dispersive surface waves and spectral analysis methods is new to the nondestructive testing of airfield pavements. Such tests are to determine whether mechanical properties of the individual layers of a pavement structure are deteriorating. The problem becomes one of the proper analysis of test results and of the need for careful examination of the problem before applying common techniques of data processing. Dispersion analysis based on conventional methods of spectral analysis and data processing is shown to be inadequate. Special signal processing is required, and a new partial transform.

Introduction

An airfield pavement is an inverted geologic structure. The values of the elastic constants of each successively deeper layer are lower than those of the layer above. This particular inversion, frequently with concrete as the top layer, then a layer of crushed stone, then soil, has ruled out analysis based on seismic reflection or refraction methods.

Surface waves are dispersive in a layered medium. Shortest wavelengths recognize only the concrete surface layer. Longer wavelengths depend into lower layers, wavespeeds reduced by the lessened values of the elastic constants. Analysis of the dispersion provides a measure of the elastic constants for the several layers. The background of this work is described by Jones (1962), and Vidale (1964).

The development of this method is stimulated by (a) desire for a method that is based on fundamental quantities, rather than depending for success upon use of a comparative method to be conducted by very experienced personnel; (b) the need for a method that does not depend upon prior knowledge of the local pavement; (c) need for equipment of light weight that is easily transported, even by backpack if necessary.

The Nature of a Test

A test to obtain the data from which to determine dispersion seems a simple matter. The procedure, to produce a "thump", may be no more than manually lifting, then dropping a weight onto a pad on the surface. Two accelerometers, quick-bonded on the slab, establish a gage length.

The weight is dropped. The mechanical signal travels away from the thump, passing one accelerometer, then the other. The two signals are recorded and the test is over.

[1]Professor, Department of Civil Engineering, North Carolina State University, Raleigh, N.C., 27695-7908.
[2]Research Assistant, Department of Civil Engineering, North Carolina State University, Raleigh, N.C.

Customary Signal Processing

The hallmark of dispersion is, that the shape of the second signal will have changed. This is shown in Figure 1, two successive time-signals from a single thump.

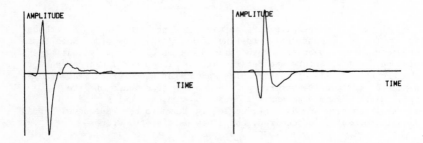

Figure 1.   Signal Shapes, First and Second Gages

Customary signal analysis is to subject the data for each gage to a Fourier transform. The output of primary interest in this work is phase vs. frequency, rather than magnitude vs. frequency. The expression for phase, equation (1), is central to subsequent computations.

$$\text{phase of } \phi = \tan^{-1}[F_{im}(\phi)/F_{re}(\phi)] \tag{1}$$

where ϕ is a particular frequency and $F_{re}(\phi)$ and $F_{im}(\phi)$ are the real and imaginary parts of the transform at that frequency.

On the basis of the frequency and the gagelength, a set of possible wavelengths and corresponding wavespeeds are computed from the phase differences as described by Yew and Chen (1980). The consequence

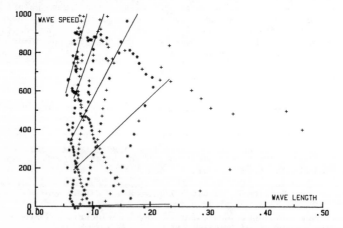

Figure 2. A Computed and Plotted Dispersion Field

of all of the computations over all the frequencies involved, is shown
in Figure 2, the dispersion field, plotting wavespeed (as ordinate)
versus wavelength.

The computations on which such a graph is based are outwardly
simple. The gagelength GL is the difference in the radial distances of
the two gages from the thump. The phase difference for any one
frequency is a measure of the advance of a sinusoid of that frequency
over that distance, so that wavelength λ is given by

$$\lambda = GL/[\Delta\phi/360] \tag{2}$$

Since period T is simply the inverse of the frequency, the wavespeed C_R is

$$C_R = \lambda f = [GL \cdot f]/[\Delta\phi/360] \tag{3}$$

The snake in the computational garden begins to become visible at
this point, by the ambiguity in equation (3). It is not clear that a
phase difference of 20°, actually is 20°. The true phase difference
could be that value plus multiples of 360 degrees. Other variations
appear when the leads and lags are considered. An expert system cannot,
yet, accurately select, at each frequency, the single proper value.
It is best to plot, at each frequency, the possible values and then
identify the appropriate related points.

Difficulties in Spectral Analysis Caused by the Physical Problem

One problem is associated with the surface layer. A concrete
surface layer is jointed, rarely more than 25 ft. (7.6m) on a side.
The shear wave velocity is customarily 6000-8000 ft/sec (1800-
2200 m/sec) so "clean time" before a reflection can return from a
boundary and influence the two records is less than .004 second.
Figure 3 shows signals containing reflections from the sides of a slab.

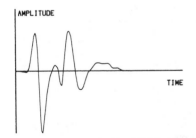 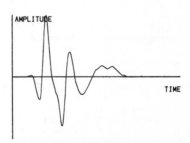

Figure 3. First and Second Signals Containing Reflections

Referring to equation (3), the computation of phase is based on
the summed contributions of the sine and of the cosine terms over the
entire signal, at each frequency considered. The reflections never
pass the gages from the same aspect. The reflections "see" a smaller

gagelength and contributions to phase will be unlike the part of the
wave that passes the gages directly.

The next problem stems from the analytical solution for a disturbance on the surface of a layered halfspace. Jones (1962) and by
Vidale (1964) show the dispersion field that must result, Figure 4.
A number of branches appear, associated with different kinds of behavior, and different layers. At the top of the field, for instance, two
curves depart in different directions from a single value. They are
the principal branches of the Lamb solutions for plate waves,
associated with the surface layer itself.

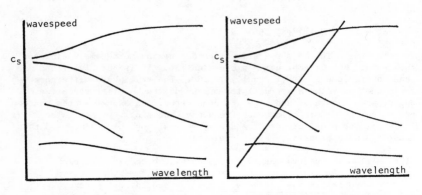

Figure 4. Elements of a   Figure 5. Line of Constant
     Dispersion Field   Frequency in the Field

The problem that arises from this multi-leaved solution is shown
in Figure 5. Lines of constant frequency in wavespeed – wavelength
space are straight lines radiating outward from the origin. A line of
constant frequency may intersect several of the different leaves of the
solution and thus should display, from experimental results, several
different sets of wavespeed/wavelength combinations. Conventional
processing displays only one set, a homogenized result of the several
sets of actual solutions. The problems described may be overcome by
adjustments to experimental technique or to the signal processing. The
influence of the reflection on phase may be overcome by replacing the
broadband input by medium-band input. The reflection is more distinct
and it may be possible to cut the reflection out of the signal.

The really troublesome problem is that the Fourier transform only
provides one phase value for each frequency. We must process each
frequency of the transform by cross-correlation means to identify
different contributions in separate regimes of the signals.

Nature of the Solution

The nature of the solution to the physical problem governs the
nature of the processing involved. The Fourier transform has been
employed successfully to study experimental data from a wide range of
problems. This is confirmation that many problems have solutions that

can be expressed adequately in terms of sine and cosine expansions. It
is not a proof of the validity of universal application of these
methods.

The problem at hand illustrates such a difficulty. A solution for
the problem of an impulse on the surface of a layered halfspace is
described by Ewing, Jardetsky, and Press (1957). That solution,
obtained by the method of separation of variables, is not a general
solution; it yields nodal points and thus the forms of standing waves.
It has been applied successfully to problems involving traveling waves
since the large argument description of the Bessel functions is in
terms of sinusoids. Here, the nearfield is involved. Long wavelengths,
low wavespeeds, and limited record length permit only a few zeroes of
the Bessel functions by the end of an acceleration record. Adequate
description of travelling waves cannot be accomplished under these
conditions.

A Discrete One-Sided Hankel Transform for Experimental Data

The wavelengths in this application cover nearly two orders of
magnitude. The work of Weinstock (1970) and Tasi (1973) suggested to
us that a useful approach would be to continue analysis by modified
Fourier methods for all but the longer wavelengths, then use a transform involving the Bessel function J_0 for arguments only out to
several zeroes.

The consequences of using Fourier methods for a solution described
by Bessel functions is shown by synthesizing acceler ometer records for
first and second gages, then performing the conventional analysis for
dispersion based on the Fourier transform. Figure 6 shows the result.
The dispersion field was computed from signals based on the Bessel
function J_0 with only a few arguments. It looks remarkably like
ordinary solutions.

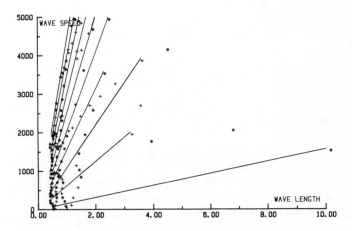

Figure 6. Dispersion Field Based on Synthetic Signals

Use of this J- side of a Hankel transform requires computational protocols. The Fourier transform involves a single element of a cross-correlation computation and includes wraparound of the sinusoid. The protocol established by us was to perform the more extensive cross-correlation of J_0 with the accelerometer signal, but without wraparound. To insure "fairness" in performing the cross-correlations, no comparison receiving more value than another, a truncated Bessel function is used, never extending beyond the end of the data. A limit was set on the extent of the cross-correlation computation so that each involved the sum of the same number of products of ordinates.

The principal drawback to this transform lies in increased computational time. The Fourier transform is beautifully economical of computation, compared with other methods of correlation.

An advantage that does accrue is that the choice among multiple solutions is reduced, from a large number of choices, to a few. Wavespeed is found by dividing gage length by transit time, which is the difference between the locations of the maxima of the cross-correlations for the records from two gages.

Acknowledgements

The work described has been supported by the United States Air Force Engineering & Services Center, Tyndall Air Force Base, Florida.

Appendix. -- References

1. Douglas, R. A. and Eller, G. L., "Soil Properties at Depth from Surface Measurements," Proc. Sym. on Interaction of Non-Nuclear Munitions with Structures, Panama City Beach, FL, April, 1985.

2. Ewing, W. M., Jardetsky, W. S., and Press, F., Elastic Waves in Layered Media, McGraw-Hill, 1957.

3. Jones, R., "Surface Wave Technique for Measuring the Elastic Properties and Thickness of Roads: Theoretical Development," Brit. J. Appl. Phys., 13, 21, 1962.

4. Tasi, J., "Far-Field Analysis of Nonlinear Shock Waves in a Lattice," J. Appl. Phys., 44, 4569, 1973.

5. Vidale, R. F., The Dispersion of Stress Waves in Layered Media Overlying a Half-Space of Lesser Acoustic Rigidity, Ph.D. Dissertation, U. of Mich., 1964.

6. Weinstock, R., "Propagation of a Longitudinal Disturbance on a One-Dimension Lattice," Amer. J. of Phys. 38,11289, 1970.

7. Yew, C. H., and Chen, C. S., "Study of Linear Wave Motions Using FFT and its Potential Application to Nondestructive Testing," Int. J. Eng. Sci. 18, 8, 1027, 1980.

STIFFNESS PARAMETERS IDENTIFICATION OF FRAMES

Morteza A.M. Torkamani*, M.ASCE
Ahmad K. Ahmadi**

ABSTRACT

Four widely used methods of structural system identification are presented. A new formula and its derivation, based on the Bayesian estimation theory, is given. Time domain responses of a frame when subjected to the ground motion of El-Centro earthquake are computed and transformed to the frequency domain. Frequencies and mode shapes of the frame are extracted from Fourier Spectra. Using these frequencies and mode shapes, a parametric study is conducted, and the methods are compared. The advantages, disadvantages, and problems of each method are discussed.

INTRODUCTION

In the traditional analysis of buildings, only the structural elements are considered. Non-structural elements, i.e., stairs, claddings, partition walls, mechanical systems, etc., are ignored. Alone these non-structural elements have negligible effect on the behavior of a structure; however, the cumulative effect of these elements may be considerable.

Full-scale measurements of buildings show that analytical models do not give mode shapes and frequencies which concur with test results. In order to close the gap between the results of analytical and experimental analysis, improved mathematical model of buildings is needed. The purpose of this paper is to present a new method of stiffness parameters identification, which incorporate non-structural and structural elements into mathematical models and updates the stiffness matrix for shear buildings.

Eigenvalues and eigenvectors of linear mathematical models are used as response quantities. These mode shapes and frequencies may be obtained from earthquake input or from data processing of measured displacements, velocities and/or accelerations of full-scale or model testing. Since it is assumed that a modal superposition method with predefined modal damping ratio is used in dynamic analysis, identification of damping coefficients is not considered in this paper.

Parameter identification methods, basically, fall into two categories. They either attempt to construct a structural model using only the measured excitation and responses or they seek to verify or

*Associate Professor, Department of Civil Engineering, University of Pittsburgh, Pittsburgh, Pennsylvania 15261.

**Graduate Student, Department of Civil Engineering, University of Pittsburgh, Pittsburgh, Pennsylvania 15261.

modify a pre-established analytical model using available experimental data. The method in this paper takes the latter approach. Measured test data (independent of any constraints on the location of the sensors) is used to identify the mode shapes and frequencies. A second set of mode shapes and frequencies is calculated considering the analytical stiffness matrix. A system identification technique is then used to estimate equivalent stiffness coefficients.

FORMULATION OF THE PROBLEM

The following discussion is restricted to shear buildings, which are emphasized in this paper; however, the theory may be applied to any planar frame building with joint rotations and axial column displacements provided the dynamic degrees of freedom are lateral displacements. If so, the total stiffness matrix of the frame is superposition of two stiffness matrices: 1. constant stiffness matrix, which represent the structural as well as non-structural elements; 2. variable banded stiffness matrix, which represent the variable structural parameters. Although, the total stiffness matrix of the frame is the sum of these two stiffness matrices, parameter identification is applied to the second stiffness matrix only.

Undamped free vibration equation of motion for a n-story shear building is

$$[M]\{\ddot{X}\} + [K]\{X\} = \{0\}, \tag{1}$$

where,

$\{\ddot{X}\}, \{X\}$ n x 1 vector of physical accelerations and displacements, respectively.

$[M]$ n x n structural mass matrix.

$[K]$ n x n structural stiffness matrix.

n is the number of physical degrees of freedom.

The mass and stiffness matrices are both functions of the structural parameters of the system. Therefore, if we denote the structural parameters of interest by $P_1, P_2, \ldots, P_m$ then the eigenvalues and eigenvectors of the structure will also be functions of $P_1, P_2, \ldots, P_m$, i.e.,

$$[[K] - \lambda_j(p)[M]]\{\phi(p)\}_j = \{0\} \tag{2}$$

where,

$\lambda_j(p)$ j^{th} eigenvalue of structure

$\{\phi(p)\}_j$ n x 1 eigenvector (mode shape) for the jth mode

The eigenvalues and eigenvectors (mode shapes) of structures, like any other mathematically well-behaved, smooth function, can be expanded by a Taylor's series.

$$\left\{ \frac{\{\lambda(p)\}}{\{\phi(p)\}} \right\}_{\{p\}=\{p\}_{i+1}} = \left\{ \frac{\{\lambda(p)\}}{\{\phi(p)\}} \right\}_{\{p\}=\{p\}_{i}} + [S][\{p\}_{i+1} - \{p\}_{i}] + \{0^{+}\} \tag{3}$$

where,

$\{\lambda\}$ 1 x 1 structural eigenvalue vector $1 \le 1 \le n$

$\{\phi(p)\}$ n_1 x 1 structural eigenvector $1 \le n_1 \le n^2$

$\{P\}$ m x 1 structural parameters vector

$[S] = \left[\frac{[\partial\lambda/\partial p]}{[\partial\phi/\partial p]} \right]$ n_2 x m sensitivity matrix $n_2 = 1 + n_1$

rearranging Eq. 3 we obtain Eq. 4.

$$\left\{ \frac{\{\Delta\lambda(p)\}}{\{\Delta\phi(p)\}} \right\} = [S] \{\Delta p\} \tag{4}$$

In Eq. 4 the unknowns are the elements of $\{\Delta P\}$. If $[S]^{-1}$ exists, the solution to the problem is trivial. In other words, if the number of measured eigenvalues and independent elements of eigenvectors entered in the left-hand side of Eq. 3 is equal to the number of parameters, then $[S]$ is a square matrix, i.e., $n_2 = m$. From Eq. 4, it immediately follows that

$$\{\Delta p\} = [S]^{-1} \left\{ \frac{\{\Delta\lambda(p)\}}{\{\Delta\phi(p)\}} \right\} \tag{5}$$

However, this is seldom the case. For the situation where the number of parameters exceeds the number of measured eigenvalues and eigenvectors entered in the left hand side of Eq. 3 (under-determined systems), special mathematical techniques may be employed to calculate an estimation of the parameters. On the other hand, there are cases that the number of measured eigenvalues and eigenvectors exceed the number of parameters (over-determined systems). Since these measured quantities are an estimation of the eigenvalues and eigenvectors, all of them should be considered in the estimation of the parameters. Then, the resulting model has characteristics which simulate the system dynamic response over the frequency range of the measured normal mode data. Therefore, a mapping (estimator) matrix $[W]$ is sought such that the transformation, or mapping, equation may be expressed as:

$$\{\Delta p\} = [W] \left\{ \frac{\{\Delta\lambda(p)\}}{\{\Delta\phi(p)\}} \right\} \tag{6}$$

Since there are in general more unknowns than measurements or vise-versa, the choice of $[W]$ is not unique and one must therefore establish criteria which will be used in the selection of an appropriate $[W]$.

The use of any statistical method depends upon the degree of randomness of variable parameters, measured frequencies and mode shapes. Statistical concepts and estimation theory form the basis of these methods. The most commonly used methods are: 1. Least Squares Structural Identification, which is completely deterministic and does not recognize the possibility that either the test data or the analytical structural model may be random variables (i.e., nonrepeatable); 2. the Weighted Least Squares Structural Identification, considers the structural model to be deterministic but recognizes that test data are random; 3. the Statistical Structural Identification, considers both the structural model and the test data to be random and 4. the Statistical Structural Identification Using an Altered Error Function [2].

The estimator matrix, [W], resulting from the first three methods is shown in Table 1.

Table 1.: Estimator Matrix, [W], for Different Methods

Method		$[W]\{\Delta R\}$
Least Squares Method	Under determined	$[S]^T([S][S]^T)^{-1}\{\Delta R\}$
	Over determined	$([S]^T[S])^{-1}[S]^T\{\Delta R\}$
Weighted Least Squares Method	Under determined	$[S]^T([S][S]^T)^{-1}\{\Delta R\}$
	Over determined	$([S]^T[D_r][S])^{-1}[S]^T[D_r]\{\Delta R\}$
Statistical System Identification		$([C_p]^{-1} + [S]^T[C_r]^{-1}[S])^{-1}[S]^T[C_r]^{-1}\{\Delta R\}$
	or	$[C_p][S]^T([S][C_p][S]^T + [C_r])^{-1}\{\Delta R\}$

Where $[C_p]$ and $[C_r]$ are the diagonal covariance matrices of the errors on prior parameters and measured responses, respectively. $[D_p]$ and $[D_r]$ are the weighting (confidence) matrices for the errors on prior parameters and measured responses which can be assumed $[D_p] = [C_p]^{-1}$ and $[D_r] = [C_r]^{-1}$. Note that in Weighted Least Squares method, under-determined case, the weighting factor is canceled; this indicates that the statistical properties of the measured data do not affect the final result.

The writers have used the Statistical System Identification formula extensively [1, 3]. Experience indicates that when the number of parameters is large, the rate of convergence is very slow. However, in some instances the estimation parameters diverge. In order to resolve this problem, Ref. [2] suggests a new error function in the Bayesian estimation theory for the derivation of the system identification formula. This is

$$E(P) = (\{R^*\} - \{R\}_{n+1})^T[D_r](\{R^*\} - \{R\}_{n+1}) + (\{P^*\} - \{P\}_0)^T[D_p](\{P^*\} - \{P\}_0) \tag{7}$$

where,

$\{P^*\}$ final paramters

$\{P\}_0$ initial parameters

$\{R^*\}$ experimentally measured response

$\{R\}_{n+1}$ predicted response at step n + 1

$[D_r]$ confidence matrix of measured responses

$[D_p]$ confidence matrix of analytical parameters

Considering Taylor series expansion and substituting Eq. 3 into Eq. 7, differentiating the resulting error function with respect to the parameters, and setting the results equal to zero, the following equation is given in Ref. [2]

$$\{p^*\} = \{p\}_{n+1} = \{p\}_n + ([D_p] + [S]_n^T[D_r][S]_n)^{-1}([D_p](\{p\}_0 - \{p\}_n)$$
$$+ [S]_n^T[D_r](\{R^*\} - \{R\}_n)) \tag{8}$$

A NEW FORMULA

We suggest that the error function, Eq. 7, be modified to the following form

$$E(P) = (\{R^*\} - \{R\}_{n+1})^T[D_r](\{R^*\} - \{R\}_{n+1}) + (\{P^*\} - \{P\}_{n+1})^T[D_p](\{P^*\} - \{P\}_{n+1}) \tag{9}$$

The second term on the right-hand side of Eq. 9 will inforce the estimated parameters at each iteration, $\{P\}_{n+1}$, to converge to the desired parameters which are not known, but are estimated. Substituting Eq. 3 into Eq. 9, differentiating the resulting error function with respect to the parameters, and setting the results equal to zero, gives the following

$$\{p\}_{n+1} = \{p\}_n + ([D_p] + [S]_n^T[D_r][S]_n)^{-1}([D_p](\{p^*\} - \{p\}_n) + [S]_n^T[D_r](\{R^*\} - \{R\}_n)) \tag{10}$$

Equation 10 is expressed in terms of the confidence matrices $[D_r]$ and $[D_p]$. Equation 10 may be expressed in terms of covariance matrices $[C_r]$ and $[C_p]$ in the following form

$$\{p\}_{n+1} = [W]_n(\{\Delta R^*\} - [S]_n(\{p^*\} - \{p\}_n)) + \{p^*\} \tag{11}$$

where
$$[W]_n = [C_p][S]_n^T([S]_n[C_p][S]_n^T + [C_r])^{-1} \tag{12}$$

and
$$\{\Delta R^*\} = \{R^*\} - \{R\}_n \tag{13}$$

The covariance matrix of vector $\{\Delta P\}_{n+1}$ may be expressed in the following form

$$Cov(\{\Delta p\}_{n+1}) = [C_p]_n - [C_p]_n[S]_n^T([S]_n[C_p]_n[S]_n^T + [C_r]_n)^{-1}[S]_n[C_p]_n \tag{14}$$

The new structural identification formula is recommended strongly for the cases where the elements of the analytical parameters and measured response have different confidence coefficients, or covariances.

EXAMPLES

In order to demonstrate capability, advantages and disadvantages of the different system identification methods, a three story shear type frame is considered as an example. Parameters to be identified are the stiffness coefficients of the columns. Masses are assumed to be known based on the assumption that calculations of the lumped masses are more accurate than stiffness coefficients. See Fig. 1.

To simulate the time domain responses of this hypothetical frame, (Fig. 1a), N-S earthquake accelerogram of the El-Centro, 1940 is applied to the base of the frame and velocities at each floor are computed. Frequencies and mode shapes of the frame are extracted from FFT of time domain responses. See Table 1.

The measured responses are frequencies and mode shapes, which are presented in Table 1. The exact stiffness parameters of the frame are given on Fig. 1a.

Two analytical models are considered in this example. Example 1 Analytical Model 1 (E1AM1) and Example 1 Analytical Model 2 (E1AM2). Figures 1.b and 1.c show E1AM1 and E1AM2, respectively. All prior column stiffness coefficients of E1AM1 are smaller than the real parameters and in E1AM2 some of the prior column stiffness coefficients are smaller while others are larger than the real parameters.

The covariance matrix of the errors on prior parameters, $[C_p]$, is selected based on the experience and judgement of the analyst as:

$$[C_p] = \begin{bmatrix} 100 & & \\ & 81 & \\ & & 64 \end{bmatrix}$$

Measured eigenvalues have variance of .0441 and elements of the measured eigenvectors have variance of .0004. A typical case, which is the application of the New Statistical System Identification formula (NSSI) in the identification of the stiffness coefficients of the columns is presented here. The cases of two measured eigenvalues are considered. Figure 2 illustrates the results. In the calculation of revised parameters, one has to estimate a value for $\{P^*\}$. In this study $\{P^*\}$ is estimated as the result of the fifth iteration in the Least

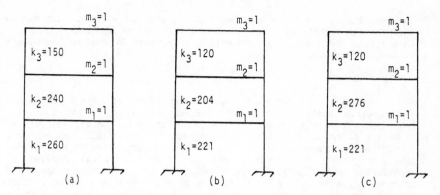

Fig. 1. Three Story Frame: (a) Exact Structure with Real
Parameters: (b) Analytical Model 1: (c) Analytical
Model 2

Table 1: Measured Responses

Frequencies			Mode Shapes		
1	2	3	1	2	3
			0.3656	−1.0023	4.5245
1.0742	2.6856	4.2480	0.6915	−0.8849	−3.6988
			1.0000	1.0000	1.0000

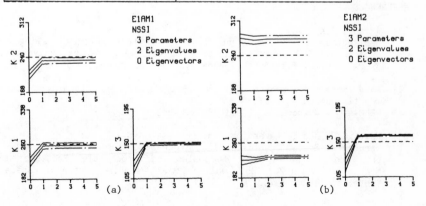

Fig. 2. Parameter Estimation of Three Story Frame Using NSSI Method

Squares method.   One may say that this method is a double estimation theory.   The results of this method are the same as LSM except that for this method the statistical properties of revised parameters, which are valuable information, are known.

CONCLUSIONS

Five methods of structural system identification are presented and a parametric study is conducted.   Based on Ref. [1] the following conclusions have been drawn: 1.   A prior analytical model has a very important effect on the rate of convergence of the estimated parameters. A model with a lower value of prior analytical parameters is recommended over the models of mixed values of prior analytical models; 2. Considering only eigenvalues as a measured response, if the number of measured eigenvalues are equal to the parameters of the system, least squares and weighted least squares will provide unique parameters which are exact; the other method will provide an estimate of parameters.   The accuracy of the estimated parameters depends on the number of iterations and covariances of the errors on measured responses and prior analytical parameters; 3.   If the number of measured eigenvalues is smaller than the parameters, all methods will provide an estimate of real parameters; however, the least squares method will provide more accurate estimated parameters than any statistical identification methods; 4.   Considering the combined use of measured eigenvalues and eigenvectors, one eigenvalue and one eigenvector are used as measured responses; least squares and weighted least squares will converge to exact parameters; other statistical methods will provide estimates of revised parameters. By increasing the number of measured eigenvalues and eigenvectors the rate of convergence will improve and revised parameters would be more accurate.

ACKNOWLEDGEMENT

The writers are grateful for partial support from National Science Foundation Grant CEE-8206909 and to Dr. John B. Scalzi, program director.

REFERENCES

[1]   Ahmadi, A.K.
      Application of System Identification in Mathematical Modeling of Buildings.   PhD thesis, Department of Civil Engineering, University of Pittsburgh, Pittsburgh, Pennsylvania, November, 1985.

[2]   ASTRO/Move; A Finite Element Program for Performing System Identification Analysis
      GELMANCO, [Culver City, California, 1983.]

[3]   Hart, G.C. and Torkamani, M.A.M.
      Structural System Identification, Stochastic Problems in Mechanics. In Proceedings of the Symposium on Stochastic Problems in Mechanics, pages 207-228.   University of Waterloo, Ontario, Canada, September 24-26, 1973.

Optimal Collection of Data for
Parameter Identification

F.E. Udwadia*

INTRODUCTION

The identification of parameters in dynamic models for building
structural systems is a field that is rapidly gaining importance [1-4].
In this paper, we attempt to study the optimal spacing of measurements
for a structural system modelled by a single-degree-of-freedom
oscillator so that the variance of one or more of the parameters being
identified is minimized.

We start with a linear oscillator, and using Fourier transforms,
derive a set of linear algebraic equations. The condition on the
measurement frequencies so that the estimated variance (from noisy
measurement data) of either the mass parameter, the stiffness parameter
or the damping parameter is minimal is derived analytically. It is
also found that there may exist a set of frequencies ω, at which no
additional information on that parameter is available, yielding no
reduction in its estimated variance.

The determination of the optimal measurement frequencies depend
solely on the nature of the forcing function used in the identification
procedure and are invariant with respect to the actual values of the
parameters being estimated. Several of the proofs of the results
obtained have been omitted here due to lack of space. The interested
reader may refer to [5].

Though the results obtained at this time are purely analytical, it is
anticipated that they will help in the design of experiments, especially
where data handling and reduction are a major cost concern.

PROBLEM STATEMENT

Consider a structure modelled by a single-degree-of-freedom system
oscillator subjected to an excitation force q(t). If x is the
displacement of such an oscillator, then its equation of motion is

$$m\ddot{x} + c\dot{x} + kx = q(t), \quad x(0) = 0, \quad \dot{x}(0) = 0 \tag{1}$$

where the parameters m, c and k denote the mass, damping and stiffness
respectively and are assumed to be real numbers. Taking Fourier
transforms, this yields

$$-m\omega^2 X(\omega) + ic\omega X(\omega) + kX(\omega) = Q(\omega) \tag{2}$$

*Professor, Department of Civil Engineering, DRB 394, University of
Southern California, Los Angeles, CA 90089-1114.

535

so that

$$\frac{1}{X(\omega)} = \frac{-m\omega^2}{Q(\omega)} + \frac{ic\omega}{Q(\omega)} + \frac{k}{Q(\omega)} , \quad \omega\varepsilon I_\Omega \quad (3)$$

where we shall assume that the division on both sides of equation (2) is possible i.e., there exists an open interval, I_Ω, such that for all $\omega\varepsilon I_\Omega$, $X(\omega)$ and $Q(\omega)$ are not identically zero.

We shall assume that the parameters m, k and c need to be identified, and shall direct our interest to finding if at all there exist a set of frequencies ω_i, i=1, 2, ... such that the variance of any one of the desired parameters (i.e., m or k or c) can be minimized by using the data (i.e., $X(\omega)$ and $Q(\omega)$)at those specific frequencies.

Relation (3) can be rewritten, after separating the real and imaginary parts, as,

$$\begin{bmatrix} U(\omega) \\ V(\omega) \end{bmatrix} = \begin{bmatrix} -\omega^2 f(\omega) & -\omega g(\omega) & f(\omega) \\ -\omega^2 g(\omega) & \omega f(\omega) & g(\omega) \end{bmatrix} \begin{bmatrix} m \\ c \\ k \end{bmatrix} + \begin{bmatrix} \varepsilon_U(\omega) \\ \varepsilon_V(\omega) \end{bmatrix} \quad (4)$$

where we have represented

$$\left.\begin{aligned} \frac{1}{X(\omega)} &= U(\omega) + i\ V(\omega), \\ \frac{1}{Q(\omega)} &= f(\omega) + i\ g(\omega), \end{aligned}\right\} \quad (5)$$

and the measurements $U(\omega)$ and $V(\omega)$ are corrupted by measurement noise $\varepsilon_U(\omega)$ and $\varepsilon_V(\omega)$. Defining,

$$[a_j(\omega)] \triangleq \underline{a}(\omega) \triangleq [-\omega^2 f(\omega) \quad -\omega g(\omega) \quad f(\omega)]^T, \quad \text{and}$$

$$[b_j(\omega)] \triangleq \underline{b}(\omega) \triangleq [-\omega^2 g(\omega) \quad \omega f(\omega) \quad g(\omega)]^T \quad (6)$$

for each value of $\omega=\omega_i$, i=1, 2, ... N, equation (4) can be written and the BLUE estimator obtained. This can be expressed by the relation

$$\underline{z} = H\ \underline{\theta} + \underline{\varepsilon} \quad (7)$$

where

$$\underline{z} = [U(\omega_1)V(\omega_1)\ U(\omega_2)\ V(\omega_2)\ ...\ U(\omega_n)\ V(\omega_n)]^T$$

$$H = [\underline{a}(\omega_1)\ \underline{b}(\omega_1)\ \underline{a}(\omega_2)\ \underline{b}(\omega_2)\ ...\ \underline{a}(\omega_n)\ \underline{b}(\omega_n)]^T$$

$$\underline{\varepsilon} = [\varepsilon_U(\omega_1)\ \varepsilon_V(\omega_1)\ \varepsilon_U(\omega_2)\ \varepsilon_V(\omega_2)\ ...\ \varepsilon_U(\omega_n)\ \varepsilon_V(\omega_n)]^T$$

$$\underline{\theta} = [m\ c\ k]^T \quad (8)$$

Assume that the error vector $\underline{\varepsilon}$ has the statistic

$$E[\underline{\varepsilon}] = 0, \quad \text{and} \quad R = E(\underline{\varepsilon}\ \underline{\varepsilon}^T) = \text{diag}\ (\sigma_1^2\ ,\ \sigma_1^2\ ,\ \sigma_2^2\ ,\ ...\ \sigma_n^2\ ,\ \sigma_n^2) \quad (9)$$

where $\sigma_i = \sigma(\omega_i)$, i = 1, 2, ..., n.

The covariance of the BLUE estimate of the vector $\underline{\theta}$ becomes [5]

$$P = (H^T R^{-1} H)^{-1} \tag{10}$$

where P is a 3×3 matrix whose diagonal elements P_{11}, P_{22} and P_{33} are the variances in the estimate of m, c and k respectively.

OPTIMAL CHOICE OF FREQUENCIES, ω_k

We shall now attempt to choose the frequencies $\omega_k \, \varepsilon I_\Omega$ in such a way that the i-th element of P, P_{ii}, is minimized. To that end we first differentiate relation (10) with respect to ω_k, to yield

$$\frac{\partial P_{rr}}{\partial \omega_k} = -\frac{2}{\sigma_k^2} \left\{ P[\underline{\dot{a}}(\omega_k)\underline{a}^T(\omega_k) + \underline{\dot{b}}(\omega_k)\underline{b}^T(\omega_k)]P \right\}_{rr}$$

$$+ \frac{2\dot{\sigma}(\omega_k)}{\sigma^3(\omega_k)} \left\{ P[\underline{a}(\omega_k)\underline{a}^T(\omega_k) + \underline{b}(\omega_k)\underline{b}^T(\omega_k)]P \right\}_{rr}. \tag{11}$$

Thus the condition that P_{rr} be extremal then yields

$$\left[\sum_j P_{rj}\dot{a}_j(\omega_k)\right] \left[\sum_s P_{sr}a_s(\omega_k)\right] + \left[\sum_j P_{rj}\dot{b}_j(\omega_k)\right] \left[\sum_s P_{sr}b_s(\omega_k)\right]$$

$$- \frac{\dot{\sigma}(\omega_k)}{\sigma(\omega_k)} \left[\left(\sum_j P_{rj}a_j\right)^2 + \left(\sum_j P_{rj}b_j\right)^2\right] = 0. \tag{12}$$

We next present two results, the proofs of which are omitted (see [5]).

Lemma 1: If a real $\omega_k \, \varepsilon I_\Omega$ exists such that $f(\omega_k)$ and $g(\omega_k)$ are not zero, and for any $r \, \varepsilon [1,2,3]$,

$$\sum_s P_{sr} \, a_s(\omega_k) = 0 , \tag{13}$$

then

$$\sum_s P_{sr} \, b_s(\omega_k) = 0 \text{ for that value of r,} \tag{14}$$

and vice-versa.

Lemma 2:

$$\frac{\sum P_{rj} \, b_j}{\sum P_{rj} \, a_j} = \frac{g(\omega_k)}{f(\omega_k)} , \qquad \sum P_{rj} \, a_j \neq 0, \quad r = 1, 3 \tag{15}$$

$$\frac{\sum P_{rj} b_j}{\sum P_{rj} a_j} = -\frac{f(\omega_k)}{g(\omega_k)} , \quad \sum P_{rj} a_j \neq 0, \quad r = 2 . \tag{16}$$

Theorem 1: For a given forcing function $Q(\omega)$, and any $r \in [1,2,3]$, there exist frequencies, ω_k, such that the inclusion of data at those frequencies does not yield any improvement in the variance P_{rr} of our estimate of parameter r. Specifically, when $\omega_k \in I_\Omega$ satisfies equation (13), $\omega = \omega_k$ is such a frequency.

Proof: Let us imagine that the measurements at the frequencies ω_1, ω_2, ... ω_{k-1} have been made and that with each measurement, the covariance matrix P is updated. After making the kth measurement at $\omega = \omega_k$, the updated covariance matrix becomes [6]

$$P^+ = P^- - P^- H_k^T [R_k + H_k P^- H_k^T]^{-1} H_k P^- . \tag{17}$$

where

P^- denotes the covariance before the measurement at ω_k

P^+ denotes the covariance after the measurement at ω_k

H_k denotes $[\underline{a}, \underline{b}]^T$ evaluated at ω_k

and

$$R_k = \sigma_k^2 \text{ diag. } (1, 1) .$$

Relation (17) can be rewritten using the notation

$$L \triangleq P^- H_k^T \tag{18}$$

as,

$$P^+ = P^- - L[R_k + H_k P^- H_k^T]^{-1} L^T . \tag{19}$$

Using relations (6) and (8), we have,

$$L = \begin{bmatrix} \sum P^-_{1j} a_j & \sum P^-_{1j} b_j \\ \sum P^-_{2j} a_j & \sum P^-_{2j} b_j \\ \sum P^-_{3j} a_j & \sum P^-_{3j} b_j \end{bmatrix}$$

If relation (13) is valid, for some r, then

$$\sum P^-_{rj} a_j(\omega_k) = \sum P^-_{rj} b_j(\omega_k) = 0 \tag{20}$$

so that

$$L_{rj} = 0, \quad j = 1, 2, \text{ for that } r. \tag{21}$$

Consequently, from equation (19) we find

$$P^+_{rj} = P^-_{rj}, \quad j = 1, 2, 3. \tag{22}$$

We note in passing using relations (20) and (22),

$$\sum P^+_{rj} a_j = \sum P^+_{rj} b_j = 0. \square \tag{23}$$

Theorem 2: The optimal locations for measurements $\omega = \omega_k$, $\omega_k \in I_\Omega$, (if at all they exist), for minimizing the variance in the estimates, P_{rr}, satisfy the following relations:

$$\sum P_{rj} \dot{a}_j (\omega_k) + \frac{g(\omega_k)}{f(\omega_k)} \sum P_{sr} \dot{b}_s (\omega_k) - \frac{\dot{\sigma}(\omega_k)}{\sigma(\omega_k)} \left[1 + \left(\frac{g(\omega_k)}{f(\omega_k)} \right)^2 \right] \sum P_{rj} a_j = 0$$

$$r = 1, 3 \tag{24}$$

when $f(\omega_k) \neq 0$.

When $f(\omega_k) = 0$, they satisfy the relations

$$\sum P_{sr} \dot{b}_s (\omega_k) - \frac{\dot{\sigma}(\omega_k)}{\sigma(\omega_k)} \sum P_{sr} b_s (\omega_k) = 0, \quad r = 1, 3 \tag{25}$$

Proof: Using equation (12) and lemma 2, the result follows. A similar result can be written for the cases when $g(\omega_k) \neq 0$ and $g(\omega_k) = 0$ respectively.

Equations (24) and (25) express the criteria for finding observation points ω_k such that the mass m or the stiffness, k, can be optimally identified. We note that the optimal location, ω_k, of the kth observation point depends in general upon the location of all the previous observation points as contained in P_{rj} and P_{sr}.

Theorem 3: The optimal locations for the measurements $\omega = \omega_k$, $\omega_k \in I_\Omega$, if at all they exist, for minimizing the variance of the damping parameter c, satisfy the relation

$$\left[1 - \frac{\omega_k \dot{\sigma}(\omega_k)}{\sigma(\omega_k)} \right] \left[g^2 (\omega_k) + f^2 (\omega_k) \right] = - \frac{\omega_k}{2} \frac{d}{d\omega} [g^2 (\omega) + f^2 (\omega)] \big|_{\omega_k} \tag{26}$$

Proof: Noting relation (12) and lemma 2, the result follows. We
assume that the covariance matrix is strictly positive definite.□

Theorem 3 states that to optimally locate the kth measurement, relation
(26) needs to be satisfied. We note that in this case the optimal
location of the kth measurement does not depend on the locations of
the preceeding measurements and is purely controlled by the nature of
the graphs of $f(\omega)$ and $g(\omega)$.

CONCLUSIONS

In this paper we have tried to understand the optimal measurement
strategy for identifying the parameters of a single degree of freedom
dynamic system. We have shown that in general, data acquired at all
frequencies, in the interval $\omega_k \in I_\Omega$, does not equally enhance our
knowledge of the parameters being estimated. Specifically, one can
often, given a data stream collected at frequencies ω_i, i = 1, 2, ...
k-1, forecast the next frequency at which data collection would be
maximally beneficial to obtaining a more confident estimate of any one
of the desired parameters. Likewise, one can predict the frequency at
which data collection would have no influence on improving the uncertainty in our estimate of any desired parameter. It is shown that for
$\{\omega \in I_\Omega | f(\omega), g(\omega) \neq 0\}$, data at all the frequencies carry information
about the damping parameter, c. Also as opposed to the optimal
measurement location, ω_k, for identification of m and k, which do
depend on the previous measurement locations ω_1, ω_2, ...ω_{k-1}, the
optimal locations for the identification of c, do not depend on the
locations of the measurement stream. They are solely controlled by the
nature of the forcing functions used in the identification procedure.

Irrespective of which parameter is being identified, the optimal
measurement locations do not depend on the values of the parameters.
It is noted that the solution of equations (24) and (26) may not exist
for any ω belonging to the open interval I_Ω. For such situations the
optimal locations would have to be chosen as the end point(s) of the
interval. Also, it is observed that the optimal measurement locations
can be calculated a priori to obtaining the measurement stream. A
numerical example has been included to illustrate the analytical
results obtained.

The results of this paper, it is hoped, will shed light on the
manner in which experimentation can be performed so that the amount of
data handling and reduction required could perhaps be significantly
decreased in the dynamic testing of structural and mechanical systems.

REFERENCES

1. D.E. Hudson, "Resonance Testing of Full-Scale Structure," Proc. ASCE, J. Engg. Mech., June 1964.

2. F.E. Udwadia and D.K. Sharma, "Some Uniqueness Results in Building Structural Identification," SIAM Journ. of Applied Math, 34, 1, January 1978.

3. W.O. Keightley, "Vibrational Characteristics of Earth Dams," Bull. Seismological Soc. Amer., 56, 6, December 1966.

4. P.C. Shah and F.E. Udwadia, "Optimal Sensor Locations in Building Structural Identification," Journ. of Appl. Mech., Vol 45, March 1978.

5. F.E. Udwadia, "Some Results on the Optimal Spacing of Measurements in the Identification of Structural Systems," Quarterly of Applied Mathematics, Volume 43, October 1985.

6. Sage, A.P. and Melsa, J.L., "Estimation Theory with Applications to Communications and Control," McGraw Hill, 1971.

RESEARCH TOPICS IN STRUCTURAL IDENTIFICATION

H. G. Natke[1] and J. T. P. Yao[2], F.ASCE

Abstract

Authors examined a variety of possible research topics including identification and detection of nonlinearities, model improvement approaches, damage identification, power estimation for nonlinearity representations, and effect of data processing techniques on structural identification. The purpose of this paper is to present our viewpoints on these research topics. An overview of the subject area is given along with a description of various aspects of structural identification. Possible approaches are suggested. It is hoped that the discussion of these practically oriented research topics will further stimulate interest in this important and timely subject area.

INTRODUCTION

Background

Structural identification refers to the application of systems identification techniques to structural engineering in general [Rodeman and Yao (1973), Natke (1975 and 1976)]. In addition to the identification of the order and parameters of equations of motion, it is desirable for structural engineers to know the damage state, residual strength, safety level, and reliability of existing structures [Hart and Yao (1977), Liu and Yao (1978), Moses and Yao (1984), and Yao (1985)].

The state of the art of identifying linear structural systems is given by Natke (1982, 1983) and Cottin (1983). In principle, the following three kinds of parameters can be chosen: (i) modal parameters, (ii) physical parameters such as flexibility influence coefficients and (iii) design variables (e.g. linearized). Difficulties can arise in the choice of the damping model and the order of the associated discretized model.

Relevant literature on nonlinear mechanical systems are reviewed by Natke and Gawronski (1984). Toussi and Yao (1983) discussed the problem of identifying nonlinear behavior of existing structures and estimated hysteretic load-deformation relations of small-scale reinforced concrete frames under simulated earthquake loads. Stephens (1985) extended this work to include (a) the estimation of nonlinear load-deformation relations of a full-size building during an actual earthquake and (b) the estimation of structural damage. Stephens, Brady, and Yao (1985)

---

1 Prof. Dr. rer. nat., and Director, Curt-Risch-Institute of Dynamics, Accoustics, and Measurements, University of Hannover, D-3000 Hannover 1, West Germany.

2 Professor of Civil Engineering, Purdue University, West Lafayette, IN 47907.

(Supported in part by NATO Research Grant No. 625/84)

reviewed several techniques for data processing recently.

Because of their common interest in structural identification, both authors decided to collaborate with each other on a collaborative research project (NATO Research Grant No. 625/84), which was awarded to Professor Natke at the University of Hannover. As a part of this collaboration, Professor Natke visited the School of Civil Engineering at Purdue University during the week of 29 July 1985. Professor Yao visited the Institute of Dynamics, Acoustics, and Measurements at the University of Hannover in December 1985.

Objective and Scope

The collaborative project is entitled "Identification of relevant characteristics and mathematical representation of existing systems". The objective of this project is to review methods for order determination of linear and nonlinear systems including those in control theory. An attempt will be made to apply these methods to identify nonlinear structural systems. In addition, several approximate methods will be investigated regarding their application in structural identification. Meanwhile, problems of reduction of test expenditure (measuring only accelerations and to obtain displacements with small measurement errors, etc.), choice of time sub-intervals, dependency on initial conditions must be considered for practical purposes.

In this paper, general aspects of identification as a part of modeling discretized mechanical and structural systems are discussed. Several methods for the identification of nonlinear structures are reviewed. In addition, new methods for power estimation within a polynomial description of the nonlinearities and of nonlinear separable systems are presented. Problems and solutions concerning data processing are outlined along with other problems in structural identification. Finally, several concluding remarks are made.

GENERAL ASPECTS OF IDENTIFICATION

The interrelationship among an existing structure in the real world, its laboratory model(s) in an experimental study, and its mathematical model(s) in an analysis is shown schematically in Figure 1. In the real world, an existing structure is subjected to various loading and environmental conditions (including special test loads). The loads as well as structural response may be measured and recorded through the use of suitable instrumentation. Moreover, the structure may be inspected. At times, appropriate laboratory models may be fabricated using dimensional analysis. Test results may be obtained through the use of instrumentation, and observations may be made and recorded. Meanwhile, the existing structure may be idealized and generalized with a mathematical model. Results of inspection, tests, observation, and calculations are subject to interpretation and decision-making process with due consideration to economical and sociological requirements as shown in Figure 1. As a result, the damage (or safety) level of the structure may be estimated. The results of such studies may also be used to obtain improved mathematical models and/or repair and maintenance strategies.

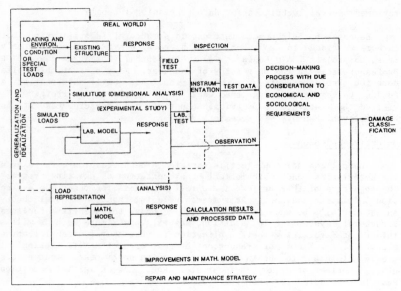

FIG. 1 INTERRELATIONSHIP OF REAL-WORLD, EXPERIMENTAL STUDY, AND ANALYSIS.

Some Comments on Parametrical/Nonparametrical Identification

One of the first steps before applying identification procedures is to look for linear/nonlinear behavior of the structure under test. Difficulties may arise for linear models in the choice of their orders (what are the effective degrees-of-freedom within the frequency interval considered?). It is intended to summarize methods for order determination by the staff of CRI[*]. The detection of nonlinearities must be accompanied by finding their characteristics which result in structure identification (of the model).

Model Improvement Approaches Developed by CRI

References concerning this subject start with Natke et al. (1974) and are summarized in Natke (1982, and mainly in 1983). Additional work is published by Cottin et al. (1984), Natke and Cottin (1985) and Cottin and Natke (to be published). The improvement to the mathematical model is obtained with an assumed structure and with the use of measured data. The procedures can take into account the Bayesian approach and are discussed on the basis of weighted least squares. The residuals cover modal quantities [partial errors, Natke (1983)] and equation error [Natke (1983) and Cottin, Natke (to be published)], input error and output error [Natke (1983), Cottin et al. (1984)]. The latter procedures include dynamic stiffness matrix formulation and frequency response (transfer) matrix formulation [Natke (1984)].

---

[*] Curt-Risch-Institut fur Dynamik, Schall-und Messtechnik, Universität Hannover, FRG

One of the advantages to be mentioned here is the use of subsystem formulation [Natke et al. (1974)]. It results in an economic procedure and the introduced parameters to be improved can be interpreted physically by linearized design variables. On the other hand, there exists only limited experience with the improvement methods [Natke and Cottin (1985)] concerning a few degrees-of-freedom and parameters to be estimated. In most cases, the inertias are assumed to be known a priori.

The present work concerns optimum input design for improvement of mathematical models. Further identification research work will be done in the time domain by Gawronski and Natke ([GN1–6] 1984,1985).

Damage Identification and Evaluation

The state of the art for the condition evaluation and interpretation of existing concrete buildings was reviewed by Yao, Bresler, and Hanson (1984). The current practice of structural assessment requires examination of available building documentation, visual inspection, field testing, laboratory testing, computational analysis, and in-depth study of structural systems and components. Results of these investigations are usually summarized and interpreted by experienced engineers. A proposed methodology as shown schematically in Figure 2 is an expert system, which consists of a knowledge base and an inference machine. The expert system may be used to aid experienced engineers in making their decisions. For demonstration purposes, several pilot expert systems such as SPERIL-I [Ishizuka, Fu, and Yao, (1983)] and SPERIL-II [Ogawa, Fu, and Yao, (1984)] have been developed. However, practical implementation of these systems would require further investigation with close cooperation of experienced engineers [Yao, Bresler, and Hanson (1984)].

AFTER YAO, BRESLER, AND HANSON (1984)

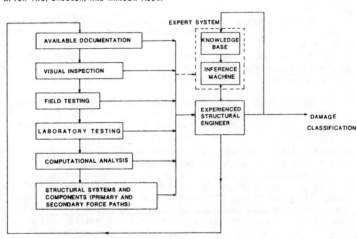

FIG. 2   VARIOUS SOURCES OF INFORMATION FOR
DAMAGE CLASSIFICATION.

As one input of SPERIL, results of nonlinear and hysteretic system identification may be used. Toussi and Yao (1983), Cifuentes (1984) and Stephens (1985) developed methods for the estimation of hysteretic load-deformation relations using records of structural response to earthquake excitations. In addition, Stephens (1985) calculated several damage indices using such results. Because of uncertainties in data processing, material properties, structural complexity, and idealized damage models, the accuracy and reliability of these damage indices are subject to interpretation and should be studied further.

REVIEW OF KNOWN METHODS FOR IDENTIFYING NONLINEAR STRUCTURES

As mentioned earlier, there exist brief reviews in some of the authors' papers. In addition, the book by Eykhoff (1977) is an excellent reference as well as the papers of Unbehauen and that of Billings etc. [see Refs. in the summaries of Natke and Gawronski (1984)]. Further important approaches are mentioned in Natke and Gawronski (1984). The validation of the model (a posteriori) can be proven by parameter significance both physically and statistically. Another proof is, for example, given by the final response test.

Without considering the dispersion function method [Rajbman and Cadeev (1980) and Rajbman (1981)], the situation can be described in such a way that, besides the linearization methods, a parametric model is generally assumed, whereby some of the parameters have no physical meaning. Therefore, the structure identification problem is shifted to parameter estimation. It is an identification in steps, and the transition from one step to the next step (from one structured model to another model) is generally not done algorithmically.

POWER ESTIMATION OF A POLYNOMIAL-ASSUMED MODEL [Natke and Gawronski (1984), Natke (1984)]

The nonlinearity under consideration may be modeled within the class of polynomials. These nonlinearities deal with the restoring and damping forces. Instead of the usual parameter estimation of the polynomials of a priori assumed powers, the powers of the polynomials ought to be determined along with their coefficients. (Recursive determination of the power will not be discussed herein.)

Starting with a very special ansatz (for polynomials up to a power of 5), it is extended to a factorial correction of a prior polynomial approximation. It seems, when looking at the first results, that the method can be used for indicating (as a criterion) the power of the desired equivalent model. The influence of randomly varying measurement errors have yet to be investigated.

Another approach deals with the "economization method" for polynomials known in classical approximation theory using Tshebysheff polynomials. This formula is taken for power estimation with simulated test results in a discrete (considering various points of the interval) and in an integral manner. The results are reasonable even taking into account the simulated measuring errors. Further investigations will be carried out.

The structure (of the model) construction of separable nonlinear time independent functions by equivalent amplification function would be rejected temporarily. The pseudo-linearization (substitution of the powers x^i by variables ξ_i) in order to perform linear regression analysis will not be pursued at present because of additional complications resulting from interdependency among these new variables.

A further proposal (not published up to now) discussed takes pattern from the before-mentioned methods of improvement.

CALCULATION OF DISPLACEMENTS AND VELOCITIES FROM MEASURED ACCELERATIONS

Stephens, Brady, and Yao (1985) studied several techniques for processing noise-polluted data of structural acceleration response to seismic excitation. These techniques include (a) median filtering, (b) least-squares linear trend analysis, (c) frequency filtering, and (d) polynomial trend identification and removal. Structural response data thus treated include those from (a) shake-table tests of small-size multi-story reinforced concrete frames from the University of Illinois and (b) a full-size building during the Imperial Valley earthquake of 15 October 1979. Three different displacement histories of the full-size building were obtained from integrating acceleration data using acceptable data processing techniques. The displacement response thus obtained using a low frequency cut-off of the magnitude resulting from shake-table investigations was rejected by Professor M. E. Sozen of the University of Illinois, who is an expert of dynamic testing and behavior of reinforced concrete structures. Based on his inspection results and damage observations of the building immediately following the earthquake, Professor Sozen concluded that the estimated motions were much larger than those possible. Although reasonable results of data processing may be obtained at present, available information is usually insufficient and incomplete for the conclusive determination of the true but unknown displacement response, which may be important in structural damage assessment.

CONCLUDING REMARKS

The first author had many years of practical experience with the aircraft industry (mainly in aeroelasticity and structural testing) before commencing his teaching and research career in civil engineering. The second author was educated as an experimentalist during the fifties, and is primarily interested in studying problems related to structural safety and reliability. Their common interest in structural identification brought them together in this joint study.

This report is the result of several intense discussions during the week of 29 July 1985, when Professor Natke visited Purdue University. Professor Yao visited the University of Hannover in December 1985. They plan to extend their collaborative research project. Meanwhile, it is hoped that topics as discussed herein will be of interest to other engineers and researchers. If sufficient interest exists or may be generated in the structural engineering profession, an international workshop may be organized to discuss these and other research topics in structural identification.

REFERENCES

Cifuentes, A.O. (1984), System Identification of Hysteretic Structures, Ph.D. Thesis, California Institute of Technology, Pasadena, California

Cottin, N. (1983), Parameterschatzung mit Hilfe des Bayesschen Ansatzes bei linearen elastomechanischen Systemen, Mitteilung des Curt-Risch-Institutes der Universität Hannover, CRI-F-2/1983

Cottin, N., Felgenhauer, H.-P., and Natke, H.G. (1984), On the Parameter Identification of Elastomechanical Systems Using Input and Output Residuals, Ingenieur-Archiv 54, pp. 378-387

Cottin, N., and Natke, H.G. (1985), On the Parameter Identification of Elastomechanical Systems Using Weighted Input and Modal Residuals; (to appear in Ing.-Archiv)

Eykhoff, P. (1977), System Identification-Parameter and State Estimation, John Wiley & Sons, Ltd. London, New York, Sydney, Toronto

Gawronski, W., and Natke, H.G., (GN):
1. Lattice Filters in the Identification of Linear Systems, Curt-Risch-Institut, Universität Hannover, Report CRI-B-1185

2. The Identification of Linear Vibrating Systems by ARMA Modelling (Review), Curt-Risch-Institut, Universität Hannover, Report CRI-B-2185

3. Lattice Filters and Non-Toeplitz Systems of Equations, 13. Internat. Conference on Modelling and Simulation, Lugano (Switzerland), June 1985

4. On ARMA Models for Vibrating Systems, Submitted for publication in Probabilistic Engineering Mechanics

5. On uniformly Balanced Linear Systems, Submitted for publication in International Journal of Systems Science

6. On the Realizations of Transfer Function Matrix, Submitted for publication in International Journal of Systems Science

Hart, G. C., and Yao, J. T. P. (1977), System Identification in Structural Dynamics, Journal of the Engineering Mechanics Div., ASCE Vol. 103, No. EM6, December 1977, pp. 1089-1104

Ishizuka, M., Fu, K. S., and Yao, J. T. P. (1983), "Rule-Based Damage Assessment System for Existing Structures", Solid Mechanics Archives, Vol. 8, pp. 99-118

Liu, S. C., and Yao, J. T. P. (1978), Structural Identification Concept, Journal of the Structural Div., ASCE, Vol. 104, No. ST12, December 1978, pp. 1845-1858

Moses, F., and Yao, J. T. P. (1984), Safety Evaluation of Buildings and Bridges, The Role of Design, Inspection, and Redundancy in Marine Structural Reliability, Edited by D. Faulkner, M. Shinozuka, R. R. Fiebrandt, and I. C. Franck, Committee on Marine Structures, Washington, D.C., 1984, pp. 549-572

Natke, H. G., Collmann, D., and Zimmermann, H. (1974), Beitrag zur Korrektur des Rechenmodells eines elastomechanischen Systems an Hand von Versuchsergebnissen, VDI-Schwingungstagung, Darmstadt, 3.u.4. Oktober 1974, in: VDI-Berichte Nr. 221 (1975), pp. 23-32

Natke, H. G. (1975), Probleme der Strukturidentifikation - Teilubersicht uber Stand- und Flugschwingungsversuchsverfahren, Z. Flugwiss. 23, Heft 4, pp. 116-125 (1975)

Natke, H. G. (1976), Survey of European Ground and Flight Vibration Test Methods, SAE paper 760828, Aerospace Engineering and Manufacturing Meeting, San Diego, Nov. 1976

Natke, H. G. (Editor) (1982), Identification of Vibrating Structures, CISM Courses and Lectures, No. 272, Springer-Verlag Wien, New York

Natke, H. G. (1983), Einfuhrung in Theorie und Praxis der Zeitreihen- und Modalanalyse, Fried. Vieweg & Sohn, Braunschweig/Wiesbaden

Natke, H. G. (1984), Comments on the Difference when Improving the Computational Model via Measured Dynamic Responses and Transfer Functions, Curt-Risch-Institut, Universität Hannover, Report CRI-B-4/84

Natke, H. G., and Gawronski, W. (1984), On Structure Identification within Non-linear Structural Systems, Curt-Risch-Institut, Universität Hannover, Report CRI-B-2/84

Natke, H. G. (1984), On Structure Identification within Non-linear Structural Systems, Addendum 1, Curt-Risch-Institut, Universität Hannover, Report CRI-B-2. 1/84

Natke, H. G. (1984), On Structure Identification within Non-linear Structural Systems, Addendum 2, Curt-Risch-Institut, Universität Hannover, Report CRI-B-2. 2/84

Natke, H. G., and Cottin, N. (1985), Updating Mathematical Models on the Basis of Vibration and Modal Test Results - A Review of Experience, Proceedings, International Symposium on Aeroelasticity, 1-3 April 1985, Aachen

Ogawa, H., Fu, K. S., and Yao, J. T. P. (1984), An Expert System for Damage Assessment of Existing Structures, Proceedings, First IEEE-AAAI Conference on Artificial Intelligence, Denver, Colorado

Rajbman, N. S. (1981), Extensions to Non-linear and Min-max Approaches, Chapter 7 of Trends and Progress in System Identification, ed. by P. Eykhoff, Pergamon Press, Oxford, New York, Toronto, Sydney, Paris, Frankfurt, pp. 185-237

Rajbman, N. S., and Cadeev, V. M. (1980), _Identifikation_ - _Modellierung industrieller Prozesse_, VEB-Verlag Technik, Berlin

Stephens, J. E., (1985), _Structural Damage Assessment Using Response Measurements_, Ph.D. Thesis, School of Civil Engineering, Purdue University, West Lafayette, IN

Stephens, J. E., Brady, G., and Yao, J. T. P., (1985), Techniques of Data Processing in Earthquake Engineering", manuscript, August 1985

Toussi, S., Yao, J. T. P., (1983), "Hysteresis Identification of Existing Structures", _J. Engrg. Mech._, Vol. 109, No. 5, pp. 1189-1202 Existing Structures", J. Engrg. Mech., Vol. 109, No. 5, pp. 1189-1202

Yao, J. T. P., (1985) _Safety and Reliability of Existing Structures_, Pitman Advanced Publishing Program, Boston, London, Melbourne

Yao, J. T. P., Bresler, B., and Hanson, J. M., (1984), "Condition Evaluation and Interpretation for Existing Concrete Buildings", presented at the ACI Spring Convention, Phoenix, AZ (to appear in a special technical publication)

A Discrete Model for the Estimation of the
Elastic Abutment Stiffness

Emmanuel Maragakis*

Abstract

The purpose of this paper is the presentation of a simple method
for the estimation of the abutment stiffness. In this method, the
abutment is represented as a uniform rigid plate with soil deposits on
both sides. Each soil deposit is divided into smaller discrete soil
layers with independently determined constants. The abutment stiffness
is evaluated by using the static equation of equilibrium of the system
in a displaced position.

Introduction

One of the most important parameters for the examination of the
seismic interaction between the bridge deck and the abutments during
earthquakes is the abutment stiffness. The abutment stiffness, in
this case, is defined as the force by which the abutment reacts to a
unit displacement imposed by the bridge deck. The calculation of a
precise value of the abutment stiffness would involve complex three-
dimensional geometry, many degrees of freedom, and the complicated
constitutive relations for the soil. Therefore, a simplified method by
which an approximate value for the abutment stiffness can be found
appears to be very attractive for practical applications (Maragakis,
1984).

The purpose of this paper is to present a simple method by which
the elastic stiffness of the abutment can be estimated. In this
method, the abutment is assumed to behave as a uniform rigid plate with
soil deposits on both sides. Each soil deposit is divided into n
discrete soil layers. Each soil layer is represented by an elastic
translational spring. The required abutment stiffness is the constant
of proportionality, which relates the force imposed by the bridge deck
on the abutment to the displacement of the abutment at the point of the
application of this force. This constant will be evaluated by writing
the equations of equilibrium of the abutment at the displaced position.

Abutment Types

The abutments of a bridge support the ends of the span and retain
the earth behind them. For highway bridges, there are several types of
abutments depending on the material of construction (plain concrete,

---

* Assistant Professor of Civil Engineering, University of Nevada-Reno

reinforced concrete, stone) and on their function (full height abutment, stub or semi-stub abutment, open abutment). The method which follows deals with abutments whose profile can be approximated by the two-dimensional configuration shown in Fig. 1.

Statement of the Problem

The problem to be solved can be briefly summarized as follows: Let W_a be the deflection imposed by the bridge deck on the soil through the abutment and let P_t equal the reaction of the soil on the bridge. The problem is to find an equivalent linear stiffness k_{ab} such that

$$P_t = k_{ab}W_a$$

Basic Assumptions

The approach presented below is based on the following simplifying assumptions.

(a) The problem to be solved is static; consequently, no inertia forces are included in the analysis.

(b) The abutment is assumed to behave as a uniform rigid plate, i.e., deformations due to bending and shear are neglected.

(c) Each soil deposit is represented by n discrete translational springs with independently determined constants. Each spring represents the resistance of a soil layer to the translational movement of the abutment and is assumed to behave only within the elastic range. The yielding of the springs will not be considered.

(d) The resistance of the soil at the bottom of the abutment will be modeled by a torsional spring which resists the rigid body rotation of the abutment.

(e) A soil deposit is considered to be incapable of assuming tensile stresses. The deposit is said to be tensioned only in the sense that its initial compressive stresses are decreased.

(f) The contact between the abutment and the soil is assumed to be frictionless.

Evaluation of the Abutment Stiffness

Consider a strip of the abutment of unit width loaded by a load P per unit width applied at a distance a from the top (see Fig. 2a). Let the displaced position of the abutment be that shown in Fig. 2b, and let W_0 and W_1 be the displacements of the top and bottom of the abutment respectively. If $W(Z_1)$ is the displacement at a depth Z_1, then:

$$W(Z_1) = W_0 - \frac{W_0 - W_1}{\ell} Z_1 \qquad (1)$$

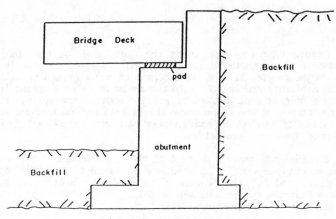

Figure 1. Abutment Profile

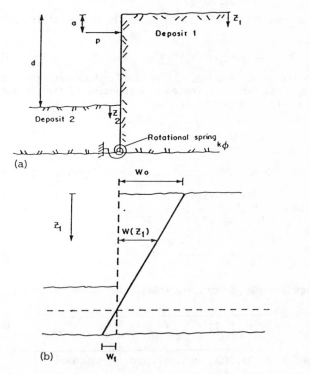

Figure 2. (a) Initial Position of the Abutment with the Soil
Deposits and the Load Applied by the Bridge Deck
(b) Displaced Position of the Abutment

$$\phi = \frac{W_0 - W_1}{\ell} \tag{2}$$

Assume that the deposit on the right side of the abutment is divided in n_r segments, while the deposit on the left side is divided in n_ℓ segments (see Fig. 3). The depths of the segments on either side can be arbitrary and unequal. In the analysis, a soil spring is placed at the middle of each segment of every deposit; the springs represent the resistance of the segments to the lateral movement of the abutment. The values of the spring stiffnesses can be assigned arbitrarily or estimated from soil properties.

Consider now the ith segment of the right deposit. Let the middle point of this segment be located at a distance Z_r^i from the top of the abutment, and let the length of the segment be δZ_r^i (see Fig. 3). Suppose that the spring constant of this segment is $k_{w,r}^i$, then the resisting force of the segment will be:

$$f_r^i = k_{w,r}^i \delta Z_r^i W_r^i \tag{3}$$

The moment of the force f_r^i about the bottom of the abutment will be:

$$m_r^i = k_{w,r}^i \delta Z_r^i W_r^i (\ell - Z_r^i) \tag{4}$$

In the above relations, W_r^i is the displacement of the abutment at depth Z_r^i ; it can be expressed as a function of the displacements W_0 and W_1 from the relation (1). Thus:

$$W_r^i = W_0 - \frac{W_0 - W_1}{\ell} Z_r^i \tag{5}$$

Similarly, for the jth segment of the left deposit, one gets:

$$f_\ell^j = k_{w,}^j \delta Z_\ell^j W_\ell^j \tag{6}$$

$$m_\ell^j = k_{w,}^j \delta Z_\ell^j W_\ell^j (\ell - Z_\ell^j) \tag{7}$$

$$W_\ell^j = W_0 - \frac{W_0 - W_1}{\ell} Z_\ell^j \tag{8}$$

Next, application of the force equilibrium gives:

$$P = \sum_{i=1}^{n_r} f_r^i + \sum_{j=1}^{n} f_\ell^j \tag{9}$$

Combination of (3), (5), (6), (8), and (9) produces

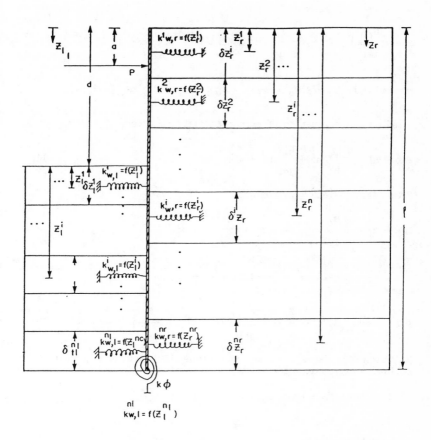

Figure 3. Modeling of the Soil Deposits with Discrete Springs

$$P = \sum_{i=1}^{n_r} k_{w,r}^i \delta z_r^i (W_0 - \frac{W_0 - W_1}{\ell} z_r^i) + \sum_{j=1}^{n_\ell} k_{w,\ell}^j \delta z_\ell^j (W_0 - \frac{W_0 - W_1}{\ell} z_\ell^j$$

or

$$P = (\sum_{i=1}^{n_r} k_{w,r}^i \delta z_r^i - \frac{1}{\ell} \sum_{i=1}^{n_r} k_{w,r}^i \delta z_r^i z_r^i + \sum_{j=1}^{n_\ell} k_{w,\ell}^j \delta z_\ell^j$$

$$- \frac{1}{\ell} \sum_{j=1}^{n_\ell} k_{w,\ell}^j \delta z_\ell^j z_\ell^j) W_0 + \frac{1}{\ell} (\sum_{j=1}^{n_r} k_{w,r}^i z_r^i \delta z_r^i + \sum_{j=1}^{n_\ell} k_{w,\ell}^j z_\ell^j \delta z_\ell^j) W_1$$

(10)

Similarly, the equation of moment equilibrium gives:

$$P(\ell - a) = \sum_{i=1}^{n_r} m_r^i + \sum_{j=1}^{n_\ell} m_\ell^j + k_\phi \phi \tag{11}$$

From (4), (5), and (7), one gets:

$$P(\ell - a) = [\sum_{i=1}^{n_r} k_{w,r}^i [\ell - 2z_r^i + \frac{(z_r^i)^2}{\ell}] \delta z_r^i$$

$$+ \sum_{j=1}^{n_\ell} k_{w,\ell}^j [\ell - 2z_\ell^j + \frac{(z_\ell^j)^2}{\ell}] \delta z_\ell^j + \frac{k_\phi}{\ell}] W_0 +$$

$$[\sum_{i=1}^{n_r} \delta z_r^i k_{w,r}^i z_r^i (1 - \frac{z_r}{\ell}) + \sum_{j=1}^{n_\ell} \delta z_\ell^j k_{w,\ell}^j z_\ell^j (1 - \frac{z_\ell^j}{\ell}) - \frac{k_\phi}{\ell}] W_1 \tag{12}$$

Solution of the system of equations (10) and (12) provides the following expressions for W_0, W_1:

$$W_0 = PA_0$$

$$W_1 = PA_1$$

$$A_0 = \frac{R_1(\ell - a) - T_1}{R_1 T_0 - R_0 T_1} \qquad A_1 = \frac{R_0(\ell - a) - T_0}{R_0 T_1 - R_1 T_0}$$

(13)

$$R_0 = \sum_{i=1}^{n_r} k_{w,r}^i \delta z_r^i + \sum_{j=1}^{n_\ell} k_{w,\ell}^j \delta z_\ell^j -$$

$$\frac{1}{\ell} [\sum_{i=1}^{n_r} k_{w,r}^i \delta z_r^i z_r^i + \sum_{j=1}^{n_\ell} k_{w,\ell}^j \delta z_\ell^j z_\ell^j]$$

$$R_1 = \frac{1}{\ell}[\sum_{i=1}^{n_r} k_{w,r}^i z_r^i \delta z_r^i + \sum_{j=1}^{n_\ell} k_{w,\ell}^j z_\ell^j \delta z_\ell^j]$$

$$T_o = \sum_{i=1}^{n_r} k_{w,r}^i [\ell - 2z_r^i + \frac{(z_r^i)^2}{\ell}] \delta z_r^i +$$

$$\sum_{j=1}^{n_\ell} k_{w,\ell}^j [\ell - 2z_\ell^j + \frac{(z_\ell^j)^2}{\ell}] \delta z_\ell^j + \frac{k_\phi}{\ell} \qquad \text{(13)} \atop \text{cont.}$$

$$T_1 = \sum_{i=1}^{n_r} \delta z_r^i k_{w,r}^i z_r^i (1 - \frac{z_r^i}{\ell}) +$$

$$\sum_{j=1}^{n_\ell} \delta z_\ell^j k_{w,\ell} z_\ell^j (1 - \frac{z_\ell^j}{\ell}) - \frac{k_\phi}{\ell}$$

From equations (13) and (1)

$$W(Z_1) = P(A_o - \frac{A_o - A_1}{\ell} Z_1) \qquad (14)$$

For $Z_1 = a$, equation (14) gives:

$$P = \frac{W_a}{A_o - \dfrac{A_o - A_1}{\ell} a} \qquad (15)$$

The total force P_t is found by multiplying by the foundation width, b. Thus,

$$P_t \frac{b}{A_o - \dfrac{A_o - A_1}{\ell} a} \qquad (16)$$

So, the desired stiffness coefficient is

$$k = \frac{b}{A_o - \dfrac{A_o - A_1}{\ell} a} \qquad (17)$$

Conclusions

A simple method has been presented for the estimation of the elastic abutment stiffness. The method is based on the assumption that the abutment behaves as a uniform rigid plate and as the discrete modeling of the soil deposits on both sides of the abutment.

The discrete modeling of the soil deposits allows the independent evaluation of the properties of each soil segment; and, thus, the method does not depend on any particular assumption regarding the evaluation of the soil properties. The discrete formulation is also particularly convenient for evaluation by small computers and programmable calculators.

References

1. Maragakis, E.A., "A Model for the Rigid Body Motions of Skew Bridges," EERL 85-02, California Technical Institute, Pasadena, California, 1985.

DYNAMIC SOIL-STRUCTURE INTERACTION FOR RC LIFELINES

T. Krauthammer[1], M. ASCE, and Y. Chen[1]

INTRODUCTION

The behavior of buried cylinders merits attention because of the impact of lifeline survivability under earthquake effects, and failures may adversely affect many communities that employ such systems. It has been ascertained that the soil has significant effect on the response of such structures, and vice versa. This interaction effect is mainly arised from the relative flexibility between the soil and structure. The interaction effects on buried cylinders have been extensively studied for the past two decades, but most of them were addressed for circular geometries in the linear domain of behavior.

The prevailing analytical approache for the interaction problems utilizes the so-called half space theory in which the structure is usually assumed to be perfectly rigid. The half space approach limited to linear problems genarally requires the analyst to make sound engineering judgements which are difficult to be achieved because of the lack of useful guidelines from proponents so far. However, in view of the complexity of the problem itself and possible behavioral nonlinearities, the anlytical approach may be conceptually and mathematically difficult for a general dynamic soil-structure interaction (SSI) problem. Another approach could be experimental where full or reduced models would be tested under simulated earthquakes. Here, the problems are not simpler especially since one would attempt to test these systems in the nonlinear domain and the associated costs are expected to be extremely high. Naturally, before such tests are proposed, one would need adequate predictions of the system behavior without which such studies should not be planned. Therefore, use of numerical techniques for the solution of the problem is desirable for the general purpose of this study. Among various available numerical techniques, the finite element method (FEM) has been shown to be more advantageous because of the ease of handling the complicated boundary, various material properties and damping, interface behavior and construction sequence. A brief literature review of its appication on the soil-structure interaction for cylinders is presented as follows.

Brown (8) developed an analytical procedure first and then employed finite element analysis to determine the static pressure distribution on buried culverts under various soil backfills, in which the structure was assumed to be rigid and linearly elastic, and the soil was assumed linearly elastic. Ferritto (10) performed a dynamic interaction analysis to evaluate the structural response of a semi-circular horizontal cylinder covered with a soil berm and subjected to air-blast waves, in which the soil was treated

----

1 Department of Civil and Mineral Engineering, University of Minnesota, Minneapolis, MN 55455

nonlinearly, while the structure was modelled linearly and the material damping was not considered. Anand (1) and Krizek et al. (13) studied the behavior of shallow-buried pipes under various static surface loads and the assumption of linear soil model. Anand et al. (2) studied the response of shallow buried reinforced concrete circular cylinders under simulated earthquake conditions, and compared the response to that induced by static surface loads. The entire analysis was performed in the linear domain on a single geometrical configuration, and there was a separation between the free field response and the structural performance. As a result, it is unlikely that one can learn about the soil-structure interaction effects on the dynamic nonlinear response of rectangular cylinders from that study. Hwang and Lysmer (11) evaluated the seismic responses of underground tunnels in the frequency domain by using a special type of two-dimensional finite elements and making the assumption that that the geometry and material properties were identical to the axis of tunnel structures. Tamura et al. (16) emphasized the importance of twisting deformation for tunnel structures buried in a soft ground with drastically varying characteristics and subjected to earthquakes. However, the adequacy of their approach is restricted to model tunnel structures.

Based on the study of available literature, it reveals that the present understading of SSI effects for rectangular reinforced concrete (RC) cylinders could be inadequate, and the related experimental work is very limited. The main purpose of this work is to obtain an overall understanding of the behavior of the rectangular RC cylinders under earthquake effects through the use of numerical method.

PROBLEM DESCRIPTION

The problem under investigation is a rectangular RC cylinder embeded to the soil medium with a various depth and subjected to the 1940 El Centro earthquake (NS-component). To see the influence of embedment depth on the structural response, three typical values of d/D: 0 (no embedment), 0.5 (partial embedment) and 2.0 (full embedment) were considered, as shown in Figure 1. The general finite element computer program ADINA (3) was used to execute the computations.

In previous studies, researchers decided to separate the structure from the free field in order to reduce considerably the required computer time, similarly to the approach employed by Anand et al. (2), in some publications this procedure is termed the "soil-island" approach. This approach however may not assure that the numerical soil-structure interaction will accurately represent the physical behavior under earthquake conditions. Therefore, it was decided in this study to adopt some of the recommendations in a recent workshop on dynamic soil-structure interaction (7) and to perform the analysis on the entire system.

Also, several assumptions were employed during this study, as follows: 1. The soil medium can be simplified as a single layer with its base resting on a rigid boundary. 2. For this phase of the study, the structure is fully bonded to its surrounding soil during the entire earthquake motion. 3. The cylindrical structure can be idealized as a two-dimensional plane strain model.

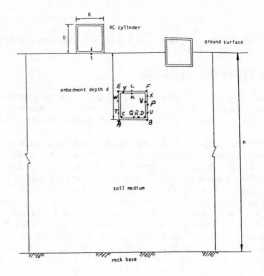

Figure 1. Problem Configuration

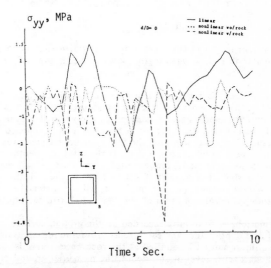

Figure 2. Stress Time Histories

METHODOLOGY

The dynamic equilibrium equations of a finite element system can be expressed by the well-known matrix form

$$[M] \{U\} + [C] \{U\} + [K] \{U\} = \{F(t)\} \qquad (1)$$

where $\{U\}$, $\{U\}$ and $\{U\}$= displacement vector, velocity vector and acceleration vector, respectively. $[M]$, $[C]$ and $[K]$= total mass matrix, toatl damping matrix and total stiffness matrix, respectively, and $\{F(t)\}$= loading vector.

To save considerable computer time and storage while at a small reduction of accuracy, the lumped mass approximation was adopted. The construction of $[C]$ is based on the viscous damping concept. The Lysmer and Kuhlemeyer (15) viscous conditions were imposed on the infinitive (vertical) boundaries to eliminate the artificial wave reflections arised from the finite domain effect. $[K]$ should be reformed in the nonlinear analysis, and the ground displacement was used here instead of the conventional ground acceleration for the input motion $\{F(t)\}$. Because of the various damping and the significant nonlinearities expected in the soil-structure system, a direct step-by-step time-domain integration scheme is needed for the solution of equation (1). Among many available integration schemes, the implicit Newmark constant-average-acceleration method, as discussed in (4) was selected since it is generally better for dynamic problems.

Finite element mesh

The following empirical rule (12) can be used as a thumb of rule to discretize the continuum for a wave propagation problem. Namely

$$L \leq L_w /8, \quad b \leq L_w /2 \quad \text{and} \quad L_w = c/f_{co} \qquad (2)$$

where L= the element length in the wave-propagation direction, b= the element length perpendicular to L, L_w= wave legth, c= wave speed, and f_{co}= cut-off frequency

However, much finer discretization is needed in the structure due to the abrupt change of geometry and the interest in structural response. Since nonreflecting boundary conditions are imposed on the vertical boundaries, the lateral extent of the finite element mesh can be significantly reduced without affecting the results. The total width of the finite element system was therefore chosen to be 30m which is the same as the total soil depth. For the structure, the plain concrete model implemented in ADINA (5) was employed with the consideration of steel confinement effects, while for the soil, the Drucker-Prager with cap model (9) was employed. All the material properties were based on the geological condition of the 1940 El Centro earthquake.

Earthquake motions were usually recorded on the ground surface or at a shallow level of soil depth. Since the soil-structure interaction analysis requires that the input motion be applied at the rock base, where is located certainly at a much deeper level of soil depth, the measured ground motion therefore has to be modified to be used as input excitation at the base. Unlike in conventional analyses, the modification of input motion here was achieved in the time domain whoose basic procedure can be described as the following: (1) perform the free-field (downward) analysis by applying the known motion on the ground surface first, (2) compare the computed response at the base with the known motion and thus obtain a modification factor, and (3) perform the free-field (upward) analysis by applying the modified ground motion (i.e. multiplying the known ground motion by a modification factor) at the base and compare the computed surface motion with the known one to verify the suitability of the modification factor. The modification factor for linear analysis can be simply obtained by performing this procedure once, however for nonlinear analysis repetition of the procedure is possible. As a result, the modification factor of 1.3 and 65 were found to be adequate for linear analysis and nonlinear analysis, respectively. The high magnification of the input for the nonlinear analysis demonstrates the significant energy dissipation which occurs in the system. Such dissipation results from the irreversible deformations in the soil material, combined with material damping.

The determination of time step size t for a nonlinear dynamic analysis is not a trivial task. Some guidelines for it are available (6). For the integration of the dynamic equilibrium equation (1), where the unconditionally stable and implicit Newmark method was employed, t must be small enough to ensure the stability as well as accuracy. Here, it was decided that 5 msec (T_1 /24) was sufficient for the linear analysis, where T_1= the natural period of the first mode, and 0.1 msec for the nonlinear analysis, which is based on Courant's stability criterion. It is noted here that the time step size t can be reasonably increased if only nodal responses (displacement, velocity and acceleration) are of interest, since they are less sensitive to t than stresses.

EXAMPLES

Three typical cases, i.e. no embedment (d/D= 0), partial embedment (d/D= 0.5) and full embedment (d/D= 2) were analyzed in the linear domain. The obtained peak responses along with the corresponding occurence times and locations for the structure in the linear domain are summarized in Tables 1 and 2. The maximum stresses at a number of typical locations in the structure for the linear analysis are presented in Table 3. To study the influence of the rigidity at the base of soil medium on the structural response, a thin layer (6m) of stiff material overlying on the base was considered in the nonlinear analysis for the no-embedment case. The $_{yy}$ time histories from different analyses at the lower right corner of the structure for the no-embedment case is shown in Figure 2.

In order to investigate the possible tensile cracking, crushing and shear slip, the maximum permissible tensile stress 7 f_c', maximum permissible compressive stress f_c' and maximum permissible shear stress 10 f_c' were used for the reinforced concrete structure, which had a 1% longitudinal reinforcement ratio. For the interface, the tension cut-off limit (0.07 MPa) assigned in the Drucker-Prager soil model was used as the maximum permissible tensile stress, the maximum permissible shear stress (0.7 MPa) was obtained by multiplying the shear modulus with the maximum average shear strain reached, and the maximum permissible compressive stress (1.0 MPa) was directly computed from the Mohr-Coulomb friction law. Also in order to evaluate the degree of structural damage, the velocity criteria, as presented in (14) was adopted.

Table 1 Peak Resultant Responses Linear Domain

d/D	PEAK RESULTANT RESPONSES				
	displ [m]	vel [m/s]	acc [m/s^2]	σ_1 [MPa]	σ_2 [MPa]
0	0.240 * pt. E 7.2 sec	0.23 pt. L 4.4 sec	1.64 pt. K 2.8 sec	2.88 El. 1 pt. C 4.3 sec	-2.88 El. 2 pt. D 4.3 sec
0.5	0.240 pt. L 7.1 sec	0.25 pt. E 4.4 sec	9.04 pt. F 2.8 sec	15.37 El. 1 pt. C 4.4 sec	-15.42 El. 2 pt. D 4.4 sec
2	0.241 pt. B 6.9 sec	0.21 pt. B 4.3 sec	1.56 pt. R 2.8 sec	47.79 El. 1 pt. C 4.4 sec	-47.96 El. 2 pt. D 4.4 sec

Table 2 Peak Nodal Responses Linear Domain

d/D	PEAK HORIZ. NODAL RESPONSES			PEAK VERT. NODAL RESPONSES		
	displ [m]	vel [m/s]	acc [m/s^2]	displ [m]	vel [m/s]	acc [m/s^2]
0	0.240 pt. E 7.2 sec	0.23 pt. L 4.4 sec	1.64 pt. K 2.8 sec	0.0078 pt. P 4.3 sec	-0.017 pt. P 1.6 sec	-0.44 pt. V 2.2 sec
0.5	0.240 pt. L 7.1 sec	0.24 pt. E 4.4 sec	-9.04 pt. F 2.8 sec	0.0073 pt. U 4.3 sec	-0.016 pt. W 2.9 sec	1.17 pt. Y 2.8 sec
2	0.241 pt. B 6.9 sec	0.21 pt. B 4.3 sec	1.55 pt. Q 2.8 sec	-0.0056 pt. T 4.3 sec	0.014 pt. X 2.8 sec	-0.46 pt. Y 1.5 sec

Table 3 Peak Element Responses Linear Domain

d/D	PEAK ELEMENT RESPONSES			
	σ_{yy} [MPa]	σ_{zz} [MPa]	τ_{yz} [MPa]	σ_{xx} [MPa]
0	2.32 El. 1 pt A 4.4 sec	-2.66 El. 2 pt D 4.3 sec	-0.66 El. 2 pt D 4.3 sec	-0.73 El. 2 pt D 4.3 sec
0.5	-12.82 El. 2 pt. D 4.4 sec	11.75 El. 1 pt. C 4.4 sec	-5.41 El. 1 pt. C 4.4 sec	-3.85 El. 2 pt. D 4.4 sec
2	-39.27 El. 2 pt. D 4.4 sec	37.77 El. 1 pt. C 4.4 sec	-16.02 El. 1 pt. C 4.4 sec	-11.98 El. 2 pt. D 4.4 sec

* The first line indicates the magnitude of response, the second line referring to Fig. 1 indicates the point for the corresponding response, and the third line provides the occurence time for the corresponding response. This order is maintained for all tables in this paper, unless stated otherwise.

CONCLUSIONS

Based on these numerical results, the following conclusions can be made:

(1) For the simple elastic analysis, the peak stresses of the structure occurred at the lower corners, and increased substantially as the embedment depth was increased. While the peak nodal responses of the structure were not sensitive to the embedment depth.

(2) The peak stresses in the nonlinear domain for the no-embedment case were generally higher than those in the linear domain. This is especially evident when the rock layer was introduced, and the reaseon for it is that the stiffer material causes more energy reach the structure. This general trend is expected to be also true for the partial- and full-embedment cases.

(3) As far as the structural damage is concerened, based on the velocity criteria, it seems tolerable at this stage since no major cracks were observed. However, attention must be paid to the post-event structural integrity due to corrosion of steel reinforcement since the concrete cover might be badly damaged particularly in the full-embedment case.

(4) The interface was expected to exhibit relative soil-structure motions such as sliding and separation, as indicated by stress levels obtained in the present study, especially in the full-embedment case, and thus it may have a serious effect on the system.

(5) As the embedment is increased, additional reinforcement should be properly provided in the corner joints of the structure, especially at the lower corners, to prevent the noted failures.

APPENDIX I - REFERENCES

1. Anand, S.C., "Stress Distributions around Shallow Buried Rigid Pipes", J. struc. Div., ASCE, ST1, Jan. 1974, pp. 161-174.

2. Anand, S.C., Parekh, J.C. and Elling, R.E., "Seismic Loads on Urban Area Tunnel Linings", Proc. 7th European Conf. Earthq. Eng., Sep. 20-25 1982, pp. 329-335.

3. Bathe, K.J., "ADINA- A Finite Element Program for Automatic Dynamic Incremental Nonlinear Analysis", Report No. 82448-1, Acoustics and Vibration Lab., Dept. Mechanical Eng., M.I.T., 1975.

4. Bathe, K.J. and Wilson, E.L., Numerical Methods in Finite Element Analysis, Prentice-Hall, 1976.

5. Bathe, K.J. and Ramaswamy, S., "On Three-Dimensional Nonlinear Analysis of Concrete Structures", Nucl. Eng. Des., 52, 1979, pp. 385-409.

6. Bathe, K.J., Finite Element Procedures in Engineering Analysis, Prentice-Hall, 1982.

7. Beskos, D.E., Krauthammer, T. and Vardoulakis, I., "International Workshop on Dynamic Soil-Structure Interaction, State-of-the-Art and

Research Needs", Report to the National Science Foundation, Univ. of Minnesota, Nov. 1985.

8. Brown, C.B., "Forces on Rigid Culverts under High Fills", J. Struc. Div., ASCE, ST5, Oct. 1967, pp. 195-215.

9. DiMaggio, F.L. and Sandler, I.S., "Material Model for Granular Soils", J. Eng. Mech. Div., ASCE, EM3, June 1971, pp. 935-950.

10. Ferritto, J.M., "Dynamic Analysis of a Cylinder Buried in an Earth Berm", in Appl. FEM Geo. Eng., Vol. III, ed. by C.S. Desai, Proc. held at Vicksburg, Mississippi, May 1972, pp. 913-941.

11. Hwang, R.N. and Lysmer, J., "Response of Buried Structures to Traveling Waves", J. Geo. Eng. Div., ASCE, GT2, Feb. 1981, pp. 183-200.

12. Kuhlemeyer, R.L. and Lysmer, J. "Finite Element Method Accuracy for Wave Propagation Problems", J. Soil Mech. Found. Div., ASCE, 99, SM5, May 1973, pp. 421-427.

13. Krizek, R.J. and McQuade, P.V., "Behavior of Buried Concrete Pipe", J. Geo. Eng. Div., ASCE, GT7, July 1978, pp. 815-836.

14. Langefors, V. and Kihlstrom, B., The Modern Techniques of Rock Blasting, Johnn-Wiley, 3rd edition, 1978.

15. Lysmer, J. and Kuhlemeyer, R.L., "Finite Dynamic Model for Infinite Media", J. Eng. Mech. Div., ASCE, EM4, Aug. 1969, pp. 859-877.

16. Tamura, C. Okamoto, S. and Kato, K., "On Twisting Deformations of a Tunnel in Soft Ground during Earthquakes", VII Sym. Earthq. Eng., Vol. I, U. of Roorke, Nov. 1982, pp. 485-490.

GREEN'S FUNCTION IMPLEMENTATION FOR PILE ANALYSIS

by

Rajan Sen[*], M.ASCE and Prasanta K. Banerjee[**], M.ASCE

Abstract

This paper describes the implementation of a new Green's function based on a layer stiffness approach in a boundary element analysis of harmonically loaded pile groups embedded in a non-homogeneous medium. Two separate problems are addressed : the optimum layer discretization appropriate for the analysis of pile problems and the depth of soil stratum required for simulating a half space. Validity is established by comparisons with previously published static and dynamic solutions.

Introduction

Soil-structure interaction problems are complicated by the need to model the infinitely extending medium surrounding a structure. An approach that is increasingly finding favor, is the application of the boundary element method. By judiciously superposing available fundamental solutions (also referred as Green's functions) of the governing differential equations, the boundary element method can automatically satisfy the boundary conditions for unbounded domains. Consequently, only the structure-soil interfaces have to be modeled and the problem size is greatly reduced.

A major obstacle in the application of the boundary element method in the analysis of harmonically loaded piles is the absence of fundamental solutions of the governing elastodynamic equations, i.e. the dynamic counterpart of Mindlin's solution for the static case. Recourse, has therefore been made to approximate Green's function such as those used by Kaynia & Kausel (1982) or Sen, Davies and Banerjee (1983) in the boundary element analysis of piles embedded in homogeneous soil. Recently, Kausel & Peek (1982) developed an explicit solution for a Green's function corresponding to periodic loads in the interior of a layered stratum. This Green's function has the potential for solving many problems of practical interest since it can represent the variation in soil modulus that occurs in many natural deposits. The implementation of this Green's function for determining the response of pile foundations embedded in a non-homogeneous medium in which the modulus varies linearly with depth is the subject of this paper.

---

[*] Assistant Professor, Department of Civil Engineering and Mechanics, University of South Florida, Tampa, Florida.
[**] Professor of Civil Engineering, State University of New York, Buffalo, NY.

Method of Analysis

A complete description of the hybrid boundary element formulation used for the dynamic analysis of pile foundations in a non-homogeneous medium may be found in Sen, Kausel & Banerjee (1985). Only an outline of the formulation is given here.

The boundary element analysis of the pile-soil response essentially involves coupling of an integral representation of the soil domain with the equations of motion of the piles. Following Banerjee and Butterfield (1981), the integral representation for the displacement field $u_i(\underline{\xi},\omega)$ at any point within the soil domain resulting from surface tractions $\phi_j(\underline{x},\omega)$ at the pile-soil interface can be represented as :

$$u_i(\underline{\xi},\omega) = \int_S G(\underline{x},\underline{\xi},\omega) \, \phi_j(\underline{x},\omega) \, dS(\underline{x}) \qquad (1)$$

where S = soil-pile interface,
      G = dynamic Green's function,
      ω = circular frequency of vibration,
      $\underline{\xi},\underline{x}$ = spatial coordinates

If the piles are represented as compressible columns or flexible beams, the equations of motion for periodic axial and lateral excitation are given by :

$$m\omega^2 u_z + E_p A_p \frac{d^2 u_z}{dz^2} = -\pi d\phi_z \qquad (2)$$

for the axial response and

$$m\omega^2 u_x - E_p I_p \frac{d^4 u}{dz^4} = d\phi_x \qquad (3)$$

for the lateral response where:

    A_p = cross-sectional area of the pile,
    E_p = Young's modulus of the pile material,
    I_p = second moment of area of the pile cross-section,
    d = pile diameter,
    m = mass of pile per unit length.

Equations (2) and (3) are essentially the dynamic counterparts of those derived by Banerjee & Driscoll (1976) for the static behavior of piles and include an additional term representing the inertia forces in the equation of motion.

The integral equation for the soil domain (equation 1) may be solved by discretizing the pile-soil interface into boundary elements and performing the indicated integrations numerically with respect to the nodal values of displacements and tractions, resulting in a linear set of equations:

$$\{u_s\} = [G]\,\{\phi_s\} \qquad\qquad (4)$$

where the subscript s indicates that the equations (4) have been derived from the consideration of the soil response alone.

Since both displacements and tractions are unknown at the pile-soil interface, equation (4) must be coupled with an equivalent set describing the motion of the piles. These are obtained by solving the differential equation of motion for the piles (equations 2 and 3) and substituting the pile head boundary conditions which yield :

$$\{u_p\} = [D]\,\{\phi_p\} + \{b_p\} \qquad\qquad (5)$$

where the subscript p indicates that the quantities refer to the pile domain and $\{b_p\}$ the vector of pile head boundary conditions.

The unknown tractions and displacements are determined by satisfying equilibrium and compatibility at the pile-soil interface assuming no slippage. The forces in the piles can then be obtained from equilibrium. By summing the individual loads and moments acting on the piles in a group for unit vertical, horizontal and rotational harmonic displacements, a 3 x 3 overall group impedance [S] can be determined. This quantity is complex, i.e.

$$S_{ij} = K_{ij} + iC_{ij} \qquad\qquad (6)$$

where K_{ij} is the real part of the impedance representing the stiffness of the system and C_{ij} the imaginary part which represents the damping in the system.

Green's Function for Non-Homogeneous Soils

The entire analysis is of course dependent on the availability of the fundamental solution for a harmonic point force, i.e. $G\,(\underline{x},\underline{\xi},\omega)$ in equation (1). No exact solution for a periodic point force embedded within a layered half space is available in the published literature. An approximate Green's function for harmonic loads applied in the interior of a layered stratum has recently been developed by Kausel & Peek (1982). Variations in soil modulus with depth may be easily incorporated by representing the medium as a layered stratum and specifying a different modulus for each layer.

Details of the formulation and the derivation of explicit solutions for different loading including horizontal and vertical ring and disk loads used in this study are given in Kausel & Peek cited earlier. However, the following restrictions are implicit in their analysis:

a) The soil layers are underlain by rigid rock;

  b) The displacements within each layer are assumed to vary
     linearly;
  c) The solution tends to diverge if the layer thickness is
     greater than one fourth the dominant wavelength excited.

To implement this Green's function in the analysis of floating piles,
it is necessary to devise criterion for simulating a half-space and
also a layer discretization scheme that ensures that the solution does
not diverge.

                         Simulation of Half-Space

     In simulating a half space by a finite stratum, the main concern
is to ensure that propagating waves are not trapped artificially by the
bedrock underlying the soil layer. For the static case, this problem
has been solved by trial and error. Thus, Randolph & Wroth (1979)
found that using a soil layer that was two and a half times the pile
length, finite element results were within 5% of a half space boundary
element solution.
     In this study, the stratum depth is determined from a
consideration of the attenuation distance. The aim is to make the
stratum large enough so that the waves decay before reaching the
bedrock. Since the attenuation distance is a function of the
wavelength, the following approximate expression is used :

$$H = 2.5\lambda + L \qquad\qquad\qquad (7)$$

where H = stratum depth for simulating a half space,
      λ = the wavelength of the shear wave excited,
      L = pile length.

Examination of displacements obtained from the Green's function
indicated that displacements decayed substantially over a distance
2.5λ, a distance that Lysmer (1966) [quoted in Richart et al (1970) on
p 91] had defined as a 'large' distance. Only the shear waves are
considered since the pile analysis is concerned with the immediate
(undrained) response. As the soil is incompressible under this
condition, P-waves contribute little to the soil deformations.
     The validity of (7) was tested by examining two different stratum
depths (H/L = 2 and 4) and comparing the real and imaginary parts of
the vertical pile impedance (K_{11} and C_{11} in equation 6) for the
static and dynamic case. The requirements of (7) were only met for the
dynamic case. A summary of the results for both compressible ($K=E_p/E_s =$
10^2) and rigid ($K=10^4$) piles normalised with respect to those in the
shallower stratum, e.g. $K_{11}^* = K_{11}/K_{11}$ $_{(H/L=2)}$, is shown in Table 1.
It may be seen that the results in Table 1 bear out the predictions
made by (7). For the dynamic case satisfying (7) the effect of
increasing the stratum depth has no effect on the stiffness whereas it
has some effect on the static results that do not meet the criterion.

TABLE 1   Effect of stratum depth on axial static and dynamic pile stiffness

H/L	Static		Dynamic			
	$K=10^2$	$K=10^4$	$K=10^2$		$K=10^4$	
	K^*_{11}	K^*_{11}	K^*_{11}	C^*_{11}	K^*_{11}	C^*_{11}
2	1.0	1.0	1.0	1.0	1.0	1.0
4	0.98	0.96	1.0	1.0	1.0	1.0

Table 1 also illustrates an important conclusion of (7) namely, larger depths are needed for simulating the static and low frequency response of floating piles than the high frequency dynamic response.

Layer Discretization

As the underlying assumption in the derivation of the dynamic Green's function is the linearization of the transcendental functions in the layer stiffness equations, the thickness of the sub-layer (h) should be small in comparison to the wavelength (λ) of the shear wave at the highest frequency considered. The minimum requirement for this is given by Kausel & Peek as:

$$\lambda \geq 4h \tag{8}$$

If the layer thickness does not meet this requirement, the solution diverges. Lysmer & Waas (1972), who pioneered the layer stiffness approach, advocate a very fine mesh near the loaded zone and a coarser one away from it. Their criterion is $\lambda = 12h$, though in examples they actually used $\lambda = 15h$.

Sensitivity studies were conducted to determine the effect of layer discretization on the results. Two different schemes were compared one having 18 layers and the other 9. For the latter case, the criterion in (8) was not satisfied away from the loaded segment. The results indicated that (8) was only critical for the loaded segment and the region close to it.

The discretization scheme adopted, nevertheless complied with (8) for all regions although the mesh near the load point was made finer. In a typical example for the sub-division for the top load point distance d from the ground surface, the following discretization was used : 8 sublayers @ 0.25d (loaded segment) followed by 4 sublayers @ 0.5d, 4 sublayers @ d, 8 sublayers @ 1.5d making 24 sub-divisions over a depth 20d.

Validation

The accuracy of the implementation of the Green's function was assessed by comparisons with previously published solutions for the static and dynamic case.

Comparison with previous static results indicated that earlier boundary element models (Banerjee & Davies, 1978 & Davies, 1979) tended to overestimate the stiffness of the pile-soil system. However, very close agreement was obtained with the finite element analyses reported by Randolph & Wroth, 1978 & Randolph, 1981.

Comparison with previous dynamic results (Krishnan, Gazetas and Velez, 1983) showed excellent agreement. Thus, the comparisons confirm the validity of the implementation used and the Green's function.

Discussion

The implementation of the Green's function in the solution of the pile problem is somewhat laborious since the analysis requires n+1 different discretization schemes, one each for the n shaft segments and an additional one for the base. In a parametric study, where the effect of different frequencies is investigated, each of these n+1 discretizations have to be analysed for each frequency. This can lead to a lot of output, and therefore implementation of the Green's function requires that some thought be given to the organization that is adopted. In this study, separate output files were created in which the results of each of the n+1 discretizations were stored. These files were subsequently combined and given suitable names identifying the frequency and the degree of non-homogeneity in the soil. As the files may be re-used the overall analysis is quite economical.

Conclusions

The implementation of a new Green's function in a hybrid boundary element analysis to determine the response of single piles and pile groups is described. This Green's function allows the soil to be represented as layered stratum in which the modulus in each layer may be varied. In the study, the soil modulus is taken to vary linearly with depth.

To simulate piles in a half space, approximate criterion were developed. Investigations were also conducted to determine a discretization that was appropriate for pile analysis. Comparisons with previously obtained solutions confirmed the validity of the method used to implement the Green's function.

The implementation of the Green's function in a pile analysis is somewhat laborious because of the need to link two separate analyses. The studies carried out only examined fairly small (2 x 2) groups. Further studies are needed to ascertain the accuracy of the Green's function in applications to larger groups. However, the initial experience suggests that this Green's function has considerable potential in the solution of many practical problems.

Acknowledgement

The work reported in this paper was funded by the National Science Foundation under grant no CEE-80-18402. The authors wish to record their indebtedness to Dr Eduardo Kausel of the Massachusetts Institute of Technology for providing the dynamic Green's function used in this study.

References

1. Banerjee, P.K. and Butterfield, R. (1981) Boundary Elements Methods in Engineering Science. McGraw-Hill, London and New York.
2. Banerjee, P.K. and Driscoll, R.M.C. (1976) Three-dimensional analysis of raked pile groups. Proceedings Institute of Civil Engineers, Part 2, Vol 61, 653-671.
3. Banerjee, P.K. and Davies, T.G. (1978) The behaviour of axially and laterally loaded piles embedded in non-homogeneous soils. Geotechnique, Vol 28, 309-326.
4. Davies, T.G. (1979) Linear and non-linear analysis of pile groups. Ph.D Thesis, University of Wales.
5. Kaynia, A.M. and Kausel, E. (1982) Dynamic behavior of pile groups. 2nd International Conference on Numerical Methods in Offshore Piling, Austin, Texas.
6. Kausel, E. and Peek, R. (1982) Dynamic loads in the interior of a layered stratum: an explicit solution. Bulletin of the Seismolog. Society of America, Vol 72, No 5.
7. Krishnan, R., Gazetas, G. and Velez, A. (1983) Static and dynamic lateral deflection of piles in non-homogeneous soil stratum. Geotechnique, Vol 33, No 3, 307-326.
8. Lysmer, J. and Waas, G. (1972) Shear waves in plane infinite structures. J Engng Mechanics Div, ASCE, Vol 98, EM1, 85-105.
9. Randolph, M.F. and Wroth, C. (1978) Analysis of deformation of vertically loaded piles. J. Geotech Engng Div, Vol 104, GT12, 1465-1488.
10. Randolph, M.F. (1981) The response of flexible piles to lateral loading. Geotechnique, Vol 31, No 2, 247-260.
11. Richart, F.E., Hall, J.R. and Woods, R.D. (1970) Vibrations of Soils and Foundations, Prenctice-Hall, Englewood Cliffs, NJ.
12. Sen, R., Davies, T.G. and Banerjee, P.K. (1983) Dynamic behavior of axially and laterally loaded piles and pile groups embedded in homogeneous soil. Report No GT/1983/2, Department of Civil Engineering, State University of New York, Buffalo.
13. Sen, R., Davies, T.G. and Banerjee, P.K. (1983) Dynamic behavior of axially and laterally loaded piles and pile groups embedded in non-homogeneous soil. Report No GT/1983/3, Department of Civil Engineering, State University of New York, Buffalo.
14. Sen, R., Kausel, E. and Banerjee, P.K. (1985) Dynamic Analysis of Piles and Pile Groups Embedded in Non-Homogeneous Soils. To appear in International Journal of Numerical Methods in Geomechanics.

SOIL-STRUCTURE INTERACTION ON STRUCTURAL DYNAMICS

Bulent A. Ovunc* M. ASCE

ABSTRACT

The dynamic analysis of structures under the effect of soil-structure interaction and the effect of member axial force is based on the
continuous mass matrix method, in which the equations of motion are satisfied at any arbitrary point of the structure not only at nodal points.
The member stiffness matrices are developed by considering the actual
mass distribution and the effect of member axial force. For the members
embedded in the soil, the soil reactions and the skin frictions are also
considered as continuously varying over the members. The functions describing the variation of the soil characteristics with the depth are
introduced in the analysis in order to take properly into account the
properties of granular and cohesive soils. The dynamic stiffness matrix
of the members is very much dependent on the ratio of the member characteristics to the soil characteristics. Even beyond a limit ratio of
the characteristics the nature of the dynamic member stiffness matrix
changes entirely. The dynamic member stiffness matrix thus developed is
introduced in the general purpose computer program STDYNL for the practical applications. The modal analysis is used if all the externally applied dynamic forces are proportional to the same temporal function.
Otherwise, the analysis is performed by means of a numerical scheme. It
is noticed that the natural circular frequencies and the corresponding
modal shapes are very sensitive to the foundation model of the structures. Depending on the soil characteristics the sequence of the excitation of the modal shapes changes and some additional modal shapes may
appear. The buckling mode of the structures is obtained as the magnitude
of the static loads increases to tend the natural circular frequency of
that mode to zero.

Introduction

The steel structures compared to reinforced concrete structures are
more flexible systems. Under dynamic loads, the effect of the member axial force and the effect of soil-structure interaction on the vibration
of steel structures are more significant. Viscous dashpots were introduced between the floors of a nine story steel frame building to represent the nonstructural damping (10). Nonlinear dynamic analysis of steel
frames using computer graphics which is based on concise formulation of
tangent stiffness matrix provides a reasonably fast procedure (7). The
justification of the theoretical approach for the wave propagation in
the piles and soil has been given through the results obtained from the

*Professor of Civil Engineering, University of Southwestern Louisiana,
USL 40172, Lafayette, Louisiana 70504-0172

measurements made on the site (2). Most of the actual studies concerning the dynamic carrying capacity of the piles are based on the discretized models (9). The rigid cylindrical inclusion partially embedded in an isotropic homogeneous elastic halfspace and subjected to lateral forces and moments has been modelled by a field of distributed forces (21). Dynamic stress-strain relationships for soils, soil dynamics and it's application to foundation engineering are are included in a state-of-art report (23). The result of a state-of-art study has shown that the parameters which provide the best correlation for piles in sand are the of the depth to diameter and the sand friction angle (8). The nonlinearity due to high strains around the piles and the lack of bond between soil and sand have been studied (1). The most important factors affecting the magnitude of the lateral movement and the bending movement in the pile are the relative flexibility of the pile, the boundary conditions at the head and tip of the pile and the distribution of the soil movement along the pile (20). In some analysis, the soil-structure interaction is taken into account by considering that the interface surface is so smooth that no shearing stress could be transmitted or that is so rough that the soil and structure would be a perfect continuum (4). The unit tip and side resistances showed pronounced increase with the increase in the ratio of depth to diameter (3).

Herein, the dynamic analysis is based on the continuous mass matrix method (11). The soil-structure interaction is taken into account as the deformations of the soil caused by the motion of the structure which in turn modifies the response the response of the structures (12), (16). The friction between the pile embe ded in soil and the soil is assumed such that the geometrical continuity along the interface of the two materials holds till the friction stress along the interface reaches it's maximum value, then the slippage starts (18). The effect of the axial force within the individual members are also introduced in the formulations (17). The general purpose computer program STDYNL is modified to include the recently developed dynamic member stiffness matrix (15), (17). The vibration of the structure may be due to externally applied dynamic loads or due to support motion. The variation of the external dynamic loads by the time may be continuous or discontinuous. The support motion may be determined from the eartquake ground motion (14). If the temporal part of all the external loads are proportional to the same time function, the forced vibration of the structure is performed by modal analysis. Otherwise, Otherwise, the forced vibration is carried out by a numerical integration scheme (13).

Procedure of analysis

The structure is assumed to be linear, elastic whereas the soil characteristics may vary with the depth of the ground (22).

For cohesive soils, the distribution of the soil subgrade coefficient C_s is considered as follows,

$$C_s = k_L (Z/L)^n$$

where, k_L is the value of C_s at the tip of pile ($Z = L$) and n is an empirical constant that depends upon the type of soil.

For sand, the subgrade coefficient C_s varies linearly with depth,

$$C_s = k.Z$$

As skin friction in case of cohesive soils is directly proportional to the surface area of the pile, the axial force is assumed to decrease linearly with the depth. In case of cohensionless soils, as skin friction in general varies linearly with the depth, the axial force distribution is assumed parabolic. For any other variations of soil modules and axial force, the parameters of the corresponding functions are taken into account as data.

The dynamic stiffness matrix of a member is developed by considering the vibration of an infinitesimal element of the member. An arbitrary vibration is obtained by combining the four independent vibrations due to: axial displacements, torsional rotations and bending in two orthogonal planes.

The vibration of a member under bending in Oyz plane can be obtained from the differential equation of motion, as follows (Fig. 1),

$$EI_x \frac{\partial^4 w}{\partial y^4} + \frac{p \partial^2 w}{\partial y^2} + C\frac{\partial w}{\partial t} + m\frac{\partial^2 w}{\partial t^2} + C_s p' = 0 \tag{1}$$

where P, C_s, p' and m are the axial force in member, soil subgrade coefficient, the projection of the cross section and the mass per unit length of the member,

$$m = \frac{A\rho + q}{g}$$

Assuming that the deflection function can be written in separable variable form,

$$w(y,t) = Z(y) \, f(t)$$

and substituting in the equation (1), one has,

$$\frac{d\,Z}{dy^4} + 2k^2\frac{d^2 Z}{dy^2} - \beta^4 Z = 0 \tag{2}$$

where, $k^2 = \dfrac{P}{2EI_x}$ and

$$\beta^4 = \frac{1}{EI_x} \left(\omega^2 m - C_s p' \right)$$

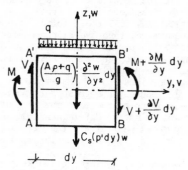

Fig. I - Bending element.

The temporal part f(t) is same for all four independent vibrations.

The general solution for the differential equation 1 is,

$$Z(y) = \{\psi(y)\}^T [L]^{-1} \{d\} \tag{3}$$

The nature of the shape function $\{\psi(y)\}$ depends on the sign of the parameters β^4 and k^2. In the derivations the compressive member axial force P is considered as positive. Thus for members in compression one has,

i. for $\beta^4 + k^4 > 0$, $\alpha_1^2 = ((\beta^4 + k^4)^{\frac{1}{2}} - k^2)$, $\alpha_2^2 = ((\beta^4 + k^4)^{\frac{1}{2}} + k^2)$

$\{\psi(y)\}^T = (\sin\alpha_1 y \quad \cos\alpha_1 y \quad \sinh\alpha_2 y \quad \cosh\alpha_2 y)$

ii. for $\beta^4 + k^4 > 0$, $\{\psi(y)\}^T = (\sin ky \quad \cos ky \quad y \quad 1)$

iii. for $\beta^4 + k^4 > 0$, $\alpha_1^2 = .5(\beta^2 - k^2)$, $\alpha_2^2 = .5(\beta^2 + k^2)$

$\{\psi(y)\}^T = (\cosh\alpha_1 y \cos\alpha_2 y \quad \cosh\alpha_1 y \sin\alpha_2 y \quad \sinh\alpha_1 y \cos\alpha_2 y \quad \sinh\alpha_1 y \sin\alpha_2 y)$

The case ii corresponds to consistent mass matrix approch. The shape functions $\{\psi(y)\}^T[L]^{-1}$ for the remaining three independent cases are determined by following the same steps as in the case of bending in Oyz plane (Fig. 2).

The dynamic stiffness matrix [k] for members are obtained by expressing the kinetic energy K, the strain energy U_i and the damping energy W_c in terms of the displacements u, v and w along x, y and z directions, respectively, then substituting them into the Lagrangian dynamic equation,

$$\frac{d}{dt} \{ \frac{\partial K}{\partial q_j} \} + \{ \frac{\partial U_i}{\partial q_j} \} + \{ \frac{\partial W_c}{\partial q_j} \} = \{ \frac{\partial W_e}{\partial q_j} \}$$

in which the work done by the external dynamic forces W_e, is considered as zero. The dynamic member stiffness matrix [k], thus obtained is function of the natural circular frequency ω, axial force P in member, and if the member is in soil, soil subgrade coefficient C_s and soil friction coefficient C_f. The dynamic member stiffness matrix is incorporated in the general purpose computer program STDYNL for the practical applications (15), (19).

Practical applications.

In the practical applications, the static loads on the members vibrating within the plane of the externally applied dynamic forces are considered as distributed on these members, whereas those acting on transversal members are lumped at the ends of these transversal members. Load factors m and l are introduces on the distributed and lumped loads ,respectively. When the distributed load factor m, is zero the results yield to those obtained by lumped mass matrix method. The case where the factor of lumped loads l, is zero constitutes the continuous mass matrix method.

The parameters considered in the two following examples are:

C_s: soil subgrade coefficients, the skin friction coefficient C_f is expressed as a percentage of C_s at the tip of piles.

r_a: load factor is the ratio of increased member axial force to the actual one.

r_I: moment inertia factor is the ratio of the section characteristics to the actual ones for the member embedded in soil.

Moreover two different types of boundary conditions are taken into account at the tips of the piles: fixed tip and free tip.

The ratio of the rigidity of the beams to columns, the embedded length of the piles may have been selected as additional parameters. The variation of the frequencies with respect to the above mentioned parameters are plotted in terms of the frequency ratio α_{ri}, where,

$$\alpha_{ri} = \frac{\omega_{ri}}{\omega_{oi}}$$

is the ratio of the i'th natural circular frequency with the effect of the parameter r to the one without it.

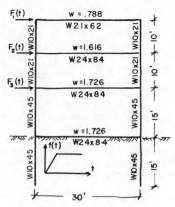

Fig.2-Three story frame

The first illustrative example is a three story steel building (Fig. 2) which has been in previous investigations (11), (14), (17). Two different boundary conditions , free and fixed, are considered at the bottom tip of the piles. The ratio of a particular natural circular frequency of the frame supported on piles to the corresponding natural circular frequency of the three frame fixed at the base are plotted with respect to the varia- of the soil characteristics. The variation of the characteristics of the members embedded in the soil are also taken into account in the graph (Fig. 3). For frames supported by piles fixed at the bottom, the variation of a natural circular frequency is limited by the corresponding natural circular frequencies of three and four story frames with fixed

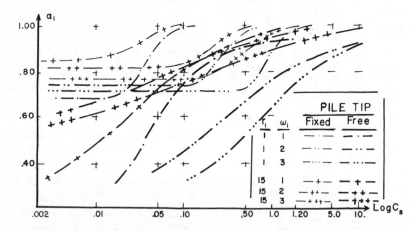

Fig.3-Variation of natural circular frequencies.

bases at ground level. For fixed base frames the variations of natural circular frequencies remain within a limited range of the variation of soil modulus, whereas the variations of the natural circular frequencies for frames with free tips cover a large range of variation of soil modulus. For the free tip frames, when the magnitude of the soil characteristic decreases beyond a certain limit, the frame vibrating in a mode starts to undergo rigid body motion. The natural circular frequency of the frame undergoing a rigid body motion tends to zero. Moreover, for frames with free tip piles, additional modes corresponding to the combination of sway and

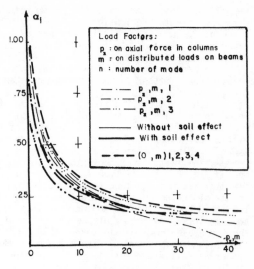

Fig. 4-Variation of natural circular frequencies.

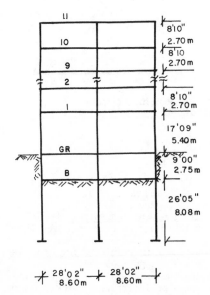

Fig.5 - Twelve story frame.

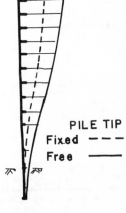

Fig.6- Lateral sway.

support settlement are excited. The magnitudes of these additional modes decrease with the decreasing soil modulus. The variation of the natural circular frequencies including and excluding the soil effect is plotted graphically with respect to the variation of the static loads acting on the structure (Fig. 4).

The second example consists of a twelve story steel frame supported on piles and similar to the Muir Medical Center building in California (Fig. 5), (5). The building is assumed to be subjected to a constant force with a finite rise time as in the first example (Fig. 2). Compared to the three story building, the column loads at lower floors are much larger and the ratio of the height of the embedded piles to the overall height of the building is much smaller that those of the three story building. Therefore, as reflected by the results of the computataions, the twelve story building is more susceptible to the rigid body motion. In the case of free tip piles, for the lower modes the natural circular frequencies tend to zero below some comparatively high limiting values of the soil characteristics, ascertaining the manifestation of the rigid body motions. The increase in the rigidity of the embedded members increases the magnitude of the natural circular frequencies of the frame. The horizontal floor displacements of the frame under the external loads are shown in Figure 6. The frame is considered under two different soil characteristics. The absolute horizontal displacements are larger for frame in soft soil compared to those of the frame in hard soil. But the relative displacements of the columns for the frame on soft soil and for the frame on hard soil are very close to each other above the level of

frist floor. The variation of the natural circular frequency with respect to the variation of soil characteristics is plotted in Figure 7.

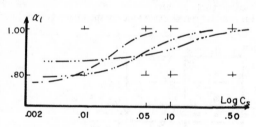

Fig.7-First three natural circular frequencies.

References

1. Aboul-ella, F. and Novak, M., "Dynamic Response of Pile-Supported Frame Foundations", ASCE, Vol. 106, No. SM6, Nov. 1980.
2. Boelle, J.L., Meunier, J., Le Houedec, D. and Levacher, D., "Caractérisation des Phenomènes de Transmission dans les sols et les Pieux des Ebranlements Liés au Battage ou Vibrofonçage ", Proc., CANCAM'85 June 2-7, 1985, pp. A.317-A.318.
3. Coyle, H.M. and Castello, R.R, "New Design Correlations for Piles in Sand ", ASCE, Vol. 107, No. GT7, July 1981, pp. 965-986.
4. Gangsheng, F. and Zaidao, Y., "Earthquake Analysis of Soil-Structure System Including Relative Displacements on Interface ", Proc., Third Intern. Earthquake Microzonation Conf., Uni. of Washington, Seattle, June 28 - July 1, 1982, pp. 969-979.

5. Gates, W.E., "Muir Medical Center ", San Fernando, California, Earthquake of February 9, 1971, Vol 1, Part B, pp. 481-508, U.S. Depart. of Commerce, National Oceanic and Atmospheric Administration, Washington, D.C., 1973

6. Goel, S.C. and Berg, G.V., "Inelastic Earthquake Response of Tall Steel Frames ", ASCE Vol. 94, No. ST8, Aug. 1968, pp. 1907-1934.

7. Hilmi, I.S. and Abel, J.F., "Nonlinear Dynamic Analysis of Steel Frames Using Computer Graphics ", Compt. & Struct. Vol. 21, No.4, 1985, pp. 825-840.

8. Ito, M.Y. and England, R.H., "Computational Method for Soil-Structure Interaction Problems ", Compt. & Struct. Vol. 13, 1981 pp. 157-162.

9. Le Houedec, D. and Riou, R., "Analytical Response of a Semi-Infinite Soil Mass to Road Traffic Vibrations", Proc., Conf. on Soil Dynamics and Earthquake Engineering, July 13-15, 1982, Southampton.

10. Nielsen, N.N., "Vibration Tests of a Nine-Story Steel Frame Building ", ASCE, Vol. 92, No. EM1, Feb. 1966, pp. 81-110.

11. Ovunc, B.A., "Dynamics of Frameworks by Continuous Mass Matrix Method ", Compt. & Struct. Vol. 4, Oct. 1974, pp. 1061-1089.

12. Ovunc, B.A., "Analysis of Buildings on Elastic Medium under Dynamic or Seismic Loads ", Proc., Finite Element Methods in Engineering, Univ. South Wales, Australia, Aug. 1974, pp. 639-659.

13. Ovunc, B.A., "Dynamic Response Time Histories of Continuous Mass Frameworks ", Abst. Research in Progress, 16th Midwestern Mechanics Conf., Kansas State Univ., Sept. 19-21, 1979, pp. 49-50.

14. Ovunc, B.A., "Effect of Axial Force on Framework Dynamics ", Compt. & Struct. Vol. 11, No. 5, 1980, pp. 389-395.

15. Ovunc, B.A., "STDYNL Computer Program for Structures ", Handbook, 3rd Intern. Seminar on Finite Element Systems, Southampton Univ. CML Publications, 1981, pp. 434-445.

16. Ovunc, B.A., "Effect of Soil-Structure Interaction on the Dynamics of Structures ", Proc., Third Intern. Earthquake Microzonation Conf. Univ. of Washington, Seattle, June 28 - July 1, 1982, pp. 821-830.

17. Ovunc, B.A., "Soil-Structure Interaction and Effect of Axial Force on the Dynamics of Offshore Structures ", Compt. & Struct. Vol. 21, No. 4, 1985, pp. 629-637.

18. Ovunc, B.A., "Dynamic Response of Buried Pipelines ", Proc., Intern. Conf. on Advances in Underground Pipeline Engineering, ASCE, Madison, Wisconsin, Aug. 27-29, 1985.

19. Ovunc, B.A., "STDYNL, A Code for Structural Systems ", STRUCTURAL ANALYSIS SYSTEMS, The Intern. Guidebook, Edited by Niku-Lari, Pergamon Press. Vol. III, 1985, pp. 225-238.

20. Poulos, H.G., "Analysis of Piles in Soil Undergoing Lateral Movement ", ASCE, Vol. 99, No. SM5, May 1973, pp. 391-406.

21. Rajapakse R.K.N.D. and Selvadurai, A.P.S., "Laterally Loaded Rigid Cylinder Embedded in Elastic Halfspace ", CANCAM'85, June 2-7, 1985, pp. A.329-A.330.

22. Reddy, A.S. and Valsanghar, A.J., "Buckling of Fully and Partially Embedded Piles ", ASCE, Vol. 107, No. SM6, Nov. 1970, pp. 1951-1965.

23. Richart, R.E. Jr., "Dynamic Stress-Strain Relationships for Soils, State-of-Art Report, Soil Dynamics and it's applications to foundation Engineering ", Proc., 19th ICSMFE, Vol. 2 Tokyo, 1977, pp. 605-612.

COMPARATIVE SEISMIC RESPONSE OF DAMPED BRACED FRAMES

P. Baktash[1], C. Marsh[2] M.ASCE

ABSTRACT

The behaviours of braced steel building frames with friction joints and with eccentric bracing, under seismic forces, are compared. Nonlinear time history dynamic analysis is used. Friction damping is shown to be of particular merit.

INTRODUCTION

The design of structures subjected to seismic forces is receiving increasing attention, not only in order to reduce the risk to life but also to reduce the secondary damage by controlling the peak accelerations and deflections.

The most popular structural systems in steel, designed to withstand lateral forces, are the moment resisting frame and the braced frame with or without moment resistance, which have been widely used in seismic regions. These systems have disadvantages and the need for improved systems has been perceived in recent years. Of the new concepts that have been introduced, two are compared in this paper.

The first is the eccentric braced frame (EBF), in which the line of action of the brace is deliberately offset from a common joint (Roeder and Popov 1977; Popov and Roeder 1978). The eccentricity is introduced so that, under the action of severe earthquakes, energy is dissipated through plastic distortions as the beams yield in shear.

The second system is a friction damped braced frame (FDBF), which utilizes a friction device in the bracing to dissipate energy mechanically (Pall and Marsh 1982). It is designed not to slip during normal service loads, and moderate earthquakes. It will slip during high seismic excitation at a predetermined load, before yielding or buckling occurs in any of the structural elements. Shake table tests (Baktash et al 1983) have shown the beneficial effects of mechanical

---

(1) Research Assistant, Centre for Building Studies, Concordia University, Montreal, Quebec, Canada.

(2) Professor of Engineering, Centre for Building Studies, Concordia University, Montreal, Quebec, Canada.

energy dissipation devices in improving the seismic response of a structure.

Comparison of the behaviour of these framing systems during severe earthquakes is based on nonlinear time-history dynamic analysis.

STRUCTURAL SYSTEMS

a) Moment Resisting Frame:

These structures have stable ductile hysteretic behaviour under reversed cyclic loading, and energy dissipation is obtained from cyclic plastic bending of the beams. However, they tend to be relatively flexible, and it is usually uneconomical to control the story drift by the moment resistance of the frame alone to prevent secondary damage. Moreover, because of their greater deflection, their structural stability is affected by the P-Δ effect, which can be significant.

b) Braced Frame:

Braced frames resist lateral forces with a greater economy of material than do moment resistant frames.

However, these structures have a limited capacity for energy dissipation, due to their pinched or deteriorating hysteresis loops. The capacity of a member under tension-compression loading falls rapidly after a few cycles. Pure tension braces have even lower energy dissipation capacity since under high lateral loadings they will yield only in half a cycle, and on the next application of load in the same direction, this elongated brace is not effective until it is again taut. As a result, energy dissipation degrades very quickly.

c) Novel Structural Sytems:

Attention has been given to combining the stiffness of braced frames with the ductility of moment resisting frames, as in braced moment resisting frames, in which the braces carry only a portion of the lateral loads during earthquakes. Two structures of this type are discussed here: the friction damped braced frame (FDBF) Fig. 1 and the eccentrically braced frame (EBF) Fig. 2(b).

In the eccentrically braced frame, the energy dissipation is obtained by the beams yielding in shear at eccentric joints. This is achieved by offsetting the centre lines of the diagonal bracing from the columns. The braces are usually designed to withstand 1.5 times the force required to yield the beam in shear. This is more than the capacity required to resist lateral wind forces. After a severe earthquake, major repairs may be required because of the large permanent deformation of the beams.

In a friction damped braced frame, the energy input from earth-quakes is dissipated mechanically by sliding friction devices. In this system, each brace is provided with a connection which, in major earthquakes will slip before exceeding the yield stress in any component. The value of the force to cause slipping exceeds that due to wind action.

GENERAL CONSIDERATIONS

To compare the performance of the structural systems, a study was conducted on the two 10 story frames shown in Fig. 2, (a) Friction damped braced frames (FDBF) and (b) Eccentric braced frames (EBF).

The frames chosen for the analysis had the same properties as those utilized by Workman (1969), except that for the eccentric braced frame stronger braces were used in order not to buckle before shear yielding occurred in the eccentric beam joints (Roeder and Popov 1977). For the friction damped braced frame, the braces were designed not to buckle before slip occurred.

Equal mass was assigned to all the floors of the frames, and zero viscous damping was assumed. The earthquake record of El-Centro 1940 (N.S. component) was used. It was scaled by factors of 0.5, 1.0 and 2.0 to give peak ground accelerations of .16g, .32g and .64g, respectively.

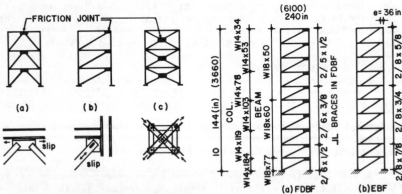

FIG. I. ALTERNATIVE ARRANGEMENTS OF
FRICTION JOISTS.
(a) and (b) Braces act in tension and
compression (c) Braces act in tension only

FIG. 2. STUDIED FRAMING SYSTEMS

FDBF:

For an ideal "elasto-plastic" behaviour of a friction damped braced frame, as shown in Fig. 1(b), the optimum slip or yield force may be obtained as follows. If the horizontal deflection, due to slipping, of each story is Δ, the work done against friction in each story is:

$$W_f = V_1 \, \Delta$$

in which V_1 is the horizontal component of the slip force in the diagonals.

The deflection of the moment resisting frames is related to the portion of the horizontal shear carried by the frame action. If the stiffness of each story is K, then:

$$\Delta = \frac{(V - V_1)}{K}$$

in which V is the lateral shear force at that story.

Thus the energy dissipated per story by friction becomes:

$$W_f = \frac{V_1 \, (V - V_1)}{K}$$

The maximum is obtained when $\frac{dW_f}{dV_1} = 0$, giving:

$$V_1 = \frac{V}{2}$$

This represents a condition in which the shear force is shared equally between the columns and the diagonals, thus the shear force causing the braces to slip is equal to the shear force causing the rigid frame to yield, giving a slip force equal to:

$$P_s = \frac{2M_p}{h \, \cos\alpha}$$

in which M_p is the plastic moment capacity of the beam, h is the floor height, and α is the brace angle.

No equivalent simple optimizing procedure appears to be available for the EBF.

EBF:

In eccentric braced frames, the braces are designed not to buckle before the shear yielding of the links, which is dictated by the beam size. The force in the braces is obtained from:

$$P = \frac{V_b}{\sin\alpha}$$

Plastic hinges are expected to form at both ends of the "link" in the beam, between the connection of the diagonal and the column, shortly after the shear yielding. A satisfactory relationship is obtained when (Popov and Roeder 1978):

$$1.1 \ V_b \leqslant \frac{2M^*_p}{e \ \sin\alpha} \leqslant 1.3 \ V_b$$

in which M^*_p is the plastic moment contribution by the flanges only.

Comparing these forces, it can be seen that for the same moment resisting frame, the bracing in the friction damped braced frame is lighter. For the present analysis, the steel required for the bracing of the FDBF was only 50% of that required for EBF.

RESULTS OF THE ANALYSIS

Nonlinear dynamic analysis was carried out using the computer programme "Drain-2D", developed at the University of California, Berkeley (Kannan and Powell 1973). This programme provides all the subroutines required for the different structural systems used in the analysis. It carries out a step-by-step integration of the dynamic equilibrium using a constant acceleration at each time step.

The effectiveness of the friction device is demonstrated in the comparisons of the results obtained, shown in Fig. 3.

Peak accelerations at the top were:

Factor of El-Centro	0.5	1.0	2.0
FDBF	0.33g	0.40g	0.43g
EBF	0.9g	1.07g	1.16g

Structural damage of the different framing systems after being subjected to an earthquake is shown in Fig. 4. For 0.5 times El-Centro both frames remained elastic, but 80% of the braces in the FDBF slipped. For 1.0 El-Centro excitation, 60% of the eccentric joints

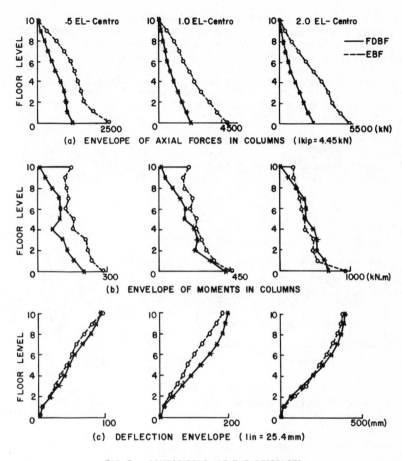

(a) ENVELOPE OF AXIAL FORCES IN COLUMNS (Ikip = 4.45 kN)

(b) ENVELOPE OF MOMENTS IN COLUMNS

(c) DEFLECTION ENVELOPE (I in = 25.4mm)

FIG. 3. COMPARISONS OF THE RESPONSES

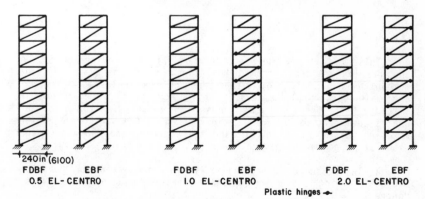

FIG. 4. STRUCTURAL DAMAGE AFTER EARTHQUAKE

yielded in EBF, while 100% of the braces slipped in the FDBF with no yielding in the structural members. For 2.0 times El-Centro, 90% of the joints yielded in the EBF, while 70% of the beams yielded in FDBF, all braces having slipped.

Fig. 5 shows the time histories of the damping elements for EBF and FDBF for El-Centro. The maximum vertical displacement of the "link" for EBF was found to be 20 mm (0.7 in.) for El-Centro, and 70 mm (2.7 in.) for 2.0 times El-Centro.

Time histories of the top deflections of the two systems are shown in Fig. 6. The results shown are for the first seven seconds of El-Centro followed by two seconds of zero acceleration. It can be seen that the two system have comparable residual displacement.

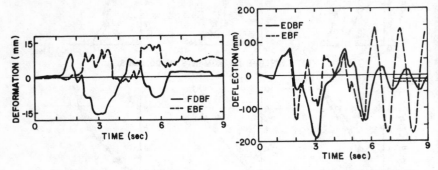

**FIG. 5 . TIME HISTORIES OF DAMPING
ELEMENTS (1in = 25.4mm)**

**FIG. 6. TIME HISTORIES OF THE TOP
FLOOR DEFLECTIONS (1in = 25.4mm)**

CONCLUSION

This study shows the efficiency of friction devices when used in the bracing of steel framed buildings. The energy is dissipated mechanically by friction, rather than by the inelastic behaviour of the main structural members. While eccentric braced frames have sufficient stiffness and energy absorbing capacity, after major earthquakes permanent damage will be introduced to the structural elements. Friction joints on the other hand are cheaper, and can absorb energy with little permanent damage.

A large scale steel model frame with friction devices is presently being tested on a shake table at Concordia University, Montreal.

An independent study (Austin and Pister 1983) has been reported, assessing the usefulness of friction damped braced frames by using the computer aided design environment DELIGHT.STRUT (Balling and Pister 1981). It also concluded that the use of friction devices will reduce damage in the event of a major earthquake, while providing savings in material costs.

REFERENCES

1. Austin, M.A., Pister, K.S., 1983, "Optimal Design of Friction Braced Frames under Seismic Loading", Report No. UCB/EERC-83/10, Earthquake Engineering Research Centre, University of California, Berkeley, California, June.

2. Baktash, P., Marsh, C., Pall, A., 1983, "Seismic Tests on a Model Shear Wall with Friction Joints", Canadian Journal of Civil Engineering, March, Vol. 10, pp. 52-59.

3. Balling, T.J., Pister, K.S., Polak, E., 1981, "Delight Struct. A Computer Aided Design Environment for Structural Engineering", Report No. EERC81-19, Earthquake Engineering Research Centre, University of California, Berkeley, California, December.

4. Kannan, A.E., Powell, G.H., 1973, "Drain-2D, A General Purpose Computer Program for Dynamic Analysis of Inelastic Plane Structures", College of Engineering, University of California, Berkeley, California.

5. Pall, A.S., Marsh, C., 1982, "Seismic Response of Friction Damped Braced Frames", American Society of Civil Engineers, Journal of Structural Division, ASCE, No. ST6, Proc. Paper 17175, June, pp. 1313-1323.

6. Popov, E.P., and Roeder, C.W., 1978, "Design of Eccentrically Braced Steel Frame", AISC Engineering Journal, Vol. 15, No. 3, pp. 77-81.

7. Roeder, C.W., and Popov, E.P., 1977, "Inelastic Behaviour of Eccentrically Braced Steel Frames under Cycling Loadings", EERC Report No. 77-18, Earthquake Engineering Research Center, University of California, August.

8. Workman, G.H., 1969, "The Inelastic Behaviour of Multistory Braced Frame Structures Subjected to Earthquake Excitation", Ph.D. Thesis submitted to the University of Michigan, Ann Arbor, Michigan.

VIBRATIONS OF SUBMERGED SOLIDS IN A VIBRATING CONTAINER

Medhat A. Haroun[1], M. ASCE and Farhad Bashardoust[2]

ABSTRACT

Dynamic forces on a vibrating submerged solid cylinder in both closed and open-top rectangular containers are evaluated. The submerged cylinder is attached to the base of a dynamically-excited tank by a spring of finite and infinite stiffnesses in the horizontal and the vertical directions, respectively. The distribution of impulsive and convective accelerations of the liquid in a vibrating tank without the submerged solid is presented. The equation of motion of the oscillating cylinder in the vibrating container is formulated, and the effective dynamic force on the submerged structure is defined.

INTRODUCTION

Practical problems which occur in a seismic design of nuclear power plant structures may involve submerged elements vibrating in liquid storage containers. A basic step in any dynamic analysis is concerned with the definition of an effective force vector. It was suggested in [1] that the effective force vector for submerged elements be taken proportional to their buoyant mass. The present paper covers the fundamental physical concepts and the associated governing equations which characterize the dynamic forces on a vibrating submerged solid in both closed and open-top rectangular containers. At first, the forces exerted on an isolated, circular, cylindrical solid fully submerged in a moving liquid, and those associated with a vibrating solid in a stationary tank are reviewed. The velocity distribution in a seismically excited tank without the submerged object are summarized. Distinction is made between the impulsive velocity of the liquid, which is directly proportional to the velocity of the container, and the velocity of the liquid due to sloshing; the excitation for the latter case must however be limited to a sinusoidal motion. Finally, the equation of motion of the oscillating submerged object in the vibrating container is formulated, and the effective dynamic force on the submerged cylinder is defined.

STRUCTURAL SYSTEM

A vibrating solid cylinder in a seismically-excited rigid rectangular tank is shown in Fig. (1). The submerged cylinder has a

---

[1]Associate Professor and [2]Graduate Research Assistant, Civil Engineering Department, University of California, Irvine, CA 92717

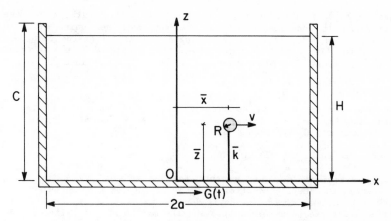

Fig. (1). Tank Geometry and Coordinate System.

radius R, and is attached to the base of the tank by a spring of a stiffness $\bar{k}$ in the horizontal direction, and of infinite stiffness in the vertical direction. The container has dimensions of 2a, 2b and C, and is filled with liquid of density ρ_ℓ to a height H. A Cartesian coordinate system (x,y,z) is used and the cylinder is assumed to vibrate in the x-direction with a displacement v(t) relative to the base displacement G(t).

ASSUMPTIONS

It is assumed that the liquid is incompressible and that the wall of the tank is infinitely rigid. The velocity of the liquid is confined to the x-z plane, and is perpendicular to the longitudinal axis of the cylinder. It is further assumed that the radius of the cylinder is relatively small in comparison with the tank dimensions so that its effect on the velocity of the liquid is negligible.

FORCES EXERTED BY INVISCID LIQUIDS

In a plane flow, a net liquid force per unit length, q, acts on the submerged element. For an unseparated, unsteady, ideal flow, an inertia coefficient, C_M, is defined to yield the force per unit length, q_I, required to hold a rigid cylinder stationary in a liquid of uniform free stream acceleration [7]

$$q_I = C_M \rho_\ell \pi R^2 \ddot{u} \tag{1}$$

If a rigid cylinder, of a mass per unit length of m_0, undergoes a translational displacement v(t) in a liquid at rest except for the presence of the cylinder, then the force per unit length required to accelerate the cylinder can be written as

$$q_I = (m_0 + C_A \rho_\ell \pi R^2) \ddot{v} = m \ddot{v} \tag{2}$$

Equation (2) defines a virtual mass per unit length, m, of the cylinder. It consists of the actual body mass per unit length in vacuum and an added mass per unit length resulting from the fact that some of the liquid particles are permanently displaced by the movement of the cylinder. The added mass coefficient, C_A, is defined as

$$C_A = C_M - 1 \tag{3}$$

It should be noted that, as the cylinder length becomes much larger than its diameter, the value of C_M approaches a theoretical limit of 2, and as such, C_A approaches unity.

FORCES EXERTED BY VISCOUS LIQUIDS

In real liquids, measurements have shown that viscosity often influences the inertia coefficient C_M which becomes time-dependent. A flow separation behind the cylinder occurs and is accompanied by differential pressure forces opposing the motion of the cylinder. In applications, a root-mean-square measured average of C_M and C_A is used in Eqs. (1) and (3).

In addition to the change in the inertia loading, a viscous, drag force, q_D, is also exerted on the submerged body. Such a force is defined as the force per unit length necessary to hold a cylinder at rest in a stream of constant velocity, $\dot{u}$; and it is given by

$$q_D = C_D \rho_\ell R \left| \dot{u} \right| \dot{u} \tag{4}$$

The use of the absolute value sign on one of the velocity terms guarantees that q_D will always oppose the direction of $\dot{u}$.

For a stationary cylinder in a plane flow field in which the free stream velocity, $u(t)$, is arbitary, one should superimpose the inertia and drag loadings given by Eqs. (1) and (4) to obtain the total time-varying load per unit length on the cylinder. This approach was first proposed by Morison. It should be noted, based on experimental observations, that C_D and C_M are not simple constants, and that some observed phenomena are neglected herein such as cavitation, liquid compressibility, three-dimensional flow and proximity effects.

VIBRATING CYLINDER

Consider a liquid with a horizontal velocity $\dot{u}(t)$ in line with the translational velocity $\dot{v}(t)$ of a rigid circular cylinder. Let the constant $\bar{c}$ characterizes the linear viscous damping of the structural material and the constant $\bar{k}$ represents the linear elastic stiffness coefficient. Thus, opposing the motion of the cylinder are the corresponding forces per unit length, $\bar{c}v$ and $\bar{k}v$. As suggested in [7], Morison's equation should be modified to obtain the effective loading because the drag term depends on the relative velocity, $(\dot{u} - \dot{v})$, between the liquid and the cylinder; and the inertia term involves the absolute acceleration of the liquid. If Newton's second law is applied to a free body sketch of the cylinder, and with the aid of Eqs. (1), (2) and (4), one obtains

$$(m_0 + C_A \rho_\ell \pi R^2) \; \ddot{v} + \bar{c}\dot{v} + \bar{k}v = C_D \rho_\ell R \; |\dot{u} - \dot{v}|(\dot{u} - \dot{v}) + C_M \rho_\ell \pi R^2 \; \ddot{u} \qquad (5)$$

It is observed that Eq. (5) is nonlinear because of the drag force term. For small amplitude motion of the cylinder, one may approximate this term by defining a new drag coefficient, C_D', such that

$$C_D' = C_D |\dot{u} - \dot{v}| \simeq \text{constant} \qquad (6)$$

and therefore, Eq. (5) becomes

$$(m_0 + C_A \rho_\ell \pi R^2) \; \ddot{v} + (\bar{c} + C_D' \rho_\ell R) \; \dot{v} + \bar{k}v = C_D' \rho_\ell R \; \dot{u} + C_M \rho_\ell \pi R^2 \; \ddot{u} \qquad (7)$$

Equation (7) clearly shows that the damping of a moving cylinder is increased due to the liquid drag force.

VIBRATION OF A CYLINDER IN A CLOSED TANK

Consider a completely filled, closed top container. If the tank is subjected to a horizontal ground acceleration, then the absolute liquid acceleration in the x-direction is equal to the ground acceleration everywhere. Thus Eq. (7) is reduced to

$$(m_0 + C_A \rho_\ell \pi R^2) \; \ddot{v} + (\bar{c} + C_D' \rho_\ell R) \; \dot{v} + \bar{k}v = - (m_0 + C_A \rho_\ell \pi R^2) \; \ddot{G} + C_M \rho_\ell \pi R^2 \ddot{G} \qquad (8)$$

and therefore, the effective seismic force becomes

$$F_{eff} = - (m_0 - (C_M - C_A) \; \rho_\ell \pi R^2) \; \ddot{G} \qquad (9)$$

and by using the definition of Eq. (3) and the fact that $m_0 = \rho_s \pi R^2$, Eq. (9) is further reduced to

$$F_{eff} = -(\rho_s - \rho_\ell) \; \pi R^2 \; \ddot{G} = -m_b \; \ddot{G} \qquad (10)$$

where m_b is the buoyant mass of the cylinder.

LIQUID ACCELERATION IN A SEISMICALLY EXCITED OPEN-TOP RIGID CONTAINER

To determine the acceleration distribution of a liquid in a seismically excited tank, and eventually the seismic response of the vibrating cylinder, one must solve for the velocity potential function $\phi(x,y,z,t)$ which satisfies the Laplace equation and the appropriate boundary conditions. Two different idealizations are used herein. At first, only the impulsive velocity of the liquid, which is directly proportional to the velocity of the container, is computed. The velocity due to liquid sloshing is then introduced; however, the excitation in this case must be limited to a sinusoidal motion.

IMPULSIVE RESPONSE

Noting that the velocity vector is the gradient of the velocity potential function, one can write the following boundary conditions

$$\frac{\partial \phi}{\partial x} (\pm a, y, z, t) = \dot{G}(t) \qquad (11a)$$

$$\frac{\partial \phi}{\partial y} (x, \pm b, z, t) = 0 \tag{11b}$$

$$\frac{\partial \phi}{\partial z} (x, y, 0, t) = 0 \tag{11c}$$

Neglecting sloshing modes at the free surface yields

$$\frac{\partial \phi}{\partial t} (x, y, H, t) = 0 \tag{12}$$

The solution $\phi(x, y, z, t)$ is given by

$$\phi = \sum_{i=1}^{\infty} A_i(t) \cos(\alpha_i z) \sinh(\alpha_i x) \tag{13}$$

where

$$\alpha_i = (2i-1)\pi/2H \tag{14a}$$

and

$$A_i(t) = \frac{2 (-1)^{i+1}}{H \alpha_i^2 \cosh(\alpha_i a)} \dot{G}(t) \tag{14b}$$

and therefore, the liquid acceleration in the x-direction is given by

$$\ddot{u} = \frac{\partial}{\partial t}(\frac{\partial \phi}{\partial x}) = \sum_{i=1}^{\infty} \frac{2 (-1)^{i+1}}{H \alpha_i} \frac{\cosh(\alpha_i x)}{\cosh(\alpha_i a)} \cos(\alpha_i z) \ddot{G}(t) \tag{15}$$

CONVECTIVE RESPONSE

If the sloshing of the liquid is considered, the boundary condition given by Eq. (12) must be replaced by

$$\frac{\partial^2 \phi}{\partial t^2} (x, y, H, t) + g \frac{\partial \phi}{\partial z} (x, y, H, t) = 0 \tag{16}$$

where g is the acceleration of gravity. When the excitation is in the x-direction, only asymmetric modes of x can be excited, and hence, the natural frequencies of sloshing are given by

$$\omega_j^2 = g\beta_j \tanh(\beta_j H) \tag{17}$$

and the corresponding mode shapes are

$$\phi_j(x, y, z) = \sin(\beta_j x) \cosh(\beta_j z) \tag{18}$$

where

$$\beta_j = (2j-1)\pi/2a \tag{19}$$

The liquid acceleration in the x-direction under a sinusoidal ground displacement of the form

$$G(t) = \bar{G} \sin(\bar{\omega}t) \tag{20}$$

can be written as

$$\ddot{u} = -\bar{\omega}^2 \bar{G} \sin(\bar{\omega}t) \left\{ 1 + \sum_{j=1}^{\infty} \frac{4(-1)^{j+1}}{(2j-1)\pi\cosh(\beta_j H)} \frac{\Delta_j^2}{1 - \Delta_j^2} \cos(\beta_j x) \cosh(\beta_j z) \right\} \tag{21}$$

where

$$\Delta_j = \frac{\bar{\omega}}{\omega_j} \tag{22}$$

This distribution of liquid acceleration is essential for the evaluation of liquid forces on the vibrating submerged cylinder.

NUMERICAL EXAMPLES

A computer program was developed to obtain the distribution of liquid accelerations in the tank, and subsequently, to evaluate the coefficient C_ρ which is defined as

$$C_\rho = (1 - 2\ \ddot{u}/\ddot{G}) \tag{23}$$

This coefficient is used to compute an effective mass density which determines the effective seismic force on the submerged cylinder via

$$F_{eff} = -\rho_{eff}\pi R^2 \ddot{G} = -(\rho_s + C_\rho\ \rho_\ell)\ \pi R^2 \ddot{G} \tag{24}$$

If the acceleration of the liquid adjacent to the submerged cylinder is equal to the acceleration of the ground motion, then C_ρ becomes equal to (-1), and therefore, the expression of the effective seismic force is reduced to that given by Eq. (10); i.e., it becomes proportional to the buoyant mass of the cylinder.

Example (1): At first, only the impulsive acceleration of the liquid is considered. For a given value of (H/a) ratio, the liquid region inside the tank is divided to a mesh of rectangular elements and the liquid acceleration is computed at each node. The coefficient C_ρ is then computed and its values are displayed in Fig. (2) for selected ratios of (H/a) of 0.5, 1.0, 1.5 and 2.0. Inspection of the figure shows that the effective force on the submerged cylinder becomes proportional to the buoyant mass of the cylinder only near the vertical boundary of the tank where the acceleration of the liquid is equal to the ground acceleration. It is also clear that the rate of change of the values of the coefficient C_ρ is slow for elevations at or near the bottom of the tank. As expected, a rapid variation of C_ρ is observed from the mid depth of the liquid up to its free surface, and also for lower elevations in shallow tanks. For these cases, the sloshing of the liquid influences the distribution of liquid acceleration. It is also clear that the values of C_ρ ranges from -1 to 1 since liquid acceleration varies from a maximum near the wall to a minimum of zero. It should be noted that C_ρ approaches a value of (-1) if (H/a) becomes very large. On the other hand, if one considers very small values of (H/a), then C_ρ approaches a value of (1) near the center of the tank, and consequently, the effective seismic force becomes proportional to the virtual mass of the cylinder; i.e., the mass of the cylinder in vaccum plus an added mass equal to $C_A\rho_\ell\pi R^2$. In this case, the acceleration of the tank wall produces a negligibly small acceleration of the liquid at the center of the tank, and hence, the cylinder accelerates, relative to the liquid, with an acceleration equals to the acceleration of the ground.

Example (2): The effect of liquid sloshing is considered by using the expression of liquid acceleration given by Eq. (21). For selected values of (H/a) ratio, three different excitation frequencies are used, namely, $(\omega/\omega_1) = 2$, 5 and 10 where ω_1 is the fundamental

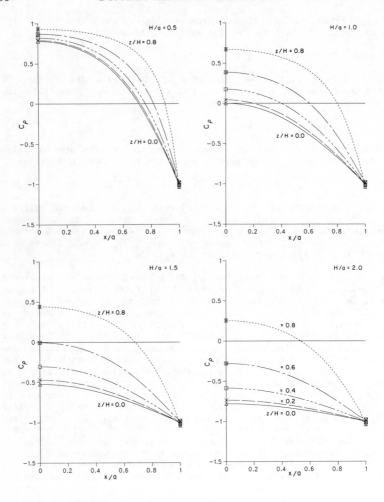

Fig. (2). Values of C_ρ neglecting sloshing.

frequency of sloshing. As expected, the largest value of $(\bar{\omega}/\omega_1)$ for large (H/a) ratio yielded results similar to the case of impulsive motions since sloshing effects in such a case are negligibly small. However, for the lower frequency ratios, liquid acceleration near the top of the tank may be out of phase with ground acceleration resulting in larger values of C_ρ that exceed unity. Sample plots of C_ρ for two different frequency ratios, and for a representative value of (H/a) of 2, are displayed in Fig. (3).

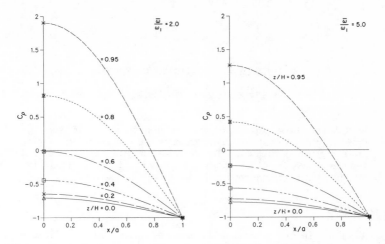

Fig. (3). Values of C_ρ considering sloshing.

CONCLUSION

The present simple analysis showed that the use of the buoyant mass for the computation of the effective force on a vibrating submerged cylinder in a seismically excited open-top container is unconservative. It is demonstrated that the effective mass, for the purpose of calculating the effective force, is larger than the buoyant mass and can exceed the mass of the cylinder in vacuum. For a closed top tank, the effective mass is equal to the buoyant mass.

REFERENCES

1. American Society of Civil Engineers, <u>Structural Analysis and Design of Nuclear Plant Facilities,</u> draft report, 1976.

2. Dong, R.G., <u>Effective Mass and Damping of Submerged Structures,</u> UCRL-52342, Lawrence Livermore Laboratory, April 1978.

3. Fritz, R.J., "The Effects of Liquids on the Dynamic Motions of Immersed Solids," <u>J. of Engineering for Industry,</u> ASME, Feb. 1972.

4. Hadjian, A.H., personal communications, Bechtel Power Corporation, Norwalk, California.

5. Haroun, M. A., "Stress Analysis of Rectangular Walls under Seismically Induced Hydrodynamic Loads," <u>Bulletin of the Seismological Society of America,</u> Vol. 74, No. 3, June 1984.

6. Housner, G. W., "Dynamic Pressures on Accelerated Fluid Containers," <u>Bull. Seis. Soc. America,</u> Vol. 47, No. 1, Jan. 1957.

7. Wilson, J. F., editor, <u>Dynamics of Offshore Structures,</u> John Wiley & Sons, 1984.

MODIFIED DYNAMIC CONDENSATION METHOD

By Mario Paz, M.[1] ASCE

ABSTRACT: This paper presents a modification of the Dynamic Condensation Method for the reduction or condensation of problems in structural dynamics. The proposed modification substantially reduces the total numbers of numerical operations in the application of the method.

INTRODUCTION

A survey in the published literature shows that the first major step toward a method of reducing the size of a structural dynamic problem appeared in the paper published by Guyan (1965). Since this initial publication, numerous investigators have proposed improvements or variations of this original work. The well-known method of Guyan is based on the static condensation of the unwanted or dependent coordinates. The relation between dependent and independent degrees of freedom is thus determined by establishing the static relation between them. In order to reduce the mass matrix, the same static relation is assumed to remain valid for the dynamic problem. Shortly after the publication of Guyan's paper, Irons (1965) presented an algorithm to reduce the stiffness and mass matrices in which the reduction is performed for each degree of freedom sequentially. A number of investigators have applied Guyan's reduction method using a variety of computational algorithms. The method proposed by Anderson, Irons and Zienkiewicz (1968) consists of eliminating those coordinates that do not appreciably contribute to the kinetic energy of the system. Ramsden and Stoker (1969) presented a technique which reduces the dynamic problem using the flexibility matrix.

Several general computer programs such as NASTRAN (1970) and SUPERB (1978) incorporate provisions for the reduction or condensation of coordinates. Some authors have developed formulas for estimating the errors introduced by the application of Guyan's reduction in structural dynamics. Also, some have presented new algorithms based on iterative schemes in an attempt to decrease errors introduced in the process of reduction. Wright and Miles (1971) have approached the problem of matrix reduction by considering the exact problem, establishing the exact relation between dependent and independent coordinates followed by a series expansion in which terms of up to the second order are retained. The net effect of retaining the second

---

[1]Professor, University of Louisville, Louisville, KY 40292

order terms is an eigenvalue problem of twice the dimensions of the reduced eigenvalue problem. This approach cannot be recommended since it is only an approximation to the original eigenvalue problem without a substantial reduction in dimensions. Kidder (1973) also followed the series expansion approach to obtain the reduced stiffness and mass matrices after neglecting higher-order terms. However, the exact relation between dependent and independent coordinates is recommended by Kidder for the back-transformation to obtain the modal shape in terms of all the coordinates in the original modeling of the structure. This transformation, which is affected by the errors introduced in the calculation of the eigenvalues of the reduced system, as was pointed out by Flax (1975), may even introduce larger errors than the simple Guyan's transformation. The method recommended by Kidder (1975) will give more accurate results if the selection of coordinates to be reduced is such that the series expansion to produce the reduced stiffness and mass matrices is convergent.

In an effort to reduce the error inherent in the approximations of the Guyan method, Miller (1980, 1981) introduced a modification of Guyan reduction. Unfortunately this modification is presented as requiring the inversion of matrices of dimensions equal to the number secondary coordinates to be condensed. Furthermore, the proposed modification does not improve the igenvalues which are calulated by Guyan's method; thus, it can only attempt to improve the eigenvectors of the reduced system.

DYNAMICS CONDENSATION METHOD

The method of dynamic condensation, recently proposed by Paz (1984, 1985), may be explained by considering the generalized eigenproblem expressed in the following partitioned matrix:

$$
\left[
\begin{array}{c|c}
[K_{ss}] - \omega^2 [M_{ss}] & [K_{sp}] - \omega^2 [M_{sp}] \\
\hline
[K_{ps}] - \omega^2 [M_{ps}] & [K_{pp}] - \omega^2 [M_{pp}]
\end{array}
\right]
\left\{
\begin{array}{c}
\{Y_s\} \\
\{Y_p\}
\end{array}
\right\}
=
\left\{
\begin{array}{c}
\{0\} \\
\{0\}
\end{array}
\right\}
\tag{1}
$$

where [K] and [M] are respectively the stiffness and mass matrices. $\{Y_s\}$ is the displacement vector corresponding to the s degrees of freedom to be reduced, $\{Y_p\}$ is the vector corresponding to the remaining p independent degrees of freedom and ω^2 is the eigenvalue determined approximately at each step of the calculations.

The process starts by assigning to ω^2 any approximate or zero value for the first eigenvalue; it is followed by elementary operations applied to eliminate the first s displacements in Eq.(1). At this stage of the elimination process, Eq.(1) may be written as

$$
\left[
\begin{array}{c|c}
[I] & [\bar{T}] \\
\hline
[0] & [\bar{D}]
\end{array}
\right]
\left\{
\begin{array}{c}
\{Y_s\} \\
\{Y_p\}
\end{array}
\right\}
=
\left[
\begin{array}{c}
\{0\} \\
\{0\}
\end{array}
\right]
\tag{2}
$$

in which $[\overline{T}]$ is the transformation matrix defined in the following relation:

$$[Y] = \begin{bmatrix} \{Y_s\} \\ \hline \{Y_p\} \end{bmatrix} = \begin{bmatrix} [\overline{T}] \\ \hline [I] \end{bmatrix} \{Y_p\}$$

(3)

and $[\overline{D}]$ the dynamic matrix which satisfies the relation

$$[\overline{K}] = [\overline{D}] + \omega^2 [\overline{M}]$$

(4)

where $[\overline{K}]$ and $[\overline{M}]$ are respectively the stiffness and mass matrices of the reduced system. The reduced mass matrix is calculated by

$$[\overline{M}] = [T^T][M][T]$$

(5)

where

$$[T] = \begin{bmatrix} [\overline{T}] \\ \hline [I] \end{bmatrix}$$

(6)

while the reduced stiffness matrix is determined using Eq.(4).

The resulting reduced eigenproblem

$$\left[[\overline{K}] - \omega^2 [\overline{M}] \right] \{Y_p\} = \{0\}$$

(7)

is solved to obtain a virtually exact eigenvalue and corresponding eigenvector for the first mode, and an approximate eigenvalue for the second mode. The calculations continue by substituting into Eq.(1) this approximation of the eigenvalue for the second mode followed by application of Eqs. (2) through (7) to obtain a virtually exact second eigenvalue and an approximation to the third eigenvalue. The continuation of this solution process in which at each step a virtually exact eigenvalue and an approximation to the next eigenvalues are computed, will result in virtually exact solution for the lowest p eigenvalues and corresponding p eigenvectors of the original system.

MODIFIED DYNAMIC CONDENSATION

The dynamic condensation method essentially requires the application of elementary operations as it is routinely used for the solution linear system of algebraic equations. The elementary operations are required to transform Eq.(1) to the form given by Eq.(2). However, the method also requires the calculation of the reduced mass matrix by Eq.(5). This last equation involves the multiplication of three matrices of dimensions equal to the total number coordinates in the system. Thus, for a system with many degrees of freedom, the calculation of the reduced mass matrix [M] requires a large number of numerical operations.

In this paper, a modification of the dynamic condensation method which substantially decreases the required number of operations is proposed. This modification consists in calculating the reduced

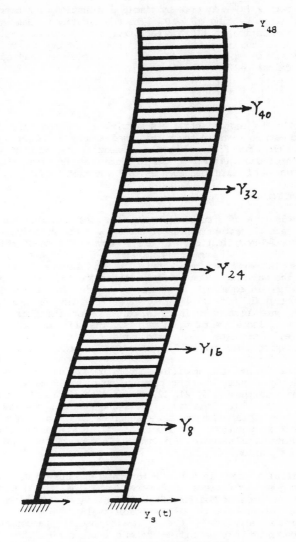

Fig. 1 Shear Building Model

stiffness matrix $\overline{[K]}$ only once by simple elimination of s displacements in Eq.(1) after setting $\underline{\omega}^2 = 0$. The modifications, thus avoids the repeated calculation of $[K]$ for each mode using Eq.[4]. Furthermore, it also avoids the long time assumed in calculating the reduced mass matrix $\overline{[M]}$ using Eq.(5). In the modified algorithm the reduced mass matrix is calculated from Eq.(4) as

$$\left[\overline{M}\right] = -\frac{1}{\omega^2}\left[\left[\overline{K}\right] - \left[\overline{D}\right]\right]$$

(8)

where $\overline{[K]}$ is the reduced stiffness matrix, already calculated and $\overline{[D]}$ is dynamic matrix given in the partitioned matrix in Eq.(2). As it can be seen, the modified algorithm essentially requires only the application of Gauss-Jordan method to eliminate the first s unknown in a linear system of equations such as the system in Eq.(1)

NUMERICAL EXAMPLES

A simple type of structure, known as a shear building was selected to demonstrate the effectiveness of the Dynamic Condensation Method and of the proposed modification. Figure 1 shows a sketch of this type of structure. Under the assumptions imposed in defining a shear building, the system is permitted only horizontal displacements at the floor levels of the building. Consequently, the number of degrees of freedom of this system is equal to the number of stories in the building. The results obtained for a 48-story shear-building reduced to six coordinates are presented in this paper. For this structure, the masses at the floors are of equal values. The stiffness of the stories are, as well, of equal values adjusted to the condition that the fundamental period for the buildings be equal to one second.

Results, using the modified algorithm, applied to a 48-story shear-building reduced to six primary coordinates selected as the displacements in the 8, 16, 24, 32, 40 and 48th floors of the building, are given in Table 1. This table shows and compares results with the exact solution (obtained using all 48 degrees of freedom in the system) for the following methods of reduction: (1) Static Condensation, (2) Modified Static Condensation, (3) Dynamic Condensation and (4) Modified Dynamic Condensation.

The results shown in Table 1 for this example of 48 coordinates reduced to just six degrees of freedom, indicates the following: (1) Static Condensation or Modified Statics Condensation produced only the fundamental frequency with an acceptable value. Higher frequencies have errors in excess of 5% and are in general unacceptable, (2) Dyanmic condensation or Modified Dynamic Condesnation gave virtually exact natural frequencies for all the modes in the reduced problem.

CONCLUSIONS

A modification for the Dynamic Condensation Method has been proposed. This modification substantially decreases the number of numerical operations required on the application of the method. Further investigation to evaluate the proposed modification as well as other possible alternatives is warranted.

TABLE 1   Natural Frequencies (cps) for a 48-Story Shear-Building.
(System modeled with 48 coordinates condensed to 6 coordinates)

Mode	Exact Frequency	METHOD OF CONDENSATION			
		Static Frequency % Error	Modified* Static Frequency % Error	Dynamic** Frequency % Error	Modified** Dynamic Frequency % Error
1	1.0000	1.0061 0.61%	1.0061 0.61%	1.0000 0.0%	1.0000 0.0%
2	2.9990	3.1641 5.5%	3.1641 5.50%	2.9990 0.0%	2.9990 0.0%
3	4.9948	5.7032 14.18%	5.7032 14.18%	4.9948 0.0%	4.9948 0.0%
4	6.9858	8.1709 16.97%	8.1709 16.97%	6.9929 0.1%	6.9967 0.1%
5	8.9686	10.5183 17.27%	10.5183 17.27%	8.9686 0.0%	9.2048 0.6%
6	10.9424	12.5690 14.86%	12.5690 14.86%	10.9456 0.0%	11.1548 1.9%

* Same values as static condensation, though the eigenvectors, are
  improved.
**One iteration.

APPENDIX - REFERENCES

Anderson, R. G., Irons, B. M., and Zienkiewicz, O. C., "Vibration and Stability of Plates Using Finite Elements," Int. Journal of Solids and Structures, Vol. 4, 1968, pp. 1031-1055.

Flax, A. H., Comment on "Reduction of Structural Frequency Equations," AIAA Journal, Vol. 13, No. 5, 1975, pp. 701-702.

Guyan, R. J., "Reduction of Stiffness and Mass Matrices," AIAA Journal, Vol. 3, No. 2, 1965, p. 380.

Irons, B. M., "Structural Eigenvalue Problems: Elimination of Unwanted Variables," AIAA Journal, Vol. 3, No. 5, 1965, pp. 961-962.

Kidder, R. L., "Reduction of Structural Frequency Equations," AIAA Journal, Vol. 11, No. 6, 1973, p. 892.

Kidder, R. L., "Reply by Author to A. H. Flax," AIAA Journal, Vol. 13, No. 5, 1975, pp. 702-703.

Miller, C. A., "Dynamic Reduction of Structural Models," Journal of the Structural Division ASCE, 1980, pp. 2097-2108.

Miller, C. A., "Errors Resulting from Dynamic Reduction,"Proceedings of the First National Conference on Computers in Civil Engineering, May 1981, pp. 58-71.

NASTRAN Theoretical Manual, Cosmic, University of Georgia, Athens, GA 1970.

Paz, M., "Dynamic Condensation Method," Journal of AIAA, Vol. 22, No. 5, May 1984, pp. 724-727.

Paz, M., Structural Dynamics: Theory and Computation, Van Nostrand Reinhold Company, 1985.

Ramsden, J. N. and Stoker, J. R., "Mass Condensation: A Semi-automatic Method for Reducing the Size of Vibration Problems," Int. Journal for Numerical Methods in Engineering, Vol. 1, 1969, pp. 333-349.

SUPERB, Structural Dynamics Research Corporation, Cincinnati, OH, 1978.

Wright, G. C. and Miles, G. A., "An Economical Method for Determining the Smallest Eigenvalues of Large Linear Systems," Int. Journal for Numerical Methods in Engineering, Vol. 3, 1971, pp. 25-34.

DIFFRACTION FORCES ON AXISYMMETRIC BODIES

Y.-C. Hsu,[1] D. Syriopoulou,[1]
J.M. Roesset,[2] and J.L. Tassoulas[3]

Introduction

The diffraction of water waves by structures has been a subject of intensive investigation during the last two decades. Analytical solutions are available in cases involving simple geometries. In other situations, more frequently encountered in practice, numerical techniques become necessary. Integral equation and finite element formulations are most commonly in use today. Comprehensive reviews of these techniques are given by Mei (8) and Sarpkaya and Isaacson (10).

Use of a Green function satisfying the governing differential equation and all boundary conditions except those on the boundary of the structure yields an integral equation, when these boundary conditions are imposed. A system of algebraic equations can then be obtained by discretizing the boundary of the structure ("direct" or "indirect" boundary element methods).

Finite element techniques are not as straightforward because the water domain is not bounded and an artificial boundary must be introduced at some distance from the structure. The question then is what boundary conditions must be imposed on the artificial boundary to satisfy the radiation condition. An approximate solution to this problem was suggested by Bettess and Zienkiewicz (1), using the so-called infinite elements. A more accurate formulation, expressing the solution in the unbounded water domain as a linear combination of modes of wave motion and using finite elements in the bounded region around the structure, was proposed by Chen and Mei (4) and Yue, Chen and Mei (14). The analytical solution for the far field and the discrete solution in the core region are coupled by imposing the stationarity condition on an appropriate functional thus specifying continuity of pressure and normal velocity across the common boundary. The resulting system of algebraic equations involves the values of the velocity potential at the nodes of the finite element mesh in the core region and the participation factors of the wave modes in the far field.

The technique developed by Yue, Chen and Mei (14) is applicable to the analysis of diffraction of water waves by structures of arbitrary, three-dimensional geometry. It is, in a certain way, a generalization of an approach presented previously by Waas (13) for the solution of soil-structure interaction problems. In Waas' formulation, a consistent transmitting boundary, modelling the far field, is combined directly with the finite elements in the discretized region around the structure without the need to introduce additional variables (modal participation factors). The boundary matrix, relating nodal values of the velocity potential to consistent nodal values of the flux at the nodes on the boundary, is obtained in terms of semi-discrete modes of wave motion in the unbounded region (discrete with respect to the vertical direction, continuous with respect to any horizontal direction).

In this paper, the derivation of the consistent boundary matrix for a fluid domain is presented. The formulation has been implemented in a computer program and has been applied to the study of linear diffraction forces on rigid and flexible axisymmetric bodies of otherwise arbitrary geometry, fixed or movable, and to the approximate solution of nonlinear diffraction problems.

---

[1]Graduate Student, Department of Civil Engineering, The University of Texas, Austin, Texas.
[2]Paul D. and Mary Robertson Meek Centennial Professor of Civil Engineering, The University of Texas, Austin, Texas.
[3]Assistant Professor, Department of Civil Engineering, The University of Texas, Austin, Texas.

Numerical Technique

Eigenvalue Problem. Time-harmonic irrotational waves of frequency ω are considered in water of constant depth d. The water is assumed to be an inviscid fluid. The free surface is the plane $z = 0$ and the bottom is at $z = d$. Let

$$\phi(r, \theta, z) \, exp(i\omega t)$$

be the velocity potential. The governing differential equation is Laplace's equation:

$$\frac{\partial^2 \phi}{\partial r^2} + \frac{1}{r}\frac{\partial \phi}{\partial r} + \frac{1}{r^2}\frac{\partial^2 \phi}{\partial \theta^2} + \frac{\partial^2 \phi}{\partial z^2} = 0 \tag{1}$$

At the surface, the linearized boundary condition is:

$$\omega^2 \phi(r, \theta, 0) + g\frac{\partial \phi}{\partial z}\bigg|_{z=0} = 0 \tag{2a}$$

g being the acceleration of gravity. The boundary condition at the bottom is:

$$\frac{\partial \phi}{\partial z}\bigg|_{z=d} = 0 \tag{2b}$$

Solutions of Eq. 1 satisfying the homogeneous boundary conditions (Eqs. 2a,b) can be written as linear combinations of modes given by:

$$\begin{Bmatrix} \phi^s(r, \theta, z) \\ \phi^a(r, \theta, z) \end{Bmatrix} = \Phi(z) C_n(kr) \begin{Bmatrix} cos(n\theta) \\ sin(n\theta) \end{Bmatrix} \quad n = 0, 1, 2, \cdots \tag{3}$$

The superscripts s, a indicate symmetric and antisymmetric modes respectively (symmetry or antisymmetry of the velocity field with respect to the plane $\theta = 0$). C_n is any solution of Bessel's equation of order n. The eigenfunction Φ and the eigenvalue k must satisfy the equation:

$$\frac{d^2 \Phi}{dz^2} - k^2 \Phi = 0 \tag{4}$$

and the boundary conditions:

$$\omega^2 \Phi(0) + g\frac{d\Phi}{dz}\bigg|_{z=0} = 0 \tag{5a}$$

$$\frac{d\Phi}{dz}\bigg|_{z=d} \tag{5b}$$

It is easily found that

$$\Phi(z) = cosh[k(z-d)] \tag{6}$$

satisfies Eqs. 4 and 5, if the dispersion relation (frequency equation):

$$\omega^2 = gk \, tanh(kd) \tag{7}$$

holds. Clearly, if k is an eigenvalue with eigenfunction Φ, $-k$ is another eigenvalue with the same eigenfunction. There is one and only one positive eigenvalue k_0 and $-k_0$ is the only negative eigenvalue. The other eigenvalues are imaginary. Let Φ_0 be the eigenfunction corresponding to k_0.

For any frequency ω, Eq. 7 can be solved for the eigenvalues k by a numerical method, e.g., Newton's method. Alternatively, it is straightforward to obtain discrete eigenfunctions and the corresponding eigenvalues by the finite element method. The interval $[0, d]$ is divided into N subintervals (finite elements) $[z_j, z_{j+1}]$, $1 \leq j \leq N$, with

$$0 = z_1 < z_2 < \ldots < z_j < z_{j+1} < \ldots < z_{N+1} = d$$

Each subinterval corresponds to a water sublayer. The depth of sublayer j is $h_j = z_{j+1} - z_j$. Assuming linear interpolation of Φ in each sublayer (any other interpolation can be employed as well), the finite element method applied to the problem defined by the differential equation 4 and the boundary conditions 5a,b yields:

$$[k^2 A + G - \omega^2 M] V = 0 \qquad (8)$$

The components of the eigenvector V are:

$$V_j = \Phi(z_j), \quad 1 \le j \le N+1$$

A, G, M are symmetric tridiagonal matrices assembled from element matrices A^j, G^j, M^j, $1 \le j \le N$, respectively, given by:

$$A^j = h_j \begin{bmatrix} 1/3 & 1/6 \\ 1/6 & 1/3 \end{bmatrix} \quad G^j = \frac{1}{h_j} \begin{bmatrix} 1 & -1 \\ -1 & 1 \end{bmatrix} \quad 1 \le j \le N$$

$$M^1 = \begin{bmatrix} 1/g & 0 \\ 0 & 0 \end{bmatrix} \quad M^j = \begin{bmatrix} 0 & 0 \\ 0 & 0 \end{bmatrix} \quad 2 \le j \le N$$

The eigenvalues k are the roots of the polynomial of degree $2N+2$:

$$det[k^2 A + G - \omega^2 M] = 0 \qquad (10)$$

If k is an eigenvalue of the algebraic eigenvalue problem (Eq. 8) with eigenvector V, $-k$ is another eigenvalue with the same eigenvector. It can be shown that, for any ω, N, there is one and only one positive eigenvalue k_0, one negative eigenvalue $-k_0$ and the other eigenvalues are imaginary. Let V^0 be the eigenvector corresponding to k_0. It is convenient to choose the N eigenvalues k_j, $1 \le j \le N$, with corresponding eigenvectors V^j, such that $Im[k_j] < 0$. A modal matrix can be defined as:

$$X = [V^0, V^1, V^2, ..., V^N] \qquad (11)$$

Since the eigenvectors are orthogonal with respect to A, they can be normalized so that

$$X^T A X = I \qquad (12)$$

Thus, for each eigenvalue k with corresponding eigenvector V, semidiscrete modes are obtained:

$$V C_n(kr) \begin{Bmatrix} cos(n\theta) \\ sin(n\theta) \end{Bmatrix} \qquad (13)$$

Consistent Transmitting Boundary. Let us consider the region $r \ge R$, $0 \le \theta \le 2\pi$, $0 \le z \le d$. Radiation and boundedness conditions are imposed on waves in this region. An appropriate solution of Bessel's equation of order n is the Hankel function of the second kind, of order n, with asymptotic behavior:

$$H_n^{(2)}(kr) \sim \left[\frac{2}{\pi kr} \right]^{1/2} exp\left(-ikr + i\frac{n\pi}{2} + \frac{\pi}{4}\right), \; r \to \infty \qquad (14)$$

The semidiscrete modes are:

$$V H_n^{(2)}(kr) \begin{Bmatrix} cos(n\theta) \\ sin(n\theta) \end{Bmatrix} \qquad (15)$$

Considering the asymptotic behavior (Eq. 14), it is seen that the radiation and boundedness conditions are satisfied, if the modes with eigenvalues k_0, k_j, $1 \le j \le N$, $k_0 > 0$, $Im[k_j] < 0$ are chosen.

For the region under consideration, the nodal values of the outward flux through $r = R$, consistent with the interpolation of the velocity potential in the sublayers, are calculated by integrating, on this boundary, $-\partial\phi/\partial r$ multiplied by the nodal interpolation functions. The vector of consistent nodal values of the flux corresponding to a semidiscrete mode given by Eq. 15 can be found by straightforward integration:

$$-\pi R \begin{Bmatrix} \varepsilon_n^s \\ \varepsilon_n^a \end{Bmatrix} k\, H_n^{(2)\prime}(kR)\mathbf{A}\mathbf{V} \tag{16}$$

with $\varepsilon_0^s = 2$, $\varepsilon_0^a = 0$, $\varepsilon_n^s = 1$, $\varepsilon_n^a = 1$, $n \geq 1$. $\mathbf{A}$ is the same matrix as in the algebraic eigenvalue problem (Eq. 8) and the prime indicates differentiation with respect to the argument. Considering, for any value of n, symmetric or antisymmetric waves, the vector of consistent nodal values of the flux at $r = R$ can be written as a linear combination of the modes:

$$\begin{Bmatrix} \mathbf{U}_n^s \\ \mathbf{U}_n^a \end{Bmatrix} = -\pi R \begin{Bmatrix} \varepsilon_n^s \\ \varepsilon_n^a \end{Bmatrix} \mathbf{A}\mathbf{X}\, diag\,[k_l H_n^{(2)\prime}(k_l R), l = 0, 1, ..., N]\,\Gamma \tag{17}$$

Γ being the vector of modal participation factors. $\mathbf{X}$ is the modal matrix defined earlier (Eq. 11). The vector of nodal values of the velocity potential at $r = R$ is given by the same linear combination of the modes (Eq. 15):

$$\mathbf{P} = \mathbf{X}\, diag\,[H_n^{(2)}(k_l R), l = 0, 1, ..., N]\,\Gamma \tag{18}$$

Clearly, from Eq. 12,

$$\mathbf{X}^{-1} = \mathbf{X}^T \mathbf{A} \tag{19}$$

Elimination of Γ from Eq. 17, using Eq. 18, yields:

$$\begin{Bmatrix} \mathbf{U}_n^s \\ \mathbf{U}_n^a \end{Bmatrix} = \begin{Bmatrix} \mathbf{C}_n^s \\ \mathbf{C}_n^a \end{Bmatrix}\mathbf{P} \tag{20}$$

with

$$\begin{Bmatrix} \mathbf{C}_n^s \\ \mathbf{C}_n^a \end{Bmatrix} = -\pi R \begin{Bmatrix} \varepsilon_n^s \\ \varepsilon_n^a \end{Bmatrix} \mathbf{A}\mathbf{X}\mathbf{K}\mathbf{X}^T \mathbf{A} \tag{21}$$

in which

$$\mathbf{K} = diag\,[k_l H_n^{(2)\prime}(k_l R)/H_n^{(2)}(k_l R), l = 0, 1, ..., N] \tag{22}$$

The surface $r = R$, $0 \leq \theta \leq 2\pi$, $0 \leq z \leq d$ may be referred to as a consistent transmitting boundary. In the derivation of the matrices $\mathbf{C}_n^s$, $\mathbf{C}_n^a$, corresponding to symmetric and antisymmetric waves respectively, radiation and boundedness conditions were imposed and, furthermore, these matrices are consistent with the assumed interpolation of the velocity potential in the sublayers. It should be noted that $\mathbf{C}_n^s$, $\mathbf{C}_n^a$ are symmetric matrices. Moreover, the modal participation factors Γ have been eliminated and the matrices relate the consistent nodal values of the flux to the nodal values of the velocity potential. Thus, the consistent transmitting boundary can be combined directly with conventional fluid finite elements. In fact, it can be demonstrated, as in similar developments for layered soil strata (12), that the boundary matrices obtained above are the limits of matrices corresponding to a finite element mesh in the region $r \geq R$, $0 \leq z \leq d$ with the same vertical spacing of the nodes as in the consistent transmitting boundary and the values of the velocity potential at nodes beyond $r = R$ eliminated, as the maximum radial spacing of the nodes goes to zero. This should be expected since the modes used in the calculation of the matrices are continuous with respect to the radial direction. It further suggests that the consistent transmitting boundary can be placed as close to the structure as the geometry permits (Fig. 1).

Applications

In order to assess the accuracy of the numerical results and the effect of the number of sublayers, the case of a rigid, vertical, circular cylinder was considered first. The velocity potential of the incident plane wave is:

$$A\frac{g}{i\,\omega\Phi_0(0)}\,\Phi_0(z)\,exp\,(-ik_0 x)\,exp\,(i\,\omega t)$$

which, in semidiscrete form, is given by:

$$A \frac{g}{i \omega V_1^0} \mathbf{V}^0 \left[J_0(k_0 r) + \sum_{n=1}^{\infty} 2i^{-n} J_n(k_0 r) \cos(n\theta) \right] exp(i \omega t)$$

where A denotes the amplitude of the vertical displacement at the free surface. It should be noticed that only the first term, corresponding to $n = 0$, contributes to the resultant vertical force on an axisymmetric body and only the term $n = 1$ contributes to the resultant horizontal force and moment.

The analytical solution to this problem was presented by MacCamy and Fuchs (7). For the present formulation, the boundary can be placed directly at the surface of the cylinder. Let $\mathbf{P}^0$ and $\mathbf{U}^0$ be the vectors of nodal potentials and consistent nodal values of the flux at $r = R$:

$$\mathbf{P}^0 = -2A \frac{g}{\omega V_1^0} J_1(k_0 R) \mathbf{V}^0$$

$$\mathbf{U}^0 = 2A \frac{g}{\omega V_1^0} \pi R k_0 J_1{}'(k_0 R) \mathbf{A} \mathbf{V}^0$$

The appropriate boundary matrix is $\mathbf{C}_1^s$ ($n = 1$, symmetric waves, Eq. 21). Imposing the condition of no flow at the surface of the cylinder, the vector of velocity potentials is obtained as:

$$\mathbf{P} = -(\mathbf{C}_1^s)^{-1} \mathbf{U}^0 + \mathbf{P}^0 = 2A \frac{g}{\omega V_1^0} \left[J_1{}'(k_0 R)/H_1^{(2)}{}'(k_0 R) \right] H_1^{(2)}(k_0 R) \mathbf{V}^0 + \mathbf{P}^0$$

The pressures on the cylinder can then be computed and combined to obtain the total horizontal force and moment. Fig. 2 shows a comparison of the results with the analytical solution. It can be seen that the differences for 16 sublayers are very small for the range of $k_0 R$ studied.

In the case of an axisymmetric body of arbitrary geometry, it is necessary to consider a core region with finite elements. Letting $\mathbf{S}$ be the matrix relating velocity potentials to consistent nodal values of the flux in the finite element region and using the subscript 2 for the nodes on the transmitting boundary and 1 for the remaining nodes, the final equations become:

$$\begin{bmatrix} \mathbf{S}_{11} & \mathbf{S}_{12} \\ \mathbf{S}_{21} & \mathbf{S}_{22} + \mathbf{C} \end{bmatrix} \begin{bmatrix} \mathbf{P}_1 \\ \mathbf{P}_2 \end{bmatrix} = \begin{bmatrix} 0 \\ -\mathbf{U}^0 + \mathbf{C} \mathbf{P}^0 \end{bmatrix}$$

where $\mathbf{C}$ denotes the appropriate boundary matrix. Fig. 3 shows the results for a rigid conical body, fixed at the base, with an angle of 30° and three values of the ratio of the radius to water depth. These results are in excellent agreement with those presented by Isaacson (9).

Most solutions of linear diffraction problems presented to date assume a rigid, immovable body. The finite element formulation allows, however, to consider a rigid movable body as well as flexible structures. The degrees of freedom associated with a rigid mass or a flexible structure are displacements and forces which are related for a steady state harmonic motion by a dynamic stiffness matrix of the form $\mathbf{K} - \omega^2 \mathbf{M}$. In order to couple the equations for the structure with those of the fluid domain, the displacements can be converted to velocities via multiplication by $i\omega$ and to equivalent nodal values of the flux by applying the same transformation matrix used for the fluid. The forces can be related in a consistent way to nodal values of the pressure which are, again, related to the values of the velocity potential since $p = -i\omega\rho\phi$ (ρ is the mass density of water). Thus, it is possible to modify the dynamic stiffness matrix of the structure, or the equations of motion of a rigid mass, expressing the degrees of freedom in terms of the nodal values of the potential and the flux. These equations can then be coupled directly with the transmitting boundary matrix in the case of a flexible cylinder or with the appropriate submatrix of the fluid finite element mesh. Fig. 4 shows the resultant moment on a fixed circular dock and an identical movable one with mass of 10^6 slugs (14.593×10^6 Kgs). The solution for the fixed dock is in excellent agreement with the results presented by Garrett (5). The solution for the movable dock assumes that the displacements are small.

Chen and Hudspeth (3) have suggested an approximate procedure to determine diffraction forces on rigid cylinders due to nonlinear waves. In their approach, the free field kinematics are computed using a nonlinear wave theory (Stokes' second order theory) while the reflected waves are determined using a linear formulation. This procedure can be easily implemented within the present formulation, the only change occurring in the expressions for

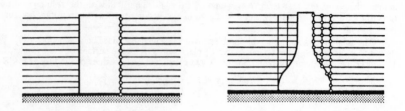

Figure 1: Discretization of the water region around axisymmetric bodies using finite elements in a core region and modelling the far field by means of a consistent transmitting boundary.

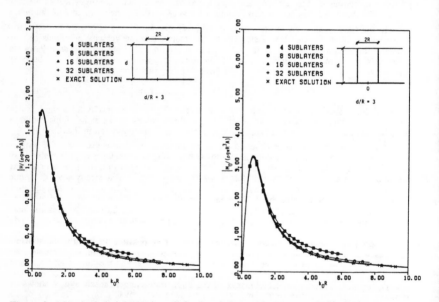

Figure 2: Magnitudes of the horizontal force and the moment on a vertical cylinder.

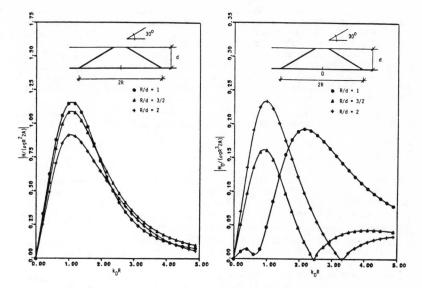

Figure 3: Magnitudes of the horizontal force and the moment on a circular cone.

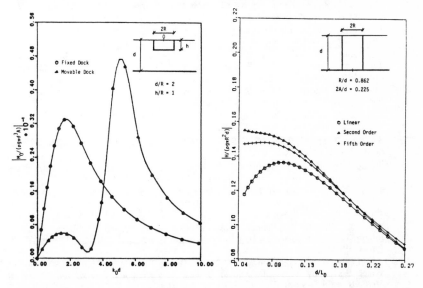

Figure 4: Magnitude of the moment on a fixed and an identical but movable circular dock.

Figure 5: Magnitude of the horizontal force on a vertical cylinder for linear, second order and fifth order Stokes waves.

the vectors U^0 and P^0 which will involve several terms depending on the wave theory used. Fig. 5 shows the resultant horizontal force on a rigid circular cylinder computed using linear, second order and fifth order Stokes waves. The solution for second order waves coincides with that of Chen and Hudspeth over the entire range of values of d/L_0 for which they presented results (d is the water depth and $L_0 = 2\pi g/\omega^2$, the wavelength in water of infinite depth). The solution for fifth order waves is in very good agreement with approximate results obtained by Chakrabarti (2). The solutions coincide for values of d/L_0 greater than 0.12. For smaller values of d/L_0, the present solution gives results which are slightly below those of Chakrabarti with a maximum difference of about 2% at $d/L_0 \approx 0.05$.

Additional results for fixed, movable and flexible bodies of various geometries using linear and approximate nonlinear diffraction theory can be found in Refs. 6 and 11.

References

1. Bettess, T. and Zienkiewicz, O.C., "Diffraction and Refraction of Surface Waves Using Finite and Infinite Elements," *International Journal for Numerical Methods in Engineering*, Vol. 11, 1977, pp. 1271-1290.

2. Chakrabarti, S.K., "Nonlinear Wave Forces on a Vertical Cylinder," *Journal of the Hydraulic Division*, ASCE, Vol. 98, No. HY11, November 1972.

3. Chen, M.-C. and Hudspeth, R.T., "Nonlinear Diffraction by Eigenfunction Expansions," *Journal of the Waterway, Port, Coastal and Ocean Division*, ASCE, Vol. 108, No. WW3, August 1982.

4. Chen, H.S. and Mei, C.C., "Oscillations and Wave Forces in a Man-Made Harbor in the Open Sea," in *Proceedings, Tenth Symposium on Naval Hydrodynamics*, pp. 573-596, Cambridge, Massachusetts, 1974.

5. Garrett, C.J.R., "Wave Forces on a Circular Dock," *Journal of Fluid Mechanics*, Vol. 46, Part 1, 1971, pp. 129-139.

6. Hsu, Y.C., "Diffraction Analysis for Large Axisymmetric Offshore Structures," *Ph.D. Dissertation*, The University of Texas, Austin, Texas, December 1984.

7. MacCamy, R.C. and Fuchs, R.A., "Wave Forces on Piles: A Diffraction Theory," *Technical Memorandum No. 69*, Beach Erosion Board, U.S. Army Corps of Engineers, December 1954.

8. Mei, C.C., "Numerical Methods in Water-Wave Diffraction and Radiation," *Annual Reviews of Fluid Mechanics*, Vol. 10, 1978, pp. 393-416.

9. Isaacson, M. de St. Q., "Nonlinear Wave Effects on Fixed and Floating Bodies," *Journal of Fluid Mechanics*, Vol. 120, 1982, pp. 267-281.

10. Sarpkaya, T. and Isaacson, M. de St. Q., *Mechanics of Wave Forces on Offshore Structures*, Van Nostrand Reinhold Co., 1981.

11. Syriopoulou, D., "A Numerical Technique for the Analysis of Water-Wave Diffraction," *M.S. Thesis*, The University of Texas, Austin, Texas, May 1983.

12. Tassoulas, J.L. and Kausel, E., "Elements for the Numerical Analysis of Wave Motion in Layered Strata," *International Journal for Numerical Methods in Engineering*, Vol. 19, 1983, pp. 1005-1032.

13. Waas, G., "Linear Two-Dimensional Analysis of Soil Dynamics Problems in Semi-Infinite Layered Media," *Ph.D. Dissertation*, University of California, Berkeley, California, 1972.

14. Yue, D.K.P., Chen, H.S., and Mei, C.C., "A Hybrid Element Method for Diffraction of Water Waves by Three-Dimensional Bodies," *International Journal for Numerical Methods in Engineering*, Vol. 12, 1978, pp. 245-266.

EFFECT OF SEISMIC WAVE INCLINATION ON STRUCTURAL RESPONSE

M. Tabatabaie,[1] A.M. ASCE, N. Abrahamson,[2]
and J. P. Singh,[3] M. ASCE

ABSTRACT

A new procedure is developed using the computer program SASSI to evaluate the effects of inclined propagating shear waves on three-dimensional (3-D) seismic soil-structure interaction (SSI) response. An example of an offshore gravity structure is used to illustrate the procedure. Using the recorded motions from the two-dimensional (2-D) dense strong motion array in Taiwan (SMART-1), the angle of incidence of incoming shear waves is estimated at 20 degrees. SSI analyses are performed for the same input motion using vertical as well as inclined wave propagation. The results show an increase of up to 50 percent in the horizontal response of the structure using the inclined wave assumption as opposed to the vertically propagating wave assumption.

INTRODUCTION

In the past, seismic SSI analyses of nuclear power plants, off-shore platforms, and other important structures have generally been performed using seismic wave fields consisting primarily of vertically propagating body waves (p-, SV- and SH-waves). This assumption may not be valid if, for example, the structure is sited close to a potential seismic source or if it is supported on a sediment-filled basin. During an earthquake, such structures will likely be subjected to strong ground motions due to inclined propagating body waves and surface waves (R- and L-waves). A comprehensive seismic SSI analysis must, therefore, consider appropriate angle of incidence for the seismic waves.

While current methodology for SSI analysis can handle complex seismic wave patterns, little information exists regarding the composition of seismic wave fields. This has resulted from a limited understanding of the effects on strong ground motions of seismic source mechanism and transmission path. Recent recordings of strong ground motion obtained from dense arrays have significantly enhanced our understanding of the properties of seismic wave fields.

We have developed procedures to study the effects of inclined propagating shear waves (SV- and SH-waves) on the three-dimensional seismic SSI response. An example of an offshore gravity platform is used to illustrate this problem. Estimates of the angle of incidence for inclined shear waves were obtained by analyzing the recorded motions from SMART-1, which provided the first recordings of strong

---

[1] Associate Engineer, [2]Senior Seismologist, [3]Director, New Technology Group, Harding Lawson Associates, San Francisco, California

ground motion made by a dense 2-D array (1). This was done by using
the array recordings to measure the slowness of the seismic waves and
then applying an understanding of the local velocity structure below
the array to estimate the angles of incidence of various wave types.

Comparing the results obtained using the assumption of inclined
propagating shear waves with those obtained using the assumption of
vertically propagating shear waves, we see a considerable increase in
the predicted response of the example offshore structure when the
inclined wave assumption is used.

METHOD OF ANALYSIS

The computer program SASSI (4) was used to perform the 3-D SSI
analysis. This program uses the flexible volume method, a general
finite element substructuring procedure. The method permits true 3-D
modeling of soil-structure systems which are subjected to ground
motions resulting from inclined propagating body waves and surface
waves. The method also allows modeling of structures with multiple
foundations of arbitrary shape supported on or embedded in horizontal
viscoelastic layers on top of a rigid base rock or a halfspace.

In the flexible volume method, the complete soil-structure system
is subdivided into two substructures, namely the foundation and the
structure. The foundation is solved first, and the impedance and
scattering properties are established at the soil-structure interface.
These properties are then input as boundary conditions and load vectors
in the dynamic finite element analysis of the structure. In performing the partitioning, the mass and stiffness of the structure are reduced by the corresponding properties of the volume of soil excavated,
but are retained within the halfspace. The interaction is thus assumed
to occur over a volume rather than at boundaries of the foundation.
This greatly simplifies the analysis procedure in that the impedance
problem is reduced to a series of axisymmetric solutions for the
response of a horizontally layered site to point loads, and the scattering problem is reduced to a free-field problem.

THE CONDEEP PLATFORM

The procedures developed are applied to study the response of the
Condeep platform (Figure 1). The structure is a gravity-type platform
designed for water depth of 240 meters. The structure consists of a
deck, a tower, three legs, and three foundation pads. Its height from
seabed to the helicopter deck is 284 m. The foundation pads are approximately 50 m in diameter and 180 m from center to center; each pad
includes 12 concrete cells, approximately 15 m in height. The pads
are connected to each other slightly above the foundation level.

Site Conditions

The soil profile at the platform site consists of 120 m of dense
fine silty sand overlying rock. The sand has total unit weight of
2 tons per cubic foot. Figure 2 shows the variation of low-strain
shear modulus of the sand with depth. Poisson's ratio was assumed to
be 0.45 for the profile.

Figure 1. General View of the
Condeep Platform

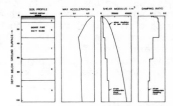

Figure 2. Strain Compatible Soil Properties from
Free Field Column Study

Seismic Criteria

Free-field input motion to the SSI analysis was obtained by
analyzing the strong ground motion recordings of the January 29, 1981
Taiwan earthquake from the SMART-1 array (Figure 3). The free-field
ground motions recorded at central station C-00 were analyzed, and the
resulting SH- and SV-wave components with recorded peak ground accelerations of 0.1 and 0.11g were selected as the x- and y-components of
the free-field control motion at ground surface for SSI analysis, respectively. The peak horizontal ground acceleration of both components
was scaled to 0.25g for analysis. Figure 4 shows the response spectrum of the two selected control motions.

Figure 3. Smart-1 Array Location
and Configuration

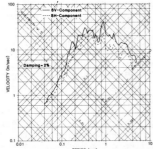

Figure 4. Horizontal Components of
Control Motion

The angle of incidence of the shear waves was estimated using
frequency-wavenumber analysis (2). Figure 5 shows the S-wave power
contours in wavenumber space at 1 Hz. The center of the plot is at
zero slowness (infinite velocity) and corresponds to vertical incidence. Moving away from the center of the plot, the slowness increases
(apparent velocity decreases), which corresponds to increasing angle
of incidence. In Figure 5, the spectral peak labeled "A" indicates
that the plane wave is equivalent to an apparent velocity of 3.6
km/sec. Based on measurements of the velocity structure below the
array, the apparent velocity of 3.6 km/sec corresponds to an angle of
incidence of about 20 degrees. This angle was used as the angle of
incidence of shear waves in the underlying rock for the SSI analysis.

The analyses previously performed on this gravity platform indicate that frequencies above 3 Hz are of minor importance for the SSI response of the structure (3). Hence, to save computer time and storage, it was decided to modify the original time history for the interaction analysis by removing all frequencies above 3 Hz. The response spectrum of the modified control motion, SV- and SH-wave components, are shown in Figures 9(a) and 9(c), respectively. The maximum acceleration of the motion did not change significantly for the SV-wave component but dropped about 50 percent for the SH-wave component. P-waves and surface waves were not considered in this study.

Site Response Analysis

A one-dimensional shear wave propagation analysis was performed using the computer program SHAKE (5) to obtain strain-compatible soil properties shown in Figure 2. The y-component of the control motion was used as input motion into the profile.

Three-Dimensional SASSI Analysis

We used the computer program SASSI to perform the SSI analysis. The SASSI procedure assumes that the site is horizontally layered. This assumption may not be appropriate for the region immediately under foundation pads. That is, on one hand, the confining pressure due to the static weight of the structure tends to increase the stiffness of the sand, while on the other hand, high shear strains due to interaction tend to decrease the effective stiffness. Both effects can be approximated by including in the structural model a number of solid elements which span the region over which the irregularities occur. Using this procedure, it is possible to perform the 3-D SASSI analysis at the actual induced soil-strain level (one iteration) and reduce the computational effort.

The increase in confining pressure due to the weight of the structure was determined by static analysis. In making this calculation, it was assumed that the three pads act independently; the stresses were added to those in the free field, and the resulting mean effective confining pressure was used to estimate the low-strain dynamic shear modulus under each pad, as shown in Figure 6. These low-strain dynamic moduli were then adjusted for strain effects due to interaction using a simple 2-D model. The 2-D model includes a single pad and a single damped oscillator with one-third the mass of, and the same fundamental frequency as, the structure. Assuming vertically propagating shear waves with the SV-component of the control motion specified at mud line level, the analysis yielded the approximate strain-compatible shear moduli and damping values for the soil region below each pad, as shown in Figure 7.

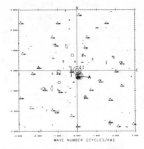

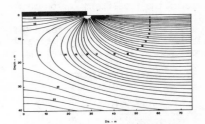

Figure 5.   Frequency-Wavenumber
Spectrum at 1 Hz

Figure 6.   Variation of Low Strain Shear Modulus
Beneath a Pad (G_{max}-*10^3 t/m^2)

The 3-D structural finite element SASSI model is shown in Figure 8.
The superstructure was idealized using beam elements and lumped masses.
The pads and a region of soil beneath each pad were modeled using
8-node solid elements.   The latter region was estimated from Figure 7
and was included to account for the weight of the structure and the
nonlinear effects in this region, as described previously.   The hydrodynamic inertia effects were included in the analysis by using an
added mass coefficient, and hydrodynamic damping was added to the
structural damping by specifying a constant damping ratio to represent
both damping effects.   The details of the structural finite element
and properties of the model are given in (3).   The inclined shear wave
assumption considered the waves to propagate in the y-z plane.

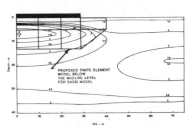

Figure 7.   Variation of Strain Compatible Shear Modulus
Beneath a Pad (G_{max} - *10^3 t/m^2)

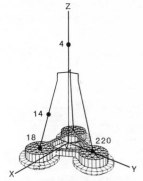

Figure 8.   View of Finite Element Model for
SASSI Analysis.

ANALYSES, RESULTS, AND DISCUSSION

The results of the 3-D SSI analysis using vertically and inclined
propagating shear waves are computed at nodes 4, 14, 18, and 220 as
shown on Figure 8.   The results in terms of 2 percent response spectra
are compared in Figures 9 through 10.   Figures 9(a) and (b) show the
response spectra corresponding to the motions calculated in the
x-direction at nodes 18 and 220 on the pads, respectively.   Figures

9(c) and (d) show the same results calculated in the y-direction at nodes 18 and 220, respectively. Superimposed on these figures are the spectra of the free-field motion in the x- and y-directions corresponding to the SH- and SV-wave components of the control motion at the mud line level, respectively. Comparing the results obtained assuming vertically and inclined propagating shear waves, we find that both assumptions result in horizontal pad motions which are essentially the

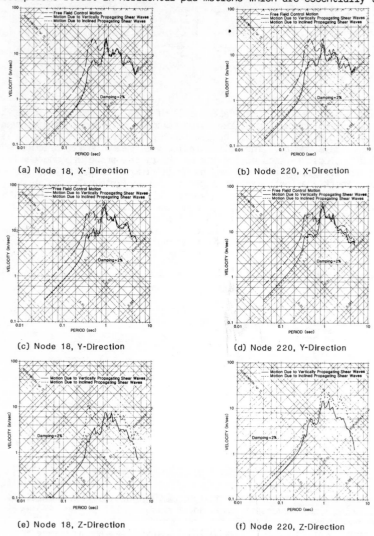

(a) Node 18, X-Direction

(b) Node 220, X-Direction

(c) Node 18, Y-Direction

(d) Node 220, Y-Direction

(e) Node 18, Z-Direction

(f) Node 220, Z-Direction

Figure 9. Comparison of Calculated Motions at the Base of the Structure

same but are considerably lower than those of the free-field response. The decrease in the structural base motion compared to that of the free-field is on the order of 60 percent at frequencies above 1.2 Hz for both vertically and inclined propagating shear waves; this is attributed to the interaction between the soil and the platform structure. Figures 9(e) and (f) compare the vertical motion of the pads calculated at nodes 18 and 220, respectively. Significant asymmetric rocking of the platform can be seen. The magnitude of rocking, however, is twice as great for the inclined propagating waves as for the vertically propagating waves.

Comparisons of the motions calculated on the superstructure are shown on Figures 10(a) and (b) for node 14 and on Figures 10(c) and (d) for node 4. The acceleration responses calculated near the top of the structure at node 4 in the x- and y-directions are about 10 percent lower at frequencies above 1.2 Hz assuming inclined propagating shear waves, as compared to the responses calculated for vertically propagating shear waves. As shown in Figure 10(b), this difference is more marked on the leg of the platform, where the calculated acceleration response at node 14 is about 50 percent greater than that calculated for vertically propagating shear waves. This considerable increase in the motion is attributed primarily to the higher rocking effects in the structure due to inclined wave propagation and the special geometry of the structural support.

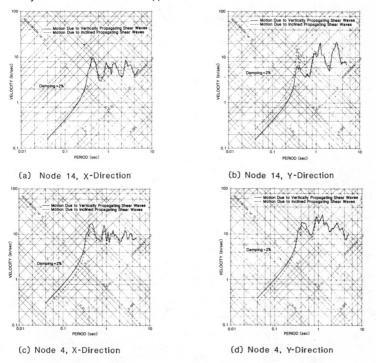

(a) Node 14, X-Direction

(b) Node 14, Y-Direction

(c) Node 4, X-Direction

(d) Node 4, Y-Direction

Figure 10. Comparison of Calculated Motions on the Superstructure

SUMMARY AND CONCLUSIONS

A new procedure is developed using the computer program SASSI and seismic wave-field information obtained from the SMART-1 array to analyze the 3-D SSI response of the Condeep platform subjected to inclined propagating shear waves. Using an angle of incidence of 20 degrees for inclined propagating shear waves in rock, we calculated the SSI response of the platform at foundation and superstructure levels and compared it to the response calculated for vertically progagating shear waves. Comparison of the results leads to the following conclusions.

1. A considerable decrease in the horizontal structural base response results from interaction between the soil and structure using both vertically and inclined wave propagation assumptions.
2. Considerable rocking in the structure results from SSI effects. The magnitude of rocking was approximately twice as great for the inclined as compared to the vertically propagating waves.
3. The computed response of the tower was basically the same for both wave propagation assumptions (difference less than 10 percent). However, the predicted response of the platform legs increased about 50 percent when inclined rather than vertically propagating waves were assumed.

ACKNOWLEDGMENTS

The authors gratefully acknowledge the support for this study provided by Harding Lawson Associates. The data for the Condeep platform were obtained from a previous study sponsored by the Norwegian Geotechnical Institute, Oslo, Norway, performed at the University of California, Berkeley. The SMART-1 array recordings were provided by the seismographic station at the University of California, Berkeley, and the Institute of Earth Sciences, Taiwan National University.

REFERENCES

1. Bolt, B. A., Loh, C. H., Penzien, J., Tsai, Y. B. and Yeh, Y. T., "Preliminary Report on the SMART-1 Strong Motion Array in Taiwan." Earthquake Engineering Research Center Report UCB/EERC-82/13, Berkeley, California, 1982.
2. Capon, J., "Signal Processing and Frequency Wavenumber Spectrum Analysis for a Large Aperture Seismic Array." Methods of Computational Physics, 1973.
3. Lysmer, J. and Tabatabaie, M., "Three-Dimensional Soil-Structure Interaction Analysis of the Condeep T300-Mono-Tower Platform." Report submitted to Norwegian Geotechnical Institute, Oslo, Norway by the University of California, Berkeley, March 1981.
4. Lysmer, J., Tabatabaie, M., Tajirian, F., Vahdani, S. and Ostadan, F., "SASSI - A System for Analysis of Soil-Structure Interaction." Report UCB/GT/81-02, Department of Civil Engineering, University of California, Berkeley, April 1981.
5. Schnabel, P. B., Lysmer, J. and Seed, H. B., "SHAKE - A Computer Program for Earthquake Response Analysis of Horizontally Layered Sites." Report EERC/70-10, Earthquake Engineering Research Center, University of California, Berkeley, December 1972.

APPROXIMATE SOLUTIONS FOR LAYERED FOUNDATIONS

A. J. Philippacopoulos [*]

A simple approximation to the dynamic response of foundations resting on layered soil profiles is presented. The approximation is based on an equivalent radiation damping which is numerically evaluated for different foundation configurations. Comparisons between approximate and rigorously computed foundation response are presented.

INTRODUCTION

Soil-structure interaction analysis involving general type of layered sites is usually performed by finite element computer programs. When, however, horizontal layering can be justified, then the supporting medium is idealized by mathematical models such as the stratum or more generally, the layered halfspace. Thus, using the stratum or the layered halfspace impedance functions, one can pefrorm the analysis by the substructure method, provided that deformations in the linear range are of interest.

Foundation impedances for foundations on uniform halfspace are readily available. For practical applications the numerical data published by Veletsos (13) for the circular and Wong and Luco (17) for the rectangular foundations may be used. With respect to the case of foundations resting on the elastic and viscoelastic stratum, analytical solutions, parametric studies using finite element models as well as approximate formulas have been reported (1,2,3,6,7,11,14). In the latter studies, impedances for circular(1,11,14), rectangular(6,7) and strip foundations (2,3) were investigated. Finally, the most known analytical investigations of the dynamic stiffness associated with the layered halfspace foundation model are those published by Kashio(5), Wei (15) and Luco (8,9).

Obviously, the above list of references is not intended to be exhaustive, in view of the fact that a substantial amount of literature on the subject of dynamically excited uniform halfspace, stratum and multilayered half space is available. It may be noted, however, that in many recent soil-structure interaction evaluations, numerical results for foundation impedances associated with layered sites are obtained by application of the general procedure decribed by Wong and Luco (16). In the latter paper, results were presented for foundations resting on uniform halfspace. Emphasis was placed on the ability of the method to handle foundations with arbitrary geometry. The procedure, however, is equally applied to generate the dynamic stiffness of foundations resting on layered halfspaces by simply replacing the Green's functions of the uniform halspace with those of the layered halfspace. The numerical routine remains otherwise the same. The procedure has been implemented into the CLASSI computer program.

Although, analytical and numerical methods exist today for the evaluation of the dynamic stiffness of foundations resting on the uniform and layered halfspace models, it is desirable to have simplified formulas which give reasonable approximation to these type of problems. Such approximations can be used in preliminary design as well as for checking the overall results from computer codes. Approximate formulas have been proposed especially for the cases of the uniform halfspace and stratum. A review can be found in a recent paper by Pais and Kausel (10).

[*] Civil Engineer, Brokhaven National Laboratory, Upton, New York, 11973

Generally speaking, it is more difficult to derive simple approximations for the dynamic stiffness of foundations on layered soil profiles than on uniform ones. (It is well known that the wave propagation is dispersive in the former and non-dispersive in the latter case). An interesting simplified approximation to this probem has been reported by Kashio (5). According to the latter, approximate impedance functions for a circular foundation supported by a layer-halfspace medium were derived considering a conical bar and a conical shear beam model. Later on, based on a similiar idea, Gazetas (4) proposed a simple model to compute radiation damping for foundations resting on the uniform and on some type of nonhomogeneous halfspace.

More recently, a simple approximation of the dynamic stiffness of foundations resting on a layer-halfspace medium has been developed by the author (12). The accuracy of the approximation has been investigated for massless foundations as well as foundations with mass. The present paper extents the applicability of the approximation to the case of flexible structures.

FOUNDATION RESPONSE

The foundation considered here, consists of a horizontal layer with constant thickness H underlain by a very deep alluvial which is treated as a uniform half space. The velocity of dilatational and shear waves propagating in the material of the layer are α_L and β_L respectively. The coresponding wave velocities in the halfspace are α_H and β_H respectively. It is assumed that the halfspace is acoustically stiffer than the layer.

Let $H_i(\omega)$ be the transfer function between applied force and response displacement of a rigid massless disk having radius R, which is placed at the surface of the layer-halfspace system. Typically, this function is referred to as foundation compliance. The foundation response to a general forcing function is obtained in the frequency domain by the relation

$$\mathcal{F}_{u_i}(\omega) = H_i(\omega) \, \mathcal{F}_{P_i}(\omega) \tag{1}$$

where:
H_i : transfer function
P_i : forcing function
u_i : displacement response function

The symbol $\mathcal{F}$ denotes Fourier Transform and the subscript i is used in place of the symbols H and R to denote horizontal and rocking mode of vibration. Since the foundation is assumed to be massless, $H_i(\omega)$ is given by the expression

$$H_i(\omega) = \frac{1}{K_i(\omega) + i\omega C_i(\omega)} \tag{2}$$

where K_i, C_i are the real and imaginary parts of the foundation complex stiffness. The latter are obtained numerically by finite element computer programs and analytically by rigorous solutions of a mixed boundary-value problem for the wave equation. In the latter case, displacements are prescribed at the contact area between foundation and soil, the tractions vanish at the free surface and typically "welded" contact is assumed at the interface between the layer and the halfspace. Finally, boundary integral methods have been used to find K_i, C_i. Assuming a contact stress distribution, enables one to obtain dynamic stiffness functions without having to formulate the question as a boundary value problem.

An analytical expresion was proposed by ths author (12) to approximate $H_i(\omega)$. Based on a purely numerical study, it was shown that $H_i(\omega)$ can be approximated by the equation

$$H_i^a(\omega) = \frac{1}{K_i^0} \quad \frac{1}{-\lambda^2\eta^2 + i\gamma\lambda\eta + 1} \tag{3}$$

where
K^1_0 = foundation static stiffness

$\eta = H/\Lambda$

$\lambda = \{8(2-\gamma^2)/A_f\}^{1/2}$

$A_f = 4Hf_{max}/\alpha_L,\beta_L$

Λ = wavelength of shear or dilatational waves in the layer

EQUIVALENT DAMPING

A little thought shows that the approximation expressed by Eq. 3 is a one-mode type of approximation. The basic parameter involved, is the gama (γ) factor which measures the equivalent damping of the vibration of the foundation. Specifically, if ζ^e represents the equivalent damping, then $\zeta^e = \gamma/2$.

A procedure has been applied succesfully to obtain equivalent damping values associated with the vibration of foundations resting on layered soil profiles (12). According to this procedure, the equivalent damping is evaluated by matching of the foundation compliances. This was accomplished by a purely numerical process. An analytical justification is underway and equivalent damping values for representative layering configurations will be given in a future publication.

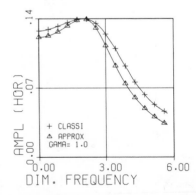 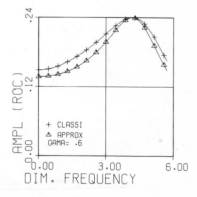

Figure 1. Compliance Functions. Layering Case: β_L/β_H=0.60, H/R=0.50

Application of the procedure are shown in Figs. 1 and 2 for β_L/β_H equal to 0.60 and 0.30 respectively. In these figures, the moduli of the compliance functions computed from Eq. 3 are plotted versus those obtained by the CLASSI program. The equivalent damping is shown to be 50% in translation and 30% in rocking for the case $\beta_L/\beta_H = 0.60$ (Fig. 1). The corresponding damping values for the $\beta_L/\beta_H = 0.30$ case are shown to be 30% and 20% respectively (Fig. 2). Thus for higher shear wave velocity ratios the equivalent damping value increased. This is reasonable, since as β_L/β_H approaches one, the problem is reduced to the case of the uniform halfspace in which more energy is radiated away. At lower β_L/β_H values, the energy remains close to the layer-halfspace interface and thus not radiated away in the supporting halfspace.

BUILDING RESPONSE

The method was applied to evaluate structural responses for building-foundation systems. In the latter, the foundation medium is represented by a layer over a halfspace. The superstructure is considered to vibrate according to its fundamental fixed-base mode having frequency f_s (cps). Based on this assumption, the structural mass m is lumped at the c.g. which is located at a height h above the ground surface. Hysteretic damping ζ^s is assumed for the fixed-base motion of the structure. This type of simplified superstructure models have been often used in soil-structure interaction studies.

The approximation presented in this paper, was applied to computations of the harmonic as well as the transient response of the soil-structure system. The results described below were obtained for a fixed-base frequency equal to 5 cps and a hysteretic structural damping ratio equal to 4%. The structural mass and the height h were treated as free parameters in order to adjust the soil-structure system frequencies.

<u>Harmonic Response</u>

The approximation of the harmonic response of the soil-structure system was evaluated in

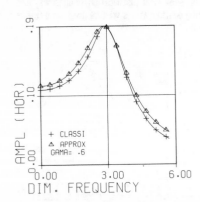
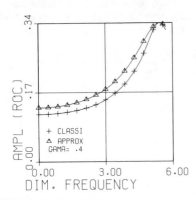

Figure 2. Moduli of Compliance Functions. Case: $\beta_L/\beta_H = 0.30$, H/R=0.50

terms of transfer functions. Plots of the latter are shown in Figs. 3 and 4 for ratios β_L/β_H equal to 0.6 and 0.3 respectively. The transfer functions shown in these figures are associated with the structural motion relative to the base. Results for layer depth-to-radius (H/R) ratios equal to 0.5 and 1.0 are presented. From the plots given in Figs. 3, 4 it can be seen that the transfer functions computed by the approximation presented here, compare very well with those obtained using CLASSI generated impedance functions.

<u>Earthquake Response</u>
A synthetic acceleration time history representing the horizontal component of the free-field motion was used to compute absolute accelerations of the structure. Soil-structure interaction evaluations were performed in the frequency domain using Fourier Transformations. Based on the later, the absolute acceleration of the structure was computed for the foundation layering cases employed in the harmonic response computations discussed above. In Fig. 5 response waveforms (absolute accelerations) are plotted for a 10 sec duration for the layering case: β_L/β_H=0.3 , H/R=1.0. The agreement shown is excellent.

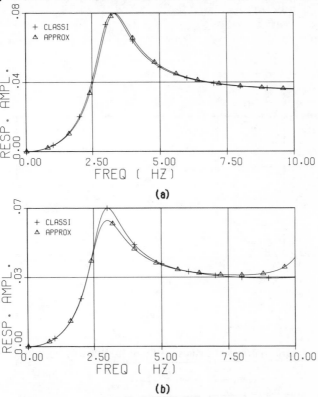

Figure 3. Transfer Functions. β_L/β_H=0.60. H/R=0.50 (a), 1.0 (b)

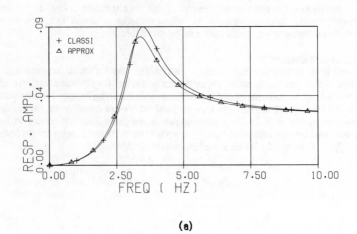

(a)

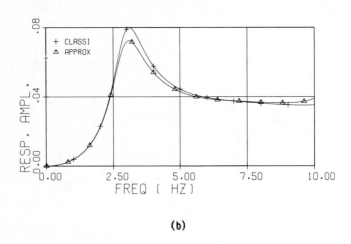

(b)

Figure 4. Transfer Functions. β_L/β_H=0.30. H/R=0.50 (a), 1.0 (b)

DISCUSSION

The simplicity of form of the approximation presented in the previous sections is striking. It should be kept in mind, however, that the nature of the problem which is approximated here is generally very complex. Accuracy limitations associated with this approximation are discussed next.

Structural accelerations computed using this approach, showed generally poor agreement with the CLASSI results for large layer depth-to-radius ratios (i.e., H/R=3). In general, the accuracy deteriorated when the frequencies at the maximum moduli of the compliance functions are in the low dimensionless frequency range (smaller than 0.5). This is due to the fact that under such conditions, the layered foundation approaches the behavior of the uniform halfspace. Numerical data generated for larger layer depth-to-radius ratios, indicated that the agreement between the moduli of the compliance functions was not satisfactory. Specifically, these moduli matched only over a narrow frequency band. As expected, this was found to be more pronounced for the horizontal than the rocking motion.

In the high frequency range, (highest dimensionless frequency considered in the study is equal to 6) application of the procedure produced less accurate results. Although, the moduli of the compiance functions are predicted well at high frequencies, their phase angles are predicted with less accuracy (12). Plots of the corresponding impedance functions (inverse of compliance functions), showed that the real part of the latter demonstrates unusual high frequency oscillations. The imaginary part, however, compared well with the results from the CLASSI code. This observation suggests that some refinement of the phase angle will improve the accuracy of the approximation.

Despite the above limitations, predictions made for the response of the rigid foundations as well as flexible structures are in remarkable agreement with rigorous solutions.

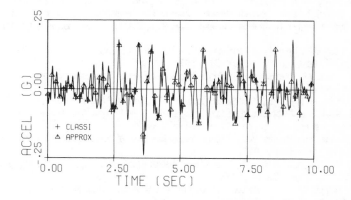

Figure 5. Absolute Structural Acceleration Time History

CONCLUSION

An approximate procedure to compute response of structures on layered soil profiles has been presented. This procedure is based on an equivalent damping associated with the steady-state motion of the foundation. Numerical data were presented in order to demonstrate the accuracy of the approximation. Based on these results it is concluded that the response of soil-structure systems is predicted with very good accuracy.

ACKNOWLEDGEMENT

The work reported here was sponsored by the United States Nuclear Regulatory Commission. Mr. Herman Graves is the Technical Monitor and his many contributions to the study are gratefully acknowledged. This paper reflects the views of the author and not necessarily the views of the sponsoring organization.

APPENDIX- REFERENCES

1. Bycroft, G.N., "Forced Vibrations of a Rigid Circural Plate on a Semi-Infinite Elastic Space or on an Elastic Stratum," Phil. Trans., Royal Soc. of London, Vol. 248, pp. 327-368, 1956.

2. Chang, L.V., "Dynamic Response of Structures in Layered Soils," Report R74-10, Dept. of Civil Engineering, MIT, 1974.

3. Gazetas, G., Roesset, J.M., "Forced Vibrations of Strip Footings on Layered Soils," National Structures Engineering Conference, ASCE, Madison, Wisconsin, 1976.

4. Gazetas, G., Dobry, R., "Simple Radiation Damping MOdel for Piles and Footings," J. of Engrg. Mech., ASCE, Vol. 110, No. 6, pp. 937-956, 1984.

5. Kashio, J., "Steady-State Response of A Circular Disk Resting on A Layered Medium," PhD Dissertation, Rice, Univ., 1971.

6. Kobori, T., et. al, "Dynamic Ground Compliance of REctangular Foundation on an Elastic Stratum," Proc. 2nd Japan Nat. Symp. on Earthq. Engrg., Japan, pp. 261-266, 1966.

7. Kobori, T., et al., "Dynamic Ground Compliance of Rectangular Foundationon an Elastic Stratum Over a Semi-Infinite Rigid Medium," Ann. Rept. Dis. Prev. Res. Inst. of Kyoto Univ., 1967.

8. Luco, J.E., "Impedance Functions for a Rigid Disk Foundation on a Layered Medium," J. Nucl. Engrg. and Des., Vol. 31, pp. 204-217, 1974.

9. Luco, J.E, "Vibrations of a Rigid Disc on a Layered Viscoelastic Medium," J. Nucl. Engrg. and Des., Vol. 36, pp.325-340, 1976.

10. Pais, A., Kausel,E., "Stochastic Response of Foundations," Report R85-6, MIT, 1985.

11. Paul, H.S., "Vibration of a Rigid Circular Disk on an Infinite Isotropic Elastic Plate," J. of Acous. Sos. of Am., Vol. 42, pp. 412-416, 1967.

12. Philippacopoulos, A.J., "Approximate Response of Some Layered Foundations," (submitted for publication), 1985.

13. Veletsos, A.S., Verbic, B., "Vibration of Viscoelastic Foundations," J. Earth. Engrg. and Struct. Dyn., Vol. 2, pp. 87-102, 1973.

14. Warburton, G.B., "Forced Vibration of a Body on an Elastic Stratum," J. of Appl. Mech., ASME, Vol. 24, pp. 55-58., 1957.

15. Wei, Y-T, "Steady-State Response of Certain Foundation Systems," PhD Dissertation, Rice Univ., 1971.

16. Wong, H.L., Luco, J.E., "Dynamic Response of Rigid Foundations of Arbitrary Shape," J. of Earthq. Engrg. and Struc. Dyn., Vol. 4, pp. 579-587, 1976.

17. Wong, H.L., Luco, J.E., "Tables of Impedance Functions and Input Motion for Rectangular Foundations," Rept. CE78-15, Univ. of Southern Calif., 1978.

Soil-Structure Interaction
at the Waterfront

J. M. Ferritto, MASCE*

Introduction

The Navy has $25 billion worth of facilities in seismically active regions. Each year $200 million of new facilities are added in those seismically active areas. The Navy, because of its mission, must locate at the waterfront often on marginal land with a high watertable. Seismically induced liquefaction is a major threat to the Navy. To understand the significance of liquefaction, it is important to note the damage caused in recent earthquakes.

Recent earthquakes, particularly those in Alaska, Japan, and Chile, have emphasized the high damage threat the soil liquefaction phenomenon poses to waterfront structures. In the 1960 Chilean earthquake (magnitude 7) quaywalls, sheet piles, and sea walls were damaged by liquefaction of loose, fine, sandy soils. In the 1964 Alaskan earthquake (magnitude 8.4) severe damage to Anchorage, Cordova, and Valdez occurred, including large-scale land slides, as a result of liquefaction. Japanese earthquakes (Niigata, 1964, magnitude 7.3; Miyagi-Ken-Oki, 1978, magnitude 7.4) caused severe waterfront damage to wharfs, bulkheads, quaywalls, piers, and conventional structures. The majority of the damage sustained in waterfront areas was primarily from liquefaction of loose, cohesionless sands (1).

As can be seen, liquefaction has played a major role in waterfront damage; most of the time being the single cause of widespread losses. Fortunately the United States has not suffered a devastating earthquake in recent years. However, the seismic risk is great, particulary in the West where, for example, it is estimated that there is a 5% annual probability of a major even in Southern California which could affect a number of Naval bases. The experience noted in recent earthquakes is that liquefaction greatly increased the amount of damage observed in waterfront facilities. Particular problems exist with sheet piles, quaywalls, wharfs, and embedded structures. Conventional buildings also suffer severe damage.

Present engineering practice does not consider the presence of the structure in determining the potential for liquefaction. Further, once liquefaction is evaluated, the resulting deformation state is not determined. Clearly this is inadequate to determine the likelihood of liquefaction and the consequences for a major structure. The finite

---

*Senior Research Engineer, Facility Systems Division, Naval Civil Engineering Laboratory, Port Hueneme, CA 93043.

element technique offers the potential for detailed analysis of soil structure problems and has been used extensively. Development of a constitutive model to predict pore pressure and characterize soils in terms of effective stresses (stress on soil grains causing deformation) rather than simple total stress offers the potential capability to analyze complex structures where liquefaction is possible. The finite element method has in recent years been a useful tool in analyzing structures, structures on soil, and soil-structures. An emphasis has been placed on use of nonlinear plasticity soil models to more accurately capture the soil response. Recent Navy research work has focused on effective stress analysis -- the ability to not only calculate the soil stress but also to calculate the pore fluid pressure.

An effective stress model can be used for analysis of ocean floor soils and nearshore and offshore structures, for seismic analysis, and for static consolidation. Oscillations in loading, whether from wave action or seismic shaking, produce a dynamic loading that can induce significant increases in pore pressure. The increase in pore pressure can reduce allowable soil capacities and increase deformations by a reduction in effective confining stress. Under extreme conditions flow slides and liquefaction occur. Although liquefaction has been identified as a phenomenon for 20 years, soil mechanics is just beginning to understand the interaction of stress confinement and drainage path, which occurs in the field, such as under a foundation. For example, common engineering practice in the evaluation of seismically induced soil liquefaction considers level ground conditions away from the structure. Shear stresses from the structure are not considered.

Princeton University Effective Stress Model

In 1981 the Naval Civil Engineering Laboratory (NCEL) undertook a review of available material models. Comparison was made between test data and model predictions (2). Based on this study, it was concluded that the material model under development at Princeton University by Professor J.H. Prevost was able to predict the behavior of cohesive and cohesionless materials, and that its development was well advanced into the implementation stage in a finite element code (3).

The Princeton University Soil Model represents soil as a two-component system: the soil skeleton and the pore fluid. The hysteretic stress-strain behavior of the soil skeleton is modeled by using the effective-stress elastic-plastic model reported in Reference 4. The model is an extension of the simple multi-surface J_2-plasticity theory and uses conical yield surfaces. The model has been developed to retain the extreme versatility and accuracy of the multi-surface J_2-theory in describing observed shear nonlinear hysteretic behavior, shear stress-induced anisotropy effects, and to reflect the strong dependency of the shear dilatancy on the effective stress ratio in granular cohesionless soils. The model is applicable to general three-dimensional stress-strain conditions, but its parameters can be derived entirely from the results of conventional triaxial soil tests. The yield function is selected of the following form:

$$f = \frac{3}{2} (\underset{\sim}{s} - p \underset{\sim}{\alpha}) \cdot (\underset{\sim}{s} - p \underset{\sim}{\alpha}) - m^2 p^2 = 0$$

where $\underset{\sim}{s} = \underset{\sim}{\sigma} - p\underset{\sim}{\delta}$ = deviatoric stress tensor; $p = 1/3$ tr $\underset{\sim}{\delta}$ = effective mean normal stress; $\underset{\sim}{\delta}$ = effective stress tensor; $\underset{\sim}{\alpha}$ = kinematic deviatoric tensor defining the coordinates of the yield surface center in deviatoric stress subspace; m = material parameter. The yield function plots as a conical yield surface (Drucker-Prager type) in stress space, with its apex at the origin, as shown in Figure 1. Unless $\underset{\sim}{\alpha} = \underset{\sim}{0}$, the axis of the cone does not coincide with the space diagonal. The cross section of the yield surface by any deviatoric plane (p = constant) is circular. Its center does not generally coincide with the origin, but is shifted by the amount p $\underset{\sim}{\alpha}$. This is illustrated in Figure 1 in the principal stress space. The plastic potential is selected such that the deviatoric plastic flow be associative. However, a non-associative flow rule is used for its dilatational component.

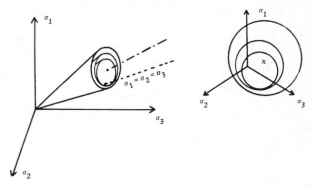

Figure 1. Yield surfaces in principal stress space.

In order to allow for the adjustment of the plastic hardening rule to any kind of experimental data, a collection of nested yield surfaces is used. The yield surfaces are all similar conical surfaces and a purely kinematic hardening rule is adopted. Upon contact, the yield surfaces are translated by the stress point and the direction of translation is selected such that no overlappings of the surfaces can take place. A plastic modulus is associated with each yield surface. Portions of the development of the research at Princeton University were under Navy sponsorship.

Validation - Laboratory Test Data

Soil test data required to develop model material parameters was obtained from a series of drained and undrained triaxial compression and extension tests conducted on a number of sands by universities and private geotechnical firms under NCEL sponsorship (5). Cyclic undrained tests and proportional loading tests were also conducted. With detailed soil test data available, NCEL began the task of validating the soil model by using the drained test data to determine model parameters and comparing predicted undrained monotonic and cyclic behavior with actual test data. In general the material model was capable of giving excellent representation of drained test data under a variety of loading conditions using parameters based on drained triaxial compression and extension. The model is capable of giving good agreement with undrained monotonic test data and can track the occurrence of liquefaction in cyclic tests in approximately the same number of cycles. Figures 2, 3, 4, and 5 give typical results.

Validation Centrifuge Model

The actual measurement of pore water pressure in the field during an earthquake would serve as the best source of data upon which to validate an effective stress soil model. Unfortunately these data are not available. The centrifuge has been used to study models of undrained soil deposits under seismic type excitation. This can serve as an approximate comparison given that the test data itself is only an indirect measure of field behavior and has errors associated with it.

A model study from Reference 6 was selected for analysis. The soil was Leighton-Buzzard 120/200 sand. The soil was deposited in a stacked-ring apparatus by pulviating the sand in layers into water and the rodding to achieve the desired density. A brass weight simulating a footing was placed on top of the saturated sand deposit. The test was conducted in a centrifuge under centrifugal acceleration of 80g. The deposit was then subjected to sinusoidal base acceleration input motion. The soil was discretized by using 240 elements and the brass footing by using two rows of 10 elements each. The brass footing is modelled as a one-phase elastic solid and a static pressure was applied to the top of the footing to achieve the static bearing pressure. The water table was located at the ground surface. No drainage of the pore fluid was allowed to take place through the rigid bottom boundary or through the impermeable sides. A ground shaking was applied as a horizontal sinusoidal input acceleration at the bottom boundary nodes, with a maximum acceleration = 0.17g and a frequency of 1 Hertz for a duration of 10 seconds (10 cycles).

The stacked-ring apparatus controlled the side boundaries in the test. Test boundary conditions were simulated by assigning the same equation number both to each nodal degree of freedom on the same horizontal plane for both side boundaries.

An analysis of the computed acceleration time histories shows slight attenuation of base motion at the top of the footing. The computed acceleration time histories at the corners of the footings show rocking motions are imported to the footing. Settlement beneath the footing increases continuously and almost linearly while shaking occurs. Further no additional significant settlements are computed to occur after the shaking stops as noted in the test. As observed in the test, in the "free-field" close to the sides, the pore-water pressure rises quickly. Directly under the structure, the pore-pressure increase is slower and always remains smaller than the pore-pressure in the free-field at the same elevation. Immediately following shaking, the excess pore-pressures dissipate rapidly and reach their steady state conditions 5 seconds after the end of the shaking. This is further illustrated in Figure 6 which shows time histories for the vertical effective stress and excess pore-water pressures. Also shown in Figure 6 is a comparison of computed pore-pressure in comparison with the measured values. The computer simulation gives qualitatively good agreement showing liquefaction to occur in the same regions as observed in the test and also gives comparable pore-pressure ratios.

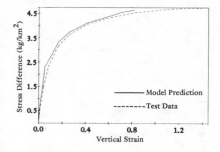

Figure 2. Drained compression test.

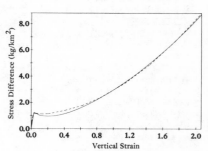

Figure 3. Undrained compression test.

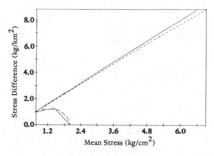

Figure 4. Undrained stress path.

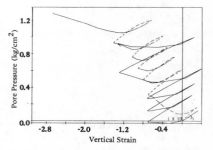

Figure 5. Cyclic test results.

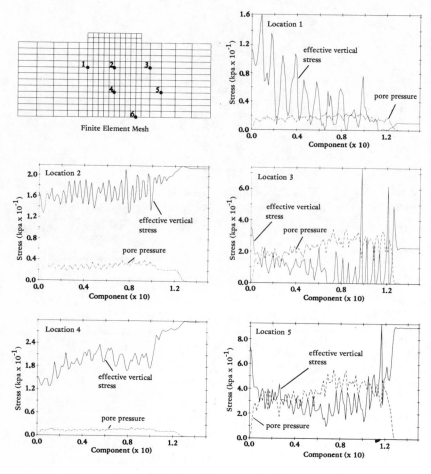

Peak Pore Pressure (kpa)

Location	Computer	Measured
1	20	20
2	30	30
3	30	25
4	15	12
5	50	65
6	40	60

Figure 6. Computed stress and pore pressure.

Conclusion

Work is underway validating the Princeton University Effective Stress Soil Model. When the effective stress soil model is validated, it will be an important tool in the analysis of waterfront construction. To obtain basic soil behavior, the Navy is participating with the Natural Science Foundation in sponsoring a test program to measure insitu pore water pressures during an earthquake. The Navy is also sponsoring, with the National Science Foundation, research to develop the pressuremeter into a tool capable of being used to obtain effective stress material model parameters directly eliminating the need for extensive triaxial testing.

Acknowledgment

The analytical studies presented in this paper were performed by Mr. Robert Slyh, Research Civil Engineer, Naval Civil Engineering Laboratory under the direction of Professor J.H. Prevost, Princeton University. Funding for this task was part of the Navy's Earthquake Mitigation Program.

Appendix

References

1. Werner, S.D. and Hung. "Seismic Response of Port and Harbor Facilities." Azbabian Associates. Report No. R-8122-5395, Oct 1982.

2. Ferritto, J.M. "The Effective Stress Soil Model." Naval Civil Engineering Laboratory, Technical Memorandum M-51-81-12. Port Hueneme, Calif., Aug 1981.

3. Prevost, J.H. "DYNAFLOW: A Nonlinear Transient Finite Element Analysis Program," Princeton University, Department of Civil Engineering. Princeton, N.J., 1981.

4. Prevost, J.H. "A Simple Plasticity Theory for Frictional Cohesionless Soils," International Journal of Soil Dynamics and Earthquakes Engineering, vol 4, no. 1, 1985.

5. Ferritto, J.M., et al. "Utilization of the Prevost Soil Model for Soil-Structure Problems." Naval Civil Engineering Laboratory, Technical Memorandum M-51-83-09. Port Hueneme, Calif., June 1983.

6. Whitman, R.V. and P.C. Lambe, "Liquefaction: Consequences for a Structure," Proceedings of the Soil Dynamics and Earthquake Engineering Conference, Southhampton, U.K., 1982.

INTERACTION EFFECTS IN VERTICALLY EXCITED STEEL TANKS

A. S. Veletsos, M.ASCE and Yu Tang[*]

ABSTRACT: Following a brief review of the hydrodynamic effects induced by a vertical component of ground shaking in liquid-filled, circular, cylindrical, steel tanks that are rigidly supported at the base, the effects of soil-structure interaction on the response of such systems are examined, and numerical data are presented with which the critical design forces for both rigidly and flexibly supported tanks may be evaluated readily.

INTRODUCTION

Recent studies of the dynamic response of liquid-storage tanks have shown that the hydrodynamic effects induced in flexible tanks by a vertical component of ground shaking may be significantly larger than those induced in rigid tanks of the same dimensions, and that the peak values of these effects may be of the order of the hydrostatic effects for a ground motion with a peak acceleration of one-third the acceleration due to gravity (e.g., Refs. 1,3,4).

These studies referred to rigidly supported tanks for which the motion experienced at the tank base is the same as the free-field ground motion. Although it has been recognized that, because of the high radiational damping capacity of vertically excited foundations, soil-structure interaction may significantly reduce the design forces for tanks supported on soft soils, the consequences of such action have not been adequately addressed to date.

The objectives of this paper are: (1) to review briefly the nature of the hydrodynamic effects induced in tanks that are rigidly supported at the base, paying special attention to the influence of the flexibility of the tank wall; (2) to highlight the principal consequences of soil-structure interaction; and (3) to provide information with which the critical design forces for both rigidly and flexibly supported steel tanks may be evaluated readily. The material is presented in physically motivated terms with minimum reference to the underlying mathematics. The mathematical details, along with complementary numerical data, are given in Ref. 5.

SYSTEM CONSIDERED

A ground-supported, upright, circular, cylindrical, steel tank of radius, a, height, H, and constant wall thickness, h, is considered. The tank is supported through a rigid base at the surface of a homogeneous elastic halfspace and is filled with liquid of mass density, ρ_ℓ. The liquid is considered to be incompressible, inviscid and free

---

[*]Brown & Root Professor and Graduate Student, respectively, Department of Civil Engineering, Rice University, Houston, TX 77251.

at its upper surface. The tank wall is presumed to be clamped at the base; its mass density is denoted by ρ; the modulus of elasticity and Poisson's ratio for the tank material are denoted by E and ν, respectively; and the corresponding quantities for the supporting soil are denoted by ρ_s, E_s and ν_s. The excitation is a vertical component of ground shaking with a free-field acceleration $\ddot{x}_g(t)$, in which t denotes time.

Points for the tank and the contained liquid are specified by the radial and axial components of a cylindrical coordinate system, r and z, the origin of which is taken at the tank base. The ground displacement and its derivatives are considered to be positive in the upward direction.

RIGIDLY SUPPORTED TANKS

Let p(z,t) be the hydrodynamic pressure exerted against the tank wall, positive when acting radially outward. For the conditions considered, this pressure is uniformly distributed in the circumferential direction and its axial distribution depends on the flexibility of the tank wall. For a completely rigid wall, the pressure increases linearly from top to bottom and may be expressed as

$$p(z,t) = [1 - \frac{z}{H}]\rho_\ell H \ddot{x}_g(t) \tag{1}$$

Note that the temporal variation of this pressure is the same as that of the acceleration of the input motion, and that its peak value at the junction of the wall and the base is $\rho_\ell H \ddot{x}_g$, in which $\ddot{x}_g$ = the maximum value of $\ddot{x}_g(t)$.

Tank flexibility affects both the magnitude, axial distribution, and temporal variation of the hydrodynamic wall pressure and of the associated tank forces. However, as is true of the impulsive components of the hydrodynamic effects for laterally excited tanks (Ref. 3), a reasonable, generally conservative, approximation to the resulting hydrodynamic effects may be obtained on the assumption that the axial distribution of p(z,t) is independent of the tank wall flexibility and the same as that for a rigid tank.

The hydrodynamic wall pressure under this assumption may be determined from Eq. 1 merely by replacing $\ddot{x}_g(t)$ with A(t). The latter quantity represents the pseudoacceleration of a similarly excited, single-degree-of-freedom oscillator, for which the natural frequency, f, and percentage of critical damping, ζ, are equal to those of the fundamental, axisymmetric, breathing mode of vibration of the tank-liquid system. The instantaneous value of the pressure is then given by

$$p(z,t) = [1 - \frac{z}{H}]\rho_\ell H A(t) \tag{2}$$

and its maximum value at any point is obtained by replacing A(t) with its maximum or spectral value, A. The latter value may be determined from the response spectrum for the prescribed ground motion using the values of f and ζ referred to above. Inasmuch as A may be much greater than $\ddot{x}_g$, it should be clear that the peak values of p(z,t) for a flexible tank may be significantly greater than for the associated rigid tank.

With the magnitude and distribution of the maximum values of the hydrodynamic wall pressure established, the corresponding values of the displacements of the tank wall and of its internal forces may be determined by static analysis. Since the distributions of the hydrodynamic and hydrostatic pressures are identical for the assumption made, the maximum effects of the two pressures will be proportional, the proportionality ratio being A/g, in which g is the gravitational acceleration. Although the contribution of the radial inertia of the flexible tank wall is not considered explicitly in this approach, it is provided for implicitly in the computation of p(z,t), as indicated in Refs. 4 and 5.

The linear variation of the hydrodynamic wall pressure represented by Eq. 2 is strictly valid only for very stiff tanks. For the tank proportions normally encountered in practice, an improved approximation is given by the expression,

$$p(z,t) = 0.8 \left[\cos \frac{\pi}{2} \frac{z}{H} \right] \rho_\ell H A(t) \tag{3}$$

which leads to essentially the same total wall force per unit of circumferential length as Eq. 2. However, this refinement is hardly warranted in practice because the tank forces and displacements for the pressure defined by Eq. 3 cannot be evaluated as readily as those for Eq. 2, and because these effects are not particularly sensitive to variations in the distribution of the wall pressure.

Natural Frequency of Tank-Liquid System

The fundamental natural frequency of axisymmetric vibration of the tank-liquid system may conveniently be expressed in the form

$$f = \frac{C_\ell}{2\pi} \frac{1}{H} \sqrt{\frac{E}{\rho}} \tag{4}$$

in which f is in cycles per unit of time, and C_ℓ is a dimensionless coefficient that depends on the values of H/a, h/a, ρ_ℓ/ρ and ν.

The values of C_ℓ for steel tanks filled with water (i.e., $\nu = 0.3$ and $\rho_\ell/\rho = 0.127$) are listed in Table 1 for several different values of H/a and h/a. Identified with the symbol C_w, these results were obtained by application of the Galerkin method using an appropriate configuration for the radial displacement of the tank wall, as indicated in Ref. 5. One observes that C_w is approximately proportional to the square root of h/a, and that for values of H/a in excess of unity, it is relatively insensitive to variations in H/a.

With the values of C_w established, the values of C_ℓ for tanks filled with liquid of arbitrary mass density, ρ_ℓ, may be determined from

$$C_\ell = C_w \sqrt{\frac{\rho_w}{\rho_\ell}} \tag{5}$$

Table 1. Values of Frequency Coefficient, $C_\ell = C_w$, for Tanks Filled with Water; $\nu = 0.3$, $\rho_\ell/\rho = 0.127$

H/a \ h/a	0.0005	0.001	0.002	0.003	0.005
0.3	0.0427	0.0611	0.0875	0.1078	0.1399
0.4	0.0478	0.0682	0.0974	0.1197	0.1549
0.5	0.0518	0.0738	0.1053	0.1294	0.1675
0.6	0.0550	0.0783	0.1116	0.1372	0.1776
0.7	0.0576	0.0819	0.1167	0.1434	0.1857
0.8	0.0596	0.0848	0.1207	0.1483	0.1922
0.9	0.0612	0.0870	0.1239	0.1522	0.1973
1.0	0.0625	0.0889	0.1264	0.1554	0.2014
1.2	0.0644	0.0915	0.1301	0.1599	0.2073
1.4	0.0657	0.0933	0.1326	0.1630	0.2113
1.6	0.0666	0.0946	0.1344	0.1651	0.2140
1.8	0.0673	0.0955	0.1356	0.1665	0.2158
2.0	0.0678	0.0961	0.1365	0.1676	0.2172
2.2	0.0681	0.0966	0.1371	0.1684	0.2182
2.4	0.0684	0.0970	0.1376	0.1689	0.2189
2.6	0.0686	0.0973	0.1380	0.1694	0.2194
2.8	0.0688	0.0975	0.1383	0.1697	0.2198
3.0	0.0689	0.0977	0.1385	0.1700	0.2202

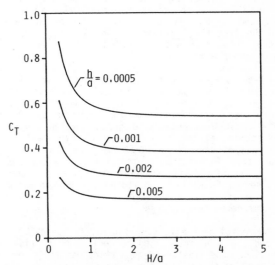

FIG. 1 Values of Coefficient C_T in Expression for Fundamental Natural Period, T, of Rigidly Supported Steel Tanks Filled with Water

in which ρ_w = the mass density of the water. Equation 5 is based on the assumption that the mass of the tank itself is negligible in comparison to that of the liquid, a condition normally satisfied in practice.

For the steel tanks examined here, $\sqrt{E/\rho}$ = 16,860 ft/sec, and the fundamental natural period of the tank-liquid system, $T = 1/f$, may be written simply as

$$T = C_T \frac{H}{100} \tag{6}$$

in which H is expressed in feet. The values of C_T corresponding to the values of C_w presented in Table 1 are displayed in Fig. 1.

FLEXIBLY SUPPORTED TANKS

For the rigidly supported systems examined so far, the vertical motion of the tank base is the same as the free-field ground motion. By contrast, for an elastically supported tank, the two motions are different, as are the corresponding tank responses. Two factors are responsible for these differences. First, the elastically supported system has one additional degree of freedom than the rigidly supported system, and hence different fundamental natural frequency and mode of vibration. Second, a substantial part of the vibrational energy of the elastically supported system may be dissipated into the supporting medium by radiation of waves and by hysteretic action in the soil itself. There is, of course, no counterpart of this mechanism of energy dissipation in a rigidly supported tank. The term soil-structure interaction is normally used to express the difference in the responses of the tank computed by: (a) assuming the motion of the tank base to be the same as the free-field ground motion; and (b) considering the modified or actual foundation motion, including the energy dissipating capacity of the supporting medium.

Studies of laterally excited simple structures (e.g., Ref. 2) have shown that a reasonable approximation to the interaction effects for such structures may be obtained by modifying the dynamic properties of the superstructure and evaluating the response of the modified structure to the prescribed free-field ground motion, considering it to be rigidly supported at the base. The requisite modifications involve changes in the fundamental natural frequency and damping of the fixed-base structure. This approach, which has provided the basis of the design provisions for soil-structure interaction recommended by the Applied Technology Council for buildings, has been shown in Ref. 5 also to yield satisfactory results for the vertically excited tank-liquid systems considered herein.

Let $\tilde{f}$ be the modified or effective natural frequency of the interacting system, and $\tilde{\zeta}$ be the associated percentage of critical damping. Further, let $\tilde{A}(t)$ be the instantaneous value of the corresponding pseudoacceleration for the postulated free-field ground motion, and $\tilde{A}$ be its maximum or spectral value. The hydrodynamic wall pressure for the elastically supported tank may then be determined from Eq. 2 or Eq. 3 by replacing $A(t)$ with $\tilde{A}(t)$; and the maximum values of this pressure may be determined by replacing $\tilde{A}(t)$ with $\tilde{A}$.

The damping factor of the flexibly supported system, $\bar{\zeta}$, may be expressed in the form

$$\bar{\zeta} = \zeta_f + (\bar{f}/f)^3 \zeta \tag{7}$$

in which the first term on the right represents the contribution of the foundation damping, including radiation and soil material damping; and the second term represents the contribution of the structural damping. Note that ζ is not directly additive to ζ_f.

The quantities $\bar{f}/f$ and $\bar{\zeta}$ depend on the properties of both the structure and the supporting medium, and may be determined from Fig. 2. The results are plotted as a function of H/a for two different values of h/a and several different rigidities for the supporting medium; the latter quantity is expressed in terms of the velocity of shear wave propagation for the medium, $v_s = \sqrt{G_s/\rho_s}$, in which G_s = the shear modulus of elasticity for the soil. Note that the foundation damping may be quite substantial for the softer soils and the stiffer tanks (tanks with the larger values of h/a), particularly when H/a is close to unity. A large increase in system damping, with no substantial change in natural frequency, would naturally be expected to lead to a significant reduction in response.

The mass of the foundation mat was presumed to be negligible for the solutions presented in Fig. 2, and no provision was made for the effect of soil material damping. Consideration of the first factor would reduce the overall system damping, whereas consideration of the second factor would increase it. However, the net effect is expected to be inconsequential for design purposes. Other assumptions made in the development of this information are that the radius of the foundation is the same as that of the tank, and that $v_s = 1/3$ and $\rho_s = 1.8\rho_\ell$. The frequency dependence of the foundation compliance function was duly accounted for.

ILLUSTRATIVE EXAMPLE

As an illustration of the extent to which the response of a tank may be influenced by the flexibilities of its wall and of the supporting soil, consider a steel tank with $H/a = 1$ and $h/a = 0.001$ that is full of water and is subjected to the vertical component of an earthquake-induced ground motion, for which $\ddot{x}_g = g/3$. The velocity of shear wave propagation for the soil is taken as $v_s = 700$ ft/sec.

If both the tank and the supporting medium are infinitely rigid, the hydrodynamic wall pressure would increase linearly from zero at the still liquid surface to a maximum at the base, and its absolute maximum value would be

$$p_{max} = \rho_\ell H \ddot{x}_g = \frac{1}{3}\rho_\ell H g = \frac{1}{3}\gamma_\ell H$$

in which $\gamma_\ell = \rho_\ell g$ = the unit weight of the contained liquid.

The effects of tank and foundation flexibilities are evaluated on the assumption that the axial distribution of the hydrodynamic wall pressure remains linear. If, as is likely to be the case, the fixed-

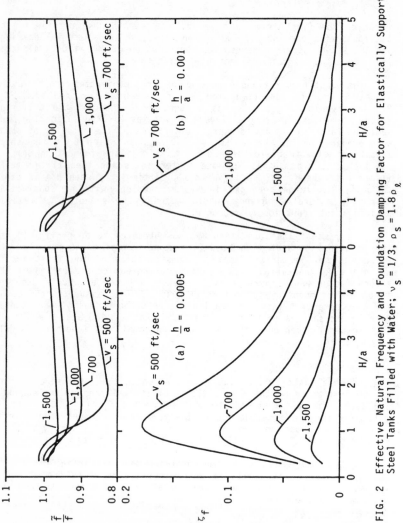

FIG. 2   Effective Natural Frequency and Foundation Damping Factor for Elastically Supported Steel Tanks Filled with Water; $\nu_s = 1/3$, $\rho_s = 1.8\rho_\ell$

base fundamental natural frequency of axisymmetric vibration of the tank-liquid system falls within the amplified, constant pseudoacceleration region of the design response spectrum; and if the damping factor for that mode is taken, as is reasonable to do, as $\zeta = 0.02$; then, the value of A would be of the order $2.8\ddot{x}_g = 0.93g$, and the peak value of the hydrodynamic pressure at the junction of the tank wall and the base would be

$$p_{max} = \rho_\ell H A = 0.93\rho_\ell H g = 0.93\gamma_\ell H.$$

In assessing the effect of foundation flexibility, it is further assumed that the effective natural frequency of the elastically supported tank remains within the amplified, constant pseudoacceleration region of the response spectrum. The foundation damping factor for the system is determined from Fig. 2 to be $\zeta_f = 0.177$, and on neglecting the contribution of structural damping, one obtains $\tilde{\zeta} = \zeta_f = 0.177$. The value of $\tilde{A}$ under these conditions is approximately $1.3\ddot{x}_g = 0.43g$. This leads to a peak wall pressure

$$p_{max} = \rho_\ell H \tilde{A} = 0.43\rho_\ell H g = 0.43\gamma_\ell H$$

which is less than one-half of that computed without regard for soil-structure interaction, and only 30 percent in excess of that computed considering both the tank wall and the foundation soil to be rigid.

CONCLUSION

It has been shown that soil-structure interaction may significantly reduce the hydrodynamic effects in vertically excited, circular, cylindrical, steel tanks, and that the consequences of such interaction may be evaluated readily with the information presented herein.

APPENDIX.— REFERENCES

1. Haroun, M. A. and Tayel, M. A., "Response of Tanks to Vertical Seismic Excitations," Journal of Earthquake Engineering and Structural Dynamics, Vol. 13, 1985, pp. 583-595.

2. Veletsos, A. S., "Dynamics of Structure-Foundation Systems," Structural and Geotechnical Mechanics, A Volume Honoring N. M. Newmark (W. J. Hall, Editor), Prentice-Hall, Inc., Englewood Cliffs, NJ, 1977, pp. 333-361.

3. Veletsos, A. S., "Seismic Response and Design of Liquid Storage Tanks," Guidelines for the Seismic Design of Oil and Gas Pipeline Systems, ASCE Technical Council on Lifeline Earthquake Engineering, ASCE, NY, 1984, Chapter 7, pp. 255-370 and 443-461.

4. Veletsos, A. S. and Kumar, A., "Dynamic Response of Vertically Excited Liquid Storage Tanks," Proc. Eighth World Conference on Earthquake Engineering, San Francisco, CA, 1984, Vol. VII, pp. 453-460.

5. Veletsos, A. S. and Tang, Y., "Dynamics of Vertically Excited Liquid Storage Tanks," to appear in Journal of Structural Engineering, ASCE.

For additional references, see lists in items 3 and 5.

NONLINEAR RESPONSE COMBINATION IN SEISMIC ANALYSIS

Armen Der Kiureghian[1] and Chih-Dao Wung[2]

Simple, approximate methods are developed to evaluate the maxima of response measures which are nonlinear functions of modal responses to seismic input. Examples for such response quantities include principal stresses and deviatoric stress measures. The results are given in the form of combination rules involving linear response components, which in turn are evaluated in terms of the modal properties of the structure and the input ground response spectrum. Comparisons with Monte Carlo simulations are used to examine the relative accuracy of the proposed and other methods.

Introduction

In seismic analysis or design of linear structures, one is often interested in critical response measures which are expressed as nonlinear functions of stress, force, or displacement components. A good example is the set of principal stresses in a plane stress state

$$\sigma_1, \sigma_2 = \frac{\sigma_x + \sigma_y}{2} \pm \left[\left[\frac{\sigma_x + \sigma_y}{2} \right]^2 + \tau_{xy}^2 \right]^{1/2} \tag{1}$$

which are nonlinear functions of stress components σ_x, σ_y, and τ_{xy} in a selected coordinate system. Another example is the deviatoric stress measure

$$\sigma_0^2 = (\sigma_x - \sigma_y)^2 + (\sigma_y - \sigma_z)^2 + (\sigma_z - \sigma_x)^2 + 6(\tau_{xy}^2 + \tau_{yz}^2 + \tau_{zz}^2) \tag{2}$$

used in the von Mises yield criterion. Other examples can be found in ordinary structural design codes.

One popular and effective method for seismic analysis of linear structures is the response spectrum method. If properly used, this method can provide good approximation to the maximum response by a simple combination of individual modal responses. With recent improvements, this method can be used for classically or nonclassically damped systems with arbitrary spacing of natural frequencies and subjected to wide-band or narrow-band ground motions [4,5]. All existing response spectrum methods, however, are limited to response quantities that can be expressed as linear functions of the modal responses. For instance, in the above two examples, the response spectrum method can be used to compute the maximum values of the stress components σ_x, σ_y, τ_{xy}, etc., but not the principal stresses σ_1 and σ_2 or the deviatoric stress measure σ_0. This is because the latter response quantities cannot be expressed as linear superposition of individual modal responses.

The problem of nonlinear response combination can be formulated as the first outcrossing of a vector process from a nonlinear safe domain [8]. The exact solution of the associated probability is not yet available and

[1]Prof., Dept. of Civ. Engrg., Univ. of California, Berkeley, Calif.
[2]Grad. Student, Dept. of Civ. Engrg., Univ. of California, Berkeley, Calif.

approximate solutions in that context (e.g, 6,8) require extensive computations. The objective in this paper is to present simple, approximate rules which can be used in conjunction with the response spectrum method to obtain good approximations of the maximum of nonlinear response expressions under seismic excitation. These are in the form of combination rules for response components in manners similar to the existing rules of modal combination for linear response components. By necessity, the rules presented are rather crude approximations based heuristic grounds. Nevertheless, they exhibit significant improvement over conventional methods, as evidenced by comparisons with Monte Carlo simulation results. It is emphasized, however, that the results presented in this paper are preliminary in nature and are under continuing refinement and investigation by the authors.

In the following sections, after describing the basic assumptions and limitations of the study, two methods of nonlinear response combination, denoted *the peak factor method* and the *extended Turkstra method*, are developed. Using Monte Carlo simulations, these methods are compared with two currently used methods in the last section of the paper, thus demonstrating the relative accuracies of the proposed methods.

Assumptions and Limitations

It is assumed that the modal responses of the structure, and hence response components expressed as their linear combinations, are Gaussian processes. In formulating the combination rules, the processes are assumed to be stationary. However, this assumption is relaxed through an interpretation of the modal responses in terms of the conventional mean response spectrum (see Appendix I). To account for the effect of ever-present gravity loads, the response components are considered to be non-zero mean processes. Finally, the second-moment statistics of the response components are assumed to be known. These are easily computed in terms of the modal responses and the input ground response spectrum, as summarized in Appendix I.

Nonlinear response expressions considered in this study are of the general form

$$R(t) = \sum_i R_i(t) \tag{3}$$

where $R_i(t)$ are called *members* of the combination. $R_i(t)$ is called a *simple* member if it takes on one of the forms $X_i(t)$, $|X_i(t)|$, $X_i^2(t)$, or $|X_i(t)|^{1/2}$, where $X_i(t)$ represents a linear response component, i.e., one which can be expressed as a linear superposition of modal responses and, therefore, is Gaussian. A combination involving only simple members is called a *simple* combination. $R_i(t)$ is called a *compound* member when it is expressed as the square root of a simple combination. A combination of the form in Eq. 3 which involves one or more compound members is called a *compound* combination. These forms of nonlinear response combination are sufficient for analysis of the combinations in Eqs. 1 and 2, as well as other nonlinear combinations that occur in conventional design codes. For example, the combination in Eq. 2 is a simple combination involving the squares of the stress components $(\sigma_x - \sigma_y)$, $(\sigma_y - \sigma_z)$, $(\sigma_z - \sigma_x)$, τ_{xy}, τ_{xy}, and τ_{xy} and the combination in Eq. 1 is a compound combination involving the simple member $(\sigma_x + \sigma_y)/2$ and a compound member defined as the square root of the simple combination $[(\sigma_x + \sigma_y)/2]^2 + \tau_{xy}^2$.

Throughout this work, the quantity of interest is the random variable

$$R_\tau = \max_\tau R(t) \tag{4}$$

where τ is a specified time interval. In particular, the attention here is focused on the mean value, μ_{R_τ}, as this measure provides adequate information for most practical applications. Where possible, expressions for the standard deviation, σ_{R_τ}, are also given.

The Peak Factor Method

This method is based on the earlier work of the first author on linear combination of stochastic loads [2]. It is based on the following fundamental relations for a stationary process $X(t)$

$$\mu_{X_\tau} = \mu_X + p_X \sigma_X \tag{5}$$

$$\sigma_{X_\tau} = q_X \sigma_X \tag{6}$$

where μ_X and σ_X are the instantaneous mean and standard deviation, respectively, and μ_{X_τ} and σ_{X_τ} are the mean and standard deviation of the maximum $X_\tau = \max_\tau X(t)$, respectively. The coefficients p_X and q_X are known as *peak factors* and are functions of the characteristics of the process and the duration τ.

When $X(t)$ is Gaussian, one set of approximate expressions of the peak factors that accounts for the bandwidth of the process is given by [3]

$$p_X = p_X(\nu\tau,\delta) = \sqrt{2\ln\nu_e\,\tau} + \frac{0.5772}{\sqrt{2\ln\nu_e\,\tau}} \tag{7}$$

$$q_X = q_X(\nu\tau,\delta) = \frac{1.2}{\sqrt{2\ln\nu_e\,\tau}} - \frac{5.4}{13 + (2\ln\nu_e\,\tau)^{3.2}} \tag{8}$$

where

$$\nu_e\,\tau = \max\left\{2.1, \begin{bmatrix} 3.56\,\delta\,\nu\,\tau & \text{for } 0 < \delta \le 0.056 \\ (2.11\,\delta^{0.45} - 0.38)\nu\tau & \text{for } 0.056 < \delta \le 0.387 \\ \nu\tau & \text{for } 0.387 < \delta \le 1.00 \end{bmatrix}\right\} \tag{9}$$

in which $\nu = (\lambda_2/\lambda_0)^{1/2}/(2\pi)$ is the mean upcrossing rate of the mean level and $\delta = (1 - \lambda_1^2/(\lambda_0\lambda_2))^{1/2}$ is a bandwidth parameter, where λ_m are the spectral moments defined in Appendix I. These results are obtained from Ref. 3 by proper modification to account for the one-sided definition of the maximum of the process. Table 1 lists the instantaneous means, standard deviations, and peak factors of the processes $X(t)$, $|X(t)|$, $X^2(t)$, and $|X(t)|^{1/2}$, which are derived based on the Gaussian assumption for $X(t)$. The expressions for q factors for $X^2(t)$ and $|X(t)|^{1/2}$ are extensive and are not given here. Simple approximations of these quantities are currently being developed.

Now consider a simple nonlinear response combination $R(t)$ in the manner defined earlier in Eq. 3, where each $R_i(t)$ takes on one of the forms listed in Table 1. In the peak factor method, the mean and standard deviation of R_τ are computed from the fundamental expressions in Eqs. 5-6 using the instantaneous values

$$\mu_R = \sum_i \mu_{R_i} \tag{10}$$

$$\sigma_R^2 = \sum_i \sum_j \rho_{R_i R_j} \sigma_{R_i} \sigma_{R_j} \tag{11}$$

where $\rho_{R_i R_j}$ is the instantaneous correlation coefficient between $R_i(t)$ and $R_j(t)$ and is computed from

$$\rho_{R_i R_j} = \frac{1}{\sigma_{R_i} \sigma_{R_j}} \int_{-\infty}^{+\infty} \int_{-\infty}^{+\infty} r_i\,r_j\,\varphi_{ij}\,dx_i\,dx_j - \mu_{R_i}\mu_{R_j} \tag{12}$$

where φ_{ij} is the bivariate normal density with means μ_{X_i} and μ_{X_j}, standard deviations σ_{X_i} and σ_{X_j}, and correlation coefficient $\rho_{X_iX_j}$. The associated peak factors are obtained from the approximate rules:

$$p_R^2 \approx \frac{\sum_i \sum_j p_{R_i} p_{R_j} \rho_{R_iR_j} \sigma_{R_i} \sigma_{R_j}}{\sum_i \sum_j \rho_{R_iR_j} \sigma_{R_i} \sigma_{R_j}} \tag{13}$$

$$q_R^2 \approx \frac{\sum_i \sum_j q_{R_i} q_{R_j} \rho_{R_iR_j} \sigma_{R_i} \sigma_{R_j}}{\sum_i \sum_j \rho_{R_iR_j} \sigma_{R_i} \sigma_{R_j}} \tag{14}$$

These are extensions of rules previously examined in Ref. 2. They are based on the simple idea that the peak factors p_R and q_R should approach the peak factors for any $R_i(t)$ whose variance dominates, and, hence a weighted average of the peak factors provides a good approximation. In situations where the correlation coefficient is not a good measure of dependence between the members of a combination, such as when $R_i(t) = X_i(t)$ and $R_j(t) = |X_i(t)|$ where $X_i(t)$ is a zero-mean process, a modified value of the correlation coefficient should be used in the above expressions to obtain meaningful results. For example, for the case just mentioned, a value close to unity rather than zero should be used, since the maximum values of the two members are likely to occur at the same time. This modification is applied in conjunction with the numerical examples later in this paper.

The above method is easily extended to compound combinations. Let $R_i(t)$ denote a compound member of the combination in Eq. 3, defined as the absolute value or square root of a simple combination. In order to use the above combination rule, all that is necessary are the instantaneous mean and standard deviation, μ_{R_i} and σ_{R_i}, the correlation coefficients $\rho_{R_iR_j}$ with other members of the combination, and the peak factors p_{R_i} and q_{R_i}. The mean, the standard deviation, and the correlation coefficients may be computed either exactly by integration (similar to that in Eq. 12), or approximately by first-order or second-order Taylor-series expansion [1]. The peak factors may be computed from Eqs. 13-14 in terms of the properties of the constituent members of $R_i(t)$. Thus, in this case, Eqs. 13-14 are used twice: Once in computing the peak factors for compound members, such as $R_i(t)$, and once in computing the peak factors for the compound combination $R(t)$.

The Extended Turkstra Method

This method is based on an approximation proposed by Turkstra [7] for linear combination of stochastic loads and has been used in the context of code development. For the combination in Eq. 1, the method is based on the following bounding relation

$$R_\tau \geq \max_i \left[R_{i\tau} + \sum_{j \neq i} R_j(t_i) \right] \tag{15}$$

where t_i denotes the time at which $R_i(t)$ takes on its maximum value, $R_{i\tau}$. Replacing variables with their expectations, an approximation of the mean is

$$\mu_{R_\tau} \approx \max_i \{ \mu_{R_{i\tau}} + \sum_{j \neq i} E[R_j \mid R_i = \mu_{R_{i\tau}}] \} \tag{16}$$

When R_i are jointly Gaussian, the conditional expectation is given by

$$E[R_j \mid R_i = \mu_{R_{i\tau}}] = \mu_{R_j} + \rho_{R_iR_j} \frac{\sigma_{R_j}}{\sigma_{R_i}} (\mu_{R_{i\tau}} - \mu_{R_i})$$

$$= \mu_{R_j} + \rho_{R_i R_j} p_{R_i} \sigma_{R_j}$$

$$= \mu_{R_j} + c_{ij} p_{R_j} \sigma_{R_j} \tag{17}$$

where $c_{ij} = \rho_{R_i R_j} p_{R_i} / p_{R_j}$. With this result, the approximation of the mean is

$$\mu_{R_\tau} \approx \max_i \left[\mu_{R_i \tau} + \sum_{j \neq i} (\mu_{R_j} + c_{ij} p_{R_j} \sigma_{R_j}) \right] \tag{18}$$

The above approximation of the mean might be unconservative, as the maximum of $R(t)$ may occur at a time when no $R_i(t)$ takes on a maximum value. To account for this, the expression for c_{ij} is altered so that Eq. 18 gives the correct result for linear combination of two Gaussian components. The result is

$$c_{ij} = \sqrt{\alpha_{ij}^2 + 1 + 2\rho_{R_i R_j} \alpha_{ij}} - \alpha_{ij} \tag{19}$$

in which $\alpha_{ij} = (p_{R_i} \sigma_{R_i}) / (p_{R_j} \sigma_{R_j})$.

Although the above result is derived when each $R_i(t)$ is a Gaussian process, it is in a form that is directly applicable, in an approximate sense, to simple nonlinear combinations defined earlier. For that application, of course, one will need to use the appropriate means, standard deviations, correlation coefficients, and peak factors of $R_i(t)$, as defined in Table 1.

For compound combinations, two approaches are possible: One approach is to compute the properties of each compound member $R_i(t)$ in the manner described in the last paragraph of the previous section, and then proceed as before. The other, more general approach is to apply the Turkstra's rule directly to the nonlinear combination. Let the combination be expressed in terms of simple members in the general form

$$R(t) = R[R_1(t), \cdots, R_i(t), R_{i+1}(t), \cdots] \tag{20}$$

Then, the approximation of the mean of R_τ is

$$\mu_{R_\tau} \approx \max_i R[\mu_{R_1} + c_{i1} p_{R_1} \sigma_{R_1}, \cdots, \mu_{R_i \tau}, \mu_{R_{i+1}} + c_{i,i+1} p_{R_{i+1}} \sigma_{R_{i+1}}, \cdots] \tag{21}$$

The second method is easier to apply, since it does not require the intermediate computation of the properties of the compound members. However, the expression for c_{ij} in Eq. 19 for this application is less rigorous, as it is based on a linear combination.

Numerical Comparisons

To examine the relative accuracy of the proposed methods, a set of 20 artificial accelerograms from Ref. 4 are employed. A structure with two modes having natural frequencies 10 and 20 rad/sec and with 5 percent damping in each mode is considered. The modal responses, $S_k(t)$, of the structure and a set of responses components expressed as linear superpositions of the modal responses, $X_i(t) = \mu_{X_i} + \psi_{i,1} S_1(t) + \psi_{i,2} S_2(t)$, are computed by step-by-step time-history analysis. The response components considered for this analysis are shown in the first two columns of Table 2. The response time histories are used in selected nonlinear response expressions and through a step-by-step analysis the maximum of the nonlinear response for each input accelerogram is determined. The ensemble mean values of these maxima are considered to be the "exact" values.

The ensemble mean of the modal maxima, $S_{k\tau}$, obtained from the time-history analysis are used as the ordinates of the response spectrum for the two modes. These spectral ordinates are used to compute the second-order statistics of the response components shown in Table 2 in the manner described in Appendix I. The approximate results from the peak factor (PF) and the extended Turkstra (ET) methods are obtained by using these properties of the response components. The ratios of the approximate to "exact" mean maxima of the nonlinear response results are reported in Tables 3 and 4.

The first example is a simple combination in the form $R(t) = X_1^2(t) + X_2^2(t)$. This is the combination in Eq. 2, when all but σ_x and τ_{xy} stress components are zero. The results for two cases, one with zero mean and one with $\mu_{X_1} = \mu_{X_2} = 5$, are shown in Table 3. In this case, the PF and ET methods for two variables give identical results. The symbols MAX and SRSS in this table refer to two methods conventionally used in practice. In the MAX method, μ_{R_τ} is computed by using the maximum μ_{R_τ} of each member in the combination, regardless of the correlation between the members, i.e., $\mu_{R_\tau} \approx \mu_{R_{1\tau}^2} + \mu_{R_{2\tau}^2}$. In the SRSS method, μ_{R_τ} is computed as the square-root-of-sum-of-squares of the maxima of the members, regardless of the correlation between them, i.e., $\mu_{R_\tau} \approx (\mu_{R_{1\tau}^2} + \mu_{R_{2\tau}^2})^{1/2}$.

The second example considered is a compound combination in the form $R(t) = X_1(t) + [X_1^2(t) + X_2^2(t)]^{1/2}$. This, clearly, is of the form in Eq. 1. The results are listed in Table 4. In this table, ET1 and ET2 refer to the two ET methods in the order described in the preceding section. The MAX and SRSS methods, again, refer to the use of member maxima or their square-root-of-sum-of-squares in computing the maximum of the compound member and of the combination. In the PF and ET methods, in computing the correlation coefficient between the simple and compound members of the combination, the absolute value $|X_1(t)|$ rather than $X_1(t)$ is used, as, for the mean values considered, it provides a better measure of the dependence.

The results in Tables 3 and 4 indicate that the PF and ET methods are more consistently accurate than the MAX or SRSS methods. This is because the PF and ET methods, in an approximate way, account for the dependence between the members of the combination. It is strongly emphasized that the results reported here are preliminary in nature and final conclusions should await the culmination of further improvements and refinements that are currently underway.

Appendix I.

Let $X_i(t) = \mu_{X_i} + \hat{X}_i(t)$ denote a linear response component, in which μ_{X_i} is the mean value resulting from gravity loads and $\hat{X}_i(t)$ is the fluctuating part resulting from the earthquake loading. Using the response spectrum method in Ref. 4, the spectral moments of the process $\hat{X}_i(t)$ are given by

$$\lambda_{i,m} \approx \sum_k \sum_l \psi_{i,k} \psi_{i,l} \rho_{m,kl} \frac{w_{m,k} w_{m,l}}{p_k p_l} S_{k\tau} S_{l\tau} \qquad m = 0,1,2 \qquad (22)$$

in which the summations are over the modes of the structure, $\psi_{i,l}$ are the modal effective participation factors, $S_{k\tau} = E[\max_\tau |S_k(t)|]$ is the mean response spectrum ordinate associated with mode k, and $\rho_{m,kl}$, $w_{m,k}$, and p_k are modal properties given as functions of the natural frequencies and modal damping ratios. Note, in particular, that λ_0 is the variance of the response, $\sigma_{X_i^2}$. The other spectral moments are needed in computing the peak factors according to Eqs. 7-9.

The correlation coefficient between any two linear response components $X_i(t)$ and $X_j(t)$ is obtained from

$$\rho_{X_i X_j} \approx \frac{1}{\sigma_{X_i} \sigma_{X_j}} \sum_k \sum_l \psi_{i,k} \psi_{j,l} \rho_{0,kl} \frac{1}{p_k p_l} S_{k\tau} S_{l\tau} \tag{23}$$

References

1. Ang, A. H-S., and Wilson, T. H., *Probability Concepts in Engineering Planning and Decision*, Volume I, John Wiley, New York, N.Y., 1975.

2. Der Kiureghian, A., "Second-Moment Combination of Stochastic Loads," *Journal of the Structural Division*, ASCE, Vol. 104, No. ST10, October 1978, pp. 1551-1567.

3. Der Kiureghian, A., "Structural Response to Stationary Excitation," *Journal of the Engineering Mechanics Division*, ASCE, Vol. 106, No. EM6, December 1980, pp. 1195-1213.

4. Der Kiureghian, A., "A Response Spectrum Method for Random Vibration Analysis of MDOF Systems," *Earthquake Engineering and Structural Dynamics*, Vol. 9, 1981, pp. 419-435.

5. Igusa, T., and Der Kiureghian, A., "Response Spectrum Method for Systems with Nonclassical Damping'" *Proceedings*, ASCE-EMD Specialty Conference, West Lafayette, Indiana, May 1983, pp. 380-384.

6. Madsen, H., "Extreme-Value Statistics for Nonlinear Stress Combination," *Journal of Engineering Mechanics*, ASCE, Vol. 111, No. 9, September 1985, pp. 1121-1129.

7. Turkstra, C. J., "Theory of Structural Design Decision," *Solid Mechanics Study*, No. 2, University of Waterloo, Waterloo, Ontario, Canada, 1970.

8. Veneziano, D., Grigoriu, M., and Cornell, C. A., "Vector-Process Models for System Reliability," *Journal of the Engineering Mechanics Division*, ASCE, Vol. 103, No. EM3, June 1977, pp. 441-460.

Table 1. Properties of Simple Response Components

$R(t)$	μ_R	σ_R	p_R	q_R
$X(t)$	μ_X	σ_X	$p(\nu\tau,\delta)$	$q(\nu\tau,\delta)$
$\lvert X(t)\rvert$	$\mu_{\lvert X\rvert}$	$(\mu_X^2+\sigma_X^2-\mu_{\lvert X\rvert}^2)^{1/2}$	$\dfrac{\mu_X+p^*\sigma_X-\mu_{\lvert X\rvert}}{\sigma_{\lvert X\rvert}}$	$q_X\dfrac{\sigma_X}{\sigma_{\lvert X\rvert}}$
$X^2(t)$	$\mu_X^2+\sigma_X^2$	$\sigma_X(2\sigma_X^2+4\mu_X^2)^{1/2}$	$\dfrac{2p^*\mu_X+\sigma_X(p^{*2}+q^{*2}-1)}{(2\sigma_X^2+4\mu_X^2)^{1/2}}$	-
$\lvert X(t)\rvert^{1/2}$	$\sqrt{\sigma_X}[F(a)+F(-a)]$	$(\mu_{\lvert X\rvert}-\mu_{\lvert X\rvert^{1/2}}^2)^{1/2}$	$\dfrac{(\mu_X+p^*\sigma_X)^{1/2}-\mu_{\lvert X\rvert^{1/2}}^2}{\sigma_{\lvert X\rvert^{1/2}}^2}$	-

Note:

$$a = \frac{\mu_X}{\sigma_X}, \qquad \mu_{\lvert X\rvert} = \sigma_X[\sqrt{2/\pi}\exp(-a)+2a\,\Phi(a)-a\,],$$

$$\Phi(a) = \frac{1}{\sqrt{2\pi}}\int_{-\infty}^{a}\exp(-\frac{u^2}{2})\,du, \qquad F(a) = \frac{1}{\sqrt{2\pi}}\int_{-a}^{\infty}\sqrt{a+u}\,\exp(-\frac{u^2}{2})\,du$$

$$p^* = p(\nu^*\tau,\delta^*), \qquad \nu^* \approx \nu[1+\exp(-5\lvert a\rvert)], \qquad \delta^* \approx \delta/[1+\exp(-5\lvert a\rvert)]$$

Table 2. Statistics of Response Components

Case	$X_1(t)$	$X_2(t)$	$\rho_{X_1 X_2}$	σ_{X_1}	σ_{X_2}
1	$S_1 + 3S_2$	$S_1 + 3S_2$	1.00	5.26	5.26
2	$S_1 + 1.5S_2$	$0.5S_1 + 3S_2$	0.80	4.29	4.04
3	$S_1 + S_2$	$0.33S_1 + 3S_2$	0.59	4.08	3.77
4	$S_1 + 0.6S_2$	$0.2S_1 + 3S_2$	0.38	3.98	3.63
5	$S_1 + 0.3S_2$	$0.1S_1 + 3S_2$	0.19	3.94	3.57
6	S_1	$3S_2$	−0.01	3.92	3.55
7	$S_1 - 0.3S_2$	$-0.1S_1 + 3S_2$	−0.21	3.94	3.58
8	$S_1 - 0.6S_2$	$-0.2S_1 + 3S_2$	−0.40	4.00	3.65
9	$S_1 - 1.2S_2$	$-0.4S_1 + 3S_2$	−0.70	4.19	3.90
10	$S_1 - 1.8S_2$	$-0.6S_1 + 3S_2$	−0.89	4.49	4.29
11	$S_1 - 3S_2$	$-S_1 + 3S_2$	−1.00	5.33	5.33

Table 3. Ratios of Approximate to Exact Values for
$$R(t) = X_1^2(t) + X_2^2(t)$$

Case	$\mu_{X_1} = \mu_{X_2} = 0$			$\mu_{X_1} = \mu_{X_2} = 5$		
	PF and ET	MAX	SRSS	PF and ET	MAX	SRSS
1	0.99	0.96	0.68	1.08	1.12	0.80
2	0.98	1.04	0.74	1.07	1.18	0.83
3	0.98	1.13	0.80	1.07	1.24	0.88
4	1.00	1.24	0.87	1.07	1.31	0.93
5	1.05	1.34	0.95	1.08	1.39	0.98
6	1.11	1.44	1.02	1.10	1.49	1.06
7	1.12	1.43	1.01	1.14	1.64	1.16
8	1.08	1.32	0.93	1.17	1.80	1.27
9	1.03	1.15	0.81	1.12	1.88	1.33
10	1.03	1.06	0.75	1.04	1.77	1.25
11	1.03	1.01	0.71	1.02	1.61	1.14

Table 4. Ratios of Approximate to Exact Values for
$$R(t) = X_1(t) + [X_1^2(t) + X_2^2(t)]^{1/2}$$

Case	$\mu_{X_1} = \mu_{X_2} = 0$				$\mu_{X_1} = \mu_{X_2} = 5$			
	PF and ET1	ET2	MAX	SRSS	PF and ET1	ET2	MAX	SRSS
1	1.07	1.07	1.08	0.69	1.03	1.03	1.06	0.73
2	1.05	1.09	1.12	0.72	1.03	1.05	1.08	0.75
3	1.05	1.12	1.15	0.74	1.03	1.07	1.10	0.77
4	1.05	1.14	1.19	0.76	1.02	1.09	1.13	0.79
5	1.04	1.16	1.20	0.77	1.02	1.11	1.15	0.81
6	1.04	1.16	1.21	0.78	1.02	1.13	1.18	0.83
7	1.06	1.17	1.21	0.78	1.02	1.15	1.21	0.85
8	1.08	1.18	1.22	0.79	1.01	1.17	1.24	0.87
9	1.10	1.16	1.19	0.77	1.01	1.21	1.29	0.90
10	1.11	1.14	1.16	0.74	0.99	1.20	1.28	0.89
11	1.12	1.12	1.13	0.73	0.99	1.17	1.26	0.87

MODAL ANALYSIS OF RANDOM VIBRATION WITH HYSTERESIS

Thomas T. Baber, M., ASCE

Univ. of Virginia, Charlottesville, VA

INTRODUCTION

Under extreme dynamic loadings, structures may undergo numerous cycles of inelastic response. Individual structural elements may be damaged, leading to losses in element stiffness and strength. The response can be modeled by degrading hysteresis at the damage sites, while response away from damage locations remains elastic (4,5).

If the excitation is highly unpredictable, e.g., earthquake ground motion, stochastic process models may be used to characterize the excitation and the structural response. For many nonlinear systems, exact determination of response statistics is difficult. Hence, numerous analytical appoximations have been developed. Of the techniques available, equivalent linearization has been perhaps the most widely used for multidegree of freedom (MDOF) systems (2,3,4,5, 6,8,9,11,12,13). The theoretical basis for the method is well developed for both zero mean and non-zero mean responses.

Modal analysis is well known to structural dynamics researchers as a means of decoupling equations of motion, and reducing active degrees of freedom. Modal analysis has also been used in nonlinear dynamics under deterministic or random loading (5,7,8,10). Often, only a single mode is used.

HYSTERETIC SYSTEM MODELS

Baber and Wen (5) and Baber (4) have developed an approach for direct stiffness analysis of hysteretic plane frames under zero-mean and non-zero mean random excitation. Simple frames and their idealizations are shown in figures 1 and 6. The governing equations can be written in the form

$$M \ddot{v}_T + C \dot{v}_T + Q_3 v_T + Q_4 r = f(t) \qquad [1]$$

$$v_R = K_{RR}^{-1} [K_{RT} v_T + K_{RH} r] \qquad [2]$$

$$H = Q_1 v_T + Q_2 r \qquad [3]$$

$$Q_1 \dot{v}_T + Q_2 \dot{r} = g(H, \dot{r}, \ldots) \qquad [4]$$

where v_T and v_R are the nodal translations and rotations respectively, r is the hysteretic hinge element rotation vector, and g (H, $\dot{r}$,...) is the vector of rate functions describing the hysteretic elements (11)

If equations [1] – [4] are used directly in random vibration analysis of planar frames, the variable set to be integrated is

$$y = [v_t^T \; v_T^T \; r^T] \qquad [5]$$

If y is an N vector, and first and second order response statistics are required, the number of unknowns is on the order of $N(N+3)/2$. As N increases, the computational problem quickly becomes unmanageable. Modal decomposition offers a means of reducing the computational labor in Monte Carlo simulation, or linearization analysis. Associated with eqn. [1] is the eigenvalue problem

$$M \; \ddot{\phi}_i \; + \; \lambda_i Q_3 \; \phi_i \; = \; 0 \qquad [6]$$

If the eigenvalues and eigenvectors associated with this problem are λ_i and ϕ_i, i=1, n, then v_T may be rewritten in terms of the ϕ_i as

$$v_T \; = \; \phi \; a \qquad [7]$$

where

$$\phi \; = \; [\phi_1 \; \phi_2 \; \phi_3 \; \cdots \; \phi_p \;] \qquad [8]$$

Substitution of this change of variables for v_T in eqns. [1] – [4] leads to a reduction in the number of variables, provided that only p eigenvectors, p < n, is used, where n is the number of translational variables, and leads to a set of reduced equations of the form

$$\ddot{a} \; + \; C^m \dot{a} \; + \; \Lambda \; a \; + \; Q_4^m \; r \; = \; f^m(t) \qquad [9]$$

$$v_R \; = \; K_{RR}^{-1} [K_{RT}^m a \; + \; K_{RH} r] \qquad [10]$$

$$H \; = \; Q_1^m \; a \; + \; Q_2 \; r \qquad [11]$$

$$Q_1^m \; \dot{a} \; + \; Q_2 \; \dot{r} \; = \; g(H, \; \dot{r}, \ldots) \qquad [12]$$

The superscript 'm' indicates that the matrix is modified by the modal transformation. Construction of typical discrete hinge models for frames indicates that the vectors v_T and r tend to be about the same size. Hence, reducing the translational variable set by half typically reduces the set of equations by about 1/3, and the second moment equations by more than half. No hysteretic element rotations are eliminated by the transformation.

RANDOM VIBRATION ANALYSIS

It is relatively straightforward to compute the transformations described in the above analysis. The result of the transformation is two-fold. A reduction in the number of equations is achieved, and the matrix Q_3 is replaced by the diagonal matrix Λ of the p eigenvalues. The response of the system under random excitation can be approached in several ways. Direct integration by Monte Carlo simulation can be used to obtain sample functions for ergodic averaging, or statistical ensembles. An associated Fokker-Planck equation could be constructed for the probability transition density functions, and used to provide equations for moments of arbitrary order. Finally, various analytical approximations could be utilized on the model itself. In the studies reported herein, the method of equivalent linearization, together with modal decomposition was used to analyze discrete hinge models of two frames.

A three story, single bay frame, shown together with its discrete hinge idealization in figure 1, was analyzed for stationary random response under Gaussian white noise excitation. Detailed specification of frame properties is given in reference 5. One, three, four and five mode models were considered. Monte Carlo simulation with ergodic averaging was used to estimate the response statistics for comparison. The RMS lateral displacements and velocities are shown in figures 3 and 4, while hysteretic hinge element mean energy dissipation rates are shown in figure 5. The Monte Carlo estimates are shown as discrete points in the figures. A nodal response estimate could not be obtained, because of the instability of the algebraic equation solving routine used to solve the Liapunov matrix equations for the nodal formulation. The first mode dominated the response. However, it was necessary to include at least 4 modes to obtain reasonable response estimates. Notably, the first mode response does not decouple from the higher modes, because of the system nonlinearity. This was particularly evident in the hinge element energy dissipation rate predictions, which changed dramatically as more modes were added to the solution. The fourth mode produced non-negligible changes in energy dissipation rates, even though the modal displacements of that mode were on the order of 1/10000 of mode 1.

The transient response of a single story single bay frame, shown in figure 6, was computed under a white noise with power spectral density 422.4, in the presence of a non-zero mean gravity acceleration. It is known (1) that gravity can significantly modify the responses, including the yield mechanism, of a frame. Transient responses were calculated since it has been shown that a unique stationary solution may not exist in the presence of a non-zero mean loading (4). Linearized system RMS lateral displacement and velocity estimates, together with 100 sample Monte Carlo simulation ensemble estimates, are shown in Figure 7 and 8, respectively. Mean vertical translation of the beam third point hinge elements is shown in Figure 9. Mean and RMS estimates of corner and third point hinge element hysteretic restoring forces are shown in Figure 10 and 11, respectively. Mean energy dissipation for those elements is shown in Figure 12. In Figures 7-12, 6ML is six mode linearization, 4MS is four mode simulation, FL is full (nodal) linearization, etc. The first two modes, shown in Figure 6, dominate the response. However, it was not possible to obtain convergent results using less than 4 modes for the non-zero mean, transient case.

SUMMARY AND CONCLUSIONS

A model for the zero or non-zero mean random vibration response of hysteretic plane frames has been summarized, in which the plane frame is idealized as a series of elastic sub-elements connected by discrete hinge sub-elements. Lumped masses are located at the nodal interfaces.

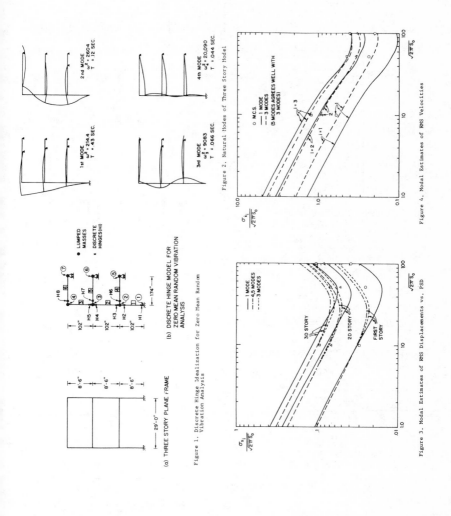

Figure 1. Discrete Hinge Idealization for Zero Mean Random Vibration Analysis

(a) THREE STORY PLANE FRAME

(b) DISCRETE HINGE MODEL FOR ZERO MEAN RANDOM VIBRATION ANALYSIS

Figure 2. Natural Modes of Three Story Model

1st MODE $\omega_1^2 = 214.4$ T = .43 SEC.

2nd MODE $\omega_2^2 = 2604$ T = .12 SEC.

3rd MODE $\omega_3^2 = 9083$ T = .066 SEC.

4th MODE $\omega_4^2 = 20,090$ T = .044 SEC.

Figure 3. Modal Estimates of RMS Displacements vs. PSD

Figure 4. Modal Estimates of RMS Velocities

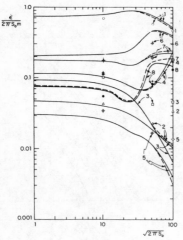

Figure 5. Modal Estimates of Hinge Element
Energy Dissipation (—— 3 modes;
--- 4 modes)

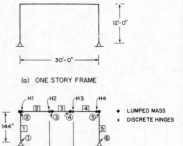

(a) ONE STORY FRAME

(b) DISCRETE HINGE IDEALIZATION

Figure 6. Discrete Hinge Model for Non-zero Mean
Random Vibration Analysis

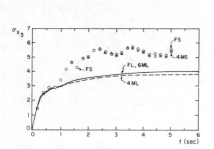

Figure 7. RMS Lateral Displacement of Node 3 of Single
Story Frame

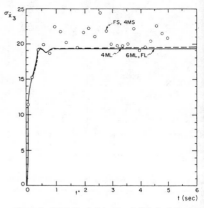

Figure 8. RMS Lateral Velocity of Node 3 of Single
Story Frame

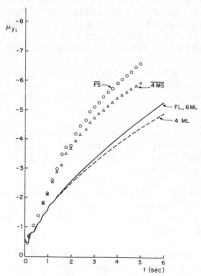

Figure 9. Mean Vertical Translation of Beam
Third Point Hinges

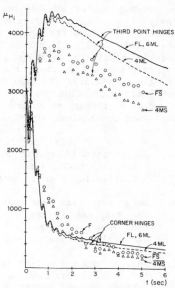

Figure 10. Mean Hysteretic Restoring Forces of
Single Story Frame

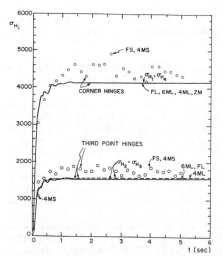

Figure 11. RMS Hinge Hysteretic Restoring Forces
Single Story Frame

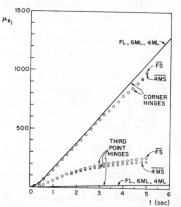

Figure 12. Mean Hysteretic Subelement Energy
Dissipations

Using equivalent linearization as a tool for the approximate analysis of random vibration leads to a set of response variables which increases as the square of the number of displacement variables. This leads to significant computational expense, and can destabilize certain algebraic solution procedures. Since a unique stationary solution may not exist in the non-zero mean case, transient analysis is often necessary. To reduce the number of displacement variables, a modal transformation was outlined, which uses the mode shapes of an associated linear structure. Two structures were analyzed. The following conclusions may be drawn:

1.   The use of modal coordinates provides significant reduction of computational expense by decreasing the number of variables. Greater computational savings are obtained in equivalent linearization than in Monte Carlo simulation.

2.   Under zero-mean base excitation, the first lateral mode dominated the response in the structure considered. Under gravity load plus base excitation, the first lateral mode and the first vertical mode are the dominant modes.

3.   Several modes in addition to the dominant modes must be included to obtain reasonable response estimates, because of the interaction between modes due to system nonlinearity. Response predictions for the dominant modes generally increased when additional modes were included.

4.   In stationary zero mean random vibration analysis, the elimination of superfluous higher modes improves the convergence of the algebraic iterative solution.

5.   In transient non-zero mean analysis, elimination of too many modes caused the iteration at each time step to diverge.

6.   The equivalent linearization analysis tends to overestimate system stiffness as compared with Monte Carlo simulation, but does correctly predict significant response trends. This situation appears to be more pronounced for non-zero mean response predictions.

7.   Under combined gravity and lateral base excitation, a shakedown response of beams can occur unless the system can transfer the gravity loads to regions not affected by the lateral excitation.

8.   Additional computation savings may be achieved by increasing the time step size when higher modes are eliminated, a situation previously recognized in deterministic vibration analysis.

9.   The modal decomposition has considerable promise for application to more general stochastic finite element models of nonlinear systems.

REFERENCES

1. Anderson, J.C., and Bertero, V. V., "Seismic Behavior of Multistory Frames Designed by Different Philosophies," Report No. EERC-69-11, Earthquake Engineering Research Center, University of California, Berkeley, CA, Oct. 1969.

2. Atalik, T. S., and Utku, S., "Stochastic Linearization of Multi-degree of Freedom Nonlinear Systems," Earthquake Engineering and Structural Dynamics, Vol. 4, pp. 411-420, 1976.

3. Baber, T. T., "Nonzero Mean Random Vibration of Hysteretic Systems," Journal of Engineering Mechanics, ASCE, Vol. 110, No. 7, pp. 1036-1049, July, 1984.

4. Baber, T. T., "Nonzero Mean Random Vibration of Hysteretic Frames," Computers and Structures, to appear.

5. Baber, T. T., "Modal Analysis for Random Vibration of Hysteretic Frames," Earthquake Engineering and Structural Dynamics, to appear.

6. Baber, T. T., and Wen, Y. K., "Stochastic Response of Multistory Yielding Frames," Earthquake Engineering and Structural Dynamics, Vol. 10, 1982, pp. 403-416.

7. Eastep, F. E., and McIntosh, S. C. Jr., "Analysis of Nonlinear Panel Flutter and Response under Random Excitation of Nonlinear Aerodynamic Loading," Journal AIAA, Vol. 10, No. 3, pp. 276-281, 1972.

8. Iwan, W. D., and Krousgrill, C. M., Jr., "Equivalent Linearization for Continuous Dynamical Systems," Journal of Applied Mechanics, Transactions, ASME, Vol. 50., pp. 415-420, 1983.

9. Iwan, W. D., and Mason, A. B., Jr., "Equivalent Linearization for Systems Subjected to Non-stationary Random Excitation," International Journal of Non-linear Mechanics, Vol. 15, pp. 71-82, 1980.

10. Lin, Y. K., "Response of a Nonlinear Flat Panel to Periodic and Randomly Varying Loading," Journal of the Aerospace Sciences, Vol. 29, pp. 1029-1033, 1962.

11. Noori, M. N., and Baber, T. T., "Random Vibration of Degrading Systems with General Hysteresis," Report No. UVA/526378/CE85/104, Department of Civil Engineering, University of Virginia, September 1984.

12. Singh, M. P., and Ghafary-Ashtiany, M., "Seismically Induced Stresses and Stability of Soil Media," Soil Dynamics and Earthquake Engineering, Vol. 1, No. 4, pp. 167-177, 1978.

13. Spanos, P-T. D., "Formulation of Stochastic Linearization for Symmetric or Asymmetric MDOF NOnlinear Systems," Journal of Applied Mechanics, Transations, ASME, Vol. 47, No. 1, p. 209, March 1980.

Random Vibration of Mass-Loaded Rectangular Plates

by Lawrence A. Bergman[*], M. ASCE and James W. Nicholson[**]

Abstract:

The classical method of separation of variables in conjunction
with the Green's function of the vibrating plate is used to obtain the
solution of the free vibration problem of the mass-loaded rectangular
plate. The stationary response of the system to stationary Gaussian
excitation is determined by the normal mode method.

Introduction

The free and deterministic forced vibration analysis of linear
combined and/or constrained dynamical systems utilizing the method of
separation of variables in conjunction with Green's functions of the
point-connected component vibrating substructures is now well
documented. In particular, the authors have previously examined the
torsional vibration of multi-disk shaft systems [1], transverse
vibration of beam-oscillator systems [2,9], transverse vibration of
thin rectangular plate-oscillator systems [10], and mass-loaded
rectangular Mindlin plates [11]. The interested reader can consult
these papers for details and for additional references addressing
these and similar problems by various other methods.
The advantages of the Green's function method are apparent,
leading to the exact natural frequencies, natural modes and the
orthogonality relation for the full system. The latter, of course,
facilitates solution of the forced vibration problem by classical
modal analysis.
In a previous paper [3], the authors have given the solution to
the problem of stationary response of a simple beam-oscillator system
to stationary excitation, for both proportional and nonproportional
damping. This existance of the orthogonality relation leads to a
succinct relationship between the modal force cross-spectral density
function and the cross-spectral densities of the applied forces,
distributed on the beam and concentrated on the individual
oscillators.
In this paper, the stationary response of a mass-loaded thin
rectangular plate, simply supported on four edges, subjected to
stationary Gaussian excitation will be considered. A similar problem
has previously been considered by Chirkov [5] using a component mode
approach, which is discussed in a recent book by Bolotin [4]. Here,

* Associate Profesor, Aeronautical and Astronautical Engineering
   Department, University of Illinois, Urbana, IL 61801
** Technical Staff, Shell Development Co., Houston, TX 77001

the plate will be viscously damped, and the masses will be arbitrarily disposed on the mid-plane surface so that rotatory inertia effects can be ignored.

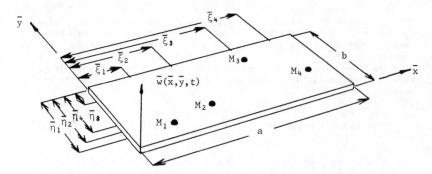

Figure 1. Mass-Loaded Simply Supported Rectangular Plate

Undamped Free Vibration

Following [10], the nondimensional equation of motion of the system, shown in Fig. 1, is

$$\nabla^4 w(x,y,\tau) + \ddot{w}(x,y,\tau) + \varepsilon_p \dot{w}(x,y,\tau) = - \sum_{i=1}^{R} \mu_i \ddot{w}(\xi_i,\eta_i,\tau)\delta(x-\xi_i)\delta(y-\eta_i)$$
$$+ \gamma \, f(x,y,\tau) \tag{1}$$

subject to the boundary conditions

$$w(0,y) = w(1,y) = w(x,0) = w(x,\lambda) = 0 \tag{2a}$$

and

$$\frac{\partial^2 w(0,y)}{\partial x^2} = \frac{\partial^2 w(1,y)}{\partial x^2} = \frac{\partial^2 w(x,0)}{\partial y^2} = \frac{\partial^2 w(x,\lambda)}{\partial y^2} = 0 \, . \tag{2b}$$

Here, overdot denotes differentiation with respect to nondimensional time τ, ∇^4 is the nondimensional biharmonic operator, and the normalized parameters are given, in terms of dimensional quantities, by

$$x = \bar{x}/a \, , \quad y = \bar{y}/a \, , \quad w = \bar{w}/a \, , \quad \xi_i = \bar{\xi}_i/a \, , \quad \eta_i = \bar{\eta}_i/a$$

$$\lambda = b/a \ , \ \Omega^2 = D/\rho a^4 \ , \ \tau = \Omega t \ , \ \delta(\ ) = a\bar{\delta}(\ )$$

$$\mu_i = M_i/\rho a^2 \ , \ \gamma = \rho g a^3/D \ , \ \varepsilon_p = C_p/\rho\Omega \ , \ f = \bar{f}/\rho g \qquad (3)$$

where ρ is the mass density per unit area, D is the bending stiffness, C_p is the plate viscous damping coefficient, M_i is the ith concentrated mass acting at $(\bar{\xi}_i, \bar{n}_i)$ and g is the acceleration due to gravity.

The plate damping and external force are set equal to zero, and separation of variables leads to the generalized partial differential equation

$$\nabla^4 W(x,y) - \alpha^4 W(x,y) = \sum_{i=1}^{R} \mu_i \alpha^4 W(\xi_i, n_i) \ \delta(x-\xi_i) \ \delta(y-n_i) \ . \qquad (4)$$

Here α^4 is the separation constant and α^2 is a frequency. The solution of Eq. 4 with the homogeneous boundary conditions (Eq. 2) gives the eigenvalues α_n and eigenfunctions $W_n(x,y)$ of the full system and is obtained using the Green's function of the vibrating plate. The Green's function $G(x,y,\xi,n;\alpha)$ satisfies

$$\nabla^4 G - \alpha^4 G = \delta(x-\xi) \ \delta(y-n) \qquad (5)$$

plus the homogeneous boundary conditions (Eq. 2). It can be obtained as a single, rapidly converging, series using the Levy method, by a less rapidly converging modal series, or by a Navier double series, all of which are discussed in [8].

The solution of the generalized differential equation, Eq. 4, is then

$$W(x,y) = \sum_{i=1}^{R} \mu_i \alpha^4 \ G(\xi_i, n_i, x, y; \alpha) \ W(\xi_i, n_i) \qquad (6)$$

To determine the unknown displacements $W(\xi_i, n_i)$, let $x \to \xi_j$, $y \to n_j$ for $j=1,2,\ldots, R$, which results in the system of homogeneous equations

$$\sum_{i=1}^{R} [\delta_{ij} - \mu_i \alpha^4 \ G(\xi_i, n_i, \xi_j, n_j; \alpha)] W(\xi_i, n_i) = 0; \ j=1,2,\ldots, R \qquad (7)$$

where δ_{ij} is the Kronecker delta. For nontrivial solution $W(\xi_i, n_i)$ to Eqs. 7 and thus $W(x,y)$, the determinant of the coefficient matrix in Eqs. 7 must be zero, defining the characteristic equation for the eigenvalues α_n. The corresponding nontrivial solution to Eqs. 7 for $\alpha = \alpha_n$ determines the system natural mode through Eq. 6

$$W_n(x,y) = \sum_{i=1}^{R} \mu_i \alpha_n^4 \ G(\xi_i, n_i, x, y; \alpha_n) W_n(\xi_i, n_i) \qquad (8)$$

where the $W_n(\xi_i, \eta_i)$ are known to within an arbitrary constant. The orthogonality relation, following [9], is given by

$$\int_0^\lambda \int_0^1 \{1 + \sum_{i=1}^R \mu_i \delta(x-\xi_i)\delta(y-\eta_i)\} W_m(x,y)W_n(x,y) \, dxdy = d_n \delta_{mn} \tag{9}$$

where d_n is an orthogonality constant or modal mass.

Forced Response of the Damped System

Assuming $f(x,y,\tau)$ to be an arbitrary deterministic function, solution to Eq. 1 is sought by the method of modal analysis. Assume

$$w(x,y,\tau) = \sum_{n=1}^\infty a_n(\tau) \, W_n(x,y) \tag{10}$$

where the generalized coordinates $a_n(\tau)$ are to be determined. Substitution of Eq. 10 into Eq. 1 and subsequent operations including use of the orthogonality relation, Eq. 9, leads to the infinite dimensional system of equations given by

$$[M]\ddot{\underset{\sim}{a}} + [C]\dot{\underset{\sim}{a}} + [K]\underset{\sim}{a} = \underset{\sim}{f}(\tau) \tag{11}$$

where

$$M_{mn} = d_n \, \delta_{mn} \tag{12a}$$

$$C_{mn} = \varepsilon_p \left[d_n \delta_{mn} - \sum_{i=1}^R \mu_i W_m(\xi_i, \eta_i) W_n(\xi_i, \eta_i) \right] \tag{12b}$$

$$K_{mn} = \alpha_m^4 \, d_n \, \delta_{mn} \tag{12c}$$

$$f_m(\tau) = \gamma \int_0^\lambda \int_0^1 W_m(x,y)f(x,y,\tau)dxdy = \gamma g_m(\tau) \tag{12d}$$

Examination of Eq. 12 reveals that, with the exception of the damping matrix, the system decouples. However, the damping matrix will diagonalize <u>only</u> when $W_m(\xi_i, \eta_i)$ and/or $W_n(\xi_i, \eta_i)$ vanish, for all i, m, and n. Physically this corresponds to the impossible case of locating each mass on a nodal line of every plate mode (i.e., modes of the plate without considering added masses). Thus, mass loading the plate acts to make the system damping inherently nonproportional.

Solution proceeds, therefore, as in [2]. The system, Eq. 11, is truncated to N modes and recast into a coupled symmetric 2N dimensional system of first order equations. A discrete eigenproblem is solved numerically for 2N eigenvalues p_m and associated eigenvectors ϕ_m. The modal matrix $[\Phi]$ is constructed and is used to decouple the first order system in the usual manner. Each equation is solved independently, and the 2N responses are transformed back to the original coordinates, resulting in

$$a_m(\tau) = \gamma \sum_{n=1}^{2N} \sum_{i=1}^{N} \left\{ \phi_{mn} \phi_{in} \int_{-\infty}^{\infty} g_i(\sigma) h_n(\tau-\sigma) d\sigma \right\}; \quad m=1,2,\ldots,N \qquad (13)$$

where

$$h_n(\tau) = \begin{cases} 0, & t < 0 \\ e^{-p_n \tau}, & \tau > 0 \end{cases} \qquad (14)$$

Eqs. 8, 13 and 14 substituted into Eq. 10 provide a closed form solution to the problem, approximate due to truncation, but convergent as N grows large.

Stationary Response to Stationary Excitation

The first and second moments of the applied force, assumed wide-sense stationary in time are

$$E[f(x,y,\tau)] = m_f(x,y,\tau) = m_f(x,y) \qquad (15)$$

$$E[f(x_1,y_1,\tau_1) f(x_2,y_2,\tau_2)] = R_{ff}(x_1,y_1,x_2,y_2;\tau_1,\tau_2)$$

$$= R_{ff}(x_1,y_1,x_2,y_2;T) \qquad (16)$$

where $T = \tau_2 - \tau_1$, m denotes a mean function and R a correlation function.

The generalized forces and coordinates are represented by their Fourier transform pairs

$$g_m(\tau) = \int_{-\infty}^{\infty} G_m(\omega)e^{i\omega\tau} d\omega , \quad G_m(\omega) = \frac{1}{2\pi} \int_{-\infty}^{\infty} g_m(\tau)e^{-i\omega\tau} d\tau \qquad (17a,b)$$

$$a_m(\tau) = \int_{-\infty}^{\infty} A_m(\omega)e^{i\omega\tau} d\omega , \quad A_m(\omega) = \frac{1}{2\pi} \int_{-\infty}^{\infty} a_m(\tau)e^{-i\omega\tau} d\tau \qquad (18a,b)$$

where $\omega = \alpha^2$ is the frequency. Expressing the applied force in terms of the system eigenfunctions and generalized forces results in

$$f(x,y,\tau) = \sum_{m=1}^{\infty} W_m(x,y)g_m(\tau) = \sum_{m=1}^{\infty} W_m(x,y) \int_{-\infty}^{\infty} G_m(\omega)e^{i\omega\tau} d\omega \qquad (19)$$

which, when substituted into Eq. 16, gives

$$R_{ff}(x_1,y_1,x_2,y_2;\tau_1,\tau_2) = \sum_{m=1}^{\infty} \sum_{n=1}^{\infty} W_m(x_1,y_1)W_n(x_2,y_2) U_{mn} \qquad (20)$$

where

$$U_{mm} = \int\limits_{-\infty}^{\infty} \int\limits_{-\infty}^{\infty} E\left[G_m^*(\omega_1) \; G_n(\omega_2)\right] e^{i(\omega_2\tau_2 - \omega_1\tau_1)} \, d\omega_1 d\omega_2 \qquad (21)$$

and * denotes complex conjugation. Asserting that [6]

$$E\left[G_m^*(\omega_1)G_n(\omega_2)\right] = S_{G_m G_n}(\omega_1)\delta(\omega_2-\omega_1) \qquad (22)$$

and implementing the Wiener-Khintchine relations [6] leads to the spectral density of the applied force in terms of the cross-spectral density of the generalized forces

$$S_{ff}(x_1,y_1,x_2,y_2;\omega) = \sum_{m=1}^{\infty} \sum_{n=1}^{\infty} W_m(x_1,y_1)W_n(x_2,y_2)S_{G_m G_n}(\omega) \qquad (23)$$

The inverse relation, obtained as a consequence of the orthogonality relation, Eq. 9, is

$$S_{G_m G_n}(\omega) = \frac{1}{d_m d_n} \int\limits_0^\lambda \int\limits_0^1 \int\limits_0^\lambda \int\limits_0^1 W_m(x_1,y_1)W_n(x_2,y_2)S_{ff}(x_1,y_1,x_2,y_2;\omega)$$

$$dx_1 dy_1 dx_2 dy_2 \qquad (24)$$

In a similar fashion, the response is expanded in terms of the system natural modes and generalized coordinates using Eqs. 10 and 18 to obtain

$$w(x,y,\tau) = \sum_{m=1}^{\infty} W_m(x,y)a_m(\tau) = \sum_{m=1}^{\infty} W_m(x,y) \int\limits_{-\infty}^{\infty} A_m(\omega)e^{i\omega\tau}d\omega \qquad (25)$$

The first and second moments of the response, also assumed wide-sense stationary in time, are

$$E\left[w(x,y,\tau)\right] = m_w(x,y,\tau) = m_w(x,y) \qquad (26)$$

$$E\left[w(x_1,y_1,\tau_1) \; w(x_2,y_2,\tau_2)\right] = R_{ww}(x_1,y_1,x_2,y_2;\tau_1,\tau_2)$$

$$= R_{ww}(x_1,y_1,x_2,y_2;T) \qquad (27)$$

The relationship, in the frequency domain, between the N generalized forces and N generalized coordinates for the nonproportionally damped system is given by the Fourier transform of Eq. 13,

$$A_m(\omega) = \gamma \sum_{n=1}^{2N} \sum_{i=1}^{N} \Phi_{mn}\Phi_{in}H_n(\omega) \; G_i(\omega); \quad m=1,2,\ldots,N \qquad (28)$$

where $H_n(\omega)$ is the nth modal transfer function given by the Fourier transform of $h_n(\tau)$,

$$H_n(\omega) = \frac{1}{j\omega + p_n} \tag{29}$$

and $j = \sqrt{-1}$. Substitution of Eqs. 25, 28, and 29 into Eq. 27 leads to the second moment of the response process

$$R_{ww}(x_1,y_1,x_2,y_2;\tau_1,\tau_2) = \sum_{r=1}^{N} \sum_{s=1}^{N} \sum_{m=1}^{2N} \sum_{n=1}^{2N} \sum_{k=1}^{N} \sum_{\ell=1}^{N} W_r(x_1,y_1)W_s(x_2,y_2)$$

$$\cdot V_{rsmnk\ell} \tag{30}$$

where

$$V_{rsmnk\ell} = \gamma^2 \, \phi_{rm}^* \phi_{km}^* \phi_{sn} \phi_{\ell n} \int_{-\infty}^{\infty} \int_{-\infty}^{\infty} H_m^*(\omega_1)H_n(\omega_2)E\left[G_k^*(\omega_1)G_\ell(\omega_2)\right]$$

$$\cdot e^{i(\omega_2\tau_2 - \omega_1\tau_1)} \, d\omega_1 d\omega_2 \tag{31}$$

Again, substituting Eq. 22 and the Wiener-Khintchine relations, the response spectral density is

$$S_{ww}(x_1,y_1,x_2,y_2;\omega) = \gamma^2 \sum_{r=1}^{N} \sum_{s=1}^{N} \sum_{m=1}^{2N} \sum_{n=1}^{2N} \sum_{k=1}^{N} \sum_{\ell=1}^{N} W_r(x_1,y_1)W_s(x_2,y_2)$$

$$\cdot \phi_{rm}^* \phi_{km}^* \phi_{\ell n} \phi_{sn} H_m^*(\omega) \, H_n(\omega) \, S_{G_r G_s} \tag{32}$$

and, finally, the mean square response $D_w^2(x,y)$ is

$$D_w^2(x,y) = \int_{-\infty}^{\infty} S_{ww}(x,y,x,y;\omega) \, d\omega \tag{33}$$

Evaluation of the integral in Eq. 33 poses no special difficulties, and for the case where $S_{ff}(x_1,y_1,x_2,y_2;\omega)$ is separable, with a constant frequency component, integration is confined to products of transfer functions given by Eq. 29.

Summary and Conclusions

A concise, closed form expression for the stationary response of a mass-loaded, simply supported thin rectangular plate subjected to stationary Gaussian white noise has been given. It has been demonstrated that the mass-loaded plate is inherently a nonproportionally damped structure. As the response is expressed in terms of a series in the undamped natural modes of the full system, the expression for the damped response is necessarily approximate due to truncation. However, in practice, convergence of the series has been robust.

The system natural modes are expressed in terms of the Green's function of the vibrating simply supported thin plate. Alternate methods of obtaining the Green's function have been discussed; however, the Levy series is the method of choice when two opposite edges of the plate are simply supported. We note that the Green's function can be constructed for any vibrating thin rectangular plate for which the natural frequencies and natural modes of the plate alone can be obtained. Therefore, the method discussed herein is entirely general and applies equally well to other linear elastic distributed systems.

The problem discussed herein is one of practical interest, particularly for experimentalists seeking optimal sensor locations as well as spectral changes due to the addition of the sensors.

Appendix. - References

1. Bergman, L. A. and Nicholson, J. W., "On the Free and Forced Vibration of Multi-Disc Shaft Systems," <u>Proceedings of the 26th Structures, Structural Dynamics and Materials Conference (AIAA, ASCE, AHS)</u>, Vol. 1, 1985, pp. 515-521.

2. Bergman, L. A. and Nicholson, J. W., "Forced Vibration of a Damped Linear Combined System," <u>ASME Journal of Vibration, Acoustics, Stress and Reliability in Design</u>, Vol. 107, 1985, pp. 275-281.

3. Bergman, L. A. and Nicholson, J. W., "Stationary Response of Combined Linear Dynamic Systems to Stationary Excitation," <u>Journal of Sound and Vibration</u>, in press.

4. Bolotin, V. V., <u>Random Vibrations of Elastic Systems</u>, Martinus Nijhoff Publishers, The Netherlands, 1984.

5. Chirkov, V. P., Random Vibrations in Thin Wall Structures Carrying Concentrated Masses," <u>Izvestia AN USSR: Izvestia Mekhanika Tverdogo Tela</u>, Vol. 10, 1975, pp. 141-145.

6. Elishakoff, I., <u>Probabilistic Methods in the Theory of Structures</u>, New York, Wiley, 1983.

7. Marek, E. L., Bergman, L. A. and Nicholson, J. W., "Solution of the Nonlinear Eigenproblem for the Free Vibration of Linear Combined Dynamical Systems," Report UILU-ENG-85-0507, University of Illinois at Urbana-Champaign, 1985.

8. Nicholson, J. W. and Bergman, L. A., "On the Efficacy of the Modal Series Representation for the Green Function of Vibrating Continuous Structures," <u>Journal of Sound and Vibration</u>, Vol. 98, 1985, pp. 299-304.

9. Nicholson, J. W. and Bergman, L. A., "Free Vibration of Combined Dynamical Systems," <u>ASCE Journal of Engineering Mechanics</u>, in press.

10. Nicholson, J. W. and Bergman, L. A., "Vibration of Damped Plate-Oscillator Systems," <u>ASCE Journal of Engineering Mechanics</u>, in press.

11. Nicholson, J. W. and Bergman, L. A., "Vibration of Thick Plates Carrying Concentrated Masses," <u>Journal of Sound and Vibration</u>, in press.

Dynamic Response Moments of Random Parametered Structures with Random Excitation[*]

Linda J. Branstetter, Assoc. Member ASCE[**]
Thomas L. Paez, Member ASCE[***]

First and second order statistical response moments are formulated
for linear multi-degree of freedom dynamical systems having both
random load and random structural characteristics. A matrix, ordinary
differential equation describing system response is discretized in
time, integrated, and then rewritten in the form of a Taylor series
expansion of each random element about its mean value. This equation
is then used to form response moment expressions. Evaluation of the
moments is made possible by the development of useful recursive
formulas. Results are verified with analytic solutions. The method
is useful for the analysis of static or dynamic random response.

Introduction

Conventional methods of structural analysis usually do not incorporate randomness in both applied loads and structural characteristics. Considerable research has been devoted to the study of deterministic structures as they respond to random excitations. In these
applications it is assumed that variation in structural response is
due largely to uncertainty in the applied loads, and that variations
in the characteristics of the structure do not affect the response.

Randomness does exist in all real structures to some degree. Under
certain conditions it is necessary to account for all the features of
a structure and its excitation, including uncertainty. Ignoring this
uncertainty may lead to errors in analytical response predictions, or
to overconfidence in the reliability of a single deterministic
prediction. It may also lead to misinterpretation of test results.
One application where this could easily occur is in damage assessment
through experimental determination of the fundamental frequencies of a
system at various times during its useful life. Such a test would be
a convenient means of detecting internal damage levels without direct
visual observation. However, some variation in the measured frequency

[*] This work was performed at Sandia National Laboratories and was
supported by the U.S. Department of Energy under contract number
DE-AC04-76DP00789 for the U.S. Department of Energy.

[**] Division 1524, Sandia National Laboratories, Albuquerque, NM 87185

[***] Division 7542, Sandia National Laboratories, Albuquerque, NM 87185

is expected for nominally identical structures, and also for the same structure with a constant damage level when tested several times. The amount of expected variation in measured frequency due to inherent randomness must be quantified before meaningful statements can be made relating measured frequency to structural damage.

There are many other areas where meaningful analysis must include structural randomness (Branstetter, 1985). Using the method developed in this paper, we consider systems having random stiffness characteristics. This is a common problem in the analysis of, for example, space vehicles where a finite number of relatively massive components are attached to each other with connections of uncertain stiffness. The method could be extended to include uncertain damping within structures as well.

Other investigators have studied this same general problem, usually considering more specialized applications, and not always proceeding with the numerical evaluation of their results. Some of the problems that have been considered and the analytical approaches used to develop solutions are given in (Soong, 1973). A few authors have written computer programs in a finite element framework to establish response moments on a step-by-step basis, for linear or nonlinear systems (e.g. Chang, 1985).

In the present investigation we also develop a computer code for evaluation of the response moments. However, our approach avoids some of the approximating assumptions made in previous studies. In addition, our formulation has the numerical advantage that not all elements of the stiffness matrix and the nodal load vector must be random. Our computer code was written to use this to advantage in keeping required input to a minimum. The size of many arrays is also minimized by this feature. These advantages will be transparent to readers of this paper. Due to space limitations only the response of two very simple systems is illustrated. Validity of the response moment equations was verified by the authors using the computer code to analyze several example systems.

Derivation of the Response Moment Equations

The n-degree of freedom systems considered in this paper may be represented by a matrix, ordinary differential equation including time independent mass, damping, and stiffness matrices of size n x n ($\tilde{M}$, $\tilde{C}$, and $\tilde{K}$), and a time dependent nodal loading vector of size n x 1 ($\tilde{L}$),

$$\tilde{M}\ddot{\tilde{x}} + \tilde{C}\dot{\tilde{x}} + \tilde{K}\tilde{x} = \tilde{L} \qquad (1)$$

In Equation (1), $\tilde{x}$ is an n x 1 displacement vector, and dots denote differentiation with respect to time. The mass and damping matrices are deterministic, while any or all of the elements of the stiffness matrix may be random variables having known first and second order moments (means, variances, and covariances). Any or all of the elements comprising the nodal load vector may be random processes whose first and second order moments are known as a function of time.

The response moments which we desire are the first order moments (mean values) of the nodal displacements and velocities as a function of time, and the second order moments (covariances) between all pairs of these displacements and velocities as a function of time. It is most convenient to find these moments by casting Equation (1) into the form of a first order differential equation of dimension 2n. If the global system response vector is defined as

$$\widetilde{X} = [\ \widetilde{\dot{x}}^T\ \widetilde{x}^T\ ]^T \tag{2}$$

then the system equation becomes

$$\widetilde{\dot{X}} + \widetilde{A}\widetilde{X} = \widetilde{P} \tag{3}$$

where $\widetilde{A}$ depends on $\widetilde{M}$, $\widetilde{C}$, and $\widetilde{K}$, and $\widetilde{P}$ depends on $\widetilde{M}$ and $\widetilde{L}$.

The general solution of Equation (3) may be written in the form of a convolution integral added to an exponential term involving initial conditions:

$$\widetilde{X}(t) = \exp\left(-\widetilde{A}(t-t_0)\right)\widetilde{X}(t_0) + \int_{t_0}^{t} \exp\left(-\widetilde{A}(t-\tau)\right)\widetilde{P}(\tau)d\tau \tag{4}$$

where t_0 is the initial time. Equation (4) may be discretized in time and then solved to obtain the response at time $t_{j+1}=(j+1)\Delta t$, where $\Delta t=t_{j+1}-t_j$. Deterministic initial conditions are assumed, and it is assumed that the load vector does not change over the length of a time step. When the exponentials are approximated up through quadratic terms and then integrated, the result is

$$\widetilde{X}_{j+1} = (\widetilde{I} - \Delta t \cdot \widetilde{A} + \frac{\Delta t^2}{2} \cdot \widetilde{A} \cdot \widetilde{A})\widetilde{X}_j + (\Delta t \cdot \widetilde{I} - \frac{\Delta t^2}{2} \cdot \widetilde{A} + \frac{\Delta t^3}{6} \cdot \widetilde{A} \cdot \widetilde{A})\widetilde{P}_j \tag{5}$$

where $\widetilde{I}$ is a 2n x 2n identity matrix. Equation (5) is accurate as long as Δt is small compared to the smallest period of the system, and the load changes very little over Δt.

We assume that ℓ of the nodal loads contained within $\widetilde{L}$ ($\ell \leq n$) are random processes (the remaining nodal loads are deterministic), and that all 2n elements of the response $\widetilde{X}$ are random processes. The random load elements at time step j are denoted γ_{kj}, k=1,...,ℓ. The random response elements at time step j are denoted β_{kj}, k=1,...,2n. Further, we assume that m elements of the stiffness matrix $\widetilde{K}$ ($m \leq (n^2+n)/2$) are random variables. (Recall that $\widetilde{K}$ is symmetric, i.e. at most $(n^2+n)/2$ entries are independent). These random elements are denoted α_k, k=1,...,m. We assume that all random quantities are normally distributed and that stiffness is independent of applied load. We assume knowledge of the mean values of all the elements of $\widetilde{A}$, the covariance matrix for $\widetilde{A}$, as well as the mean values and the covariance matrix for the nodal loads at all time.

The response of Equation (5) can be written as a Taylor series approximation involving the random terms expanded about their mean values:

$$
\tilde{X}_{j+1} = \tilde{G}_1(\mu) + \tilde{G}_2(\mu) + \sum_{i=1}^{m} (\alpha_i - \mu_{\alpha_i}) \frac{\partial(\tilde{G}_1 + \tilde{G}_2)}{\partial \alpha_i}\bigg|_{\mu} + \sum_{k=1}^{2n} (\beta_{kj} - \mu_{\beta_{kj}}) \frac{\partial \tilde{G}_1}{\partial \beta_{kj}}\bigg|_{\mu}
$$

$$
+ \sum_{k=1}^{\ell} (\gamma_{kj} - \mu_{\gamma_{kj}}) \frac{\partial \tilde{G}_2}{\partial \gamma_{kj}}\bigg|_{\mu} + \frac{1}{2} \sum_{i=1}^{m} \sum_{k=1}^{m} (\alpha_i - \mu_{\alpha_i})(\alpha_k - \mu_{\alpha_k}) \frac{\partial^2(\tilde{G}_1 + \tilde{G}_2)}{\partial \alpha_i \partial \alpha_k}\bigg|_{\mu}
$$

$$
+ \frac{1}{2} \sum_{i=1}^{m} \sum_{k=1}^{2n} (\alpha_i - \mu_{\alpha_i})(\beta_{kj} - \mu_{\beta_{kj}}) \frac{\partial^2 \tilde{G}_1}{\partial \alpha_i \partial \beta_{kj}}\bigg|_{\mu}
$$

$$
+ \frac{1}{2} \sum_{i=1}^{m} \sum_{k=1}^{\ell} (\alpha_i - \mu_{\alpha_i})(\gamma_{kj} - \mu_{\gamma_{kj}}) \frac{\partial^2 \tilde{G}_2}{\partial \alpha_i \partial \gamma_{kj}}\bigg|_{\mu} \tag{6}
$$

In Equation (6), $\tilde{G}_1$ and $\tilde{G}_2$ are the first and second terms of Equation (5), respectively, and μ denotes mean value. $\tilde{G}_1(\mu)$ and $\tilde{G}_2(\mu)$ denote the 2n x 1 vectors obtained by evaluating $\tilde{G}_1$ and $\tilde{G}_2$ when all random elements assume their mean values as of time step j. Likewise, the indicated 2n x 1 partial derivatives are evaluated at the point where all random quantities assume their time step j mean values. Note in Equation (5) that $\tilde{G}_1$ and $\tilde{G}_2$ are linear in some underlying variables.

Therefore, quadratic terms in these variables do not occur in Equation (6). Numerical evaluation of the partial derivatives of Equation (6) is done using a lumped mass representation of the system, and Rayleigh damping, where the damping matrix is a linear combination of the mass and mean stiffness matrices

The response mean results from taking the expected value of Equation (6):

$$
\tilde{\mu}_{X_{j+1}} = \tilde{G}_1(\mu) + \tilde{G}_2(\mu) + \frac{1}{2} \sum_{i=1}^{m} \sum_{k=1}^{2n} \rho_{\alpha_i \beta_{kj}} \sigma_{\alpha_i} \sigma_{\beta_{kj}} \frac{\partial^2 \tilde{G}_1}{\partial \alpha_i \partial \beta_{kj}}\bigg|_{\mu} \tag{7}
$$

where ρ is the correlation coefficient between the subscripted quantities. The required covariances in Equation (7) between each random stiffness element and each element of the response vector from time step j may be calculated from a recursive formula. Subtracting the response mean from the total response of Equation (6), premultiplying this difference by the difference between any single random stiffness element and its mean, and taking the expected value of the result, produces a 2n x 1 vector whose recursively calculated elements are the required covariances between this random stiffness element and all elements of the time step j response vector. If [(p)] denotes the pth element of the indicated 2n x 1 vector of partial derivatives, then

$$
\rho_{\alpha_r \beta_{pj}} \sigma_{\alpha_r} \sigma_{\beta_{pj}} = \sum_{i=1}^{m} \rho_{\alpha_r \alpha_i} \sigma_{\alpha_r} \sigma_{\alpha_i} \left(\left. \frac{\partial \widetilde{G}_1}{\partial \alpha_i} \right|_\mu (p) + \left. \frac{\partial \widetilde{G}_2}{\partial \alpha_i} \right|_\mu (p) \right)
$$

$$
+ \sum_{k=1}^{2n} \rho_{\alpha_r \beta_{k(j-1)}} \sigma_{\alpha_r} \sigma_{\beta_{k(j-1)}} \left. \frac{\partial \widetilde{G}_1}{\partial \beta_{kj}} \right|_\mu (p) \qquad \begin{array}{l} (r=1, \ldots m \; ; \\ p=1, \ldots, 2n) \end{array} \qquad (8)
$$

We assume that all random stiffness and response elements are uncorrelated at time zero.

The response covariance matrix is evaluated by subtracting the response mean vector from the response vector of Equation (6), multiplying by the transpose of the resulting vector to form a 2n x 2n matrix, and taking the expected value. The response covariance matrix is found to be the sum of matrices $\widetilde{T}_i$, $i=1,\ldots,8$, along with the transposes of a few of the $\widetilde{T}_i$:

$$
\widetilde{S}_{X_{j+1}} = (\widetilde{T}_1) + (\widetilde{T}_2) + (\widetilde{T}_3) + (\widetilde{T}_4 + \widetilde{T}_4^T) + (\widetilde{T}_5 + \widetilde{T}_5^T) + (\widetilde{T}_6) + (\widetilde{T}_7 + \widetilde{T}_7^T) + (\widetilde{T}_8) \qquad (9)
$$

Each quantity enclosed in parentheses in Equation (9) is a 2n x 2n symmetric matrix. Each of the matrices $\widetilde{T}_i$ is the expected value of a term in Equation (6) multiplied by its transpose or the transpose of another term in Equation (6). The $(r,p)^{th}$ element of the complete response covariance matrix is

$$
\widetilde{S}_{X_{j+1}}(r,p) = \rho_{\beta_{r,j+1} \beta_{p,j+1}} \sigma_{\beta_{r,j+1}} \sigma_{\beta_{p,j+1}} \qquad \begin{array}{l} (r=1, \ldots, 2n \; ; \\ p=1, \ldots, 2n) \end{array} \qquad (10)
$$

The constituent matrices are listed in (Branstetter, 1985), and, for example, $\widetilde{T}_1$ is:

$$
\widetilde{T}_1 = \sum_{i=1}^{m} \sum_{g=1}^{m} \rho_{\alpha_i \alpha_g} \sigma_{\alpha_i} \sigma_{\alpha_g} \left(\left. \frac{\partial \widetilde{G}_1}{\partial \alpha_i} \right|_\mu + \left. \frac{\partial \widetilde{G}_2}{\partial \alpha_i} \right|_\mu \right) \left(\left. \frac{\partial \widetilde{G}_1}{\partial \alpha_g} \right|_\mu^T + \left. \frac{\partial \widetilde{G}_2}{\partial \alpha_g} \right|_\mu^T \right) \qquad (11)
$$

In order to completely evaluate the response covariance matrix, covariances between all random load elements at time step j and all response elements at time step j are required. Using the same method which resulted in Equation (8), a recursive relation for these quantities is:

$$
\rho_{\gamma_{rj} \beta_{pj}} \sigma_{\gamma_{rj}} \sigma_{\beta_{pj}} = \sum_{k=1}^{2n} \rho_{\gamma_{rj} \beta_{k,j-1}} \sigma_{\gamma_{rj}} \sigma_{\beta_{k,j-1}} \left. \frac{\partial \widetilde{G}_1}{\partial \beta_{kj}} \right|_\mu (p)
$$

$$
+ \sum_{k=1}^{\ell} \rho_{\gamma_{rj} \gamma_{k,j-1}} \sigma_{\gamma_{rj}} \sigma_{\gamma_{k,j-1}} \left. \frac{\partial \widetilde{G}_2}{\partial \gamma_{kj}} \right|_\mu (p) \qquad \begin{array}{l} (r=1, \ldots, \ell; \\ p=1, \ldots, 2n) \end{array} \qquad (12)
$$

Equation (12) may be specialized for various types of stochastic
loading. For a white noise loading random process, the structural
response results from the value of the applied load at the current and
all previous time steps. It is independent of future load, however,
because in a white noise random process the load is uncorrelated with
itself from step to step. Therefore both sums in Equation (12) become
zero. For single step memory loading, where the load is correlated
with itself only at the present step and at the preceding step, it may
be shown that only the first sum in Equation (12) becomes zero.
Higher step memory loading further complicates the recursive expression but may be evaluated (in theory).

Numerical Results

An analytic expression for the transient
mean—square response (the displacement
variance) of the system of Figure 1 when
excited by white noise may be found in
(Lin, 1967, p. 127). The analytic solution
is valid for deterministic stiffness and
zero initial conditions of displacement
and velocity. Dynamic system behavior for
these conditions predicted by the response
moment equations of this report was found
to agree with Lin.

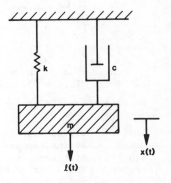

Figure 1.

The authors used several different
single degree of freedom systems having
widely varying parameters to investigate
the effect of random stiffness on system
behavior. The systems were all subjected
to white noise excitation. Comparisons were made between systems
having random stiffness of known mean and variance, and systems having
a deterministic stiffness matching the mean value of the random
system. It was found that the displacement and velocity variances are
not dependent on initial conditions when the stiffness is
deterministic. However, when the stiffness is a random variable, the
early time variances are very sensitive to initial conditions.

Off-diagonal terms in the response covariance matrix are expected
to be small. For this system, this is the covariance between displacement and velocity of the mass. As expected, larger magnitudes
are seen at early time when both response quantities show greater
dependence on initial conditions. The effect of random stiffness on
the early time covariance is pronounced. This effect is shown in
Figure 2 for one of the example systems, where the solid line is for
the deterministic system and the dashed line is the corresponding
random system. The units are not discussed since only the general
trend is of interest here.

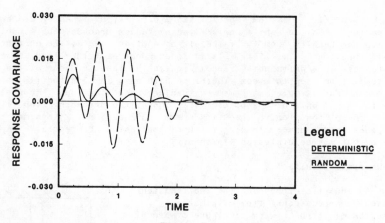

Figure 2.

Consider the undamped 2-degree of freedom system shown schematically in Figure 3. The lower mass is excited with an ideal white noise input. The system natural frequencies are $0.618(k/m)^{0.5}$ rad/sec and $1.618(k/m)^{0.5}$ rad/sec. Unit mass m and stiffness k apply. Analytic expressions were found for the displacement variances of nodes 1 (at the top mass) and 2 (at the bottom mass), assuming deterministic stiffness and zero initial displacement and velocity of both masses. This was done by finding a matrix of mass-normalized orthonormal modes which decoupled the system equation of motion. The deterministic system was then analyzed using the response moment equations developed in this report. The displacement variance histories of both masses computed using the response moment equations were identical to the analytic predictions.

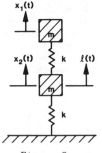

Figure 3.

The system was then analyzed considering the stiffness to be random. It was assumed that the two springs of Figure 3 were manufactured from a single bar of metal, and that the mean and variance of the batch of bars it was taken from was known. The mean of k is 1.0 lb/in (175.1 N/m), and the variance of k is 0.0625 lb/in (10.9 N/m). Consideration of the stiffness matrix of this system shows that the elements (and thus their statistical properties) are related. This was accounted for in the computer solution. An initial displacement of 1.0 in (0.0254 m) was given to the lower (excited) mass, and both masses had zero initial velocity. Solutions were performed for the case of zero stiffness variance (deterministic) as well as the nonzero variance (random) case stated above. Mean displacement behavior of the upper mass is shown in Figure 4, where displacement is in inches and time is in seconds. The displacement of this mass in the system with random stiffness is lower than that of the same system having deterministic stiffness. This was observed in all systems studied by the authors.

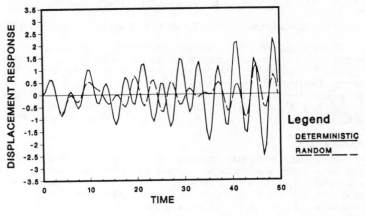

Figure 4.

Summary

Expressions are formulated for the first and second order response moments of linear multi-degree of freedom dynamical systems having both random loading and random stiffness characteristics. Recursive relations are developed which allow convenient numerical solution of the moment relations. Some simple numerical results are shown.

References:

Branstetter, L. J., and Paez, T. L., 1985, Dynamic Response of Random Parametered Structures with Random Excitation, Sandia National Laboratories, SAND85-1175 (to be published).

Chang, F., 1985, Probabilistic Dynamics of Elasto-Plastic Structures, PhD thesis, University of New Mexico.

Lin, Y. K., 1967, Probabilistic Theory of Structural Dynamics, McGraw-Hill, New York.

Soong, T. T., 1973, Random Differential Equations in Science and Engineering, Academic Press, New York.

Stochastic Response for Tertiary Systems

Takeru Igusa,[1] A.M. ASCE and Armen Der Kiureghian,[2] M. ASCE

A modal superposition method is presented for obtaining the response of a single-degree-of-freedom tertiary subsystem mounted on a primary-secondary system. The method is direct in that the response of the tertiary subsystem is obtained in terms of the properties of the input excitation without need to generate intermediate floor response spectra for the primary-secondary system. The method is also general in that any method in random vibrations based on modal superposition can be applied to obtain the tertiary subsystem response.

Introduction

Many engineering structures contain vital components that can be characterized as tertiary subsystems. In such structures, a relatively heavy primary subsystem supports a light secondary subsystem, which in turn supports an even lighter tertiary subsystem. Examples include measurement and control devices mounted on piping in nuclear power plants.

A primary-secondary-tertiary (P-S-T) system is a complex composite system possessing dynamic characteristics uncommon in ordinary structures. In previous work (7,8), the dynamic characteristics of composite primary-secondary (P-S) systems have been investigated and have been identified to be: *Tuning* -- which is the coincidence of the frequencies of the primary and secondary subsystems and gives rise to closely-spaced frequencies of the composite system; *Interaction* -- which is the feedback effect between motions of the two subsystems and is significant in cases of tuning; *Non-Classical Damping* -- which occurs when the damping characteristics of the two subsystems are different and results in complex-valued mode shapes of the composite system; and *Spatial Coupling* -- which is the effect of multiple support excitations of the secondary subsystem and is influenced by the locations of the support points and the stiffnesses of the connecting elements. In a P-S-T system, the above characteristics are also present, albeit in a more complex manner. For example, tuning might occur between the primary and the secondary, the primary and the tertiary, the secondary and the tertiary, or between all three subsystems. The first three forms of tuning are called *first degree tunings*, whereas the latter form is called *second degree tuning*. Interaction effects might be significant between any two subsystems, or between all three subsystems, depending on the conditions of tuning.

There is a need in the industry for a relatively simple, yet accurate method of determining the response of tertiary subsystems. The authors are currently developing an efficient method for obtaining the tertiary subsystem response which accounts for all four dynamic characteristics described above, yet avoids the solution of large eigenvalue problems or other iterative numerical schemes. This method is based on previous results for P-S systems (6,7,8). The method consists of two steps: (a) synthesis of the modal properties of the composite P-S-T system in terms of the known properties of the individual subsystems, and

---

[1] Asst. Research Engr., Dept. of Civ. Engrg., Univ. of California, Berkeley, Calif.
[2] Prof., Dept. of Civ. Engrg., Univ. of California, Berkeley, Calif.

(b) dynamic response analysis by modal superposition employing the derived properties of the composite system. Using perturbation techniques, closed-form solutions for the modal properties of the composite system are obtained and numerical solution of the eigenvalue problem for the composite system is avoided. The effects of interaction, tuning, non-classical damping, and spatial coupling are implicitly included in the derived modal properties. Once these properties are obtained, the response to any prescribed input excitation can be evaluated using one of many modal superposition methods available in the literature.

In this short paper, detailed results are presented for the case when the tertiary subsystem is a single-degree-of-freedom (SDOF) oscillator which is sufficiently light so that it does not interact with its supporting primary and secondary subsystems. If the mean peak responses are obtained for a set of oscillators with varying frequencies and damping ratios, the resulting curves are commonly known as floor response spectra. In the current practice, such spectra are commonly associated with P-S systems. To distinguish the spectra developed in this paper for P-S-T systems, the term *tertiary floor response spectra* (TFRS) is used. Of special interest here is computation of TFRS when the input is a seismic base motion specified in terms of the ground response spectrum.

The method developed here is direct in the sense that TRFS are generated directly in terms of the design input response spectrum specified at the base without the need to obtain intermediate floor response spectra for the secondary subsystem. The method is efficient since neither large eigenvalue analyses nor time-history integrations are needed, and can be easily coded for computer implementation. Finally, the method is accurate since the formulations are based on precise mathematical (perturbation) analysis of the composite P-S-T system, and account for all important dynamic characteristics of the system.

An example is presented to illustrate the method, investigate its accuracy, and demonstrate the important characteristics of the tertiary response.

Equations for the Response of the Tertiary Subsystem

Let $\mathbf{M}$, $\mathbf{C}$, and $\mathbf{K}$ be the mass, damping, and stiffness matrices of an N degree-of-freedom (DOF) viscously damped linear system. For a base acceleration $\ddot{x}_g(t)$, the equation for the system relative displacement response $\mathbf{x}(t)$ is

$$\mathbf{M}\ddot{\mathbf{x}} + \mathbf{C}\dot{\mathbf{x}} + \mathbf{K}\mathbf{x} = -\mathbf{M}\mathbf{r}\ddot{x}_g(t) \tag{1}$$

where $\mathbf{r}$ is the influence vector that couples the ground motion to the degrees of freedom of the structure. The response quantity of interest $v(t)$ is usually expressed as a linear combination of the components of the displacement vector $v(t) = \mathbf{q}^T\mathbf{x}$. For example, if the relative displacement between nodes 1 and 2 is of interest, then $\mathbf{q} = [-1\ 1\ 0\ \cdots\ 0]^T$. The standard method to solve for the response is to obtain the mode shapes Φ_i, frequencies, ω_i, and damping ratios, ζ_i, of the system, calculate the participation factors, and express the response by modal superposition. For classically damped systems, where the mode shapes diagonalize the damping matrix, the modal participation factors are given by $\Gamma_i = (\Phi_i^T\mathbf{M}\mathbf{r})/(\Phi_i^T\mathbf{M}\Phi_i)$ and the effective participation factors, relating the modal responses to the response quantity $v(t)$, are given by $\psi_i = \mathbf{q}^T\Phi_i\Gamma_i$. The equation for the response is given by modal superposition

$$v(t) = \sum_{i=1}^{n} \psi_i h_i(t) \tag{2}$$

where $h_i(t)$ is the well-known Duhamel integral for mode i

$$h_i(t) = \frac{1}{\omega_i \sqrt{1-\zeta_i^2}} \int_0^t \ddot{x}_g(\tau) \exp[-\zeta_i \omega_i(t-\tau)] \sin[\omega_i \sqrt{1-\zeta_i^2}(t-\tau)] d\tau \qquad (3)$$

For non-classically damped systems, the mode shapes are complex-valued (4); however, modal superposition can still be applied. In this case, the equation for the response is given in terms of the Duhamel integral and its derivative (9)

$$v(t) = \sum_i [\psi_i^{(1)} h_i(t) + \psi_i^{(2)} \dot{h}_i(t)] \qquad (4)$$

The coefficients $\psi_i^{(1)}$ and $\psi_i^{(2)}$ are real-valued, generalized effective participation factors given by

$$\psi_i^{(1)} = -2\mathrm{Re}[b_i s_i], \qquad\qquad \psi_i^{(2)} = -2\mathrm{Re} b_i \qquad (5)$$

where s_i is the complex form of the frequencies

$$s_i = \omega_i(-\zeta_i + i\sqrt{1-\zeta_i}) \qquad (6)$$

and b_i are the complex effective participation factors

$$b_i = \frac{\mathbf{q}^T \Phi_i \Gamma_i}{2i\omega_i \sqrt{1-\zeta_i}}, \qquad\qquad \Gamma_i = \frac{2i\omega_i \sqrt{1-\zeta_i}(\Phi_i^T \mathbf{M}\mathbf{r})}{s_i^{-1}\Phi_i^T \mathbf{K}\Phi_i - s_i \Phi_i^T \mathbf{M}\Phi_i} \qquad (7)$$

where Γ_i are the complex modal participation factors. The second term in the summation in Eq. 4 is due to the phase shift caused by the non-classical damping of the system and arises from the imaginary components of the mode shapes. For systems with classical damping, the modal participation factors Γ_i defined in Eq.7 are real-valued, coinciding with the previous definition for Γ_i, the first effective participation factor $\psi_i^{(1)}$ coincides with the previous definition for ψ_i, the second participation factor $\psi_i^{(2)} = 0$, and the summation Eq. 4 reduces to Eq. 2. Thus, Eq. 4 is a general formulation for the response $v(t)$ valid for both classically damped and non-classically damped systems.

The properties of the Duhamel integral and its derivative have been studied extensively in random vibration theory. Thus, once Eqs. 2 or 4 are obtained, the response $v(t)$ can be analyzed using a wide variety of methods. For this reason, most linear systems are analyzed using modal superposition methods.

Next, consider a structure-oscillator system consisting of a SDOF oscillator supported on a N-DOF structure, where the mass of the oscillator is much lighter than that of the structure and the support of the oscillator coincides with the k-th DOF of the structure (see Fig. 1). The composite structure-oscillator system is a P-S system and has special dynamic properties not found in ordinary structures, as stated in the introduction. However, the equations of motion are still in the form of Eq. 1. In the past, the equations have been altered to ignore interaction between the oscillator and the structure. In this approach, the response of the structure is used as input to the oscillator, and is commonly referred to as the "cascaded" method of analysis (2). By ignoring interaction, however, the resulting equations of motion are no longer symmetric and a straightforward modal analysis of the system is not possible. Thus, alternative expressions for the response which are more complicated than Eqs. 2 and 4 must be used.

The authors have developed a method wherein the interaction between the two subsystems is included, and the modal properties of the $N+1$ DOF system are derived in closed form. Once the modal properties are obtained, the modal superposition rules based on Eqs. 2 and 4 can be immediately applied. In the following, this modal approach to the structure-oscillator problem is described, and its extension to P-S-T systems is developed.

Let $\omega_i{}^*$, $\zeta_i{}^*$, $\Phi_i{}^*$, $\Gamma_i{}^*$ be the frequencies, damping ratios, mode shapes, and modal participation factors for the structure for $i = 1,\ldots,n^* \le N$ and ω_0 and ζ_0 be the frequency and damping ratio of the oscillator. The composite $N+1$ DOF structure-oscillator system also possesses frequencies, damping ratios, mode shapes, and effective participation factors which are denoted ω_j, ζ_j, Φ_j, and b_j for $j = 1,\ldots,n^*+1$. Using perturbation theory and modal synthesis techniques, the authors have derived expressions for ω_j, ζ_j, Φ_j, and b_j directly in terms of the six set of basic subsystem parameters $\omega_i{}^*$, $\zeta_i{}^*$, $\Phi_i{}^*$, $\Gamma_i{}^*$ ω_0, and ζ_0 when the structure is classically damped (7). These results have now been extended for the case where the structure is non-classically damped. For the sake of brevity, in this short paper the final results for the case when interaction can be neglected are presented. More general results including the effect of interaction is forthcoming. Once the modal properties for the composite structure-oscillator system are obtained, the response $v(t)$ is given by Eq. 4.

For the special, yet important case where the oscillator is very light in comparison with the structure modal masses and the interaction effect can be ignored, the expressions for the modal properties of the composite structure-oscillator system reduce to the following simple formulas

$$\omega_i = \omega_i{}^*, \qquad\qquad \zeta_i = \zeta_i{}^*, \qquad\qquad i = 1, \ldots, n^* \tag{8a}$$

$$\omega_{n+1} = \omega_0, \qquad\qquad \zeta_{n+1} = \zeta_0 \tag{8b}$$

$$b_i = \frac{i\Gamma_i{}^*}{2\omega_i{}^*}\left[q_i\varphi_{ik}{}^* - \frac{q_{n+1}\omega_0^2}{\omega_i{}^{*2} - \omega_0^2 + 2i\omega_i{}^*(\omega_i{}^*\zeta_i{}^* - \omega_0\zeta_0)}\right], \; i = 1, \ldots, n^* \tag{9a}$$

$$b_{n+1} = \frac{iq_{n+1}}{2\omega_0}\left[\Gamma_{n+1}{}^* + \sum_{i=1}^{n}\frac{\Gamma_{n+1}{}^*\omega_0^2}{\omega_i{}^{*2} - \omega_0^2 + 2i\omega_i{}^*(\omega_i{}^*\zeta_i{}^* - \omega_0\zeta_0)}\right] \tag{9b}$$

where $\varphi_{ik}{}^*$ is the k-th component of the mode shape $\Phi_i{}^*$. The expressions for the mode shapes Φ_i are intermediate results used to find b_i and are not needed to obtain the response $v(t)$. The above expressions, used in conjunction with Eq. 4, yield the equations for the response of an structure-oscillator system with negligible interaction. Next, the above analysis is expanded to the study of tertiary subsystems.

Consider an M-DOF secondary subsystem supported at multiple attachment points by an N-DOF primary subsystem (see Fig. 2). Assume both subsystems are viscously and classically damped and possess normal modes. Let the frequencies, modal damping ratios, modal masses, mode shapes, and modal participation factors be denoted by ω_{pi}, ζ_{pi}, m_{pi}, Φ_{pi}, and Γ_{pi} $i = 1, \ldots, n \le N$, respectively, for the primary subsystem, and by ω_{sj}, ζ_{sj}, m_{sj}, Φ_{sj}, and Γ_{sj}, $j = 1, \ldots, m \le M$, respectively, for the secondary subsystem. Attached to one of the DOF of the secondary subsystem is a SDOF tertiary subsystem represented as an oscillator with frequency ω_0, damping ratio ζ_0, and a negligible mass (see Fig. 2).

Considering only the primary and secondary subsystems and setting the tertiary subsystem temporarily aside, the composite P-S system has frequencies, damping ratios, mode shapes, and modal participation factors given by $\omega_i{}^*$, $\zeta_i{}^*$, $\Phi_i{}^*$, and $\Gamma_i{}^*$, $i = 1,\ldots,n+m$. In Ref. 7, the authors have developed expressions for these composite system modal properties in terms of the subsystem properties using perturbation methods and modal synthesis techniques. In general, this composite system is non-classically damped and the mode shapes and participation factors are complex-valued.

In the next step of the analysis, the P-S system is considered as the "structure" with $n^* = n+m$ modes and the tertiary subsystem is considered as the "oscillator" of the previous analysis (see Fig. 2). Using the modal properties of the structure, the response of the oscillator (tertiary subsystem) is given by Eqs. 4-9.

A special and important case is where the mass of the secondary subsystem is also very light in comparison with the primary subsystem. As for the structure-oscillator system, the expressions for the modal properties of the P-S system reduce to simple forms

$$\omega_i^* = \omega_{pi}, \qquad\qquad \zeta_i^* = \zeta_{pi}, \qquad i = 1, \ldots, n \qquad (10a)$$

$$\omega_{n+j}^* = \omega_{sj}, \qquad\qquad \zeta_j^* = \zeta_{sj}, \qquad j = 1, \ldots, m \qquad (10b)$$

$$\omega_{n+m+1}^* = \omega_0, \qquad\qquad \zeta_{n+m+1}^* = \zeta_0 \qquad (10c)$$

$$\varphi_{ik}^* = -\sum_{j=1}^{m} \frac{(\Phi_{pi}^T \mathbf{K}_{sp} \Phi_{sj})\varphi_{sjk}}{m_{sj}[\omega_{pi}^2 - \omega_{sj}^2 + 2i\omega_{pi}(\omega_{pi}\zeta_{pi} - \omega_{sj}\zeta_{sj})]}, \qquad i = 1, \ldots, n \qquad (11a)$$

$$\varphi_{n+j,k}^* = \varphi_{sjk}, \qquad j = 1, \ldots, m \qquad (11b)$$

$$\Gamma_i^* = \Gamma_{pi}, \qquad i = 1, \ldots, n \qquad (12a)$$

$$\Gamma_{n+j}^* = \Gamma_{sj} - \sum_{i=1}^{n} \frac{(\Phi_{pi}^T \mathbf{K}_{sp} \Phi_{sj})\Gamma_{pi}}{m_{pi}[\omega_{pi}^2 - \omega_{sj}^2 + 2i\omega_{pi}(\omega_{pi}\zeta_{pi} - \omega_{sj}\zeta_{sj})]}, \qquad j = 1, \ldots, m \qquad (12b)$$

where $\mathbf{K}_{sp}$ is the matrix of stiffnesses between the primary and secondary subsystems.

Response to Stochastic Input

Using the modal superposition rule given by Eq. 4, probabilistic description of the response $v(t)$ can be obtained for several types of stochastic input using a wide variety of methods. In the following, the widely used state vector formulation for the response and one response spectrum rule is given.

Introducing the state vector $\mathbf{V} = [v(t) \ \dot{v}(t)]^T$, the response cross-correlation matrix may be written as

$$E[\mathbf{V}\mathbf{V}^T] = \sum_i \sum_j \mathbf{A}_i E[\mathbf{X}_i \mathbf{X}_j^T] \mathbf{A}_j^T \qquad (13)$$

where

$$\mathbf{A}_i = \begin{bmatrix} \psi_i^{(1)} & \psi_i^{(2)} \\ -\omega_i^2 \psi_i^{(2)} & \psi_i^{(1)} - 2\zeta_i \omega_i \psi_i^{(2)} \end{bmatrix}, \qquad \mathbf{X}_i = \begin{Bmatrix} h_i(t) \\ h_i(t) \end{Bmatrix} \qquad (14)$$

The above formulation is similar to that used for stochastic analysis of classically damped systems. The only difference is in matrix $\mathbf{A}_i$: For classically damped systems $\mathbf{A}_i = \text{diag}\{\psi_i, \ \psi_i\}$. It follows that existing procedures for stochastic response analysis of classically damped systems are directly applicable to non-classically damped systems, provided matrix $\mathbf{A}_i$ is properly modified. The procedure to be used in computing the cross-correlation matrix depends on the form of specification of the input excitation. If $\ddot{x}_g(t)$ is specified in the frequency domain through an evolutionary power spectral density, then a procedure such as in Ref. 3 can be used, and if the input is specified in the time domain by modulation, then a procedure such as in Refs. 1 and 10 can be used.

If the input is specified by the mean response spectrum, one expression for the mean peak response $\bar{V}$ is given by (5)

$$\bar{V} = \left[\sum_i \sum_j \left[C_{ij}\rho_{0,ij} - D_{ij}\eta_{1,ij}w_{1,ij} + E_{ij}\rho_{2,ij}w_{2,ij} \right] \frac{p^2}{p_i p_j} S_i S_j \right]^{1/2} \qquad (15)$$

where C_{ij}, D_{ij}, E_{ij}, are in terms of the effective participation factors; $\rho_{0,ij}$, $\eta_{1,ij}$,

$\rho_{2,ij}$, $w_{1,ij}$, $w_{2,ij}$ are in terms of the frequencies and modal damping ratios; S_i is the response spectrum ordinate corresponding to the i-th mode of the system; p_i is the i-th modal peak factor; and p is the peak factor for the response process.

Example

As example applications, floor response spectra for tertiary systems are calculated for two 7-DOF P-S systems. The systems consist of a 4-DOF secondary subsystem attached at two points to a 3-DOF primary subsystem (Fig. 3). Two secondary subsystems are considered, one which is tuned to the first and third frequencies of the primary subsystem, and one which is detuned with all of the modes of the primary subsystem. The properties of the individual subsystems are shown in Fig. 3 and Table 1. The base input, specified by its mean response spectrum, is described in Ref. 7.

For each DOF of the secondary subsystems, the TFRS is calculated for 0.02 tertiary subsystem damping and are plotted in Figs. 4 and 5 for the tuned and detuned P-S systems, respectively. For the tertiary subsystem attached to DOF x_5, a numerical eigenvalue analysis of the composite P-S-T system is used to compute the modal properties and the corresponding TFRS are plotted if Figs. 4 and 5 (using squares). The figures indicate that the above analytical method is in close agreement with the numerical eigenvalue analysis results.

The effects of first and second-degree tunings can be observed in the TFRS. For the tuned P-S system, a very high peak is observed at the second-degree tuning frequencies near 1.4 Hz, where all three systems are tuned to each other. For the detuned P-S system, peaks can be observed corresponding to each first-degree tuning. For example, at 0.87 Hz the tertiary system is tuned to the secondary subsystem and at 1.4 Hz, the tertiary system is tuned to the primary subsystem. It is observed that the TFRS peaks for first-degree tuning is significantly less than that for second-degree tuning.

The shape of the TFRS are considerably different according to the location of the tertiary subsystem on the secondary subsystem. These shapes are influenced by the spatial coupling effect between the primary and secondary subsystems, which is discussed in detail in Ref. 7, as well as the modal deflections of the secondary subsystem.

Acknowledgment

This research work is supported by the National Science Foundation under Grant No. ECE-8416518, which support is gratefully acknowledged.

References

1. D. A. Gasparini and A. DebChaudhury, "Dynamic Response to Nonstationary Nonwhite Excitation," *Journal of the Engineering Mechanics Division*, vol. 106, no. EM6, December, 1980.

2. D. A. Gasparini, A. Shah, G. Tsiatas, S. L. Shien, and W.-J. Sun, "Random Vibration of Cascaded Secondary Systems," Case Institute of Technology, Dept. of Civil Engineering Research Report 83-1, January 1983.

3. J. K. Hammond, "On the Response of Single and Multidegree of Freedom Systems to Non-Stationary Random Excitations," *Journal of Sound and Vibration*, vol. 7, no. 3, pp. 393-416, 1968.

4. W. C. Hurty and M. F. Rubinstein, *Dynamics of Structures*, Prentice Hall, Inc., Englewood Cliffs, N.J., 1964.

5.   T. Igusa and A. Der Kiureghian, "Response Spectrum Method for Systems with Nonclassical Damping," in *Proceedings*, ASCE-EMD Specialty Conference, pp. 380-384, West Lafayette, Indiana, May 23-25, 1983.

6.   T. Igusa and A. Der Kiureghian, "Dynamic Analysis of Multiply Tuned and Arbitrarily Supported Secondary Systems," Report No. UCB/EERC-83/07, Earthquake Engineering Research Center, University of California, Berkeley, Calif., July 1983.

7.   T. Igusa and A. Der Kiureghian, "Dynamic Response of Multiply Supported MDOF Secondary Systems," *Journal of Engineering Mechanics*, vol. 111, no. 1, pp. 20-41, January 1985.

8.   T. Igusa and A. Der Kiureghian, "Dynamic Characterization of Two-Degree-of-Freedom Equipment-Structure Systems," *Journal of Engineering Mechanics*, vol. 111, no. 1, pp. 1-19, January 1985.

9.   T. Igusa, A. Der Kiureghian, and J.L. Sackman, "Modal Decomposition Method for Stationary Response of Nonclassically Damped Systems," *Earthquake Engineering and Structural Dynamics*, vol. 12, no. 1, pp. 121-136, 1984.

10.   S. F. Masri, "Response of a Multi-Degree-of Freedom System to Nonstationary Random Excitation," *Journal of Applied Mechanics*, vol. 45, no. 3, pp. 649-656, September 1978.

Table 1. Properties of Example Primary and Secondary Subsystems

Subsystem	Mode	Frequency (Hz)
Primary: $k_p/m_p = 400$ rad²/ sec² $m_p = 10,000$ slugs	1 2 3	1.42 3.98 5.73
Tuned Secondary: $k_s/m_s = 1,180$ rad²/ sec² $EI/L^3m_s = 13.2$ rad²/ sec² $m_s = 100$ slugs	1 2 3 4	1.37 3.11 5.67 5.83
Detuned Secondary: $k_s/m_s = 484$ rad²/ sec² $EI/L^3m_s = 5.38$ rad²/ sec² $m_s = 100$ slugs	1 2 3 4	0.87 1.99 3.63 3.73

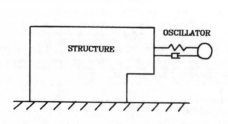

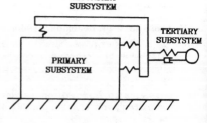

Figure 1.   **Oscillator-Structure System**

Figure 2.

Primary-Secondary-Tertiary System

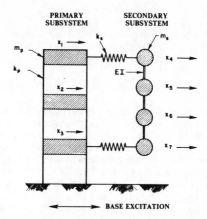

Figure 3.   Example Primary-Secondary System

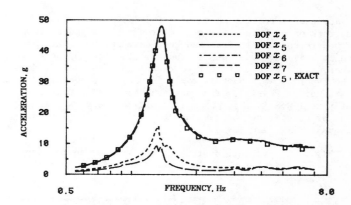

Figure 5.   Mean Tertiary Floor Response Spectra:
Tuned Primary-Secondary System

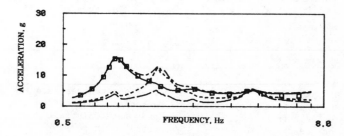

Figure 4.   Mean Tertiary Floor Response Spectra:
Detuned Primary-Secondary System

Rigid Body Random Rocking Prediction
*Pol D. Spanos and Aik-Siong Koh
Member ASCE

ABSTRACT

The stochastic rocking of a rigid block due to shaking
of its foundation is examined. The foundation is flexible
(Winkler) and its independent springs and dashpots separate
from the block base when tension is incipient. Modulated
white noise is used as a model of horizontal acceleration of
the foundation. The statistics of the rocking response are
found by an analytical procedure which involves a combination
of static condensation and stochastic linearization. The
analytical solution is computationally efficient and compares
well with pertinent data obtained by numerical simulations.

INTRODUCTION

Structures resting on shaking foundations are likely to
sway, rock or do both. The swaying response is generally
well understood and the related knowledge has been incorporated in building codes and textbooks. The rocking behavior,
however, has proved difficult to treat quantitatively. Most
of the reported work on this topic focuses on structural impulse
and sinusoidal shaking. The pertinent body of literature is
briefly reviewed in Ref. [2]. In addition there is a limited
number of studies on the response to random excitations and
earthquakes such as in Ref. [4].

This paper presents a study of the stochastic rocking
response of rigid blocks on horizontally excited flexible
foundations.

Most structures are designed to avoid uplifting of the
base from the foundation. Nevertheless, several studies such as
in Ref. [1],suggest that uplifting can be beneficial in reducing
the dynamic stresses of the structure during ground motions.
This phenomenon may be exploited in future design of structures
situated in seismically active regions. For this class of

*Pol D. Spanos
Professor of Civil Engineering
Rice University
Houston, Texas 77251

Aik-Siong Koh, Assistant Professor
Mechanical Engineering
Texas Tech University
Lubbock, Texas 79409

structures it is important that the method of analysis incorporates the effects of uplifting. The exact rocking response of these structures can presently be obtained by direct numerical integration of the corresponding equations of motion. This approach, however, is computationally costly. Therefore, approximate but more efficient methods are needed to predict the response of an uplifting structure under dynamic loading.

RIGID BLOCK-FLEXIBLE FOUNDATION MODEL

Fig. 1 shows a two-dimensional block which can rock and bounce on a foundation of independent springs and dashpots (Winkler foundation). The block has mass, m, and polar moment of inertia, I_{yy}, about the center of mass, cm. The center of mass is located at height, H, above the base of width, 2B, while θ_{cr} is the critical tilt-angle when the center of mass is directly over the supporting base corner. To simplify the ensuing analysis, the center of the block base, cb, is confined to vertical motions only. Two generalized coordinates, z_{cb} and θ, are employed to define the position of the block with respect to the foundation. The first defines the upward vertical displacement of the base center from its position of static equilibrium. The Winkler foundation involves massless springs and dashpots whose lines of reaction are directed vertically. Its stiffness coefficient k, has units of force per unit width of base per unit vertical deformation. Further, its damping coefficient c has units of force per unit width of base per unit vertical deformation velocity. The foundation can be exposed to horizontal acceleration, $\ddot{x}_g$, and vertical acceleration, z_g, where a dot above a variable denotes differentiation with respect to time. Note that the entire system is under the influence of gravity, g, which acts vertically downwards.

Soil materials rarely exhibit significant tensile strength, and debounding or uplifting has been reported between the base of a structure and its foundation. To account for this behavior the springs and dashpots in the dynamic model can separate from the base of the block when tension is incipient. Uplifting of the block base is allowed and it occurs when either of the base corners rises above the undeformed surface of the foundation.

It is useful at this stage to introduce the results of a dimensional analysis of the problem. The non-dimensional π-terms found are $\pi_H = H/B$, $\pi_I = I_{yy}/(mB^2)$, $\pi_k = kB^2/(mg)$, $\pi_c = c^2B^3/(m^2g)$, and $\pi_t = t\sqrt{g}/B$.

It is noted that real structures do not meet perfectly the assumptions of the model; they can only be approximated by a judicious choice of the model parameters. For example, Ref. [3] provides expressions for the effective values of k and c to represent the dynamic behavior of an identical block rocking on an elastic half space.

EQUATIONS OF MOTION

Two equations of motion describe the rocking and bouncing of the block on the Winkler foundation. These equations are piecewise nonlinear, and they change fundamentally in character depending on whether there is uplifting of the base or not. In principle, complete separation of the block from its foundation is possible, but for realistic applications such extreme levels of response are rare.

Applying the principles of dynamics, the following equations of motion for the block before uplifting are obtained

$$(mH^2 + I_{yy})\ddot{\theta} + (2B^3c/3)\dot{\theta}\cos^2\theta$$
$$+ [(2/3)B^3k \cos \theta - mH(g + \ddot{z}_{cb})]\sin \theta$$
$$+ mh(\ddot{x}_g\cos \theta - \ddot{z}_g\sin \theta) = 0, \tag{1}$$

$$m\ddot{z}_{cb} + 2Bc\dot{z}_{cb} + 2Bkz_{cb} + m(g + \ddot{z}_g)$$
$$- mH(\dot{\theta}^2\cos \theta + \ddot{\theta} \sin \theta) = 0. \tag{2}$$

Similar equations of motion while uplifting occurs are found

$$(mH^2 + I_{yy})\ddot{\theta} + (1/3)c(B^3 - z_{cb}^3\text{sgn } \theta/\sin^3\theta)\dot{\theta}\cos^2\theta$$
$$+ [(1/3)B^3k\cos \theta - mgH]\sin \theta$$
$$+ mh[\ddot{x}_g\cos \theta - (\ddot{z}_g + \ddot{z}_{cb})\sin \theta]$$
$$+ (1/2)c\dot{z}_{cb}(z_{cb}^2/\sin^2\theta - B^2)\cos \theta \text{ sgn}\theta$$
$$+ kz_{cb}[(1/6)z_{cb}^2/\sin^2\theta - (1/2)B^2\text{sgn } \theta]\cos \theta = 0, \tag{3}$$

$$m\ddot{z}_{cb} + c(B - z_{cb}\text{sgn } \theta/\sin \theta)\dot{z}_{cb}$$
$$+ [Bk + (1/2)(z_{cb}\text{sgn } \theta/\sin \theta)(c\dot{\theta} \cos \theta/\sin \theta - k)]z_{cb}$$
$$- mH(\ddot{\theta} \sin \theta + \dot{\theta}^2\cos \theta) - (1/2)B^2(c\dot{\theta}\cos \theta + k\sin \theta)\text{sgn } \theta$$
$$+ m(g + \ddot{z}_g) = 0, \tag{4}$$

where sgn($\cdot$) denotes the signum function which equals +1 or −1 when its argument is positive or negative, respectively.

Equations 1 to 4 are coupled, nonlinear with no obvious

solutions. Good approximations can be obtained, however, by assuming $\sin\theta \cong \theta$, $\cos\theta \cong 1$, and by neglecting all terms having order greater than one; this is acceptable, since θ and z_{cb} are small for most cases of interest. For added simplicity the ground excitation is assumed to be horizontal, i.e. $\ddot{z}_g = 0$. Due to these assumptions the equations for no uplifting become

$$(mH^2 + I_{yy})\ddot{\theta} + (2/3)B^3c\dot{\theta} + [2B^3k/3 - mgH]\theta + mH\ddot{x}_g = 0, \quad (5)$$

$$m\ddot{z}_{cb} + 2Bc\dot{z}_{cb} + 2Bkz_{cb} + mg = 0. \quad (6)$$

Equation 5 governs a linear single degree of freedom system with mass $m_{sdof} = (mH^2 + I_{yy})$, damping coefficient $c_{sdof} = 2B^3c/3$, and stiffness $k_{sdof} = (2B^3k/3 - mgH)$. Similarly, the equations of motion with uplifting become

$$(mH^2 + I_{yy})\ddot{\theta} + (1/3)c[B^3 - (z_{cb}\mathrm{sgn}\,\theta/\theta)^3]\dot{\theta}$$

$$+ [B^3k/3 - mgH]\theta + (1/2)c[(z_{cb}/\theta)^2 - B^2]\dot{z}_{cb}\mathrm{sgn}\theta$$

$$+ (1/6)k[(z_{cb}/\theta)^2 - 3B^2]z_{cb}\mathrm{sgn}\theta + mH\ddot{x}_g = 0, \quad (7)$$

$$m\ddot{z}_{cb} + c(B - z_{cb}\mathrm{sgn}\,\theta/\theta)\dot{z}_{cb} + (1/2)k(2B - z_{cb}\mathrm{sgn}\theta/\theta)z_{cb}$$

$$+ (1/2)c[(z_{cb}/\theta)^2 - B^2]\dot{\theta}\mathrm{sgn}\theta + (1/2)B^2k\theta\mathrm{sgn}\theta + mg = 0. \quad (8)$$

The method of static condensation is applied to Equations 5–8 in order to eliminate their dependence on z_{cb}. This method seeks θ and z_{cb} that satisfy Equation 6 and 8 under static conditions. Thus, z_{cb} is determined by the equation

$$z_{cb} = z_{st} = \frac{-mg}{2Bk}, \text{ for no uplifting and} \quad (9)$$

$$z_{cb} = \frac{Bk\theta\mathrm{sgn}\theta - \sqrt{2mgk\theta}\mathrm{sgn}\theta}{k} \text{ for uplifting.} \quad (10)$$

It is noted that under static conditions uplifting occurs when $|\theta| > \theta_{ul}$, where

$$\theta_{ul} = \frac{|z_{st}|}{B} = \frac{mg}{2B^2K}. \quad (11)$$

Substituting for z_{cb}, $\dot{z}_{cb}$, and $\ddot{z}_{cb}$, Equation 5 remains unchanged, but Equation 7 becomes

$$(mH^2 + I_{yy})\ddot{\theta} + \sqrt{2mgk\theta}\mathrm{sgn}\theta\ mgc\dot{\theta}/(6k^2\theta^2)$$

$$+ mg[B\mathrm{sgn}\theta - H\theta - 2\sqrt{2mgk\theta}\mathrm{sgn}\theta/(6k\theta)] + mH\ddot{x}_g = 0 \quad (12)$$

Equations 5 and 12 may be rewritten in the form

$$m_{sdof}\ddot{\theta} + C(\theta)\dot{\theta} + K(\theta)\theta = -mH\ddot{x}_g, \quad (13)$$

where the variable damping and stiffness terms are

$$C(\theta) = c_{sdof}, \qquad\qquad |\theta| \leqslant \theta_{ul},$$

$$\doteq \sqrt{2mgk\theta}sgn\theta \ mgc/(6k^2\theta^2) \qquad |\theta| > \theta_{ul}, \qquad (14)$$

$$K(\theta) = k_{sdof}, \qquad\qquad |\theta| \leqslant \theta_{ul},$$

$$= mg \ [sgn\theta B/\theta - H - 2\sqrt{2mgk\theta}sgn\theta/(6k\theta^2)], \ |\theta| > \theta_{ul}. \qquad (15)$$

The horizontal ground acceleration is assumed to be a modulated Gaussian white noise. Such a process can be modeled by

$$\ddot{x}_g/g = q(t)w(t) \qquad\qquad\qquad\qquad (16)$$

where $q(t)$ is a deterministic function, and $w(t)$ is a white noise process which has doubled-sided spectral density S_w. The purpose of introducing the modulating function is to capture the intensity variation of typical earthquake acceleration records; it rises rapidly from zero to a peak of unity and then decays exponentially to zero - see Fig. 2.

ANALYTIC APPROXIMATION

The method of stochastic linearization is applied to predict rigid block response to moderate level random ground shaking. To do this conveniently, analytic expressions are used to approximate the piecewise functions of the spring restoring moment, $M(\theta) = K(\theta)\theta$, and the damping coefficient, $C(\theta)$. The spring restoring moment is approximated by the expression

$$M_{app}(\theta) = K_1 arctan(K_2\theta) - K_3\theta, \qquad\qquad (17)$$

where K_1, K_2 and K_3 are constants. The conditions used for the matching are

$$\frac{dM(0)}{d\theta}_{app} = \frac{dM(0)}{d\theta} \ , \ and \qquad\qquad\qquad (18)$$

$$M(\theta)_{app} = M(\theta) = mg(R - H\theta) \ for \ \theta \to \infty \ . \qquad (19)$$

Hence, the constants become $K_1 = 2mgR/\pi$, $K_2 = \pi(k_{sdof} + mgH)/(2mgR)$, and $K_3 = mgH$. The damping coefficient in Equation 14 is approximated by the equation

$$C_{app}(\theta) = C_1/(1 + C_2\theta^2), \qquad\qquad\qquad (20)$$

where C_1 and C_2 are constants. The conditions used for the matching are $C(0)_{app} = C(0)$, and $C(1.5\theta_{ul})_{app} = C(1.5\theta_{ul})$. The above conditions yield $C_1 = C_{sdof}$, and

$$C_2' = \frac{1}{(1.5\theta_{u1})^2}\left[\frac{C_1}{C(1.5\theta_{u1})} - 1\right].\tag{21}$$

Fig. 3 and 4 show the analytic approximations of the spring restoring moment and damping coefficients compared with the exact expressions. Thus, the equation of motion of the rocking block with and without uplifting is approximated as follows

$$\ddot\theta + \frac{C_1\dot\theta}{m_{sdof}(1 + C_2\theta^2)} + \frac{K_1\arctan(K_2\theta)}{m_{sdof}} - \frac{K_3\theta}{m_{sdof}} = -\frac{m\bar{l}\ddot{x}_g}{m_{sdof}}.\tag{22}$$

The method of stochastic linearization introduces a linear structure

$$\ddot\theta + c_e\dot\theta + k_e\theta = -m\bar{l}\ddot{x}_g/m_{sdof},\tag{23}$$

where the damping coefficient, c_e, and stiffness, k_e, are chosen based on an optimality criterion. Specifically given θ, and $\dot\theta$, the expectation of the square of the difference of left-hand sides of Equations 22 and 23 is minimized. The equivalent damping and stiffness coefficient are determined by the equation

$$c_e = E\left[\frac{\partial g(\theta,\dot\theta)}{\partial\dot\theta}\right], \text{ and}\tag{24}$$

$$k_e = E\left[\frac{\partial g(\theta,\dot\theta)}{\partial\theta}\right],\tag{25}$$

where $E[\cdot]$ is the operator of mathematical expectation,

$$g(\theta,\dot\theta) = \frac{C_1\theta}{m_{sdof}(1 + C_2\theta^2)} + \frac{K_1\arctan(K_2\theta)}{m_{sdof}} - \frac{K_3\theta}{m_{sdof}},\tag{26}$$

and θ, $\dot\theta$ are approximated by Gaussian processes. That is the approximate joint probability density of θ and $\dot\theta$ is given by the equation

$$P_{\theta\dot\theta} = \frac{1}{2\pi\sigma_\theta\sigma_{\dot\theta}\sqrt{1 - \rho^2}}\exp\left[\frac{-(\sigma_{\dot\theta}^2\theta^2 - 2\sigma_\theta\sigma_{\dot\theta}\rho\theta\dot\theta + \sigma_\theta^2\dot\theta^2)}{2\sigma_\theta^2\sigma_{\dot\theta}^2(1 - \rho^2)}\right],\tag{27}$$

where σ_θ and $\sigma_{\dot\theta}$ are the standard deviations of θ and $\dot\theta$ respectively, while ρ is their correlation coefficient. Evaluating the expectations in Equations 24 and 25 the equivalent damping and stiffness coefficients are found

$$c_e = \frac{C_1}{m_{sdof}\sigma_\theta\sqrt{2C_2}}\frac{\sqrt{\pi}}{\ }\exp\left[\frac{1}{2C_2\sigma_\theta^2}\right]\text{erfc}\left[\frac{1}{\sqrt{2C_2}\sigma_\theta}\right]\tag{28}$$

$$k_e' = - \frac{C_1 \rho \, \dot{\sigma}_\theta \sqrt{\pi}}{m_{sdof} \sigma_\theta^2 \sqrt{2} C_2} \exp\left(\frac{1}{2 C_2 \sigma_\theta^2}\right) \text{erfc}\left(\frac{1}{\sqrt{2} C_2 \sigma_\theta}\right)\left(1 + \frac{1}{C_2 \sigma_\theta^2}\right)$$

$$+ \frac{C_1 \rho \, \dot{\sigma}_\theta}{m_{sdof} C_2 \sigma_\theta^3} + \frac{K_1 \sqrt{\pi}}{m_{sdof} \sigma_\theta \sqrt{2}} \exp\left(\frac{1}{2 K_2^2 \sigma_\theta^2}\right) \text{erfc}\left(\frac{1}{\sqrt{2} K_2 \sigma_\theta}\right)$$

$$- \frac{K_3}{m_{sdof}} \tag{29}$$

NUMERICAL RESULTS

The preceding procedure is applied to a particular model for which $\Pi_H = 4.0$, $\Pi_I = 5.6$, $\Pi_k = 215.9$, $\Pi_C = 251.3$ and $\sqrt{g/B} = 3.6/s$. The particular modulation function used is shown in Fig. 2 and the response statistics of the model are shown in Fig. 3 and 4.

To examine the reliability of the proposed procedure, numerical simulations of the rigid block-flexible foundation model subjected to modulated white noise accelerations are performed. A comparison of the approximate solution and the simulated results shows a reasonable agreement, but the former systematically under-estimates the magnitude of the tilt-angle. This observation suggests that the approximate method assumes a stiffer model than the original one. The stiffening is associated with the static condensation, and the analytic approximation of the spring restoring moment.

Note that the preceding procedure is applicable even when the seismic excitation model involves a filtered white noise process. In this case the equation of motion of the structure can be combined with the differential operator representing the filtered white noise to yield an augmented dynamic system under modulated white noise excitation. The latter problem has been addressed in this article.

The results of the present study should be deemed as strictly preliminary. Clearly, if the excitation is strong enough to the extent that a non-negligible fraction of its realizations induces toppling, the problem should be formulated on a first-passage basis. Furthermore, despite its mathematical convenience, the Gaussian approximation of the response should be rejected, if the response standard deviation is not quite smaller than the threshold of toppling.

REFERENCES

1. Meek, J. W., "Effects of Foundation Tipping on Dynamic Response," Journal of the Structural Division, ASCE, Vol. 101, No. ST7, July 1975, pp. 1297-1311.

2. Spanos, P-T.D. and Koh, A. S., "Rocking of Rigid Blocks Due to Harmonic Shaking," Journal of the Engineering Mechanics Division, ASCE, Vol. 110, No. EM11, Nov. 1984.

3. Veletsos, A. S. and Wei, Y. T., "Lateral and Rocking Vibration of Footings," Journal of the Soil Mechanics and Foundation Division, ASCE, Vol. 97, No. SM9, Sept. 1971, pp. 1227-1248.

4. Yim, C. S., Chopra, A. K. and Penzien, J., "Rocking Response of Rigid Blocks to Earthquakes," Earthquake Engineering and Structural Dynamics, Vol. 8, No. 6, 1980, pp. 565-587.

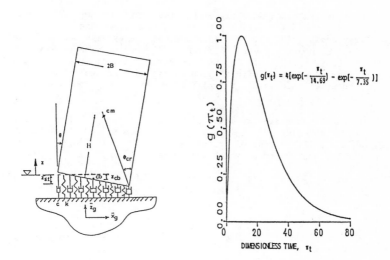

Fig. 1 Rocking Block on Winkler   Fig. 2 Excitation Modulating Function
Foundation

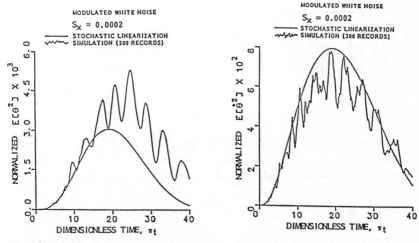

Fig. 3 Tilt-Angle Variance   Fig. 4 Tilt-Angle Velocity Variance

Dynamic Response of Systems with Uncertain Properties

F.E. Udwadia*

INTRODUCTION

The structural analyst/designer has two basic sources of uncertainty to contend with, while striving to obtain economical and safer designs. The first deals with uncertain knowledge of the dynamic loading time histories that may be brought to bear on the structure. The second deals with the basic uncertainty in the structural model of the system itself. While a large amount of effort has been expended in dealing with the first source of uncertainty very little research has been done dealing with the latter source. :

This paper addresses the methodology for determining the response of a structure modelled as single-degree-of-freedom system whose mass, damping and stiffness parameters are known only approximately. Consider a system defined by the differential equation

$$m\ddot{x} + c\dot{x} + kx = f(t), \quad x(0) = x_o, \quad \dot{x}(0) = \dot{x}_o \tag{1}$$

where,

$$\left. \begin{array}{l} 0 < m_1 \leq m \leq m_2 < \infty \\ 0 < k_1 \leq k \leq k_2 < \infty \\ 0 < c_1 \leq c \leq c_2 < \infty \end{array} \right\} \tag{2}$$

We assume that the ranges of variation $\{m_i, k_i, c_i\}$; $i = 1, 2$ are known apriori. We shall assume that the variables m, k and c are independent random variables. Then proceeding from the viewpoint of information theory, these variables may be thought of as being uniformly distributed. We shall assume that the parameters though unknown cause the system to respond in an oscillatory mode. The function $f(t)$ is assumed to be known as are also the initial conditions.

The random system may be thought of as a point in parameter space where each point represents the triplet $\{m_i, k_j, c_\ell\}$ (see Figure 1).

Statistics of Undamped Natural Frequency: Let m_0, k_0 and c_0 denote the means of the distribution of m, k and c respectively (Figure 2). The probability density function (pdf) of the normalized natural frequency, ω/ω_0, can be easily obtained in closed form.

*Professor, Department of Civil Engineering, DRB 394, University of Southern California, Los Angeles, CA 90089-1114.

Using the notation

$$\xi_{ij} = \frac{\sqrt{k_i/m_j}}{\omega_o} , \quad m_1 = m_o(1 - n_1), \quad k_1 = k_o(1 - n_2),$$
$$\overline{n} = n_1/n_2 , \quad \omega_o = \sqrt{k_o/m_o} , \tag{3}$$

(a) for $\overline{n} \leq 1$,

$$P_{\omega/\omega_o}(x) = \begin{cases} 0 & , x \leq \xi_{12} \\[2ex] \frac{1}{4n_1 n_2} [(1+n_1)^2 x - (1-n_2)^2 \frac{1}{x^3}], & \xi_{12} < x \leq \xi_{11} \\[2ex] \frac{1}{n_2} \cdot x & , \xi_{11} < x < \xi_{22} \\[2ex] \frac{1}{4n_1 n_2} [(1+n_2)^2 \frac{1}{x^3} - (1-n_1)^2 x], & \xi_{22} \leq x < \xi_{21} \\[2ex] 0 & , \xi_{21} \leq x , \end{cases} \tag{4}$$

and,

(b) for $\overline{n} \geq 1$

$$P_{\omega/\omega_o}(x) = \begin{cases} 0 & , x \leq \xi_{12} \\[2ex] \frac{1}{4n_1 n_2} [(1+n_1)^2 x - (1-n_2)^2 \frac{1}{x^3}], & \xi_{12} < x \leq \xi_{22} \\[2ex] \frac{1}{n_1 x^3} & , \xi_{22} < x < \xi_{11} \\[2ex] \frac{1}{4n_1 n_2} [(1+n_2)^2 \frac{1}{x^3} - (1-n_1)^2 x], & \xi_{11} \leq x \leq \xi_{21} \\[2ex] 0 & , \xi_{21} \leq x . \end{cases} \tag{5}$$

The distribution (see Figure 3), while widely different in the central regions, surprisingly, have exactly the same mean value and the variance given by

$$E[\omega/\omega_o] = \frac{4}{3} \left\{ \frac{2 + \sqrt{1 - \eta_2^2}}{[(1+\eta_1)^{\frac{1}{2}} + (1-\eta_1)^{\frac{1}{2}}][(1+\eta_2)^{\frac{1}{2}} + (1-\eta_2)^{\frac{1}{2}}]} \right\} \tag{6}$$

and

$$\sigma[\omega/\omega_o] = \left\{ \frac{1}{2\eta_1} \ln\left(\frac{1+\eta_1}{1-\eta_1}\right) - E^2[\omega/\omega_o] \right\}^{\frac{1}{2}} \tag{7}$$

As η_2 increases the $E[\omega/\omega_o]$ decreases, and as η_1 increases, $E[\omega/\omega_o]$ increases. For small values of η_1 and η_2, the frequency of the system having the mean properties (i.e. $\sqrt{k_o/m_o}$) is an overestimate of the expected value of the natural frequency of the random system as long as $\eta_2 > \sqrt{3}\,\eta_1$. Analytical results obtained for the damped system while more complex show similar features for values of $\xi_o \triangleq k_o/2\sqrt{c_o m_o}$ < 10% [see reference 1].

Response to Dynamic Loads:

Consider the system represented by equation (1) with zero initial conditions. Let

$$X(\omega) = \int_0^\infty e^{-i\omega t} x(t)dt , \quad i = \sqrt{-1} \tag{8}$$

Then relation (1) transforms to

$$X(\omega) = [k - m\omega^2 + ic\omega]^{-1} F(\omega)$$

The transfer function $H_f(\omega)$ can be defined as

$$H_f(\omega) = [k - m\omega^2 + ic\omega]^{-1} . \tag{9}$$

Should the oscillator be subjected to a base acceleration $\ddot{z}(t)$, the transfer function $H_e(\omega)$ corresponding to its response relative to the base would be given by

$$H_e(\omega) = -m\,H_f(\omega) .$$

Noting the m, k and c are uniformly distributed indepedent random variables, we obtain

$$E[H_f(\omega)] = \frac{1}{V} \sum_{s_\ell=1}^{2} \sum_{s_j=1}^{2} \sum_{s_i=1}^{2} \varepsilon_{s_i s_j s_\ell} \cdot h_f(m_{s_i}, k_{s_j}, c_{s_\ell}) \tag{10}$$

where,

$$\varepsilon_{pqr} = \begin{cases} (-1)^{p+q+r} & , \text{ when } p \neq q \neq r \\ 0 & , \text{ otherwise} \end{cases}$$

$$V = [(m_2 - m_1)(k_2 - k_1)(c_2 - c_1)]^{-1} ,$$

$$h_f(m_s, k_j, c_\ell) = -\frac{i}{2\omega^2}\left[\frac{y^2}{2} - y^2 \ell n\, y\right] ,$$

with $y = (k_j - m_s\omega^2 + i\omega c_\ell)$, and

$\ell n\, y = \ell n|y| + i\theta$, $0 \leq \theta \leq 2\pi$.

$$(11)$$

The amplitude of the steady state harmonic response, A_f, of the oscillator subjected to a forcing function $e^{i\omega t}$ is obtained as

$$E[A_f] = \sum_{s_\ell=1}^{2} \sum_{s_j=1}^{2} \sum_{s_i=1}^{2} \frac{1}{V}\, \varepsilon_{s_i s_j s_\ell}\, a_f(m_{s_i}, k_{s_j}, c_{s_\ell}) \qquad (12)$$

where $a_f(m_s, k_j, c_\ell) = \dfrac{1}{2\omega^3}\left\{ p\, \sqrt{p^2 + q^2} - q^2 \ell n[p + \sqrt{p^2 + q^2}] \right.$

$$\left. - \frac{pq}{\omega^3}\, \sinh^{-1}\left(\frac{q}{p}\right) \right\} \qquad (13)$$

where ε and V are as defined above and

$$p = c_\ell\omega , \qquad q = (k_j - m_s\omega^2)$$

Expressions for the variance of the response can also be obtained in closed form as also the corresponding expressions when the oscillator is subjected to a harmonic base acceleration [see reference 1].

Figure 4(a) shows the expected value of the amplitude of the transfer function for the case $m_0 = 1$, $k_0 = 100$, $c_0 = 0.5$, $\eta_1 = 0.05$, $\eta_2 = 0.15$, $\eta_3 = 0.6$ ($c_1 \triangleq c_0(1-\eta_3)$). The analytically obtained curves for expected values and the 1σ band are compared with Montecarlo simulations using a sample size of 20,000. The transfer function amplitude, A_f^0, for the system with the mean properties $\{k_0, c_0, m_0\}$ is also shown. We note that the expected transfer function curve is broader showing higher "effective" damping. The normalized amplitude, $k_0 A_f$, is found to be a function only of η_1, η_2, η_3 and ξ_0. The effect of ξ_0 on the expected value of the transfer function amplitude ($c_0 = 1$) is shown in Figure 4(b).

The expected response of the system to a transient forcing function, f(t), can be obtained as

$$E[x(t)] = \frac{1}{2\pi} \int_{-\infty}^{\infty} E[H_f(\omega)] \, F(\omega) d\omega$$

where the expected value on the right hand side is obtained from relations (10) and (11). The variance of the response of the system can be similarly obtained as also the corresponding expressions for the case of base accelerations. Figures 5(a) and 5(b) show the pdfs of the percentage of critical damping and the damped natural frequency for the random system of Figure 4(b). Figure 6(a) shows the expected value of the response of this system to the base acceleration shown in Figure 6(b), together with the computed 1-σ bounds on the expected response (shown shaded). The response of the system with the mean properties is also shown. The expected value of the response shows a higher effective damping as observed earlier.

CONCLUSIONS

We have in this paper developed a methodology for obtaining the response of a random system to steady state harmonic and transient excitations. Analytical expressions have been obtained to yield these responses. Comparisons with Montecarlo simulations are shown to be very good.

REFERENCES

F.E. Udwadia, "Dynamics of Uncertain Systems," Report No. FEU-01-86, University of Southern California, Los Angeles, CA 90089-1114.

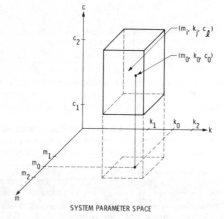

SYSTEM PARAMETER SPACE

Figure 1.

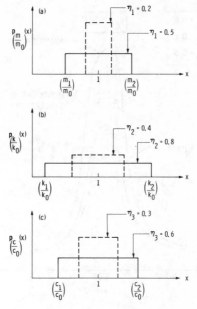

Figure 2.   Apriori Probability Density
Functions for $\frac{m}{m_0}$, $\frac{k}{k_0}$, $\frac{c}{c_0}$

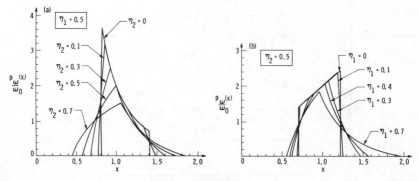

Figure 3.

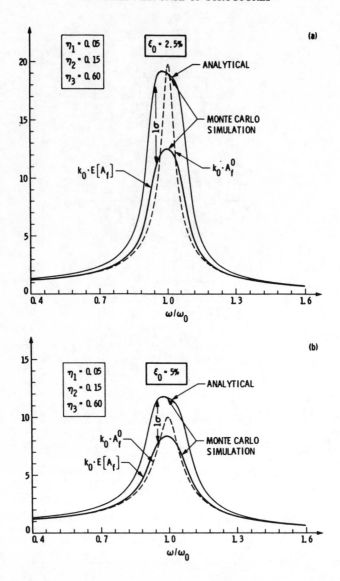

Figure 4.

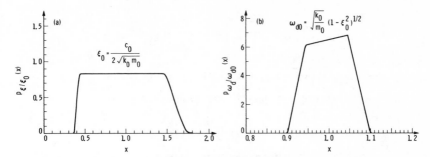

Figure 5. Analytically Obtained pdfs of Random Oscillator with
$m_1 = 0.95$, $m_2 = 1.05$, $k_1 = 85$, $k_2 = 115$, $c_1 = 0.4$, $c_2 = 0.6$

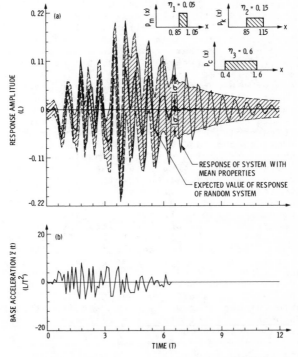

Figure 6. Expected Response of Random System of Figure 5 and the
$1-\sigma$ Band

STOCHASTIC RESPONSE OF STRUCTURES
WITH APPENDAGES CONTAINING FLUIDS

WEI-JOE SUN*, A. MASCE, AND

AHSAN KAREEM** MASCE

ABSTRACT

The influence of appendages containing fluids on the dynamic response of multi-degree-of-freedom systems subjected to stochastic environmental loads, e.g., earthquakes, waves, and winds, is investigated. The modal properties of a system comprising of an appendage containing a fluid attached to a multi-degree-of-freedom system are expressed in terms of individual dynamic properties of the primary and secondary systems. The primary system is modeled as a lumped mass multi-degree-of-freedom system. An equivalent lumped mass model of the sloshing fluid is used to represent the secondary system. All the frequencies of the secondary system that are less than or equal to the fundamental natural frequency of the primary system are included in the dynamic analysis of the combined system. The input to the system can be a stationary or a nonstationary white or filtered white noise vector-valued excitation. The nonstationary response was computed at any time interval ultizing the modal impulse-response function by approximating the envelope intensity function with a staircase unit impulse function. This noval formulation facilitates the computational procedure by reducing the multiplicity of the integrals. The covariance matrices of the response components of the combined system were computed using this formulation. The peak response values at any level on the structure are obtained following the evolutionary distribution of the extreme values. An important feature of the combined system is that the response of the primary system is suppressed when

*   Research Associate

**   Director Structural Aerodynamics and Ocean System Modeling Laboratory Dept. of Civil Engineering, University of Houston, TX 77004

one of the sloshing modes of the secondary fluid appendage is tuned with the fundamental mode of the primary system.

INTRODUCTION

The dynamic behavior of liquids in moving containers has been investigated by many researchers (1,2,3). Theoretical and experimental investigations of the dynamic behavior of ground-supported, deformable, cylinderical liquid storage tanks have been described in reference 4. An investigation of the properties and associated dynamic behavior of a fluid containing appendage attached at any level to a multi-degree-of-freedom system subjected to stochastic environmental load is reported here. Some typical examples are a tall building with a water tank, a tall refinary structure with fluid containers and an offshore platform with water tanks etc. In this study the dynamic properties of the combined system are expressed in terms of the dynamic properties of the primary and secondary appendage separately. This formulation helps to avoid computational difficulties that may arise as a result of difference in the magnitude of various elements in the mass and stiffness matrices. The present approach is based on a perturbation technique used earlier by Sackman, et. al (5) for the dynamic analysis of a light equipment, represented by a single-degree-of-freedom system, attached to a structure.

MODELING OF PRIMARY AND SECONDARY SYSTEMS

The primary system is modeled as a discrete lumped mass system with n lumped masses. The secondary system represented by an appendage containing a fluid is attached to any level of the primary system. The sloshing motion of the fluid in the appendage induces hydrodynamic forces on the primary system. The boundaries of the appendage are assumed to be rigid, which precludes any dynamic fluid-structure interaction between the fluid and the container. The dynamic behavior of the fluid can be mathematically derived following a potential flow theory (1,6). Alternatively, the lateral sloshing can be represented by an equivalent mechanical model in which the fluid is replaced by lumped fluid masses, springs and dashpots. Such a representation allows possible inclusion of the fluid damping as well. The mechanical characteristics of the equivalent system are established following the dynamic similitude of the sloshing fluid or by

m_ℓ ; $\ell = 1,\ldots.J$ are given by

$$m = \frac{8\,Tanh\,[(2\ell-1)]\pi r]M_L}{\pi^3 r\,(2\ell-1)^3} \qquad --(1)$$

and $k_\ell = m_\ell \omega_\ell^2$ for $\ell = 1,2\ldots J$.

The equivalent mechanical model of sloshing is shown in Fig. 1a. The combined primary and secondary, fluid containing appendage, systems are shown in Fig 1b, in which the secondary system may be attached at any level of the primary system. The overall system is excited by a unidirectional base acceleration, $\ddot{x}_g$.

The equations of motion for the combined primary and secondary systems are given by

$$\underline{M}^s \ddot{\underline{X}}^s + \underline{C}^s \dot{\underline{X}}^s + \underline{K}^s \underline{X}^s = -\underline{M}^s \underline{I}\,\ddot{x}_g \qquad --(2)$$

$$\underline{X}^s = [\underline{X} \vdots \underline{x}]^T$$

The frequency and the mode shapes of the combined system are obtained from the following equations:

$$[\underline{K}^s - \underline{M}^s \omega_s^2]\,\phi_i^s = 0 \qquad ---(3)$$

in which ϕ_i^s = ith mode shape of the combined system and it is expressed as

$$\phi_i^s = \left\{ \frac{\phi_{Ni}^s}{\phi_{Ji}^s} \right\} = \left[\phi_{1i}^s \,----\, \phi_{pi}^s \,-----\, \phi_{Ni}^s \vdots \phi_{p1i}^s \,-----\, \phi_{pJi}^s \right]^T \qquad ---(4)$$

In Eq. superscript, s, represents the combined system and subscript, p, denotes the level at which the secondary system is attached. The first part of the vector, ϕ_{Ni} ,represents the ith mode shape corresponding to the primary system and the second part, ϕ_{Ji}^s ,describes the ith mode shape of the secondary system. It is assumed here that the mass of the secondary system is small, therefore, the major contribution to ϕ_i^s comes from ϕ_i the ith mode shape of the primary system. Rearranging Eq. 3 gives

$$\underline{K}\,\phi_{Ni} - \omega_{si}^2\,\underline{M}\,\phi_{Ni} + \left\{ (\overset{0}{\underset{\ell}{\textstyle\sum^J}} k_\ell - m_0 \omega_{si}^2)\phi_{pi} - \overset{}{\underset{\ell}{\textstyle\sum^J}} k_\ell\,\phi_{p\ell i}^s \right\} = 0$$

and

$$k_\ell\,\phi_{pi} + (k_\ell - m_\ell\,\omega_{si}^2)\,\phi_{p\ell i}^s = 0$$

$$\ell = 1,2\cdots J \qquad ---(5)$$

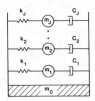

Fig. 1a. Equivalent Mechanical Model of Sloshing

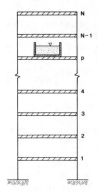

Fig. 1b. Combined Primary and Secondary Systems

From Eq. 5

$$\alpha_{li} = \frac{\phi_{pli}^{s}}{\phi_{pi}} = \frac{\omega_l^2}{\omega_l^2 - \omega_{si}^2} \qquad \qquad l = 1, 2 \cdots J \qquad ---(6)$$

in which α_{li} is the modal influence factor for the secondary system. Substituting Eqs. 6 in Eq. 5 and following the orthogonality of modes one obtains

$$\omega_i^2 - (1 + m_o \phi_{pi}^2) \omega_{si}^2 + \sum_l \frac{m_l \omega_l^2 - \omega_{si}^2}{\omega_{si}^2 - \omega_l^2} \phi_{pi}^2 = 0 \qquad ---(7)$$

The solution of the above equation is obtained

$$\left(\frac{\omega_{si}}{\omega_i}\right)^2 = \frac{\left(1 + \frac{\gamma_i + \theta_{li}}{2}\right) \pm \left[\left(1 + \frac{\gamma_i + \theta_{li}}{2}\right)^2 - (1 + \gamma_{oi})(1 + \theta_{li})\right]^{1/2}}{(1 + \gamma_{oi})(1 + \theta_{li})}$$

in which

$$\gamma_{oi} = m_o \phi_{pi}^2 \qquad ; \qquad \gamma_i = \gamma_{oi} + \gamma_{li}$$

$$\gamma_{li} = m_l \phi_{pi}^2 \qquad ; \qquad \theta_{li} = (\omega_i^2 - \omega_l^2)/\omega_l^2 \qquad --(8)$$

The factor θ_{li} and γ_{li} can be referred as the detuning parameter and the effective mass at each node of the secondary system, respectively. These parameters have significant influence on the dynamics of the problem under consideration. The modal influence factor of the secondary system with respect to the location of the attachements is given by

$$\alpha_{li} = \frac{1}{1 - \left(\frac{\omega_{si}}{\omega_i}\right)^2 (1 + \theta_{li})} \qquad ---(9)$$

Now the mode shape of the secondary system as an attachement to the primary system are given by

$$\phi_{pli}^{s} = \alpha_{li} \phi_{pi} \qquad ---(10)$$

The new modes resulting from the addition of the secondary system are denoted by

$$\underset{\sim}{\phi}_l^{s} = \left\{ \begin{array}{c} \underset{\sim}{\phi}_{Nl}^{s} \\ ----- \\ \underset{\sim}{\phi}_{Jl}^{s} \end{array} \right\} = \left[\underset{1l}{\phi}^{s} ----- \underset{Nl}{\phi}^{s} \Big| \underset{pl}{\phi}^{s} ----- \underset{PJl}{\phi}^{s} \right]^T$$

$$---(11)$$

The derivation of the new mode shapes and associated frequencies is available in reference 7.

In the case of perfect tuning of a secondary system mode to a primary system mode, the previous results are no longer valid. Following the previous procedure, the mode shapes associated with the combined system can be obtained (7).

DYNAMIC RESPONSE ANALYSIS

The dynamic response of the combine multi-degree-of-freedom primary and fluid containing appendage is computed using modal superposition approach. The mode acceleration approach can be used to efficiently truncate higher modes of response (7). A computationally efficient procedure for evaluating the covariance of response components of a system subjected to a nonstationary random excitation due to an earthquake or gusts in a hurricane is presented by Sun and Kareem in Ref. 8. This method is restricted to separable nonstationary stochastic models of excitation. The nonstationary response is computed at anytime interval by the modal impulse-response function by approximating the envelope intensity function with a staircase unit impulse function(8).

EXAMPLE:

An example is presented to illustrate the methodology developed here. A 10-story uniform building with a water tank at the top floor is considered. The frequencies and the mode shapes of the system were obtained using the perturbation technique and the direct analysis of the eigenvalue problem. In Table 1 the comparison between the methodology presented here and the exact results of the first 10 frequencies of the combined system are given. These results show a good agreement with the exact values obtained from a direct eigenvalue analysis. Similarly, good agreement with the exact values was obtained for the mode shapes (7). A water tank located either at the 5th floor or the top of the building was used to illustrate the procedure for computing the response. The building was excited by an earthquake which was represented by the Kanai-Tajimi power spectrum. The filter characteristics used in this study are: $\omega_f = 5\pi$ and $\zeta_f = 0.6$. The nonstationarity feature was introduced by a trapezoidal envelope function. The modal frequency of the water tank depends on the tank dimension, a, in the direction of motion. This dimension was assumed as 120 in, which results in tuning the second tank mode with the fundamental building mode. The water level and the tank width were arbitrarily chosen. In Table II the RMS displacement response, for stationary excitation, at the top floor of the building is summerized for various water levels. The results suggest that there is indeed significant reduction in the response of the primary system when the fluid containing appendage is attached.The response estimates and the statistics of their exceedance for nonstationary excitation were available in reference 7.

CONCLUSIONS

Following a perturbation technique a procedure is developed to compute the modal properties of a system consisting of a fluid containing appendage attached to a multi-degree-of-freedom system. The formulation includes the effect of interaction between the secondary and the primary systems that is particularly noteworthy when a modal frequency of the sloshing fluid is tuned to a natural frequency of the structure. The methodology is illustrated using a numerical example of a 10-story building with a water tank as the secondary appendage. The procedure for computing the eigenvalues and mode shapes of the combined system shows good agreement with the exact values. The dynamic response analysis suggests that an optimally tuned tank acts as a tuned mass damper in reducing the response of the building. The percentage of reduction is proportional to the mass ratio of the water tank to the building. The fluid damping also influences the reduction in response.

REFERENCES

1. Abramson, H.N "The Dynamic Behavior of Liquid in Moving Containers,", NASA, SP-106, 1966.

2. Housner, G.W., "Dynamic Pressures on Accelerated Fluid Containers", Bulletin of the Seimological Society of America, vol. 47, No. 1. Jan., 1957

3. Housner, G.W., "The Dynamic Behavior of Water Tanks," Bulletin of the Seimological Society of America. Vol. 53. No. 1. Feb., 1963.

4. Haroun, M.A. "Vibration Studies and Tests of Liquid Storage Tanks," Earthquake Engineering and Structural Dynamics, Vol. 11, 1983.

5. Sackman, J.L. and A.D. Kiureghian and B. Nour-Omid "Dynamic Analysis of Light Equipment in Structures: Modal Properties of the combined System", J. of the Eng. Mech., ASCE, Vol. 109, No. 1, Feb 1983.

6. Lamb, Sir Horace, "Hydrodynamics," Dover publications, New York.

7. Sun, W.J., and A. Kareem,"Stochastic Response of Structures with Appendages Containing Fluids," Tech Report, Civil Engineering Department, UHCE86-1

8. Sun, W.J., and A. Kareem, "Response of Torsionally Coupled Multi-story Buildings to Random Excitation", Technical Report, Civil Engineering, UHCE86-2

ACKNOWLEDGMENTS

Financial assistance for this research was provided by the National Science Foundation Grant CEE 8352223 and matching funds from several industrial groups.

TABLE I

Comparison of frequencies with exact values

Mode	Method Developed	Exact Solution
1	2.55	2.55
2	5.41	5.42
3	6.03	6.06
4	7.11	7.11
5	16.72	16.73
6	27.18	27.19
7	17.16	37.29
8	46.84	46.87
9	55.36	55.40
10	61.76	61.79

TABLE II

RMS displacement response at top floor
for stationary excitation
(Water damping ratio = 0.05)

Water level (in.)	Tank at top fl. (in.)	Reduction %	Tank at 5th floor (in.)	Reduction %	mass ratio%
0	2.76	0	2.76	0	0
30	2.29	17.03	2.46	10.87	2.68
60	2.00	27.54	2.26	18.12	3.58
90	1.86	32.61	2.06	25.36	3.69
150	1.97	28.62	1.87	32.25	3.51
200	2.09	24.28	1.93	30.00	3.30

RESPONSE OF EMBEDDED FOUNDATIONS:

A HYBRID APPROACH

Akira Mita and J. Enrique Luco[*]

ABSTRACT

A hybrid approach to study the dynamic through-the-soil coupling between adjacent foundations embedded in a uniform soil model when subjected to harmonic external forces and seismic excitation is presented. The hybrid approach is based on the use of the Green's functions for a uniform half-space combined with a finite element model of the finite volume of soil removed for the foundations. The formulation employs Green's functions for sources located on the surface of the half-space and receivers located at a constant depth below the free-surface. This formulation represents a variation of the substructure deletion approach proposed by Dasgupta(1980) which involves both sources and receivers located on the soil surface. Applications of an iterative scheme to obtain the coupling between foundations and of a reciprocity theorem to obtain the response of the foundations to seismic waves are presented. The use of the hybrid approach is illustrated by numerical results for a two-dimensional antiplane shear model involving two identical foundations.

INTRODUCTION

Dynamic analysis of linear soil-structure interaction effects under seismic excitation can be particularly important in the preliminary design stage of large scale structures such as nuclear power plants and high rise buildings. Finite element and boundary integral equation methods have been used extensively to solve the mixed boundary-value problems associated with the interaction between the foundations and the soil. Direct applications of the classical finite element method to interaction problems are limited by the overall finiteness of the model. Even though several methods have been proposed to overcome this shortcoming by introducing special boundaries or infinite elements, some limitations still remain. Boundary integral equation methods based on the use of Green's functions for a half-space do not suffer from this problem. For three-dimensional foundations embedded in a layered half-space, the calculation of the Green's functions involving a large number of combinations of sources and receivers can become extremely costly. The alternative substructure deletion approach proposed by Dasgupta(1980) which combines the continuum and the finite element approaches, can dramatically reduce the number of Green's function calculations and, therefore, appears to be promising. However, extensive numerical tests of the substructure deletion approach have not been presented. In addition, the approach involves the calculation of Green's functions for sources and receivers located on the ground surface and requires some special consideration for the singularity which occurs when source and receiver points coincide.

The hybrid approach proposed here corresponds to a variance of the substructure deletion method in which the singularities are eliminated by moving the receivers from the free-surface to a constant depth below the sources. In addition, the approach is extended to include multiple foundations as well as the calculation of the response to seismic excitation. The response of foundations to seismic excitation is obtained by use of a reciprocity theorem involving the free-field ground motion and the surface sources appearing in the calculation of impedance

[*] Department of Applied Mechanics and Engineering Sciences, University of California, San Diego, La Jolla, California 92093

functions. In this method only the free-field motion on the ground surface is involved and no deconvolution is required.

Numerical results for a two-dimensional anti-plane shear model show that the proposed hybrid approach works well for shallow embedded foundations but not for deeply embedded foundations. These results suggest that surface sources may not provide a complete representation of the displacement field for deeply embedded foundations.

FORMULATION OF A HYBRID APPROACH

The substructure approach proposed by Dusgupta(1980) and the variance described here are based on establishing a relation between the tractions and displacements on a surface C internal to an unindented half-space for a distribution of loads acting on the surface of the half-space(Fig. 1). This relation which is obtained by combining the Green's functions for the half-space with a finite element discretization of the finite volume of soil enclosed within the surface C is also assumed to hold for the indented half-space. This relation can be used to obtain the force-displacement relationship for rigid foundations interfacing with the soil along the surface C. The following discussion assumes harmonic time dependence and, for brevity, the factor $e^{i\omega t}$ is omitted.

Isolated foundations

The relation between forces applied at the source points S and displacements at receiver points R (Fig. 1) in the unindented half-space is assumed to be available in the form

$$\{u_R^l\} = [C_{RS}^{ll}]\{f_S^l\} \tag{1}$$

where $\{u_R^l\}$ is the displacement vector at n_R receiver points and $\{f_S^l\}$ is the force vector at n_S source points. The number of source and receiver points is assumed to be equal ($n_R = n_S$). The superscript l corresponds to the lth foundation which, in this first part, is considered isolated from other foundations. The matrix $[C_{RS}^{ll}]$ represents the Green's function matrix. The portion of the soil to be occupied by the foundation can be modeled by displacement-based finite elements as follows

$$\begin{bmatrix} K_{SS}^l & K_{SR}^l & K_{SC}^l \\ K_{RS}^l & K_{RR}^l & K_{RC}^l \\ K_{CS}^l & K_{CR}^l & K_{CC}^l \end{bmatrix} \begin{Bmatrix} u_S^l \\ u_R^l \\ u_C^l \end{Bmatrix} = \begin{Bmatrix} f_S^l \\ 0 \\ -f_C^l \end{Bmatrix} \tag{2}$$

in which each submatrix is the dynamic stiffness involving stiffness and mass matrices. Nodes other than the source nodes S, the receiver nodes R or the nodes on the connecting surface C have been condensed out. In Eq. (2), $\{f_C^l\}$ represents the nodal force vector that the finite soil region exerts on the surrounding unbounded soil. Under the assumption that the displacement at nodes R can be described by Eq. (1), substitution from Eq. (1) into Eq. (2) leads to

$$[H_{RS}^{ll}]\{f_S^l\} = -[K_{RC}^{l*}]\{u_C^l\} \tag{3}$$

where H_{iS}^{lm} and K_{iC}^{l*} are defined by

$$H_{iS}^{lm} = (K_{iR}^l - K_{iS}^l K_{SS}^{l-1} K_{SR}^l)C_{RS}^{lm} + \delta_{lm} K_{iS}^l K_{SS}^{l-1} \quad, \quad (l,m=1,2,..\ ;\ i=R,C) \tag{4}$$

$$K_{iC}^{l*} = K_{iC}^l - K_{iS}^l K_{SS}^{l-1} K_{SC}^l \quad, \quad (l=1,2,..\ ;\ i=R,C) \tag{5}$$

in which δ_{lm} is the Kronecker delta function. The force vector $\{f_C^l\}$ can be obtained from $\{f_S^l\}$ and $\{u_C^l\}$ in the form

$$\{f_C^l\} = -[H_{CS}^{ll}]\{f_S^l\} - [K_{CC}^{l*}]\{u_C^l\} \ . \tag{6}$$

Substitution from Eq. (3) into Eq. (6) gives

$$\{f_C^l\} = \left[[H_{CS}^{ll}][H_{RS}^{ll}]^{-1}[K_{RC}^{l*}] - [K_{CC}^{l*}] \right]\{u_C^l\} \tag{7}$$

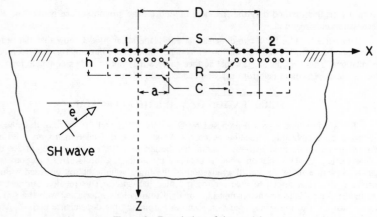

Figure 1. Description of the model

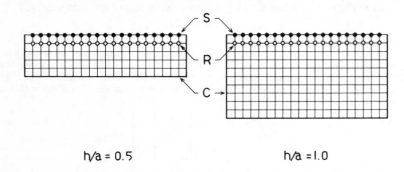

$h/a = 0.5$ $h/a = 1.0$

Figure 2. Finite element model of the excavated soil

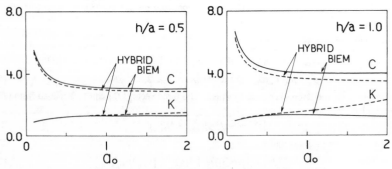

Figure 3. Impedance functions for an isolated foundation

which corresponds to the relation between the displacement $\{u_C^l\}$ and the force $\{f_C^l\}$ on the internal surface C of a half-space for sources applied on the free surface S. The assumption is made that Eq. (7) also provides an estimate of the force-displacement relation for the indented half-space after the excavation for the foundation has been made. The conditions for the validity of this assumption will be discussed elsewhere.

For a rigid foundation, the displacement vector at C is given by

$$\{u_C^l\} = [\alpha^l]\{U_{oC}^{l*}\} \tag{8}$$

where $\{U_{oC}^{l*}\} = (\Delta_{ox}^{l*}, \Delta_{oy}^{l*}, \Delta_{oz}^{l*}, \theta_{ox}^{l*}, \theta_{oy}^{l*}, \theta_{oz}^{l*})^T$ represents the generalized displacement response of the foundation and $[\alpha^l]$ is a rigid-body-motion influence matrix defined by

$$[\alpha^l] = \begin{bmatrix} \alpha_1^l \\ \alpha_2^l \\ . \\ \alpha_{n_C}^l \end{bmatrix}, \alpha_i^l = \begin{bmatrix} 1 & 0 & 0 & 0 & (z_i^l - z_o^l) & -(y_i^l - y_o^l) \\ 0 & 1 & 0 & -(z_i^l - z_o^l) & 0 & (x_i^l - x_o^l) \\ 0 & 0 & 1 & (y_i^l - y_o^l) & -(x_i^l - x_o^l) & 0 \end{bmatrix} \tag{9}$$

where (x_i^l, y_i^l, z_i^l) represents the coordinates of the ith node on C and (x_o^l, y_o^l, z_o^l) represents a reference point. The impedance matrix for the rigid foundation is given by

$$[\overline{K}] = [\alpha^l]^T \Big[[H_{CS}^{ll}][H_{RS}^{ll}]^{-1}[K_{RC}^{l*}] - [K_{CC}^{l*}] \Big] [\alpha^l] \tag{10}$$

and it defines the relation between the resulting forces and moments acting on the soil and the motion of the foundation.

Multiple foundation system

The approach described above can be also applied to a multiple foundation system. For a system of N foundations the relation between forces at source points and displacements at receiver points in the unexcavated half-space is assumed to be given in the form

$$\begin{Bmatrix} u_R^1 \\ u_R^2 \\ . \\ u_R^N \end{Bmatrix} = \begin{bmatrix} C_{RS}^{11} & C_{RS}^{12} & \cdot & C_{RS}^{1N} \\ C_{RS}^{21} & C_{RS}^{22} & \cdot & C_{RS}^{2N} \\ . & . & . & . \\ C_{RS}^{N1} & C_{RS}^{N2} & \cdot & C_{RS}^{NN} \end{bmatrix} \begin{Bmatrix} f_S^1 \\ f_S^2 \\ . \\ f_S^N \end{Bmatrix} \tag{11}$$

where superscripts $1, 2, .., N$ denote the foundation numbers. The soil regions to be occupied by the foundations are represented by finite element models and the resulting force-displacement relations in each region are given by Eq. (2). A procedure similar to that used to obtain Eqs. (3) and (7) leads to

$$\{f_C\} = \Big[[H_{CS}][H_{RS}]^{-1}[K_{RC}^*] - [K_{CC}^*] \Big] \{u_C\} \tag{12}$$

where $\{f_C\} = (f_C^1, f_C^2, \ldots, f_C^N)^T$ and $\{u_C\} = (u_C^1, u_C^2, \ldots, u_C^N)^T$. Eq. (12) is also assumed to hold for the excavated half-space. The $6N \times 6N$ impedance matrix for rigid body motion of the N foundations is given by

$$[\overline{K}] = [\alpha]^T \Big[[H_{CS}][H_{RS}]^{-1}[K_{RC}^*] - [K_{CC}^*] \Big] [\alpha] \tag{13}$$

where

$$[\alpha] = \begin{bmatrix} \alpha^1 & 0 & \cdot & 0 \\ 0 & \alpha^2 & \cdot & 0 \\ . & . & . & . \\ 0 & 0 & \cdot & \alpha^N \end{bmatrix} \tag{14}$$

and

$$
\left[H_{iS}\right] = \begin{bmatrix} H_{iS}^{11} & H_{iS}^{12} & \cdot & H_{iS}^{1N} \\ H_{iS}^{21} & H_{iS}^{22} & \cdot & H_{iS}^{2N} \\ \cdot & \cdot & \cdot & \cdot \\ H_{iS}^{N1} & H_{iS}^{N2} & \cdot & H_{iS}^{NN} \end{bmatrix} , \quad \left[K_{ij}^{*}\right] = \begin{bmatrix} K_{ij}^{1^{*}} & 0 & \cdot & 0 \\ 0 & K_{ij}^{2^{*}} & \cdot & 0 \\ \cdot & \cdot & \cdot & \cdot \\ 0 & 0 & \cdot & K_{ij}^{N^{*}} \end{bmatrix} , \quad (i=R,C\,;\,j=S,C) \quad (15)
$$

Iterative approach

For a foundation system involving a large number of foundations the inversion of the matrix $[H_{RS}]$ can become costly. To circumvent this problem Wong and Luco(1985) have proposed an iterative scheme. The extension of that approach to embedded foundations is obtained by writing the matrix $[H_{RS}]$ in the form

$$
[H_{RS}] = [D] - [B] = [D]\Big[[I] - [D]^{-1}[B]\Big] \quad (16)
$$

where $[I]$ is an identity matrix and

$$
[D] = \begin{bmatrix} H_{RS}^{11} & 0 & \cdot & 0 \\ 0 & H_{RS}^{22} & \cdot & 0 \\ \cdot & \cdot & \cdot & \cdot \\ 0 & 0 & \cdot & H_{RS}^{NN} \end{bmatrix} , \quad [B] = - \begin{bmatrix} 0 & H_{RS}^{12} & \cdot & H_{RS}^{1N} \\ H_{RS}^{21} & 0 & \cdot & H_{RS}^{2N} \\ \cdot & \cdot & \cdot & \cdot \\ H_{RS}^{N1} & H_{RS}^{N2} & \cdot & 0 \end{bmatrix} \quad (17)
$$

The inverse of $[H_{RS}]$ can be expressed in the series form

$$
[H_{RS}]^{-1} = \sum_{n=0}^{\infty} \Big[[D]^{-1}[B]\Big]^{n} [D]^{-1} \quad (18)
$$

in which succesive terms can be obtained by iteration.

Response of foundations subjected to seismic waves

Luco and Wong(1985) have shown that the response of massless foundations to seismic waves can be obtained, by use of a reciprocity theorem, in terms of the free-field ground motion of the half-space(prior to excavation) and of forces $[\lambda\,(\overline{x})]$ distributed over a surface S' internal to the foundation. This distribution of forces is involved in the calculation of the impedance matrix. The seismic response of the foundations is described by the $6N \times 1$ vector $\{U_o^*\}$ given by

$$
\{U_o^*\} = [\overline{K}]^{-1} \int_{S'} [\lambda\,(\overline{x})]^T \{u_g\,(\overline{x})\}\, dS'(\overline{x}) \quad (19)
$$

which can be rewritten in discretized form

$$
\{U_o^*\} = [\overline{K}]^{-1}[\Lambda]^T\{U_g\} \quad (20)
$$

where the matrix $[\overline{K}]$ represents the impedance matrix defined by Eq. (13) and $\{U_g\}$ represents the free-field ground motion. A component U_{gj}^{li} of $\{U_g\}$ represents the i-component of displacement at the jth surface node of the lth foundation. The matrix $[\Lambda]$ contains forces acting at the source points and is given by

$$
[\Lambda] = -[H_{RS}]^{-1}[K_{RC}^*][\alpha] \quad (21)
$$

where $[H_{RS}]$, $[K_{RC}^*]$ and $[\alpha]$ are defined by Eqs. (14) and (15). One of the advantages of this procedure to calculate the seismic response of foundations is that it involves only the free-field ground motion on the soil-surface and, consequently, no deconvolution is required.

NUMERICAL EXAMPLES FOR
TWO-DIMENSIONAL ANTI-PLANE MODEL

To test the validity of the proposed approach and to illustrate its application, the two-dimensional anti-plane shear vibrations of infinitely long, rigid, rectangular foundations embedded in a uniform elastic half-space are considered. The elastic half-space is characterized by the shear modulus μ and the shear wave velocity β. The cases of a single rigid foundation of width $2a$ embedded in the soil to a depth h and of two identical rigid foundations separated by a distance $(D-2a)$ are considered(Fig. 1). The Green's function for anti-plane shear vibrations of the half-space is given by

$$C_{RS}(\overline{x},\overline{x}_S) = -\frac{i}{4\pi\mu}\Big[H_o^{(2)}(k_s|\overline{x}-\overline{x}_s|)+H_o^{(2)}(k_s|\overline{x}-\overline{x}_s^*|)\Big] \tag{22}$$

where $H_o^{(2)}(\cdot)$ represents the Hankel function of the second kind and order zero, $\overline{x}$ and $\overline{x}_S$ denote the coordinates of the receiver and the source, $\overline{x}_S^*$ gives the coordinates of the mirror image of the source with respect to the plane defined by $z = 0$. The wave number k_s is defined by $k_s = \omega/\beta$ where ω is the frequency. The finite element models of the excavated soil regions are shown in Fig. 2 in which the number of sources in each foundation was chosen to be nineteen.

Impedance functions

Impedance functions for a single foundation have been calculated for two values of embedment ratio h/a of 0.5 and 1.0. The real and imaginary parts of the impedance function $\overline{K}_{yy}$, which is defined as the y-component force needed to produce a unit harmonic vibration in the y-direction, are normalized in the form

$$k = Real(\overline{K}_{yy}/\mu) \quad , \quad c = Imag(\overline{K}_{yy}/\mu)/a_o \tag{23}$$

where a_o is a dimensionless frequency defined by

$$a_o = \omega a/\beta \quad . \tag{24}$$

The terms k and c are the apparent stiffness and damping coefficients respectively. Obviously the damping effects are due to the radiation of waves. The results calculated are compared in Fig. 3 with those obtained by the boundary integral equation method[Kobori et.al.(1982)]. It is apparent that the hybrid method works well for the shallow foundation(h/a=0.5) but fails for the deeply embedded foundation(h/a=1.0). At extremely low frequencies the hybrid approach seems to give reasonable rsults even for the deeply embedded foundation. However, when two deeply embedded foundations are considered the estimates are poor even in the low frequency range. This fact suggests that surface sources may be insufficient to represent completely the displacement field for deeply embedded foundations.

Impedance functions for two identical foundations have been obtained for three values of the separation D/a of 2.5, 4.0 and ∞. Results for the embedment ratio h/a=0.5 are shown in Fig. 4, in which k^{11} and c^{11} are associated with the force needed to vibrate the first foundation with unit amplitude while k^{12} and c^{12} are associated with the force required to keep the second foundation at rest. Comparison of these results with those obtained by the boundary integral equation method shows good agreement. All results shown in Fig. 4 have been obtained through the direct inversion of $[H_{RS}]$. The effects of the soil column between the two foundations produces an oscillation as a function of a_o.

The convergence of the iterative approach for two embedded foundations with embedment ratio h/a=0.5 and separations D/a=2.5 and 4.0 has been studied by comparison of the results for 2, 6 and 10 iterations with those obtained by direct inversion. The results shown in Fig. 5 indicate that the closer the two foundations the more iterations are required to reach a stable solution. Ten or more iterations are required when the separation D/a is 2.5. The iterative scheme will be more advantageous for three-dimensional problems and for a system of more than two foundations.

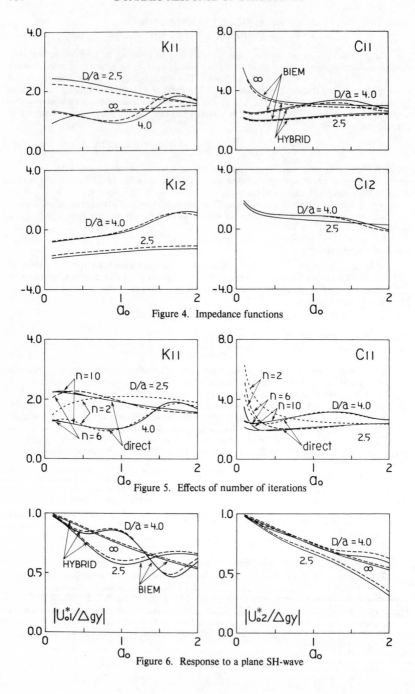

Figure 4. Impedance functions

Figure 5. Effects of number of iterations

Figure 6. Response to a plane SH-wave

Seismic response of foundations

Once the impedance functions have been calculated the response of the massless foundations to seismic excitation can be readily obtained by Eq. (20) which involves only the free-field ground motion on the free-surface. The free-field ground motion on the surface at $\bar{x} = (x,0)$ for a plane SH-wave with angle of incidence e is given by

$$u_{gy}(x,0) = \Delta_{gy} \exp(-ik_s x \cos e) \tag{25}$$

where Δ_{gy} represents the amplitude of the free-field displacement on the ground surface. The numerical results for two foundations with embedment ratios $h/a = 0.5$ excited by a plane SH-wave with angle of incidence $e = 45°$ are shown in Fig. 6(left panel corresponds to the foundation on the left in Fig. 1). The results obtained by the hybrid approach are in good agreement with those obtained by the boundary integral equation method.

CONCLUSIONS

A hybrid approach based on the use of Green's functions for the half-space combined with a finite element model of the soil volume removed for the foundations has been presented. This approach corresponds to a variation of the substructure deletion approach advanced by Dasgupta(1980) for a single foundation. The approach has been extended to multiple foundations and to the calculation of the seismic response. One advantage of the hybrid approach vis-a-vis the boundary integral equation method is that the Green's functions need to be calculated only for a small number of source-receiver depth combinations. A second advantage is that singularities, which occur when source and receiver points coincide, are avoided by locating the receivers below the sources. A third advantage is that since sources are located only on the surface no deconvolution of the free-field ground motion is needed to obtain the response of the foundations to seismic excitation. One disadvantage of the approach is that by considering only sources on the surface it is not possible to obtain a complete representation of the displacement field for deeply embedded foundations. For deeper foundations the procedure may have to be extended to include sources at depth.

ACKNOWLEDGMENTS

The stay of A. Mita at University of California, San Diego while on leave from Ohsaki Research Institute has been supported by Shimizu Construction Co., Ltd., Tokyo, Japan. Partial support from the National Science Foundation Grant ECE-8312441 is also acknowledged.

REFERENCES

Dasgupta, G. (1980). Foundation Impedance Matrices by Substructure Deletion, *J. of the Engrg. Mech. Div., ASCE*, 103, 517-523.

Kobori, T., Y. Shinozaki and A. Mita (1982). Soil-Structure Interaction Analysis by Boundary Integral Equation Method, *Proc. of 6th Japan Earthq. Engrg. Symp.*, 1737-1744.

Luco, J. E. and H. L. Wong (1985). Seismic Response of Foundations Embedded in a Layered Half-Space, *Int. J. of Earthq. Engrg. and Struct. Dyn.*, submitted.

Wong, H. L. and J. E. Luco (1985). Dynamic Interaction Between Rigid Foundations on a Layered Half-Space, *Int. J. of Soil Dyn. and Earthq. Engrg.*, submitted.

MODELING FOR FINITE-ELEMENT SOIL-STRUCTURE INTERACTION ANALYSIS

A. H. Hadjian[1], M. ASCE, S. T. Lin[2] and J. E. Luco[3], M. ASCE

Abstract

The modeling for finite-element soil-structure interaction analysis for embedded structures is evaluated in an attempt to resolve the issue related to the elevation at which design ground motion should be specified. Based on the premise that the design ground motion in large excavations should tend to become identical to that on the free-surface, a finite-element model is recommended in which the rigidity of the basemat must be incorporated and which has a contoured bottom that reflects the excavation at the surface. With this model the design ground motion should be specified at the free-surface.

Introduction

Design ground motion is usually specified at the free-surface mainly because that is where most of the recorded earthquake motions have been obtained. The use of this free-surface design motion for structures founded on the ground surface is obviously considered appropriate. It is only when embedded structures are to be analyzed for soil-structure interaction that a long standing controversy seems to persist: Should the design ground motion be applied at the free-surface or in the free-field at the foundation base level? Clearly it does not make sense to relocate the design ground motion as a function of the structure being analyzed. In addition, a unique definition of design ground motion is desirable where a number of structures are considered in close proximity but at different embedment levels. The controversy is primarily due to a misconception relative to the variation of ground motion amplitudes with depth. Since the amplitude of upcoming waves is theoretically doubled upon reflection at the free ground surface, motion at depth (of engineering interest) and within the medium itself should have lower amplitudes relative to the free-surface amplitude. This however should not imply that at the top of the foundation of embedded structures, where free-surface conditions exist, significant amplitude reductions, simply due to depth and exclusive of interaction effects, should take place. Nevertheless, within the medium itself and at the foundation base level significant amplitude reductions should be expected. These observations have been substantiated at the Fukushima Nuclear Power Plant during the Magnitude 7.4 Miyagi-ken-Oki, Japan Earthquake of June 12, 1978.

The thrust of this study is on the soil modeling requirements when the design ground motion is specified at the free-surface.

1 Principal Engineer, Bechtel Power Corp., Los Angeles, CA.
2 Member of the Technical Staff, Rockwell International, Downey, CA.
3 Professor, University of California, San Diego, La Jolla, CA.

Basis for Study

If the soil-structure interaction analysis of embedded structures were performed using a model that starts with the source mechanism of the earthquake and motion proceeds towards the embedded structure then no special precautions would be required to assure proper response calculations. This type of source-to-structure modeling is of course not practical and hence artifacts must be used to achieve the expectations of motion at the foundation.

In the past research on this issue, these expectations were evaluated invariably by the use of the response of embedded structures, thus mixing the effects of the variation of ground motion with depth and inertial interaction. A more appropriate investigation of this problem would be through the elimination of inertial soil-structure interaction effects, i.e. no structure will be included in the response calculations. The adequacy of the soil modeling could then be judged by comparing the motions in large shallow "excavations" to those specified at the free-surface. Given the relatively non-flat surfaces where ground motions have been recorded, it is to be expected that ground motion characteristics would not, and should not at least from the viewpoint of design, be dissimilar in and out of large relatively shallow depressions. Even though ground motions in close proximity could be different, from the view point of design, two structures one in and the other out of a large shallow depression, should not be designed for different motions. Our knowledge of detailed microzonation is still miniscule. However, as the lateral size of the "excavation" is reduced, it is to be expected that the ground motion in the "excavation" would tend towards the free-field motion at the same elevation within the medium. In fact, in the limit, as the lateral "excavation" size goes to zero, the two above motions would be identical. On the other hand, as the lateral "excavation" size becomes larger, the motion in the excavation should become identical to that on the free-surface. These characteristics are the basis by which the appropriateness of the soil modeling is judged.

Model Description

The basic study model is shown in Fig. 1a. The model is expected to represent three lateral excavation sizes in an elastic half-space ranging from small (H = 30 ft) to intermediate (H = 150 ft) to large (H = 300 ft). In all cases the depth of the excavation is fixed at 30 ft. Fig. 1c provides a visual appreciation of the size of the excavation in relation to its depth. The following two variations of the basic model of Fig. 1a are also used either separately or together: a) the top layer in the excavation is replaced with concrete to simulate the presence of a rigid basemat and b) the bottom boundary of the model is contoured to follow the surface excavation as shown in Fig. 1b. Table 1 summarizes the seven FLUSH models used in this study.

In the study the medium is assumed to be uniform, and is represented by strain-independent solid elements having the following material properties: weight density γ = 120 pcf, Poisson's ratio μ = 1/3, shear wave velocity v_s = 1000 ft/sec, damping ratio β = 3%. In those cases where a concrete basemat is placed in the excavation area or the contoured rigid base is to be simulated, the material properties of the

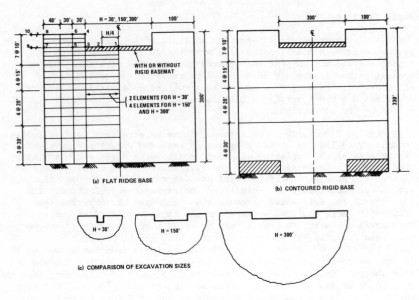

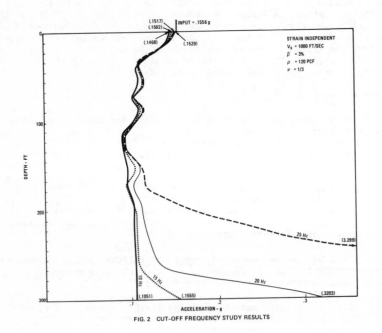

solid elements (shaded area in Fig. 1) are replaced by: weight density
γ = 150 pcf, Poisson's ratio μ = 0.17, shear wave velocity v_s =
8000 ft/sec, damping ratio β = 0.01%.

At top layers where the response is of interest, the vertical dimension
of the elements is chosen not to exceed one fifth (1/5) of the shear wave
length for frequencies below 20 Hz. This dimension is increased gradually
with depth as shown in Fig. 1 in order to reduce the cost of the
computations. The horizontal dimension of solid elements is also shown in
Fig. 1. In the excavation area four equal widths are used for the small
excavation, and eight equal widths for the intermediate and large size
models. A transmitting boundary is specified at the edge of the model to
simulate the semi-infinite soil system beyond the finite element region.
Only a half model is used in the analysis with vertical and rotational
degrees of freedom being constrained on the line of symmetry for
horizontal excitation. Although the Nyquist frequency of the control
motion is 25 Hz, a cut-off frequency of 15 Hz is used throughout the
analyses. This decision is based on results from a SHAKE analysis (see
Model Validation).

The Taft earthquake record (N21E) is used as the input control motion.
The control motion is specified at the top of the free surface. The
excitation is in the horizontal direction only. Horizontal response
accelerations are computed at eight selected nodes as denoted by solid
dots in Fig. 1a. Then the response spectra for 1% damping is calculated
for each response time-history acceleration. Since the objective is to
compare the motions in the excavations to those on the free-field, the
ratios of response spectra in the excavation and its vicinity to the
corresponding free-field response spectra are calculated. These
free-field responses are designated as "nodes" 9 and 10 in Fig. 1a.

Model Validation

Prior to the FLUSH analyses, several soil-column studies are carried
out using SHAKE to assure that the models of Fig. 1 are adequate for this
study. The mesh size (layer thickness in the medium) is checked with the
control motion being cut off at 10 Hz, 15 Hz, 20 Hz, and 25 Hz. The
resulting peak acceleration profiles are graphically shown in Fig. 2. For
cut-off frequencies of 20 Hz and 25 Hz, the control motion cannot be
deconvolved to the rigid base by the soil profile shown in Fig. 1.
Consequently, 15 Hz is taken to be the appropriate cut-off frequency for
the subject analyses.

The two dimensional finite element model, without any excavation, that
is comparable to the SHAKE model is also analyzed. The purpose of this
analysis is to check the mesh size, the transmitting boundary at the edge
of the model, and the boundary specified on the line of symmetry for the
half soil model. Response at the free surface and at 30 ft below the
surface are compared to the corresponding SHAKE model results. The
results are found to be satisfactory.

In order to check the material properties that are used to simulate the
contoured rigid base, a SHAKE model with a 30 ft layer thickness of the
rigid material is added to the bottom of the soil-column. Excellent
agreement in the peak acceleration profiles of both models is observed.

Discussion of Results

Fig. 3 is representative of the complete results for one model. Because of space limitations only selected results from the remaining six models will be presented. The conclusions however will be based on the complete available results. Response of Model 3 with 150 feet wide excavation will not be discussed simply because it represents intermediate results between the large and small excavations. Its usefulness lies in the fact that it substantiates the trends from the large to the small excavation models, and thus would be considered as a contribution to model validation. As discussed earlier the results are presented as ratios of response spectra. Ratios greater than unity indicate that the response in question is higher than the corresponding free-field response. Each frame of Fig. 3 and subsequent figures as well carry a designation in the top right hand side corner. The first number refers to the node as defined in Fig. 1 and the second number to the free-field motion: 9 refers to the free-field motion at 30 feet below the free-surface and 10 refers to the free-surface motion.

The results in Fig. 3 indicate that the response ratio in the excavation relative to the surface (node 10) is, in general, larger than unity. Additionally, the response in the excavation is not uniform. The centerline response (node 1) is larger than the response at both the quarter point (node 2) and the edge of the excavation (node 3). As the size of the excavation is reduced to 150 feet and 30 feet the response ratio in the excavation relative to the surface (node 10) becomes less than unity. This is to be expected since as the excavation width tends to zero the response ratio with respect to node 9 rather than node 10 should tend to unity. The response ratios at nodes 7 and 8 are not unity simply because these locations are not yet far enough away from the excavation. For the 30 feet excavation models the response ratios at nodes 7 and 8 do get to be very nearly unity. Because of the use of transmitting boundaries it is not necessary to have a large finite-element region to simulate free-field conditions.

The above observations lead to the conclusion that the model of Fig. 1a is inadequate for the purpose of carrying out soil-structure interaction analyses. From the viewpoint of design it would be necessary to assure that the input motion be at least the same along the base of the structure. One possibility to achieve this objective is to connect the nodes at the surface of the excavation by rigid links. However, since heavy embedded structures usually have relatively rigid basemats, the properties of the top layer elements in the excavation are instead modified to represent concrete. The results of this modification are compared in Fig. 4 to the corresponding results without the inclusion of a rigid basemat. Significant improvements are to be noted. Given the basic acceptance criteria attention should be focused on the comparisons of Figs. 4b and 4c. Figs. 4a and 4d are provided solely to accentuate the thrust of the earlier discussion.

The relatively rigid concrete basemat affects the results in the excavation in two ways: the response becomes uniform across the excavation and the response ratios are reduced, particularly for the large excavation. Thus the maximum ratio of 2.7 is reduced to 1.4 in Fig. 4c. However, it also enhances underestimates across the frequency spectrum.

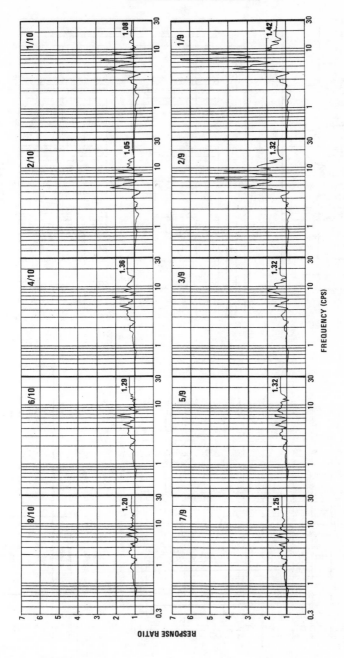

FIG. 3   RESPONSE RATIOS AT NODES 1 THROUGH 8 FOR MODEL 4

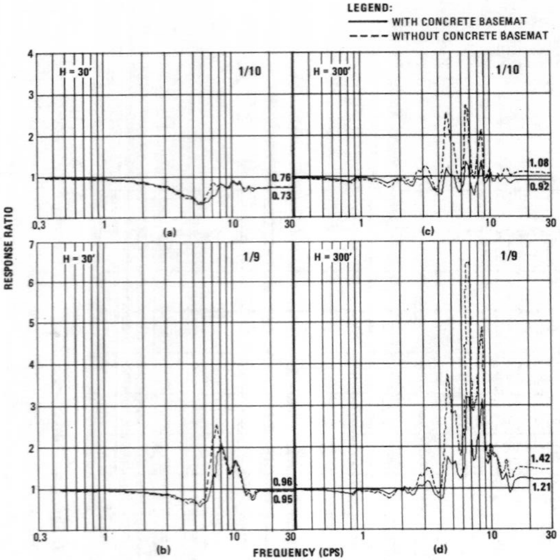

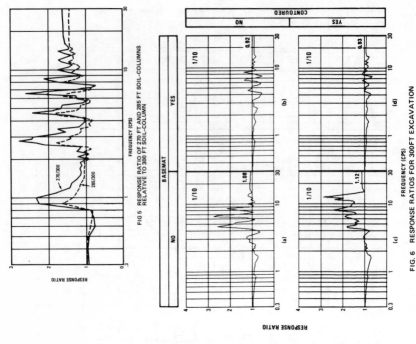

FIG 5   RESPONSE RATIO OF 270 FT AND 285 FT SOIL-COLUMNS
RELATIVE TO 300 FT SOIL-COLUMN

FIG. 6   RESPONSE RATIOS FOR 300FT EXCAVATION

FIG. 4   COMPARISON OF RESPONSE RATIOS AT CENTER OF EXCAVATION

Table 1.   FLUSH Analyses

FLUSH Model	Excavation Width	Presence of Basemat	Shape of Rigid Base
1	30 ft	No	Flat
2	30 ft	Yes	Flat
3	150 ft	No	Flat
4	300 ft	No	Flat
5	300 ft	Yes	Flat
6	300 ft	No	Contoured
7	300 ft	Yes	Contoured

Additionally (not shown) it enhances the response off of the excavation implying that buildings influence the response of adjacent structures simply by the rigidity they impart to the foundation. It is interesting to note that whereas the large excavation response ratio (relative to node 10, Fig. 4c) fluctuates from 0.5 to 1.4, the response ratio for the small excavation (relative to node 9, Fig. 4b) fluctuates from 0.6 to 2.1). Given the acceptance criteria, the modeling is not yet acceptable even though the situation has been significantly improved.

In order to justify the next level of model improvement the following side study is carried out. A one-dimensional 300 feet soil-column, representative of the soil profile without an excavation, is used to define the motion at the soil-column base (using SHAKE). This motion is then prescribed under a soil-column 270 feet long, representative of the large excavation. The response ratio at the top of the two columns is compared in Fig. 5 (solid line). It is obvious that the length of the column is a very sensitive parameter and since in a two-dimensional model with vertically propagating waves the motion propagates essentially in a one-dimensional fashion (compare response ratio of Fig. 5 with frame 1/10 of Fig. 3), the acceptance criteria would then require a length of 300 feet below the lateral extent of the excavation. This argument then leads to the contoured model of Fig. 1b. (To check against a fortuitous selection of soil-column lengths relative to the time history input motion, the response ratio of a second intermediate column length of 285 ft. to the 300 ft soil-column is also shown in Fig. 5 with dashed lines. As to be expected the results are lower all across the frequency spectrum --- in the limit the response ratio becomes unity -- and a shift of about 5% in the frequencies of the peaks occurs which is due to the change in the soil-column length, from 270 ft. to 285 ft.).

Fig. 6 shows the response ratios at the center of the excavation of all four models for the 300 feet excavation: a) Model 4, b) Model 5, c) Model 6 and d) Model 7. Although taken separately the basemat concrete does a better job than contouring in bringing the response ratios closer to unity, the combined effect of these two modeling artifacts does achieve, in a remarkable fashion, the objectives of the study. For all practical purposes the response ratio in Fig. 6d fluctuates, only slightly, about 0.93. It is expected that as the excavation becomes larger, the average response ratio will tend towards unity with minor fluctuations. As shown in Fig. 6c it is interesting to note that contouring all by itself is not a very useful modeling artifact for the present objective.

Conclusions

Based on the premise that ground motion characteristics should not, at least from the viewpoint of design, be dissimilar in and out of large shallow depressions, the following modeling requirements have been established for finite-element analyses using methodologies similar to the FLUSH code: a) The design ground motion should be specified at the free-surface. b) The foundation basemat should be modeled such as to incorporate its large stiffness. c) The bottom of the soil model should be contoured to reflect the excavation at the surface.

The Simulation of Soil-Concrete Interfaces
with a Non-local Constitutive Model

Z. Chen[1] and H.L. Schreyer[1], M. ASCE

This investigation explores a method for predicting deformation
in the vicinity of interfaces of dissimilar materials such as concrete and soil. Based upon the basic features of the deformation
field, a new nonlocal approach for simulating soil-concrete interfaces is proposed. It is assumed that the stress field is a
function of both strain and strain gradients, and that no relative
slip occurs at the contact surface between soil and concrete.
Typical response features for the region adjacent to the interface
can be obtained by adjusting material parameters. Numerical results
for static and dynamic cases show that softening and localization
are handled in a stable manner independent of mesh size.

Introduction

In the past, many models for representing interfaces were based
on Coulomb's law for which no slip between two bodies will occur
unless a critical tangential stress is reached, and the relative
motion is constrained to be along the contact surface if the critical state is reached. For dissimilar materials, however, Coulomb's
law may not be applicable since observed zones with large strain
gradients can not be predicted automatically. Nevertheless, the
simple friction approach has been utilized with either slip lines or
special interface elements to obtain computational procedures, which
in turn, introduce questions of stability and computational
complexity.

Among the investigators dealing with soil-concrete interface
phenomena, Huck et al. (1973, 1981) were unique in that both experimental and analytical investigations were performed. Vermeer
(1983) investigated the interface response in powder compaction by
means of a non-associated perfect plasticity model. The idea of
sliding on a single plane was considered much too restrictive for
representing physical data, and therefore, the new idea was employed
of small slipping motions on an infinite number of closely spaced
parallel surfaces, with the Mohr-Coulomb criterion being used. In
order to calculate the stress and density distributions at several
stages of the process of loading, three approaches were suggested.
The first is the use of a non-associated constitutive model and a
fine mesh near the walls of the die so that a thin shear band can

[1] Research Assistant and Professor, respectively, Dept. of
Mechanical Engineering and New Mexico Engineering Research
Institute, University of New Mexico, Albuquerque, NM 87131.

develop.   The second consists of special thin interface elements to
model   the   shear band along the wall.   Frictional sliding is implemented directly as a mixed boundary condition in the third approach.
Drumm and Desai (1983), and Oden and Pires (1983), described the
constitutive behavior of an interface by using frictional principles
with some modifications to include either elasticity or a perfectly-
plastic relation.

    From the research efforts mentioned above, it is found that
assumptions based upon frictional theories are commonly adopted, and
numerical methods are under active development.   So far, these
simple frictional theories have not proven useful for representing
interfaces involving materials with disparate properties such as
concrete and soil.   In addition, there are some unpleasant features
associated with numerical procedures for frictional models.   As
examples,   the   response may depend on mesh size and often there are
computational difficulties associated with slideline algorithms.

    For   large   static and dynamic loads, the stress in the softer
material next to an interface may exceed the limit state and softening may be initiated.   Some experimental stress-strain data beyond
the peak stress has been documented for concrete, rocks and soils.
One approach is to model the post-peak behavior by using plasticity
and a reduction of stress with strain.   However, serious theoretical
and computational difficulties caused by strain softening have
arisen,   and they have been intensely debated at recent conferences.
Valanis (1985), Sandler and Wright (1983), and Read and Hegemier
(1984) discussed in detail the various arguments concerning strain
softening.   Bazant et al. (1984) incorporated strain softening into
a constitutive model by using a nonlocal continuum theory.   Schreyer
and Chen (1986) investigated the effects of strain softening and
localization in a structure by assuming that stress is a function of
both strain and strain gradient.   It is concluded that current
experimental techniques can not rule out strain softening, and that
the various nonlocal approaches have overcome the theoretical objec-
tions to strain softening.

    It is suggested that the material in the vicinity of an inter-
face be modeled as a nonlocal continuum which means that stress is
considered to be a function of strain gradient as well as strain.
The effect of the interface can be incorporated by assuming no slip
but with a reduced limit stress for the weaker of the two adjacent
materials.   Then with the use of conventional, constitutive-equation
algorithms and finite element methods, static and dynamic response
features in the vicinity of an interface can be obtained with a
simple and convenient calculation.

Assumptions

    There are not enough experimental data to provide a good
qualitative description of the behavior at soil-concrete interfaces.
For example, it is still not known at what stage of the loading path
that macroscopic slip between concrete and soil occurs if it occurs

at all. As a consequence of this uncertainty, the following assumptions are made to construct a new nonlocal continuum model, which is different from the usual Mohr–Coulomb model:

(1)   Slip does not occur at the contact surface between concrete and soil under shear loading. Instead, with an increase in load, a small zone with a large strain gradient will form due to dispersed microcracking. This zone represents a localization of deformation that is accompanied by softening.

(2)   The imperfection of the soil–concrete interface reduces the strength of the soil, so that with large enough loads the softening is initiated in the region of soil adjacent to the interface. With this assumption, the roughness of the concrete surface is simulated through the strength property of the adjacent material rather than through a friction coefficient.

(3)   A nonlocal constitutive theory is required to represent the localized behavior of softening in soil. This theory provides a means for controlling the size of the localization.

(4)   A suitable nonlocal theory is obtained if the limit stress depends on the strain gradient. Read and Hegemier (8) indicate that strain softening is not a true material property but is simply a manifestation of the effects of progressively increasing inhomogeneity of deformation. An alternate interpretation that explains equally well the observed phenomena, however, is that softening occurs only in the presence of strain gradients. The assumption that the limit stress depends upon the strain gradient is therefore a simple mathematical representation of the observation.

(5)   The concrete is assumed to be rigid because the stiffness of concrete is usually higher by an order of magnitude than that of soil.

(6)   The deformation of the soil is assumed to be small so that the total strain field can be divided into elastic and inelastic parts.

Constitutive Equation For Soil

To illustrate the nonlocal aspect of the theory, a modified form of the Drucker–Prager formulation is used for the yield surface. In addition to the linear dependence on mean pressure, the yield surface incorporates strain hardening and strain softening, and dependence on the inelastic strain gradient as follows:

$$F = \bar{\sigma} - (1+a_1 P)GH \qquad (1)$$

in which P is the mean pressure and $\bar{\sigma}$ is the second invariant of deviatoric stress

$$P = -\frac{1}{3} \text{ tr } \underset{\approx}{\sigma} \qquad \underset{\approx}{\sigma}^d = \underset{\approx}{\sigma} + P\underset{\approx}{I}$$

$$\bar{\sigma} = [\frac{3}{2} \text{ tr } (\underline{g}^d \cdot \underline{g}^d)]^{1/2} \tag{2}$$

Strain hardening and strain softening is incorporated in G through additional parameters as follows:

$$G = \begin{cases} G_0 & e^i = 0 \\ G_0 + (G_L - G_0) \sin [\frac{\pi}{2} (\frac{e^i}{e^i_L})^n] & 0 < e^i \le e^i_L \\ (G_L - G_a) [1 + a_5 e^*] e^{-a_5 e^*} + G_a & e^i > e^i_L \end{cases} \tag{3}$$

in which e^i is an inelastic strain invariant formed from increments in inelastic strain

$$e^i = [\frac{2}{3} \text{ tr } (d\underline{e}^i \cdot d\underline{e}^i)]^{1/2} \tag{4}$$

e^i_L is the value of e^i at the limit state, and

$$e^* = \frac{e^i - e^i_L}{e^i_L} \tag{5}$$

The values of G at initial yield and the limit state are G_0 and G_L, respectively, while G asymptotically approaches G_a for large values of e^i. The shape of a stress-strain curve in the hardening regime is controlled through the parameter n and the parameter a_2 performs a similar role for softening.

The dependence of F on strain gradients is taken to be the simple expression

$$H = (1-a_2) e^{-a_3 g} + a_2 \tag{6}$$

where a_2 and a_3 are material parameters and

$$g = (\nabla e^i \cdot \nabla e^i)^{1/2} \tag{7}$$

Here it has been assumed that H decreases with a suitable norm of the inelastic strain gradient, and H asymptotically approaches a_2 for large values of the product $a_3 g$. The parameter a_3 has the dimensions of length which presumeably can control the size of the localization zone.

Inelastic strain increments are obtained from a nonassociated flow rule

$$d\underset{\approx}{e}^i = d\lambda \frac{\partial \phi}{\partial \underline{\sigma}} \qquad \phi = \bar{\sigma} - G \qquad (8)$$

so that inelastic compaction or dilatation is not incorporated in the model.

The constitutive relation is completed by assuming that the total strain is a sum of inelastic and elastic parts, and that the elastic part is governed by the usual linear relation for isotropic materials.

To illustrate the model, consider a case of pure shear and the following hypothetical material parameters:

$$G_0 = 50\sqrt{3} \qquad G_L = 100\sqrt{3} \qquad G_a = 10\sqrt{3}$$

$$\qquad (9)$$

$$G = 2 \qquad e_L^i = 10 \qquad n = 0.5$$

$$a_2 = 0.1$$

Also, additional parameters were set as follows:

$$P = 100 \qquad g = 10 \qquad (10)$$

Figure 1 shows the effect of the parameter a_5 in the softening regime for $a_1 = 0$ and $a_3 = 0$. The material is perfectly plastic when $a_5 = 0$. The effect of the strain gradient is shown in Figure 2 for $a_1 = 0$ and $a_5 = 0$. It is assumed that the strength of the material is reduced with an increase in strain gradient.

Results for a Static Problem

To illustrate the attractive features of a nonlocal plasticity problem, consider the one-dimensional model problem shown in Figure 3. The domain $0 \leq x \leq L$ is separated into a number of elements with linear shape functions. At $x = L$ a transverse force f is applied and a no slip condition is prescribed at $x = 0$. To make the static problem stable in the strain-softening regime, the prescribed force condition is replaced with a prescribed displacement.

Static results were obtained for the following material parameters:

$$G_0 = 0.5\sqrt{3} \qquad G_L = \sqrt{3} \qquad G_a = 0.1\sqrt{3}$$

$$\qquad (11)$$

$$G = 2 \qquad e_L^i = 0.16 \qquad n = 0.4$$

$$a_1 = 0 \qquad a_3 = 0.08 \qquad a_5 = 0.6$$

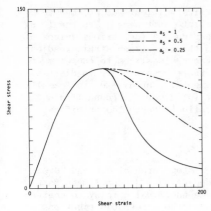

Figure 1. The effect of the softening parameter on stress-strain response.

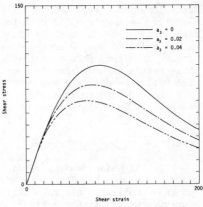

Figure 2. The effect of the strain gradient parameter on stress-strain response.

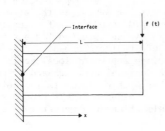

Figure 3. One-dimensional model problem.

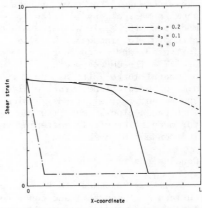

Figure 4. The effect of the strain-gradient parameter on the static response of the soil.

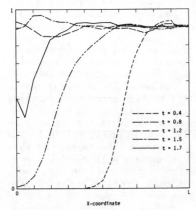

Figure 5. Spatial distribution of stress at various times for dynamic problem.

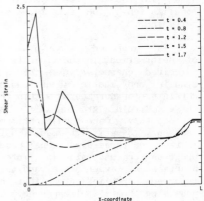

Figure 6. Spatial distribution of strain at various times for dynamic problem.

Ten elements were used and the first element was assumed to be
slightly weaker than the rest to initiate softening. The resulting
distribution of strain is shown in Figure 4 for various values of
a_3. A value of $a_3 = 0$ corresponds to the case of no strain-gradient
effect and the result is that softening localizes to one element.
This result is undesireable because the stress and strain distribution will depend on the number of elements. For $a_3 \neq 0$, the results
will not depend on mesh size except for numerical roundoff and the
degree of localization can be controlled through the specification
of a_3.

Results For a Dynamic Problem

For a mass density of unity, a step force of 0.9 and 20 elements, dynamic results were obtained using a diagonal mass matrix
and an explicit integrator with some numerical damping to control
spurious oscillations. Figure 5 shows the stress distributions at
various times. The material parameters were chosen so that an
elastic wave traverses the bar in unit time. For $t < 1$ the results
show numerical and inelastic dispersion and dissipation which
modifies the basic step feature of the elastic wave. As the stress
attempts to increase because of the rigid boundary, softening is
initiated and a considerable decrease in stress is noted for a time
of 1.7. The corresponding distribution of strain is shown in Figure
6 where large strains near the interface represent localization.
Similar predictions using an alternative nonlocal approach are given
by Bazant et al. (1984), which together with our results indicate
that a nonlocal constitutive equation provides a stable procedure
for predicting strain softening and localization in a manner that is
independent of mesh size

Conclusions

It is assumed that no relative slip occurs at the contact
surface between soil and concrete. Instead, for sufficiently large
loads a shear band appears adjacent to the interface where the shear
band is just the physical representation of softening and
localization. The interface serves as an initiator of strain
softening.

The proposed new nonlocal constitutive model provides features
of deformation in the vicinity of the interface that represent the
perceived characteristics of a soil-concrete interface for both
dynamic and static shear loading. Because of the difficulties in
obtaining measurements within the localization zone which contains
large strain gradients, existing experimental data are not
satisfactory. Once good experimental data are available, it will be
necessary to verify the model, not only in one-dimensional, but also
in two- and three-dimensional cases with curved interfaces. Further
progress on understanding interface phenomena will require a combination of experimental and theoretical investigations. The work
reported here indicates that a nonlocal constitutive model is a
promising theoretical approach.

References

Bazant, Z.P., Belytschko, T.B., and Chang, T., 1984, "Continuum Theory for Strain Softening," Journal of Engineering Mechanics, Vol. 110, No. 12.

Drumm, E.C., and Desai, C.S., 1983, Testing and Constitutive Modeling for Interface Behavior in Dynamic Soil-Structure Interaction, Report to National Science Foundation, Washington, D.C.

Huck, P.J., Liber, T., Chiapetta, R.L., Thomopoulas, N.T., and Singh, N.M., 1973, Dynamic Response of Soil/Concrete Interfaces at High Pressure, Technical Report No. AFWL-TR-73-264, Air Force Weapons Laboratory.

Huck, P.J., and Saxena, S.K., 1981, "Response of Soil-Concrete Interface at High Pressure" Proc. Xth Int. Conf. Soil Mech. Found. Engng., Stockholm, Sweden.

Oden, J.T., and Pires, E.B., 1983, "Numerical Analysis of Certain Contact Problems in Elasticity with Non-classical Friction Laws," Computers and Structures, Vol. 16, No. 1-44, pp. 481-485.

Read, H.E., and Hegemier, G.A., 1984, "Strain Softening of Rock, Soil and Concrete," Mechanics of Materials, Vol. 3, pp. 271-294.

Sandler, I., and Wright, J., 1983, "Summary of Strain-Softening," Theoretical Foundations for Large-Scale Computions of Nonlinear Material Behavior, Preprints, DARPA-NSF Workshop, S. Nemat-Nasser, ed., Northwestern University, Evanston, IL, Oct. 1983, pp. 225-241.

Schreyer, H.L., and Chen, Z., 1986, "One Dimensional Softening with Localization," Journal of Applied Mechanics, (to appear).

Valanis, I.C., 1985, "On the Uniqueness of Solution of the Initial Value Problem in Strain Softening Materials," Journal of Applied Mechanics, Vol. 52, No. 3, pp. 649-653.

Vermeer, P.A., 1983, "Frictional Slip and Non-associated Plasticity," Scandinavian Journal of Metallurgy, Vol. 12, pp. 168-276.

Acknowledgment

This work was supported by the Air Force Office of Scientific Research.

BOUNDARY ZONE SUPERPOSITION METHOD FOR LINEAR AND NONLINEAR DYNAMIC ANALYSIS OF INFINITE DOMAIN PROBLEMS

by

Richard B. Nelson*
Civil Engineering Department
University of California, Los Angeles
Los Angeles, CA 90024

and

Yoshio Muki**
California Research and Technology, Inc.
Chatsworth, CA 91311

ABSTRACT

A boundary zone superposition method is presented for treating radiation boundary conditions in linear and nonlinear transient dynamic finite element analysis of structures in an infinite medium. The method is applicable to problems where input loads are imposed from within the finite element model or to problems where disturbances outside the finite element model generate input data on the boundary. The method is general and not restricted by any special properties of the medium transmitting stress waves, nor to any frequency characteristics beyond those inherent in the definition of the finite element model. The method is valid for problems with confined nonlinearities, with linear behavior required only in the boundary zone surrounding the confined region. Several examples are presented to demonstrate the method.

INTRODUCTION

The approach of using limited size finite element or finite difference models to investigate transient dynamic response of structure/media problems introduces artificial boundaries not present in the infinite medium. These artificial boundaries are known to be a major source of numerical error brought about by spurious reflections and/or refractions of dynamic stress waves impinging upon the artificial boundaries. As a result, results for response to dynamic loads are usually valid only for the period of time before reflections off the

---

* Professor of Engineering and Applied Science, Consultant,
  California Research and Technology, Inc.

**Research Engineer, Former Graduate Student,
  Civil Engineering Department.

surrounding artificial boundary return to the area of interest and con-
taminate the solution.

The need for a finite element or finite difference model capable of
giving accurate numerical results over arbitrarily long periods of time
has lead to the development of a number of different procedures for
treating the artificial boundary [1-9]. In one way or another all these
methods attempt to produce conditions at the boundary which will eliminate the unwanted spurious reflections.

Several important characteristics of any method for treating the
radiating boundary should be discussed. First, the basic formulation
should be independent of the frequency of the input signal. Otherwise
transient dynamic loadings cannot be easily processed, and then only if
the problem is linear so that the superposition principle can be used to
construct transient dynamic response.

The method also should be independent or at least very insensitive
to the angle of incidence of waves impinging on the radiating boundary.
Otherwise, the method can only be used effectively given prior knowledge
of the angle of incidence of all the waves encountering the boundary
which, in general, will be variable. This situation becomes hopeless in
a situation such as an inclusion induced wave scattering problem, or in
a multilayered system with significantly different characteristics
between the layers.

Time step requirements on the direct time integration algorithm
should not be more restrictive than that dictated by the requirements
inherent in the numerical grid. For example, if the integration algorithm is of the unconditionally stable implicit type, then the time step
should be set to accurately capture important wave effects below a
characteristic frequency (i.e., periods greater than a specified
minimum).

In addition, the method should allow excitation from outside the
bounds of the finite element model in order to analyse structure/media
interaction to far-field excitations, such as response to seismic waves.
The method also should require only near-boundary information to establish all information required for its implementation. This requirement
is important in controlling the bandwidth of the information in the
model, that is, in controlling the numerical efficiency in the numerical
model.

Finally, recognizing that a host of approximations are in the finite element (difference) model itself, a radiating boundary condition
should be able to tolerate, i.e., be insensitive, to various types of
numerical behavior in the interior of the numerical grid, since important aspects of the response may be nonlinear, as for example slip,
opening and closing at the structural media interface.

Several earlier techniques for treating the radiating boundary
problem should be briefly discussed. One well known technique is to use
"mapped infinite elements", actually far-field displacement functions
[1-2]. The primary difficulty with this approach is that infinitely
many different dynamic motions of the exterior infinite domain can
develop and unless the proper exterior (infinite domain) solutions are
employed, numerical reflections will occur on the finite
element/infinite element interface.

A so-called consistent boundary method [3] has recently been proposed, which appears to have promise for problems formulated in the frequency domain. The method involves considerable matrix manipulation, including a substructuring operation based on a spatial eigensolution. One additional drawback is that the direction of "infiniteness" is assumed to be horizontal and the finite element grid is assumed to be stratified with some rigid layer at the bottom of the grid. This may be a fair assumption for certain geological applications if a rigid basement is not too remote to be included in a numerical model.

The use of nodal springs, dampers and or masses has been used for years to attempt to eliminate spurious reflections off the boundary of the numerical grid. For example, viscous dampers have been employed to absorb energy from outgoing waves [3]. The damping coefficients are tuned so the impendance of the interior materials match the damping coefficients on the grid boundary. Typically, the boundary nodes are attached to dampers aligned normal and tangential to the surface of the boundary, with impedances chosen to match dilatational and shear wave impedances, respectively. This approach can be shown to effectively cancel both dilatational and shear waves provided their angle of incidence to the grid boundary is normal. If such is not the case, the amplitude reflected waves become quite large, (in fact, equal to the amplitude of the incident wave when the angle of incident is 90° away from the normal case.) Research has been done on the application of viscous boundaries to non-normal incidence angles [4] using coefficients in the impedance relationships, but little improvement is seen over the standard viscous boundaries.

An innovative attempt to treat the radiating boundary problem has appeared in an approach termed the paraxial boundary method [5,6]. In this approach the hyperbolic wave equation is modified to give a second order partial differential equation which admits only waves in one direction. Actually, the paraxial boundary is very similar to the viscous boundary [7]. However, the method is substantially more involved, in fact, for problems with Poisson's ratio >1/3 the equations for this boundary condition can lead to instabilities. The paraxial equations also lead to nonsymmetric element arrays, a computational difficulty for programs configured to process symmetric matrices. In addition, special hybrid elements must be developed to generate an interface between paraxial elements and the nominal finite element grid. Since the paraxial finite element is similar to the viscous boundary condition it is not surprising that both methods are similar in their performance. Both are frequency independent and both have a strong dependence on angle of incidence.

BOUNDARY ZONE SUPERPOSITION METHOD

The boundary zone superposition method is in essence a computationally efficient implementation of the simple and elegant superposition approach to canceling boundary reflections first presented by Smith [7].

Smith's method can be easily visualized using the classic one dimensional rod problem, see Figure 1. Given two identical rods, one end fixed, (Dirichlet type boundary value problem, the other free, Neumann type boundary value problem), identical stress waves can be sent down the rods to the boundaries. The fixed rod will reflect a stress wave of equal magnitude and same sign, whereas the free rod will reflect a stress wave of equal magnitude but of opposite sign. If the two solutions were superimposed, the reflection waves will cancel each other out, while leaving the incident (outgoing) waves undisturbed. This simple example forms the basis of the present method.

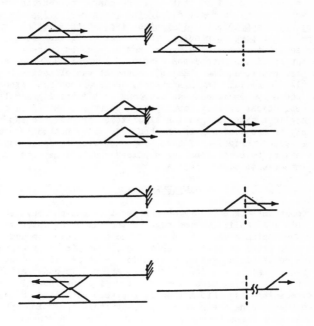

Figure 1   Motion of a wave at the boundaries of the finite element model, and the effective cancellation of the reflection wave. Figures on left are the response to fixed and free boundary conditions. Figures on the right are averaged to give the infinite domain solution.

The superposition method as presented by Smith required that the interior region be linear as the superposition was applied over the entire solution grid. The major flaw in Smith's approach of constant superposition is that when a wave inpinges off a <u>second</u> boundary the reflected waves which were of opposite sign, again reflect, but this time with the same sign, thereby generating real reflection. As a result Smith required more solutions to superimpose in order that the second reflection will also be canceled, and in fact, for n number of reflections that occur, required 2^n separate solutions.

In terms of computer costs, for an explicit solution, this approach was just as costly for a single grid made large enough to accommodate the solution time frame, since a comparable number of equations must be solved.

Smith's approach reveals an important characteristic of the super-position process, namely, that the boundary reflection waves and incident waves can be segregated. If the first reflection were to be segregated and completely eliminated from both solutions the second reflections would pose no problem. The number of grids needed for an unlimited solution time window would then be reduced to two.

The superposition method was developed further by Kunar and Marti [7] who demonstrated that Smith's approach was effective for general waves in elastic media. In this method, the shortcoming introduced by multiple boundary incidence was overcome by eliminating completely the boundary reflections as they developed. Kunar and Marti accomplished this for explicit time integration algorithms applying the superposition process only in the vicinity of the boundary. The method presented in this paper is a variation of their approach with extended capabilities to handle a variety of engineering applications, and may be used with any time integration algorithm.

IMPLEMENTATION

To implement the boundary zone superposition method, the finite element grid is split at the edge of the region of interest into two separate overlapping grids, the boundary zones, leading to complementary boundary conditions at the outermost edge. That is, at the boundaries, one grid is constrained against motion in the normal direction, and free in the tangential direction. The other grid is constrained against motion in the tangential direction, and free in the normal direction. The fixity condition in this method is replaced with large masses, large enough, so that during the regular cycle by cycle solution phase, the masses act as a fixed boundary condition. (This is convenient because the degrees of freedom needs to be maintained in order to easily modify the displacements at these degree of freedom.) In the overlap region, the two grids each have half the stiffness and density characteristics of the material just interior to the interface, therefore maintaining impedance and stiffness continuity at the interface, see Figure 2.

As each wave passes into the overlap region, the wave splits into equal parts in each of the overlapping grids. As the wave reflects off the boundary, the complementary boundary conditions in the two grids will cause the reflection waves of the two grids to be of opposite sign and equal magnitude.

Before the reflection waves enter the interior region of the model they are eliminated completely without altering the outgoing wave by averaging the solution for the two grids at corresponding locations and reassigning the resulting solution (the incident wave) back to the two grids. The only remaining task is to retain the previous boundary values. At the outer boundary of the overlap region, the averaged dis-placement and velocity information is assigned to unrestrained degrees of freedom, and the average displacement is assigned for the fixed (massed) degrees of freedom. In this way the same boundary values are retained at the outer boundary as before the averaging process. Since the outgoing waves both have the same sign and magnitude in both grids the averaging process will leave the incident wave undisturbed while eliminating all reflections.

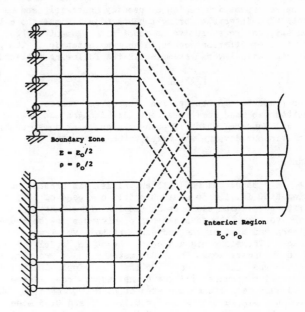

Figure 2   Grid Configuration used for the Boundary Zone Superposition Method.   (Dirichlet-type Boundary Conditions are replaced by lumped masses, sufficiently large to maintain fixity during cycle by cycle solution phase.)

By repeating this method at time intervals slightly shorter than the shortest transit-time across the overlap region, the finite element solution can be carried on indefinitely, without any reflections contaminating the primary interior region. It is important to recognize that the superposition only occurs in the overlapping region. Thus the interior region may be linear or non-linear.

Since either of the grids for the interior/Neumann or

interior/Dirichlet problems are nearly identical, they have closely
matching dispersion effects so that only very slight inconsistencies or
incompatibility should exist. Thus, direct time integration may be
expected to produce well behaved dynamically compatible results.

Finally, it is important to realize that implementation of the
current method requires no major modifications or additions to a finite
element code. The method requires no new element formulations, only
that the stresses, and nodal (displacement, velocity and acceleration)
information in the superposition region by readily accessible and alter-
able at discrete time intervals during the solution. Since the method
does not rely on any feature exclusive to the finite element method, the
application to finite difference codes is an easy extension. The procedure is illustrated with several examples in the following section.

EXAMPLES

In examples presented in the section Newmark's constant average
acceleration method was used for the direct time integration due to its
absence of numerical damping, which if present might mask some problems
with the method. Viscous damping was taken to be zero.

Central Disturbance in Rod

A infinite one dimensional rod with an initial disturbance at x=0
was analyzed using a 50 in. finite element model, composed of 60 1.0 in.
elements, with five additional elements on the last five inches of each
end of the bar to form the double group of elements in the boundary
zone. The stiffness and mass of the rod were set to give a wave speed
of 1×10^5 in./sec. Therefore signals cross the 50 in. model 0.5 msec,
and cross the 5 in. boundary zones in 0.05 msec. As a result, the
averaging technique was applied every 0.05 msec. A simple initial dis-
placement $U(x,t=0) = 1 + \cos\pi x/4$; $|x| < 4.0$ was used, see Figure 3a.
The Newmark method was used with a time step of $\Delta t = .001$ msec to obtain
the response for t>0. Displacements at t = 0.1, 0.25 and 0.35 msec are
shown in Figures 3b-3d, respectively. Note the different behavior of
the fixed boundary model compared to the free boundary case, see Figure
3c, and the outgoing wave produced by superposition. By 0.35 msec the
wave has passed and the grid is virtual motionless, except for some
imperceptible high frequency noise due to the slight differences in the
dispersive characteristic between the finite element models with free
and fixed boundary conditions.

Boundary Input on Rod

The same problem discussed in the first example was again
analyzed, but in this case, the loading was a sinusoidal input at x =
-25 in., to simulate the motion of a harmonic wave traveling in the +x
direction in an infinite one dimensional rod. The boundary conditions
at the edge of the "fixed" portion of the left boundary zone were taken
to be specified velocity and on the "free" portion, applied stresses
associated with the input stress wave. That is, the Dirichlet and Neumann conditions were specified nonhomogeneous functions of time. At

times when superposition was used to reset the data in the boundary zone the adjustment in velocities and displacements was made exactly the same way as in the previous example.

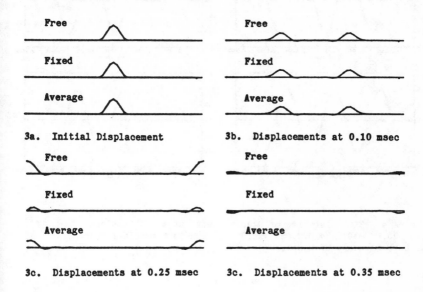

3a. Initial Displacement

3b. Displacements at 0.10 msec

3c. Displacements at 0.25 msec

3c. Displacements at 0.35 msec

Figure 3   Wave Propagation in Rod.

Results for this example are shown in Figures 4a-4d, which gives displacement histories at x = -20.0, -10.0, +10.0, and +20.0 in. respectively. The technique captures the traveling wave very accurately, with very little feedback from either the upstream or downstream boundaries.

An additional example was considered in [9] exactly identical to the second example except for the addition of a concentrated mass at x = 0. It was found that the boundary superposition method had no difficulty in tracking the partial scattering of the sinusoidal input wave by the concentrated mass. Thus, the method allows for input signals on boundaries of the model but cancels reflections generated by internal waves.

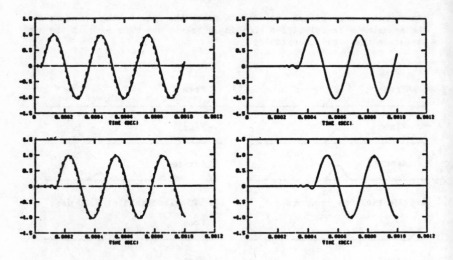

Figure 4   Time Histories of Displacements at Axial Stations
x = -20.0, -10.0, +10.0 and +20.0 in. Due to Simusoidal Input
on Left Boundary of Model.

Half-space Subjected to Surface Load

The one-dimensional examples were simple in that only a single
wave speed and wave form was present in the medium. For two and three-
dimensional media multiple wave forms and wave speeds exist, with
dispersive characteristics attendant with boundary conditions at free
surfaces. Several two dimensional problems were investigated in [9],
one of which, the case of a homogeneous elastic half space subjected to
a normal impulsive loading on the surface, will be presented here. This
problem is axisymmetric and a relatively simple finite element grid may
be used to represent the nearfield portion of the infinite medium, see
Figure 5.

The grid has a radius of 13 inches and is composed of 170 four node continuum elements. The radial dimensions of the elements increases from an initial value of 0.5 inches, expanding by ~ 5% per row, an effective scheme for suppressing artificial impedance due to mismatch of elements [10]. Material properties were taken to be $E = 10.0 \times 10^5 \text{lb/in}^2$, $= 0.3$ and $\rho = 1.0 \times 10^{-3} \text{lb.sec}^2/\text{in}^4$. One remarkable result of the analysis was that the superposition approach did not perform well for the axisymmetric problem when lumped mass was used. Consistent mass was therefore used and the method worked properly.

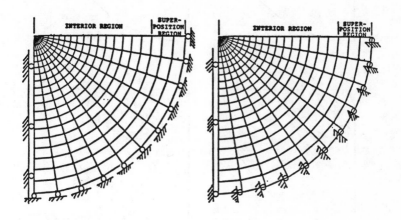

Figure 5 Finite Element Model for Two Dimensional Axisymmetric Impact on Half Space.

While a tremendous amount of data could be presented for this analysis only a series of velocity field plots are shown to indicate graphically the nature of the outward wave and the relative absence of reflections, see Figures 6a-d. These plots are not drawn at the same scale, but rather to depict the nature of the motion. The results here are not as clean as for the one dimensional rod, but nevertheless contain only a very small amount of noise compared to the strength of the initial signal.

For more complex loadings more noise may be expected. However, in the authors' experience, this approach has been superior to other approaches described earlier. Clearly, if any damping were introduced, either directly through, say, proportional damping, or indirectly through the time integration algorithm then the highest frequency noise

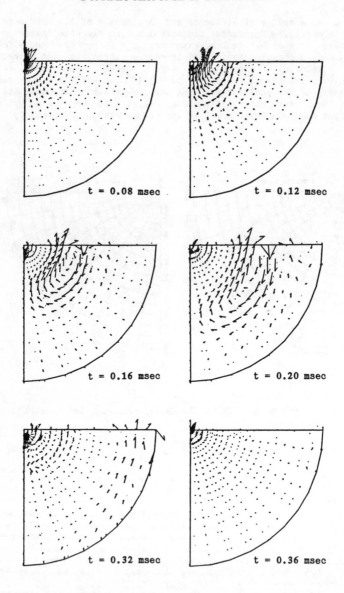

Figure 6   Velocity Plots (Not to Common Scale) Showing Outward Motion and Residual Noise.

would be quickly eliminated, a procedure often employed to reduce or eliminate undesirable numerical "noise".

CONCLUDING REMARKS

The boundary zone superposition method appears to hold great potential for providing the silent boundary needed when using finite size numerical models to investigate transient dynamic response of structure media problems. While the work presented here and in [9] is regarded as preliminary the approach appears to be very promising. A major bonus is the ease with which the technique can be implemented in existing computer programs.

If the method proves to be valid for general problems, it will provide a means for investigating many types of problems ranging from (nonlinear) seismic response of structures, to the analysis of impact stresses in very large space structures, to crack detection and nondestructive test evaluation.

ACKNOWLEDGEMENT

The authors wish to express their gratitude to California Research and Technology, Inc., for providing significant technical support, and Dr. Y.M. Ito and Dr. R.H. England with whom the authors held many helpful discussions.

REFERENCES

1. Zienkiewicz, O.C., Bettess, P., Chiam, T.C. and Emson, C. "Numerical Methods for Unbounded Field Problems and a New Infinite Element Formulation" Computational Methods for Infinite Domain Media Structure Interaction ASME, pp. 115-148, 1982.

2. Tassoulas, J.L., Kausel, E., and Roesset, M. "Consistent Boundaries for Semi-Infinite Problems Computational Methods for Infinite Domain Media-Structure Interaction, ASME, pp. 146-166, 1982.

3. Lysmer, J., and Kuhlemeyer, R.L. "Finite Dynamic Model for Infinite Media" J. Eng. Mechanics, Div., ASCE, pp. 859-877, Aug. 1969.

4. Cohen, M., and Jennings, P.C. "Silent Boundary Methods for Analysis," Computational Methods for Transient Analysis ASME, pp. 301-360, 1982.

5. Cohen, M. "An Evaluation of the Paraxial Boundary" Computational Methods for Infinite Domain Media-Structure Interaction ASME, pp. 167-182, 1982.

6. Cohen, M., and Jennings, P.C., "Silent Boundary Methods for Transient Analysis", Chapter 7, pp. 301-360, Computational Methods for Transient Analysis, T. Belytschko and T.J.R. Hughes, Ed., Elsevier Sci. Publ. 1983.

7. Smith, W.D. "A Nonreflecting Plane Boundary for Wave Propagation Problems," Journal of Computational Physics, Vol. 15, pp. 492-503, 1974.

8. Kunar, R.R., and Marti, J., "A Non-Reflecting Boundary for Explicit Calculations" Computational Methods for Infinite Domain Media-Structure Interaction ASME, pp. 183-503, 1982.

9. Muki, Y., "A Radiating Boundary Method for Linear and Nonlinear Transient Finite Element Analysis," M.S. Thesis, UCLA School of Engineering and Applied Science, 1983.

10. Day, S.M. "Finite Element Analysis of Seismic Scattering Problems" Ph.D. Thesis U.C.S.D., 1977.

ON ANALYSIS OF DYNAMIC STRUCTURE/MEDIA INTERACTION

V. Avanessian[1], S.B.Dong[2], M. ASCE and R. Muki[2]

Abstract

A version of the Global—Local Finite Element Method (GLFEM) is applied to the analysis of soil—structure interaction under steady-state excitation. The soil medium is considered to be homogeneous, isotropic and of semi—infinite extent. The structure is axisymmetric and may lie on the surface of the half—space or be partially or fully buried. Both axisymmetric and asymmetric excitations are discussed. GLFEM is a numerical analysis procedure which combines contemporary finite element modeling with classical Ritz functions to produce a more effective and efficient mathematical model. Herein, the Ritz functions are the spherical harmonics which are the eigensolutions of the entire space problem. The solution procedure enforces (1) both traction and displacement continuity at the interface between the finite element mesh and the outer region and (2) the traction free surface conditions on the half—space surface beyond the finite element mesh. Herein, the advantages of this approach are demonstrated by a series of examples.

Introduction

A Global—Local Finite Element Method (GLFEM) is used herein for the evaluation of steady—state structure/medium interaction. An axisymmetric structure in contact with a homogeneous, isotropic elastic medium of semi—infinite extent subjected to axisymmetric and asymmetric inputs is considered. GLFEM refers to a class of methods where contemporary finite elements and classical Ritz functions are used in the modeling of the system. Finite elements are used for the structure and a portion of its surroundings, i.e., the inner region. Finite element modeling allows arbitrary elastic structures under any form of contact conditions with the medium to be investigated, i.e., it may lie on the surface or may be partially or fully submerged. Beyond the finite element mesh, outgoing spherical harmonics serve as Ritz functions (termed global functions herein) for representing the radiated and/or scattered waves. Full traction and displacement continuity are enforced at the interface between the finite element model and the outer region. A sequence of zero—energy integrals is prescribed to meet the traction—free conditions in the outer region's free surface.

---

(1) Visiting Researcher, Ohsaki Research Institute, Tokyo, Japan

(2) Professor, Dept. of Civil Engineering, University of California, Los Angeles, CA 90024

The state-of-the-art can be summarized in four categories of methods: (1) energy absorbing boundaries, (2) non-reflecting boundaries, (3) infinite elements, and (4) consistent boundaries. Reviews of previous works on these approaches were conducted in earlier papers, Refs. [1,2]. The present GLFEM approach obviates a host of difficulties inherent in all methods to date. The most troublesome issue is an inability to capture the true mesh interface compliance conditions in a given problem. This compliance depends on the excitation (its frequency and the directionality of the input), the size and shape of the structure, and its contact conditions with the surrounding medium. While each method to date enjoys certain advantages, all suffer from the limitation regarding the compliance. Restating this shortcoming in slightly different terms, none are valid over all frequency ranges and are not amenable to an error analysis. All these methods are ad hoc in nature and their reliability in a new problem always poses a question.

In presenting the GLFEM approach herein, only the essence of the method will be given. The mathematical details, especailly those on manipulation of the spherical harmonics in connection with mesh interface continuity and traction-free surface requirements are covered in Refs. [1,2,3] and will not repeated here for brevity's sake. After summerizing the method, a set of example will be given to illustrate the range of possible applications with this approach.

Global-Local Modeling

Consider an elastic axisymmetric structure in contact with a homogeneous, isotropic elastic half-space as shown in Fig. 1. The structure may be excited by two types of steady-state input with frequency ω:

(1)   a forced excitation applied directly to the structure.

(2)   a seismic wave arriving from the outer region with an arbitrary angle of incidence.

Because of rotational symmetry, the analysis can be conducted with all variables expanded in terms of trigonometric series with respect to ϕ of the cylindrical coordinates (R,ϕ,z). The dynamic inputs are also expanded in a similar series. Each mode in this expansion can be analyzed separately and the complete solution would then be the synthesis of a participating modes.

In the modeling scheme, the structure and some portion of the adjoining half-space are represented by finite elements with the mesh boundary as a hemispherical surface of radius a_o. In the authors' finite element code, the model employed eight-node toroidal elements (Serendipity quad) with quadratic interpolation functions. Cylindrically orthotropy can be accommodated with no complication. Since the finite element aspects of this method involve only standard methodology, further elaboration is not necessary. The crucial point to recall is that the modeling accuracy is chained to the discretization, a maxim for the finite element method. Thus, frequency levels, which are related to wave lengths in the model, in part dictate the fineness of the discretization.

The assembled stiffness and mass matrices for the finite element model for a given circumferential mode number m may be written as

$$[D(\omega,m)]\{U\} = \Big[[K(m)] - \omega^2[M] \Big]\{U\} = \{F\} \tag{1}$$

with $[K(m)]$ and $[M]$ as the assembled stiffness and mass matrices and $\{U\}$ as the array of finite element nodal displacements. Without causing confusion, the steady-state time factor $\exp(-i\omega t)$ has been omitted and will remained suppressed throughout our discussion. It is convenient to partition Eq. (1) according to nodal degrees of freedom in the interior and on the mesh boundary.

$$\begin{bmatrix} D_{ii}(\omega,m) & D_{ib}(\omega,m) \\ D_{bi}(\omega,m) & D_{bb}(\omega,m) \end{bmatrix} \begin{bmatrix} U_i \\ U_b \end{bmatrix} = \begin{bmatrix} F_i \\ F_b \end{bmatrix} \tag{2}$$

The steady-state structure/medium interaction problem is viewed as a forced vibration problem, whose forcing functions are consistent load vectors appearing in F_i and F_b. Dynamic inputs applied directly to the structure or an incident seismic wave are represented as known consistent load vectors whose modal magnitudes and distributions are completely prescribed. Forced oscillation involves terms entering into F_i, while an incident wave gives consistent loads in F_b in Eq. (2). The scattered or radiated waves are the unknowns. The stresses associated with them that act on the mesh interface will also be perceived as consistent loads that will appear in F_b in Eq. (2). The essential strategy is the manner in which these unknown consistent load vectors are determined.

The outer region in our formulation is the half-space with a hemispherical dent of radius a_0. This region is occupied by a homogeneous, isotropic material with Lame constants λ, μ, and density ρ. To capture the scattered waves in this outer region, spherical harmonics are proposed. It is noted that this system of global functions, being a complete set, has the mathematical capability for representing any outgoing wave from the mesh boundary to any given accuracy.

For the mathematical representation of these spherical harmonics, we referred to the forms given by Pao and Mow [5], who expressed them in terms of three displacement potentials Φ, χ, Υ in spherical coordinates (r, Θ, ϕ):

$$\Phi(r,\Theta,\phi,t) = \sum_{m=o}^{\infty} \sum_{n=o}^{\infty} A_n^m(t) h_n^{(1)}(k_p r) P_n^m(\cos\Theta)\cos m\phi$$

$$\chi(r,\Theta,\phi,t) = \sum_{m=o}^{\infty} \sum_{n=o}^{\infty} B_n^m(t) h_n^{(1)}(k_s r) P_n^m(\cos\Theta)\cos m\phi \tag{3}$$

$$\underline{Y}(r,\circ,\underline{Y},t) = \sum_{m=o}^{\infty} \sum_{n=o}^{\infty} D_n^m(t)h_n^{(1)}(k_s r)P_n^m(\cos\theta)\sin m\phi$$

where $h_n^{(1)}$ is the spherical Hankel function of the first kind, P_n^m the associated Legendre function of the first kind of degree n and order m, and k_p, k_s are longitudinal and shear wave numbers. The A_n^m, B_n^m, D_n^m are the global coefficients which are the unknowns in this problem. These functions obey the equations of motion, the Sommerfeld radiation conditions, but not the traction-free surface conditions. Spherical displacement components $\underline{u} = (u_r, u_\theta, u_Y)$ may be derived from Eq. (3) according to the differentiation:

$$\underline{u} = \nabla\Phi + k_s\nabla \ (e_r r \ \chi) + \nabla x\nabla x(e_r r\underline{Y}) \tag{4}$$

Strains may be obtained from $\underline{u}$ by differentiation and the stresses are given by an isotropic constitutive law. With the adoption of spherical harmonics, the displacement and stress distributions for each component are known and only the coefficients, A_n^m, B_n^m, D_n^m, remain undetermined. As will be shown, the problem of determining the scattered field is transformed to one of finding the unknowns A_n^m, B_n^m, D_n^m rather than actual distributions of displacements and tractions in numerical form.

It is noted that an incident wave may be represented by incoming spherical harmonics. In the analysis of a general incident wave or some non-axisymmetric forced vibration, these inputs in terms of the trigonometric series representation would be known in the analysis of each mode.

Solution Procedure

The three main requirements of the solution are: (1) traction-continuity at the mesh interface between the inner and outer regions, (2) displacement continuity at the mesh interface between the inner and outer regions, and (3) traction-free surface conditions in the outer region.

To satisfy traction-continuity at the mesh interface, consistent load vectors due to global function tractions must be formed. They are obtained by differentiating the globasl displacements according to the strain-displacement relations, using the stress-strain law and integrating these stress distributions with the finite element quadratic displacement interpolations over the surface of each element on the mesh boundary. The finite element solution due to these consistent loads represents the inner region displacement distributions arising from traction continuity. For an incident seismic wave, both the magnitude and distribution are known. For the scattered or radiated field, only their distributions are known and their strengths are as yet undetermined.

The result of meeting traction continuity yields two sets of displacements at the mesh interface, a global function representation and a finite element representation. For conceptual purposes, it is useful to regard the former as the limiting displacement values on the mesh

interface from the exterior and finite element solution as the counterparts from the interior. For a given circumferential mode m, we will
denote the unknown and known displacements on the mesh interface of
these two representations, respectively, as

$$\{U_{gb}^m\} = \Sigma\, C_n^m \{U_{gs}^{n,m}\} + \{U_{gi}^m\} \; ; \; \{U_{FEb}\} = \Sigma\, C_n^m \{U_{FEs}^{n,m}\} + \{U_{FEi}^m\} \qquad (5)$$

with C_n^m as an ordered concatenation of the coefficients
A_n^m, B_n^m, D_n^m and $\{U_{gi}^m\}$ and $\{U_{FEi}^m\}$ are the known inputs to the problem. To
enforce displacement continuity at the mesh interface, equate
$\{U_{bg}^m\}$ $\{U_{FEb}^m\}$ to give:

$$[M_1(m)]\{C_n^m\} = \{B_1^m\} \qquad (6)$$

where

$$[M_1(m)] = \{U_{gs}^{n,m}\} - \{U_{FEs}^{n,m}\} \; ; \; \{B_1^m\} = \{U_{FEi}^m\} - \{U_{gi}^m\} \qquad (7)$$

In general, $[M_1(m)]$ will be a rectangular matrix of dimensions (pxq),
where q represents the number of global functions used and p is the
number of finite element degrees of freedom on the mesh interface. The
solution to Eq. (10) reflects displacement continuity at the mesh interface. We will solve this equation simultaneously with that which satisfies traction-free surface conditions in the outer region.

Although the global functions do not meet the traction-free surface conditions on a pointwise basis, we can insist that they be met on
an integral basis. On a modal basis, let the spherical harmonic tractions at $\theta = \pi/2$ be integrated with a sequence of displacements of the
form $r^{-\ell}$. The integrals are of the form

$$\int_{a_o}^{\infty} r^{-\ell}\, C_n^m\, \sigma_{rr}^{n,m}(r,\pi/2)\, dr = 0 \qquad \int_{a_o}^{\infty} r^{-\ell}\, C_n^m\, \sigma_{\theta\theta}^{n,m}(r,\pi/2)\, dr = 0$$

$$\int_{a_o}^{\infty} r^{-\ell}\, C_n^m\, \sigma_{\theta\Upsilon}^{n,m}(r,\pi/2)\, dr = 0 \qquad (\ell = 0,1,2,\ldots)$$

$$\qquad (8)$$

and each one expresses a zero-net work condition due to the surface
tractions in the outer region. These integrals are tantamount to
zero consistent loads on the outer region free surface, and are
equivalent to the weaker statement used in a finite element sense.
Equation (8) may be cast into the form

$$[M_2(m)]\{C_n^m\} = 0 \qquad (9)$$

where $[M_2(m)]$ is a (3sxq) where s is the number of integral constraints.
The mathemtical details going into forming $[M_2(m)]$ may be found in Refs.
[1,2,3]. The minimum number of such constraints is problem-dependent.

However, for the calculations reported herein, an over-abundant of integrals were used so that this minimum number issue was avoided.

Equations (6) and (9) must be solved simultaneously. Combining them into one euqations gives

$$[M(m)]\{C_n^m\} = \{B^m\} \qquad (10)$$

where

$$[M(m)]^T = \left[[M_1(m)]^T \ [M_2(m)]^T \right] \ ; \ \{B^m\}^T = \left[\{B_1^m\}^T \ 0 \right] \qquad (11)$$

The solution of Eq. (10) is obtained on a least squares basis which can be written as

$$\{C_n^m\} = \left[[M(m)]^T [M(m)] \right]^{-1} [M(m)]\{B^m\} \qquad (12)$$

The uniqueness of the results is discussed in Ref. [1]. From the solution for the global coefficients, displacements and stresses both in the inner and outer regions may be calculated for the harmonic under consideration. Scattered cross-sections, which depict the directional nature of the radiated energy by the structure, can also be calculated. The expressions for the scattered cross-sections may be found in Refs. [1,2,3].

Examples

Some examples of the GLFEM method are given below. Due to space limitations, only the results are presented. References are cited in which the details are covered more thoroughly.

1. Forced Oscillations of a Rigid Circular Plate on an Elastic Half-Space

In the first example, the problem of a rigid circular plate in smooth contact with an elastic half-space under normal force and rocking excitations were considered. These problems were solved by Krenk and Schmidt [4,6]. The GLFEM results for normal excitation are shown in Fig. 2 and that for rocking motions in Fig. 3. Excellent agreement between the analytical and GLFEM results may be seen in these figures. More details are given in Refs. [1,3].

In another example, the plate that is fully bonded to the half-space with different embedment conditions was considered, see Fig. 4. Results for normal and horizontal force oscillations are shown in Figs. 5,6 and 7. Greater details on this problem may be found in Refs. [1,3].

2. Elastic Wave Scattering by a Rigid Plate from Plane Compressional Wave

The GLFEM results for a plate subjected to a steady compressional wave with normal incidence are shown in Figs. 8 and 9. Scattering cross-sections, which illustrate the directional nature of the energy radiated from a scatterer, are shown in Figs. 10 and 11 for two wave numbers. The details to this problem may be found in Ref. [2].

References

1.   Avanessian, V., Muki, R. and Dong, S.B., "Forced Oscillations of an Axisymmetric Structure in Contact with an Elastic Half-Space by a Version of Global Local Finite Elements," Journal of Sound and Vibration (in press) 1986.

2.   Avanessian, V., Muki, R. and Dong, S.B., "Axisymmetric Soil-Structure Interaction by Global-Local Finite Elements," Earthquake Engineering and Structural Dynamics (in press) 1986.

3.   Avanessian, V., Muki, Dong, S.B. and Muki, R., "Asymmetric Forced Vibration of an Axisymmetric Body in Contact with an Elastic Half-Space - A Global-Local Finite Element Approach," (in preparation).

4.   Krenk, S. and Schmidt, H., "Vibration of an Elastic Circular Plate on an Elastic Half-Space - A Direct Approach," Journal of Applied Mechanics, ASME, Vol. 48, pp. 161-168, 1981.

5.   Pao, Y.H. and Mow, C.C., Diffraction of Elastic Waves and Dynamic Stress Concentration, Crane and Russak, Inc., New York, 1973.

6.   Schmidt, H. and Krenk, S., "Asymmetric Vibrations of a Circular Elastic Plate on an Elastic Half-Space," International Journal of Solids and Structures, Vol. 18, pp. 91-105, 1982.

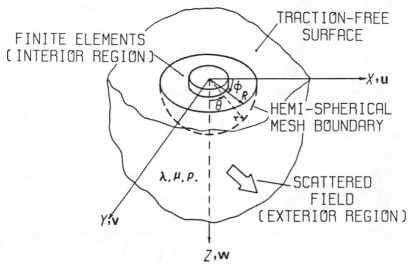

FIGURE 1

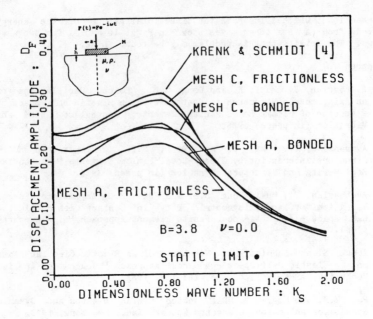

FIGURE 2

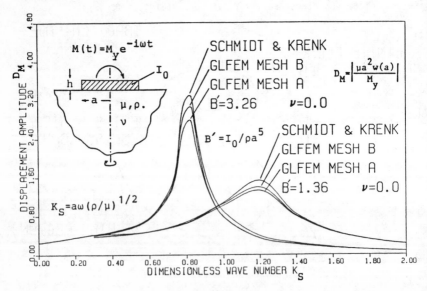

FIGURE 3

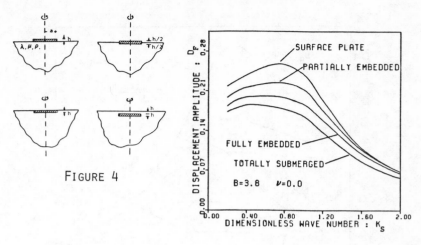

FIGURE 4

FIGURE 5

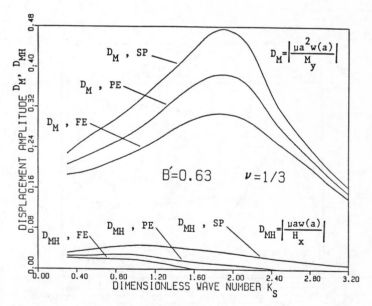

FIGURE 6

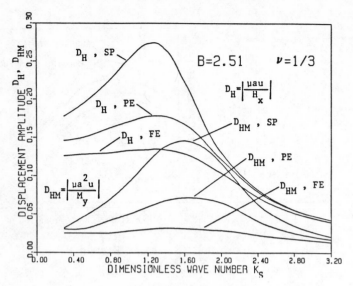

FIGURE 7

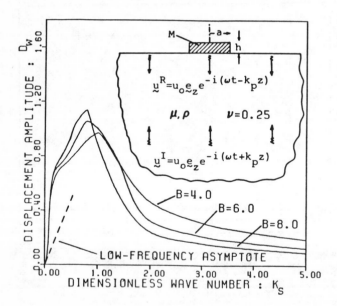

FIGURE 8

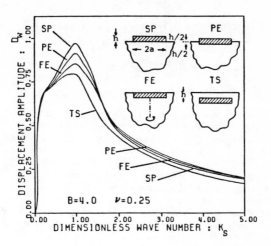

FIGURE 9

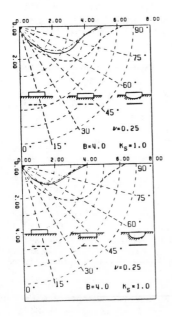

FIGURE 10

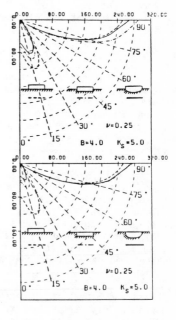

FIGURE 11

Active Control of Distributed Parameters Structures by Pointwise Control Actions

Mohamed Abdel-Rohman[1], M. ASCE

Abstract

In structural control, one is faced with distributed parameter structures which involve infinite degrees of freedom. To control a structure properly one has to use an infinite number of sensors and control actions. This is not feasible to implement in civil structures. Usually, the proposed control mechanism generates a finite number of control actions. Several methods have been offered to control a structure. These methods are the optimal control, the pole assignment, the independent modal space control, and the direct velocity feedback. This paper shows the advantages and disadvantages of each method and recommending the suitable method to be used in the active control of civil structures. The control of a simple span beam is used as illustrated example.

Introduction

Modern flexible structures are modeled by infinite degrees of freedom systems. To control completely the response of these structures against dynamic loads, one must use an infinite number of sensors and actuators. In other words, the control actions and observations must also be continuously distributed. The feasibility of implementing this control is hard to achieve for civil structures. Usually, the control mechanisms which are developed for civil structures such as tuned mass dampers, tendons, or aerodynamic appendages, create finite number of control actions. Therefore, one should be aware that the structure could not be controlled completely in the presence of pointwise control actions. In the presence of this control and dealing with structures of infinite vibrational modes, one faces some numerical difficulties, as pointed out in Ref. [6].

Several methods have been used to provide the control law to control the flexible structures. The optimal control method [3,11] and pole assignment method [2,4] are based on controlling the assumed dominant vibrational modes. However, the designed control actions excite the neglected modes or may even destabilize any of these modes. The independent modal space control [8-10] is considered as an approximation of the distributed control law. By a distributed control and observation one is able to control all the infinite modes

---

[1] Visiting Scientist, Department of Engineering Science and Mechanics, Virginia Polytechnic Institute and State University, Blacksburg, VA 24061, and Associate Professor, Civil Engineering Department, Kuwait University, Kuwait.

independently. Therefore, by using a number of discrete control actions one can control independently the same number of the vibrational modes. A better control is obtained when increasing the number of control actions. The direct velocity feedback method was suggested [6,7] in order to overcome the problems of spillover and the instability of the risidual modes due to the feedback control, but it only allows velocity feedback.

The purpose of this paper is to show the advantages and disadvantages of each of the above methods and recommending the suitable method which could be used safely to control the civil engineering structures. The control of a simple span beam is used as an illustrative example.

Equations of Motion

The equation of motion of a simple span beam subjected to distributed loads and control is given by

$$m\ddot{w} + EI\Delta^2 w = q(x,t) - u(x,t) \tag{1}$$

in which m is the mass per unit length, EI is the flexural rigidity, $\Delta = \partial^2/\partial x^2$ is Laplace operator, $q(x,t)$ is a distributed load, and $u(x,t)$ is the distributed control.

Solution of Eq. 1 is assumed as

$$w(x,t) = \sum_{i=1}^{\infty} \phi_i(x)A_i(t) \tag{2}$$

in which $\phi_i(x)$ is the ith eigenfunction, and $A_i(t)$ is the ith generalized coordinate.

For self-adjoint systems, Eq. 1 can be transformed into an infinite number of ordinary differential equations. Substituting Eq. 2 into Eq. 1 and applying the integral transformation one obtains for ith mode

$$\ddot{A}_i + \omega_i^2 A_i = q_i(t) - u_i(t) \tag{3}$$

in which ω_i, $q_i(t)$, and $u_i(t)$ are obtained from

$$\omega_i^2 = \frac{i^4 \pi^4}{L^4} \frac{EI}{m} \tag{4}$$

$$q_i = \frac{2}{mL} \int_0^L q(x,t)\phi_i(x)dx \tag{5}$$

$$u_i = \frac{2}{mL} \int_0^L u(x,t)\phi_i(x)dx \tag{6}$$

in which L is the beam span.

The simple form of Eq. 3 was obtained based on the orthogonality between vibrational modes, and that the control is continuous with respect to x and t. If the continuous feedback control is linear in terms of $w(x,t)$ and $\dot{w}(x,t)$, Eq. 3 remain uncoupled. In this case, the design of a continuous control is very simple as pointed out in Ref. [6]. However, due to the difficulties of implementing such control law, the control by pointwise actions is usually considered.

The equation of motion under pointwise control actions reads

$$m\ddot{w} + EI\Delta^2 w = q(x,t) - \sum_{i=1}^{N} \bar{U}_i(t)\delta(x - x_i) \qquad (7)$$

in which $\bar{U}_i(t)$ is the ith control action applied at $x = x_i$; and δ is the Dirac delta function. It follows that the ordinary differential equation of jth mode becomes

$$\ddot{A}_j + \omega_j^2 A_j = q_j(t) - \sum_{i=1}^{N} U_i \phi_j(x_i) \qquad (8)$$

in which $U_i = 2\bar{U}_i/mL$.

If the feedback control actions $U_i(t)$ are linear functions of pointwise measurements $y_i(t)$, they then involve the response of all infinite vibrational modes. For example, the measured deflection at $x = x_i$ and the measured velocity at $x = x_j$ are, respectively, given by

$$y_i(t) = w(x_i,t) = \sum_{k=1}^{\infty} \phi_k(x_i)A_k(t) \qquad (9)$$

$$y_j(t) = \dot{w}(x_j,t) = \sum_{k=1}^{\infty} \phi_k(x_j)\dot{A}_k(t) \qquad (10)$$

It follows that Eq. 8 is coupled with all other modes. In the following it is shown how each of the aforementioned control methods can be applied and find out a reliable method to be used in structural control.

The Optimal Control Method

In this method, the structure is modeled by a certain number of dominant modes. Suppose that the simple beam is modeled by the first three vibrational modes. The state equation in this case is given by

$$\dot{\underline{X}} = \underline{A}\,\underline{X} + \underline{B}\,\underline{U} + \underline{d}; \; x_0 \qquad (11)$$

in which

$$\underline{A} = \begin{bmatrix} 0 & 1 & 0 & 0 & 0 & 0 \\ -\omega_1^2 & 0 & 0 & 0 & 0 & 0 \\ 0 & 0 & 0 & 1 & 0 & 0 \\ 0 & 0 & -\omega_2^2 & 0 & 0 & 0 \\ 0 & 0 & 0 & 0 & 0 & 1 \\ 0 & 0 & 0 & 0 & -\omega_3^2 & 0 \end{bmatrix}_{6\times6} \qquad (12)$$

$$\underline{X}^T = [A_1 \dot{A}_1 A_2 \dot{A}_2 A_3 \dot{A}_3]. \qquad (13)$$

If the beam is controlled by r control actions applied at $x = x_i$, $i = 1, 2, \ldots, r$, respectively, then, matrix $\underline{B}$ is obtained from Eq. 8 as

$$\underline{B} = \begin{bmatrix} 0 & 0 & \cdots & 0 \\ -\phi_1(x_1) & -\phi_1(x_2) & \cdots & -\phi_1(x_r) \\ 0 & 0 & \cdots & 0 \\ -\phi_2(x_1) & \cdots & \cdots & -\phi_2(x_r) \\ 0 & 0 & \cdots & 0 \\ -\phi_3(x_1) & \cdots & \cdots & -\phi_3(x_r) \end{bmatrix}_{6xr} \qquad (14)$$

The control actions are determined to minimize an objective function in the form

$$J = \frac{1}{2} \int_0^T (\underline{X}^T \underline{Q} \ \underline{X} + \underline{U}^T \underline{R} \ \underline{U}) dt \qquad (15)$$

in which $\underline{Q}$ and $\underline{R}$ are specified weighting matrices; and T is a final control time. A solution to Eq. 11 subjected to Eq. 15 is given by

$$\underline{U}^* = - \underline{R}^{-1} \underline{B}^T \underline{P} \ \underline{X} \qquad (16)$$

in which $\underline{P}$ is the solution of the Riccati equation,

$$- \underline{\dot{P}} = \underline{P} \ \underline{A} + \underline{A}^T \underline{P} - \underline{P} \ \underline{B} \ \underline{R}^{-1} \underline{B}^T \underline{P}, \ \underline{P}(T) = 0 \qquad (17)$$

which contains 21 first order simultaneous nonlinear differential equations for a structure modeled by three modes. The optimal control is obtained by trials, assuming various values for the weighting matrices $\underline{Q}$ and $\underline{R}$ [3]. In each trial, however, the compilation of $\underline{P}$ is expensive especially when the structure is modeled by a large number of modes.

The control in Eq. 16 is evaluated upon estimating the state variables from the available measurements. The measurements, are also assumed to be comprised from the dominant modes only. If the number of measurements is less than the number of state variables, one has to design a filter to estimate the state variables from the available measurements. The disadvantage of this method lies in designing a control law only for the assumed dominant modes. There is a chance that the designed control spillover energy to the neglected modes or may even destabilize any of them. However, the method can be used for linear time-varying systems, and for determining tracking control for both deterministic and random disturbances [3,5].

The Pole Assignment Method

This method is only applicable for linear time-invariant systems. The control law for a system of order n is given by

$$\underline{U} = - \underline{K} \ \underline{X} \qquad (18)$$

in which $\underline{K}$ is a constant gain matrix of dimension r x n.

Substituting Eq. 18 into Eq. 11, the eigenvalues are found from

$$\Delta(\lambda) = |\lambda \underline{I}_n - \underline{A} + \underline{B} \ \underline{K}| = 0 \tag{19}$$

in which λ is the eigenfunction parameter; $\underline{I}_n$ is an identity matrix of order n; and $\Delta(\lambda)$ is the polynomial in terms of λ.

Equation 19 can be written as [4]

$$\Delta(\lambda) = |\lambda \underline{I}_n - \underline{A}| \ |\underline{I}_n + \underline{\phi}(\lambda)\underline{B} \ \underline{K}|$$

$$= |\lambda \underline{I}_n - \underline{A}| \ |\underline{I}_r + \underline{K}\underline{\phi}(\lambda)\underline{B}| = 0 \tag{20}$$

in which $\underline{I}_r$ is an identity matrix of order r, and $\underline{\phi}(\lambda)$ is the transition matrix of n x n.

For each one of the n specified eigenvalues, Eq. 20 is satisfied by forcing a column of the matrix $[\underline{I}_r + \underline{K}\underline{\phi}(\lambda)\underline{B}]$ to be zero. When r is greater than one, there is multiple choice for the column. However, when r is one, a unique column must be chosen. Although it is possible to determine the optimal gain matrix from all possible gain matrices [4], it is much easier and faster to use the method when r equals one. Moreover, the considered number of modes whatever how large it is does not cause any numerical difficulties [2]. In this method too, the control is based on estimating the state variables when the number of the measurements is less than the assumed order of the system. The main advantage of this method is that it provides an easy and fast solution for the gain matrix, whatever is the order of the system, when the structure is controlled by one control action.

The Independent Modal Space Control

This method provides the control for individual modes independently. For controlling three modes in the beam using three control actions applied at x_1, x_2, and x_3, Eq. 8 can be expressed as

$$\begin{bmatrix} \ddot{A}_1 \\ \ddot{A}_2 \\ \ddot{A}_3 \end{bmatrix} + \begin{bmatrix} \omega_1^2 & 0 & 0 \\ 0 & \omega_2^2 & 0 \\ 0 & 0 & \omega_3^2 \end{bmatrix} \begin{bmatrix} A_1 \\ A_2 \\ A_3 \end{bmatrix} = \begin{bmatrix} q_1 \\ q_2 \\ q_3 \end{bmatrix} - \begin{bmatrix} \phi_1(x_1) & \phi_1(x_2) & \phi_1(x_3) \\ \phi_2(x_1) & \phi_2(x_2) & \phi_2(x_3) \\ \phi_3(x_1) & \phi_3(x_2) & \phi_3(x_3) \end{bmatrix} \begin{bmatrix} u_1 \\ u_2 \\ u_3 \end{bmatrix}$$

$$\tag{21}$$

By modal control, each mode in Eq. 21 can be expressed independently as

$$\ddot{A}_r + \omega_r^2 A_r = q_r - f_r \tag{22}$$

in which r is 1, 2, or 3; and $f_r = C_{1r}A_r + C_{2r}\dot{A}_r$ is the modal control, and C_{1r} and C_{2r} are gain factors of rth mode.

The independent modal control f_r could be determined either by specifying new poles for each mode or using the optimal control theory and specifying an objective function [10]. Upon finding f_1, f_2, and f_3, the actual control actions are obtained from

$$
\begin{bmatrix} U_1 \\ U_2 \\ U_3 \end{bmatrix} = \begin{bmatrix} \phi_1(x_1) & \phi_1(x_2) & \phi_1(x_3) \\ \phi_2(x_1) & \phi_2(x_2) & \phi_2(x_3) \\ \phi_3(x_1) & \phi_3(x_2) & \phi_3(x_3) \end{bmatrix}^{-1} \begin{bmatrix} f_1 \\ f_2 \\ f_3 \end{bmatrix} \qquad (23)
$$

which indicates that U_r is a linear combination of $A_1, \dot{A}_1, A_2, \dot{A}_2$, A_3, and $\dot{A}_3$. Let's now consider the mode no. 4. Its equation is

$$
\ddot{A}_4 + \omega_4^2 A_4 = q_4 - \phi_4(x_1)U_1 - \phi_4(x_2)U_2 - \phi_4(x_3)U_3 \qquad (24)
$$

which indicates that whereas the first three modes are uncoupled and controlled, the remaining modes are excited by the modal control forces of the first three modes. However, the eigenvalues of the high order modes remain the same as before applying the control.

In order to use IMSC, however, one has to estimate, independently, the coordinates of the controlled modes. This is only possible if one can collect distributed measurements. For example, if $w(x,t)$ is available, then the generalized coordinate $A_i(t)$ can be found from

$$
A_i(t) = \int_0^L \phi_i(x)w(x,t)dx \ / \ \int_0^L \phi_i^2(x)dx \qquad (25)
$$

However, the observation of $w(x,t)$ is not so easy to obtain. The best that one can do is to fit the collected pointwise observations with a certain interpolation function. Let the resulting function be denoted by $\bar{w}(x,t)$. The generalized coordinate $A_i(t)$ is obtained from

$$
A_i(t) \simeq \int_0^L \bar{w}(x,t)\phi_i(x)dx \ / \ \int_0^L \phi_i^2(x)dx \qquad (26)
$$

The main disadvantage of the method lies in estimating the generalized coordinates of the structure from pointwise observations. The number of these observations must be quite large in order to enable accurate estimates. It turns out that when $\bar{w}(x,t)$ is not exactly like $w(x,t)$, then the estimated coordinate $A_i(t)$ shall be contaminated by other coordinates. It follows that the designed modal control is not totally independent but contaminated by other coordinates. Therefore, there is a possibility of instability here as well. This conclusion can be witnessed when comparing the modal control using a certain number of measurements with another using different number of measurements.

Another disadvantage for this method exists when the structure is controlled by one or two control forces. In this case the excitation provided by the modal control forces on the higher order modes may affect, badly, the controlled response of the structure.

The Direct Velocity Feedback

This method requires that the number of control actions must equal the number of the velocity sensors. Moreover, the controlled actions and the sensors must be collocated [1,7]. The measurements at $x = x_j$ is given by

$$y_j = \sum_{i=1}^{\infty} \phi_i(x_j)\dot{A}_i(t) \tag{27}$$

Considering three control actions and velocity sensors, and using Eqs. 8 and 27 one obtains the following relations:

$$\begin{bmatrix} \ddot{A}_1 \\ \ddot{A}_2 \\ \vdots \\ \ddot{A}_\infty \end{bmatrix} + \begin{bmatrix} \omega_1^2 & 0 & 0 & 0 \\ 0 & \omega_2^2 & 0 & 0 \\ \vdots & \vdots & \ddots & \\ \cdots & \cdots & & \omega_\infty^2 \end{bmatrix} \begin{bmatrix} A_1 \\ A_2 \\ \vdots \\ A_\infty \end{bmatrix} = \begin{bmatrix} q_1 \\ q_2 \\ \vdots \\ q_\infty \end{bmatrix} - \begin{bmatrix} \phi_1(x_1) & \phi_1(x_2) & \phi_1(x_3) \\ \phi_2(x_1) & \phi_2(x_2) & \phi_2(x_3) \\ \vdots & \vdots & \vdots \\ \phi_\infty(x_1) & \phi_\infty(x_2) & \phi_\infty(x_3) \end{bmatrix} \begin{bmatrix} U_1 \\ U_2 \\ U_3 \end{bmatrix} \tag{28}$$

$$\begin{bmatrix} y_1 \\ y_2 \\ y_3 \end{bmatrix} = \begin{bmatrix} \phi_1(x_1) & \phi_2(x_1) \cdots \phi_\infty(x_1) \\ \phi_1(x_2) & \phi_2(x_2) \cdots \phi_\infty(x_2) \\ \phi_1(x_3) & \phi_2(x_3) \cdots \phi_\infty(x_3) \end{bmatrix} \begin{bmatrix} \dot{A}_1 \\ \dot{A}_2 \\ \vdots \\ \dot{A}_\infty \end{bmatrix} \tag{29}$$

Equation 29 can be expressed in a short form by

$$\underline{Y}_{3\times 1} = \underline{B}^T_{3\times\infty} \dot{\underline{A}}_{\infty\times 1} \tag{30}$$

The relationship between the control actions and the collocated measurements can be expressed as

$$\underline{U} = \underline{K}\ \underline{Y} \tag{31}$$

in which $\underline{U} = [U_1\ U_2\ U_3]$ transposed; and $\underline{K}$ is of dimension 3×3. Equation 28, then, reads in a short form

$$\underline{A} + \underline{\Lambda}\underline{A} = g - \underline{B}\ \underline{K}\ \underline{B}^T \dot{\underline{A}} \tag{32}$$

in which $\underline{\Lambda}$ is a diagonal matrix of general dimension $\infty \times \infty$.

Balas [7] has shown that the designed control always stabilizes the system whenever the matrix $[\underline{B}\ \underline{K}\ \underline{B}^T]$ is non-negative definite. The matrix K could be symmetric matrix provided that $[\underline{B}\ \underline{K}\ \underline{B}^T]$ is non-negative definite. An easy way for designing K is to presume $[\underline{B}\ \underline{K}\ \underline{B}^T]$ is a diagonal matrix and specify the desirable damping ratios for a number of modes equals the number of the diagonal elements.

Conclusion

It is obvious that the direct velocity feedback method is the recommended method. It does not need filtering and recognizes the distribution of the structure. Since it only provides active damping, one may compensate for the need of more stiffness by introducing passive stiffness. In the second place stands the pole assignment method. The considered number of modes does not cause any numerical difficulties especially when the structure is controlled by one control action. The optimal control method can be used in stochastic and tracking control provided checking the stability of the structure. The independent modal space control method could only be used when the estimated generalized coordinates obtained from certain number of observations are the same when using greater number of observations. Otherwise, the independent control is not really independent but contaminated with all other modes. Moreover, good control necessitates that the number of control actions must be greater than two or three.

APPENDIX I - References

1. Abdel-Rohman, M., and Leipholz, H. H., "Active Control of Flexible Structures", Journal of Structural Division, ASCE, Vol. 104, ST8, Aug. 1978, pp. 1251-1266.
2. Abdel-Rohman, M., and Leipholz, H. H., "Structural Control by Pole Assignment Method", Journal of Engineering Mechanics, ASCE, Vol. 104, EM5, Oct. 1978, pp. 1159-1175.
3. Abdel-Rohman, M., Quintana, V. H., and Leipholz, H. H., "Optimal Control of Civil Engineering Structures", Journal of the Engineering Mechanics, ASCE, Vol. 106, EM1, Feb. 1980, pp. 57-73.
4. Abdel-Rohman, M., "Active Control of Large Structures", Journal of Engineering Mechanics, ASCE, Vol. 108, Oct. 1982, pp. 719-730.
5. Abdel-Rohman, M., "Stochastic Control of Tall Buildings Against Wind", Journal of Wind Engineering and Industrial Aerodynamics, Vol. 17, 1984, pp. 251-264.
6. Abdel-Rohman, M., and Leipholz, H. H., "Optimal Feedback Control of Elastic, Distributed Parameter Structures", Journal of Computers & Structures, Vol. 19, No. 5/6, 1984, pp. 801-805.
7. Balas, M. J., "Direct Velocity Feedback Control of Large Space Structures", Journal of Guidance, Control and Dynamics, 1979, pp. 252-253.
8. Meirovitch, L. and Öz, H., "Modal Space Control of Distributed Gyroscopic Systems", Journal of Guidance and Control, 3(2) 1980, pp. 140-150.
9. Meirovitch, L., and Baruh, H., "Control of Self-Adjoint Distributed Parameter Systems", Journal of Guidance and Control, 5(1), 1982, pp. 60-66.
10. Meirovitch, L., and Silverberg, L. "Globally Optimal Control of Self-Adjoint Distributed Systems", Optimal Control Applications and Methods, John Wiley & Sons, Vol. 4, 1983, pp. 365-386.
11. Yang, J. N., "Application of Optimal Control Theory to Civil Engineering Structures", Journal of the Engineering Mechanics, ASCE, Vol. 101, EM6, Dec. 1975, pp. 819-838.

OPTIMAL CONTROL OF SEISMIC STRUCTURES

By F.Y. Cheng[1], M.ASCE, and C.P. Pantelides[2]

ABSTRACT: This paper presents the algorithm for optimal design of structural systems with active tendon controllers and/or active mass dampers. The structure is subjected to earthquake excitation modelled as a stationary random process. Using the approach one may economically resize structural members and effectively determine the optimal level of the control forces and control parameters.

INTRODUCTION

The application of active control theory to civil engineering structures has been studied by previous researchers, (1,2,3,5,6,8). Two main approaches have been developed in the analysis of buildings by using active control systems. The first approach (2), emphasizes on an optimal control problem with deterministic loads, by minimizing a quadratic performance index. The second approach uses the transfer matrix to analyze the stochastic response of buildings (6), for which the control force is not optimal. For both approaches, the structural systems are not optimized.

This paper presents the application of active control in structural optimization by a simultaneous integrated design of the structure and the control parameters. The objective function is structural weight; the constraints are the floor displacements and control forces and the design variables are the floor stiffness and the control parameters. The constraints are expressed in terms of the standard deviations of floor displacements and control forces.

The active tendon and active mass damper control systems are considered for general building systems subjected to an earthquake acceleration modelled as a stationary random process with zero mean (4,7). The implementation of standard deviation expressed in the constraints is in the sense that for a given maximum allowable displacement and a probability of not exceeding this value, the standard deviation of the displacement can be obtained. A Gaussian probability distribution is assumed.

FORMULATION

Let us consider an N-story shear building shown in Fig. 1 with tendon controllers in some floors and a mass damper on the top floor acting simultaneously; the equations of motion for a typical floor can be derived as

$$Y_j = Y_{j-1} + M_j \ddot{X}_j + C_j \dot{X}_j \tag{1a}$$

---

[1]Prof., [2]Res. Asst., Dept. of Civil Eng., Univ. of MO-Rolla, 65401

$$Y_{j-1} = K_j(X_j - X_{j-1}) + f_i(t) \qquad 1 \leq j \leq N-1 \tag{1b}$$

$$Y_N = k_d(X_{N+1} - X_N) + c_d(\dot{X}_{N+1} - \dot{X}_N) \tag{2a}$$

$$f_d(t) = m_d \ddot{X}_{N+1} + Y_N \tag{2b}$$

$$Y_N = Y_{N-1} + M_N \ddot{X}_N + C_N \dot{X}_N + f_d(t) \tag{2c}$$

$$Y_{N-1} = K_N(X_N - X_{N-1}) + f_{tN}(t) \tag{2d}$$

in which X_j and Y_j are the displacement and shear of the jth floor; M_j, C_j and K_j signify respectively the mass, damping and stiffness of the jth floor; and $f_{ti}(t)$ and $f_d(t)$ are the control forces of the ith tendon and mass damper respectively. X_{N+1} is the mass damper displacement, Y_N the force exerted on the mass damper (m_d) from the spring (k_d) and dashpot (c_d). Taking Fourier transforms of Eq. 1 yields Eq. 3 and Eq. 2 yields Eq. 4

$$\{Z\}_L = [A(L)]\{Z\}_0 \qquad 1 \leq L \leq N-1 \tag{3}$$

$$\{Z\}_N = [A(N)]\{Z\}_{N-1} + \left\{ \overline{\frac{0}{grd(\omega)\overline{X}_N}} \right\} \tag{4a}$$

$$\{Z\}_{N+1} = [T_d]\{Z\}_N - \left\{ \overline{\frac{0}{grd(\omega)\overline{X}_N}} \right\} \tag{4b}$$

in which an over-bar denotes the Fourier transform of a quantity. In Eqs. 3 and 4, $\{Z\}_j = \{\overline{X}_j(\omega), \overline{Y}_j(\omega)\}^T$ is the state vector.

$$[A(N)] = [A]_N [A]_{N-1} \cdots [A]_1 \tag{5}$$

where $[A]_j$ is the transfer matrix (8) of the jth floor, given by

$$\{Z\}_j = [A]_j \{Z\}_{j-1} \tag{6a}$$

$$[A]_j = \begin{bmatrix} 1 & K_c^{-1} \\ -M\omega^2 + i\omega C & 1 + K_c^{-1}(-M\omega^2 + i\omega C) \end{bmatrix}_j \tag{6b}$$

where $i = \sqrt{-1}$, $K_c = K + grt(\omega)$ and

$$\overline{f}_{ti}(\omega) = grt(\omega)[\overline{X}_i(\omega) - \overline{X}_{i-1}(\omega)] \tag{7a}$$

If a certain floor is without a tendon controller then $grt(\omega) = 0$ and $K_c = K$; $grt(\omega)$ is described later. In Eq. 4 $\{Z\}_{N+1} = \{\overline{X}_{N+1}, 0\}^T$, $\{Z\}_N = \{\overline{X}_N, \overline{Y}_N\}^T$, $grd(\omega)$ will be described in Eq. 23, and

$$[T_d] = \begin{bmatrix} 1 & (k_d + i\omega c_d)^{-1} \\ -m_d\omega^2 & 1 - m_d\omega^2(k_d + i\omega c_d)^{-1} \end{bmatrix} \tag{7b}$$

Applying a delta function $X_0 = \delta(t)$ as an earthquake displacement, its Fourier transform is $\overline{X}_0 = 1$, and $\{Z\}_j$ become the frequency response functions to the ground displacement. The boundary conditions at the top and at the base are

$$\{Z\}_N = \{\overline{X}_N, 0\}^T, \qquad \{Z\}_0 = \{1, \overline{Y}_0\}^T \qquad (8a,b)$$

Repeated applications of Eq. 6a to Eq. 4a give the relationship between $\{Z\}_N$ and $\{Z\}_0$

$$\{Z\}_N = [A(N)]\{Z\}_0 + \left\{ \frac{Q}{grd(\omega)\overline{X}_N} \right\} \qquad (9)$$

in which $\{Z\}_0$ is given by Eq. 8b. Solving Eq. 4b and Eq. 9 gives four equations for the four unknowns $\overline{Y}_0$, $\overline{X}_N$, $\overline{Y}_N$ and $\overline{X}_{N+1}$ as

$$\overline{Y}_0 = - \frac{[D21 + TD21*TD12*A11(N)*grd(\omega)]}{[D22 + TD21*TD12*A12(N)*grd(\omega)]} \qquad (9a)$$

$$\overline{X}_N = A11(N) + A12(N)*\overline{Y}_0 \qquad (9b)$$

$$\overline{Y}_N = A21(N)+grd(\omega)*A11(N)+\overline{Y}_0*[A22(N)+grd(\omega)*A12(N)] \qquad (9c)$$

$$\overline{X}_{N+1} = \overline{X}_N + TD12 * \overline{Y}_N \qquad (9d)$$

in which $TD12=(k_d+i\omega c_d)^{-1}$, $TD21=-m_d\omega^2$, $AIJ(N)$ are the elements of $[A(N)]$ in Eq. 5, and D21, D22 are elements of the following matrix $[D]$ based on $[T_d]$ and $[A(N)]$ as

$$[D] = \begin{bmatrix} D11 & D12 \\ D21 & D22 \end{bmatrix} = [T_d][A(N)] \qquad (10)$$

$[T_d]$ and $[A(N)]$ are shown in Eqs. 7b and 5 respectively. The state equations for floors $j=1$ to $j=N-1$ are given by Eq. 3 in which $\overline{Y}_0$ is obtained from Eq. 9a. In the case where one has only floor tendons, then

$$\overline{Y}_0 = - \frac{A21(N)}{A22(N)}, \qquad \overline{X}_N = A11(N) + A12(N)\overline{Y}_0 \qquad (11a,b)$$

The state equations at any floor are given by Eq. 3 where L is now defined as $1 \leq L \leq N$ and $\overline{Y}_0$ is obtained from Eq. 11a.

The earthquake ground acceleration of (4,7) is used with a power spectral density $\Phi\ddot{x}_0\ddot{x}_0(\omega)$. The structural response and the control forces are also stationary random processes with zero mean. The mean square response vector is defined as

$$\sigma_i^2 = \int_{-\infty}^{\infty} |\{Z\}_i|^2 \ \omega^{-4} \ \Phi\ddot{x}_0\ddot{x}_0(\omega)d\omega, \qquad i = 1,\ldots,N \qquad (12)$$

in which $|\{Z\}_i|^2 = \{|\overline{X}_i|^2, |\overline{Y}_i|^2\}^T$. The standard deviation is given by $\sqrt{\sigma_i^2}$ for the quantity of interest. The mean square value of the control force from either the tendon or mass damper is

$$\sigma_f^2 = \int_{-\infty}^{\infty} |\overline{f}(\omega)|^2 \ \omega^{-4} \ \Phi\ddot{x}_0\ddot{x}_0(\omega)d\omega \qquad (13)$$

in which $\overline{f}(\omega)$ is given by Eq. 7a or Eq. 23a.

CONTROL FORCES

The controller considered is an electrohydraulic servomechanism similar to that described in (8). For the ith tendon controller the motions are obtained by sensors placed on the floor above and below it.

$$X_i^r = \frac{d^r X_i}{dt^r}, \qquad X_i^r = \frac{d^r X_{i-1}}{dt^r} \qquad (14)$$

For the current study r=1 was used for the tendon controllers and r=2 for the mass damper. The sensed motions are transmitted to the controller through the voltage V(t),

$$V(t) = pr \ (X_i^r - X_{i-1}^r) \qquad (15)$$

in which pr is a proportionality constant. The displacement of the hydraulic rams of the servomechanism U(t), is regulated by the feedback voltage V(t) through

$$\dot{U}(t) + R_1 \ U(t) = \frac{R_1 \ V(t)}{R_0} \qquad (16)$$

where R_1 = loop gain and R_0^{-1} = feedback gain. In Fig. 2 the electrohydraulic mechanism is shown in detail. The definitions of R_1 and R_0^{-1} are:

$$R_1 = \frac{U}{R} = \frac{G}{1+GH} \qquad (17)$$

$$R_0^{-1} = H \qquad (18)$$

The tendon control force consists of two parts; the elongation of the tendon and U(t), the hydraulic ram displacement.

$$f_{ti}(t) = k_t[(X_i - X_{i-1}) \cos\theta + U(t)] \cos\theta \qquad (19)$$

in which k_t is the tendon stiffness and θ is the angle the tendon makes with the horizontal (Fig. 1). Taking the Fourier transform of Eqs. 15, 16 and 19

$$\bar{f}_{ti}(\omega) = k_t \cos\theta \ \left[\cos\theta + \left(\frac{i\omega}{\omega_1}\right) \frac{\tau\varepsilon}{\varepsilon + \left(\frac{i\omega}{\omega_1}\right)}\right](\bar{X}_i - \bar{X}_{i-1}) \qquad (20)$$

in which ω_1=the fundamental frequency of the building without control; ε is the normalized loop gain and τ is the normalized feedback gain

$$\varepsilon = \frac{R_1}{\omega_1}, \qquad \tau = \frac{pr\omega_1}{R_0} \qquad (21)$$

The expression grt(ω) appearing in Eq. 7a can now be established from Eq. 20 as

$$grt(\omega) = k_t \cos\theta \left[\cos\theta + \left(\frac{i\omega}{\omega_1}\right) \frac{\tau\varepsilon}{\varepsilon + \left(\frac{i\omega}{\omega_1}\right)} \right] \tag{22}$$

For the mass damper, one acceleration sensor is placed on the top floor and the voltage $V(t)$ is now proportional to $\ddot{X}_N$, i.e. $V(t) = pr\ \ddot{X}_N$. Analogous to Eq. 20 with $\theta = 0$, the Fourier transform of the damper control force is

$$\bar{f}_d = grd(\omega)\bar{X}_N, \qquad grd(\omega) = k_{td}\left(\frac{i\omega}{\omega_1}\right)^2 \frac{\tau d\ \varepsilon d}{\varepsilon d + \left(\frac{i\omega}{\omega_1}\right)} \tag{23a,b}$$

where K_{td} = a proportionality constant, $\bar{X}_N$ is given by Eq. 9b and

$$\varepsilon d = \frac{R_1}{\omega_1}, \qquad \tau d = \frac{pr\omega_i^2}{R_o} \tag{24}$$

OPTIMIZATION

The optimization problem is formulated as follows
Find K_j, τ_i and ε_i such that

$$Min.\ W = a + \sum_{j=1}^{N} b_j K_j \tag{25}$$

subject to

$$\sigma\delta j \leq \sigma\delta max_j, \qquad\qquad j=1,\ldots,N \tag{26a}$$

$$\sigma cf_i \leq \sigma cfmax, \qquad\qquad i=1,\ldots,M \tag{26b}$$

$$K_j \geq Kmin, \qquad\qquad j=1,\ldots,N \tag{26c}$$

$$\tau_i \leq \tau max, \qquad\qquad i=1,\ldots,M \tag{26d}$$

$$\varepsilon_i \leq \varepsilon max, \qquad\qquad i=1,\ldots,M \tag{26e}$$

in which W = structural weight of the building; a and b_j are constants; K_j and $Kmin$ are the elastic floor stiffness and the minimum stiffness; $\sigma\delta j$ and $\sigma\delta max_j$ are the standard deviation of displacement of the jth floor and the allowable; σcf_i and $\sigma cfmax$ are the standard deviation of the control force of the ith controller and the allowable; τ_i and τmax are the normalized feedback gain of the ith controller and its upper bound; ε_i and εmax are the normalized loop gain of the ith controller and its upper bound, and N=number of floors, M=number of controllers including mass damper.

SAMPLE EXAMPLES AND OBSERVATIONS

The optimization procedure was applied to a two-story building for three case studies and the results are shown in Figs. 3-7. In cases A and B the building has two tendon controllers. The difference between A and B is in the parameter k_t given in Eq. 20. In case A k_{ti} is allowed to vary with K_i as $k_{ti} = 0.05\ k_i$. In case B, k_{ti} is kept constant at 40 kips/in. In case C, a mass damper is included in addition to the two tendons; k_{ti} is allowed to vary as in case A. For all three cases the parameters of the structure are: M1=M2=2kip-sec /in., θ=25°, C1=C2=1.6 kip-sec/in.

For case C, k_d=6.11 kip/in., c_d=0.10 kip-sec/in., m_d=0.04 kip-sec^2/in., k_{td}=25 kip/in. The constraints for all three cases are: $\sigma\delta max_1$=0.035 in., $\sigma\delta max_2$=0.070 in., σcf_1=10.0 kip, σcf_2=10.0 kip, K_{min}=100.0 kip/in., τmax=ϵmax=10.0. Additional constraints are imposed on case C as $\tau_d \leq 6.0$ and $\epsilon_d \leq 6.0$.

From Figs. 3 and 4, one can see that case C gives the least weight and stiffness. Fig. 5 shows that the tendon control forces of case B require a larger τ. The ϵ values, however, reach upper bound for all three cases as revealed in Fig. 6. The same tendon control force standard deviations are obtained at the optimum for all three cases. In Fig. 7 for case C, τ_d reaches the upper bound whereas ϵ_d goes to a very small value. In all three cases the control force constraints and the displacement constraint of the second floor are active. It appears that the combined tendon and mass damper system is advantageous over the other two cases. This is especially true in the studies of tall buildings not included in this paper in which tendons can be placed on the lower floors and a mass damper on the top. In summary using the approach, one may economically resize structural members and effectively determine the optimal level of the control forces and control parameters.

ACKNOWLEDGEMENTS

Partial financial support for this work through Grant No. NSF 8403875 is gratefully acknowledged.

REFERENCES

1. Abdel-Rohman, M., and Leipholz, H.H., "Active Control of Flexible Structures," Journal of the Structural Division, ASCE, Vol. 104, No.ST8, Aug., 1978, pp. 1251-1266.

2. Abdel-Rohman, M., and Leipholz, H.H., "A General Approach to Active Structural Control," Journal of the Engineering Mechanics Division, ASCE, Vol. 105, No.EM6, Dec. 1979, pp. 1007-1023.

3. Chang, J.C.H., and Soong, T.T., "Structural Control Using Active Tuned Mass Damper," Journal of the Engineering Mechanics Division, ASCE, Vol. 106, No.EM6, Dec., 1980, pp. 1091-1098.

4. Kanai, K., "Semi-Empirical Formula for Seismic Characterization of the Ground," Bulletin of Earthquake Research Institute, University of Tokyo, Tokyo, Japan, Vol. 28, 1964, pp. 1967.

5. Liu, S.C., Yang, J.N., Samali, B., "Control of Coupled Lateral-Torsional Motion of Buildings Under Earthquake Excitation," Vol. V, Proc. of 8th World Conf. on Earthquake Eng., 1984, pp. 1023-1030.

6. Samali, B., Yang, J.N., and Liu, S.C., "Active Control of Seismic-Excited Buildings," Journal of the Structural Division, ASCE, Vol. 111, No. 10, Oct., 1985, pp. 2165-2180.

7.   Tajimi, H., "A Standard Method of Determining the Maximum Response of a Building Structure During an Earthquake," Proceedings, Second World Conference on Earthquake Engineering, Tokyo, Japan, Vol.II, July, 1960.

8.   Yang, J.N., "Control of Tall Buildings Under Earthquake Excitations," Journal of the Engineering Mechamics Division, ASCE, Vol. 108, No.EM5, Oct., 1982, pp. 833-849.

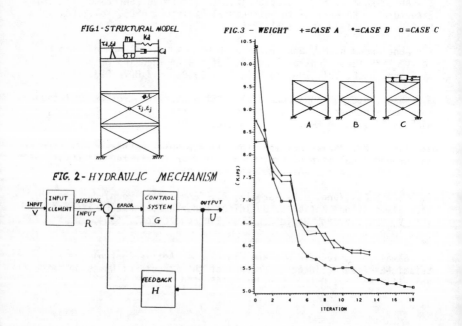

FIG.1 - STRUCTURAL MODEL

FIG. 2 - HYDRAULIC MECHANISM

FIG.3 — WEIGHT   +=CASE A   *=CASE B   □=CASE C

FIG.4 - STIFFNESS +=CASE A *=CASE B □=CASE C
—— =BOTTOM FLOOR -----=TOP FLOOR

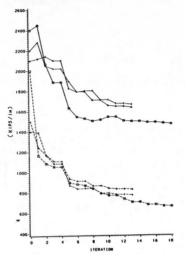

FIG.5 - TAU +=CASE A *=CASE B □=CASE C
—— =BOTTOM FLOOR -----=TOP FLOOR

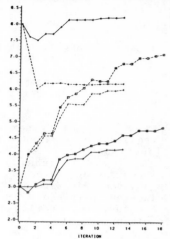

FIG.6 - EPSILON +=CASE A *=CASE B □=CASE C
—— =BOTTOM FLOOR -----=TOP FLOOR

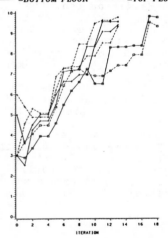

FIG.7 - CASE C +=TD □=ED OF MASS DAMPER

Application of Modern Control
Theory for Building Structures

Mohammad Ali Basharkhah*
James T. P. Yao**

Active control systems can be used to reduce the effect of strong-motion earthquakes. During the past two decades, various approaches have been applied to find the optimal control law. A technique based upon modal analysis is developed and presented herein. The present method is based upon a stochastic approach and a modal analysis for reducing displacement response of building structures to earthquakes. For the purposes of illustration, a numerical example is also given.

Introduction

The control theory has been an important component of modern engineering[1,9]. In addition to its many applications in other engineering disciplines, attempts have been made to apply the control theory to ensure comfort and/or safety of civil engineering structures in recent years. Recent developments in structural control were reviewed by authors[3]. In addition, several experimental studies have been summarized and discussed[11].

Although no well-designed tall building has collapsed to date, it is difficult to predict the magnitude of future earthquakes. Large displacement response to strong-motion earthquakes can cause severe damage to some of these structures. To reduce the structural response to strong-motion earthquakes, one can use either passive and/or active control devices. Although the concept of active control is not new, the application of control theory to reduce motions of civil engineering structures is still being developed[2,3,11].

---

* Engineer, Englekirk and Hart Consulting Engineers, Inc., 2116 Arlington Avenue, Los Angeles, CA 90018-1398
** Professor of Civil Engineering, Purdue University, W. Lafayette, IN 47907

The objective of this study is to present a technique based upon a stochastic approach and a modal analysis with the consideration of the effect of ground motions for the active control of building structures. In this method, only those modes which are disturbed are controlled.

General Approach for N Degree-of-Freedom Systems

A classical method for designing the control system is to relocate the dominant open-loop poles in the s-plane to ensure satisfactory transient response. In this paper, the design of feedback compensators is considered for linear and constant-coefficient multivariable systems. With an adequate control system, the structural response should remain within elastic limit states. It is assumed that all state variables can be used in obtaining feedback signals. In addition, it is desirable to obtain a decoupled or non-interacting system such that each input component affects only one output component (or some prescribed subset of output components).

Consider an n-degree-of-freedom system, the equation of motion can be written as

$$m \ddot{x} + c \dot{x} + k x = w \tag{1}$$

Where m, c, and k are nxn matrices representing the mass, damping, and stiffness of the structure respectively; w is a nx1 vector representing external force induced by the ground acceleration. Because the duration of an earthquake is relatively short, the effect of damping can be neglected. Then, Equation 1 may be rewritten as

$$m \ddot{x} + k x = w \tag{2}$$

It is possible to find a similarity transformation, x=Tq, such that both mass and stiffness matrices become diagonal[6]. The matrices m and k are real, symmetric and positive definite. Because the eigenvalues are real and positive, their square roots are called the natural frequencies of the system. The eigenvectors of the system are orthogonal and may be normalized. In this case, Equation 2 may be written as follows:

$$\ddot{q} + \omega^2 q = w^* \tag{3}$$

The ith equation of Equation 3 is given as follows:

$$\ddot{q}_i + \omega_i^2 q_i = w_i^* \tag{4}$$

By using the state variable approach, Equation 4 can be written as:

$$\dot{X}_i = A_i X_i + D_i W_i \tag{5}$$

Where:

$$X_i = \begin{bmatrix} q_i \\ \dot{q}_i \end{bmatrix} \qquad A_i = \begin{bmatrix} 0 & 1 \\ -\omega_i{}^2 & 0 \end{bmatrix} \qquad D_i = \begin{bmatrix} 0 \\ 1 \end{bmatrix} \tag{6}$$

To reduce the displacement response, the most critical modes of the system should be controlled. To find these modes, several different approaches may be used. The most common one is to choose the first few modes of the original system which are usually the most critical modes of the structure. This can easily be done, because it is always possible to find a coordinate transformation which diagonalizes the original system. Another method to find the most critical modes is based upon the cost analysis. Define the component cost associated with each state as follows:

$$V_i = 1/2 \lim_{t \to \infty} E \left[\frac{\sigma}{\sigma X_i} (X^T X) \star X_i \right] \tag{7}$$

The component cost in the form of a Ricatti equation usually consists of two parts: one representing the amount of energy required to generate the control force; the other representing the effectiveness of the control system. The value of V_i must be computed for all modes of the system. Then, the states which have higher component cost should be controlled. This approach may not give the same critical modes as the first approach because those modes with lowest natural frequency may not have the highest component costs. It means that those modes which are excited have higher component cost. The component cost as given by Equation 7 may be computed as follows[2]:

$$Z A^T + A Z + D \overline{W} D^T = 0 \tag{8}$$

$$V_i = t_r [Z]_{ii} \tag{9}$$

In this case, the disturbance force is assumed to be a zero-mean white-noise process, $\overline{W}$ is the intensity matrix of the process, and Z is the covariance matrix of the states which is defined as follows:

$$Z = E [X X^T] \tag{10}$$

Where E denotes the expectation operation. Note that the open-loop plant matrix of most civil engineering structures is stable. Therefore, Equation 8 will yield a solution for the covariance matrix of the states. Furthermore, it is important to note that the unstable modes of the system which are controllable or observable cannot be neglected[7] because they are the most important modes of the system and their corresponding component costs become infinite. Several optimal control laws take the form of a state feedback control and let the control force be:

$$U_i = -K_i X_i \qquad (11)$$

Where K_i is a 1x2 constant matrix. Adding the control force to the controlled modes, Equation 5 may be modified as:

$$\dot{X}_i = A_i X_i + B_i U_i + D_i W_i \qquad (12)$$

Where

$$B_i = [0 \quad 1]^T \qquad (13)$$

The elements of K_i are determined by the solution of the Riccati equation for each controlled mode. Therefore, Equation 4 can be rewritten as:

$$\ddot{q}_i + \omega_i{}^2 \dot{q}_i = w^*{}_i - \bar{c}^*{}_i \dot{q}_i - \bar{k}^*{}_i q_i \qquad (14)$$

Where $\bar{c}^*{}_i = K_{12_i}$ and $\bar{k}^*{}_i = K_{11_i}$. It can be seen that it is not necessary to control all modes. The elements of K_i for those uncontrolled modes are zero. Equation 14 can be rearranged as follows:

$$m^* \ddot{q} + \bar{c}^* \dot{q}_i + \bar{k}^* q + k^* q = w^* \qquad (15)$$

Where $\bar{c}^*$ and $\bar{k}^*$ are nxn diagonal matrices such that each one has some zero elements for those uncontrolled modes and some nonzero elements corresponding to controlled modes. By using the inverse of the similarity transformation, Equation 15 may be written as:

$$m \ddot{x} + \bar{c} \dot{x} + (k+\bar{k})x = w \qquad (16)$$

Where $\bar{c}$ and $\bar{k}+k$ are nxn equivalent damping and stiffness of the closed-loop system. In this case, the gain matrix of the control law can be found by using state space variables. To use the state space variables, Equation 16 may be written as:

$$\dot{X} = \bar{A} X + D W \qquad (17)$$

By using the same method, Equation 2 can be rearranged as:

$$\dot{X} = A\ X + D\ W \tag{18}$$

Knowing that $\bar{A} = A - BK$ the gain matrix can be found as follows:

$$K = (B^T\ B)^{-1}\ B^T\ \Delta A \tag{19}$$

Where $\Delta A = A - \bar{A}$. There is a solution for the gain matrix if the inverse of BB^T matrix exists. Otherwise, by using the least-square method, one may find the best solution for the algebraic equation $BK = \Delta A$.

Numerical Example

Consider a four-degree-of-freedom system as shown in Figure 1. The maximum displacement response of each floor without any active control system to an artificial earthquake is shown in Table 1. The natural frequencies of the system are shown in Table 2. It is found that the first two modes have highest component costs. To control the first two modes, these control laws are assumed for each mode:

$$U_1 = -[2.0 \quad 0.0] \begin{bmatrix} q_1 \\ \dot{q}_1 \end{bmatrix}$$

$$U_2 = -[1.0 \quad 0.0] \begin{bmatrix} q_2 \\ \dot{q}_2 \end{bmatrix}$$

The non-zero elements of k^* are $k^*_{11} = 2.0$ and $k^*_{22} = 1.00$. The k matrix is computed as follows:

$$\bar{k} = \begin{matrix} 1.20 & 0.76 & 0.23 & -0.03 \\ 0.76 & 0.67 & 0.50 & 0.26 \\ 0.23 & 0.50 & 0.70 & 0.53 \\ -0.03 & 0.26 & 0.53 & 0.44 \end{matrix}$$

The $\bar{c}^*$ matrix becomes a null matrix. Knowing that the control law is equal to $U = -K\ X$, the final gain matrix can be found by using Equation 19 as follows:

$$K = \begin{matrix} 1.20 & 0.0 & 0.76 & 0.0 & 0.23 & 0.0 & -0.03 & 0.0 \\ 0.76 & 0.0 & 0.67 & 0.0 & 0.50 & 0.0 & 0.26 & 0.0 \\ 0.23 & 0.0 & 0.50 & 0.0 & 0.70 & 0.0 & 0.53 & 0.0 \\ -0.03 & 0.0 & 0.26 & 0.0 & 0.53 & 0.0 & 0.44 & 0.0 \end{matrix}$$

By using this control law, the reduced maximum displacement response of each floor is computed as shown in Tabale 1. The natural corresponding frequencies of the closed-loop system are shown in Table 2. It may be seen that the two lowest

TABLE 1
MAXIMUM DISPLACEMENTS OF A FOUR-DEGREE-OF-FREEDOM
SYSTEM DUE AN ARTIFICAL EARTHQUAKE

Case	X_1(cm)	X_2(cm)	X_3(cm)	X_4(cm)
Without Control System	9.33	14.31	16.90	21.85
With Control System	6.10	9.16	12.07	12.82

TABLE 2
THE NATURAL FREQUENCIES OF THE SYSTEM

Case	Mode 1 (HZ)	Mode 2 (HZ)	Mode 3 (HZ)	Mode 4 (HZ)
Without Control System	1.00	1.89	4.42	5.43
With Control System	3.00	3.89	4.42	5.43

natural frequencies have been increased. This is the advantage of this method as compared to the pole-assignment method[2].

Concluding Remarks

In this paper, an attempt has been made to find an optimum gain matrix in the application of the active control to civil engineering structures. Because the pole-assignment method[4,5,8,10], which is based on the shifting of the open-loop poles further to the left-hand side of the s-plane, is more applicable for mechanical or electrical systems. The present method has certain advantages over the pole-assignment method. The gain matrix of the control law may be found such that those modes with higher component costs have smaller displacement than the corresponding displacement of the original system. Therefore, one may use more modes to control or use higher gain to substantially reduce the displacement response of the structure to dynamic loads.

Acknowledgement

This study was supported in part by the National Science Foundation through Grant No. CME-8018963. Authors wish to thank Drs. M. P. Gaus and J. E. Goldberg for their continued interest and encouragement.

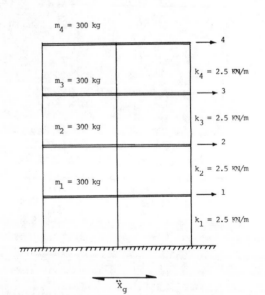

Figure 1: Four Degree-of-Freedom System.

APPENDIX A
REFERENCES

[1] Athans, M. and Falb, P.L., Optimal Control: An Introduction to the Theory and Its Applications, New York,
N.Y., McGraw-Hill Book Company, Inc., 1965.

[2] Basharkhah, M.A., Reliability Aspects of Structural
Control, Ph.D. Dissertation, School of Civil Engineering, Purdue University, W. Lafayette, IN, August
1983.

[3] Basharkhah, M.A., and Yao,J.T.P., "Some Recent Developments in Structural Control," Journal of Structural
Mechanics, Vol. II, No. 2, 1983, pp. 137-152.

[4] Chidambara, M.R., "Recursive Simplification of Pole-
Assignment Problem in Multi-Input Systems," International Journal of Control, Vol. 16, No. 2, August
1972, pp. 389-399.

[5] Davison, E.J., "On Pole Assignment in Multivariable
Linear Systems," IEEE Transactions on Automatic
Control, Vol. AC-13, No. 6, December, 1968, pp.
747-748.

[6] Falb, P.L. and Wolovich, W.A., "Decoupling in the
Design and Synthesis of Multivariable Control
Systems," IEEE Transactions on Automatic Control, Vol.
C-12, No. 6, December 1976, pp. 651-659.

[7] Gilbert, E.G., "Controlability and Observability in
Multivariable Control Systems," Journal Soc. Ind.
Appl., Math-Control Series, Series A, Vol. I, No. 2,
1963, pp. 128-151.

[8] Heymann, M., "Comments on Pole Assignments in Multi-
Input Controllable Linear Systems," IEEE Transactions
on Automatic Control, Vol. AC-13, No. 6, December
1968, pp. 748-749.

[9] Melsa, J.L. and Schultz, D..G., Linear Control
Systems, McGraw-Hill, New York, 1969.

[10] Wonham, W.M., "On-Pole Assignment in Multi-Input,
Controllable Linear Systems," IEEE Transactions on
Automatic Control, Vol. AC-12, No. 6, December 1967,
pp. 660-665.

[11] Yao, J.T.P., and Soing, T.T., "Importance of Experimental Studies in Structural Control," Preprint
84-010, ASCE Spring Convention, Atlanta, Georgia, May
14-18, 1984.

Constrained Viscoelastic Layers for Passive
Vibration Control of Structures

George Tsiatas,[1] A.M. ASCE and Dario A. Gasparini,[2] M. ASCE

Abstract

Infills with constrained viscoelastic layers for increasing the energy absorption capacity of structures are examined. The infills are expected to "work" in the linear range of response by contributing the larger portion of the overall damping. To estimate the damping of the composite systems all materials are modeled as viscoelastic and steady state response to harmonic lateral excitation is computed. Effective damping is calculated from the amplification at resonance. The approach is demonstrated for a shear frame and a tube type building.

Introduction

Current trends for new, taller structural systems and lightweight construction techniques, increase the importance of controlling the wind induced vibrations for serviceability and human comfort. Different approaches have been used to control undesirable vibrations of tall buildings. Some isolate the structure from the excitation source (Constantinou and Tadjbakhsh, 1985) or include a feedback control system which senses the vibrations and introduces corrective forces (Sae-Ung and Yao, 1978). Others incorporate mechanisms which increase the damping and therefore passively absorb the vibration energy (Gasparini et al., 1980, 1981).

Vibrations are ordinarily damped by material energy absorption during cyclic straining, friction at interfaces and ultimately by fracture. The relative importance of those factors on the overall damping depends on vibration amplitudes and construction details. For the case of wind excitation buildings are customarily designed to behave linearly, and then, material energy absorption capacity is expected to be the significant contributor to damping. Hence, addition of highly damped materials will increase the energy absorption and make serviceability design criteria easier to meet.

[1]Assistant professor, Dept. of Civil & Environmental Engineering, Washington State University, Pullman, WA.

[2]Associate professor, Dept. of Civil Engineering, Case Western Reserve University, Cleveland, OH.

Different types of highly damped composite elements can be introduced in a structure for passive vibration control. In this paper, a specific composite panel configuration is analyzed. It is a versatile panel which can be incorporated in new or even existing buildings. Applications of the approach for the case of a shear frame and a tube type building are examined.

Infill Panel

Fig. 1 illustrates a composite system with constrained viscoelastic layers. Thin layers of highly damped material are constrained between rigid panel elements. The panels are hinged at the bottom to the frame girders whereas at the top vertically slotted hinge pins eliminate any undesirable effects of vertical distortion. Some of the features of this specific configuration are:

(a) Viscoelastic layers are in nearly pure shear stress-state;

(b) Different amount of damping can be attained by changing the number and thickness of the layers;

(c) Flexibility exists in the positioning of the infills, even existing buildings could benefit with the introduction of such prefabricated panels; and

(d) Additional stiffness can be provided to the structure.

The infill panels are distributed in a building in such a way that maximum straining of the viscoelastic material takes place. There can be different design philosophies on the use of the panels as lateral load resisting elements. Fig. 2 illustrates three approaches for the case of a frame structure. In the first version (all girders simply supported) the panels resist all the lateral load and they exhibit maximum contribution to the overall damping. This results in a heavily damped system. In the second variation on the other hand, the composite panels contribute little to the lateral load resistance and since their deformation is limited the result is a less damped structure. The last scheme (midbay girders simply supported) is a hybrid between the previous two; the panels contribute a significant portion to the lateral stiffness and the structure is rather heavily damped since the hinges allow for increased straining of the viscoelastic layers. The use of a specific scheme depends upon the structural system at hand as well as on the practical feasibility of the concept. For example, the first version requires considerable stiffness and reliability from the viscoelastic material which might not be readily available. It can be anticipated that in their first applications the panels will play a small role in providing lateral stiffness.

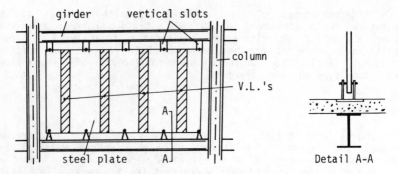

Fig. 1.   Infill Panel

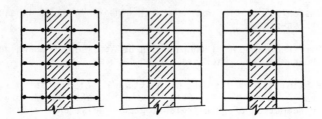

Fig. 2.   Variations of Infill Concept

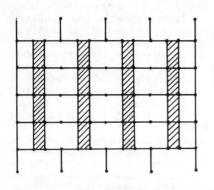

Fig. 3.   Superelement Model of
an Infill with 4 VLs

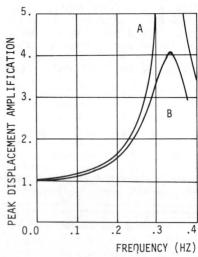

Fig. 4.   Amplification Curves
A)   Rigid 16-Story Frame
B)   Infilled Frame

Damping Inference

The infill panel consists of elements with significantly different damping characteristics. It is difficult to quantify damping of systems made up of several materials having different stiffnesses and damping properties. According to a frequently used approach the loss factor of a composite can be expressed as the weighted sum of the loss factors of the components, the weighting factors being the energies stored in each one of them (Raggett, 1975). But this method may not always give satisfactory results, especially when the damping properties of the components vary greatly (Gasparini et al., 1980). A different approach is followed in the present work.

A material's linear damping capacity may be incorporated in an analysis using viscoelastic constitutive equations. Therefore, all materials of the infilled frame are modeled as viscoelastic. Viscoelastic stress-strain relations are written

$$P_1 \, S_{ij} = Q_1 \, e_{ij} \qquad P_2 \, S = Q_2 \, e \qquad (1)$$

where S_{ij}, e_{ij} and s, e are the deviatoric and spherical components corresponding of the stress and strain. P_1, Q_1, P_2 and Q_2 are linear differential operators. Different models can be formulated by choosing appropriate operators (Flugge, 1975). For example, in the case of Kelvin-Voigt model the operators become

$$P_1 = 1 \qquad Q_1 = E_1 + K_1 \frac{d}{dt}$$

$$P_2 = 1 \qquad Q_2 = E_2 + K_2 \frac{d}{dt} \qquad (2)$$

where E_1, E_2 are spring constants and K_1, K_2 are viscosity coefficients.

A finite element program (CASE02) capable of modeling general linear viscoelastic materials and able to calculate steady-state responses under harmonic loads is used. The amplitudes of the steady state responses are functions of the effective damping of the system. By exciting the structure at different frequencies with a load distribution proportional to the first mode, the amplification A at resonance is computed and the damping is estimated by the formula

$$\zeta_c = \frac{1}{2A} \qquad (3)$$

This is exact for one degree of freedom systems or MDOF systems vibrating at only one natural frequency. For lightly damped systems resonance is defined with sufficient accuracy as the point of peak amplification. For heavily damped systems resonance is defined when the complex lateral displacements have a 90 degrees phase difference with the forces.

The complexity of the infill panel configuration as well as possible repetitions at several positions in the infilled structure, necessitate the use of a superelement representing the whole panel assembly as the only practical and accurate way for a complete finite element analysis. Fig. 3 illustrates the finite element discretization of a typical infill panel with four constrained viscoelastic layers. The assemblage shown constitutes the superelement. The nodes where it is attached on the building have 3 dof whereas all the other (internal) nodes have 2 dof. Plane stress elements are used to model the viscoelastic layers. This configuration is automatically generated by the computer. Flexibility exists in the number of viscoelastic layers as well as in the number of vertical subdivisions which is estimated for a realistic aspect ratio of the plane stress elements.

Applications

The infill concept was applied to two structural systems, a shear frame and a tube type building. For both cases the Kelvin-Voigt model of viscoelastic behavior was used. The stress-strain relation for the one-dimensional elements takes the form

$$\sigma_0 = (E + i\Omega K_s)\, \epsilon_0 \tag{4}$$

where σ_0 and ϵ_0 are the complex stress and strain amplitudes. Steady state responses should converge to the elastic static ones as the excitation frequency $\Omega \longrightarrow 0$. Hence, E is simply the modulus of elasticity of the steel. The viscosity term K_s can be computed from

$$K_s = \frac{2\zeta_s E}{2\pi f_n} \tag{5}$$

where f_n is the fundamental frequency of the system in Hz and ζ_s is the critical viscous damping coefficient for the steel which was taken for all applications to be 1%. Since the fundamental frequency is not known beforehand, an iterative procedure must be followed for the calculation of the viscosity. Eq. (5) was derived by equating the steady state response of a one dof oscillator to that of a mass on a viscoelastic support.

For the case of plane stress elements both the deviatoric and the spherical components of stress and strain must be modeled. The different operators for a Kelvin model were given in Eq. (2). For an elastic model the corresponding operators become

$$P_1 = 1.0 \qquad Q_1 = 2G \qquad P_2 = 1.0 \qquad Q_2 = \frac{E}{1 - 2\nu} \tag{6}$$

where G and ν are the shear modulus and the Poisson ratio of the material. Therefore for equivalent elastic static response as $\Omega \longrightarrow 0$, E_1 and E_2 should be

$$E_1 = 2G \qquad E_2 = \frac{2(1 + \nu)G}{1 - 2\nu} \tag{7}$$

The viscosity terms can be computed from equations equivalent to (5). The damping coefficients for the viscoelastic material were assumed $\zeta_1 = \zeta_2 = .5$. A loss factor n equal to one (n = 2ζ) can be considered to be a realistic capability of many elastomers.

Rigid Frame. Rigid frames are widely used as lateral load resisting systems in high-rise buildings. Their predominately shear behavior suggests that their dissipative characteristics could be greatly improved with the constrained viscoelastic layers. A 16-story 3-bay steel shear frame was used for an illustration of the approach. Infill panels were positioned at the center bay of the frame and the corresponding girders were simply supported. This corresponds to the third configuration of Fig. 2. The exterior bays resist most of the story shear leaving a small portion for the panels. Significant distortion of the viscoelastic layer takes place resulting in a heavily damped structure. For less computational effort the sixteen infills of the frame were divided into four groups and all the members of each individual group were modeled with the same superelement. The panels were designed so that the infilled frame has the same stiffness as the original rigid frame.

Table 1 lists the phase angles and the amplification factors for a range of frequencies. The maximum displacement did not occur at resonance indicating that the system is heavily damped. The amplification factor at resonance was found to be A = 4.10 which gives a damping ratio for the infilled frame of 12.2%. This is considerably higher than the 1% exhibited by the original rigid frame. Fig. 4 shows the dynamic amplification curves of the rigid and the infilled frame. A side benefit of the infills is that the midbay girders act primarily as axial members and their dimensions can be greatly reduced.

Tube Building. The 30-story tube type building of Fig. 5 was used. It was designed according to the current UBC regulations. The prescribed wind load was distributed linearly along the height of the building. This aimed to excite the first only mode which for tubular buildings is approximately linear. Highly damped infill panels were positioned in the two center bays of each side of the tube. Since rigid connections between beams and columns are essential for the tube behavior, no hinges were introduced at the corresponding girders as it was done for the 16-story frame. The steel tube provides all the lateral stiffness. This arrangement corresponds to the second variation of the infilled frame concept shown in Fig. 2.

Table 2 lists peak responses of the infilled tube for a range of frequencies. An exactly 90° angle was not obtained at resonance indicating that the influence of higher modes was not totally avoided. The amplification at resonance was 15.1 which corresponds to an overall damping of 3.3%. This is higher than the 1% of the unfilled tube although not as high as in the case of the shear frame. Fig. 6 shows the amplification curves for the rigid and the infilled tube. It can be seen that the infills increased the stiffness of the tube. Detailed analysis showed that the "shear lag" phenomenon was greatly reduced with the presence of the infills. Also, the viscoelastic layers were very close to a pure shear state of stress and not heavily stressed.

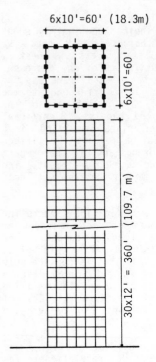

6x10'=60' (18.3m)

6x10'=60'

(109.7 m)

30x12' = 360'

Fig. 5. 30-Story Tube

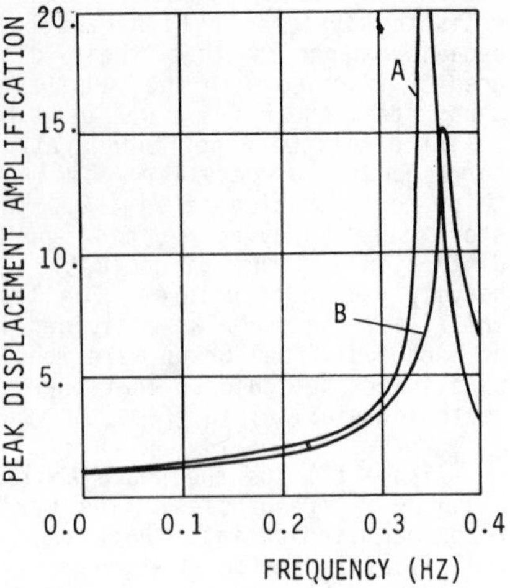

PEAK DISPLACEMENT AMPLIFICATION

20.

15.

10.

5.

0.0   0.1   0.2   0.3   0.4

FREQUENCY (HZ)

A

B

Fig. 6. Amplification Curves
A) Rigid 30-Story Tube
B) Infilled Tube

Table 1
Steady State Peak Responses
for 16-Story Infilled Frame

Ω (Hz)	ϕ (degrees)	A
0.000	0.00	1.000
0.200	13.70	1.510
0.300	48.25	3.337
0.330	82.47	4.123
0.333	83.42	4.103
0.334	86.98	4.099
0.336	81.29	4.088
0.340	80.15	4.039
0.350	72.66	3.799

Table 2
Steady State Peak Responses
for Infilled Tube

Ω (Hz)	ϕ (degrees)	A
0.000	0.00	1.00
0.340	21.04	6.34
0.350	75.71	16.52
0.351	79.64	16.58
0.352	88.33	16.31
0.353	79.51	15.45
0.360	32.82	9.42

Conclusions

The infill concept can significantly increase the damping and hence the energy absorption capacity of structures in their linear range of response. The effects were pronounced in the case of the shear frame were damping was increased from 1% to 12.2%. The primary reason for this was the incorporation of the hinges in the midbay girders which permitted additional straining of the infills. In the case of the tube, damping increase was moderate since high flexural stiffness and stiff beam-column connections do not allow much distortion in the panels. If hinges are to be introduced in the tube adequate lateral stiffness and reliability must be exhibited by the infills. Of course higher damping values could be attained using viscoelastic materials of higher damping capacity or by increasing the number of infill panels in the tube.

Acknowledgment

The support of this research by the National Science Foundation under Grant No. PFR79-21772 is gratefully acknowledged.

References

Constantinou, M. C. and Tadjbakhsh, I. G., "Hysteretic Dampers in Base Isolation: Random Approach," Journal of Structural Engineering, ASCE, Vol. 111, No. 4, April 1985, pp. 705-721.

Flugge, W., Viscoelasticity, 2nd ed., Springer-Verlag, New York, NY, 1975.

Gasparini, D. A., Curry, L. W. and DebChaudhury, A., "Damping of Frames with Viscoelastic Infill Panels," Journal of the Structural Division, ASCE, Vol. 107, No. ST5, May 1981, pp. 889-905.

Gasparini, D. A., DebChaudhury, A. and Curry, L. W., "Damping of Frames with Constrained Viscoelastic Layers," Journal of the Structural Division, ASCE, Vol. 106, No. ST1, Jan. 1980, pp. 115-131.

Raggett, J. D., "Estimating Damping of Real Structures," Journal of the Structural Division, ASCE, Vol. 101, No. ST9, Sept. 1975, pp. 1823-1835.

Sae-Ung, S. and Yao, J. T. P., "Active Control of Building Structures," Journal of the Engineering Mechanics Division, ASCE, Vol. 104, No. EM2, Apr. 1978, pp. 335-350.

BOUNDED STATE CONTROL
OF
STRUCTURES WITH UNCERTAIN PARAMETERS*

S.K. LEE,[**] F. KOZIN[***]

1.   Introduction

In general, the purpose of structural control is to maintain the
controlled variables within an allowable region determined by the
needs of structural safety and human comfort. This is a natural
extension of the optimal control problem.

The performance concept of keeping system states within a prespecified region is not new. It was mentioned, for example in [1],
as practical stability. The design of suboptimal controllers to
maintain the lateral acceleration levels within comfortable bounds on
high-speed rail vehicles was discussed in [2].

In the past decade the concept of optimal control of civil
structures, developed and studied by Yao [3], Yang [4], Soong [5],
Leipholz [6] and their co-workers has generated a significant interest
in this field. Even more recently, the renewed interest in advanced
structural design and control due to the NASA space station program
has generated even greater activity in this field.

In the civil engineering structural control literature, linear
quadratic cost functions, pole placement methods, modal control, as
well as other techniques, have been applied. However, when bounded
state constraints are incorporated, the results become highly complex
and in general cannot be easily satisfied.

Indeed, the control of linear systems in the presence of set
constrained external disturbances has been investigated maintaining
the states in a prespecified region by estimating the maximum
reachable set [7], [8], [9]. Application of this approach to
structural control requires the solution of several non-linear op-
timaztion problems. Unfortunately, the existence of a solution for
these non-linear problems has not been clearly established at this
time.

*   Partially supported under NSF Grants PFR 78-08811, CEE-8311190

** Research Engineer, Agency for Defense Development, Korea

*** Professor of Systems, Polytechnic Institute of New York,
    Route 110, Farmingdale, New York 11735, USA.

Our approach takes a somewhat different and simpler direction [11]. We essentially construct an extension of the Lyapunov function method for design of stabilizing controls.

In particular for the general linear structure

$$\dot{X} = AX + BU + Ef \tag{1.1}$$

where $X(0)\epsilon\underline{X}_o$, $X\epsilon R^{2n}$ we determine a controller u which will maintain the displacement and velocity variables within a practically allowable region X_1 based upon the following assumptions:

    (1) $f(t)\epsilon\Omega_f$, a closed bounded region, for all $t \geq 0$

    (2) $\underline{X}_o = \{X; X'P_oX \leq C_o\}$ (1.2)

    (3) $\underline{X}_1 = \{X; X'P_1X \leq C_o\}$,

where P_o, P_1 are positive definite matrices and $C_o > 0$.

We study linear feedback laws

$$U = -\Pi X \tag{1.3}$$

The problem is to determine Π, so that the invariant set , $\underline{X}(\Pi)$, generated by Π satisfies,

$$\underline{X}_o \subset \underline{X}(\Pi) \subset \underline{X}_1 \tag{1.4}$$

However, in solving the practical problem of structural control, it is necessary that the structural performance be independent of the effect of disturbances not only due to initial conditions and external excitations, but also due to uncertainties in the structural model. Because of the inherent uncertainties in the structure and the simplifications used in all mathematical modeling, it is vitally important that the control characteristics be preserved under all possible perturbations of the model parameters. The control system must be structurally stable, [1] i.e., robust in performance under parameter variations.

We are not concerned with unknown structures but, instead, are interested in a structure for which certain models are available. These models are chosen to represent what is considered to be uncertain about the structure. We will consider the uncertainties as a small perturbation in parameter coefficients of the differential equation. That is

$$\dot{X} = (A + \delta A(\theta)X + (B + \delta B(\nu))U + f(t), \; X_o \in \underline{X}_o, \text{ and}$$
$$f(t) \in \Omega_f, \; \forall \; t \geq 0, \tag{1.5}$$

_____

(1) Roughly speaking, a given system is said to be structurally stable if small perturbations of parameters do not change the essential characteristics of the system, e.g., stability.

in which $\theta(t)$ and $\nu(t)$ are uncertainty vectors contained in closed-bounded sets Θ and V respectively. The matrices $\delta A(\theta)$ and $\delta B(\nu)$ are assumed to be continuous in θ and ν, respectively.

Hence the control U must be designed so that the entire trajectory of system (1.5) will remain inside the prespecified region under all possible combinations of $X_o \in \mathfrak{X}_o$, $f \in \Omega_f$, $\theta \in \Theta$, and $\nu \in V$.

Earlier approaches, based upon Lyapunov methods have led to discontinuous or non linear type controller, even for linear time invariant systems. On the other hand recent studies, [10], have looked at the possibility of designing a linear state feedback controller by assuming matching conditions, which describe the uncertainties in lumped form. However, because of severe restrictions of the matching condition, e.g., control input matrix = external disturbance input matrix, the result is, in our opinion, not too realistic.

In this paper, we will first state a sufficient and necessary condition for the existence of linear state feedback controller, which is robust, by extending the result in [11]. We will also study the design of the linear state feedback controller by utilizing weaker matching conditions than those in [10].

II. Main Results

The external disturbance f is assumed to be modeled as

$$f = HW, \quad \dot{W} = GW, \quad W_o \in \Omega_o \tag{2.1}$$

where G is assumed to be a stable matrix.

Combining (1.5) and (2.1), we can rewrite the equation (1.5) in the augmented form

$$\dot{\hat{X}} = \hat{A}\hat{X} + \delta\hat{A}(\theta)\,\hat{X} + \hat{B}U + \delta\hat{B}(\nu)U, \quad \hat{X}(t_o) = \hat{X}_o . \tag{2.2}$$

in which $\hat{X} = \begin{bmatrix} X \\ \cdots \\ W \end{bmatrix}$, $\hat{A} = \begin{bmatrix} A & \vdots & H \\ \cdots & \vdots & \cdots \\ 0 & \vdots & G \end{bmatrix}$, $\delta\hat{A}(\theta) = \begin{bmatrix} \delta A(\theta) & O \\ O & O \end{bmatrix}$

$\hat{B} = \begin{bmatrix} B \\ \cdots \\ 0 \end{bmatrix}$, $\delta\hat{B} = \begin{bmatrix} \delta B(\nu) \\ O \end{bmatrix}$ and the uncertainty vectors θ and ν

are assumed to be contained in compact sets Θ and V, i.e., $\theta \in \Theta$ and $\nu \in V$.

Let $\hat{\mathfrak{X}}_o$ denote the set of all possible initial perturbations and $\hat{\mathfrak{X}}_1$ denote the maximum allowable region given in the augmented state space.

$$\hat{\mathbb{X}}_0 = \{\hat{X}; \ \hat{X}'P_0\hat{X} \leq C_0\}, \ \hat{\mathbb{X}}_1 = \{\hat{X}; \ \hat{X}'P_1\hat{X} \leq C_0\} \tag{2.3}$$

Hence, the control problem defined by the system (2.2) is formulated as that of choosing

$$U = -\Pi\hat{X} \text{ so that } \hat{X}(t) \text{ evolved from any } \hat{X}_0 = \begin{bmatrix} X_0 \\ W_0 \end{bmatrix} \text{ in } \hat{\mathbb{X}}_0$$

remains in $\hat{\mathbb{X}}_1$ for each $\theta\varepsilon\bigoplus, \nu\varepsilon\mathcal{V}$ and for all $t \geq 0$.

Throughout this section we will assume the following:

(i) $\bigoplus$ and $\mathcal{V}$ are compact; $\bigoplus = \{\theta: \|\theta\| \leq r\}$ and
     $\mathcal{V} = \{\nu: \|\nu\| \leq \underline{r}\}$

(ii) $\delta B(\nu) = B\pmb{\mathfrak{E}}(\nu)$.

(iii) $\max_{\{\nu\}} \|\pmb{\mathfrak{E}}(\nu)\| = \varepsilon < 1$.

(iv) $\{A, B\}$ is a controllable pair.

(v) A positive definite matrix P has been chosen so that
$$\hat{\mathbb{X}}_0 \subseteq \hat{\mathbb{X}} = \{\hat{X}; \ \hat{X}'P\hat{X} \leq C\} \subseteq \hat{X}_1.$$

Consider an ellipsoidal set $\hat{\mathbb{X}}$ defined by

$$\hat{\mathbb{X}} = \{\hat{X}; \ \hat{X}' \ P \ \hat{X} \leq C, \ C>0\},$$

in which P is an arbitrary positive definite matrix. Let $\hat{X}'P\hat{X} = V(\hat{X})$, then in order for $\hat{\mathbb{X}}$ to be an asymptotically stable invariant set, the time derivative of $V(\hat{X})$ for the system (2.2) with $U = \Pi\hat{X}$, must satisfy the condition,

$$\hat{X}'\{P(\hat{A}+\delta\hat{A}(\theta) + (\hat{B} + \delta\hat{B}(\nu))\Pi)$$
$$+ '(\hat{A} + \delta\hat{A}(\theta) + (\hat{B} + \delta\hat{B}(\nu))\Pi)'P\} \hat{X} < 0, \forall \hat{X} \neq 0,$$

for all $\theta \in \bigoplus$ and $\nu \in \mathcal{V}$. $\tag{2.4}$

The characterization of Π which satisfies (2.4) is given by the following theorem, (stated without proof).

Theorem [12]. There exists a linear state feedback control,

$$U \equiv \Pi\hat{X} = -\beta\hat{B}' \ P \ \hat{X}$$

with sufficiently large $\beta > 0$ so the $\dot{V}(\hat{X}) < 0$, if and only if

$$\hat{T}' \{P(\hat{A} + \delta\hat{A}(\theta)) + (\hat{A} + \delta\hat{A}(\theta))' \ P\}\hat{T} < 0, \tag{2.5}$$

in which Span $(\hat{T}) \equiv N(P\hat{B}\hat{B}'P)$.

The proof of this theorem depends upon the continuity of eigen-values of real symmetric matrices under perturbation.

Indeed, if Q is an nxn-real symmetric matrix and δQ is an un-certain term, also real symmetric, let $\{\lambda_i; i = 1, 2, \ldots, n\}$ be eigenvalues of Q and $\delta\lambda$ be an eigenvalue of $Q + \delta Q$.

Then $\delta\lambda$ shall be contained in at least one of the closed intervals

$$|\sigma - \lambda_1| \leq \rho, \quad i = 1, 2, \ldots, n, \tag{2.6}$$

in which $\rho \equiv \|\delta Q\|$ and σ is a real variable.

From the theorem, we find that β must satisfy

$$\beta > \beta_0 = \frac{(\lambda_Q + \rho)^2 + \lambda_{max}[Q](\delta\lambda)(\min_{j}|\lambda_j + \rho|)}{2(1-\varepsilon)\,\mu_{min}[R]\,(\min_{j}|\lambda_j + \rho|)}, \tag{2.7}$$

where $R = PBB'P$, $Q = PA + A'P$.

It is found that due to the uncertainty terms, the value β in (2.7) is greater than the corresponding β obtained in [11], where no uncertainties are present.

We now make the following assumptions:
Assumption 1: There exist continuous matrix functions $D(\cdot)$ and $E(\cdot)$ such that

(a) $\quad \delta\hat{A}(\theta) = \hat{B}\, D(\theta)$

(b) $\quad \delta\hat{B}(\nu) = \hat{B}\, \boldsymbol{\mathcal{E}}(\nu), \tag{2.8}$

where $\|\boldsymbol{\mathcal{E}}\| < 1,$ for all $\nu \in \mathcal{V}$.

The above conditions are called matching conditions. They, essentially, insure that the uncertainty does not influence the structure dynamics more than the control vector U.

These conditions are clearly weaker than those in [10], since no condition is placed upon the external disturbance matrix.

Assumption 2: There exists a scalar function $V(\cdot)$ which, for all X, possesses continuous first partial derivatives and a strictly increasing continuous function $\eta(\cdot)$ such that $\dot{V} \leq -\eta(\|\hat{X}\|)$, where $\dot{V}$ denotes a time derivative of V for the uncontrolled system,

$$\dot{\hat{X}} = \hat{A}\,\hat{X} \tag{2.9}$$

Since $\hat{A}$ is assumed to be a stable matrix, there exists a unique solution, symmetric and positive definite, of the linear matrix equation,

$$P\,\hat{A} + \hat{A}'P = -Q, \tag{2.10}$$

where Q>0. Therefore we can choose $V(\cdot)$ as $V(\hat{X}) = \hat{X}'P\,\hat{X}$ and so

$$\dot{V}(\hat{X}) \leq \eta(\|\hat{X}\|) = -|\lambda_{min}[Q]|\ \|\hat{X}\|^2 \qquad (2.11)$$

The fundamental concept of the idea is to design a control U so that the Lyapunov function of the system (2.9) can also be a Lyapunov function of the uncertain system (2.2) for all possible variations of the uncertainties. Based upon the matching conditions above, the structure of the control U, which guarantees the set of solutions $\hat{X}$ of the uncertain structure (2.2) to be an invariant set, is characterized in the following theorem, stated without proof.

Theorem [12]. The control U constructed as

$$U = -\beta \hat{B}' \nabla V(\hat{X}) \qquad (2.12)$$

will guarantee that X defined above is an asymptotically stable invariant set, if

$$\beta \geq \frac{d^2}{4(1-\delta)\alpha\lambda_{min}[Q]}\ , \qquad (2.13)$$

where $d \equiv \max_{\{\theta\}} \|D(\theta)\|$,

$$1 > \delta > e \equiv \max_{\{\nu\}} \|\boldsymbol{E}(\nu)\| < 1,$$

and $0 < \alpha < 1$.

Thus the set $\boldsymbol{X}$ for the uncertain system (2.2) will be an invariant set under all possible combinations of uncertainties with control

$$U = -2\,\beta_o\hat{B}'\,P\hat{X},$$

where

$$\beta_o = \frac{d^2}{4(1-e)\alpha\lambda_{min}[Q]}$$

for

$$V \equiv \hat{X}'P\hat{X}\ .$$

III. Conclusion

In this paper we have determined conditions for the existence of controllers which will guarantee that the overall structural response magnitudes will be constrained to lie in a preselected region. Further, it is assumed that the structural parameters are known only to lie within given bounds. Based upon a set of matching conditions, weaker than those recently appearing in the literature, a continuous state feedback controller is obtained. In numerical studies, however, it has been found that the gain parameter β is rather large due to the redundancies required by the uncertainties in the parameters. The next step, therefore, would be to investigate adaptive controller procedures for applications to structures.

Bibliography

1. J. Lasalle and S. Lefschetz, "Stability by Lyapunov's Direct Method with Applications", Academic Press, New York, New York., 1961.

2. G.N. Sarma and F. Kozin, "An Active Suspension System Design for Lateral Dynamics of a High-Speed Wheel Rail System", ASME Jour. Dynamic Systems, Measurement and Control, Vol. 93, No. 4, Dec.1971, p. 233.

3. J.T.P. Yao, "Concept of Structural Control", ASCE, J. Of Structural Div., Vol. 98, No., ST 7, Proc. Paper 9048, July 1972, pp.1567-1574.

4. J.N. Yang and F. Ginnopoulos, "Active Tendon Control of Structures" ASCE, J. of theEng. Mech. Division, Vol. 104, No. EM3, Proc. Paper 13836, June 1978, pp. 551-568.

5. T.T. Soong and C.R. Martin, "Model Control of Multistory Structures", ASCE, J. of the Eng. Mech. Division, Vol. 104, No. EM4, Proc. Paper 12321, August 1976.

6. M. Abdel Rohman and H.H.E. Leipholz, "A General Approach to Active Structural Control", Proc. of the Intl. IUTAM Symposium on Structural Control held at the University of Waterloo, Ontario, Canada, June 1979.

7. M. Delfour and S. Mitter, "Reachability of Perturbed Systems and Mini-Sup Problems", SIAM J. Control, Vol. 7, Nov. 1969.

8. J.D. Glover and F.C. Schweppe, "Control of Linear Dynamic System with Set Constrained Disturbances", IEEE Trans. on Automatic Control, Vol. AC-16, October 1975.

9. F.L. Chernousko, "Elipsoidal Bounds for Sets of Attainability and Uncertainty in Control Problems", J. Optimal Control Applications and Methods, Vol. 3, 1983.

10. B.R. Barmish, M. Corless, G. Leitmann, "A New Class of Stabilizing Controllers for Uncertain Dynamical Systems", SIAM Jour. Control and Optimization, Vol. 21, No. 2, pp. 246-255, 1983.

11. S.K. Lee, F. Kozin, "Bounded State Control of Linear Structures", proceedings 2nd Int'l. Symposium on Structural Control, Univ. of Waterloo, July 1985.

12. S.K. Lee, "Bounded State Control of Linear Systems with Application to Active Structural Control", Ph.D. Thesis in System Engineering, Dept. of Electrical Engineering, Polytechnic Institute of New York, Jan. 1985.

An Experimental Study of Active Structural Control

L.L. Chung[1], A.M. Reinhorn[2], M.ASCE, and T.T. Soong[3]

Abstract

An experimental investigation is made concerning the possible application of active control to structures under seismic excitation. The experimental facilities and structural model are carefully fabricated and calibrated. As the model is dynamically similar to real structures, it is possible to extrapolate results to illustrate the dynamic behavior of real structures under control. Methodologies for implementation of the active control system with tendons and a hydraulic actuator as controllers are developed.

Time delay in control execution, an important factor in the success of structural control, is taken into consideration in the derivation of the control algorithm. The effectiveness of structural control is shown by comparing the controlled response to the uncontrolled one. The correlation between analytical and experimental results is also studied.

Introduction

In recent years, the idea of applying active control to reduce damage to civil engineering structures caused by earthquakes has become an area of considerable interest [see, for example, (Samali, Yang and Liu, 1985)]. Various control algorithms and control systems have been proposed and investigated. However, by and large, the results have been of an analytical or simulation nature.

While some of the analytical and simulation results have been encouraging, a number of important questions need to be addressed and verified through experimental studies. These include time delay in data acquisition and on-line calculation, unsynchronized application of control force, interface between controllers and structure, and interface between sensors and structure.

The main objective of this study is to obtain experimental results using a carefully calibrated model structure under controlled experimental conditions so that they can be used to evaluate the implementability of active control to real structures. In this study, the model structure approximates a single-degree-of-freedom system and the control law studied is linear optimal closed-loop control utilizing the classical quadratic performance index. Methodologies for implementation of the control system with tendons and an electro-hydraulic actuator as controllers are studied.

---

[1]Research Assistant, [2]Assistant Professor and [3]Professor, Department of Civil Engineering, State University of New York at Buffalo, Buffalo, New York 14260.

Experimental facilities are designed so that more realistic
excitations and control forces can be realized. Hundreds of pounds of
control forces are provided by the actuator to alleviate structural
vibration under the excitation simulated through the seismic simulator
using scaled-down past accelerograms. With structural parameters
obtained from system identification, computer simulation is utilized
to obtain theoretical results which in turn are compared with
experimental data. Because the model is designed to be dynamically
similar to real structures, the applicability of structural control to
real structures can be extrapolated.

Model Structure and Experimental Facilities

The model structure is a three-story steel frame modeling a shear
building by the method of artificial mass simulation (Soong, Reinhorn
and Yang, 1985). In this study, the model is rigidly braced on the
top two floors to behave as a single-degree-of-freedom (SDOF) system.
Figure 1 shows its SDOF configuration in the laboratory. Some of its
properties are as follows: mass (m) = 16.69 lb-sec^2/in (2921 kg),
height = 8'4'' (2.53 m), natural frequency (ω_o) = 3.47 Hz, stiffness
(k) = 7934 lb/in (1390 kN/m), and damping factor (ξ) = 1.24 %.

Control forces are supplied by a hydraulic actuator and through a
system of tendons attached to the structure as shown in Fig. 2. The
tendons are pretensioned to about 500 lb (2.23 kN) to insure that they
are under tension at all times. The stiffness of the tendons is 2124
lb/in. (372 kN/m) and their angle of inclination is 36° from the
horizontal.

The model structure is bolted on a concrete block which is in
turn bolted to the center of a shaking table, which supplies the
desired base excitation. State variable measurements are made by
means of strain gage bridges installed on the columns just below the
first-floor slab. The signal from one strain gage bridge is used as
the feedback displacement signal while the signal from the second is
further passed through an analog differentiator to measure relative
velocity, which is also fed back to calculate the required control
force.

Space limitations do not allow a more detailed description of the
experimental facilities used in the study. More detailed description
on various units such as the shaking table, actuator and sensing
equipment, processing unit and data acquisition system can be found
elsewhere (Soong, Reinhorn and Yang, 1985).

Control Algorithm

Let u(t) be the based-mounted actuator displacement. The
equation of motion of the controlled system is

$$\ddot{x} + 2\xi\omega_o\dot{x} + \omega_o^2 x = -\ddot{x}_o - \frac{4k_c\cos\alpha}{m} u \tag{1}$$

where x(t) is the structural displacement relative to the base, m ,
ξ and ω_o are, respectively, the mass, damping factor and natural
frequency of the uncontrolled system, $\ddot{x}_o(t)$ is the base acceleration,

Fig. 1. The Model Structure

Fig. 2. Tendons and Hydraulic
Actuator

k_c is the tendon stiffness and α is the inclination angle of the tendons with respect to the base. As usual, the overdot indicates time derivative.

Using linear optimal control with classical quadratic performance index, the actuator displacement $u(t)$ is to be found in the form

$$u = k_1 x + k_2 \dot{x} \tag{2}$$

where k_1 and k_2 are such that the integral

$$J = \int_0^{t_f} (kx^2 + \beta k_c u^2) dt \tag{3}$$

is minimized, where t_f is the duration defined to be longer than that of the base excitation. The coefficient β in Eq. 3 is the weighting factor. As β increases, more weight or more importance is placed on the input energy and, as β decreases, more weight is placed on the relative displacement.

The problem of finding k_1 and k_2 is a standard one in optimal control theory (Sage and White, 1977). It can be shown that they have the form

$$k_{1,2} = \frac{2\cos\alpha}{\beta m} p_{21,22} \tag{4}$$

where p_{ij} is the ijth component of the 2×2 matrix P satisfying the Riccati matrix equation given by

$$\dot{P} + PA + A^T P - Pbr^{-1}b^T P + 2Q = 0 \tag{5}$$

where

$$A = \begin{bmatrix} 0 & 1 \\ -\omega_o^2 & -2\xi\omega_o \end{bmatrix} \quad , \quad b = \begin{bmatrix} 0 \\ -4k_c\cos\alpha/m \end{bmatrix} \quad , \quad r = \beta k_c \quad , \quad Q = \begin{bmatrix} k & 0 \\ 0 & 0 \end{bmatrix}$$

The substitution of Eq. 2 into Eq. 1 shows that the effect of active control is to add active stiffness and active damping to the uncontrolled system. The resulting controlled natural frequency and controlled damping factor are

$$\omega_o' = \left[\omega_o^2 + \frac{4k_c\cos\alpha}{m} k_1 \right]^{1/2} \tag{6}$$

$$\xi' = \left[2\xi\omega_o + \frac{4k_c\cos\alpha}{m} k_2 \right]/2\omega_o' \tag{7}$$

Time Delay Compensation. The system has been idealized in the derivation given above and many factors need to be modified when the control algorithm is applied to the control of a real system. Among them, time delay is particularly important in insuring the success of structural control. Therefore, it is desirable to modify the feedback gains k_1 and k_2 so that time delay can be taken into account.

It is relatively easy to compensate for time delay when the controlled structural motion can be assumed to oscillate with a dominant frequency ω. If the displacement feedback force lags displacement t_x in time while velocity feedback force lags velocity $t_{\dot{x}}$ in time, their corresponding phase lags are ωt_x and $\omega t_{\dot{x}}$, respectively. Let g_1 and g_2 be the modified feedback gains with time delay compensation. It can be shown that they are related to the ideal feedback gains k_1 and k_2 by

$$\begin{bmatrix} \cos\omega t_x & \omega\sin\omega t_{\dot{x}} \\ -\frac{1}{\omega}\sin\omega t_x & \cos\omega t_{\dot{x}} \end{bmatrix} \begin{bmatrix} g_1 \\ g_2 \end{bmatrix} = \begin{bmatrix} k_1 \\ k_2 \end{bmatrix} \tag{8}$$

Using precalculated feedback gains for the ideal system, the real system feedback gains defined by g_1 and g_2 can be obtained by solving Eq. 8.

Experimental Results

A series of structural control experiments was carried out using the SDOF model structure under the optimal closed-loop control criterion as developed above.

First, however, the time delay due to control execution must be estimated. This was accomplished by performing an uncontrolled test under a band-limited white noise excitation. Utilizing white noise as command signal input to a component of the system, the achieved signal lags the command due to time delay. Under the assumption of constant time delay, the phase shift of the transfer function is proportional to the vibration frequency. Therefore, time delay can be obtained from the phase slope. Experimentally, the observed phase shift was 2.11°/Hz due to differentiator, 3.62°/Hz due to on-line calculation and 7.21°/Hz due to actuator dynamics. Their corresponding time delays were 6, 10 and 20 msec, respectively. With known time delays, the real system feedback gains can be obtained from their values calculated for the ideal system by solving Eq. 8.

Three structural control experiments consisting of one uncontrolled test and two controlled tests were carried out using the actual El Centro accelerogram as base excitation whose intensity was scaled down to 25% to prevent the structure from exceeding elastic limit under uncontrolled test. The results are illustrated in Figs. 3-5 for three values of β ($\beta = \infty$ represents the uncontrolled case). In addition to a significant reduction in peak magnitude, the increased damping effect is clearly evident in the controlled relative displacement (Fig. 3) and controlled acceleration (Fig. 4). Since the feedback control force is a linear combination of the relative displacement and relative velocity, its time history possessed a similar pattern to those of relative displacement and relative velocity. When the control force was given less weight in the objective function (smaller β), larger control forces were required as shown in Fig. 5.

Figures 3-5 also show good agreement between analytical and experimental results obtained under the same conditions. Discrepancies do exist and they may be due to a variety of inaccuracies in the mathematical modeling of the actual system, these may include system and measurement noise, inaccuracy in structural modeling and interactions between the control system and the structure. For example, a channel beam connecting the tendons to the actuator was not rigid. Such interaction between controller and structure introduced a damping force which was a function of the relative displacement. In the uncontrolled case, the damping was small and the damping factor obtained from system identification could only reflect the average effect of the damping force. Following a large amplitude of relative displacement, the experimental result was smaller than the analytical one because of extra damping force induced by the flexible channel beam. Following a small amplitude of relative displacement, the experimental result was greater than the analytical value. However, for the controlled cases, most of the damping force was contributed by the feedback force. Therefore, the influence of the channel beam was negligible. In the two controlled tests, a smaller control force was required for $\beta = 5$. Thus, there was a better agreement between the experimental and analytical results

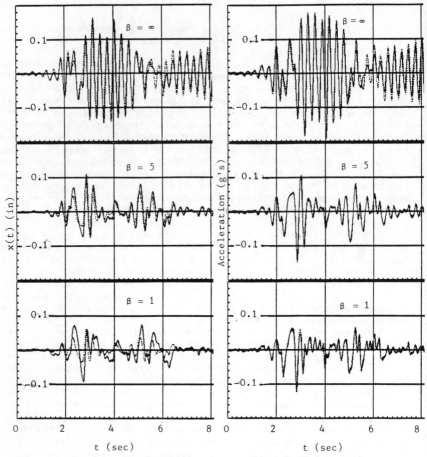

Fig. 3. Relative Displacement
     (---- Analytical, ——— Experimental)

Fig. 4. Acceleration

because the actuator performed better with a smaller rate of change of
the control force. Compared to the analytical results, a larger
control force was required but less reduction of response was obtained
experimentally because of less than 100% efficiency.

Concluding Remarks

Active structural control with a linear optimal feedback control
algorithm has been carried out experimentally using tendon control.
The experiments were carefully designed in order that a realistic
structural control situation could be affected.   Efforts made towards

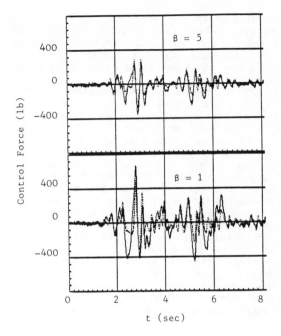

Fig. 5. Control Force (---- Analytical, —— Experimental)

this goal included making the model structure dynamically similar to a real structure, working with a carefully calibrated model, using realistic base excitations, requiring more realistic control force and taking into account such practical considerations as time delay in control execution.

As a result, these experiments permitted a realistic comparison between analytical and experimental results. Furthermore, important practical considerations involving implementability of active control to real structures could be identified and assessed.

Unavoidable time delay in control execution was given careful consideration in implementing the control algorithm. As mentioned earlier, a 30 msec displacement and 36 msec velocity delay were encountered in executing the control. Using a natural frequency of 4.0 Hz as a reference, this delay represents phase shifts of displacement and velocity feedback forces of up to 43° and 52°, respectively. It was thus a significant factor in control design and must be compensated for as was done in this experimental work. However, the magnitude of time delay is expected to decrease as more advanced control system software and hardware become available. Coupled with the fact that real structures vibrate at a lower fundamental frequency as compared with the model structural frequency, time delay effects will not be as serious in control of real structures as in this study.

Time delay effects can also be taken into account in a more direct way by being included in the problem formulation. This approach, however, would lead to the necessity of solving a system of differential-difference equations which would in turn increase real-time computation involved.

The second important practical consideration has to do with the dynamics of the control system itself. In certain cases, its interaction with the structure must be considered in the overall control problem. In this experimental study, for example, the flexibility of the channel beam modified the structural damping characteristics which could not be quantified through system identification. However, this damping effect was small in comparison with the increased damping force due to control application and thus was not considered important. Moreover, the feedback control force continuously compensated the structural response, making structural control effective even though some of the structural parameter values were not precisely known. However, massive control systems need to be used in real structural applications, and interactions between the structure and the control system must be carefully considered.

This experimental study of structural control can be readily extended in many directions. Future work includes the study of MDOF structural systems by removing structural bracing on the upper two floors. The use of other control algorithms such as optimal open-loop and suboptimal digital pulse control will also be examined.

Acknowledgement

This research was supported in part by the National Science Foundation under Grant No. CEE-8311879.

Appendix - References

1. Sage, A.P. and White, C.C., Optimal Systems Control, Second Edition, Prentice Hall, New York, 1977.

2. Samali, B., Yang, J.N. and Liu, S.C., 'Active Control of Seismic-Excited Buildings', J. Structural Engineering, ASCE, 111, 1985, pp. 2165-2180.

3. Soong, T.T., Reinhorn, A.M. and Yang, J.N., 'A Standardized Model for Structural Control Experiments and Some Experimental Results', Proc. Second International Symposium on Structural Control, Waterloo, Canada, 1985.

Dynamic Response of Highway Guardrail Post Foundations

by

Jey K. Jeyapalan[*]

ABSTRACT

The highway departments use two types of guardrail posts: a circular wood post and a steel W6 x 8.5 post. The current specifications require the steel post to be placed in a concrete footing. However, the concrete footing is not required for the wood post. Because of this requirement, the steel post guardrail systems are considered not as economical as the wood post guardrail systems. The research study reported herein was conducted to determine whether the concrete footings are necessary for the steel guardrail posts to perform satisfactorily as a traffic barrier system. An analytical model was developed to model the guardrail post as a laterally loaded drilled shaft. A series of static and dynamic impact load tests were performed to verify analytical predictions. The results of the field tests were in good agreement with the analytical predictions.

---

[*] Director, Wisconsin Hazardous Waste Management Center,
    University of Wisconsin, Madison, WI 53706

Dynamic Behavior of Full-Scale Wood Diaphragms

by

Rafik Y. Itani and Robert H. Falk[1]

Abstract

Wood diaphragms are important structural components of low-rise buildings that resist the lateral forces produced by wind and earthquakes. The main function of these paneled components is to resist in-plane shear forces and to provide stability to a structure.

Considerable research effort has been directed toward the behavior of wood building components; however, most of this has been limited to static response. Since wood buildings are known to be efficient in absorbing the energy produced by dynamic loads, an accurate understanding of their dynamic behavior is essential.

Experimental testing of full-scale wood diaphragms has been performed at Washington State University to determine basic dynamic properties. Ten diaphragms, including walls, floors, and ceilings have been structurally tested for frequency, damping, stiffness, and strength characteristics.

The experimental tests were performed on an MTS hydraulic actuator system capable of static and dynamic operation. Four 8'x24' wall diaphragms and six floor ceiling diaphragms were tested for static, free vibration, forced vibration, and ultimate static loads. Floor and ceiling diaphragm sizes were 16'x16' and 16'x28' which represent the largest diaphragms dynamically tested to date. Typical construction configurations were used and plywood and gypsum board served as sheathing material. Door and window openings were provided in two wall sections and a stairwell opening in one floor specimen to determine their effect.

Results of these structural tests will be presented. Static test results provide nonlinear stiffness characteristics, while free vibration tests yield damping ratios and natural frequencies. Diaphragm strength will result from ultimate static load tests.

These experimental test results also serve to verify a developed analytical model used to predict the response of wall, floor and ceiling diaphragms. This general finite element computer model accounts for the nonlinear behavior of nail fasteners, variable geometry of wood panels, and the effects of openings. Information provided by this model will include stiffness and damping characteristics as well as a prediction of the strength of wood diaphragms.

---

1) Professor and graduate student, respectively, Dept. of Civil Engineering and Environmental Engr., Washington State University, Pullman, WN 99164-2914

Results of this study will provide a data bank of structural properties for the modeling of three-dimensional full-scale structures and suggest recommendations for efficient methods to minimize earthquake and wind damage.

This study is part of an ongoing National Science Foundation research project on wood diaphragms.

Cladding Design for Tall Buildings

D.A.Reed[*]

Abstract

In the past twenty years, glass cladding has emerged as a popular and economic building material. Its use in tall buildings which are constructed in high-risk wind and seismic geographical regions has provided an impetus for increased research activity in the areas of fracture mechanics, wind engineering and the dynamic behavior of tall buildings. Despite recent gains, the behavior of glass cladding under various types of loading conditions is still not fully understood. The major objective of this paper will be to examine present procedures and proposed changes for glass cladding under wind pressure and seismic loading conditions.

The failure of glass cladding during wind storms is attributed to several causes, such as air-borne missile impact, extreme wind pressure loadings and in-plane loadings caused by interstory racking. Strategies for the reduction of failures due to missile impact include provisions for the reduction in potentially hazardous roofing materials, the use of metal screens or covers and the investigation of the economic feasibility of requiring laminated glass in the lower stories of buildings which are highly susceptible to impact damage. Research activities in this area include the development of numerical procedures for predicting failure and empirical studies of failure under impact loads. The assessment of in-plane loadings caused by interstory racking and torsion of the entire structure under wind loadings has not been fully investigated. The major focus of research has been in the area of failure due to extreme wind pressure loads. In this paper, recent results obtained in the development of numerical procedures for estimating damage and ultimately catastrophic failure will be examined. In particular, the dimensionless damage function will be evaluated for panels of different aspect ratios under various wind loading conditions. Relevant research from the fracture mechanics area will also be presented and discussed.

In contrast to the significant advances which have been made in predicting the behavior of glass under wind pressure loadings, relatively minor attention has been focused upon the performance of glass panels during and after an earthquake. The pioneering article in the field is that of Bouwkamp in 1961. Although deflections and rotations for panels tested under diagonal compression or buckling conditions were investigated by Bouwkamp, the stress state and fracture patterns were not studied.

Recent laboratory results obtained by Shinkai (Asahi Glass Company, Yokohama, Japan) for diagonal compression and buckling show that the fracture patterns under these conditions display numerous variabilities. It was noted by Shinkai that the crack pattern in buckling is strongly dependent upon the panel thickness. Because buckling failure is an extremely violent process, it would present a severe danger to

---

[*] Assistant Professor of Civil Engineering, University of Washington, Seattle, WN 98195

building occupants and pedestrians.

Damage to nonstructural elements for a reinforced concrete building subjected to lateral loading has been examined recently by the U.S.-Japan Joint Technical Coordinating Committee. Local attachment of glass panels was found to be the critical factor for determining failure; those panels for which the connection details allowed for greater in-plane movement and sliding did not fail. Similar results for other framing systems such as a braced steel system would be useful. Guidelines for rehabilitation to panels in existing structures and for new construction should be based upon studies on various framing types to determine critical locations for panels within the structure itself as well as an evaluation of the glass curtain wall system. A review of the studies undertaken in the behavior of panels under seismic loading conditions will be made in the paper. Recommendations for improved seismic design will be made.

Response of Underground Pipelines to Dynamic Loads

Jey K. Jeyapalan *

ABSTRACT

The dynamic response of underground pipelines to earthquake loads is of considerable interest to designers of lifeline systems. In this paper the results of an investigation of the forced response of an underground pipeline due to dynamic loads is presented. The pipe is modelled as a linearly elastic shell element and the surrounding soil is represented by both elastic and viscoelastic material models. The pipe—soil interface is modelled as a perfect bond. The stresses and the deformed shapes of the pipelines are reported for ranges of pipeline materials and properties of backfill materials. The beam on elastic foundation model is also studied for this problem and the results are compared with those obtained by the continuum model.

---

* Director, Wisconsin Hazardous Waste Management Center,
University of Wisconsin, Madison, WI 53706

Seismic Isolation of Bridge Structures

Ronald L. Mayes[1], Ian G. Buckle[2], Lindsay R. Jones[3], Trevor E. Kelly[4]

Introduction

Considerable efforts have been made in the United States in the past decade to develop improved seismic design procedures for new bridges and comprehensive retrofit guidelines for existing bridges [1, 2]. These efforts were spurred in part by the damage caused to bridges during the San Fernando earthquake of February 9, 1971. This earthquake demonstrated the potential inadequacy of past design procedures and the necessity for seismically upgrading bridges that were designed using the pre-1971 procedures.

It is generally recognized that it is uneconomic to design a structure to respond in the elastic range to the maximum credible earthquake and the current state of the art provides two alternate procedures. The first, which is embodied in most codes, is to design components, such as columns, for a much lower force level than in inherent in the elastic forces induced by a severe earthquake. Combined with this lower design force, design details are provided to ensure that the components can respond in a ductile manner, thus providing an energy dissipating mechanism for the additional force.

The second alternate is to reduce the earthquake forces by a concept called seismic isolation. The concept has been extensively developed and applied in New Zealand and is receiving increased attention in the United States. Two bridge projects have incorporated the concept and another is in the preliminary design phase. In addition to these bridge projects, four buildings have or will incorporate the concept.

Seismic Isolation Concepts

The isolation of structures from the damaging effects of earthquakes is not a new idea. The first patent for a seismic isolation scheme was taken out as early as 1909 but until recently very few structures have been built which use these principles. However, new impetus was given to the concept with the successful development of mechanical energy dissipators by the New Zealand Department of Scientific and Industrial Research [3, 4, 5, 6]. When used in combination with a flexible isolation

[1] President; [2] Director of Research & Development; [3] Vice President; [4] Project Engineer: Dynamic Isolation Systems, Inc., 2855 Telegraph Avenue, Suite 410, Berkeley, California 94705

device, an energy dissipator can control the response of the structure
by limiting displacements and forces and thereby improving seismic
performance. As a consequence, several buildings and 37 bridges have
now been designed and built in New Zealand which incorporate base
isolation and energy dissipators [5]. Many of the features of base
isolation can be used to improve the seismic performance in existing
structures. The inclusion of base isolators and energy dissipators when
upgrading a structure can significantly enhance its seismic performance.

There are two basic elements in practical base isolation systems
as follows:

1.   A flexible mounting so that the periods of vibration of the total
      system are lengthened (shifted) sufficiently to reduce acceleration
      response, and

2.   A damper or energy dissipator so that the relative deflections
      across the flexible element can be controlled.

An elastomeric bearing with a lead plug inserted into a preformed
hole is an efficient means of providing both these elements in a single
unit [6, 7]. Bridge structures have for a number of years been supported
on elastomeric bearings and thus the concept of the flexible mount is
not new. Increasing the thickness of the bearing is a natural step to
ensure adequate flexibility and period shift.

A lead—rubber bearing comprises alternate layers of rubber
vulcanized or cemented to thin steel shims, typically with a lead plug
placed in the center (Fig. 1). The internal steel shims plus the
externally applied vertical load provide confinement for the plug. The
rubber in the bearing carries the weight of the structure and the lead
plug provides energy dissipation by plastic deformation. Outer steel shims
with dowel holes are provided for transfer of lateral force from the
structure above to the bearing.

The bearing is designed to resist low levels of shear, such as wind
and braking loads, elastically with a high initial stiffness (K_u)
until a yield level determined by the characteristic strength, Q_d, is
reached. The force level, Q_d, is a function of the diameter of the lead
plug(s). The post yield stiffness K_d is kept at a minimum to ensure
good energy dissipation and low overall structure stiffness during more
severe seismic loading. The bi—linear hysteresis curve formed by these
two stiffnesses and shown in Fig. 2 has a stable form and large enclosed
area, demonstrating the energy absorption properties.

Design Application

The seismic design principles for seismic isolation are best
illustrated by Figure 3. The solid line represents a realistic elastic
ground response spectra. This is the spectra that is used to determine
actual forces and displacments for conventional design. The lowest curve
represents the design forces used for a multi—column bent in ATC—6. Also
shown is the probable overstrength of the column so designed. The force

on the isolated structure is represented by an inelastic spectrum that incorporates the damping of the isolation system. This is shown by the small dashed line of Figure 3.

In seismic isolation design, the period of the bridge is increased into the 1.5-2.5 second range. This results in a factor of 5 to 10 reduction in the actual forces to which a bridge is subjected. As a consequence of this force reduction, a factor of 2 or more reduction can be obtained in the design forces for the columns. In addition to this reduction in the design force level, the ductility demand on the column can be eliminated as shown in Figure 3. For new bridge construction this can result in up to a 10% cost savings. For the retrofit of existing bridges seismic isolation provides a solution to the three major seismic problems in that vulnerable existing bearings can be replaced, forces to the substructure are significantly reduced and displacements are reduced.

Three-Span Bridge Example

Figure 4 shows a three-span steel bridge with 5 main girders. The total number of bearings is 20 with 10 over the piers, each carrying a dead load of 104 kips and a dead plus live load of 124 kips. The abutment bearings carry one-half this load. The detailed steps for designing the lead-rubber bearings for one seismic zone are presented and a comparison of the force and displacement requirements for various seismic zones is then discussed.

1.   The maximum dead load plus live load is used with Chart 1-B [7] for circular bearings to obtain the required plan bonded dimension of 13 inches with 3/8 inch internal rubber layers. This bearing has a rated load capacity of 180 kips.

2.   The total bonded area of all bearings is then calculated and an average compressive of 588 psi over all bearings is obtained.

3.   To obtain the total rubber thickness a design chart for a specific seismic zone and soil type must be used. For this example the CalTrans 0.6g Soil Type S1 design chart is used [7]. It is within this chart that the trade-off in base isolation design between design force versus design displacement can readily be seen. The recommended rubber thickness from the chart varies from 6 to 26 inches. This encompasses a bearing displacement of 6 inches and a base shear force of 0.27W for the 6-inch rubber height where W is the weight of the superstructure. For the 26-inch rubber height, the bearing displacement is 10.4 inches and the base shear force is 0.14W. For design, a rubber height of 8 inches was selected corresponding to an effective period of 1.75 seconds, a force of 0.23W and a maximum displacement of 6.8 inches.

4.   Because the abutments carry only one-half the load that the piers carry, it was decided to use lead plugs in only one-half the abutment bearings. To retain symmetry, two plugs are used in each abutment and five in each pier for a total of 14 plugs. To obtain

a characteristic strength of 0.05W, this required a total strength
of (1560 x 0.05) 78 kips, and dividing by 14 provides a required
characteristic strength of 5.57 kips per plug. Tabulated properties
[7] show that a 2.5" plug provides 5.62 kips, very close to the
required value.

If the same bridge were to be located in a moderate seismic zone
the same bearing diameter would be used but with a lower rubber thickness.
As an example, the ATC-6 seismic design zone [1] with an Aa of 0.30g and
soil type S1 Chart 2-B-1 [7] would give an appropriate rubber thickness
of 5.00" for the same average compressive stress of 588 psi. This is only
62% the thickness selected for the CalTrans 0.6g design. In this case,
the chart provides maximum force levels of 0.14W and corresponding
displacements of 2.09". This compares with a force coefficient of 0.23W
and 6.84" displacement for the same bridge with 8" rubber thickness and
design ground motions corresponding to 0.6g CalTrans.

Table 1 illustrates the variations in force coefficients and
displacements as functions of the total rubber thickness and the level
of seismicity.

The values tabulated are for the ATC-6 spectral shape for maximum
accelerations from 0.2g to 0.4g and for the CalTrans spectra for 0.5g
and 0.6g. Therefore, they are not directly comparable for all acceleration
levels but do give an indication of the effectiveness of the isolation
for varying rubber thicknesses for varying seismic intensity.

It can be seen that the increased rubber thickness is more effective
for the higher intensity ground motions. For example, increasing T_r
from 5" to 8" reduces forces by only 10% for 0.2g spectra but by 30% when
the maximum ground motion is 0.6g.

The table also shows that bearings designed for one seismic
intensity are satisfactory for other levels of intensity. For example,
the 8" rubber thickness selected for 0.6g ground acceleration also gives
good isolation for all intensities from 0.2g to 0.6g. When the 5" rubber
thickness selected for the 0.3g region is subjected to higher intensities
the force level increases but the base isolation is still effective. In
fact, even at 0.6g, double the design earthquake, the rubber strain of
(5.73"/5.0") 1.15 is considerably less than the strain levels of 1.50
to which these bearings have been tested.

Conclusions

Clearly the seismic safety of new and existing bridges is of major
concern and recent efforts have been directed towards providing guidelines
to solving the problem. The paper has presented the application of the
seismic isolation design concept utilizing the lead-rubber bearing.

The benefits of seismic isolation for bridges can be summarized
as follows:

1. Factor of 5-10 reduction in the realistic forces to which a bridge will be subjected.

2. Elimination of the ductility demand and hence damage to the columns.

3. Control of the distribution of the seismic forces to the substructure elements with appropriate sizing of the elastomeric bearings.

4. Reduction in column design forces by a factor of approximately 2 compared to conventional design.

5. Reduction in foundation design forces by a factor of 2-3 compared to conventional design.

Failures in past earthquakes have indicated three major problem areas with existing bridges. The first is the vulnerability of existing rocker and roller type bearings and this is coupled with inadequate support lengths for superstructure girders. The third problem area is inadequate strength and ductility capacity of supporting substructures. The bearings provide a solution to the three major seismic retrofit problems in that vulnerable bearings can be replaced, resulting in decreased superstructure displacements and a force-limiting mechanism for the supporting substructure. Thus, the recently developed base isolation and energy dissipating concepts embodied in the lead-rubber bearing provide an effective and economical solution for the three major problems of seismic bridge retrofit.

References

[1] Applied Technology Council, "Seismic Design Guidelines for Highway Bridges," ATC-6 Report, Palo Alto, CA., 1981.

[2] Applied Technology Council, "Seismic Retrofitting Guidelines for Highway Bridges," ATC-6-2 Report, Palo Alto, CA., 1984.

[3] Blakeley, R.W.G., "Analysis and Design of Bridges Incorporating Mechanical Energy Dissipating Devices for Earthquake Resistance," Proceedings of a Workshop on Earthquake Resistance of Highway Bridges, Applied Technology Council, Report No. ATC-6-1, Nov. 1979.

[4] Blakeley, R.W.G., et al., "Recommendations for the Design and Construction of Base Isolated Structures," Bulletin New Zealand National Society for Earthquake Engineering, Vol. 12, No. 2, 1979.

[5] Buckle, I.G., "The Use of Energy Dissipators and Base Isolation in the Design and Retrofit of Bridges and Buildings," Proceedings of 51st Structural Engineers Association of California Convention, October, 1982.

[6] Robinson, W.H., "Lead-Rubber Hysteretic Bearings Suitable for Protecting Structures During Earthquakes," Journal of Earthquake Engineering and Structural Dynamics, Vol. 10, 1982.

[7] Dynamic Isolation Systems, Inc., "Seismic Base Isolation Using Lead-Rubber Bearings – Design Procedures for Bridges," Berkeley, CA., 1983.

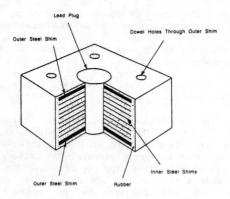

Figure 1.   Typical Lead-Rubber Bearing

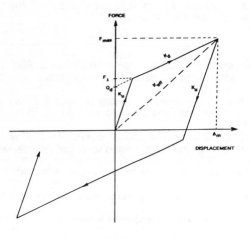

Q_d	=	Characteristic strength (kips)
F_1	=	Yield force (kips)
F_{max}	=	Maximum force (kips)
K_d	=	Post-elastic stiffness (kip/inch)
K_u	=	Elastic (unloading) stiffness (kip/inch) (=6.5K_d)
K_{eff}	=	Effective stiffness
Δ_m	=	Maximum bearing displacment

Figure 2.   Properties of Lead-Rubber Bearing

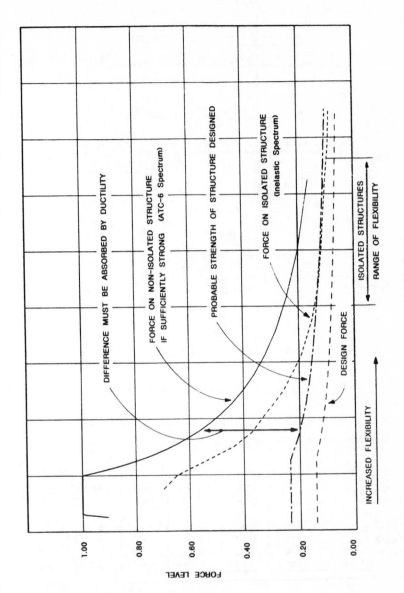

Figure 3. Earthquake Forces

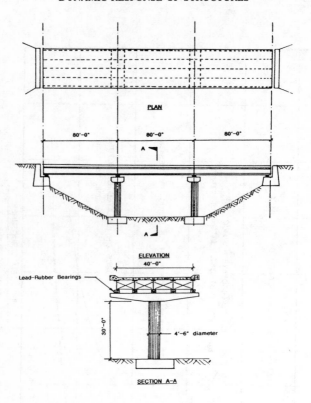

Figure 4. Bridge Example 1

MAXIMUM GROUND ACCELERATION	$T_r = 5"$		$T_r = 8"$		$T_r = 11"$	
	FORCE COEFF.	MAX. DISP.	FORCE COEFF.	MAX. DISP.	FORCE COEFF.	MAX. DISP.
0.2g–ATC-6	0.10	1.28	0.09	1.60	0.08	1.87
0.3g–ATC-6	0.14	2.09	0.11	2.35	0.10	2.70
0.4g–ATC-6	0.19	2.90	0.14	3.40	0.12	3.71
0.5g–CalTrans	0.26	4.73	0.20	5.20	0.17	5.93
0.6g–CalTrans	0.32	5.73	0.23	6.84	0.21	8.11

Table 1. Results for Different Seismic Zones

HYSTERETIC BEHAVIOR OF PRECAST
CLADDING CONNECTIONS

J.I. Craig*, B.J. Goodno*, M.J. Keister**, C. J. Fennell*

Georgia Institute of Technology
Atlanta, Georgia 30332

ABSTRACT

A preliminary experimental study is presented which investigates the behavior of connections for architectural precast concrete cladding panels. The principal objective is to quantify the lateral stiffness capacity, energy dissipation properties, and ductility of representative precast claddings designs used on several highrise buildings. Hysteretic models will be developed on the basis of test and used to predict cladding influence on overall building seismic response in the linear and nonlinear stages of behavior. Test results are presented for three 3/4 in. diameter wedge-type inserts subjected to direct pullout. The reaction frame is designed to perform moment, shear and combined shear/pullout tests. This study is expected to lead to a better understanding of precast connection behavior and performance and thereby aid in developing a more rational connection design procedure.

INTRODUCTION

Background

During the past few years, cladding has become a greater concern to engineers due to numerous cladding failures [3]. One of the reasons for these failures is pure ignorance of the behavior of cladding, for despite its widespread use, cladding behavior is still little understood. Cladding is an expensive part of a building, costing as much as 10 to 20% of a building's initial cost [3], but if cladding repairs are needed, the cost of the building's facade can become much more than originally anticipated. Until more research is done and cladding behavior is better understood, these failures will continue with great loss of money and possible risk of injury to the public.

In the design of precast cladding, most engineers follow the recommendations of the Prestressed Concrete Institute [14, 15]. The panels, according to PCI, should be designed and detailed to transfer wind and thermal forces to the supporting frame with the panel connections resisting these forces plus the panel's own weight. Sliding or flexible connections are recommended by PCI allowing movement in the plane of the panel lessening panel interaction

*Georgia Institute of Technology
**Bennett and Press, Inc., Atlanta, GA

with the supporting frame. Because of this separation between panel and frame, the panels are classified and designed as nonstructural elements apparently adding little to the lateral stiffness or lateral force resistance. Recent studies [6, 7, 10, 13-15] have shown that this disregard of precast panels to carry lateral load or add lateral stiffness is false. Not only has it been shown that cladding can add significantly to the lateral stiffness, but also it has been suggested that the panels and connections themselves could also augment the energy dissipation characteristics of the structure, especially for moderate ground motions [13].

To date, very little research has been performed on precast cladding connections. There are two parts to a connection assemblage, attachment to the precast panel and attachment to the structural frame. Both parts are important but the concern of this study is the precast panel attachment. To this end, preliminary tests have been conducted and data evaluated for a wedge-type precast insert used in the Southeastern United States. These and follow-on studies are expected to lead to a better understanding of precast connection behavior and performance.

Previous Studies

Previous research on the behavior of structures with attached cladding has been sporadic. Early work by Sherwood [16] and Gram [7] established that precast cladding could augment the lateral stiffness of highrise buildings by as much as 30% for low level motions. Later, a finite element model of a precast panel and its clip angle connections was used to show that these facade elements possessed adequate lateral stiffness potential; however the actual stiffness contribution was heavily dependent on the details of the attachment [20]. In later studies [5,12], a variety of different models for cladding were presented and used to demonstrate that cladding stiffness could significantly alter dynamic properties and linear seismic response. Torsional response in particular was susceptible to change if partial cladding failure occurred [4]. Other studies by Weidlinger [18] and Oppenheim [10] demonstrated that structural partitions, like exterior cladding, could be used effectively to control wind induced structure motions.

Several experimental investigations of cladding have also been reported. Anicic et al [1] tested large flat concrete panels for out-of-plane cyclic motions. Uchida et al [17] attached flat precast panels to a two story two bay steel frame and conducted free and forced vibration tests for in plane motions to assess the effect of the panels on frame dynamic properties and response. Finally, Aswad [2] and later Wiss, Janney, Elstner [19] conducted tests of a variety of precast concrete connections. These latter tests pertain to structural uses of precast concrete elements but may contribute to a better understanding of cladding connection behavior as well, which is the subject of the test program to be described below.

EXPERIMENTAL PROGRAM

Many of the referenced studies point to the precast panel connections as the key in understanding cladding behavior, yet very little testing has been done in this area. The present tests are designed to consider both the static as well as dynamic strength properties. It is important to note at the outset that the phrase "strength characterization" is meant to imply a characterization of the generalized connection constituitive relations which pertain directly to earthquake design. Thus, the emphasis is placed not only on simple stiffness and strength measurements but will also include assessment of ductility, hysteretic behavior, energy dissipation, and failure modes.

The tests involve only single connections but consider several load components singly and in combinations. As many as six components may act across a connection, but of these, inplane shear in the lateral and vertical directions and normal forces are of prime concern. Panel weight, environmental and earthquake forces and interaction forces (racking, interstory drift) are transferred as combinations of shear and normal forces. When the connections are eccentric, the added bending moments developed across the connection must be considered.

Since principal concern is with connection performance under earthquake loading, the measurements are carried out under cyclic loading conditions. The measurements are arranged to allow assessment of the following connection characteristics:

1. Small-deflection,linear,elastic behavior,
2. Large-deflection, nonlinear, inelastic, cyclic behavior,
3. Maximum strength and failure modes,
4. Number of large-amplitude cycles.

The first case is necessary in order to determine the cladding stiffness contribution for typical ambient conditions. Cases 2-4 assess the ductility, energy dissipation capacity, and patterns of failure for cladding under strong earthquake ground motion conditions. The load is applied for a prescribed number of cycles at which point the load-deformation plots can be examined to evaluate the stiffness degradation and hysteresis loop form and stability.

Specimen Design

The specimen design (samples A, B, C) consisted of a 3 ft. by 3 ft. by 6 in. thick slab of concrete reinforced with five #3 bars in two directions at a 3 in. depth. Following a design detail used in an existing highrise building, the connection insert was attached to two #3 bars 10 in. long as shown in Fig. 1. The insert was a Dayton Superior F-19 shelf angle insert (ASTM A 536-67 ductile iron) normally used in thin precast panels. A Dayton Superior F-7 3/4 in. askew head bolt was used. The concrete had a 28 day compressive strength above 5,000 psi and the reinforcing steel had a yield strength of 57,200 psi (ultimate strength of 84,600 psi).

Test Fixture

The test fixture consisted of a base slab, reaction frame, and hold-down device (Fig. 2). The base slab was an 8 ft. by 10 ft. by 1 ft. thick heavily reinforced concrete reaction slab capable of reacting 100 tons at any point. As shown , the intial configuration was for pull-out tests, but the fixture is designed for more complex loading cases. Instead of pull-out, moment and pull-out can easily be applied to the connection by simply employing an eccentric link in the load path. Likewise, shear and pull-out or shear and moment/pull-out can be developed with little modification. Only the initial pull-out test results will presented in this paper.

Instrumentation

The inserts were heavily instrumented with strain gages and linear variable differential transformers (LVDT's). As shown in Fig. 3, strain gages were located on various areas of the insert and reinforcing bars. All of the gages showed good survivability and yielded consistent results. The connection displacement data were measured with LVDT's placed co-planar to the slab at right angles to each other for inplane motion and attached to a collar on the actuator linkage for transverse displacement measurement.

Data Acquisition

Loading for all cases was applied servo-hydraulically by an MTS actuator through a load cell and attachment linkage. All load, displacement and strain data were acquired using an H-P 9825 calculator driving a custom assembled data acquisition and control system. The test program was also generated and controlled automatically by the same system.

EXPERIMENTAL RESULTS

All three samples failed at between 11,000 and 12,000 lbs. by fracture of the concrete in a shallow shear cone. This level of ultimate strength is three times the design load of 4,000 lbs. stated by the manufacturer. The results for the three samples tested (A, B, C) are discussed below in terms of displacement and strain response.

Displacement Response

In all tests, the LVDT's behaved well, producing displacement data with .001 in. resolution. All inserts, except experimental slab A, experienced excessive slip in the direction of the insert slot. To alleviate this problem, a steel spacer was used to block the bolt and prevent slippage out of the insert. The block was sized so as to locate the bolt in the center of the slot. The out-of-plane displacements (in the direction of load) were linear and similar in all slabs. Considering Fig. 4 as an example of vertical displacement, the displacement versus load relationship can be seen to be linear with narrow hysteresis loops. In all of the samples, failure occurred with displacement in the 0.060 to 0.070 in. range with very little change in the slope of the load versus displacement curve.

The in-plane displacement was measured by two orthogonal LVDT's but this data is suspect due to the introduction of the steel block as explained earlier. It was concluded from this data that after approximately four to five loading cycles the bearing surface on the bolt was worn smooth allowing significant slippage of as much as 0.3 in. Post-test examination of the bearing surface of the bolt confirmed this.

Strain Response

Since insert behavior was a prime interest in these tests gages 0 through 4 (see Fig. 3) located on the bottom flange of the insert were most important. Figure 5 presents behavior of gage 0 at one end of the insert and shows tension with a change to compression in successive cycles. This behavior seems to show good embedment of the insert ends producing the boundary condition of a fixed end beam. With successive cycles, the strain becomes negative indicating that cracking is occurring in the concrete near the insert end thus changing the boundary conditions of the insert end from a fixed end to a pinned end.

At the other end of the insert, gage 4 behaves the same for all three slabs. Looking at the plot for sample C (see Fig. 6), as an example, notice the change in specimen behavior. The strain starts out slightly in tension and steadily goes into compression with steadily wider hysteresis loops. This behavior along with the behavior along with the behavior of gage 0 implies that the insert behaves first as a fixed end beam and quickly changes boundary conditions to a beam with fixed and pinned ends and finally to a beam with pinned ends (see Fig. 7). To further support this beam representation, Figure 8 shows the behavior of the strain of gage 2 for specimen B. This pattern is characteristic of that observed at gage 1 and 3 for all three samples. Note that all strain is compressive thus confirming this beam representation with changing end conditions as cycles continue.

Gages 5 and 6 respond to the shear load in the insert model. Both gages behaved similarly and Figure 9 is representative of the behavior of both gages for all specimens. The data shows that the shear behavior is linear with tight hysterisis loops exhibiting little energy dissipation.

In all samples, no cracking was observed until failure occurred. When failure did occur, a loud and distinct crack or crashing sound could be heard. Concrete fracture in a very shallow circular or elliptical cone. The diameter of the cone was approximately 2 ft. with a height of 2 1/2 in.

CONCLUSIONS

The inserts experienced highly linear strain and displacement results up to 10,000 lbs. In this load range, the hysteresis loops were narrow, indicating little energy dissipation. In the load cycles from 10,000 lbs. to failure at 11,000 to 12,000 lbs., the samples experienced nonlinear behavior with changing slopes of load versus displacement and load versus strain curves. Also, in these final cycles, the hysteresis loops became very large indicating greater energy dissipation possibly due to reinforcing steel deformation.

The method of failure of the inserts was unsatisfactory for practical situations because of its suddenness and catastrophic fracturing of concrete. It was noticed after failure that the reinforcing steel underwent considerable deformation. This deformation was probably the reason the strain and deformation plots underwent pronounced changes at 10,000 lbs. with strain and displacement plots changing to wide hysteresis loops. The reinforcing steel used with the inserts was only 10 in. long and lengthening it would most likely add considerably to the ductility and survivability at loads above 10,000 lbs. To maximize survivability, it would probably be best to tie the insert into panel flexural steel instead of adding bars just for the insert alone. This would increase the insert steel's embedment providing better continuity of the insert to the panel. This continuity would induce a connection failure by a ductile structural failure of the panel instead of a sudden brittle localized fracture noticed in these tests. Specific recommendations for modifications to the insert support detail await further testing involving other loads, load combinations and panel support arrangements.

FURTHER STUDIES

With measured connection properties available from the laboratory tests, it will now be possible to refine and calibrate overall building model studies. Improved cladding models representing linear, nonlinear, and hysteretic panel-connection response characteristics will be added to the basic structure model so that earlier studies may be continued into the nonlinear, strong motion range. In particular, such considerations as, (1) energy dissipation effects due to cladding, (2) partial cladding failure and its influence on overall building torsional response, (3) progressive cladding failure under strong ground motion, and, (4) panel and connection forces for design, will be investigated.

The ultimate objective is to explore possible connection changes which will result in improved second generation designs for actual buildings. Enhanced stiffness, damping, and ductility characteristics and economy are sought. Information obtained from these Phase I experimental studies will be used in analytical models to predict overall structural behavior under moderate and strong motion loading, and it will also be used as the basis for improvements in current connection designs. At the present time, work is being performed to develop appropriate loading programs that will realistically simulate normal and strong motion connection response, and also to provide for a more logical evaluation of performance data.

REFERENCES

1. Anicic, D., Zamolo, M., and Soric, Z., "Experiments on Nonstructural Partition Walls Exposed to Seismic Forces," Proceedings, Seventh World Conference on Earthquake Engineering, Istanbul, Turkey, September 8-13, 1980, Vol. 6, pp. 144-150.

2. Aswad, A., "Selected Precast Connections: Low-cycle Behavior and Strength," _Proceedings_, Third Canadian Conference of Earthquake Engineering, Montreal, Canada, June 4-6, 1979, pp. 1201-1224.

3. "Facades: Errors Can Be Expensive," _Engineering-News Record_, Vol, 204, No. 5, January 24, 1980, pp. 30-34.

4. Goodno, B.J. and Palsson, H., "Torsional Response of Partially Clad Structures," _Proceedings_, Conference on Earthquake and Earthquake and Earthquake Engineering: The Eastern U.S., September 14-16, 1981, Knoxville, Tennessee, Vol. 2, pp. 859-877.

5. Goodno, B.J., Will, K.M., and Palsson, H., "Effect of Cladding on Building Response to Moderate Ground Motion," _Proceedings_, Seventh World Conference on Earthquake Engineering, Istanbul, Turkey, September 8-13, 1980, Vol. 7, pp. 449-456.

6. Goodno, B.J., Will, K.M., and Craig, J.I., "Analysis of Cladding on Tall Buildings," Presented at the ASCE National Convention and Exposition, Session on Developments in Methods of Structural Analysis, held in Atlanta, Georgia, October 23-25, 1979 (Meeting Preprint No. 3744).

7. Gram, K.G., "The Shear Effects of Precast Cladding Panels on Multistory Buildings," a Special Problem Report presented to the Faculty of the School of Civil Engineering, Georgia Institute of Technology, Atlanta, Georgia in June 1976, in partial fulfillment of the requirements for the degree of Master of Science.

8. Goodno, B.J. and Palsson, H., "Earthquake Performance of Building Cladding Systems," _Proceedings_, Sino-American Symposium on Bridge and Structural Engineering, Beijing, China, September 13-19, 1982.

9. Keister, M.J., "Testing and Evaluation of Wedge-Type Precast Concrete Inserts Subjected to Pullout," Master of Science Research Problem Report, School of Civil Engineering, Georgia Institute of Technology, Atlanta, GA, March 1983.

10. Oppenheim, I.J., "Dynamic Behavior of Tall Buildings with Cladding," _Proceedings_, Fifth World Conference on Earthquake Engineering, Rome, Italy, June 1973, pp. 2769-2773.

11. Palsson, H., "Cladding-Structure Interaction Case Study for a Highrise Office Tower," Student Paper Competition, AIAA/ASCE/ASME/AHS Symposium, Dayton, Ohio, October 3, 1980.

12. Palsson, H., Goodno, B.J., "A Degrading Stiffness Model for Precast Concrete Cladding," _Proceedings_, Seventh European Conference on Earthquake Engineering, Athens, Greece, September 20-25, 1982, Vol. 5, pp. 135-142.

13. Palsson, H., Goodno, B. J., Craig, J. I., and Will, K. M., Clad-
     ding Influence on Dynamic Response of Tall Buildings," Earthquake
     Engineering and Structural Dynamics Journal, Vol 12, No. 2, March-
     April, 1984, pp. 215-228.

14. PCI Manual for Structural Design of Architectural Precast Concrete,
     PCI Publication No. MNL-121-77, Prestressed Concrete Institute,
     201 N. Wells St., Chicago, Illinois 60606, 1977.

15. PCI Manual on Design of Connections for Precast Prestressed Con-
     crete, Prestressed Concrete Institute, 201 N. Wells St., Chicago,
     Illinois 60606, 1973.

16. Sherwood, G.C., "Effects of Precast Concrete Panels on the Stiff-
     ness of the 100 Colony Square Building," Master of Science Re-
     search Problem Report, School of Civil Engineering, Georgia
     Institute of Technology, Atlanta, GA, August, 1975.

17. Uchida, N., Aoygai, T., Kawamura, M. and Nakagawa, K., "Vibration
     Test of Steel Frame Having Precast Concrete Panels," Proceedings,
     Fifth World Conference on Earthquake Engineering, Rome, Italy,
     June 1973, pp. 1167-1176.

18. Weidlinger, P., "Shear Field Panel Bracing," Journal of the
     Structural Division, ASCE, Vol. 99, No. ST7, Proc. Paper 9870,
     July 1973, pp. 1615-1631.

19. Wiss, Janney, Elstner and Assoc., Inc., Cyclic and Monotonic Shear
     Tests on Connections Between Concrete Panels for the Massachusetts
     Institute of Technology, WJE No. 77578, July 30, 1981, Vol. 1,
     Wiss, Janney, Elstner and Assoc., Inc. 330 Plinsten Road,
     Northbrook, Illinois 60062.

20. Will, K.M., Goodno, B.J., and Saurer, G., "Dynamic Analysis of
     Buildings with Precast Cladding," Proceedings, ADCE Seventh
     Conference on Electronic Computation, St. Louis, Missouri, August
     6-8, 1979, pp. 251-264.

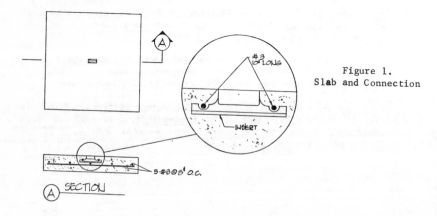

Figure 1.
Slab and Connection

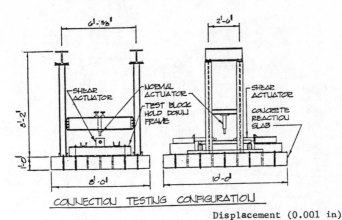

CONNECTION TESTING CONFIGURATION

Figure 2. Connection Test Frame.

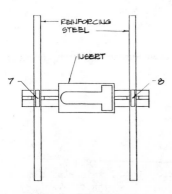

Figure 3. Wedge Insert Showing Strain Gage Locations.

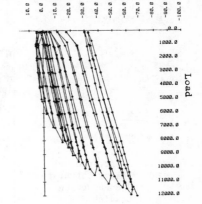

Figure 4. Load-Displacement Plot for Sample B.

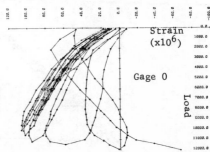

Figure 5. Load-Strain Plot/Sample B

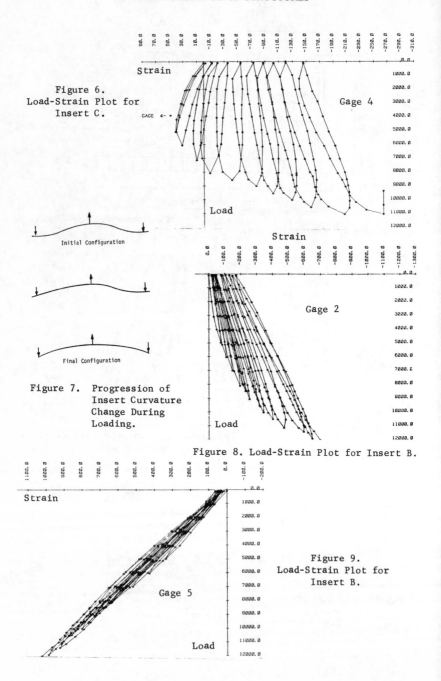

Figure 6.
Load-Strain Plot for
Insert C.

GAGE 4 - •

Gage 4

Strain

Load

Initial Configuration

Final Configuration

Figure 7.   Progression of
Insert Curvature
Change During
Loading.

Strain

Gage 2

Load

Figure 8. Load-Strain Plot for Insert B.

Strain

Gage 5

Load

Figure 9.
Load-Strain Plot for
Insert B.

Ambient Vibration Testing of the Foothill Communities Law
and Justice Building

by G. Pardoen[*] and G.C. Hart[**], and D. Nakaki[**]

This paper will present the results of two ambient vibration tests
conducted on the Foothill Communities Law and Justice Center Building.
One ambient test was performed in 1985. The second was conducted in
1986 after the first set of earthquake response records were obtained in
the building in 1985.

The paper also describes the results of a NSF funded project at
UCLA to define a periodic ambient vibration testing program for use in
evaluating future earthquake response records obtained in this building.

---
[*] Civil Engineering Dept., University of California, Irvine
[**] Civil Engineering Dept., University of California, Los Angeles

Dynamic Response Correlation Using Simplified Models

J. C. Anderson, S. F. Masri and P. Mohasseb*

Abstract

A simplified model for building response which has a Ramberg-Osgood hysteresis is developed in a discrete sense. It is then used to evaluate the response of an instrumented, bearing wall building which was subjected to a strong motion earthquake. Critical comparisons are made between recorded and calculated building response.

Introduction

The design of structures to resist extreme dynamic loads such as earthquake and blast requires an understanding of the dynamic response of the structure. For most structures, economics dictates that they be designed to respond in the inelastic range when subjected to these extreme loading conditions. This requires the development of analytical models which can be used to represent the behavior of the structure in both the elastic and inelastic range. One possibility is to represent the individual members of the structure by finite elements which can model the nonlinear behavior of the individual member. This type of solution usually requires a large number of elements and the solution time even on today's modern computers can be extensive. In many instances, modeling the nonlinear response of the individual elements can be quite difficult as is the case when considering the nonlinear response of a concrete or masonry load bearing shear wall.

An alternative solution is to consider a very simplified analytical model which has relatively few parameters. The idea is to adjust the parameters so that the simplified model represents the overall response of the entire structure in some average sense. One of the simplest of this type of model is the elasto-plastic oscillator which has been used extensively to develop response spectra for earthquake loads and shock spectra for blast loads. However, this model only has three parameters which can be altered: the elastic frequency, the viscous damping and the yield level.

Objective and Scope

This paper develops a simplified model for building response which is a discrete formulation of the mathematical model suggested by Ramberg and Osgood (1) several years ago. Application of this type of mathematical model for determining the dynamic response of

---

*Dept. of Civil Engineering, Univ. of Southern Calif., Los Angeles, CA 90089

structures has been reported by Jennings (2,3) and by Kaldjian and
Fan (4). However, their studies were limited to single degree of
freedom systems and they did not have any recorded data with which to
compare their calculations. Using a discrete, numerical formulation
of the Ramberg Osgood (R-O) model allows it to be treated as a R-O
element in the finite element sense with direct extension to multi-
degree of freedom, chain-like structures. During the last fifteen
years, a considerable amount of structural response data has been
recorded in buildings during strong motion earthquakes. This response
data can be used to both calibrate and evaluate the analytical model
under consideration. In the following sections, the development of a
single degree of freedom, R-O model is described and its application
to modeling the response of an instrumented bearing wall building is
discussed.

Ramberg-Osgood Model

The R-O model consists of a continuous mathematical function
relating force and displacement. The section of the curve starting
at zero load and continuing in either direction until reversal of load
is called the skeleton or virgin curve. The equation for this branch
of the curve has the form

$$\frac{x}{x_y} = \frac{P}{P_y} + a\left(\frac{P}{P_y}\right)^r \tag{1}$$

where x is the displacement, P is the restoring force, x_y is a characteristic displacement which is usually taken as the yield displacement, P_y is a characteristic force which is usually taken as the yield
force and a and r are two additional constants which can be used to
adjust the shape of the skeleton curve. The effect of varying a and r
can be seen in Figure 1. As limiting cases, an elastic force-displacement is obtained when a=0 and an elasto-plastic force displacement is
obtained as r and a become large.

When a yielding system is alternatively loaded and unloaded, the
response curve departs from the skeleton curve and gives rise to what
it known as a hysteresis curve. The ascending and descending branches
of the hysteresis curve are described by the following equation:

$$\frac{x-x_i}{2x_y} = \frac{P-P_i}{2P_y} + a\left(\frac{P-P_i}{2P_y}\right)^r \tag{2}$$

where x_i and P_i are the coordinates of the most recent point at which
the direction of loading has been reversed. These relationships are
shown with the skeleton curve in Figure 2. The application of these
general relationships to random earthquake excitation has been discussed in detail by Jennings (3).

Solution Procedure

The equations of motion are solved by a step-by-step integration
procedure in which the incremental stiffness during a time step is
assumed to be constant. This incremental stiffness is the tangent

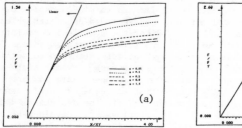
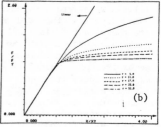

Figure 1.   Influence of Parameters on Skeleton Curve:
(a)   a varying and (b) r varying.

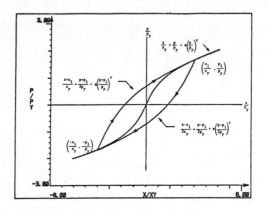

Figure 2.   Skeleton Curve and Hysteresis.

Figure 3.   View of North Hall Looking From Southwest
(Miller and Felszeghy).

stiffness to the R-O curve at the beginning of the time step. The
Newmark Beta Method with beta-1/4 (constant acceleration during the
time step) is used to determine the displacement during the time step.
With the displacement known, the restoring force at the end of the
time step can be calculated. The R-O force corresponding to this
displacement can also be determined by using a Newton-Raphson iterative procedure to solve the resulting r-th order polynomial. The
residual force, representing the difference between the numerical
solution and the R-O function is then applied as an incremental force
during the next time step.

Application

North Hall on the University of California at Santa Barbara
campus is a three story concrete and masonry, bearing wall structure
which was designed in 1960 and strengthened in 1975. It was instrumented with six strong motion recorders at the time of the 1978
Santa Barbara earthquake and produced nine excellent records of
structural response. Details of the structure and instrumentation
have been reported by Miller and Felszeghy (5). A picture of the
building is shown in Figure 3 and the location and orientation of the
strong motion accelerometers is shown in Figure 4. The peak recorded
ground acceleration was a relatively high .45g at the base but the
duration of strong motion shaking was only three seconds. Damage in
the building consisted primarily of diagonal cracking in the masonry
and concrete shear walls with no apparent yielding of the main reinforcement.

Following the earthquake a three dimensional elastic model of the
building was constructed using the ETABS program (6). Using this
model it was possible to obtain estimates of the mode shapes and
frequencies of the translational modes of vibration. These results
showed that the vibration of the building in the transverse direction
was predominantly the first mode which had a calculated period of .214
seconds. This agreed well with the value obtained from the ambient
vibration tests which was .20 seconds for the first mode. The fundamental mode shape in this direction was calculated to be <1,.55,.18>.

The first phase of this study considered the development of a
single-degree-of-freedom, Ramberg-Osgood model. The mode shape and
frequency of the first mode were used to determine the generalized
dynamic properties. The characteristic force, P_y, was taken to be
a multiple of the base shear specified in the Uniform Building Code
for a building of this type and was defined in terms of a new variable,
$M = P_y/P_{code}$. The amount of viscous damping was fixed at 2% of critical which is thought to be representative of a concrete structure
that is stressed below the yield level. The two constants, a and r
of the R-O curve were then varied along with M so as to give the best
fit of the acceleration recorded at the roof level. The best fit was
determined by calculating the root mean square deviation between the
calculated and recorded values.

The dynamic response of the following three, single-degree-of-
freedom oscillators are considered: (a) linear elastic, (b) elasto-
plastic and (c) Ramberg-Osgood. The recorded base acceleration

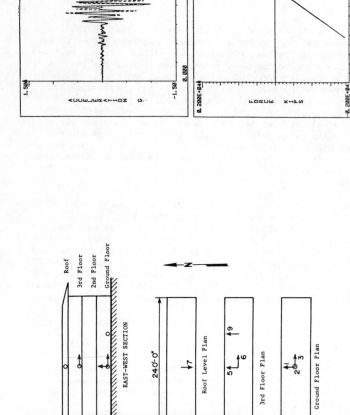

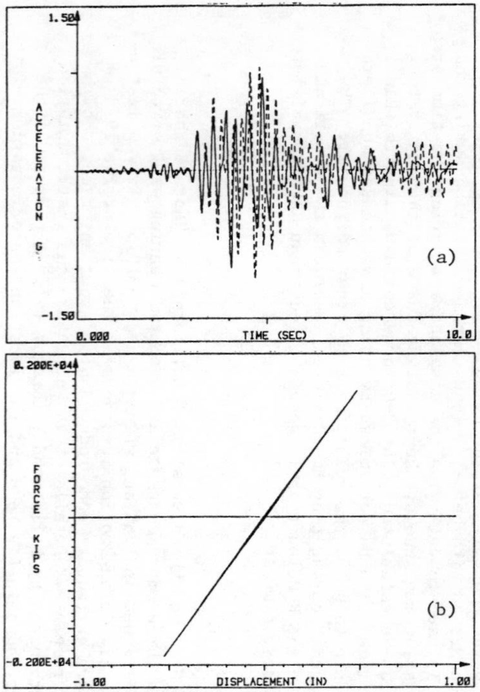

Figure 5. Linear SDOF System
           (a) Measured versus Calculated Roof
               Acceleration
           (b) Base Shear versus Displacement

Figure 4. Location of Strong Motion Instrumentation
           Within North Hall (Miller and Felszeghy).

(recording 1 in Fig.4) was used as input and the acceleration recorded at the roof (recording 7 in Fig. 4) was used as the reference output. Comparisons of the measured and calculated roof acceleration for the linear elastic model (a=0) are shown in Figure 5a. Here it can be seen that the comparison is quite good until a little over 3 seconds when the period of the recorded trace starts to lengthen indicating inelastic behavior. Beyond this point the comparison is not very good and any agreement with the recorded curve is pure chance. The force displacement curve for this condition is shown in Figure 5b as a straight line through the origin.

The acceleration time histories for the elasto-plastic model are shown in Fig. 6a. Here, the elasto-plastic force displacement has been approximated by assigning values of a=4.0 and r=100 to the R-O function. The agreement with the recorded values is better than the previous case showing good agreement until about four seconds into the record. However, the elasto-plastic model does not capture the peak accelerations and the variation between recorded and calculated values over the final 6 seconds of the time history is substantial. The resulting hysteresis curve for the elasto-plastic system is shown in Fig. 6b. Here it can be seen that there is a considerable drift in the displacement, representing a substantial permanent set.

Following these two initial studies, the parameters M, a and r were varied so as to determine the values which would result in a best fit to the recorded data in a RMS sense. These values were determined to be M=1.4, a=0.8 and r=1.6. The comparison with the recorded acceleration using these values is shown in Fig. 7a. The comparison is quite good in most areas of the time history. This is particularly true in the last six seconds which were not represented well by either of the other two models. Some of the peak acceleration values are not captured but others are represented quite well. This may be a consequence of representing the response by only a single degree of freedom and therefore neglecting any effect of higher modes. The hysteresis curve for the R-O model is shown in Fig. 7b. The behavior can be seen to be quite different from both the linear and elasto-plastic models. There is considerable energy dissipation represented by the area under the curve but there is little permanent set.

Conclusions

This paper has presented some of the initial results of a study which seeks to apply a Ramberg-Osgood force displacement relationship to the modeling of the nonlinear dynamic response of certain building systems. This initial study has been directed towards single degree of freedom systems. Calculated results have been compared to those recorded in a single building during a strong motion earthquake. Studies are currently underway to apply these same techniques to other buildings having different framing systems and to extend the application to multiple degree of freedom systems. Based on these limited results, the following conclusions are made:

(1) The Ramberg-Osgood model results in a better estimation of overall building response than does either the linear or elasto-plastic models. However, it should be noted that

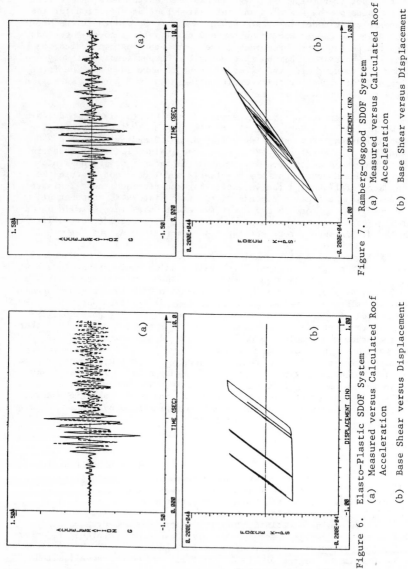

Figure 6. Elasto-Plastic SDOF System
(a) Measured versus Calculated Roof Acceleration
(b) Base Shear versus Displacement

Figure 7. Ramberg-Osgood SDOF System
(a) Measured versus Calculated Roof Acceleration
(b) Base Shear versus Displacement

at present this conclusion is limited to buildings which are not strongly nonlinear.

(2) Use of an elasto-plastic hysteresis results in a substantial drift in the displacement due to plastic deformation. This aspect of the elasto-plastic model needs to be investigated.

(3) The formulation of a discrete R-O model which can be used as an element shows considerable promise for application to multistory buildings.

Acknowledgement

This study was funded in part by a grant from the National Science Foundation.

References

1. Ramberg, W., and Osgood, W. R., "Description of Stress Strain Curves by Three Parameters," NACA, TN902, July, 1943.

2. Jennings, P. C., "Periodic Response of a General Yielding Structure," Journal of the Engineering Mechanics Division, ASCE, Vol. 90, No. EM2, April, 1964.

3. Jennings, P. C., "Earthquake Response of a Yielding Structure," Journal of the Engineering Mechanics Division, ASCE, Vol. 91, No. EM4, August, 1965.

4. Kaldjian, M. J., and Fan, W. R. R., "Earthquake Response of a Ramberg-Osgood Structure," Journal of the Structural Division, ASCE, Vol. 94, No. ST10, October, 1968.

5. Miller, R. K. and Felszeghy, ., "Engineering Features of the Santa Barbara Earthquake of August 13, 1978," University of California, Santa Barbara, UCSB-ME-78-2, Dec., 1978.

6. Wilson, E. L., Hollings, J. P. and Dovey, H. H., "ETABS, Three Dimensional Analysis of Building Systems," EERC 75-13, University of California, Berkeley, April, 1975.

DEVELOPMENT OF A NONLINEAR MODEL FOR UCSB NORTH HALL
RESPONSE UNDER EARTHQUAKE LOADS

by

J. C. Anderson[*] S. F. Masri [*] R. K. Miller [*] A. F. Saud [*]

and

T. K. Caugehy[**]

ABSTRACT

A simple and efficient nonparametric identification technique is applied to a multistory building under seismic excitation. Using the limited ammount of available excitation and response measurements, a reduced-order nonlinear nonparametric model is constructed. It is shown that the building underwent significant nonlinear nonconservative deformations, particularly in the zones involving the interstory motions between the ground and the first story, and between the second and third story.

1. INTRODUCTION

North Hall is a three-story office building constructed in 1960 and located on the campus of the University of California, Santa Barbara (UCSB). Following its rehabilitation in 1975, the building was fully instrumented by strong motion seismic recorders. Thus, the occurance of the 13 August 1978 Santa Barbara earthquake generated a large data base of strong motion measurements from the installed instruments (see Fig. 1).

This paper is concerned with the development of an appropriate not-necessarily linear building model by the use of the available response measurements of the building under earthquake ground motion. Particular attention is devoted to determining the nature, extent, and location of nonlinear building characteristics.

---

[*]Civil Engineering Department, University of Southern California, Los Angeles, CA 90089-0242

[**]Division of Engineering, California Institute of Technology, Pasadena, CA 91106

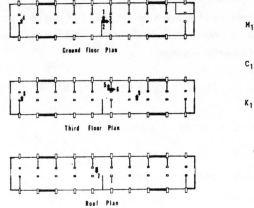

Fig. 1. Location of strong motion insturmentation (Taylor).

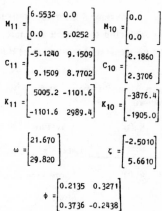

Fig. 2. Identified system matrixes and modal parameters.

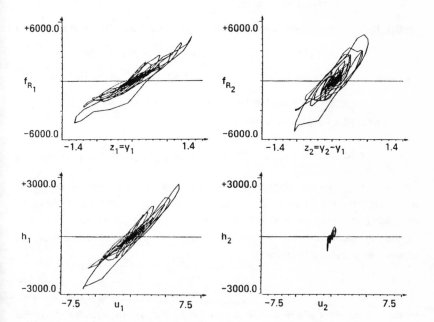

Fig. 3. Phase-plane plots of the system restoring forces.

2. FORMULATION

Consider a discrete multi-degree-of-freedom (MDOF) nonlinear system which is subjected to prescribed interface motion. The motion of this system is assumed to be generated by the set of differential equations

$$M_{11}\, \ddot{x}_1 + C_{11}\, \dot{x}_1 + K_{11}\, x_1 + M_{10}\, \ddot{x}_0 + C_{10}\, \dot{x}_0 + K_{10}\, x_0 + f_N(t) = 0 , \qquad (1)$$

where

$x_1(t)$ = System absolute displacement vector of order n_1;

           $= (x_{1_1},\, x_{1_2},\, ...,\, x_{1_{n_1}})^T$,

$x_0(t)$ = Interface displacement vector of order n_0;

           $= (x_{0_1},\, x_{0_2},\, ...,\, x_{0_{n_0}})^T$,

$M_{11},\, C_{11},\, K_{11}$ = Square Matrixes, each of order $(n_1 x n_1)$ that characterize the inertia, damping and stiffness forces associated with the system degrees of freedom,

$C_{10},\, K_{10}$ = Matrixes, each of order $(n_1 x n_0)$ that characterize the damping and stiffness forces associated with interface motions,

$f_N(t)$ = is an n_1 column vector of nonlinear nonconservative forces involving $x_1(t)$ as well as $x_0(t)$.

Let Eq. (1) be expressed in the form

$$M_{11}^{-1}C_{11}\, \dot{x}_1 + M_{11}^{-1}K_{11}\, x_1 + M_{11}^{-1}C_{10}\, \dot{x}_0 + M_{11}^{-1}K_{10}\, x_0 + M_{11}^{-1}f_N(t) = -\, I\, \ddot{x}_1 . \quad (2)$$

Consider a linearized version of Eq. (2) and assume that experimental measurements of $\ddot{x}_1(t)$ and $\ddot{x}_0(t)$ are available for a given time span. Then, by integrating the accelerations with respect to time and by taking a sufficient number of measurements of the system excitation and response, the following set of algebraic equations is obtained:

$$R\, q = b , \qquad (3)$$

where R is a matrix whose entries correspond to the system response at various observation times, q is a vector containing all the unknown influence coefficients of the system, and b is a vector containing the excitation measurements.

Using least-squares approximation methods, the parameter vector $\underset{\sim}{a}$ can be computed from

$$\underset{\sim}{a} = R^{\dagger} \underset{\sim}{b} \tag{4}$$

where $R^{\dagger}$ is the pseudoinverse of R.

With the parameters determined by Eq. (4), the residual nonlinear forces can be found from Eq. (2). Then making use of the eigenvectors associasted with the linearized version of Eq. (2), the generalized restoring force $\underset{\sim}{h}$ corresponding to the nonlinear terms in Eq. (2) can be expressed as

$$\underset{\sim}{h}(\underset{\sim}{u},\underset{\sim}{\dot{u}}) = \Phi^T M_{11}^{-1} \underset{\sim}{f}_N(\underset{\sim}{x}_1, \underset{\sim}{\dot{x}}_1, \underset{\sim}{x}_0, \underset{\sim}{\dot{x}}_0) , \tag{5}$$

where Φ is the eigenvector matrix and $\underset{\sim}{u}$ is the vector of generalized coordinates given by

$$\underset{\sim}{y} = \Phi \underset{\sim}{u} . \tag{6}$$

In the case of nonlinear dynamic systems commonly encountered in the structural field, a judicious assumption is that each component of $\underset{\sim}{h}$ can be expressed in terms of a series of the form:

$$h_i(\underset{\sim}{u}, \underset{\sim}{\dot{u}}) = \sum_{j=1}^{J_{max_i}} \hat{h}_i^{(j)} \left(v_{1_i}^{(j)}, v_{2_i}^{(j)} \right). \tag{7}$$

The approximation indicated in Eq. (7) is that each component h_i of the nonlinear generalized restoring force $\underset{\sim}{h}$ can be adequately estimated by a collection of terms $\hat{h}_i^{(j)}$ each one of which involves a pair of generalized coordinates (displacements and/or velocities). The particular choice of combinations and permutations of u_k and u_ℓ and the number of terms J_{max_i} needed for a given h_i depend on the nature and extent of the nonlinearity of the system and its effects on the specific "mode" i. Note that the formulation in Eq. (7) allows for "modal" interaction between all modal displacements and velocities, taken two at a time.

The individual terms appearing in the series expansion of Eq. (7) may be evaluated by using the least-squares approach to determine the optimum fit for the time history of each h_i. Thus, $\hat{h}_i^{(1)}$ may be expressed as a double series involving a suitable choice of basis functions,

$$\hat{h}_i^{(1)} \left(v_{1_i}^{(1)}, v_{2_i}^{(1)} \right) = \sum_k \sum_\ell {}^{(1)}C_{k\ell}^{(i)} T_k \left(v_{1_i}^{(1)} \right) T_\ell \left(v_{2_i}^{(1)} \right) . \tag{8}$$

Eq. (7) is obtained by fitting the residual error with a similar double series involving other pairs of generalized coordinates that have significant interaction with "mode" i.

Using two-dimensional orthogonal polynomials to estimate each $h_i(\underline{u},\underline{\dot{u}})$ by a series of approximating functions $\hat{h}_i^{(j)}$ of the form indicated in Eq. (8), then the numerical value of the $C_{k\ell}$ coefficients can be determined by invoking the applicable orthogonality conditions for the chosen polynomials. While there is a wide choice of suitable basis functions for least-squares application, the orthogonal nature of the Chebyshev polynomials and their "equal-ripple" characteristics make them convenient to use in the present work.

3. APPLICATION

3.1 Processing of Measured Data

Analog measurements of the earthquake ground motion and the building response were processed using modern time series analysis methods. The digitized measurements consisted of the time histories of the acceleration, velocity and displacement at different locations and directions (Porter et al, 1979).

For the purpose of this study, the building motion in the North – South direction was used. The system input was taken as the recorded ground motion corresponding to channel 1 (henceforth designated by CH_1) measured at the center of the building ground floor. The measured building response at the center of the third floor and the roof (data channels CH_5 and CH_7, respectively) were taken as the dynamic system response. Notice that the motion of the second floor was not measured during the earthquake.

3.2 Identification of Equivalent Linear Systems

Using the mass matrix determined by Taylor (1981), and following the procedure outlined in Section 2, the system matrixes appearing in Eq. (2) were identified and are given in Fig. 2. The corresponding system classical mode shapes and nonclassical modal damping parameters are also given in Fig. 2.

3.3 Identification of Nonlinear Model

Proceeding with the identification steps represented by Eqs. (5)-(6), results in the phase-plane plots of the modal state variables shown in Fig. 3. A three-dimensional representation of the surfaces associated with the measured nonlinear

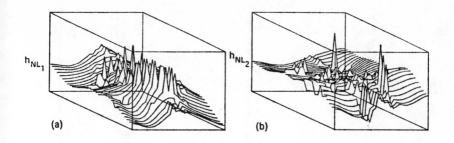

(a) (b)

$$\hat{h}_i^{(1)}\left(v_{1_i}^{(1)}, v_{2_i}^{(1)}\right) = \sum_k \sum_\ell {}^{(1)}C_{k\ell}^{(i)} T_k\left(v_{1_i}^{(1)}\right) T_\ell\left(v_{2_i}^{(1)}\right)$$

$\hat{h}_{NL_1}$

$C_{k\ell}$	0	1	2	3
0	-3.817D+01	1.367D+00	8.637D-04	-1.387D-04
1	2.045D+01	-3.588D-01	-2.875D-02	-2.671D-05
2	3.091D+01	-4.582D-01	-9.100D-03	1.132D-04
3	-1.204D+01	1.128D-01	5.106D-03	-2.140D-05
4	-1.151D+00	1.773D-02	2.647D-04	-4.139D-06
5	4.898D-01	-3.332D-03	-1.535D-04	7.388D-07

(c)

$\hat{h}_{NL_2}$

$C_{k\ell}$	0	1	2	3
0	2.649D+01	-1.421D+00	-3.238D+00	5.663D-01
1	-5.667D+02	3.685D+01	1.355D+01	-2.479D+00
2	1.162D+03	-1.605D+02	9.975D+01	-7.991D-01
3	3.587D+03	-3.994D+02	-3.297D+02	1.114D+01
4	-1.198D+04	2.105D+03	-8.949D+02	-4.383D+01
5	7.648D+03	-1.352D+03	2.079D+03	6.009D+01

(d)

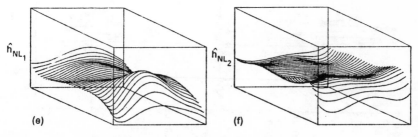

(e) (f)

Fig. 4. Measured and identified nonlinear forces.

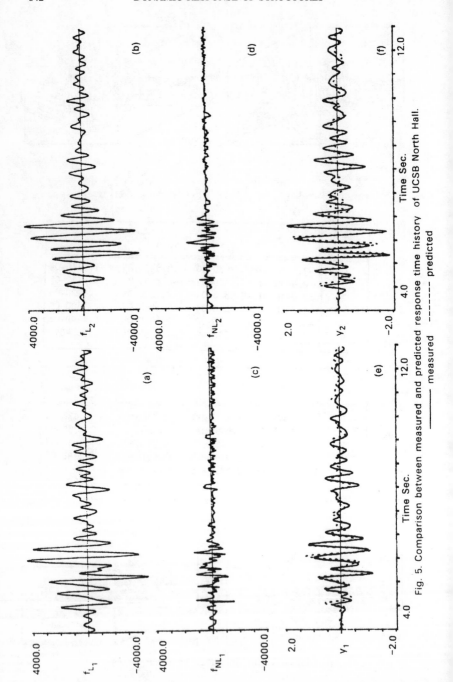

Fig. 5. Comparison between measured and predicted response time history of UCSB North Hall.
———— measured   - - - - - predicted

h_1 and h_2 are shown in Figs. 4(a) and 4(b) respectively.

Following the procedure in Section 2, the identification results shown in Fig. 4(c) and (d) are obtained. Note from the tabulated power series coefficients in Fig. 4(c) and (d) that the contribution of the nonlinear terms (i.e., those other than A_{01} and A_{10}) is not negligible. The corresponding identified (approximating) nonlinear forces $\hat{h}_1$ and $\hat{h}_2$ are shown in Fig. 4(e) and (f).

3.4 Response Estimation

By incrementally solving the reduced-order system equations and using the transformation in Eq. (6), approximate response time histories of the building's two degrees of freedom were computed (estimated) and are compared to the available measurements in Fig. 5.

Curves (a) and (b) in Fig. (5) represent the contribution of the linear terms on the left hand side of Eq. (2) that are associated with DOF number 1 and 2 (second story and roof), respectively. Similarly, curves (c) and (d) in Fig. (5) represent the contribution of the residual nonlinear forces. The plots in Figs. 5(e) and (f) , covering the strong motion part of the earthquake, show a comparison between the measured and predicted relative displacement of North Hall with respect to the first floor. Similar good agreement, both in amplitude as well as phase, is obtained for the building absolute acceleration.

The use of additional translational and rotational DOFs are expected to yield even better correlation between measured and predicted response based on simplified reduced-order nonlinear models.

ACKNOWLEDGEMENT

This study was supported in part by the National Science Foundation.

REFERENCES

1. Taylor, D. G., (1981), "Analysis of North Hall Response to 13 August 1978 Santa Barbara Earthquake Using Finite Element Structural Program (ETABS)", M.S. Thesis, Mech. Engin. Dept., University of California, Santa Barbara.

2. Porter, L. D. , Ragsdale, J. T., and McJunkin, J. D., (1979), "Processed Data From the Partial Strong-Motion Records of the Santa Barbara Earthquake of 13 August 1978, Preliminary Results," Report No. 23, Office of Strong Motion Studies, California Division of Mines and Geology, Sacramento, CA.

Large Scale Testing Facility for the Evaluation
of Structural Systems Under Critical Loads

Frieder Seible[1], Gilbert A. Hegemier[2], and Hidenori Murakami[3]

ABSTRACT

The total collapse of large buildings, as experienced recently
again during the Mexican Earthquake of September 13, 1985, emphasized
the need for a better understanding of the overall behavior of
complete structural systems, including accumulative damage and
stiffness degradation, successive three-dimensional force redistribu-
tions, and final collapse mechanisms under critical loads.

The Charles Lee Powell Structural Systems Laboratory at the
University of California, San Diego provides, with a 50 ft. (15m) high
reaction strong wall and a 120 x 50 ft. (37 x 15m) box girder test
floor, the necessary environment to test up to five story full-scale
building systems under simulated seismic loads. This new facility
allows the in-depth investigation of structural prototype behavior and
with it the necessary verification for complex nonlinear structural
computer codes.

Introduction

The complete or partial collapse of buildings, housing essential
facilities such as hospitals, schools and communication installations,
during the recent Mexican Earthquake has dramatically emphasized the
need for a better understanding of entire structural systems and their
behavior under critical loads, i.e., loads which are beyond the
working stress range and cause significant structural damage and/or
failure of individual members, subassemblages or complete structures.
While computer models play an essential role in assessing structural
behavior, the highly nonlinear constitutive laws and information on
failure modes have to be empirically obtained by experimental testing.

With component tests, subassemblage tests and scaled down model
tests not providing reliable information on time redistribution
capabilities of successively damaged, highly redundant structural
systems, the true global failure behavior can only be studied on
full-scale or prototype structures. On this basis, a recent EERI

1. Assistant Professor of Structural Engineering, Department of AMES,
   University of California, San Diego, La Jolla, California 92038.

2. Professor of Applied Mechanics, Department of AMES, University of
   California, San Diego, La Jolla, California 92038.

3. Assistant Professor of Applied Mechanics, Department of AMES,
   University of California, San Diego, La Jolla, California 92038.

workshop (EERI Report No. 84-01, 1984, EERI Report No. 84-02, 1984) on Experimental Research recommends that for Earthquake Engineering Research in the U.S., "... the understanding of structural behavior under earthquake excitations would be greatly enhanced by the construction of two or three facilities whose design is centered around a reaction wall system", and "... one reaction wall facility with the capacity to test structures with heights of at least 40 ft (12m) is needed."

To complement existing United States reaction wall facilities at the university of Texas, Austin, and at the University of Michigan, Ann Arbor, (both walls are 20 ft. (6m) high) and to keep U.S. earthquake engineering research competitive with large Japanese reaction wall facilities (Kamimura, K., et al., 1984), the Charles Lee Powell Structural Systems Laboratory at the University of California, San Diego, was built with a 50 ft (15m) reaction wall for the full-scale testing of five story buildings under simulated seismic loads. In the following the design and construction of the new facility are explained and test load capacities and dynamic characteristics of the reaction strong wall and box girder test floor are presented.

Space and Technical Requirements

Requirements for the new facility were dictated primarily by the U.S.-Japan Coordinated Program for Masonry Building Research calls in the NSF sponsored U.S. research plan (Technical Coordinating Committee for Masonry Research, 1984), for the full-scale test of a five story reinforced concrete masonry research building under simulated seismic loads. This full-scale test will provide the final verification of a coordinated systematic research effort in which different aspects of masonry behavior and design are studied by various research teams throughout the U.S. Accordingly, results from theoretical and experimental research conducted on elements, components and sub-assemblages of masonry structures will be systematically combined to predict the prototype behavior. This systems approach is then ultimately verified by the aforementioned full-scale test.

The test of such a five story prototype requires a 50 ft (15m) strong wall with sufficient capacity to react the horizontal forces necessary to simulate critical seismic loads, and a test floor which can anchor and accommodate various geometric arrangements of full-scale-test test structures. The length of the structural test floor was governed by requirements to test bridge components and bridge systems over 100 ft (30m) in length.

The complexity and variety of experiments to be performed in a large-scale structural systems laboratory poses, in addition to sophisticated high speed data acquisition and computer equipment, special mechanical prerequisites (Okamoto, S., et al., 1983). A closed loop multi-actuator environment is required to accurately simulate critical loads of complete structural systems. With a projected direct pumping capacity of 3 x 140 gpm (3 x 500 C/min) and, if necessary, gas accumulators for additional short period power boosts, the laboratory was conceived to operate in a static, quasi-static, cyclic, pseudo-dynamic or real time dynamic testing mode.

The new UCSD Structural Systems Laboratory was designed and built to meet the above requirements.

Description of the New Facility

The new UCSD Structural Systems Laboratory consists of a 4-cell box girder test floor, a reaction strong wall, a box type outside shell, and an annex containing a data acquisition/computer control room, a conference room, offices and a utility section. Two elevations of the overall outside appearance of the Laboratory Building are shown in Fig. 1. Natural lighting for the test area is provided by a continuous band of 10 ft (3m) high aluminum framed storefront windows along the top of the east and west face of the building and over the entire height of the north wall.

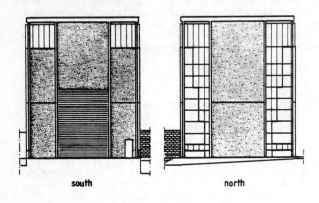

south   north

Figure 1. UCSD Structural Systems Laboratory, Elevations.

The outside building facade consists of precast concrete panels supported by a braced three-dimensional structural steel frame. Reinforced concrete masonry is used for the single story annex. Truck access to the building is provided by a 24 x 30 ft (7 x 9m) roll up door, and a loading dock is located at the west side of the building, Fig. 2. Other essential support facilities for a structures laboratory of this size such as work shops for wood and metals, machine shops, electronics shop and a materials testing laboratory will be available in a new engineering building which is currently under construction next to the new Structural Systems Facility.

The entire laboratory area is covered with a 10 ton overhead double girder crane (see Fig. 3) with top riding hoist and high spotting accuracy. The vertical clearance to the roof girders is 57 ft (17.4m) over the entire test area. All structural concrete for the new laboratory has a design strength of f'_c = 5000 psi (35 N/mm^2).

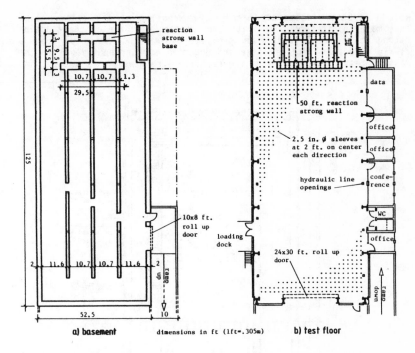

Figure 2. UCSD Structural Systems Laboratory, Plan Views

Structural Test Floor

The structural test floor is 125 ft (38m) long, 52.5 ft (16m) wide and 13 ft. (4m) deep. The top slab or test floor slab of the 4-cell box girder section is 3 ft (0.9m) thick and has 2.5 in. (∅64mm) tie down sleeves with cast iron screw caps on a 2 ft. (0.6m) module in the longitudinal and transverse directions. Plan views of the basement and structural test floor are given in Fig. 2a and b respectively, while a transverse section is shown in Fig. 3. Fork lift access to the basement is provided by a 12 1/2 percent ramp and a 10 x 8 ft (3 x 2.4m) roll up door together with an 8 ft (2.4m) minimum vertical clearance inside the 4-box girder cells. The hydraulic post-tensioning of test structures and load frames with tie down rods to the test floor will be carried out from the basement.

The base of the reaction strong wall is monolithically connected to the box girder test floor as shown in Fig. 2a and standard 3 x 4 ft (0.9 x 1.2m) manholes provide access for structural inspection.

The main reinforcedment of the box girder test floor in the longitudinal direction consists of rebars in 3 layers of No. 11 at 8 in. (∅35mm at 20 cm) at the top of the top slab and 2 layers of

No. 11 at 6 in. (ø35mm at 15 cm) at the bottom of the bottom slab. This main reinforcement is continuous through the base of the reaction strong wall, see Fig. 4b, and provides a moment capacity of 135,000 ft-kip (185 MN-m) for positive and negative moments equivalent to the maximum test load capacity of the reaction strong wall.

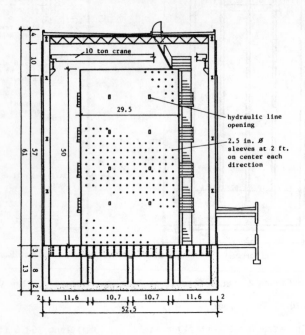

Figure 3.   UCSD Structural Systems Laboratory, Cross-Section.

Reaction Strong Wall

The 50 ft (15m) high and 29.5 ft (9m) wide reaction strong wall is positioned at the north end of the laboratory and has a two cell box girder cross-section. The front and the back wall are both 3 ft. (0.9m) thick and provide horizontal 2.5 in. ø (64 mm) sleeves on a 2 ft. (0.6m) module for the attachment of actuators or load frames. These tie through holes provide an option for hosizontal post-tensioning of the strong wall if the need for increased shear capacity rises. The detailed layout and dimensions of the reaction strong wall are shown in, Figs. 2 and 3. Metal catwalks and 3 x 4 ft (0.9 x 1.2m) manholes at every 10 ft (3m) level provide access to the back and the inside of the reaction wall.

The main reinforcement of the reaction wall and a uniform state of prestress of 850 psi (6 MN/m^2) is provided by a post-tensioning system consisting of 68 tendons with 12-1/2 in. ø 7 wire strands (grade 270). The additional nominal vertical reinforcement consists of rebars No. 6 at 24 in. (19mm ⌀ every 60 cm). The horizontal shear reinforcement in the reaction wall webs and the horizontal flexural reinforcement in the front and back wall consist of rebars No. 8 at 12 in. (25mm ⌀ every 30 cm).

The post-tensioning system utilizes fixed end loop anchors which are particularly advantageous for the vertical geometry since they can be permanently installed during the construction of the strong wall base. They are subsequently coupled to corrugated ducts for each of the reaction wall construction segments, five 10 ft (3m) lifts, and the corrosion and damage sensitive prestressing cables do not have to be installed until the actual time of post-tensioning and the subsequent grouting operation. Figure 4 shows photographs of the placement of the fixed end loop anchors (a) and the completed reaction strong wall (b).

(a)   (b)

Figure 4. Construction of Reaction Strong Wall.

The entire post-tensioning operation is conducted from the top of the reaction wall. Bond between the tendons and the reaction wall is established by subsequent grouting of the ducts with cement mortar. The ducts are grouted through vents at the bottom of the loop anchorages and continuous grouting can be assured by an overflow of grout at the high point, i.e., the air vent at the top stressing anchorage.

Test load capacities and dynamic characteristics for the reaction strong wall are discussed in the following section.

Reaction Wall and Test Floor Characteristics

The structural testing to be conducted in this new facility range from full-scale 5 story buildings loaded horizontally with up to 15 hydraulic actuators, Fig. 5., to full or large-scale tests of bridge systems and bridge components.

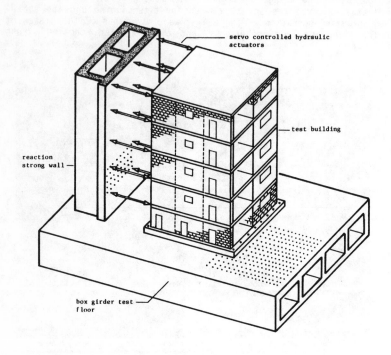

Figure 5. Schematic Sketch of 5 Story Test Building

This wide range of special load and geometric arrangements, as well as the nature of the tests, requires special design considerations over and above standard building code requirements.

While load and strength requirements are still significant design factors, particular consideration has to be given to stiffness requirements to ensure satisfactory testing boundary conditions. This requires that deformations be maintained at a minimum and consequently cracks, which are direct cause for structural stiffness deterioration, have to be eliminated or carefully controlled.

The stiffness objective was achieved by specified structural dimension not necessarily resulting from applied loads, by a finely spaced reinforcement mesh with 12 in. (30 cm) maximum spacing and by eliminating tensile stresses in the strong wall with full prestressing at the base under general test loads.

This full prestressing of the strong wall and with it the elimination of cracks under general test loads also allows for a simplified linear elastic analysis of the dynamic response characteristics of the reaction wall itself.

The first three natural frequencies of the reaction strong wall were determined, via a three-dimensional finite element analysis, to be 11.8 Hz. (flexure about the weak axis), 17.0 Hz. (flexure about the strong axis), and 25.0 Hz. (torsion). The appropriate modeshapes are depicted in Fig. 6a, b, and c, respectively.

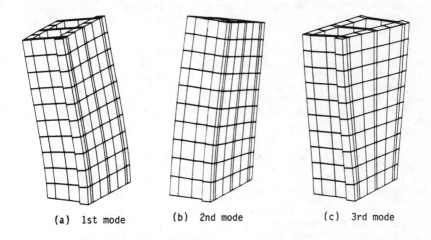

(a) 1st mode   (b) 2nd mode   (c) 3rd mode

Figure 6. Dynamic Deformation Characteristics of Reaction Strong Wall

With these dynamic strong wall characteristics, real time dynamic testing on buildings and structures with fundamental natural frequencies of up to 2 Hz. are deemed feasible.

The specific nature of the applied loads, generated with computer controlled hydraulic actuators and very accurately applied at predetermined locations, was the basis for the special safety concept. Similar to the overload concept on bridge structures where heavier loads are only allowed with a special permit, required load capacities were separated into Test Loads (TL) and Special Permit Test Loads (SPTL). Test Loads (TL) are loads and load levels which can be applied without any additional verifications as static cyclic or dynamic loads. The deterministic nature of these loads and their point of application justifies a load factor of 1.4 for the ultimate strength design.

Special permit Test Loads (SPTL) are load levels for which a reduced ultimate strength design load factor of 1.2 was applied, reflecting the fact that these high load levels will be reached only very infrequently and for very specialized tests. Since at these high load levels the type of loading, the geometry and possible mode of failure of the test structure, and redundancy in the load control and overload shut-off system became very important, the "Special Permit" term was affixed to these test loads, requiring a special structural analysis documentation of the entire test and the anticipated structural response of the strong wall and the test floor, prior to any testing.

Test Load (TL) and Special Permit Test Loads (SPTL) levels are shown in Fig. 7 for the reaction strong wall and the structural box girder test floor. The special Permit Test Loads can be looked upon as available test capacities for the structural systems testing facility.

A continuous flexural capacity of 135,000 ft-kip (185 MN-m) and a base shear capacity of 4500 kips (20 MN) pose the ultimate strength limitations for the design of any test building. A calculated stiffness value of 5400 kip/in. (950 kN/mm) for a single horizontal degree of freedom at the top of the reaction wall translates to a 0.5 in. (13 mm) theoretical horizontal displacement when the maximum horizontal load of 3 x 900 kips (3 x 4 NM) is applied at the top of the 50 ft (15m) reaction strong wall.

Point loads in tension or compression of TL = 300 kips (1.3 MN) can be applied at any location of the strong wall and TL = 150 kips (0.7 MN) can be applied at any location of the box girder test floor with a minimum spacing of 10 ft (3m), see Fig. 7.

Strategically located tie-down holes in the test floor around the perimeter of the strong wall for vertical post-tensioning, see Fig. 2b, in conjunction with additional horizontal post-tensioning of the strong wall through the horizontal sleeves can accommodate future increased requirements in strong wall capacity.

The box girder test floor with the 3 ft. (0.9m) top slab and the tie-down holes allows for the application of vertical forces to the test structure as well as for the erection of a second axis reaction wall, to accommodate three directional loading. A temporary weak axis reaction wall could be made up of prefabricated box type sections stacked on top of each other and post-tensioned together and to the test floor with temporary post-tensioning tendons or threaded rods.

With the floor and wall capacities shown in Fig. 7 and the dimensions depicted in Figs. 2 and 3, the UCSD Structural Systems Laboratory has the necessary prerequisites to become one of the leading structural research facilities world-wide.

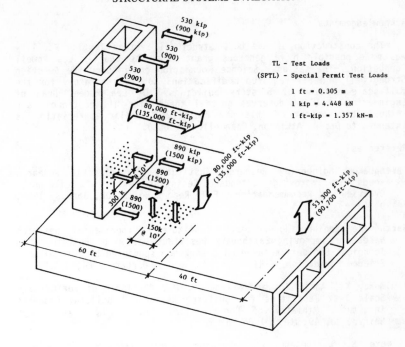

Figure 7.   Test Floor and Reaction Wall Test Load Capacities

Conclusions

The evaluation of prototype post-working stress behavior of complete structural systems and with it the validation of complex nonlinear computer models requires full-scale experimental testing. The UCSD Charles Lee Powell Structural Systems Laboratory is the only facility in the U.S. where 5 story full-scale building systems can be tested under simulated seismic loads. The laboratory dimensions and test floor and reaction strong wall capacities rank the new facility world-wide among the top structural research facilities.

The full-scale testing of structural systems will help to integrate results from experimental component and scaled down model testing with new analytical models to develop reliable structural design criteria. In addition, full-scale structural systems testing can provide invaluable information on the effectiveness of repair measures of severely damaged buildings and the effects of seismic retrofitting to existing structures.

Acknowledgements

The construction of the UCSD Structural Systems Testing Facility was made possible by a generous grant from the Charles Lee Powell Foundation. The National Science Foundation contributed the Reaction Strong Wall and all required modifications to the laboratory for the full-scale testing of 5 story buildings. Dr Lea Rudee, Dean of Engineering at UCSD, deserves special thanks for his continuous and enthusiastic support of this new facility. Similar appreciation is extended to Dr. R. Atkinson, Chancellor of UCSD.

References

Earthquake Engineering Research Institute, "Experimental Research Needs for Improving Earthquake Resistant Design of Buildings," Overview and Recommendations, EERI Report No. 84-02, January 1984, p. 43.

Earthquake Engineering Research Institute, "Experimental Research Needs for Improving Earthquake Resistant Design of Buildings," A Technical Evaluation based on a Workshop sponsored by the National Science Foundation, EERI Report NO. 84-01, January 1984, p. 154.

Kamimura, K., Aoki, Y., and Nakashima, M., "Register of World Large Scale Test Facilities for Building Research," Building Research Institute, Ministry of Construction, Japan. Kenchiku Kenkyu Shiryo, No. 49, March 1984, p. 51.

Okamoto, S., Kaminoseno, T., Nakashina, M., and Kato, H., "Techniques for Large Scale Testing at BRI Large Scale Structure Testing Laboratory," Building Research Institute, Ministry of Construction, Japan, BRI Research Paper No. 101, May 1983, p. 38.

Technical Coordinating Committee for Masonry Research, "U.S.-Japan Coordinated Program for Masonry Building Research, U.S. Research Plan," March 1984, p. 37.

A Recommended Federal Action Plan for
Reducing Earthquake Hazards of Existing Buildings

Christopher Rojahn*
Member, ASCE

ABSTRACT

In September 1985 a Five-Year Plan for reducing earthquake hazards of existing buildings was submitted to the Federal Emergency Management Agency. This plan, prepared by the Applied Technology Council, the Building Seismic Safety Council, and the Earthquake Engineering Research Institute, identifies recommended strategies, techniques, research needs, and resources required to reduce the potential for loss of life and life-threatening injuries in earthquakes. Twenty-four principal tasks associated with the following key elements are recommended: (I) Technical and Engineering Requirements, (II) Public Policy, Legal, and Financial Strategies, (III) Special Requirements of Historic Buildings, (IV) Multihazard Assessment, and (V) Information Transfer and Dissemination. The total estimated cost in federal funds for the five-year program is $40,400,000. This amount does not include funds and in-kind services expected to be provided by professional organizations, practitioners, researchers, local and state governments, and others.

INTRODUCTION

Buildings in the United States have been constructed in a variety of environments. Many (possibly many millions) located in areas of seismic risk are not adequately earthquake resistant. Thus, even moderate earthquakes threaten to cause substantial life loss and injury.

On a nationwide basis, the problem of reducing the earthquake hazards posed by existing buildings is the most pervasive--and perhaps the most complex and controversial--issue to be found in contemporary earthquake hazard reduction. Yet, more than any other hazard abatement strategy, a coordinated program for confronting the hazardous buildings problem has the greatest potential for saving lives and limiting serious injuries. The potential beneficiaries of this effort include not only the more than seventy million people at major or moderate risk in the U. S. (NSF-USGS, 1976), but also the U. S. population as a whole, since every business, industry, and sector of the economy will be affected by a major U. S. earthquake.

This paper provides an overview of a recently completed recommended action plan for reducing earthquake hazards of existing buildings (ABE, 1985b). The recommended action plan (Action Plan), written with the support of the Federal Emergency Management Agency (FEMA), was developed over a one-year period by ABE, a joint venture of the Applied Technology Council,

---
*Executive Director, Applied Technology Council, 3 Twin Dolphin Drive, Suite 275, Redwood City, California 94065.

Building Seismic Safety Council, and the Earthquake Engineering Research Institute. The intent of the Action Plan is to fulfill a task identified in the National Earthquake Hazards Reduction Program (NEHRP) Five Year Plan for Fiscal Years 1985-89 that called for the creation of a coherent, comprehensive, and detailed plan for mobilizing federal resources and influence in the support of resolving the existing earthquake hazardous buildings problem.

PLAN DEVELOPMENT PROCESS

The plan development process involved a large number of organizations and specialists in earthquake engineering, public policy, and other diciplines concerned with earthquake hazardous buildings. Nine issue papers covering the major subjects related to reducing earthquake hazards of existing buildings provided the basis for a three-day workshop in Tempe, Arizona, which was attended by approximately fifty specialists from throughout the United States (ABE, 1985a). Participants met in working groups and focused on identifying critical Action Plan tasks, estimated costs, recommended funding levels, and recommended action agents.

A draft Action Plan was developed following the workshop and submitted for review and comment to the project's Technical Advisory Committee, to all Workshop Participants, and to more than seventy organizations concerned with reducing earthquake hazards of existing buildings. These individuals and organizations were included in the Action Plan development process to insure nationwide consideration by the various groups interested in and affected by initiatives and activities related to resolving the earthquake hazardous buildings problem.

The draft Action Plan went through several stages of review and revision by the Technical Advisory Committee, and it was submitted in final form to FEMA in September 1985.

PURPOSE

The purpose of the Action Plan is to reduce the potential for loss of life and life-threatening injuries in earthquakes. To achieve this goal, the Action Plan identifies strategies, techniques, research needs, and resources required to initiate and sustain hazard abatement programs for existing seismically hazardous nonfederal buildings nationwide. The Action Plan is specifically focused on nonfederal buildings, as separate coordinated actions are being taken to evaluate and correct earthquake deficiences in buildings by federal agencies that own, lease, or are otherwise involved in the construction, use, and management of buildings. The knowledge gained and the information collected as a result of the implementation of the Action Plan, however, are expected to benefit federal agencies.

The Action Plan recommends a variety of initiatives and activities, both private and governmental, and involves two main thrusts: the need for research on a variety of subjects, and the need to accelerate the application of current knowledge, practice, and techniques. The Plan is intended to provide a focus for the activities of diverse segments of government and the private sector and to highlight the major aspects and scope of the earthquake hazardous buildings problem nationwide. It also provides a framework for coordinating actions and suggests a pattern for sharing the financing of programs by the federal agencies responsible for implementing the NEHRP.

The major goal of the Action Plan is to save lives. Although protection of property is clearly an importnat matter, and an ancillary benefit in most cases where life safety is increased, property protection is neither a specific goal, nor an objective of the Action Plan.

KEY ELEMENTS

The Action Plan recommends an initial five-year program organized under several elements as follows: (I) Technical and Engineering Requirements, (II) Public Policy, Legal, and Financial Strategies, (III) Special Requirements of Historic Buildings, (IV) Multihazard Assessment, and (V) Information Transfer and Dissemination. Each element contains a brief introduction to the work involved, the objectives to be achieved, the tasks to be accomplished, a suggested schedule, and a recommended budget. The Action Plan also contains a compendium of unresolved issues needing further attention and/or research.

Element I.--Technical and Engineering Requirements

The seven tasks recommended in this element are intended to benefit a wide variety of primary and secondary user groups including: residents and occupants of buildings; building owners; structural engineers, architects, and other design professionals; building and local regulatory officials; model code and specification writers; and the planning and education communities. These tasks are to:

1. Sponsor research to prepare a standard building inventory methodology

2. Sponsor research to provide a preliminary building evaluation methodology

3. Provide criteria for setting building strengthening priorities

4. Sponsor research to develop a detailed building evaluation methodology. Subtasks include: (a) determination of strength/ deformation characteristics of existing buildings and components; (b) evaluation and development of field condition assessments; and (c) preparation of state-of-the-art detailed building evaluation guidelines.

5. Evaluate and develop practical and effective strengthening techniques. Subtasks include: (a) development of structural performance criteria; (b) identification of practical and effective strengthening methods; (c) experimental research to determine performance characteristics of strengthening methods; (d) definition of acceptable performance criteria for strengthening buildings; and (e) preparation of state-of-the-art guidelines for seismic strengthening.

6. Develop cost estimates for alternative strengthening techniques

7. Begin development of standards of practice for seismic strengthening of existing buildings

Element II.--Public Policy, Legal, and Financial Strategies

The eight tasks recommended in this element have one purpose: to determine the potential for and to initiate change in the attitude of local governments so that programs for reducing earthquake hazards of existing buildings can either be enacted and successfully implemented, or so that existing programs can be accelerated. The tasks are to:

1. Identify major juridictions with significant hazardous building problems

2. Sponsor survey research on key individual and public attitudes in candidate jurisdictions

3. Support the initiation of ordinance development/enactment/implementation/funding processes

4. Develop financial measures to support hazardous buildings reduction programs

5. Support research on liability issues

6. Compile and evaluate existing laws, ordinances, and regulations, and prepare model legislation

7. Evaluate private sector hazard reduction initiatives

8. Support research to determine socioeconomic impacts

Element III.--Special Requirements of Historic Buildings

Historic buildings are a specialized subset of "existing buildings." They are defined as historic by their being listed in the National Register of Historic Places and require special treatment, when seismic strengthening is considered or being implemented, to preserve those aspects of the building that have made it historic. The three tasks identified specifically for historic buildings are to:

1. Evaluate and revise the "Standards of Rehabilitation" and other policy directives

2. Develop a historic building importance classification system

3. Support research to prepare annotated bibliographies on the seismic strengthening of historic buildings

Element IV.--Multihazard Assessment

Performance goals for individual hazards (e.g., earthquakes, wind, fire, tsunami, flooding, etc.), when established without consideration of other hazards, may needlessly overlap, duplicate, or even conflict with the performance goals for other hazards. Rather than consider such hazards singularly, a more rational and effective approach would be to consider performance goals for all possible hazards simultaneously. The three tasks recommended in this element are to:

1. Sponsor research on current hazard abatement measures

2. Determine feasibility of incorporating multihazard mitigation techniques in the rehabilitation process

3. Determine the fundamental relationships among hazard abatement objectives

Element V.--Information Transfer and Dissemination

The ultimate success of hazardous buildings programs will depend upon the continual transfer, implementation, and routinization of knowledge. The three tasks recommended in this element are:

1. Establish an earthquake hazardous buildings information service. Subtasks include: (a) identification and surveying of present earthquake hazards information services; (b) identification of types of information to be provided and exploration of ways to integrate such information; and (c) design, establishment and support of an Information Service.

2. Synthesize recent research and prepare commentaries on application to hazardous buildings abatement programs

3. Develop education and training programs

RECOMMENDED BUDGET

Shown in Table 1 are the estimated federal costs to conduct the tasks identified for each of the five major elements of the Action Plan. The summary budget (Table 1) identifies the estimated cost for each element on both a yearly and a total basis; Table 2 shows the five-year costs distributed by agency. Cost estimates are based upon an assumption of $100,000 per professional per work year, including allowances for overhead, clerical costs, and related incidental costs. Estimated costs include both equipment and labor.

The total estimated cost in federal funds for the five-year program is $40,400,000. In addition, professional organizations, practitioners, researchers, local and state governments, and others can be expected to contribute funds and in-kind services to help meet the objectives of the Action Plan.

Table 1.- Summary Budget

		ESTIMATED FEDERAL COSTS ($1,000)					
		Year					
		1	2	3	4	5	Total
I.	Technical and Engineering Requirements	$2,400	$5,200	$8,950	$7,700	$4,500	$28,750
II.	Public Policy, Legal, and Financial Strategies	325	1,000	2,075	2,550	1,650	7,600
III.	Special Requirements of Historic Buildings	125	275	125	75	75	675
IV.	Multihazard Assessment	100	125	250	250	200	925
V.	Information Transfer and Dissemination	250	400	600	600	600	2,450
	TOTAL COSTS	$3,200	$7,000	$12,000	$11,175	$7,025	$40,400

CONCLUDING REMARKS

The recommendations of the Action Plan must be considered as a beginning. Given the nature and extent of this problem, commitment must be thought of in terms of decades; only the foundation of this effort can be laid in five federal fiscal years.

To date, only a few jurisdictions have established hazard reduction programs for existing buildings. It is believed that more programs should be created, that local governments must continue to lead, and that the state and federal governments can do a variety of things to support local action. These include providing data, funding evaluations and inventories, providing financial incentives, funding basic or needed research, establishing minimum criteria, and helping in the sharing of information. It is the intent of the Action Plan to identify the mutual actions that must be taken and to mobilize more resources and influence.

Earth science research during the last decade has provided a much better understanding of the earthquake risk in seismically active areas of the United States. Geologists now estimate that the probability for the recurrence of a great-magnitude earthquake in California within the next 25 years is greater than 50 percent (FEMA, 1980). East of the Rocky Mountains, where the nation's largest historic earthquake occurred in the early 19th century, the earthquake risk is also substantial. The immense inventory of existing earthquake hazardous buildings in the United States constitutes a serious threat to life. It is essential that implementation of the tasks identified in the Action Plan begin soon.

Table 2.- Five-Year Costs Distributed by Federal Agency

ELEMENT	ESTIMATED COSTS ($1,000)				
	FEMA	NBS	NSF	DOI	Total
I. Technical and Engineering Requirements					
Year 1	$780	$344	$1,276	$0	$2,400
Year 2	750	842	3,608	0	5,200
Year 3	630	1,487	6,833	0	8,950
Year 4	680	1,312	5,708	0	7,700
Year 5	800	783	2,917	0	4,500
Subtotal	$3,640	$4,768	$20,342	$0	$28,750
II. Public Policy, Legal, and Financial Strategies					
Year 1	$283	$0	$42	$0	$325
Year 2	555	0	445	0	1,000
Year 3	1,255	0	820	0	2,075
Year 4	1,744	0	806	0	2,550
Year 5	1,288	0	362	0	1,650
Subtotal	$5,125	$0	$2,475	$0	$7,600
III. Special Requirements for Historic Buildings					
Year 1	$0	$0	$75	$50	$125
Year 2	0	0	75	200	275
Year 3	0	0	75	50	125
Year 4	0	0	75	0	75
Year 5	0	0	75	0	75
Subtotal	$0	$0	$375	$300	$675
IV. Multihazard Assessment					
Year 1	$0	$0	$100	$0	$100
Year 2	0	0	125	0	125
Year 3	0	0	250	0	250
Year 4	0	0	250	0	250
Year 5	0	0	200	0	200
Subtotal	$0	$0	$925	$0	$925
V. Information Transfer and Dissemination					
Year 1	$111	$0	$139	$0	$250
Year 2	178	0	222	0	400
Year 3	378	0	222	0	600
Year 4	378	0	222	0	600
Year 5	378	0	222	0	600
Subtotal	$1,423	$0	$1,027	$0	$2,450
TOTAL COSTS	$10,188	$4,768	$25,144	$300	$40,000

Key:
FEMA—Federal Emergency Mgmt. Agency   NSF—National Science Foundation
NBS—National Bureau of Standards   DOI—Department of Interior

ACKNOWLEDGMENTS

Numerous individuals contributed to the development of the Action Plan. W. G. Corley, F. R. Preece, C. Rojahn, R. E. Scholl, J. R. Smith, and A. S. Virdee served as members of the ABE Joint Venture Management Committee. R. A. Olson and C. Rojahn wrote and revised the plan. R. E. Scholl organized the Workshop and compiled the Workshop Proceedings. J. Smith served as ABE Authorized Representative. Reviews of the Action Plan were provided by the Technical Advisory Committee, consisting of the Management Committee, project consultants R. A. Olson and R. L. Sharpe, and the Issue Paper Authors--B. Bresler, W. T. Holmes, R. M. Dillon, R. K. Yin, R. S. Olson, H. C. Miller, R. D. Hanson, and R. W. Luft. J. T. Dennett served as rapportuer for the Technical Advisory Committee meetings.

The patience and significant contributions of Ugo Morelli, FEMA Project Officer, are also greatly appreciated as are the contributions of the Workshop participants and Liaison Group Representatives. The preparation of the Action Plan was supported by the Federal Emergency Management Agency under Cooperative Agreement No. EMW-84-K-1739.

REFERENCES

ABE, A Joint Venture of Applied Technology Council, Building Seismic Safety Council, and Earthquake Engineering Research Institute, 1985a, "Proceedings--Workshop on Reducing Seismic Hazrds of Existing Buildings," Federal Emergency Management Agency, Washington, D.C.

ABE, A Joint Venture of Applied Technology Council, Building Seismic Safety Council, and Earthquake Engineering Research Institute, 1985b, "An Action Plan for Reducing Earthquake Hazards of Existing Buildings," Federal Emergency Management Agency, Washington, D.C.

Federal Emergency Management Agency (FEMA), 1980, "An Assessment of the Consequences and Preparations for a Catastrophic California Earthquake: Findings and Actions Taken," Washington, D.C.

National Science Foundation and U. S. Geological Survey (NSF-USGS), 1976, "Earthquake Prediction and Hazard Mitigation Options for USGS and NSF Programs," Washington, D.C.

Performance of an Earthquake Excited Roof Diaphragm

M. Celebi, G. Brady, E. Safak and A. Converse*

Introduction

The objective of this paper is to study the earthquake performance of the roof diaphragm of the West Valley College gymnasium in Saratoga, California through a complete set of acceleration records obtained during the 24 April 1984 Morgan Hill Earthquake (M = 6.1). The gymnasium was instrumented by the California Division of Mines and Geology (CDMG) as part of its Strong Motion Instrumentation Program (SMIP). The raw and processed records have been made public by CDMG (Shakal and others, 1984; Huang and others, 1985).

The roof diaphragm of the 112'x144' (34.14m x 43.90m) rectangular, symmetric gymnasium consists of 3/8" (0.95 cm) plywood over tongue-and-groove sheathing attached to steel trusses supported by reinforced concrete columns and walls. A three-dimensional schematic of the gymnasium and the locations of the eleven channels of force balance accelerometers (connected to a central recording unit) and associated directions are provided in Figure 1. There were no free field instruments at the site. Three sensors placed in the direction of each of the axes of the diaphragm facilitate the evaluation of in-plane deformation of the diaphragm. Other sensors placed at ground level measure vertical and horizontal motion of the building floor, and consequently allow the calculation of the relative motion of the diaphragm with respect to the ground level.

Records

The recorded accelerations from the CDMG reports are shown in Figure 2. Traces 1 and 2 show vertical acceleration at ground level at each edge of one of the shear walls in the long N-S perimeter wall. Their difference would measure any rocking at the base of the shear wall, when working in the shearing direction (N-S). Excitation is very low level (peak 0.03 g) and no significant differences are apparent.

The amplitudes of trace 9, measuring acceleration in the N-S direction at ground level, are small (0.03 g), except at 12.5 seconds, when a small packet of 3.8 Hz (two cycle average) motion appears, raising the peak to 0.10 g. On the other hand, traces 6 and 8 indicate that the N-S motion at the top of the N-S walls is 1.5 times that of the ground level (peak acceleration in trace 8 at the 12.5 sec arrival of the pulse is 0.15 g). Any difference between traces 6 and 8 would indicate the existence of a torsional component about a vertical axis in the roof diaphragm response. At the levels of acceleration indicated in these traces, the differences, if any, are negligible.

The N-S motion at the center of the roof diaphragm as seen in Trace 7 clearly are magnified by a factor of 3 (peaks compared) over the mo-

* U.S. Geological Survey, 345 Middlefield Road, Menlo Park, CA 94025.

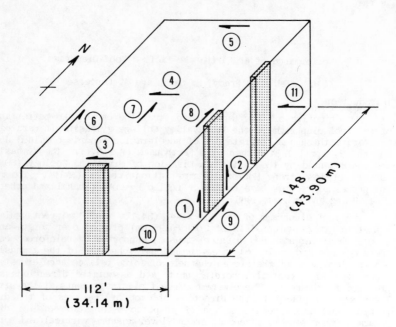

Figure 1.   Sketch of the gymnasium showing locations of the lateral
             force resisting shear walls and eleven accelerometers.
             Additional vertical load carrying columns are not shown.

tion at the edge walls. The traces show that a prominent 4.3 Hz component appears at 10 sec, and builds to a highly resonant motion with the
arrival of the 3.8 Hz pulse at the base at 12.5 sec.

The E-W motion at ground level, as depicted by the low level acceleration (0.04 g peak) of traces 10 and 11 from the north and south ends
of the structure, are similar and therefore indicate negligible torsional motion. Traces 3 and 5 from the roof diaphragm indicate similarity
(as traces 6 and 8 did in N-S direction) and therefore no significant
torsion of the diaphragm. An amplification of approximately 1.5 times
over the ground floor level amplitudes is again evident (0.06 g and 0.07
g compared with 0.04 g).

In the E-W direction there is no sudden arrival of a prominent wave
packet as was the case in the N-S direction at 12.5 sec. Consequently,
the accelerations at floor and roof levels remain at approximately constant amplitude levels from 8 sec through to the end of the record at 23
sec. The only exception is trace 4 at the center of the roof diaphragm.
This trace is the indicator of in-plane distortion of the flexible diaphragm in the E-W direction. Between 8 and 13 seconds, a partial resonance occurs with frequencies shifting slightly about 4 Hz. Intermit-
tent resonant episodes occur during the remainder of the record. The

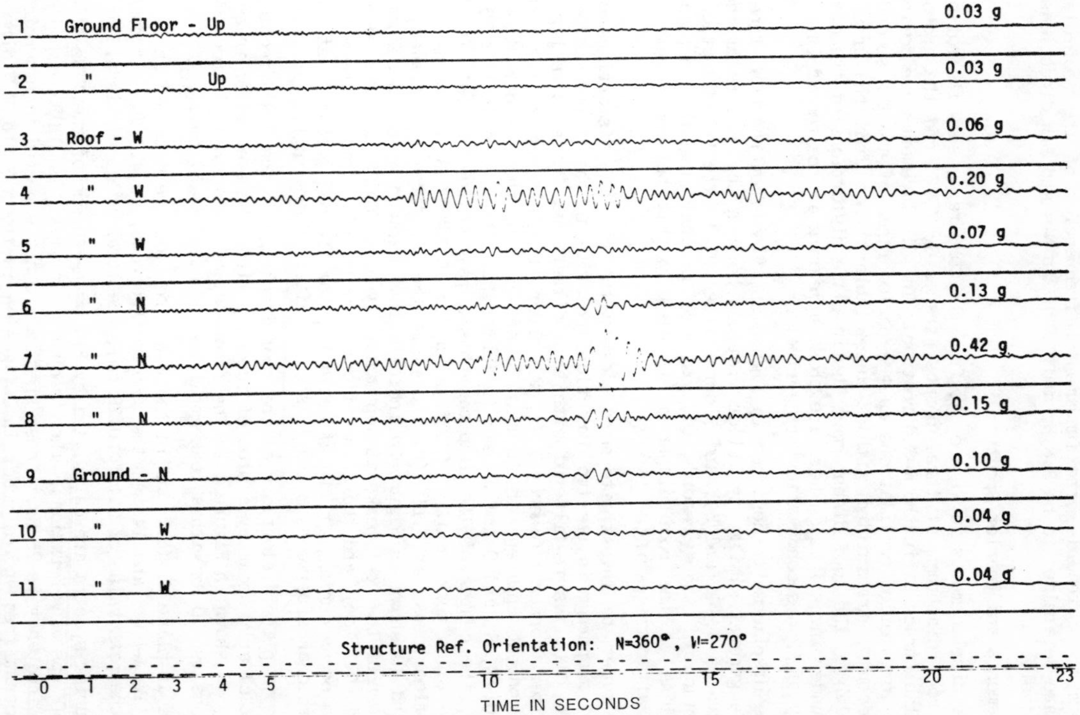

Figure 2. Original CRA-1 recording of the 11 channels at West Valley College gymnasium, Saratoga, California. Channel numbers at left correspond to locations in Figure 1. Scaled peak values at right. (Shakal and others, 1984).

amplification over the levels at the edge walls is 3 times (0.20 g compared with 1/2 (.06 g + .07 g)). This amplification is the same as the N-S case.

From these records, the fundamental frequencies of the roof diaphragm are determined to be 3.8 Hz and 3.9 Hz in N-S and E-W directions, respectively.

Displacements and Fourier Spectra

The displacements in Figure 3a, from the CDMG tape, show the motion of the roof diaphragm at three points (traces 6, 7, 8) and the ground floor slab (trace 9) in the N-S direction. The displacements are dominated by long-period oscillations in the period range of about 2 seconds to 4 seconds, transmitted without appreciable change from the ground floor slab to the roof diaphragm. The in-plane distortions of the diaphragm under investigation here are those appearing in the record in the vicinity of 12.5 seconds after triggering.

The displacements depicted in Figure 3b, also from the CDMG tape, show the general similarity at long periods between the motions of the roof (traces 3, 4, and 5), and floor (trace 10) in the E-W direction. The portion of these records of most use in analyzing the in-plane distortions of the roof diaphragm is clearly that between 9 and 14 seconds after triggering.

In order to investigate more closely the in-plane distortions of the roof diaphragm in its two directions relative to the ground motion, the CDMG data was re-filtered with a bi-directional Butterworth high-pass filter (corner frequency 1 Hz, and order 4). This removes the long period content apparent in Figures 3a and b. In the N-S direction, traces 6, 7, 8, and 9 now appear in Figure 3c. In the vicinity of 13 seconds, the center of the diaphragm has a peak motion of 0.31" (0.78 cm), the edges peak at 0.12" (0.31 cm) and at 0.14" (0.36 cm) while the ground floor peaks at 0.11" (0.28 cm). In the E-W direction, traces 3, 4, 5, and 10 shown in Figure 3d indicate that between 9 and 13 seconds the center of the roof peaks at approximately 0.14" (0.35 cm), the edges at 0.05" (0.12 cm), and the ground floor at 0.04" (0.09 cm). (The actual peaks occur during, or are affected by, low frequency content at 1.5 Hz, in resonance between 13 and 16 seconds after triggering).

The in-plane distortions of the roof diaphragm in the N-S direction are most clearly shown by a plot of trace 7, minus the average of traces 6 and 8, as shown in Figure 4a. The frequency of the high amplitude oscillations is 4.0 Hz with a peak amplitude of 0.22" (0.57 cm).

The in-plane distortions in the E-W direction are detailed in Figure 4b. Between 8 and 14 sec the average frequency is 4.0 Hz, although around the peak value of 0.11" (0.28 cm), the local frequency is 4.2 Hz.

Fourier spectra are calculated by FFT using 40 sec of the 1 Hz-filtered 50 sps accelerations; the frequency interval is 0.025 Hz; the highest frequency is 25 Hz. These spectra are not expected to pin-point the resonant frequencies under investigation in the N-S and E-W directions since the resonance occurs for a relatively short time. The linear plots of Figures 5a and b allow the approximate frequencies under investigation to be conveniently read. Trace 7 frequency, for the N-S direction, is 3.8 Hz; trace 4 frequency, E-W direction, is 3.9 Hz.

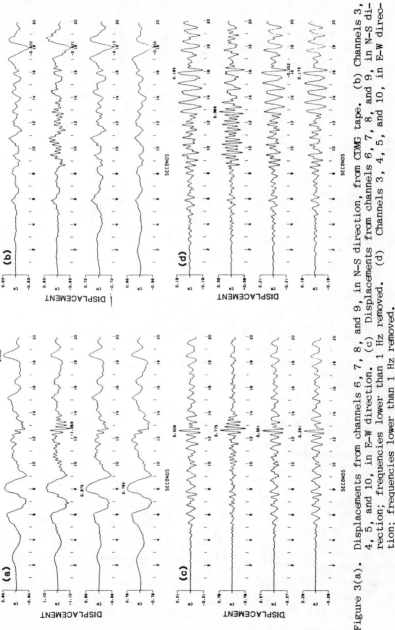

Figure 3(a). Displacements from channels 6, 7, 8, and 10, in E-W direction. (b) Channels 3, 4, 5, and 9, in N-S direction, from CDMG tape. (c) Displacements from channels 6, 7, 8, and 9, in N-S direction; frequencies lower than 1 Hz removed. (d) Channels 3, 4, 5, and 10, in E-W direction; frequencies lower than 1 Hz removed.

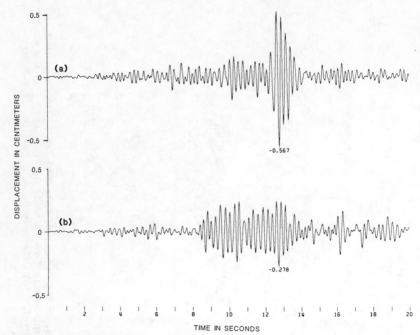

Figure 4.   Displacements at the center of the roof span relative to
the edges, (a) N–S direction, (b) E–W direction.

Analytical Considerations

In order to compare recorded displacements with analytical tools,
the diaphragm is modeled (N–S direction only) as a simple beam (incorporating bending and shear deformations) subjected to a uniformly distributed load equivalent to the product of the mass (of the diaphragm
and half the height of the walls) and the recorded average roof level
acceleration (0.28 g) as determined by the sum of a quarter of the peak
accelerations of traces 6 and 8 and half of trace 7. Cross sectional and
material properties of the plywood only are considered ($F=1\times10^{6}$ psi
(6894 MPa) and $G=110,000$ psi (758 MPa)). The calculated displacement at
center of the diaphragm in N–S direction is 0.07" (0.19 cm) of which
0.05" (0.13 cm) is due to shear deformation. However, this calculated
displacement is possibly underestimated since slips at connections are
not incorporated and only a simplified boundary condition is used. This
displacement is about 33% of the recorded peak of 0.22" (0.57 cm).

On the other hand, using ATC–7 (1981) recommended equations, a total displacement of 0.70" (1.79 cm) is calculated. Therefore, there is
no agreement among the recorded, analytical model or ATC derived
displacements.

Summary and Conclusions:

o The recorded peak accelerations in the N–S and E–W directions of
the ground floor of the building differ considerably.

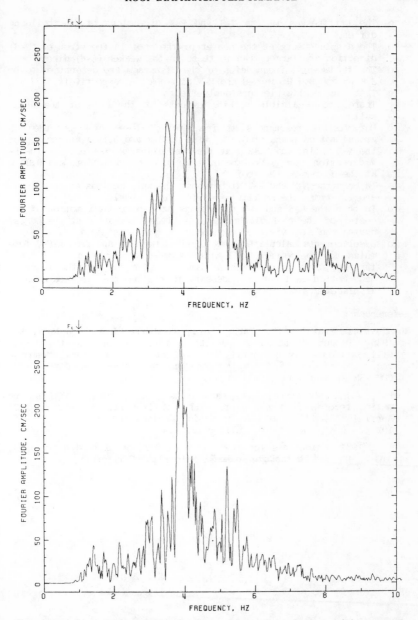

Figure 5.   Fourier amplitude for center of roof span, (a) trace 7,
filtered, in N-S direction, and (b) trace 4, filtered,
in E-W direction.

o The stiffnesses in the N-S and E-W directions of the building differ.
o The displacements at the center of the roof in the stronger (N-S) direction are larger than in those in the weaker (E-W) direction.
o The fundamental frequencies of the disphragm are determined to be 3.8 Hz and 3.9 Hz in N-S and E-W directions, respectively.
o There are negligible torsional effects.
o There are negligible rocking effects at the base of the shear walls.
o For visible resonances at frequencies close to 4 Hz, the N-S ground motion peaks at 0.1 g, with 0.13 g and 0.15 g at the top of the N-S walls, and 0.42 g at the roof diaphragm center. In the E-W direction, the peaks are 0.04 g, 0.06 g and 0.07 g, and 0.20 g at the center of the roof.
o In both the N-S and E-W directions, the wall motions at roof level are 1.5 times the amplitudes of the floor level.
o In both the N-S and E-W directions, the horizontal motion of the center of the roof diaphragm is 3 times the motion at the edges, during the same visible resonances at approximately 4 Hz.
o Displacements calculated by analytical methods and from ATC-7 formulas do not compare well with recorded displacements. Perhaps further study using a three dimensional model and the recorded base motion as input is necessary to resolve the discrepancies of displacements.

References

Huang, M.J., Shakal, A.F., Parke, D.L., Sherburne, R.W., and Nutt, R., 1985, "Processed data from the strong motion records of the Morgan Hill Earthquake of 24 April 1984 - Part II: Structural response records", *Office of Strong Motion Studies, Report No. OSMS 85-04*, CDMG-SMIP, Sacramento, California.

Shakal, A.F., Sherburne, R.W., and Parke, D.L., 1984, "CDMG strong motion records from the Morgan Hill, California, Earthquake of 24 April 1984", *Office of Strong Motion Studies Report No. OSMS 84-7*, 101 p., CDMG, Sacramento, California.

_____, 1981, Guidelines for the design of horizontal wood diaphragms, ATC-7, *Applied Technology Council*, Palo Alto, California.

AUTHOR INDEX

Page number refers to first page of paper.

875